MICROPHYSICS OF CLOUDS
AND PRECIPITATION

MICROPHYSICS OF CLOUDS AND PRECIPITATION

by

HANS R. PRUPPACHER

Department of Atmospheric Sciences,
University of California, Los Angeles

and

JAMES D. KLETT

Los Alamos Scientific Laboratory, Los Alamos, New Mexico and
Department of Physics, New Mexico Institute of Mining and Technology, Socorro

D. REIDEL PUBLISHING COMPANY

DORDRECHT : HOLLAND / BOSTON : U.S.A.
LONDON : ENGLAND

Library of Congress Cataloging in Publication Data

Pruppacher, Hans R , 1930–
 Microphysics of clouds and precipitation.

  Bibliography: p.
  Includes index.
  1. Cloud physics. 2. Precipitation (Meteorology)
 I. Klett, James D., 1940–   joint author. II. Title.
 QC921.5.P78   551.5'76   78–9686
 ISBN 90–277–0515–1

Published by D. Reidel Publishing Company,
P.O. Box 17, Dordrecht, Holland

Sold and distributed in the U.S.A., Canada and Mexico
by D. Reidel Publishing Company, Inc.
Lincoln Building, 160 Old Derby Street, Hingham,
Mass. 02043, U.S.A.

Printed in The Netherlands

TABLE OF CONTENTS

CHAPTER 14   CLOUD PARTICLE INTERACTIONS–
COLLISION, COALESCENCE, AND BREAKUP   464

CHAPTER 15   GROWTH OF CLOUD DROPS BY
COLLISION AND COALESCENCE   504

PREFACE

Cloud physics has achieved such a voluminous literature over the past few decades that a significant quantitative study of the entire field would prove unwieldy. This book concentrates on one major aspect: *cloud microphysics*, which involves the processes that lead to the formation of individual cloud and precipitation particles.

Common practice has shown that one may distinguish among the following additional major aspects: *cloud dynamics*, which is concerned with the physics responsible for the macroscopic features of clouds; *cloud electricity*, which deals with the electrical structure of clouds and the electrification processes of cloud and precipitation particles; and *cloud optics* and *radar meteorology*, which describe the effects of electromagnetic waves interacting with clouds and precipitation. Another field intimately related to cloud physics is *atmospheric chemistry*, which involves the chemical composition of the atmosphere and the life cycle and characteristics of its gaseous and particulate constituents.

In view of the natural interdependence of the various aspects of cloud physics, the subject of microphysics cannot be discussed very meaningfully out of context. Therefore, we have found it necessary to touch briefly upon a few simple and basic concepts of cloud dynamics and thermodynamics, and to provide an account of the major characteristics of atmospheric aerosol particles. We have also included a separate chapter on some of the effects of electric fields and charges on the precipitation-forming processes.

The present book grew out of a series of lectures given to upper division undergraduate and graduate students at the Department of Atmospheric Sciences of the University of California at Los Angeles (UCLA), and at the Department of Physics of the New Mexico Institute of Mining and Technology at Socorro (New Mexico Tech.). We have made no attempt to be complete in a historical sense, nor to account for all the work which has appeared in the literature on cloud microphysics. Since the subject matter involves a multitude of phenomena from numerous branches of physical science, it is impossible to make such a book truly self-contained. Nevertheless, we have considered it worthwhile to go as far as possible in that direction, hoping thereby to enhance the logical structure and usefulness of the work. In keeping with this goal, our emphasis has been on the basic concepts of the field.

This book is directed primarily to upper division and graduate level students who are interested in cloud physics or aerosol physics. Since no specialized knowledge in meteorology or any other geophysical science is presumed, the material presented should be accessible to any student of physical science who has had the more or less usual undergraduate bill of fare which includes a general background in physics, physical chemistry, and mathematics. We also hope the book will be of value to those engaged in relevant areas of teaching and research; also, we hope it will provide a source of useful information for professionals working in related fields, such as air chemistry, air pollution, and weather modification.

In the preparation of this book we have incurred many debts. One of us (HRP) is extremely grateful to his long time associate Prof. A. E. Hamielec of McMaster University at Hamilton, Canada, whose generous support provided the basis for solving many of the nydrodynamic problems reported in this book. Gratitude is also gladly expressed to the faculty and research associates at the Meteorological Institute of the Johannes Gutenberg University of Mainz and at the Max Planck Institute for Chemistry at Mainz, in particular to Profs. K. Bullrich, C. Junge, and Drs. G. Hänel, F. Herbert, R. Jaenicke, and P. Winkler for the assistance received during two stays at Mainz while on sabbatical leave from UCLA. In addition, sincere thanks are extended to the Alexander von Humboldt Foundation for a U.S. Senior Scientist Award which made possible the second extended visit at Mainz. Also, one of us (JDK) is grateful to Drs. C. S. Chiu, P. C. Chen, and D. T. Gillespie for informative discussions, and to Prof. M. Brook and Dr. S. Barr for providing time away from other duties. Appreciation is expressed also to the National Center for Atmospheric Research (NCAR) for the assistance provided during a summer visit.

A large number of figures and tables presented in this book have been adapted from the literature. The publishers involved have been most considerate in granting us the rights for this adaptation. In all cases, references to sources are made in the captions.

Our own research reported in this book has been supported over the years by the U.S. National Science Foundation. We would like to acknowledge not only this support, but also the courteous, informal, and understanding manner in which the Foundation's Officers, Drs. F. White, P. Wyckoff, E. Bierly, and F. Eden, conducted their official business with us.

Special thanks go also to the editors of the D. Reidel Publishing Company of Dordrecht, Holland, for providing a fruitful relationship with us.

Finally, we wish to express our sincere appreciation for the invaluable assistance of T. Feliciello, A. C. Rizos, and P. Sanders, who typed the manuscript, and to B. J. Gladstone who drew all the diagrams.

Los Angeles,
Los Alamos,   *March 1978*

H. R. PRUPPACHER
J. D. KLETT

Hath the rain a father? or who has begotten
the drops of dew?
Out of whose womb came the ice? and the hoary
frost of heaven who has gendered it?
Canst thou lift up thy voice to the clouds,
that abundance of waters may cover thee?
Who can number the clouds in wisdom?
or who can stay the bottles of heaven?
Canst thou send lightnings, that they may go,
and say unto thee, here we are?
Knowest thou the ordinances of heaven? canst
thou set the dominion thereof in the earth?

*Job **38**, 28–29, 33–35*

CHAPTER 1

HISTORICAL REVIEW

As one studies the meteorological literature, it soon becomes evident that cloud microphysics is a very young science. In fact, most of the quantitative information on clouds and precipitation, and the processes which are involved in producing them, has been obtained since 1940. Nevertheless, the roots of our present knowledge can be traced back much further. Although a complete account of the development of cloud physics is not available, a wealth of information on the history of meteorology in general can be found in the texts of Middleton (1965), Schneider-Carius (1962, 1955), and Humphreys (1942, 1937). Based on these and other sources we shall sketch here some of the more important events in the history of cloud physics. In so doing we shall be primarily concerned with developments between the 17th century and the 1940's, since ideas prior to that time were based more on speculation and philosophical concepts than scientific fact. As our scope here is restricted to west European and American contributions, we emphasize again that no claims for completeness are made.

It was apparently not until the 18th century that efforts were underway in Europe to give names to the characteristic forms of the clouds. Lamarck (1744–1829), who realized that the forms of clouds are not a matter of chance, was probably the first to formulate a simple cloud classification (1802); however, his efforts received little attention during his lifetime. Howard (1772–1864), who lived almost contemporaneously with Lamarck, published a cloud classification (1803) which, in striking contrast to Lamarck's, was well received and became the basis of the present classification. Hildebrandson (1838–1925) was the first to use photography in the study and classification of cloud forms (1879), and may be regarded as the first to introduce the idea of a cloud atlas (an idea beautifully realized much later by the International Cloud Atlas I, II (1956) of the World Meteorological Organization, and by the Cloud Encyclopedia of Scorer (1972), where a full description is given of the major genera, species, and varieties of atmospheric clouds).

Both Lamarck and Howard believed the clouds they studied consisted of water bubbles. The bubble idea was originated in 1672 by von Guericke (1602–1686), who called the small cloud particles he produced in a crude expansion chamber 'bullulae' (bubbles). Although he explicitly named the larger particles in his expansion chamber 'guttulae' (drops), the bubble idea, supported by the Jesuit priest Pardies (1701), prevailed for more than a century until Waller (1816–1870) reported in 1846 that the fog particles he studied did not burst on impact, as bubbles would have. Although this observation was confirmed in 1880 by Dines (1855–1927), it was left to Assmann (1845–1918) to finally end the dispute through the authority of his more comprehensive studies of cloud droplets under the microscope (1884).

The first attempt to measure the size of fog droplets with the aid of a microscope was made by Dines in 1880. Some early measurements of the size of the much larger raindrops were made by ingeniously simple and effective means. For example, in 1895 Wiesner (1838–1916) allowed raindrops to fall on filter paper impregnated with water-soluble dye and measured the resulting stains. A little later, Bentley (1904) described an arrangement in which drops fell into a layer of flour and so produced pellets whose sizes could easily be measured and related to the parent drop sizes.

The elegant geometry of solid cloud particles has no doubt attracted attention from the earliest time. A woodcut done in 1555 by Olaus Magnus, Archbishop of Upsala, represents one very early attempt to depict a snow crystal. Kepler (1571–1630) was also intrigued by the forms of snow crystals and asked the question "Cur autem sexangula?" ("But why are they six-sided?"). Descartes (1596–1650) was perhaps the first to correctly draw the shape of some typical forms of snow crystals (1635). Hooke (1635–1703) first studied the forms of snow crystals under a microscope. Scoresby (1789–1857), in his report on arctic regions (1820), presented the first detailed description of a large number of different snow crystal forms and noticed a dependence of shape on temperature. Further progress was made when Neuhaus (1855–1915) introduced microphotography as an aid in studying snow crystals. These and earlier studies on the shape of snow crystals were summarized and critically discussed by Hellmann (1854–1939). Hellmann also pointed out in 1893 that snow crystals have an internal structure, which he correctly attributed to the presence of capillary air spaces in ice. The most complete collections of snow crystal photomicrographs were gathered by Bentley in the U.S. (published by Humphreys in 1931), and during a life's work by Nakaya in Japan (published in 1954).

It was also realized early that not all ice particles have a six-fold symmetry. However, before the turn of the 18th century, interest in the large and often quite irregular shaped objects we now call hailstones was apparently restricted to their outward appearance only. Volta (1745–1827) was among the first to investigate their structure, and in 1808 he pointed out that hailstones contain a 'little snowy mass' at their center. In 1814 von Buch (1774–1853) advocated the idea that hailstones originate as snowflakes. This concept was further supported by Waller and Harting (1853), who investigated sectioned hailstones under the microscope. In addition to finding that each hailstone has a center which, from its appearance, was assumed to consist of a few closely-packed snowflakes, they discovered that hailstones also have a shell structure with alternating clear and opaque layers, due to the presence of more or less numerous air bubbles.

All known observations of cloud and precipitation particles were made at ground level until 1783, when Charles (1746–1823) undertook the first instrumented balloon flight into the atmosphere. Although frequent balloon flights were made from that time on, they were confined mostly to studies of the pressure, temperature, and humidity of the atmosphere, while clouds were generally ignored. The first comprehensive study of clouds by manned balloon was conducted by Wigand (1882–1932), who described the in-cloud shape of ice crystals and graupel particles (snow pellets or small hail) in 1903.

Attempts to provide quantitative explanations of the processes of cloud particle formation came relatively late, well into the period of detailed observations on individual particles. For example, in 1875 Coulier (1824–1890) carried out the first crude expansion chamber experiment which demonstrated the important role of air-suspended dust particles in the formation of water drops from water vapor. A few years later, Aitken (1839–1919) became the leading advocate of this new concept. He firmly concluded from his experiments with expansion chambers in 1881 that cloud drops form from water vapor only with the help of dust particles which act as *nuclei* to initiate the new phase. He categorically stated that "without the dust particles in the atmosphere there will be no haze, no fog, no clouds and therefore probably no rain." The experiments of Coulier and Aitken also showed that by progressive removal of dust particles by filtration, clouds formed in an expansion chamber became progressively thinner, and that relatively clean air would sustain appreciable vapor supersaturations before water drops appeared. The findings of Coulier and Aitken were put into a more quantitative form by Wilson (1869–1959), who showed in 1897 that moist air purified of all dust particles would sustain a supersaturation of several hundred percent before water drops formed spontaneously. This result, however, was already implicitly contained in the earlier theoretical work of W. Thomson (the later Lord Kelvin, 1824–1907), who showed that the equilibrium vapor pressure over a curved liquid surface may be substantially larger than that over a plane surface of the same liquid (1870).

As soon as experiments established the significant role of dust particles as possible initiators of cloud drops, scientists began to look closer at the nature and origin of these particles. Wilson followed up his early studies with dust-free air and discovered in 1899 that ions promote the condensation process, a result which had been predicted theoretically in 1888 by J. J. Thomson (1856–1940). However, it was soon realized that the supersaturations necessary for water drop formation on such ions were much too large for them to be responsible for the formation of atmospheric clouds. Aitken (1880), Welander (1897), and Lüdeling (1903) suggested that the oceans inject hygroscopic salt particles into the atmosphere, which may then serve as condensation nuclei. The great importance of such particles was also realized by Köhler (1888–), who pointed out that the presence of large numbers of hygroscopic particles generally should prevent large supersaturations from occurring in clouds. In addition, Köhler was the first to derive a theoretical expression for the variation of the vapor pressure over the curved surface of an aqueous solution drop (1921, 1922, 1927). His pioneering studies became the foundation of our modern condensation theory.

Although the significance of oceans as a source of condensation nuclei was by now clearly recognized, Wigand's observations (1913, 1930) suggested that the continents, and not the oceans, are the most plentiful source. Wigand's conclusions (1934) were supported by the studies of Landsberg (1906–   ) and Bossolasco (1903–   ).

Lüdeling (?) and Linke (1878–1944) were probably the first to determine the concentration of condensation nuclei in the atmosphere (1903, 1904). However, it was Wigand who, during balloon flights from 1911 to 1913, first carried out detailed studies of condensation nuclei concentrations at different levels in the atmosphere as a function of various meteorological parameters. He discovered that their concentration was related to the temperature structure in the atmosphere, and was significantly different inside and outside clouds. On comparing the concentrations of condensation nuclei and cloud drops, Wigand concluded that there are sufficient numbers of condensation nuclei in the atmosphere to account for the number of drops in clouds.

Studies during the same period brought out the fact that dust particles also play an important role in the formation of ice crystals. Thus, those researchers who ascended into clouds with instrumented balloons found ice crystals at temperatures considerably warmer than the temperatures to which Fahrenheit (1686–1736) had supercooled highly purified water in the laboratory. On the other hand, in view of the large number of condensation nuclei found in the atmosphere, they were also surprised to find that often clouds consisted largely of supercooled drops rather than ice crystals, even though they reached heights where the temperature was considerably below 0 °C. This implied that apparently only a small fraction of the dust particles present acted as ice-forming nuclei. Wegener (1880–1931) suggested that water drops form on water-soluble, hygroscopic nuclei while ice crystals form on a selected group of dust particles which must be water-insoluble. From his observations during a Greenland expedition (1912–1913), he concluded that ice crystals form as a result of the direct deposition of water vapor onto the surface of ice-forming nuclei. He therefore termed this special group of dust particles 'sublimation nuclei.' Wegener's mechanism of ice crystal formation by direct vapor deposition was also advocated by Findeisen (1909–1945). On the other hand, Wigand concluded from his balloon flights that ice crystal formation is often preceded by the formation of supercooled water drops, which subsequently freeze as a result of contact with water-insoluble dust particles. Other arguments against a sublimation mechanism for the formation of ice crystals were brought forward by Krastanow in 1936, who demonstrated theoretically that the freezing of supercooled drops is energetically favored over the formation of ice directly from the vapor.

While all these studies provided some answers concerning why and how clouds come into being, they did not provide any clues as to why some clouds precipitate and others do not. One of the first precipitation theories was formulated in 1784 by Hutton (1726–1797). He envisioned that the cloud formation requisite to precipitation is brought about by the mixing of two humid air masses of different temperatures. The microphysical details of the apportioning of the liquid phase created by this cooling process were not considered. The meteorologists Dove (1803–1879) and Fitz Roy (1805–1865) evidently were in favor of his theory, since it seemed to predict the observed location of rain at the boundary between "main currents of air" (this is now interpreted as frontal rain). Therefore, Hutton's precipitation theory persisted for almost a century. When at last given up, it was not for apparent meteorological reasons but for the physical reason that, owing to the large amount of latent heat released during the phase change of water vapor to water, Hutton's process provides far too small an amount of condensed water to explain the observed amounts of rain.

It finally became clear that only cooling by expansion of humid air during its ascent in the atmosphere would provide clouds with sufficient condensed water to account for the observed rain. The first mathematical formulation of the cooling which is experienced by a volume of expanding air was given by Poisson (1781–1840) in 1823, thus providing the basis for understanding von Guericke's 'cloud chamber' experiments carried out 150 years earlier. Soon afterwards, the idea of cooling by adiabatic expansion, according to which there is no heat exchange between the rising parcel of air and the environment, was applied to the atmosphere by Espy (1785–1860). He deduced in 1835 from experiments and theory that, for a given expansion, dry air is cooled about twice as much as air saturated with water vapor, owing to the heat released by condensing vapor. Also, Péclet (1793–1857) showed in 1843 that the rate of dry adiabatic cooling for a rising air parcel is larger than the cooling usually observed during balloon ascents in the atmosphere.

The first quantitative formulation of the 'saturation adiabatic process,' according to which the condensation products are assumed to remain inside the water-saturated air parcel, was worked out by Lord Kelvin in a paper read in 1862 and published in 1865. Meanwhile, in 1864 Reye (1838–1919) independently derived and published formulations for the same process. A mathematical description of the cooling rate of a lifted air parcel from which the condensation products are immediately removed upon formation, a 'pseudoadiabatic process,' was formulated in 1888 by von Bezold (1837–1907). In 1884 Hertz (1857–1894) further extended the thermodynamic formulation of a rising moist parcel of air. He suggested that if such a parcel rises far enough it will pass through four stages: (1) the 'dry stage' in which air is still unsaturated, (2) the 'rain stage' in which saturated water vapor and water are present, (3) the 'hail stage' in which saturated water vapor, water, and ice coexist, and (4) the 'snow stage' in which only water vapor and ice are present.

In 1866, Renou (1815–1902) first pointed out that ice crystals may play an important role in the initiation of rain. Solely on the basis of the rather restricted meteorological conditions he observed, Renou suggested that for the development of precipitation, two cloud layers are required: one consisting of supercooled drops and another at a higher altitude which feeds ice crystals into the cloud layer below.

More significant progress in the understanding of precipitation formation involving ice crystals was achieved by Wegener (1911), who showed through thermodynamic principles that at temperatures below 0 °C supercooled water drops and ice crystals cannot coexist in equilibrium. Using this result, Bergeron (1891–1977) proposed in 1933 that precipitation is due to the colloidal instability which exists in clouds containing both supercooled drops and ice crystals.

Bergeron envisioned that in such clouds the ice crystals invariably grow by vapor diffusion at the expense of the supercooled water drops until either all drops have been consumed or all ice crystals have fallen out of the cloud. Findeisen's cloud observations (1938) produced further evidence in favor of the Wegener-Bergeron precipitation mechanism.

In suggesting mechanisms for the formation of hailstones, Marcellin Ducarla Bonifas (1738–1816) proposed with considerable foresight in 1780 that "columns of air, more strongly heated than the surrounding atmosphere, may violently rise to elevations where the temperature is sufficiently low that the condensation products freeze to become little snowy globules which further grow from the vapor and by collision with supercooled water drops until they are heavy enough to fall back to Earth." Similarly, von Buch in 1814 and Maille (1802–1882) in 1853 suggested that hailstones originate on snow pellets and grow further by collision with supercooled water drops. Much later, Köhler (1927) applied the notion of collision growth to ice crystals, which he recognized might collect supercooled cloud drops. He also noted, but did not explain, his observation that both drops and crystals have to be of a minimum critical size before such growth may evolve.

The same basic idea of collisional growth, applied this time to cloud drops of different size and hence different fall velocities, was put forth independently in 1715 by Barlow (1639–1719) and by Musschenbroek (1692–1761) in 1739. Musschenbroek also proposed that drops growing by collision will not exceed a size of about 6 mm in diameter, due to the observed instability of drops larger than this size. Reynolds (1842–1912) expanded on the notion of collisional growth and showed by computation in 1877 that water drops above a certain size grow slower by vapor diffusion than by collision with other drops.

A subtle aspect of the collisional growth process was discovered by Lenard (1862–1947), who observed in 1904 that colliding drops do not always coalesce. This he attributed correctly to the difficulty of completely draining all the air from between the colliding drops. He also found (as had been noticed in 1879 by Strutt, the later Lord Rayleigh, 1842–1919) that small amounts of electric charge residing on drops could build up attractive electric forces which are sufficiently large to overcome the hydrodynamic resistance to coalescence. In agreement with the expectations of Musschenbroek, Lenard concluded from his experiments that growth by collision-coalescence continues until drops grow to a critical size, after which they become hydrodynamically unstable and break up. He suggested that the fragment drops may then continue to grow in the same manner, producing a 'chain-reaction' effect of overall rapid growth.

Despite Lenard's experimental results, the mechanism of growth by collision was given little attention for a long time, since the Wegener-Bergeron-Findeisen mechanism dominated the thinking of meteorologists, most of whom studied storm systems in the middle and higher latitudes where the ice phase is quite common. Simpson (1878–1965) attempted to revive the collision mechanism in his presidential address to the Royal Meteorological Society in 1941. On the basis of reports from airplane pilots who flew over India through precipitating clouds with tops thought to be warmer than 0 °C, and from some crude calculations made by Findeisen on the rate at which unequal size cloud drops coagulate, Simpson asserted that he found it untenable to assume that precipitation formation should be confined only to clouds which reach subzero temperature levels. However, convincing quantitative support for Simpson's position had to await the late 1940's, when radar observations and military flights finally led to a general consensus that clouds need not reach subzero temperature levels, and consequently need not contain ice crystals, for precipitation to occur.

*   *   *

In striking contrast to the rather slow development of cloud physics prior to World War II, an abrupt and accelerating increase in research and knowledge has occurred since. A confluence of several factors has brought about this dramatic change. For example, a surge of interest in cloud physics was closely tied to the military-related research in meteorology which developed during the war years (1939–1945) and produced a great number of trained workers in meteorology. Also, several new observational techniques involving aircraft, radar, and other instruments became available to scientists at a time when both the necessary funding and support personnel were also relatively abundant. In addition, interest and support was stimulated by the demonstrations of Schaefer and Langmuir in 1946 that it is possible to modify at least some clouds and affect the precipitation yield by artificial means. (They seeded supercooled stratus clouds with dry ice, which caused the formation and subsequent rapid growth of ice crystals. This induced colloidal instability led in about 20 minutes to a miniature snowfall.) Finally, the fast pace of general technological advances has had a continuing great impact on cloud physics, insuring an accelerated development by making available such important tools as computers, satellites, rockets, and accurately controlled climatic chambers and wind tunnels.

To a large extent, the rapid progress referred to above can be categorized as a fairly direct development of the ideas and discoveries which were made considerably earlier. As we shall see, the period of progress since the beginning of World War II has not been characterized by numerous conceptual breakthroughs, but rather by a series of progressively more refined quantitative theoretical and experimental studies of previously identified microphysical processes.

As we shall also see, much remains to be learned in spite of the significant advances of the past three decades. One principal continuing difficulty is that of incorporating, in a physically realistic manner, the microphysical phenomena in the broader context of the highly complex macrophysical environment of natural clouds. This problem was well expressed 20 years ago in the preface to the first edition of Mason's (1957) treatise on cloud microphysics:

Although the emphasis here is upon the *micro-physical* processes, it is important to recognize that these are largely controlled by the atmospheric motions which are manifest in clouds. These *macro-physical* features of cloud formation and growth, which might more properly be called a *dynamics*, provide a framework of environmental conditions confining the rates and duration of the micro-physical events. For example, the growth or freezing of cloud droplets is accompanied by the release of great quantities of latent heat, profoundly influencing the motion of cloudy air masses, while the motions which ultimately cause evaporation of the cloud determine its duration, and will set a limit to the size which its particles can attain. Progress in cloud physics has been hindered by a poor appreciation of these interrelations between processes ranging from nucleation phenomena on the molecular scale to the dynamics of extensive cloud systems on the scale of hundreds or thousands of kilometers.

The problem of scale which Mason refers to provides a revealing point of view for appreciating the extent of difficulties one encounters. Thus, stating the case in a very conservative manner, we are concerned in cloud microphysics with the growth of particles ranging from the characteristic sizes of condensation nuclei ($\geq 10^{-2}\,\mu$m) to precipitation particles ($\leq 10^{4}\,\mu$m for raindrops, $\leq 10^{5}\,\mu$m for hailstones). This means we must follow the evolution of the particle size spectrum, and the attendant microphysical processes of mass transfer, over about seven orders of magnitude in particle size. Similarly, the range of relevant cloud-air motions varies from the characteristic size of turbulent eddies which are small enough to decay directly through viscous dissipation ($\geq 10^{-2}\,$cm), since it is these eddies which turn out to define the characteristic shearing rates for turbulent aerosol coagulation processes, to motions on scales at least as large as the cloud itself ($> 10^{5}\,$cm). Thus, relevant interactions may occur over at least seven orders of magnitude of eddy sizes. Also, in recent years it has become increasingly clear that a strong coupling may occasionally occur between the particle growth processes, including the development of precipitation, and the growth of the cloud electric field. Since in the atmosphere field strengths range from the fair-weather value ($\leq 10^{2}\,$volts m^{-1}) to fields of breakdown value ($\sim 10^{6}\,$volts m^{-1}), to understand the formation of highly electrified clouds we must cope with about four orders of magnitude of electric field variation. At the same time, we also must be concerned with various electrostatic force effects arising from at least an eight order of magnitude range of particle charge, considering the observed presence of 1 to 10^{8} free elementary charges ($\sim 5 \times 10^{-10}$ to $\sim 5 \times 10^{-2}$ e.s.u.) on atmospheric particles. If the electrostatic contribution to the large scale cloud energetics is also considered, a much larger charge magnitude range is involved.

It is clear, therefore, that a complete in-context understanding of cloud microphysics must await some sort of grand synthesis, an elusive and distant goal even from the point of view of presently available models of cloud dynamics, microphysics, and electrification. We should emphasize that such an approach to the subject is far beyond the scope of this book. Rather, our goal has been to provide where possible a reasonably quantitative account of the most relevant, individual microphysical processes. In addition to whatever intrinsic interest and usefulness in other applications the separate case studies of this book may hold, we also hope they may help provide a useful basis for an eventual integrated treatment of overall cloud behavior. As we shall see, however, even this restricted approach to the subject must necessarily involve a degree of incompleteness, since many microphysical mechanisms are still not understood in quantitative detail. In this sense also cloud microphysics is still a developing subject, and so is characterized to some extent by inadequate knowledge as well as conflicting results and points of view.

CHAPTER 2

MICROSTRUCTURE OF ATMOSPHERIC CLOUDS AND PRECIPITATION

Before discussing the microphysical mechanisms of cloud particle formation, we shall give a brief description of the main microstructural features of clouds. Here we shall be concerned primarily with the sizes, number concentrations, and geometry of the particles comprising the visible cloud.

2.1 Microstructure of Clouds and Precipitation Consisting of Water Drops

2.1.1 CLOUDS AND FOGS

Since water readily supercools, particularly in small quantities, water clouds as well as fogs (which are clouds with ground contact sufficiently extensive to suppress vertical motions) are frequently found in the atmosphere at temperatures below 0 °C. Curves 1 and 2 of Figure 2-1 demonstrate this tendency. These curves, based on a large number of aircraft observations made by Peppler (1940) over Germany, and by Borovikov *et al.* (1963) over the ETU (European territory of the U.S.S.R.), show that supercooled clouds are quite a common occurrence in the atmosphere, especially if the cloud top temperature is warmer than −10 °C. However, with decreasing temperature the likelihood of ice increases such that at −20 °C less than 10% of clouds consist entirely of supercooled

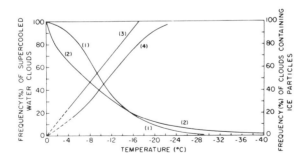

Fig. 2-1. Variation of the frequency of supercooled clouds and of clouds containing ice crystals. Curves 1 and 2 pertain to ordinate at left. Curves 3 and 4 pertain to ordinate at right. (1) Peppler (1940), Germany, all water clouds; (2) Borovikov *et al.* (1963) ETU, all water clouds; (3) Mossop *et al.* (1970), Tasmania, mixed clouds; (4) Morris and Braham (1968), Minnesota, mixed clouds.

9

drops. Only in rare cases have supercooled clouds been observed at temperatures as low as −35 °C over Germany (Weickmann, 1949), and as low as −36 °C
over the ETU (Borovikov *et al.*, 1963).

Although the relative humidity of clouds and fogs usually remains close to
100%, considerable departures from this value have been observed. Thus,
reports from different geographical locations (Pick, 1929, 1931; Neiburger and
Wurtele, 1949; Mahrous, 1954; Reiquam and Diamond, 1959; Kumai and Francis,
1962a, b) show that the relative humidity of fogs has been found to range from
100% to as low as 81%. Somewhat smaller departures from saturation are
usually observed in cloud interiors. Warner (1968a) indirectly deduced values for
the relative humidity in small to moderate cumuli from measurements of vertical
velocity and drop size. From his results (shown in Figure 2-2), we see that in
these clouds the relative humidity rarely surpasses 102% (i.e., a supersaturation
of 2%), and is rarely lower than 98%. The median of the observed supersaturations was about 0.1%. Similarly, Braham (in Hoffer, 1960) found, during
several airplane traverses through cumulus clouds, that in their outer portions the
air generally had relative humidities between 95 and 100%, dipping to as low as
70% near the cloud edges where turbulent mixing was responsible for entraining
drier air from outside the clouds. In the more interior cloud portions, the relative
humidity ranged from 100% to as high as 107%.

Most, but not all, drop size distributions measured in many different types of
clouds under a variety of meteorological conditions exhibit a characteristic
shape. Generally, the concentration rises sharply from a low value to a maximum, and then decreases gently toward larger sizes, causing the distribution to
be positively skewed with a long tail toward the larger sizes. Such a characteristic shape can be approximated reasonably well by either the lognormal or

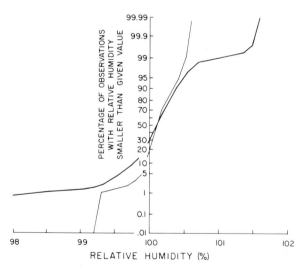

Fig. 2-2. Percentage of observations with relative humidity less than a given value for all samples
(heavy line) and for samples taken within 300 m of cloud base (thin line). (From Warner, 1968a, by
courtesy of *J. de Rech. Atmos.*, and the author.)

gamma distribution functions. Other convenient representations are the empirical size distributions developed by Best (1951a), and by Khrgian and Mazin (in Borovikov *et al.*, 1963). Best's distribution may be expressed as

$$1 - F = \exp\left[-(d/C)^k\right], \tag{2-1}$$

where F is the fraction of liquid water comprised of cloud drops with diameters smaller than $d(\mu m)$. The characteristic parameters C and k vary with the water content, the drop concentration, and the maximum drop size in the cloud. Best found $1.92 \leqslant k \leqslant 4.90$ and $12 \leqslant C \leqslant 29 \, \mu m$.

The Khrgian-Mazin drop size distribution can be expressed as

$$n(a) = A a^2 \exp(-Ba), \tag{2-2}$$

where $n(a)\,da$ is the number of drops cm^{-3} in the radius range $(a, a + da)$. The parameters A and B can be related to any two moments of the distribution. For example, in terms of the total concentration N (the zeroth moment), and the average radius $\bar{a}$ (the ratio of the first and zeroth moments) we find

$$N = \int_0^\infty n(a)\,da = \frac{2A}{B^3}, \tag{2-3}$$

and

$$\bar{a} = \frac{1}{N} \int_0^\infty a n(a)\,da = \frac{3}{B}. \tag{2-4}$$

Another related quantity of interest is the total mass concentration of liquid water. Since this often turns out to be about $10^{-6}\,g\,cm^{-3}$, one defines the cloud liquid water content, w_L, as follows:

$$w_L(g\,m^{-3}) \equiv 10^6 \left(\frac{4\pi}{3}\right) \rho_w \int_0^\infty a^3 n(a)\,da, \tag{2-5}$$

where ρ_w is the density of water in $g\,cm^{-3}$ and a is in cm. Then for the Khrgian-Mazin distribution we find

$$A \approx 1.45 \times 10^{-6} \left(\frac{w_L}{\rho_w \bar{a}^6}\right), \tag{2-6}$$

and

$$N \approx 1.07 \times 10^{-7} \left(\frac{w_L}{\rho_w \bar{a}^3}\right). \tag{2-7}$$

Of course, these analytical expressions only represent average distributions. Individual drop size spectra may be significantly different. For example, many individual spectra tend to be strongly bimodal (Figure 2-3a) (Eldridge, 1957; Durbin, 1959; Warner, 1969a). Also, they may exhibit only weak positive skewness, or may even be negatively skewed (Squires, 1958a; Durbin, 1959; Warner, 1969a).

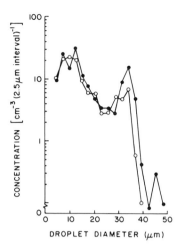

Fig. 2-3a. Variation of the cloud drop concentration with drop diameter. Adjacent samples taken 100 m apart near the top of a cloud 1400 m deep. (From Warner, 1969a; by courtesy of Amer. Meteor. Soc., and the author.)

During his flights through clouds over the east coast of Australia, Warner (1969a) observed (Figure 2-3b) that the tendency of a size distribution to be bimodal increased with height above cloud base and with decreasing stability in the cloud environment. Based on this observation, Warner suggested that bimodal drop size distributions are the result of a mixing process between the cloud and the environment. Since the drop size spectra were fairly uniform for a given level across a cloud and the bimodality was not confined to the cloud edges, Warner proposed that the mixing process producing the bimodality is due

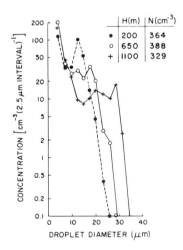

Fig. 2-3b. Variation with height of the spectrum of cloud drops (based on two clouds). The height H above cloud base at which the samples were taken and the average total drop concentration N at that height are also shown. (From Warner, 1969a; by courtesy of Amer. Meteor. Soc., and the author.)

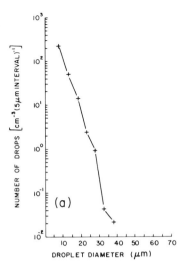

Fig. 2-4a. Mean drop size distribution in fair weather cumulus; average, total drop concentration: 293 cm⁻³. (From Battan and Reitan, 1957; by courtesy of Pergamon Press Ltd.)

mostly to entrainment of drier air at the growing cloud top, and to a lesser degree to entrainment at the cloud edges. Figure 2-3b shows further that the fraction of drops larger than 25, 30, and 35 μm diameter increased rapidly with height above cloud base, indicating that the size distribution experiences a broadening effect with increasing distance from cloud base.

The drop size distribution is also a function of the development stage of a cloud. This is exemplified by Figures 2-4a, b which show that fair weather cumuli

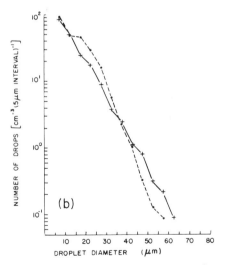

Fig. 2-4b. Mean drop size distribution in cumulus congestus over the central U.S.; average, total drop concentration in 10 clouds which developed precipitation echoes: 188 cm⁻³ (dashed line), in 21 clouds which did not develop echoes: 247 cm⁻³ (solid line). (From Battan and Reitan, 1957; by courtesy of Pergamon Press Ltd.)

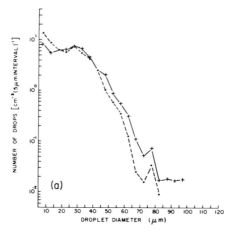

Fig. 2-5a. Mean drop size distribution in tropical cumuli over the Gulf of Mexico, near the Bahama Islands, and near Puerto Rico. Average, total drop concentration in 11 clouds with echoes: 52 cm^{-3} (dashed line), in 26 clouds without echoes: 58 cm^{-3} (solid line). (From Battan and Reitan, 1957; by courtesy of Pergamon Press Ltd.)

have relatively narrow drop size spectra, while the spectra of cumuli which have reached the more mature development stage of a cumulus congestus are considerably broader.

Significant differences are also found between spectra formed in maritime air and in continental air masses. This is illustrated in Figures 2-4b and 2-5a. Comparison of these figures shows that at the small drop size end of the spectrum, the number concentration of drops with a diameter near 10 μm is larger by about one order of magnitude in cumuli over the Central U.S. (about 100 cm^{-3} per 5 μm-size interval) than in tropical cumuli (about 10 cm^{-3} per

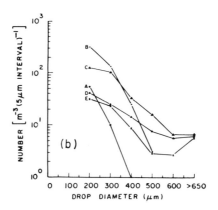

Fig. 2-5b. Drop spectrum for tropical cumuli. Total concentration ranges: (A) 200 to 3000 drops m^{-3}, $w_L = 1.0$ to 33.1×10^{-3} g m^{-3}, (B) 3000 to 20 000 drops m^{-3}, $w_L = 32.0$ to 213×10^{-3} g m^{-3}; (C) 3000 to 8000 drops m^{-3}, $w_L = 33$ to 163×10^{-3} g m^{-3}; (D) 1000 to 3000 drops m^{-3}, $w_L = 3.3$ to 116×10^{-3} g m^{-3}; (E) precipitation measurements. (From Brown and Braham, 1959; by courtesy of Amer. Meteor. Soc., and the authors.)

5 μm-size interval). Note, however, that the drop number concentration in both cloud types is similar for drops larger than about 40 μm diameter. Figure 2-5b extends the spectrum for tropical cumuli (Figure 2-5a) to larger sizes and illustrates a typical feature of cloud drop spectra in general: the concentration decreases sharply from a few tens or more per cubic centimeter at the small drop size end to between 1 ℓ^{-1} and 1 m^{-3} for the large drops with diameters >500 μm.

Squires (1958a) has carried out detailed comparative studies of the drop spectra of different types of clouds. The observational sequence shown in Figures 2-6a, b, c illustrates the dependency of spectral shape on cloud type for situations in which the nuclei on which drops form are essentially the same in type and concentration, since a given air mass has spawned all three types shown. However, we can see that even though there is little variation in liquid water content, the drops become smaller, more numerous, and more homogeneous in size as one passes from the orographic to the stratus to the cumulus cloud types. Continental cumuli appear to represent an extension of this trend, in that the spectra are even narrower, the concentrations even higher, and the average drop sizes even smaller (Figure 2-6d). For this case a different air mass type with a correspondingly different nucleus content is involved, and this largely accounts for the change from the maritime spectrum. Squires' observations clearly express the trend that high drop concentrations in clouds are associated with

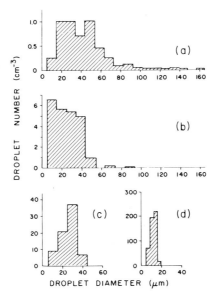

Fig. 2-6. Droplet spectra in clouds of various types. Cumulus samples are taken 2000 ft above cloud base, orographic and dark stratus values are average. Note change in ordinate scale from figure to figure. (a) Orographic cloud over Hawaii, $w_L = 0.40$ g m^{-3}; (b) dark stratus over Hilo (Hawaii), $w_L = 0.34$ g m^{-3}; (c) tradewind Cu over Pacific off the coast of Hawaii, $w_L = 0.50$ g m^{-3}; (d) continental Cu over Blue Mts. near Sidney, Australia, $w_L = 0.35$ g m^{-3}. (From *The Physics of Rain Clouds* by N. H. Fletcher, copyrighted by Cambridge University Press, 1962a, redrawn by Fletcher after Squires. 1958a.)

narrow drop size spectra and small drop sizes. This is the pattern usually encountered in continental type clouds, while in maritime clouds low drop concentrations are associated with broad size spectra and large drop sizes. This trend is also confirmed by the recent observations of Ryan *et al.* (1972) on maritime stratus off the California coast, maritime cumuli over the Gulf of Mexico, and continental cumuli near Eureka, California (Figure 2-7). Note again that above a certain level the drop size spectra of maritime cumuli and of maritime stratus tend to be bimodal.

Remote sensing with radar can also provide information on the correlation between the drop spectrum and the overall development of precipitating clouds. Most of the early work was done with 10 cm or 3 cm radars which, in general, could only detect drops larger than a few hundred microns in diameter. More recent high power 3 cm radar, and most 1 cm radar, permit the detection of drops with diameters larger than a few tens of microns (Mason, 1971; Battan, 1973).

From radar studies of various types of cumuli, Battan and Braham (1956), Morris (1957), and Battan (1963) found that the appearance of a radar echo is characteristically related to the cloud dimensions. Thus, Figure 2-8 shows that the probability of an echo developing in a cloud grows with its depth and width.

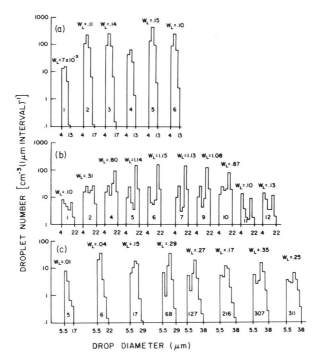

Fig. 2-7. Droplet spectra in clouds of various types. (a) Small non-precipitating continental cumulus over Northern California near Eureka, (b) small non-precipitating maritime cumulus over Gulf of Mexico near Houston, (c) shallow non-precipitating maritime stratus over Pacific off California coast near Santa Cruz. The histograms have equal widths for conservation of space; the numbers within the distributions are the elapsed times in seconds after cloud penetration; w_L in g m^{-3}. (From Ryan *et al.*, 1972; by courtesy of Amer. Meteor. Soc., and the authors.)

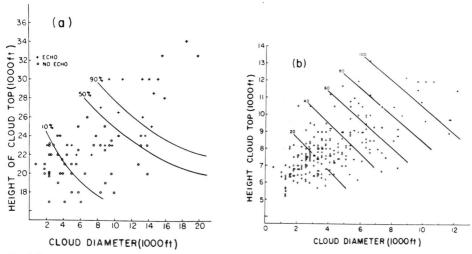

Fig. 2-8.   Occurrence (+) or non-occurrence (O) of precipitation in cumuli over Arizona as function of cloud-top height and cloud width. Lines of equal probability to find an echo are also given. (a) Clouds over Arizona (From Morris, 1957; by courtesy of Amer. Meteor. Soc., and the author.) (b) Clouds over ocean near Puerto Rico (From Battan and Braham, 1956; by courtesy of Amer. Meteor. Soc., and the authors.)

This correlation is also illustrated by Klazura's (1971) measurements shown in Figure 2-9. He further confirmed the expectation that the higher the cloud builds, the broader is the size spectrum, and the larger is the concentration of large drops. Figure 2-8 also indicates that continental type clouds need to build considerably higher than maritime clouds, and must become considerably wider before a radar echo appears. Figure 2-10 shows that the appearance of a precipitation echo is also related to the height of the visual cloud base, while Figure 2-11 supports the results shown in Figure 2-8. Thus, the higher the cloud base, the higher the level in the cloud at which a precipitation echo develops;

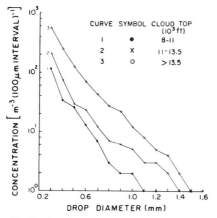

Fig. 2-9.   Mean drop-size distributions in warm cumuli over southeast Texas for drop diameters > 250 μm. Samples have been classified according to estimated cloud tops. (From Klazura, 1971; by courtesy of Amer. Meteor. Soc., and the author.)

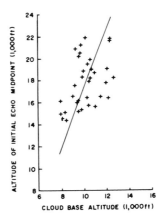

Fig. 2-10. Average altitude of the midpoint of initial echoes plotted against altitude of visual cloud base height; cumulus over Arizona; correlation coefficient for regression line: +0.48. (From Battan, 1963; by courtesy of Amer. Meteor. Soc., and the author.)

and the greater the cloud depth, the greater the probability for a precipitation echo.

Further information on the microstructure of clouds, and the relationship between their microstructure and macrostructure, can be obtained from the measurement of liquid water content. From the observations of Zaitsev (1950), Draginis (1958), Squires (1958b), Durbin (1959), Ackerman (1959, 1963), Huan Mei-Yuan (1963), Borovikov *et al.* (1963), Warner (1955, 1969a), and Vulfson *et al.* (1973) one notes that the distribution of cloud water content has four characteristic features: (1) At any given level the water content varies considerably over short distances in a manner which is closely related to the variation of the vertical air velocity (Figure 2-12). (2) Typically, on the macro-scale the cloud water content increases with height above the cloud base, assumes a maximum somewhere in the upper half of the cloud, and then

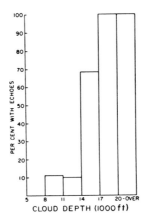

Fig. 2-11. Percentage of Arizona cumuli giving radar echo as a function of their depth. (From Morris, 1957; by courtesy of Amer. Meteor. Soc., and the author.)

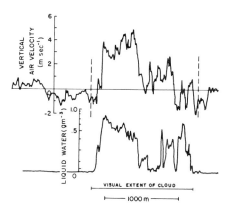

Fig. 2-12. Vertical air velocity and liquid water content as function of distance from place of entering cloud. (From Warner, 1969a; by courtesy of Amer. Meteor. Soc., and the author.)

decreases again toward the cloud top (Figures 2-13 and 2-14c). (3) The distribution of water content parallels the distribution of the drop size rather than drop concentration (Figures 2-14a, b). (4) Comparison between the observed cloud water content w_L and the water content $(w_L)_{ad}$ computed on the basis of a saturated adiabatic ascent of moist air shows that generally $w_L < (w_L)_{ad}$. This fact is illustrated by Figure 2-15 which implies that, as a cloud builds, drier air is constantly entrained and subsequently saturated at the expense of some of the water released during ascent. In most cases $w_L/(w_L)_{ad}$ is found to increase with increasing cloud width, implying that the net dilution effect by entrainment is less in larger clouds than smaller ones.

Observations by Houghton and Radford (1938), Kojima *et al.* (1952), Mahrous (1954), Reiquam and Diamond (1959), Kumai and Francis (1962a), Okita (1962a), Meszaros (1965), and Garland (1971) show that fogs, unlike clouds, are characterized by relatively low water contents (generally less than $0.1\,\mathrm{g\,m^{-3}}$), small drops (typically between $0.5\,\mu$m and a few tens of micrometers), and small number concentrations ($1\,\mathrm{cm^{-3}}$ to a few hundreds per cubic centimeter if drops of $a < 1\,\mu$m are disregarded). While in earlier observations drops were detected only if their radii were larger than $1\,\mu$m, Garland (1971) used an impactor method in conjunction with an interference contrast method to measure the size of fog drops of radii as small as $0.4\,\mu$m. Though these small drops do not add

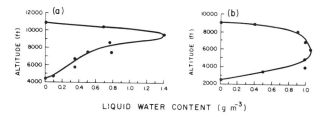

Fig. 2-13. Examples (a, b) for the variation of peak water content with height in cumulus clouds. (From Warner, 1955; by courtesy of *Tellus*.)

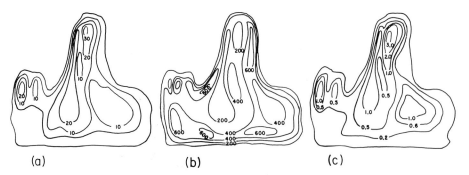

(a)   (b)   (c)

Fig. 2-14. Spatial distribution of microstructure parameters in cumulus cloud. (a) drop diameter (μm), (b) drop concentration (number cm^{-3}), (c) w_1 (g m^{-3}). (From Mei-Yuan, 1963; by courtesy of Air Force Geophys. Laboratory, Hanscom Air Force Base, Mass.)

significantly to the water content, they do dominate the number concentration which may be as large as several thousand cm^{-3} for $0.4 \leqslant a \leqslant 1 \mu$m. Some examples of fog drop size spectra are given in Figure 2-16. Note that the spectra represented in Figures 2-16d, e are strongly bimodal. Figure 2-17 indicates the spatial variation of drop concentration in fogs is often quite large, ranging up to two orders of magnitude for certain size categories.

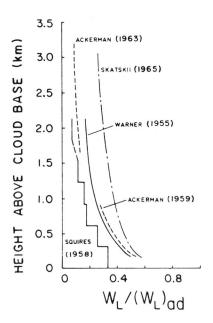

Fig. 2-15. Ratio of the observed mean liquid water content at a given height above cloud base to the adiabatic value, for presumably non-precipitating clouds. (From Warner, 1970; by courtesy of Amer. Meteor. Soc., and the author.)

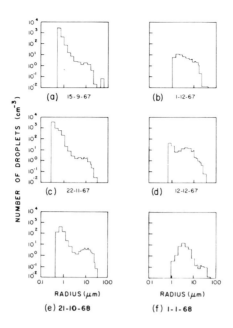

Fig. 2-16.   Fog drop size distributions observed on different days. (From Garland, 1971; by courtesy of *Quart. J. Roy. Meteorol. Soc.*)

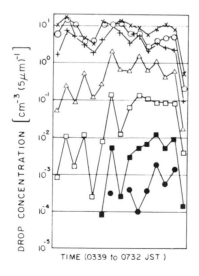

Fig. 2-17.   Variation of concentrations of fog droplets on September 11, 1957, at Asahikawa, Japan. × 5–10 μm; ○ 10–15 μm; + 15–20 μm; △ 20–25 μm; □ 25–35 μm; ■ 35–50 μm; ● 50–75 μm; JST means Japanese Standard Time. (From Okita, 1962a; by courtesy of *J. Meteor. Soc., Japan.*)

2.1.2   RAIN

A small difficulty arises in attempting to describe the spectra of rain, since raindrops are large enough to have a size-dependent shape which cannot be characterized by a single length (see Section 10.3.2). The conventional resolution, which we adopt here, is to describe rain spectra in terms of the equivalent diameter D_0 defined as the diameter of a sphere of the same volume as the deformed drop. When falling at terminal velocity, drops are nearly perfect spheres if $D_0 \lesssim 280 \, \mu$m. Larger drops are slightly deformed and resemble oblate spheroids if $280 \lesssim D_0 \lesssim 1000 \, \mu$m. For $D_0 > 1000 \, \mu$m, the deformation becomes large and the drops resemble oblate spheroids with flat bases (see Plate 1). Drops larger than about 10 mm in diameter are hydrodynamically unstable and break up, even in a laminar air stream (see Section 10.3.4).

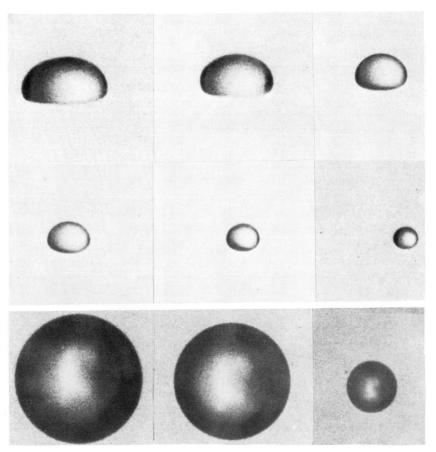

Plate 1.   Shape of cloud and raindrops as determined from wind tunnel experiments. Equivalent radius of drops. Top row, from left to right: 4.00 mm, 3.68 mm, 2.90 mm; second row, from left to right: 2.65 mm, 1.75 mm, 1.35 mm; third row, from left to right 393 μm, 354 μm, 155 μm. Drops in third row were printed comparatively large to show sphericity. (From Pruppacher and Beard, 1970; by courtesy of *Quart. J. Roy. Meteor. Soc.*)

In addition to the equivalent diameter D_0, there are three other quantities which are commonly used to characterize rain: (1) the size distribution $n(D_0)$, expressed here in terms of the number of drops per cubic meter of air per unit size interval; (2) the water content, w_L, given as

$$w_L(g\ m^{-3}) = \left(\frac{\pi}{6}\right) \times 10^{-3}\ \rho_w \int_0^\infty D_0^3 n(D_0)\ dD_0, \qquad (2\text{-}8)$$

with D_0 in mm and ρ_w in $g\ cm^{-3}$; (3) the rainfall rate or intensity, R, usually expressed in $mm\ hr^{-1}$:

$$R(mm\ hr^{-1}) = 6\pi \times 10^{-4} \int_0^\infty D_0^3 n(D_0) V_\infty(D_0)\ dD_0, \qquad (2\text{-}9)$$

with the drop terminal velocity V_∞ in $m\ sec^{-1}$.

Raindrop spectra may extend to drop diameters as large as 6 mm (Mason and Andrews, 1960; Diem, 1968; Blanchard and Spencer, 1970). Such large drops are rather rare since they are found only in very heavy rain with $R > 100\ mm\ hr^{-1}$. At smaller rainfall intensities raindrop spectra usually extend only to drop diameters of 2 to 3 mm. Larger drops tend to break up as a result of collision with other drops (see Sections 10.3.4, 14.8.2, and 15.4).

Several factors influence the spectral shape of rain at the small size end. Since rain must fall against the cloud updraft, the strength of the latter tends by itself to truncate the spectrum at some minimum size. However, this effect is largely masked by the further processing of rain after it leaves the cloud. In particular, small drops continue to be produced by breakup and evaporation. Some of these are consumed by the latter process, while others are collected by larger drops. Also, near the beginning of a rainshower, the drop spectrum at ground level may be expected to be biased toward large sizes owing to the greater fall speeds of the larger drops, and possibly toward small sizes owing to an initially high evaporation rate. The overall shaping of the spectrum is obviously quite complicated, and determined in part by such meteorological variables as temperature, relative humidity, and wind in the subcloud region. Observations show that most precipitating drops which reach the ground have $D_0 > 200\ \mu m$.

Various empirical equations have been advanced to describe the size spectra of raindrops. One often used is the size distribution proposed by Best (1950a), which has essentially the same form as (2-1):

$$1 - F = exp\ [-(D_0/A)^{2.25}], \qquad (2\text{-}10)$$

where $A = 1.30\ R^{0.232}$, with R in $mm\ hr^{-1}$ and D_0 in mm, and where F is the fraction of water comprised of raindrops with equivalent diameters smaller than D_0. Probably the most widely used description is that of Marshall and Palmer (1948), which is based on the observations of Laws and Parsons (1943). The Marshall-Palmer (MP) distribution is

$$n(D_0) = n_0 exp\ (-\Lambda D_0), \qquad (2\text{-}11)$$

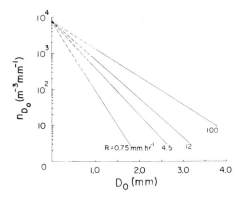

Fig. 2-18. Marshall-Palmer raindrop size distribution. (Based on Marshall and Palmer, 1948; by
courtesy of Amer. Meteor. Soc., and the authors.)

where $\Lambda = 4.1\,R^{-0.21}$ mm^{-1}, and $n_0 = 8 \times 10^3$ m^{-3} mm^{-1}. The parameter n_0 is
obtained by extrapolation and is assumed to be a constant (Figure 2-18).

Subsequent more detailed studies, including those by Blanchard (1953), Okita
(1958), Mason and Andrews (1960), Caton (1966), and Blanchard and Spencer
(1970) have demonstrated that the MP distribution is not sufficiently general to describe most observed raindrop spectra accurately. In particular, Joss *et al.* (1968),
Joss and Waldvogel (1969), Strantz (1971), Diem and Strantz (1971), Sekhorn and
Srivastava (1971), Czerwinski and Pfisterer (1972), and Waldvogel (1974) have
pointed out that n_0 cannot be considered a constant, but rather is a function of R.

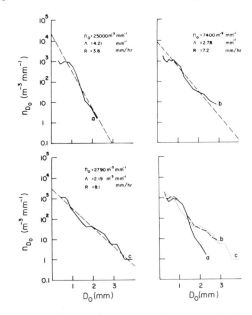

Fig. 2-19. Raindrop spectra: (a) before, (b) during, and (c) after a 'n_0-jump' measured during the
rainfall on September 18, 1969, at a station in southern Switzerland. (From Waldvogel, 1974; by
courtesy of Amer. Meteor. Soc., and the author.)

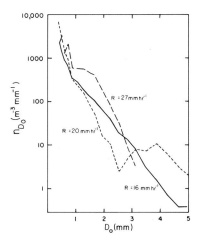

Fig. 2-20. Raindrop size distribution observed on three different occasions in southern Switzerland. (From Joos *et al.*, 1968; by courtesy of Amer. Meteor. Soc., and the authors.)

Also, the functional dependence of Λ on R varies. On analysis of 46 rain spectra obtained in southern Switzerland with $0.3 \leqslant D_0 \leqslant 5.3$ mm, Joss *et al.* (1968) and Joss and Waldvogel (1969) found $3 \times 10^3 \leqslant n_0 \leqslant 10^5$ m^{-3} mm^{-1}. Also, Waldvogel (1974) discovered that during a particular rainfall, n_0 may suddenly change. Figure 2-19 gives an example of the variation of Λ and the raindrop distribution before, during, and after a sudden change in n_0 (termed 'n_0-jump' by Waldvogel, 1974). The changes of n_0 were found to be related to changes in convective activity, i.e., air mass stability. However, 'n_0-jumps' were observed even during rainfalls of the same convective character with a continuous rainfall rate. This behavior must be attributed to changes in the microphysical processes occurring in the cloud system from which the rain fell, or in the air during the fall of the drops from cloud to ground. Examples of such 'n_0-jumps' are given in Table 2-1.

Although n_0 and Λ varied considerably within each rainfall, and from one rainfall to another, at any particular moment the rain size distributions observed over southern Switzerland could in many cases be approximated fairly well by an exponential distribution of the MP type (Figure 2-20, $R = 16$ mm hr^{-1}). Exponential type raindrop spectra were also observed by Okita (1958) in Japan

TABLE 2-1

Examples for the variation of n_0, Λ, and R during different types of rain. CET is Central European Time. (Based on data of Waldvogel, 1974.)

	Type of Rainfall	n_0 m^{-3} mm^{-1}	Λ mm^{-1}	R mm hr^{-1}
June 6, 1968,	2205–2235 CET, thunderstorm	35 000	3.7	10.2
	2235–2310 CET, thunderstorm	4 000	2.5	5.8
June 19, 1969,	0510–0540 CET, shower	16 000	3.8	4.0
	0550–0620 CET, widespread rain	8 000	2.6	8.0

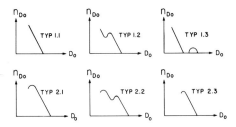

Fig. 2-21. Schematic representation of raindrop spectra according to the classification of Strantz. The plots are in terms of log n_{D_0} vs. log D_0. (From Czerwinski and Pfisterer, 1972; by courtesy of *J. de Rech. Atmos.*, and the authors.)

and by Sekhorn and Srivastava (1971) during thunderstorm rains near Cambridge, Massachusetts. In addition, the data of Sekhorn and Srivastava allowed relating n_0 and Λ in a fixed manner to the rainfall intensity R as $n_0 = 7 \times 10^3 R^{0.37}$ m^{-3} mm^{-1} and $\Lambda = 3.8 R^{-0.14}$ mm^{-1}. Similarly, Blanchard and Spencer (1970) found, from their compilation of rain data from many parts of the world, that rain spectra observed during very intense rainfalls could be well fitted to an exponential distribution for $D_0 \gtrsim 2$ mm. In other cases observed rain spectra have deviated considerably from an exponential type (see Figure 2-20). This variability has been pointed out especially by Diem (1968), Diem and Strantz (1971), Strantz (1971), and Czerwinski and Pfisterer (1972), who together

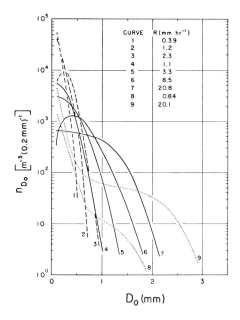

Fig. 2-22. Raindrop distributions at stations in Hawaii. Curves 1–3 are for measurements made at or near dissipating edges of non-freezing orographic clouds. Curves 4–7 represent data taken at cloud base. Curves 8–9 are for non-orographic rains. (From Blanchard, 1953; by courtesy of Amer. Meteor. Soc., and the author.)

analyzed 10,650 raindrop spectra obtained from observations at 8 locations between 80° N and 3° S in different climatological regions. However, Strantz showed that, despite the diversity of meteorological and microphysical conditions which led to observed rainfalls, most spectra could be classified within two groups of 3 types each. The 6 types are shown schematically in Figure 2-21 (note only type 1.1 conforms strictly to the MP distribution). Although a third group was needed for all spectra not fitting into groups 1 and 2, it comprised only 0.4% of the cases observed.

In addition to the parameters n_0 and Λ, the liquid water content w_L of rains can usually be expressed as a function of the rainfall intensity R according to a law of the form $w_l = AR^b$. Measurements at various locations have shown that A and b are usually quite uniform, with $0.052 \leqslant A \leqslant 0.089$ and $0.84 \leqslant b \leqslant 0.94$. However, in contrast, Blanchard (1953) found that the parameters which correlate w_L to R show marked differences between orographic rains originating in clouds warmer than 0 °C, and non-orographic rains. Spectra of orographic rains were much narrower (Figure 2-22), and their liquid water contents were considerably higher. In addition, Blanchard (1953) and Okita (1958) pointed out that w_L varies with distance from the cloud base, being higher just below the base than at the ground. They attributed this behavior to the existence of a much larger number of small drops at cloud base than at the ground, the drop depletion being caused by collision and coalescence and by evaporation.

2.2 Microstructure of Cloud and Precipitation Consisting of Ice Particles

If the temperature in a cloud decreases below 0 °C, ice particles may form. As we have discussed previously, the transformation to ice does not take place readily, the ice phase becoming frequently observed only as cloud temperatures approach −20 °C (Figure 2-1). At temperatures where supercooled water drops coexist with ice particles, the latter grow at the expense of the former. This may occur by water vapor diffusion to the ice particles, and/or by drops colliding with and freezing on the ice particles. The first mechanism is called *deposition* and the second, *riming*. Ice particles which have grown by deposition only are called *ice* or *snow crystals*. Ice crystals may also grow by collision with other crystals; this mechanism is often referred to as *clumping*. Aggregates of snow crystals are called *snowflakes*.

The terminology of ice particles formed as a result of riming is considerably less precise and has not been generally accepted. In the initial stages of riming, as long as the features of the original ice crystal are still well distinguishable, the ice particle is simply called a lightly or densely *rimed snow crystal*. When riming of an ice particle has proceeded to the stage where the features of the primary ice particle are only faintly or no longer visible, the ice particle is called a *graupel particle*, a *soft hail particle*, or a *snow pellet*. Such a particle has a white, opaque, and fluffy appearance due to the presence of a large number of air capillaries in the ice structure. It usually has a bulk density of less than $0.8\,\mathrm{g\,cm^{-3}}$ (List, 1958a, b; 1965). In the later stages of riming such particles may have a conical, rounded, or irregular shape. An ice particle is called a *small-hail*

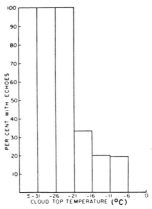

Fig. 2-23. Percentage of cumulus clouds which contained precipitation echoes; observed over semi-arid Arizona. (From Morris, 1957; by courtesy of Amer. Meteor. Soc., and the author.)

particle or *type-b ice pellet* if it has originated as a frozen drop or ice crystal and has grown by riming to an irregular or roundish, semi-transparent particle (with or without a conical tip) of bulk density 0.8 to 0.99 g cm^{-3} (List, 1958a, b; 1965). Such a particle may contain water in its capillary system. Hard, transparent, globular, or irregular ice particles consisting of frozen drops, or partially melted and subsequently refrozen snow crystals or snowflakes with bulk densities between the density of ice and 0.99 g cm^{-3} are called *type-a ice pellets* or *sleet* (List, 1958a, b; 1965). Such particles may also contain unfrozen water.

Unrimed, single ice crystals usually have maximum dimensions less than 5 mm. Snowflakes may have maximum dimensions up to several centimeters, but they are usually less than 2 cm. Rimed ice crystals, graupel particles, and ice pellets usually have maximum dimensions less than 5 mm. Ice particles grown by riming are called *hailstones* if their maximum dimensions are typically larger than 5 mm. They have a roundish, ellipsoidal, or conical shape often with lobes, knobs, or other proturerances on the surface. They are partially or completely opaque, and in cross section exhibit an onion-type layered structure with alternating opaque and clear layers caused by the presence of more or less numerous air bubbles (List, 1958a, b, 1965).

Since radar echoes indicate the presence of large cloud or precipitation size particles, and since these usually form once the temperature in a cloud is sufficiently low, one would expect the probability of a radar echo to be related to temperature. Indeed, in numerous clouds (Figure 2-23) the probability of an echo is often small as long as the cloud top temperature is warmer than or only a few degrees below 0 °C. The probability then becomes much larger once the cloud top reaches −20 °C, the temperature at which most clouds contain ice particles.

2.2.1   SHAPE, DIMENSIONS, BULK DENSITY, AND NUMBER CONCENTRATION
         OF ICE CRYSTALS

Casual observation shows that snow crystals appear in a large variety of shapes or "habits." More detailed studies, however, reveal that from a crystallographic point of view, snow crystals have one common basic shape, namely that of a six-fold symmetric (hexagonal) prism with two basal planes of type (0001) and 6 prism planes of type (10$\bar{1}$0)*. Crystal planes of the type (11$\bar{2}$0), which would contribute to a dodecagonal shape, are metastable and occur very rarely. Crystal faces of the type (10$\bar{1}$1), which would contribute to a pyramidal shape, also are metastable and rarely appear. This is also the case with faces of the type (10$\bar{1}$2). The habit of a crystal is determined by the slowest growing faces. Metastable faces, such as (11$\bar{2}$0), (10$\bar{1}$1), and (10$\bar{1}$2), grow quickly to become the crystal's edges and corners, while faces of the type (0001) and (10$\bar{1}$0) grow slowly and become the bounding faces of the crystal.

Laboratory experiments reveal that the rate of propagation of the basal faces (growth along c-axis), relative to that of prism faces (growth along the crystallo-graphic direction of type [10$\bar{1}$0], varies with temperature and supersaturation in a characteristic manner (Aufm. Kampe *et al.*, 1951; Nakaya, 1954; Mason and Shaw, 1955; Kobayashi, 1957, 1958; Hallett and Mason, 1958). The results of these studies were consolidated by Kobayashi (1961) and by Rottner and Vali (1974). The experimental findings are summarized in Figures 2-25a, b, where the variation of the ice crystal shape is given as a function of temperature and supersaturation, and as a function of temperature and water vapor density in excess of that at ice staturation. At a large vapor density excess, the ice crystal shape changes with decreasing temperature from a plate to a needle, to a column, to a sector plate, to a dendrite, back to a sector plate, and finally back to

* In Figure 2-24 the orientation of crystal lattice planes is described in terms of the intersections of the planes with the hexagonal axes a_1, a_2, a_3, and c. By convention, the reciprocals of these intersections are replaced by the smallest integers having the same ratio (the atomic structure of crystals insures that this can always be done), and the integers are placed in order within brackets to complete the orientation description. A bar over a number indicates an intersection in a negative axial direction.

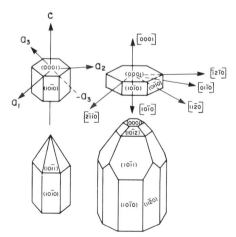

Fig. 2-24. Schematic representation of different habits of snow crystals. (Based on Wolff, 1957, with changes.)

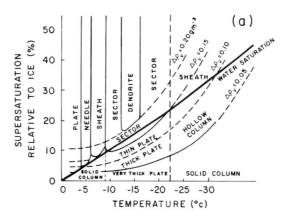

Fig. 2-25a. Variation of ice crystal habit with temperature and supersaturation. (Based on laboratory observations of Mason, 1971; Hallett and Mason, 1958; Kobayashi, 1961; and Weissweiler,
1969.)

a column. This cyclic plate-column-plate-column change in habit is due to a
cyclic change of the preferential growth direction along the crystallographic
directions of type [10$\bar{1}$0] and [0001], the changes occurring at temperatures near
$-4\,°C$, $-9\,°C$, and $-22\,°C$. While the former two transition temperatures are
rather sharply defined, the latter is diffuse, i.e., habit change may take place in a
temperature range of several degrees, centered around $-22\,°C$. In contrast, at
very low vapor density excess, the crystal shape changes between a short
column and a thick plate near $-9\,°C$ and $-22\,°C$. Close to or at ice saturation, the
ice crystal shape ceases to vary with temperature but rather assumes the
equilibrium shape, which is a thick hexagonal plate with a height to diameter
ratio of 0.81 (see Section 5.7.2).

Thus, we see that although the temperature is the principal factor, humidity
conditions in the environment also control the important growth features of

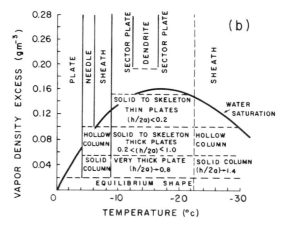

Fig. 2-25b. Variation of ice crystal habit with temperature and vapor density excess. (Based on
laboratory observations of Kobayashi, 1961; and Rottner and Vali, 1974.)

snow crystals. For example, near $-15\,°C$, the snow crystal habit varies with increasing vapor density excess from a thick plate to a thin plate, to a sector plate, and finally to a dendrite for which preferential growth is along the crystallographic direction of type $[11\bar{2}0]$. Near $-5\,°C$, the ice crystal habit varies with increasing vapor density excess from a short solid column, to a hollow column, to a needle with pronounced growth in the crystallographic direction $[0001]$. Laboratory observations of the ice crystal habit at temperatures between -22 and $-50\,°C$ have revealed no essentially new habit features (Kobayashi, 1965a, b). Long solid columns (sheaths) appear between -45 and $-50\,°C$ at low supersaturations, and change into hollow columns as the supersaturation is raised.

Although the basic shape of ice crystals is hexagonal prismatic, laboratory observations have revealed a few snow crystals with other shapes. Trigonal prismatic plates and columns, trigonal dendrites, and rhombic and scalene pentagonal ice crystals were observed by Yamashita (1969, 1971, 1973) after seeding supercooled clouds with a very cold body. Aufm. Kampe et al. (1951) and Mason (1953) also observed trigonal plates after seeding supercooled clouds with dry ice. Little is known of the detailed growth conditions of such rare ice crystal shapes. Ohtake (1970a, b) suspected that quasi-stable faces such as the pyramidal faces of type $(10\bar{1}2)$ or $(10\bar{1}1)$ may develop at rapid cooling rates when the time to complete a quasi-stable crystal face becomes comparable to the time for the completion of a stable face. Kobayashi (1965) found that at temperatures between -50 and $-90\,°C$ pyramidal faces develop at the tip of prismatic columns.

So far we have only considered ice crystals grown in the laboratory. It is essential to ask whether natural snow crystals exhibit the same characteristic changes in shape. Considerable uncertainties are involved in answering this question, due to the inherent difficulties of accurately establishing the actual temperature and humidity conditions of the locations at which the sampled snow crystals grew and acquired their shape. However, a large number of observations in different parts of the world finally have made a fairly definite conclusion possible. (Cloud observations have been made over Germany by Weickmann (1945, 1949, 1957a) and Grunow (1960); over Canada by Gold and Power (1952, 1954); over the U.S.S.R. by Bashkirova and Pershina (1956, 1964); over Japan by Magono (1960), Nakaya and Higuchi (1960), Higuchi (1962a, b, c), Lee and Magono (1967), Magono et al. (1959, 1960, 1962, 1963, 1965, 1966), and Tazawa and Magono (1973); over Australia by Ono (1970); over Colorado and the Great Lakes region by Jiusto and Weickmann (1973), and Weickmann (1972); and over the Pacific northwestern U.S. by Hobbs et al. (1971a, 1972, 1974a).)

Observations prior to 1966 have been summarized in a diagram (Figure 2-26) prepared by Magono and Lee (1966). Observations made after 1966 generally have supported the Magono-Lee diagram. Comparison between Figures 2-25a, b and Figure 2-26 shows that laboratory experiments are in basic agreement with the Magono-Lee diagram.

The outstandingly beautiful photographs of snow crystals captured at the ground by Bentley (Bentley and Humphreys, 1931, 1962), Nakaya (1954), and

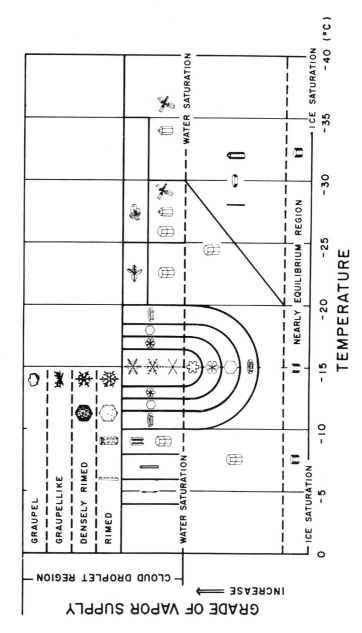

Fig. 2-26. Temperature and humidity conditions for the growth of natural snow crystals of various types. (From Magono and Lee, 1966; by courtesy of *J. Fac. Sci.*, Hokkaido University.)

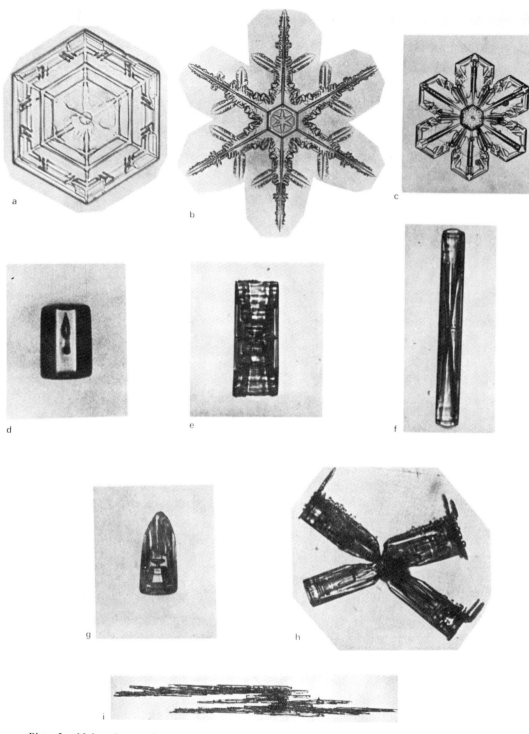

Plate 2. Major shapes of snow crystals: (a) simple plate, (b) dendrite, (c) crystal with broad branches, (d) solid column, (e) hollow column, (f) sheath, (g) bullet, (h) combination of bullets (rosette, Prismenbüschel), (i) combination of needles. (From Nakaya, 1954; by courtesy of Harvard University Press, copyright 1954 by the President and Fellows of Harvard College.)

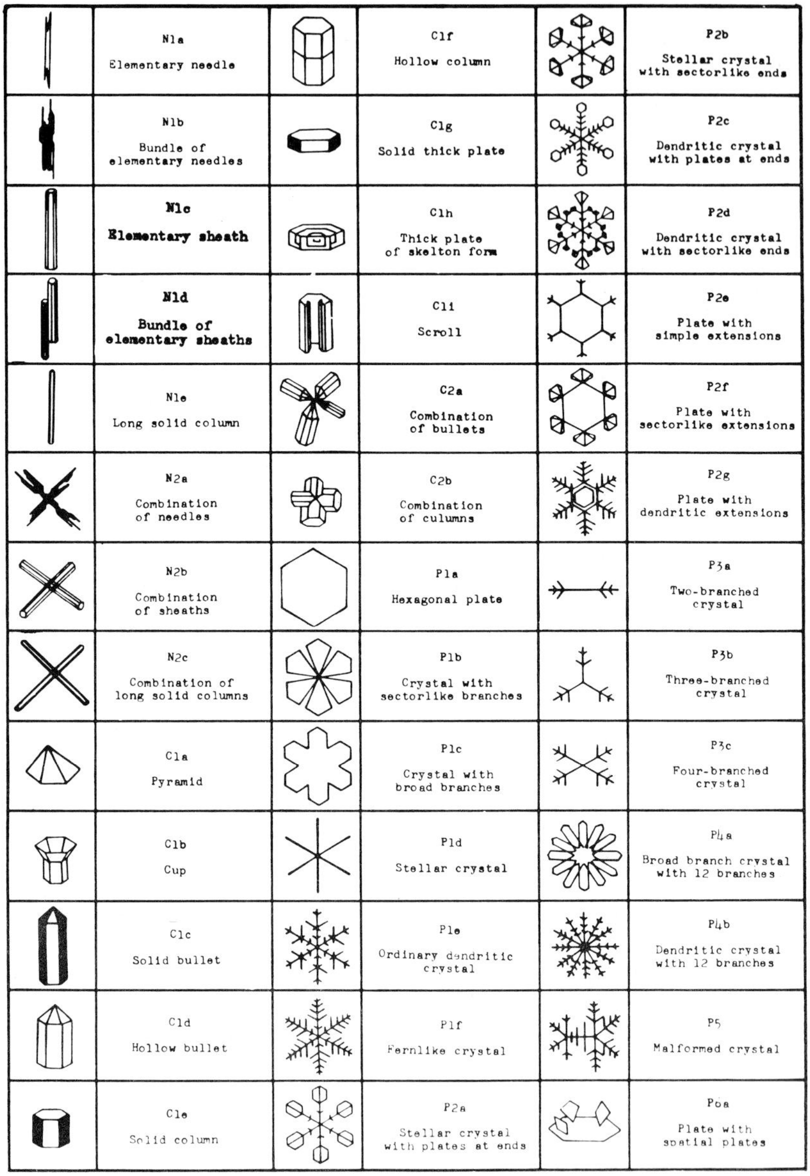

	N1a Elementary needle		**C1f** Hollow column		**P2b** Stellar crystal with sectorlike ends		
	N1b Bundle of elementary needles		**C1g** Solid thick plate		**P2c** Dendritic crystal with plates at ends		
	N1c **Elementary sheath**		**C1h** Thick plate of skelton form		**P2d** Dendritic crystal with sectorlike ends		
	N1d **Bundle of** **elementary sheaths**		**C1i** Scroll		**P2e** Plate with simple extensions		
	N1e Long solid column		**C2a** Combination of bullets		**P2f** Plate with sectorlike extensions		
	N2a Combination of needles		**C2b** Combination of columns		**P2g** Plate with dendritic extensions		
	N2b Combination of sheaths		**P1a** Hexagonal plate		**P3a** Two-branched crystal		
	N2c Combination of long solid columns		**P1b** Crystal with sectorlike branches		**P3b** Three-branched crystal		
	C1a Pyramid		**P1c** Crystal with broad branches		**P3c** Four-branched crystal		
	C1b Cup		**P1d** Stellar crystal		**P4a** Broad branch crystal with 12 branches		
	C1c Solid bullet		**P1e** Ordinary dendritic crystal		**P4b** Dendritic crystal with 12 branches		
	C1d Hollow bullet		**P1f** Fernlike crystal		**P5** Malformed crystal		
	C1e Solid column		**P2a** Stellar crystal with plates at ends		**P6a** Plate with spatial plates		

Fig. 2-27.   The Magono-Lee classification of natural snow crystals. (From Magono and Lee, 1966;
by courtesy of *J. Fac. Sci.*, Hokkaido University.)

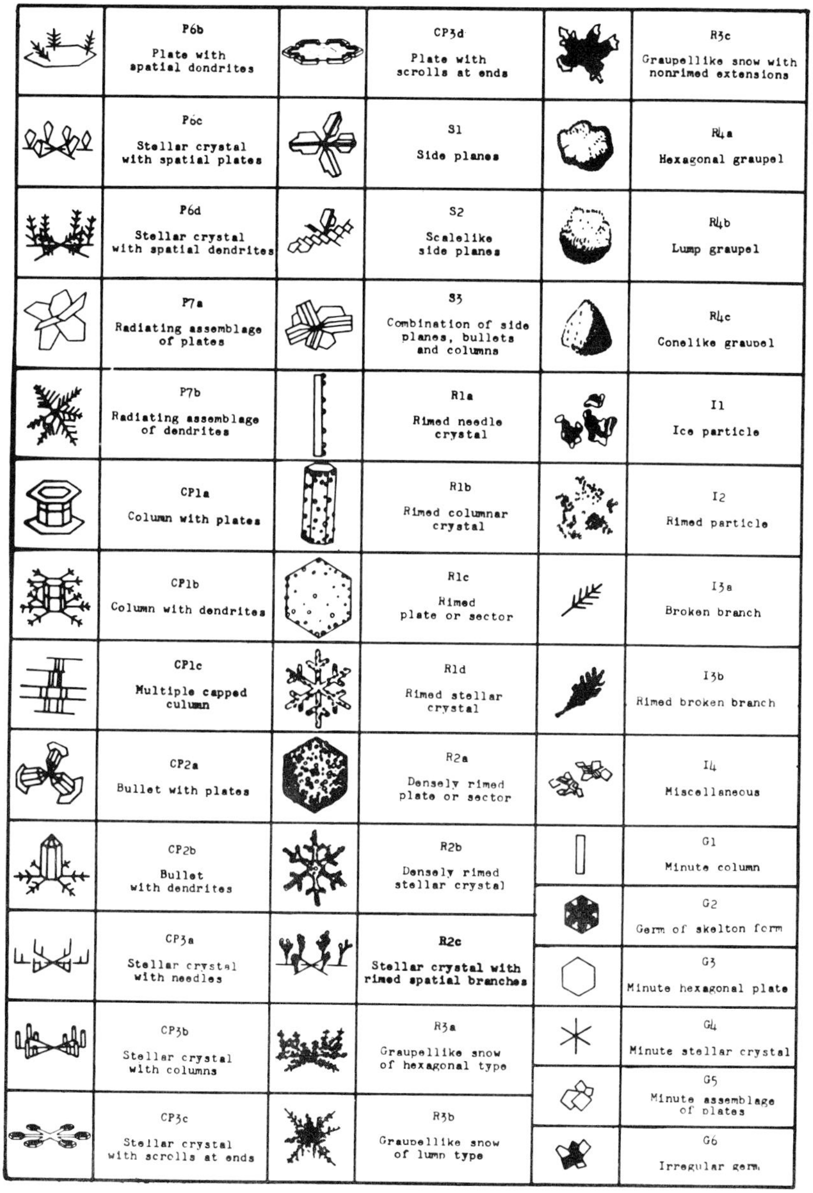

	P6b Plate with spatial dendrites		**CP3d** Plate with scrolls at ends		**R3c** Graupellike snow with nonrimed extensions
	P6c Stellar crystal with spatial plates		**S1** Side planes		**R4a** Hexagonal graupel
	P6d Stellar crystal with spatial dendrites		**S2** Scalelike side planes		**R4b** Lump graupel
	P7a Radiating assemblage of plates		**S3** Combination of side planes, bullets and columns		**R4c** Conelike graupel
	P7b Radiating assemblage of dendrites		**R1a** Rimed needle crystal		**I1** Ice particle
	CP1a Column with plates		**R1b** Rimed columnar crystal		**I2** Rimed particle
	CP1b Column with dendrites		**R1c** Rimed plate or sector		**I3a** Broken branch
	CP1c Multiple capped column		**R1d** Rimed stellar crystal		**I3b** Rimed broken branch
	CP2a Bullet with plates		**R2a** Densely rimed plate or sector		**I4** Miscellaneous
	CP2b Bullet with dendrites		**R2b** Densely rimed stellar crystal		**G1** Minute column
	CP3a Stellar crystal with needles		**R2c** Stellar crystal with rimed spatial branches		**G2** Germ of skelton form
	CP3b Stellar crystal with columns		**R3a** Graupellike snow of hexagonal type		**G3** Minute hexagonal plate
					G4 Minute stellar crystal
	CP3c Stellar crystal with scrolls at ends		**R3b** Graupellike snow of lump type		**G5** Minute assemblage of plates
					G6 Irregular germ

Fig. 2-27. (continued.).

Magono and Lee (1966), and of snow crystals captured during flights in cirrus clouds by Weickmann (1945), provide a comprehensive atlas of most snow crystal types found in atmospheric clouds. An attempt to bring order into this multiplicity of crystal forms has been made by Magono and Lee (1966). Although their classification (Figure 2-27) has not yet been formally accepted on an international basis, it has been found practical and is very widely used. Photographs of a few major snow crystal shapes are given in Plate 2.

Let us assume now that a snow crystal of one particular habit, formed by growth at a particular temperature and humidity, is suddenly moved into a new environment of different temperature and humidity where it continues to grow by vapor diffusion. Under such conditions the habit characteristic of the second temperature and humidity conditions becomes superimposed on the original habit. Thus, a columnar snow crystal suddenly surrounded by conditions characteristic of plate-like growth will develop end-plates (Figure 2-27, CP1a). A stellar crystal suddenly surrounded by conditions characteristic of needle growth will develop needles on the branches, with the needles growing perpendicular to the plane of the crystal (Figure 2-27, CP3a). Although such snow crystals appear as combinations of different shapes, from a crystallographic point of view they are still *single ice crystals* since the crystallographic orientation of the c- and a-axes is still the same throughout the crystal.

While some ice particles in clouds originate on water-insoluble aerosol particles on which water vapor is deposited as ice, others originate as frozen drops. Various observers (e.g., Koenig, 1963; Braham, 1964) have studied frozen drops in atmospheric clouds. They are irregular in shape, often with bulges and protrusions formed during the freezing process. Ice particles formed from single-crystalline frozen drops are likely to turn into two-layered crystals (Figure 2-28). The conditions for the formation of such crystals were studied by Auer (1970, 1971, 1972a), Weickmann (1972), Jiusto and Weickmann (1973), and Parungo and Weickmann (1973), who found that ice crystals with frozen drops in their centers were quite abundant. Auer's cloud studies indicated that at temperatures from −9 to −10 °C, about 19% of the total snow crystal concentration could be attributed to crystals each with a frozen drop at the center. At −15 to −16 °C this fraction reached a maximum of 48%, and decreased to about 23% at temperatures between −21 and −22 °C. The diameter of the frozen center-drop was found to range between 2.5 and 25 μm.

Drops which freeze polycrystalline and subsequently continue to grow by vapor diffusion form *spatial crystals* (Higuchi and Yoshida, 1967; Magono and Suzuki, 1967; Lee, 1972; Kikuchi and Ishimoto, 1974). Figure 2-29 illustrates how, in principle, such a crystal develops. If, for example, such a frozen drop continues to grow by vapor diffusion near −15 °C, dendritic branches will emerge from the frozen drop at various angles (Figure 2-27, P7b). Since polycrystalline drops are more likely to occur at lower temperatures, they frequently develop into a combination of columnar crystals. Note that, due to competition for water vapor, the columns may have a conical or pyramidal shape (bullet-shape) pointing towards their common growth center which is the frozen drop (Figure 2-27, C2a). A snow crystal of this form is called a *combination of*

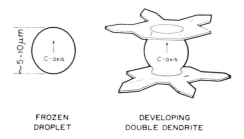

FROZEN   DEVELOPING
DROPLET   DOUBLE DENDRITE

Fig. 2-28. Schematic drawing indicating formation of double star from a single-crystalline frozen drop. Note competition for vapor causes irregular growth of opposing branches of the two crystals. (From Jiusto and Weickmann, 1973; by courtesy of Amer. Meteor. Soc., and the authors.)

bullets or *rosette*. A single column broken off a rosette is simply called a *bullet* (Figure 2-27, C1c, C1d). A rosette may consist of 2 to 9 bullets but most frequently consist of 3 to 4 bullets (Kikuchi, 1968).

Spatial crystals may also develop as a result of supercooled drops colliding with a snow crystal. At temperatures of only a few degrees below 0 °C, a drop colliding with a snow crystal turns into an ice-single-crystal with a crystallographic orientation which may be the same or different from the snow crystal it contacts. If the temperature is sufficiently cold, the colliding drop may turn into a polycrystalline mass of ice. Further growth of such polycrystalline frozen drops by vapor diffusion leads to a spatial snow crystal with two or more c-axis orientations (Figure 2-27, P6a–d, P7a).

Several peculiar snow crystal shapes not classified by Magono and Lee were encountered by Kikuchi (1970), Kikuchi and Yanai (1971), and Magono *et al.* (1971) at temperatures between −26 and −30 °C during an Antarctic expedition; by Thuman and Robinson (1954), Kumai (1965, 1966a, 1969a), and Ohtake (1967, 1968, 1970a, b) at temperatures between −30 and −55 °C during ice fog in Alaska; and by Itoo (1957) during a strong ground inversion at a station in central Mongolia. The ice crystals observed at the Antarctic station consisted mostly of combinations of bullets, columns, and side planes which very likely originated as a type of hoarfrost snow-covered surfaces. In ice fog, Thuman and Robinson observed irregular particles of 'block-shape' and polyhedral particles bounded by trapezoidal faces which were portions of a hexagonal bi-pyramid. Pyramidal planes of type (10$\bar{1}$1) and higher order planes were also observed by Itoo in 'diamond-dust' snow crystals.

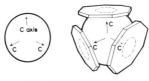

DEVELOPING SPATIAL DENDRITE

Fig. 2-29. Schematic drawing indicating formation of a spatial-dendrite from a polycrystalline frozen drop. (From Jiusto and Weickmann, 1973; by courtesy of Amer. Meteor. Soc., and the authors.)

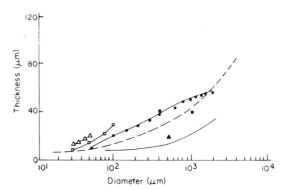

Fig. 2-30.   Dimensions of hexagonal ice plates observed in clouds. ■ Schaefer (1947), ▲ Weickmann (1949), □ Reynolds (1952), △ Mason (1953), ● Ono (1969), – – – Auer and Veal (1970), —— Hobbs *et al.* (1974a). (From Hobbs *et al.*, 1974a; by courtesy of the authors; copyrighted by American Geophysical Union.)

The size of a snow crystal can usually be characterized by two dimensions: the crystal diameter (d) and the crystal thickness (h) in the case of plate-like crystals, and the crystal length (L) and the crystal width (d) for columnar type crystals. Detailed measurements of snow crystal dimensions have been carried out in several locations (Figures 2-30, 2-31, 2-32). The length of columnar crystals and the diameter of plate-like crystals were found to range typically between 10 μm and 1 mm. Maximum dimensions reached several millimeters (Figure 2-33). The thickness to diameter ratio for simple plates, dendritic crystals, and stellar crystals was $0.01 \le (h/d) \le 0.2$, and for 'thick' plates, $0.2 \le (h/d) \le 0.5$. The width to length ratio was $0.10 \le (d/L) \le 0.77$ for columns, and $0.05 \le (d/L) \le 0.33$ for needles.

Observations have shown further that the thickness and diameter of plate-like crystals, and the length and width of columnar crystals, are characteristically

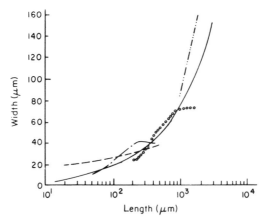

Fig. 2-31.   Dimensions of ice needles observed in clouds. –··– Magono (1954), – – – Isono (1959), –·– Ono (1969), —— Auer and Veal (1970), ○○○○ Hobbs *et al.* (1974a). (From Hobbs *et al.*, 1974a; by courtesy of the authors; copyrighted by American Geophysical Union.)

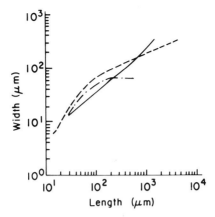

Fig. 2-32. Dimensions of warm region ice columns observed in clouds. –·– Ono (1969), – – – Auer and Veal (1970), —— Hobbs *et al.* (1974a). (From Hobbs *et al.*, 1974a; by courtesy of the authors; copyrighted by American Geophysical Union.)

related to each other. The significant fact is that a snow crystal, growing by vapor diffusion, distributes its mass in a fairly predictable manner which obeys certain dimensional relations. A comparison shows (Figures 2-30, 2-31, 2-32) that, for a particular crystal type, the dimensional relationships proposed by various authors agree reasonably well, though they were derived from observations in clouds over different parts of the world. Davis (1974) combined observations of his own with those of Auer and Veal (1970), Ono (1969, 1970), and Kajikawa (1972, 1973) to the best fit relationships given in Table 2-2a in terms of a set of power laws. Size relationships for additional snow crystal shapes are given in Table 2-2b.

Most ice crystals have a bulk density less than that of bulk ice. This is due to small amounts of air in capillary spaces, and to the tendency of snow crystals to grow in a skeletal fashion. In particular, columnar crystals often develop as hollow crystals with 'hour-glass' air spaces at either end. Heymsfield (1972), on combining his data with that of Ono (1970), determined relations between the bulk density and the crystal dimensions (Table 2-3). Somewhat lower bulk

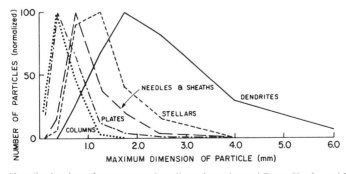

Fig. 2-33. Size distribution of snow crystals collected at Alpental Base, Hyak, and Keechelus Dam, State of Washington. (From Hobbs *et al.*, 1972; by courtesy of the authors.)

TABLE 2-2a

Dimensional relationships for various ice crystal types; d(cm), L(cm), A_B(cm^2), V_c(cm^3). The form Plc-r refers to a simple 'daisy-type' dendrite; the form Plc-s refers to a dendrite with sector-type branches; d refers to the diameter of the circle circumscribed around the snowflake. (From C. E. Davis, 1974; by courtesy of the author.)

Crystal type	Dimensional relationship, cm	Area of basal face, A_B, cm^2	Volume of crystal, V_c (cm^3)	Range of major axis μm
Pla	$h = 1.41 \times 10^{-2} d^{0.474}$	$A_B = 0.65 d^2$	$V_c = 9.17 \times 10^{-3} d^{2.475}$	10–3000
Plb	$h = 1.05 \times 10^{-2} d^{0.423}$	$A_B = 0.65 d^2$	$V_c = 6.79 \times 10^{-3} d^{2.423}$	10–40
Plb	$h = 1.05 \times 10^{-2} d^{0.423}$	$A_B = 0.55 d^{1.97}$	$V_c = 7.37 \times 10^{-3} d^{2.420}$	41–2000
Plc-r, Pld	$h = 9.96 \times 10^{-3} d^{0.415}$	$A_B = 0.65 d^2$	$V_c = 6.47 \times 10^{-3} d^{2.415}$	10–90
Plc-r, Pld	$h = 9.96 \times 10^{-3} d^{0.415}$	$A_B = 0.11 d^{1.63}$	$V_c = 1.096 \times 10^{-3} d^{2.045}$	91–1500
Plc-s	$h = 9.96 \times 10^{-3} d^{0.415}$	$A_B = 0.65 d^2$	$V_c = 6.47 \times 10^{-3} d^{2.415}$	10–100
Plc-s	$h = 9.96 \times 10^{-3} d^{0.415}$	$A_B = 0.21 d^{1.76}$	$V_c = 2.09 \times 10^{-3} d^{2.175}$	101–1000
Clg	$h = 0.138 d^{0.778}$	$A_B = 0.65 d^2$	$V_c = 8.97 \times 10^{-2} d^{2.778}$	10–1000
Cle $L/d \leqslant 2$	$d = 0.578 L^{0.958}$	$A_B = 0.65 d^2$	$V_c = 0.217 L^{2.916}$	10–1000
Cle $L/d > 2$	$d = 0.260 L^{0.927}$	$A_B = 0.65 d^2$	$V_c = 4.39 \times 10^{-2} L^{2.854}$	10–1000
Clf $L/d \leqslant 2$	$d = 0.422 L^{0.892}$	$A_B = 0.65 d^2$	$V_c = 0.116 L^{2.784}$	10–50
Clf $L/d \leqslant 2$	$d = 0.422 L^{0.892}$	$A_B = 0.65 d^2$	$V_c = 0.105 L^{2.765}$	51–1000
Clf $L/d > 2$	$d = 0.263 L^{0.930}$	$A_B = 0.65 d^2$	$V_c = 4.49 \times 10^{-2} L^{2.860}$	10–50
Clf $L/d > 2$	$d = 0.263 L^{0.930}$	$A_B = 0.65 d^2$	$V_c = 4.06 \times 10^{-2} L^{2.841}$	51–1000

TABLE 2-2b

Crystal type	Dimensional relationship	Author
Ple, Plf, P2c P2g, P3c, P4b	$h = 9.022 \times 10^{-3} d^{0.377}$	Auer and Veal (1970)
Nla	$d = 3.0487 \times 10^{-2} L^{0.61078}$	Auer and Veal (1970)
Nle	$d = 3.527 \times 10^{-2} L^{0.437}$	Jayaweera and Ohtake (1974)
Clc ($L \leqslant 0.3$ mm)	$d = 0.1526 L^{0.7856}$	Heymsfield (1972)
Cld ($L \geqslant 0.3$ mm)	$d = 0.0630 L^{0.532}$	Heymsfield (1972)

densities for columnar crystals were observed by Iwai (1973) and Jayaweera and Ohtake (1974). They found that short columns had bulk densities close to that of ice. With increasing L, however, the density decreased rapidly, reaching $\rho_c \approx 0.5$ for $L \approx 1$ mm. For needles and sheaths, they found $\rho_c \approx 0.3$ to 0.4 g cm^{-3} if $L > 1$ mm. Table 2-3 also implies that larger dimensions correlate with lower bulk densities.

Since the probability of the occurrence of the ice phase in clouds increases with decreasing temperature, we might expect a monotonic rise in the concentration of such particles with decreasing temperature. This behavior turns out to hold only in the minority of cases. More often, a rapid phase change to ice (*glaciation*) occurs, such that the ice particle concentration is not a sensitive function of further temperature lowering. We shall discuss this more fully in Sections 9.2.1 and 16.6. At this point we merely note the net effect of glaciation. Figure 2-34 summarizes measurements of Hobbs *et al.* (1974b) in clouds over the Cascade Mts. (Wash.). It is seen that at temperatures between -4 and $-25\,°$C the

TABLE 2-3
Bulk density of various snow crystals (d, L in mm). (Based on data of
Heymsfield, 1972.)

Crystal type	Bulk density, ρ_c, (g cm^{-3})
hexagonal plate	$\rho_c = 0.9$
plates with dendritic extensions	$\rho_c = 0.656 \, d^{-0.627}$ ($d \geqslant 0.7$ mm)
dendrites	$\rho_c = 0.588 \, d^{-0.377}$ ($d \geqslant 0.3$ mm)
stellar, broad arms	$\rho_c = 0.588 \, d^{-0.377}$ ($d \geqslant 0.3$ mm)
stellar, narrow arms	$\rho_c = 0.46 \, d^{-0.482}$ ($d \geqslant 0.24$ mm)
column, cold region	$\rho_c = 0.65 \, L^{-0.0915}$ (L $\geqslant 0.028$ mm)
column, warm region	$\rho_c = 0.848 \, L^{-0.014}$ (L $\geqslant 0.014$ mm)
bullet	$\rho_c = 0.78 \, L^{-0.0038}$ (L $\geqslant 0.1$ mm)

range of number concentrations of snow crystals varies little with temperature on
the average, and that the concentrations may reach values as high as 10^4 liter^{-1}.
Similar observations were made in clouds over Australia by Mossop (1970), and in
clouds over Missouri by Braham (1964) and Koenig (1963). In other cases a simple
correlation between ice nuclei and crystal concentrations holds. This is exemplified
in Figure 2-35.

The histories of ice content and particle concentrations in storm clouds have
been observed by Magono and Lee (1973) in Japan, by Borovikov and Nevzorov
(1971) in the U.S.S.R., and by Jiusto and Weickmann (1973) over the Great
Lakes region of the U.S. The case studies of Magono and Lee are exemplified in

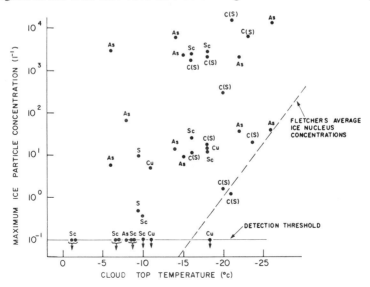

Fig. 2-34. Maximum ice particle concentrations in clouds over the Cascade Mountains (State of
Washington). Cloud types are indicated by As – altostratus, S – stratus, Sc – stratocumulus, Cu –
cumulus, C(S) – cumulus with stratified tops. (From Hobbs *et al.*, 1974b; by courtesy of the authors.)

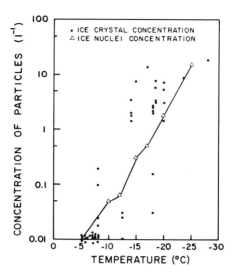

Fig. 2-35. Ice crystal concentration and average ice nuclei concentration as a function of cloud top temperature in clouds over Israel. (From Gagin, 1971; by courtesy of Amer. Meteor. Soc., and the author.)

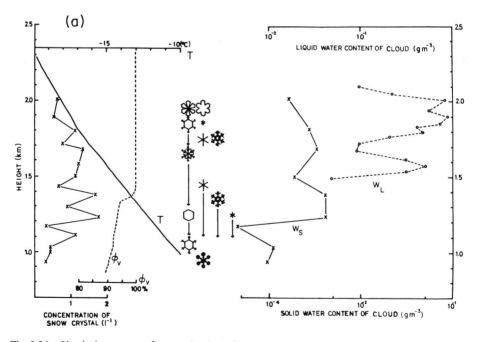

Fig. 2-36. Vertical structure of snow clouds during a storm in January, 1970, over Hokkaido, Japan: (a) early stage on Jan. 22, 1553 JST; (b) mature stage on Jan. 24, 0959 JST; (c) decaying stage on Jan. 25, 1045 JST. (From Magono and Lee, 1973; by courtesy of *J. Meteor. Soc., Japan.*)

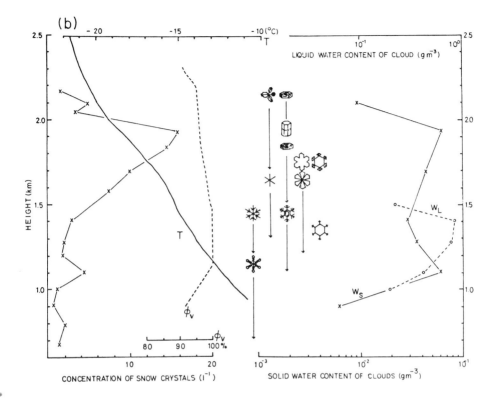

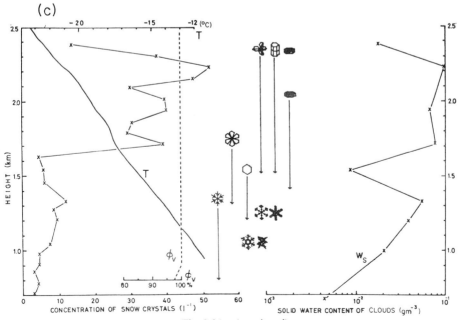

Fig. 2-36. (*continued*).

Figure 2-36. They found it useful to distinguish three stages in the development of the microstructure of cloud containing ice particles: (1) the early stage when the concentration of the snow crystals is still small and almost all portions of the cloud system are composed of supercooled drops (Figure 2-36a), (2) the mature stage when the ice content of the cloud system is comparable to its liquid water content (Figure 2-36b), and (3) the decaying stage when almost no drops are found in the cloud (Figure 2-36c). Note that the number concentration of ice crystals is highest during the decaying stage, and that the cloud ice content varies considerably with height, depending not only on the number concentration of the ice crystals but on their shape as well.

The microstructure of cirrus type clouds has been studied by Weickmann (1945, 1949), Kikuchi (1968), Heymsfield (1972, 1975a, b, c), and Heymsfield and Knollenberg (1972). These studies revealed a concentration of ice crystals which normally ranged between 10 and 50 ℓ^{-1}. Also, the crystals had mean lengths between 0.1 and 1.0 mm, and bullets, rosettes, and columns were the prevalent crystal habits. The ice water content ranged from 0.15 to 0.40 g m^{-3}.

Ice fog has a somewhat unusual microstructure. Such a fog develops during a pronounced ground inversion at very low temperatures. Most of the studies on ice fogs were carried out in Fairbanks, Alaska by Thuman and Robinson (1954), by Ohtake (1967, 1968, 1970a, b), and by Kumai (1965, 1966a, b, 1969a, b). Strong ground inversions and winter temperatures between -30 and $-55\,°C$ often develop at this location. Power plants, automobile exhausts, and exhausts from the heating systems of dwellings act as sources of moisture and dust particles. Under these conditions ice crystals stay small and develop unusual forms. At -39 to $-40\,°C$ the crystals have diameters which range from 2 to 30 μm, with most frequent diameters near 10 μm. At warmer temperatures $(-31$ to $-33\,°C)$ the size distribution broadens to diameters between 5 and 50 μm with a mode near 20 to 25 μm. The solid water content is low and ranges between 0.09 g m^{-3} (at $-40\,°C$) and 0.02 g m^{-3} (at $-30\,°C$). The number concentration of ice crystals is very high, ranging between 100 and 200 cm^{-3}. Due to this high concentration the visibility in ice clouds is severely reduced.

2.2.2 SHAPE, DIMENSIONS, BULK DENSITY, AND NUMBER CONCENTRATION OF SNOWFLAKES, GRAUPEL, AND HAILSTONES

When certain conditions prevail in a cloud, snow crystals collide to form snowflakes. Air temperature and snow crystal shape play the dominant roles in such aggregation. Hobbs *et al.* (1974b), who studied cyclonic and orographic cloud systems over the Cascade Mts. (State of Washington), and Rodgers (1974b), who studied orographic cloud systems over Elk Mountain (Wyoming), established that the probability for the occurrence of snowflakes is highest if the air temperature at the site of their formation is near 0 °C. With decreasing temperature, the probability of aggregation decreases with a secondary maximum near $-15\,°C$. Both observations show that the maximum dimensions of snowflakes are largest near 0 °C (see Figure 14-19). In addition to temperature, snowflake size is strongly affected by the shape of the component crystals. Figure

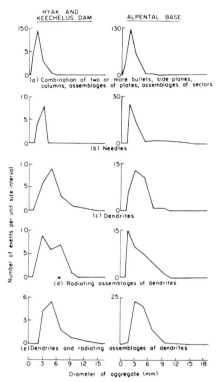

Fig. 2-37.   Size distributions of snow crystal aggregates collected at Hyak and Keechelus Dam, and Alpental Base, Cascade Mountains, State of Washington. (From Hobbs *et al.*, 1974a; by courtesy of the authors; copyrighted by American Geophysical Union.)

2-37 shows that aggregates of columns and needles tend to stay small, while aggregates of dendritic crystals tend to become large. Although maximum snowflake diameters may be as large as 15 mm, most of the snowflakes have diameters between 2 and 5 mm.

Observations of Locatelli and Hobbs (1974) in the Cascade Mountains further demonstrated that, just as with single snow crystals, snow crystal aggregates tend to follow dimensional relationships during their growth by collision with other crystals. These relations are expressible in terms of power laws of the form m = Ad^B, where A and B are constants for an aggregate of component crystals of given shape, m is the mass of the snowflake, and d is its maximum dimension (Figure 2-38). Unfortunately, at present the above power law relation is supported only by the observations in clouds over the Cascade Mts. Considering the multitude of possible snow crystal combinations in a crystal aggregate, and the large variety of bulk densities associated with each crystal type, it is not yet possible to state whether the values for A and B given by these authors will apply to clouds over other regions as well.

The number of component crystals per snowflake was examined by Hobbs *et al.* (1974b), and Rodgers (1974a). Although the results of their studies scattered greatly, they indicate the expected trend that the number of component

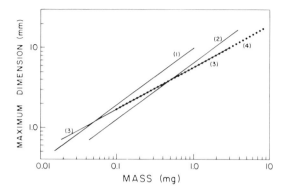

Fig. 2-38.   Best fit curves for maximum dimensions versus masses of unrimed aggregates of various types. (1) aggregates of unrimed side planes, (2) aggregates of unrimed radiating assemblages of dendrites, and aggregates of dendrites, (3) aggregates of unrimed radiating assemblages of planes and bullets, (4) aggregates of densely rimed radiating assemblages of dendrites, or of densely rimed dendrites. (From Locatelli and Hobbs, 1974; by courtesy of the authors; copyrighted by American Geophysical Union.)

crystals increases with increasing snowflake size. This correlation is more pronounced, the smaller the component crystals (Figure 2-39).

A law for the size distribution of snowflakes which is analogous to the Marshall and Palmer (1948) raindrop distribution was proposed by Gunn and Marshall (1958). From an extensive field study, these authors suggested the relation

$$n(D_0) = n_0 \exp(-\Lambda D_0), \tag{2-12}$$

where $\Lambda = 25.5\,R^{-0.48}\,\text{mm}^{-1}$, $n_0 = 3.8 \times 10^3\,R^{-0.87}\,\text{m}^{-3}\,\text{mm}^{-1}$, D_0 is the equivalent diameter of the water drop to which the ice crystal aggregate melts, and R is the

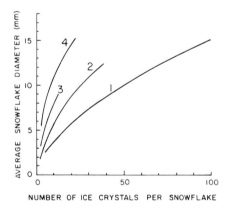

Fig. 2-39.   Best fit lines for the relation between the number of component crystals per snowflake and snowflake diameter; average diameter of component crystals: (1) <1.5 mm, (2) 1.5 to 2.5 mm, (3) 2.5 to 3.5 mm, (4) >3.5 mm. (From Rodgers, 1974a; by courtesy of Amer. Meteor. Soc., and the author.)

rate of precipitation in millimeters of water per hour (Figure 2-40). Observations by Ohtake (1969, 1970c) above and below the melting level in cloud systems over Alaska and Japan were also generally consistent with (2-12). In addition, Ohtake observed that the drop size spectrum just below the melting layer tended to be similar in shape to the size distribution of the melted snowflakes just above it. From this Ohtake concluded that snowflakes do not break up during their fall through the melting layer.

Rimed ice crystals and graupel are formed in clouds which contain both ice crystals and supercooled drops. Field studies have shown that in such clouds both snow crystals and frozen drops may serve as embryos for graupel formation. Thus, Harimaya (1976) carefully sectioned and disassembled natural graupel particles under the microscope to find both snow crystals and frozen drops as center particles. The importance of frozen drops to the formation of graupel has also been stressed by Pflaum *et al.* (1978), who experimentally studied the riming growth of frozen drops and of the crystal plates while they were freely suspended in the vertical air stream of a wind tunnel. Considerable controversy exists in the literature with regard to the type of ice particle which may serve as an embryo for conically shaped graupel. Arenberg (1941) suggested that conical graupel originate on planar snow crystals which, while falling under gravity, primarily rime on their bottom side. Under such conditions rime builds into a downward facing point, thus forming a conical graupel with its apex down. Holroyd (1964) proposed that conical graupel are the result of an aggregation of partially rimed needle crystals which continue to rime after aggregation. Nakaya (1954), List (1958a, b), and Knight and Knight (1973a), among others, advocate the ideas of Reynolds (1876), who suggested that conical graupel are the result of planar ice crystals which preferentially rime on their bottom side, the rime fanning out into the wind rather than growing into a point. Such behavior causes

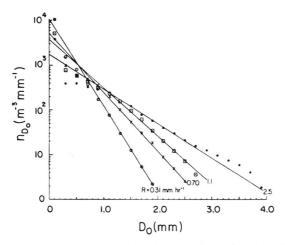

Fig. 2-40. Variation of number concentration of snowflakes with drop diameter of melted snowflake. Data fitted to exponential distribution (solid line). (From Gunn and Marshall, 1958; by courtesy of Amer. Meteor. Soc., and the authors.)

the development of a conical graupel which falls with its apex up and has its embryo near the apex. Weickmann (1953, 1964) and Takeda (1968) suggested that conical graupel can also start on frozen drops. This suggestion was experimentally verified by the wind tunnel studies of Pflaum *et al.* (1978).

In studies of the initial stages of the riming process in various types of clouds Ono (1969), Wilkins and Auer (1970), Hobbs *et al.* (1971a), Kikuchi (1972a), and Iwai (1973) found that both columnar ice crystals and ice crystal plates have to grow by diffusion to a certain critical size before they can grow by riming. Ono, Wilkins and Auer, and Hobbs concluded that riming commences on plate-like and dendritic snow crystals if $d \geq 300 \, \mu$m. The onset of riming on columnar crystals appeared to be independent of the length of the column but strongly dependent on its width. Ono, Hobbs, and Iwai found that riming commences on columnar crystals if $d \geq 50 \, \mu$m. During riming, only drops of certain sizes became attached to a snow crystal. Plate-like and dendritic crystals collected drops whose diameters had a mode near 20 μm, while on columnar crystals the

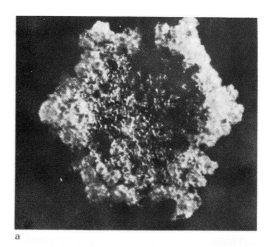

a

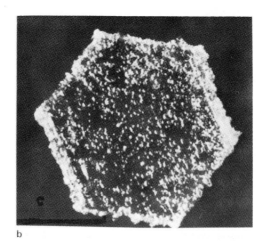

b

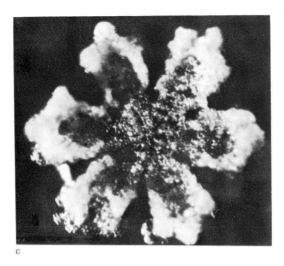

c

Plate 3. Rimed ice crystals: (a) (b) rimed simple plates, (c) rimed sector plate. (From Hobbs *et al.*, 1971; by courtesy of the authors.)

collected drops had diameters with a mode near 40 μm. The larger the crystal grows by riming, the wider is the size spectrum of the attached drops (Wilkins and Auer, 1970).

Photographs of rimed crystals (Wilkins and Auer, 1970; Zikumda and Vali, 1972; Iwai, 1973; Knight and Knight, 1973a) show that plate-like and dendritic crystals are rimed most intensely at the crystal edges, with considerably fewer frozen drops attached to the interior surface portions of the crystal (Plate 3a–c). Drops frozen onto simple columnar crystals are uniformly distributed over the crystal surface (Plate 4), and columns with end plates are most intensely rimed on the outer surface of an end plate, with few or no drops attached to the columnar stem of the crystal.

Observations show that, as in the case of snowflakes, rimed single snow crystals, rimed snow crystal aggregates, and graupel particles (Plate 5) also follow fairly definite size-mass relationships during their growth (Figure 2-41).

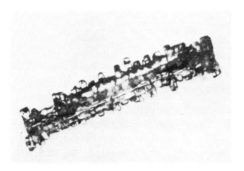

Plate 4. Rimed column. (From Iwai, 1973; by courtesy of *J. Meteor. Soc., Japan.*)

Plate 5.   Graupel particles of various sizes. Distance between lines on collection plate is 2 mm.
(From Aufdermauer, 1963; by courtesy of the author.)

The bulk density of rimed ice particles varies greatly, depending on the dense-
ness of packing of the cloud drops frozen on the ice crystal. Table 2-4 shows that
the bulk density of graupel particles ranges from about 0.05 g cm^{-3} to as high as
0.89 g cm^{-3}.

In clouds with sufficiently large updrafts, riming may continue until hailstones
are produced. We have already given some description of the various shapes
they may assume; some examples are shown in Plates 6 and 7. The various

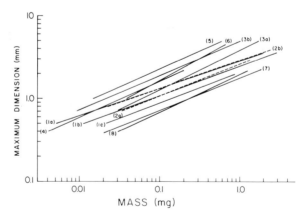

Fig. 2-41.   Best fit lines for maximum dimensions versus masses for single, solid, rimed precipitation
particles: (1a) lump graupel (ρ: 0.05 to 0.1 g cm^{-3}), (1b) lump graupel (ρ: 0.1 to 0.2 g cm^{-3}), (1c) lump
graupel (ρ: 0.2 to 0.45 g cm^{-3}), (2a) conical graupel, (2b) hexagonal graupel, (3a) graupel-like snow of
lump type, (3b) graupel-like snow of hexagonal type, (4) densely rimed columns, (5) densely rimed
dendrites, (6) densely rimed radiating assemblages of dendrites (based on best fit power laws of
Locatelli and Hobbs, 1974), (7) lump graupel (ρ: 0.25 to 0.7 g cm^{-3}), (8) conical graupel (ρ: 0.25 to
0.7 g cm^{-3}). (Based on best fit power laws by Zikmunda and Vali, 1972.)

TABLE 2-4

Densities of graupel particles.

Observer	Size range (mm)	Density (g cm^{-3})
Locatelli and Hobbs (1974), Washington	0.5–3	0.05–0.45
Zikmunda and Vali (1972), Wyoming	0.5–1	0.45–0.7
Zikmunda and Vali (1972), Wyoming	1–2	0.25–0.45
Bashkirova and Pershina (1964), USSR	0.4–3	0.08–0.35
Braham (1963), Missouri	0.5–3	0.85–0.89
List (1958a, b), Switzerland	0.5–6	0.5 –0.7
Magono (1953), Nakaya and Tereda, (1935), Japan	0.8–3	0.13

habits and surface textures of hailstones have been studied in detail by List (1958b), Hallett (1965), Browning (1966), Carte and Kidder (1966), Browning and Beimers (1967), Knight and Knight (1970a), and Mossop (1971a).

Auer (1972b) has summarized measurements reported in the literature on the size distribution of hailstones and graupel particles (Figure 2-42). The number concentration of these ice particles decreases rapidly as their size increases. Graupel particles with diameters between 0.5 and 5 mm are present in concentrations ranging between 10^3 and 1 m^{-3}, while small hailstones with diameters between 0.5 and 2.5 cm are present in concentrations ranging from 1 to 10^{-2} m^{-3}. Large hailstones of diameters between 2.5 and 8 cm are present in concentrations ranging from 10^{-6} to 10^{-2} m^{-3}.

Auer (1972b) was able to fit the average size distributions of spherical graupel and hailstones observed in storms over the High Plains surrounding Laramie to

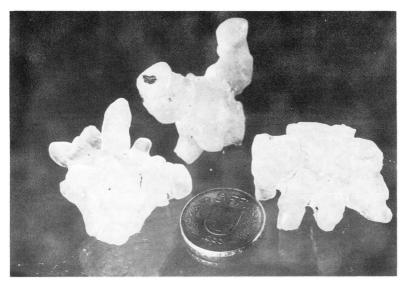

Plate 6.   Hailstones. (From Levi et al., 1970; by courtesy of Amer. Meteor. Soc., and the authors.)

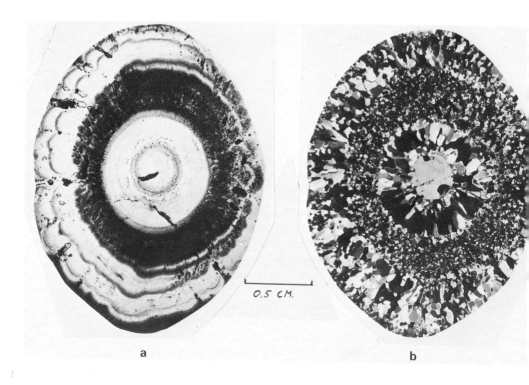

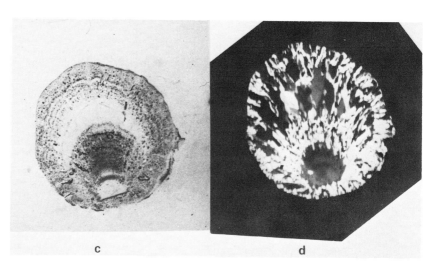

Plate 7.   Thin section of hailstones: a, c in normal transmitted light, b, d in polarized transmitted light. (Plate a, b from Knight and Knight, 1968a; by courtesy of Amer. Meteor. Soc., and the authors. Plate c, d from Federer, 1977, personal communication; max. diameter of stone 18 mm; by courtesy of B. Federer.)

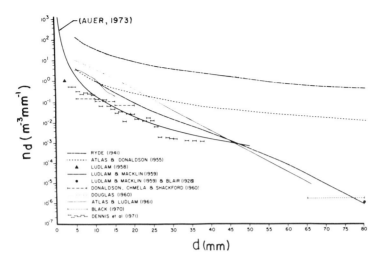

Fig. 2-42.   Number concentration of graupel particles and hailstones as a function of their diameter. (From Auer, 1972b; by courtesy of Amer. Meteor. Soc., and the author.)

the relation

$$n(d) = 561.3 \, d^{-3.4}, \qquad (2\text{-}13)$$

with $n(\text{m}^{-3}\,\text{mm}^{-1})$ and $d(\text{mm})$ for $1 \leqslant d \leqslant 50$ mm. Federer and Waldvogel (1975) found that the size distribution of spherical hailstones which fell from multicell storms over central Switzerland followed the relation

$$n(d) = n_0 \exp{(-\Lambda d)}, \qquad (2\text{-}14)$$

with $1.5 \leqslant n_0 \leqslant 52$ m^{-3} mm^{-1}, and $0.33 \leqslant \Lambda \leqslant 0.64$ mm^{-1} for $d \leqslant 22$ mm. The mean spectrum was characterized by $n_0 = 12.1$ m^{-3} mm^{-1} and $\Lambda = 0.42$. The hail fall rates (in equiv. water) ranged between 2.6 and 152 mm hr^{-1}, while the solid water content w_s ranged from 0.05 to 2.64 g m^{-3}, depending on the hail fall rate R according to the relation $w_s = 1.74 \times 10^{-2}$ R. Measurements consistent with those of Federer and Waldvogel were made by Smith *et al.* (1976), who observed exponential size distributions of the form of (2-14) for hailstones which fell over N.E. Colorado, with $0.25 \leqslant n_0 \leqslant 61$ m^{-3} mm^{-1} and $0.27 \leqslant \Lambda \leqslant 0.84$ mm^{-1}.

While hailstones usually have diameters of a few centimeters, Figure 2-42 indicates that large hailstones may have a major axis length as large as 6 to 8 cm. Hailstones of even larger sizes have been observed by Browning (1966), and by Ross (1972), who described a hailstone which weighed 766 g and had a circumference of 44 cm.

Hailstones collected at the ground are usually hard ice particles. In early studies on hail growth, this observation led to the assumption that hailstones always grow as solid ice particles. However, in more recent experimental studies during which hailstone growth was simulated in wind tunnels, List (1959a, b; 1960a, b) and Macklin (1961) discovered that hailstones are not always hard particles but, depending on the regime of growth, may also be 'soft' particles which consist of ice-water mixtures, termed *spongy ice* by List. Such ice-water mixtures are produced when the latent heat released during growth is not exchanged efficiently enough between the hailstone and its environment to allow all the water collected by the hailstone to freeze. That portion of the collected water which immediately freezes produces a skeletal framework or mesh of dendritic ice crystals (see Chapter 16), in which the unfrozen portion of the collected water is retained as in a sponge whose surface temperature is at 0 °C.

A few field studies have verified the spongy growth mode of hailstones in clouds. Thus, Summers (1968) reported on the collection of soft or slushy hailstone samples which fell during hailstorms in Alberta, Canada. Gitlin *et al.* (1968) and Browning *et al.* (1968) used calorimetric methods to analyze freshly fallen hailstones at various geographic locations. They found that water comprising up to 16% of the total mass was embedded in the ice structure of some hailstones. In a recent more definitive study, Knight and Knight (1973b) analyzed natural hailstones by the quenching technique, and concluded that while some hailstones experience spongy growth at times, this growth mode is not the rule for all hailstones.

Studies of hailstones by the thin section technique (List, 1961) reveal that usually a hailstone has one distinct central growth unit or growth embryo. Hailstones with two centers of growth exist but are extremely rare (Rogers, 1971). Considerable controversy exists in the literature concerning the nature of this central growth unit. List (1958a; 1959a, b; 1960a, b), and Knight and Knight (1976) found that about 80% of the hailstones which fell in Switzerland and in Colorado, respectively, had a graupel embryo, which, in turn, may have originated on a snow crystal or on a small frozen drop. Similarly, Mossop and Kidder (1964) and Carte and Kidder (1966) found that a large percentage of the hailstones collected during hailstorms in South Africa originated as graupel particles. In contrast, however, Macklin *et al.* (1960) found that most hailstones collected in England had clear growth centers, and they interpreted this to mean that these hailstones originated as large frozen drops. Both types of growth centers were found by Knight and Knight (1970b) and by Federer and Waldvogel (1978), who examined a large number of hailstones which fell in the U.S. and in Switzerland, respectively. They observed embryos of a few millimeters to 1 cm in diameter. Some of these were more or less opaque and had conical shapes (e.g., Plate 7c, in which we note the conical graupel developed from a frozen drop which on freezing split into two hemispheres), while others were clear or bubbly and had spherical shapes. Thus, present evidence indicates that hailstones may originate either as graupel or frozen drops.

The bulk density of hailstones tends to vary radially from surface to core, with alternating concentric layers of lower and higher density. The density of such hailstones shell has been found to vary usually between 0.8 and 0.9 g cm^{-3}, but shell densities as low as 0.7 g cm^{-3} have also been observed (List, 1958b, 1959a, b; Macklin *et al.*, 1960; Mossop and Kidder, 1961; Prodi, 1970; List *et al.*, 1970). The density variations are a reflection of varying amounts of trapped bubbles (Plate 7a, c). Many of these bubbles are quite regularly grouped within concentric layers, which alternately contain larger and smaller numbers. Hailstone shells with a large number of bubbles appear quite opaque, while shells with only a few bubbles appear as clear ice.

The size and number concentration of air bubbles in hailstones have been studied by List (1958b), Macklin *et al.* (1970), List *et al.* (1972), and List and Agnew (1973). From an examination of planar cuts through the stone centers, List and his co-workers found that the bubble size distribution was lognormal. The planar number concentration of bubbles varied across the slice surface by more than two orders of magnitude, from about 50 to 5000 cm^{-2}. Opaque shells consisted of numerous smaller bubbles, while clear, transparent shells contained fewer and larger bubbles. In hailstone layers deposited in the dry growth regime, Carras and Macklin (1975) found a volume air bubble concentration of 10^6 to 10^8 cm^{-3}, with air bubble sizes ranging from 2 to 8 μm in diameter. In layers deposited in the wet growth regime the air bubble concentration was 10^5 to 10^6 cm^{-3}, and the bubbles had diameters between 20 and 100 μm.

When thin sections of graupel particles and hailstones are studied by means of polarized light, a second interesting structural feature is revealed: One finds that the ice of the sections is polycrystalline, with large and small individual crystallites (single ice crystals) in alternating layers (Plate 7b, d). Most crystallites tend to assume preferred orientations (List, 1958a, b; 1960). Detailed studies of the size and orientation of crystallites in natural hailstones have been made by Aufdermauer *et al.* (1963), Knight and Knight (1968), List *et al.* (1970), Levi *et al.* (1970), Macklin *et al.* (1970), and Macklin and Rye (1974). These have shown that transparent layers relatively free of air bubbles tend to consist of fairly large crystallites, while opaque layers with relatively high air bubble concentrations tend to consist of numerous small crystallites. List *et al.* (1970) determined that, in the slice-plane, crystallites have surface areas which range between 1×10^{-3} and 8×10^{-2} cm^2 (0.1 to 8 mm^2). Macklin *et al.* (1970) and Rye and Macklin (1975) found that the crystallite length decreases from 8 mm to 0.25 mm and the width from 1 mm to 0.2 mm as the ambient temperature decreases from -5 to -30 °C.

Although hailstone crystallites are randomly oriented in some shells, they assume a preferred orientation in others. Aufdermauer *et al.* (1963), Knight and Knight (1968), and Levi *et al.* (1970) found that in some shells crystallites have their crystallographic *c*-axis generally either parallel or at right angles to the hailstone's radial growth direction. List *et al.* (1970) found crystallites with preferred *c*-axis orientation parallel to the radial growth direction in the clear shells of the hailstones, and with rather random orientations in opaque shells. A discussion of the reasons for the polycrystallinity of hailstones and the preferred orientation of crystallites under certain growth conditions is given in Chapter 16.

CHAPTER 3

THE STRUCTURE OF WATER SUBSTANCE

In the previous chapter we described the observed variety of shapes, sizes, and concentrations of the solid and liquid particles which comprise clouds and precipitation. The remaining chapters will be devoted to exploring how such particles come into being and how they grow. Understanding these processes depends, to a large extent, on knowledge of the physical properties of water vapor, water, ice, and, ultimately, on the physical characteristics of the water molecule itself. Therefore, as a prelude to what will follow, this chapter will describe briefly some of the relevant structural features of individual water molecules and their various combinations in water vapor, bulk water, and ice.

For a detailed study of subjects covered in this chapter, the reader may refer to the texts of Hobbs (1974), Ben-Naim (1974), Whalley *et al.* (1973), Horne (1972), Franks (1972), Fletcher (1970a), Robinson and Stokes (1970), Eisenberg and Kauzmann (1969), Riehl *et al.* (1969), Kavanau (1964), and Dorsey (1940).

3.1 Structure of an Isolated Water Molecule

Measurements of the heat capacity of water vapor at constant volume near room temperature yield a value of approximately $3\,k$ per molecule, where k is the Boltzmann constant. Since quantum statistical mechanics shows that the vibrational degrees of freedom are frozen-in at these temperatures, we must interpret the heat capacity measurements in terms of a contribution of $(1/2)\,k$ from each of the three translational degrees of freedom, and a contribution of $(1/2)\,k$ from rotation about each of the three axes for which the molecule has an appreciable moment of inertia. This interpretation implies that the water molecule cannot have its three atoms arranged in a linear fashion. The same conclusion is reached by investigating the electrical properties of the water molecule. Since such measurements reveal a large electric dipole moment of $\mu = 1.83 \times 10^{-18}$ e.s.u. cm (see Hobbs, 1974; Eisenberg and Kauzmann, 1969), a linear molecule is once again ruled out.

The geometry of the water molecule can be deduced accurately from studies of the infrared spectrum of water vapor. On the basis of such measurements, Mecke (1933) concluded that the three atoms are situated at the vertices of a triangle, the geometry of which is given in Figure 3-1. Recent experiments show that the equilibrium O—H bond length is 0.95718 Å and that the equilibrium H—O—H bond angle is 104.523° (see Fletcher, 1970a; Hobbs, 1974).

The structure of the water molecule is importantly affected by the electron

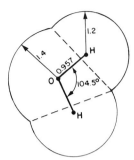

Fig. 3-1. Two dimensional geometry of a single water molecule. The O—H distance (in 10^{-8} cm) and the H—O—H angle are indicated, as are radii of the hydrogen and oxygen atoms. (From *Water and Aqueous Solutions* by A. Ben Naim, copyrighted by Plenum Press, 1974.)

configuration around the oxygen atom.* In its ground state, an oxygen atom has two electrons in the spherical 1s orbital, where they are bound tightly to the atomic nucleus, and two electrons, less tightly bound, in the spherical 2s orbital. In addition, two electrons can be considered to occupy the $2p_x$ orbital, one electron the $2p_y$ orbital, and one electron the $2p_z$ orbital. This electron configuration is illustrated in Figure 3-2. Since the $2p_y$ and $2p_z$ orbitals may contain two electrons each, these orbitals are incomplete. The electrons in these orbitals are therefore free to couple with the electrons in the 1s orbital of the two hydrogen atoms, allowing them to form two O—H bonds.

If these orbitals exactly described the O—H bond of a water molecule, one would expect water to have a bond angle of 90°. Experimentally, however, one finds that the bond angle is some 15° larger. One might try to explain this on the basis of the fact that the O—H in a water molecule is not truly covalent but is partly ionic; i.e., the electrons are not evenly shared by the oxygen atom and a hydrogen atom. Since oxygen is more electronegative than hydrogen, oxygen exerts a greater force on the shared electron pair than does the hydrogen. Consequently, the electrons spend a greater portion of their time in the outer shell of the oxygen atom than in the hydrogen shells, and so the positive charge of the hydrogen nuclei is incompletely shielded by the electrons. Electrostatic repulsion between the two hydrogen atoms must, consequently, lead to an increase of the bond angle.

* Generally, an atom's ground state electron configuration is described by specifying the number of electrons in each energy level or 'shell', characterized by the principal quantum number n and the angular momentum quantum number l. In listing these electrons, it is customary to use the spectroscopic notation in which the numbers $l = 0, 1, 2, 3$ are replaced by the respective letters s, p, d, f. The number of electrons in a shell is indicated by a superscript; e.g., $2p^6$ means there are 6 electrons in the shell characterized by $n = 2$, $l = 1$.

Each electron is said to occupy an 'orbital' corresponding to given values of n, l, and the quantum number m describing the z-component of the electron's angular momentum. By the Pauli exclusion principle, there is room in each orbital for two electrons, necessarily with opposed spins. It is the outermost, *valence* electrons in incomplete orbitals which are responsible for *covalent* and *ionic* chemical bonds. For further information, see any standard reference on quantum mechanics, e.g., Schiff (1968).

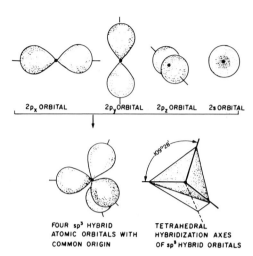

FOUR sp³ HYBRID   TETRAHEDRAL
ATOMIC ORBITALS WITH   HYBRIDIZATION AXES
COMMON ORIGIN   OF sp³ HYBRID ORBITALS

Fig. 3-2. Hybridization of orbitals of oxygen atom. (From *Organic Chemistry* by D. J. Cram and
G. S. Hammond, copyrighted by McGraw-Hill Publ. Co., 1955.)

However, Heath and Linnett (1948) showed that this repulsion is insufficient to
account for the experimentally found bond angle. They suggested that a more
significant factor in opening up the bond angle is the mixing or 'hybridization' of
the 2s orbital of the oxygen atom with its $2p_y$ and $2p_z$ orbitals, resulting in the
formation of four sp³-hybrid atomic orbitals. Two of these overlap with the
hydrogen orbitals, while the two remaining orbitals form two lobes on the side of
the oxygen atom away from the hydrogen atoms (Figure 3-2). These lobes, called
lone-pair hybrids, are symmetrically located above and below the molecular
plane and form roughly tetrahedral angles with the bond hybrids (exact tetrahedral angle, 109.467°). It is this tetrahedral character of the water molecule
which gives rise to tetrahedral coordination of water molecules in water and
ice.

Duncan and Pople (1953) and Bader and Jones (1963) have carried out
quantum mechanical ('molecular orbital theory') calculations of the electron
density distribution around a water molecule. Their results confirm the distribution shown in Figure 3-2, and show that there are four locations in the water
molecule with high electron density: close to the oxygen atom, close to each
hydrogen atom, and at the location of the lone pair orbitals which appear as an
electron density bulge 'behind' the oxygen atom.

The charge distribution around a water molecule may also be approximated by
'electrostatic' or point charge models. In these models point charges are assigned
whose sign, magnitude, and location are such that the molecule as a whole is
electrically neutral, and the electric dipole moment is equal to that experimentally measured. Such models have been worked out by Bernal and Fowler (1933),
Verwey (1941), Rowlinson (1951), Bjerrum (1951), Pople (1951), Campbell (1952),
and Cohen *et al.* (1962). Although such models are convenient in some cases,
they generally do not correctly predict the higher electric moments (Kell, 1972a).

3.2 Structure of Water Vapor

Experiments indicate that water molecules in water vapor tend to interact and form clusters, in contrast to ideal gas behavior. Dimers as well as higher order polymers are considered to be present in water vapor, though in small concentrations only. Recent experiments involving molecular beam techniques (Lin, 1973; Searcey and Fenn, 1974) suggest that in highly supersaturated water vapor, clusters of up to 180 water molecules may be present. Clusters of 21 water molecules seemed to exhibit particularly large stability. It is interesting to note that 21 water molecules can be arranged in the form of a pentagonal dodecahedron with a molecule at each corner and a single molecule in the center of the 'cage.'

However, no conclusive evidence of the actual geometric arrangement, if any, of water molecules in such clusters in vapor is available at present. Studies on the possible and more likely cluster types have been reviewed by Rao (1972) and Kell (1972a). Recent theoretical studies of the formation of water clusters have been carried out by Kistenmacher et al. (1974a, b) and Abraham (1974a). Kistenmacher et al. found two possible stable configurations for the dimers, a cyclic form and an open form which was more stable. For the trimers and tetramers, the cyclic forms seemed to be somewhat more stable than the open structures. For the large clusters the authors suggested not a single structure, but a statistical distribution of different configurations, since many configurations with significantly different geometry were found to possess nearly the same energy.

The potential energy of interaction, U, between a pair of water molecules has the general character of being strongly repulsive at very close separations and weakly attractive at longer range. One widely used and relatively simple expression for it is due to Stockmayer (1941):

$$U = -\frac{\mu^2 f}{r^3} - \frac{c}{r^6} + \frac{c\sigma^{18}}{r^{24}}, \tag{3-1}$$

where r is the separation of the molecules, μ is the dipole moment of an isolated water molecule, σ is the collision diameter (the molecular separation at which $U = 0$ if $\mu = 0$), c is an adjustable constant, and f is a known function of the mutual orientation of the two molecules.

The first term on the right side of (3-1) is just the dipole-dipole contribution to the interaction energy, and may be attractive or repulsive, depending on the dipole orientations. The second term represents contributions from: (1) the interaction energy between a permanent dipole of one molecule and the dipole it induces in the other (dipole-polarization or induction interaction), (2) the net energy arising from momentary, fluctuating dipoles interacting with the corresponding induced dipoles (polarization-polarization or dispersion interaction). Even though the time average of these dipole fluctuations may be zero, the energy contribution is proportional to their mean square, which is finite and positive. Both (1) and (2) are usually referred to as *van der Waal's interaction*, which by its nature can be seen to bring about an attractive force between the molecules. The third term in (3-1) represents the short range repulsive forces, which loosely may be ascribed to the overlap of electronic orbitals which are incompatible according to the Pauli exclusion principle.

There is little doubt that the Stockmayer potential or similar ones, such as Rowlinson's (1949, 1951) potential, portray with fair accuracy the interaction between pairs of water molecules at large separations in dilute water vapor. This is evidenced by the fact that values for the second virial coefficient computed via (3-1) can be made to fit experimental values. On the other hand, the same potential functions yield values for the third virial coefficient of water vapor which disagree substantially with experiment. Partly, this is due to the approximate nature of (3-1), and partly because three-body interactions should be included also, since other molecules in the system can significantly modify the interaction of a given pair. In particular, the Stockmayer potential is insufficiently 'directional' in character to account for the geometry of cluster formation in water vapor. A recent, more complicated potential function which has proven to be of predictive value in this respect is described briefly in Section 3.4.

3.3   Structure of Ice

At atmospheric pressures and at temperatures between about $-100°$ and $0 °C$, water substance crystallizes from its gaseous or its liquid state to form a sixfold-symmetric or hexagonal solid called ice-I_h. At different temperatures and pressures ice assumes other crystalline modifications which are discussed, for example, in Fletcher (1970a) and Hobbs (1974). We shall concern ourselves here only with ice-I_h, henceforth referred to simply as 'ice.'

X-ray diffraction studies demonstrate that in ice each oxygen atom is surrounded by four nearest neighbor oxygen atoms at a distance of about 2.76×10^{-8} cm. These four atoms form an almost regular tetrahedron. In turn, these oxygen tetrahedrons are joined together to form a hexagonal lattice (Figure 3-3). The hexagonal space group is denoted by D_{6h}^4 or $P6_3/mmc$, and is characterized by 1 sixfold axis of rotation perpendicular to 1 mirror plane, $(3+3)$ twofold axes of rotation perpendicular to $(3+3)$ mirror planes, and a center of symmetry.

Near $0° C$, any given oxygen atom in ice also has 12 second nearest neighbors at a distance of about 4.52 Å, 1 third nearest neighbor at 4.59 Å, 6 fourth nearest neighbors at 5.26 Å, 3 fifth nearest neighbors at 5.31 Å, 6 sixth nearest neighbors

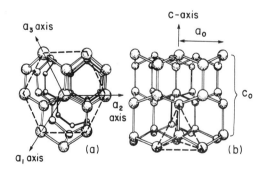

Fig. 3-3.   Position of oxygen atoms in ice-I_h: (a) view along c-axis, (b) view perpendicular to c-axis.

at 6.36 Å, 6 seventh nearest neighbors at 6.46 Å, 9 eighth nearest neighbors at 6.69 Å, 2 ninth nearest neighbors at 7.36 Å, and 18 tenth nearest neighbors at 7.81 Å (Kamb, B., 1975, personal communication).

Each water molecule is hydrogen bonded to its four nearest neighbors. (Generally, a hydrogen bond may be defined as a valence linkage joining two electronegative atoms through a hydrogen atom). This is brought about through the formation of two hydrogen (O----H—O) bonds by each water molecule, each bond being directed towards a lone electron pair of a neighboring water molecule. This manner of bonding leads to an open lattice structure, as illustrated in Figure 3-3. Perpendicular to the c-axis, the ice lattice consists of open-puckered hexagonal rings (with oxygen atoms alternately raised and lowered). Along the c-axis are vacant shafts. Comparison shows that the arrangement of oxygen atoms in ice is isomorphous with the Wurtzite structure of ZnS and the tridymite structure of SiO_2.

Each unit cell of ice, a four-sided prism set on a rhombic base, contains four water molecules and is characterized by the lattice constants a_0 and c_0 (Figure 3-4). X-ray data for a_0 and c_0 (Blackman and Lisgarten, 1957; Lonsdale, 1958; La Placa and Post, 1960; Brill and Tippe, 1967; Kumai, 1968) are summarized in Figure 3-5 as a function of temperature. These measurements show that a_0 and c_0 decrease with decreasing temperature – the rate of decrease is smaller, the lower the temperature – such that $(c_0/a_0) = 1.629$ for all temperatures. Using the values for a_0 and c_0 given in Figure 3-5, the volume of a unit cell of ice, $V_{uc} = 2(a_0^2\sqrt{3}/4)c_0$, varies from 1.305×10^{-22} cm^3 (0 °C) to 1.281×10^{-22} cm^3 (−180 °C). Thus, the number of water molecules cm^{-3} varies from 3.06×10^{22} (0 °C) to 3.12×10^{22} (−180 °C).

According to Eisenberg and Kauzmann (1960), three points of view may be taken to define the hydrogen bond energy E_H in ice. First, one may assume that E_H is given by the lattice energy E_L of one mole of ice (the difference in energy between one mole of isolated water molecules and one mole of ice, both at 0 °K and with motionless atoms), divided by the number of hydrogen bonds in a mole. Since both hydrogen atoms of a water molecule participate in one H-bond

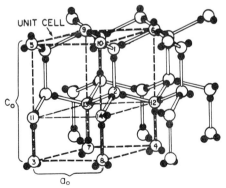

Fig. 3-4. A typical disordered arrangement of protons in the ice-I_h structure; oxygens (1) and (2) contribute 12/12 each, oxygens (3), (4), (5), (6) contribute 1/12 each, oxygens (7) to (12) contribute 2/12 each, and oxygens (13) and (14) contribute 4/12 each, for a total of 48/12 = 4 oxygens. (From Fletcher, 1970a, with changes.)

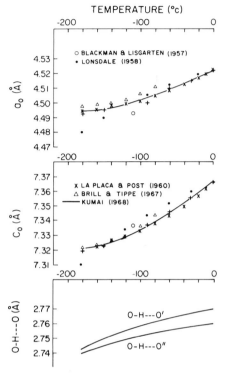

Fig. 3-5. Temperature variation of the lattice parameters of ice-I_h.

(excluding the molecules on the surface of ice), one may estimate E_H as $E_H = E_L/2 = 7.04\,\text{kcal mol}^{-1}$, based on the value for E_L found by Whalley et al. (1973). Second, one may define E_H, possibly more appropriately, in terms of the sublimation enthalpy $(\Delta H)_s$ by writing $E_H = (\Delta H)_s/2$. Since (ΔH_s) is temperature-dependent, so is E_H, and we find $E_H(0\,°\text{K}) = 5.66\,\text{kcal mol}^{-1}$ and $E_H(0\,°\text{C}) = 6.15\,\text{kcal mol}^{-1}$. In both of these definitions we ascribe the entire intermolecular energy in ice to hydrogen bonding. We therefore include in E_H the effects of dispersion and short-range repulsive forces which are present not only in ice but also in crystals of non-hydrogen bonded substances. The third definition of E_H is based on the premise that the contribution to $(\Delta H)_s$ from hydrogen bonds is distinct from that of other forces. One may therefore set $E_H = [(\Delta H)_s - E_{other}]/2$, where E_{other} represents the intermolecular energy associated with the other forces. This definition suffers from the fact that E_{other} is not an observed quantity, and cannot presently be accurately calculated.

An accurate theoretical calculation of a single hydrogen bond in ice should include at least the effects of nearest neighbors. To date, most investigators have avoided detailed computations for these effects. Generally, the approach taken has been to assume the total hydrogen bond energy is given by the sum of the four component energies (dipole-dipole, dipole-polarization, polarization-polarization, and short-range interactions), and to evaluate each of these by approximate methods for two neighboring water molecules at the relative positions found in ice. For this purpose, various models for the charge distribution in a water molecule have been assumed. For some component energies, rough estimates for the effect of neighboring molecules have also been made. The results of the most pertinent calculations on this subject have been summarized by Hobbs (1974) and Eisenberg and Kauzmann (1969). The values computed for E_H range from 4 to 8 kcal mol^{-1}. In all of these computations, the entire intermolecular energy has been attributed to hydrogen bonding.

The positions of the hydrogen atoms in ice are subject to the *Bernal-Fowler* (BF) *rules* (Bernal and Fowler, 1933). These require that: (1) each water molecule is oriented such that its two hydrogen atoms are directed approximately towards two of four oxygen atoms which surround it tetrahedrally, (2) there is only one hydrogen atom on each O—O linkage, and (3) each oxygen atom has two nearest neighboring hydrogen atoms such that the water molecule as a structural unit is preserved.

An ice structure which obeys the BF rules is termed ideal. Natural ice, however, does not behave ideally. Numerous experiments imply that a natural ice lattice contains *defects* which violate the BF rules. The following major atomaric defects are found in natural ice: stacking faults, chemical defects, molecular vacancies (Schottky defects), interstitial molecules (Frenkel defects), ionized states, and orientational defects (Bjerrum defects).

By means of regular oxygen tetrahedrons one may build up a cubic as well as a hexagonal lattice. In such a cubic arrangement of oxygen atoms there is formed a diamond-type ice lattice. *Stacking faults* occur when layers of cubic ice are intermixed in otherwise hexagonal ice. Such faults are particularly prone to occur in ice formed from vapor below about $-80\,°C$. *Chemical defects* result if foreign ions are built into the lattice during ice growth in an aqueous solution. Salt ions are either built into lattice voids or at regular lattice positions. *Molecular vacancies* denote the omission of water molecules from regular ice-lattice positions. *Interstitial molecules* are water molecules occupying irregular positions in the ice lattice. Fletcher (1970a) estimates the energy necessary for formation of a mole of vacancies to be about 12.2 kcal, whereas the energy necessary to form interstitial sites is about 14 to 15 kcal mol^{-1}. He further estimates that in natural ice at $-19\,°C$, the concentration of vacancies is about 10^{12} cm^{-3}.

Various authors (see Fletcher, 1970a; Hobbs, 1974) have shown that neither molecular nor interstitial molecules are capable of producing changes in the hydrogen configuration of ice. (These defects, therefore, cannot explain the electrical properties of ice.) Rather, such changes are produced by ionic states and orientational defects. Ice, like water, exhibits *ionized states* (H_3O^+, OH^-) in violation of the third BF rule. Such a state is created by the motion of a proton from one neutral water molecule to another. According to Jaccard (1971), the concentration of ionized states in ice at $-10\,°C$ is $C_{H^+} = C_{OH^-} \approx 3 \times 10^{11}$ cm^{-3}. He estimates the energy necessary for the formation of pairs of such states to be about 17.5 kcal mol^{-1}. Comparable figures for ionized states in water are $C_{H^+} = C_{OH^-} \approx 1.0 \times 10^{-7}$ mol liter$^{-1} = 6 \times 10^{13}$ ion pairs cm^{-3}, with a pair formation energy of 13.6 kcal mol^{-1}. *Orientational or Bjerrum defects* violate the first and second BF rules. Bjerrum defects consist either of a bond occupied by two protons instead of one (doubly occupied bond: O—H----H—O, D-defect), or of a bond which contains no proton at all (empty bond: O----O, L-defect). According to Jaccard (1971) these defects occur in concentrations of $n_D = n_L \approx 6 \times 10^{15}$ cm^{-3}, requiring an energy for pair formation of about 15.5 kcal mol^{-1}.

Pauling (1935, 1960) has pointed out that an ordered hydrogen arrangement in ideal ice would conflict with the experimental fact that ice possesses zero-point entropy. That is, from the relation $S = k \ln W$, wherein the entropy S is related to the number of distinguishable microstates W, an ordered hydrogen arrangement along with the restrictions of the BF rules would lead to $W_0 = 1$ and thus $S_0 = 0$ at $T = 0\,°K$. Consequently, Pauling proposed a disordered hydrogen arrangement, subject to the BF rules.

In Pauling's model the zero-point entropy of ice may be deduced directly by counting the allowed microstates. For this purpose we assume a perfect ice lattice which contains N_A (Avogadro's number) water molecules. There are then $2N_A$ OH----O bonds, on each of which the proton has two possible positions. This allows $(2)^{2N_A}$ possible arrangements in ice, if we assume with Pauling that all arrangements are equally probable, and if we consider the first and second BF rules. However, many of these arrangements are not consistent with the third BF rule. To account for this, let us count all the possible arrangements of the hydrogens in the immediate vicinity of a particular oxygen atom. One finds 16 such arrangements: one OH$_4^{2+}$, four OH$_3^+$, six OH$_2$, four OH$^-$, and one O^{2-}. Only 6 out of these 16 arrangements are compatible with the third BF rule. Assuming again with Pauling that all the arrangements are equally probable, the probability of a given oxygen atom having the correct arrangement around it is 6/16. Assuming further that all N_A oxygen atoms in ice are independent, the total number of possible configurations is reduced by a factor of $(6/16)^{N_A}$. Hence, $W_0 = (2^{2N_A})(3/8)^{N_A} = (3/2)^{N_A}$, from   which $S_0 = k \ln W_0 = R \ln (3/2) = 0.805$ cal mol^{-1} ($°K$)$^{-1}$. More detailed computations (Nagle, 1966) lead to S_0 (theor.) $= 0.8145 \pm 0.0002$ cal mol^{-1} ($°K$)$^{-1}$, in good agreement with the experimentally found value of S_0 (expt.) $= 0.82 \pm 0.05$ cal mol^{-1} ($°K$)$^{-1}$.

One may challenge this result on the grounds that defects are present in a real ice lattice, and energy differences exist between the various hydrogen arrangements (Gränicher *et al.*, 1957; Gränicher, 1958). As the temperature of a real ice lattice is reduced, all molecules lose energy and tend to exist in the arrangement in which the energy of the system is lowest. Now, thermodynamic equilibrium is achieved only if opportunity is provided for the molecules to have free passage to all permitted energy states. One may argue that in real ice such free passage is provided by means of the migration of atomic defects, e.g., ionized states and Bjerrum defects which alter the atomic arrangement during migration. Such a mechanism would lead to just one spatial arrangement at $0\,°K$, and thus to $S_0 = 0$, which is contrary to observation.

Gränicher *et al.* (1957) and Gränicher (1958) also supplied a way out of this dilemma. Their experimental studies on the electrical behavior of ice showed that configurational changes due to the migration of atomic defects become negligible below a temperature of about 75 °K. Below this temperature both the concentration and diffusion rate of defects, which exponentially decrease with decreasing temperature, are sufficiently small that one may consider the hydrogen configuration to be 'frozen-in.' In addition, computations by Pitzer and Polissar (1956) showed that above this freeze-in temperature, the energy differences between the various possible hydrogen arrangements in ice are small compared to the thermal energy kT. They become comparable to or larger than kT only if T < 60 °K. These results imply that real ice is disordered with respect to the hydrogen arrangement, since the hydrogen arrangement freezes-in in any of the possible configurations at temperatures where the difference between the configurational energies is still smaller than kT.

Since in the Pauling-Bernal-Fowler model for ice each hydrogen atom has two equally likely positions along a given O—O linkage, theirs may be regarded as a 'half-hydrogen' model for ice. This model has been confirmed by the neutron diffraction studies of Wollan et al. (1949) and Peterson and Levy (1957). The model reflects explicitly the idea that the structure of ice is independent of the positions of the hydrogen atoms.

Peterson and Levy also found the H—O—H valence angle to be nearly equal to the corresponding O—O—O angle. The latter is nearly tetrahedral ($\approx 109.5°$) and, therefore, about 5° larger than that of an isolated water molecule. This result was questioned by Chidambaram (1961), who argued that since the O—H---O bond is more easily bent than the H—O—H valence angle, the latter angle should not increase during solidification. On the other hand, he showed that the data of Peterson and Levy are consistent with an ice structure in which the water molecules keep the valence angle which they have in the vapor state, but in which the O—H---O bonds are slightly bent. In this structure each H is about 0.04 Å off the O---O axis. This means that the O—H---O bonds are bent by an average of 6.8°. In support of his model, Chidambaram cited the small change of frequency for the H—O—H bending mode when water vapor changes to ice, and the H—O—H angle of water molecules in hydrated crystals, which deviates very little from 104.5°. Chidambaram's views are supported by nuclear magnetic resonance studies (see Hobbs, 1974).

In closing this section we want to draw attention to an aspect of ice structure we have ignored thus far, namely the irregular motion of the molecules about their mean positions in the lattice. At 0 °C the molecules oscillate with an average period of vibration τ_V of about 2×10^{-13} sec. In addition, they undergo a random walk characterized by a translational displacement period of $\tau_D \lesssim 10^{-5}$ sec, and suffer reorientations according to a rotation period of $\tau_R \approx 10^{-5}$ sec (Eisenberg and Kauzmann, 1969).

3.4 Structure of Water and Aqueous Solutions

In ascribing 'structure' to a fluid such as water, the time periods τ_V and τ_D just mentioned become especially relevant. In fact, we must consider three different time scales: times $t \ll \tau_V$, times intermediate to τ_V and τ_D, and times $t \gg \tau_D$. Assuming we were equipped with a camera which had shutter speeds less than

τ_V, we could obtain a relatively sharp picture of the actual position of a water molecule at any given instant. This would reveal the instantaneous water structure called the I-Structure (Eisenberg and Kauzmann, 1969). If the shutter speed were between τ_V and τ_D, each molecule would complete many oscillations while the shutter was open, and the resulting somewhat blurred picture would provide information on the vibrationally averaged position of the water molecules in water, i.e., the V-Structure (Eisenberg and Kauzmann) of water. If the shutter speed were larger than τ_D, the diffusionally averaged arrangement of the water molecules, or D-Structure (Eisenberg and Kauzmann), could be found. No experimental techniques are available at present to obtain information on the I-Structure of a liquid. Experimental studies which employ infrared or Raman spectroscopy, or neutron scattering techniques, lead to information on the V-Structure, while X-ray studies determine the D-Structure.

Spectroscopic studies show that the frequency of oscillation for water molecules is slightly smaller in water than in ice, the period of vibration being $\tau_V \approx 10^{-13}$ sec. Studies on self diffusion, viscosity, dielectric relaxation, and nuclear magnetic resonance relaxation show that a water molecule in water has a characteristic displacement period near 0 °C of $\tau_D \approx 10^{-11}$ sec.

From X-ray data (Narten *et al.*, 1967; Narten and Levy, 1969, 1970, 1971, 1972) one may derive the average number $\rho(\bar{r})$ of molecules in a volume element of water which is located at a distance $\bar{r}$ from any given water molecule. Usually, however, one does not plot $\rho(\bar{r})$ but rather $g(\bar{r}) = \rho(\bar{r})/\rho_w$, where ρ_w represents the bulk density of water expressed as the number of molecules per unit volume of water. Thus, $g(\bar{r})$ is the factor by which the average local density $\rho(\bar{r})$ of water molecules differs at $\bar{r}$ from the density of water molecules in bulk water, and so at large distances from a given water molecule, $g(\bar{r}) = 1.0$. On the other hand, in the vicinity of the given molecule, the local density may differ considerably from bulk density. An example of the radial distribution function $g(\bar{r})$ for water of various temperatures is given in Figure 3-6. The first maximum near 2.9 Å must be attributed to the interactions between the oxygen atoms of nearest neighbor

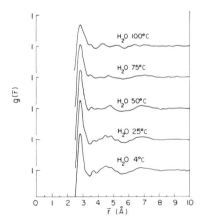

Fig. 3-6. Radial distribution functions g(r) for water at various temperatures. (From Narten *et al.*, 1967, with changes.)

water molecules. The broad maxima near 4.5 Å and 7 Å result from interactions between the oxygen atoms of second nearest and higher order nearest neighbor water molecules.

Figure 3-6 also shows that with decreasing temperature the maxima become increasingly distinct, which implies that the number of water molecules participating in interactions at the distances of the intensity peaks increases. Thus, we see that with decreasing temperature water becomes structurally more ordered. This trend continues, at temperatures below 0 °C, with the scattering intensity peaks continuing to become increasingly pronounced and shifting toward the X-ray intensity maxima observed for ice (Dorsch and Boyd, 1951). We may conclude that, although the long range order breaks down when ice melts, considerable local ordering persists in water. This implies that not all the hydrogen bonds which exist in ice become broken when ice melts. At any moment, a certain number of H-bonds are intact even though the location of the intact bonds in water rapidly fluctuates, since H-bonds break and re-form in continuous succession. (It is interesting to note the correlation between the maxima shown in Figure 3-6 and the nearest neighbor distances in ice; see Section 3.3.)

Information on the state of hydrogen bonds in water can also be obtained from infrared and Raman spectra (Walrafen, 1966, 1967, 1968a, b, 1972). Such spectra confirm that water molecules exist as entities in water. They also give evidence that some O—H groups in H$_2$O are hydrogen bonded and, therefore, point toward a free, lone electron pair of a neighboring H$_2$O molecule, while other O—H groups are nondirectionally bonded to the surrounding water molecules and hence are disoriented with respect to neighboring lone electron pairs (Kell, 1972a; Eisenberg and Kauzmann, 1969). The latter are referred to as non-hydrogen bonded or 'broken' O—H groups. Estimates of the percentage of broken H-bonds in water as a function of temperature are summarized in Figure 3-7.

These experimental findings are supported by recent studies which attempt to simulate the molecular structure of water by purely theoretical methods (Rah-

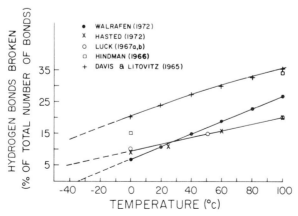

Fig. 3-7. Variation with temperature of number of hydrogen bonds broken in water.

man and Stillinger, 1971; Stillinger and Rahman, 1972; Popkie *et al.*, 1973; Kistenmacher *et al.*, 1974a, b). Stillinger and Rahman applied conventional molecular dynamics to a system of 216 water molecules which interacted via a potential function developed by Ben-Naim and Stillinger (1972) and Ben-Naim (1972). This pair potential function is considerably more complicated than the Stockmayer potential (3-1). It is based on Bjerrum's four-point charge model for a water molecule and incorporates the linear bonding tendency between neighbors in a tetrahedral pattern such as that found throughout the ice lattice, or locally around a given water molecule in water. It has been argued by Stillinger (1970) and Stillinger and Rahman (1972) that this potential function also incorporates the principal features of non-additivity; i.e., it takes into account the many-body aspect of the problem.

Recently, another pair potential function was developed by Clementi *et al.* (1973), Popkie *et al.* (1973), and Kistenmacher *et al.* (1974a, b) (see also Abraham, 1974a, and Fromm *et al.*, 1975). This function is based on an analytically fitted Hartree-Fock potential (Kern and Karplus, 1972), the Bernal and Fowler (1933) point charge model, and correlation energy corrections due to induced dipole interaction and short-range effects. It was used in conjunction with the Monte-Carlo simulation method of Barker and Watts (1969) to study a group of 125 water molecules. The computations of Clementi, Popkie, and Kistenmacher, as well as those of Stillinger and Rahman, yielded radial distribution functions for water molecules in water which are in fair agreement with X-ray results. In addition, the molecular dynamics study of Stillinger and Rahman (1972) predicted that the hydrogen bond rupture mechanism in water is characterized by an excitation energy of about $2.5 \, \text{kcal mol}^{-1}$. This is in good agreement with the Raman and infrared spectra experiments of Walrafen (1966, 1967, 1968a, b, 1972), who found that the energy necessary to rupture hydrogen bonds in water is about $2.55 \, \text{kcal mol}^{-1}$. Infrared spectra taken by Worley and Klotz (1966) imply the comparable value of $2.4 \, \text{kcal mol}^{-1}$, while Davis and Litovitz (1965) obtained $2.6 \, \text{kcal mol}^{-1}$. Finally, Bucaro and Litovitz (1971) inferred $2.5 \pm 0.1 \, \text{kcal mol}^{-1}$ from depolarized light scattering measurements.

Considerable uncertainties still exist as to how X-ray, infrared, and Raman studies should be interpreted in terms of the arrangement of the water molecules in water. Most modern theories of water assume that water has a 'structure' which can be described in terms of highly hydrogen-bonded, three dimensional configurations of molecules. We shall now describe briefly just the main features of some of the more prominent models put forward for the water structure.

In the 'quasi-crystalline model' the water structure is assumed to resemble one of several possible forms: a broken down ice-I_h structure (Bernal and Fowler, 1933; Katzoff, 1934; Morgan and Warren, 1938), a quartz structure (Bernal and Fowler, 1933), a structure of octahedrally arranged molecules (Van Eck *et al.*, 1958), or a structural mixture of molecules arranged in a tridymite structure dispersed in a denser ice-*III* structure (Jhon *et al.*, 1966). In the 'interstitial model' the water is visualized as consisting of a highly hydrogen bonded structure inside of which non-bonded or partially bonded molecules occupy interstitial structure positions (Samoilov, 1946, 1957; Forslind, 1952; Namiot, 1961; Danford and Levy, 1962;

Krestov, 1964; Gurikov, 1960, 1965). In the 'clathrate* model' water is assumed to have a structure similar to the clathrate structure of gas hydrates except that instead of a gas molecule a water molecule is held inside each cavity of a cage-like, hydrogen bonded framework of pentagonal dodecahedron cages (Pauling, 1959, 1960; Frank and Quist, 1961). The 'flickering cluster model' makes use of the partially covalent character of the hydrogen bond and assumes that H-bond formation in water is a cooperative phenomenon, in that the formation of a hydrogen bond between two water molecules reinforces the tetrahedral hybridiza-tion in the oxygen atoms. This in turn strengthens all existing bonds and promotes the formation of new bonds. Conversely, the breaking of an H-bond in water results in the almost simultaneous rupture of a whole group of bonds, thus leading to the formation and dissolution of water clusters in a 'flickering' manner (Frank and Wen, 1957; Frank, 1958). The 'mixture model' pictures water as a mixture of 0-, 1-, 2-, 3-, and 4-bonded water molecules engaging in the formation of various sized clusters (Haggis et al., 1952; Némethy and Scheraga, 1962a, b, 1964; Walrafen, 1966, 1967, 1968a, b, 1972). Finally, the 'bent-bond model' assumes that few, if any, bonds between water molecules are broken upon melting of ice, but instead become bent to various degrees (Pople, 1951).

Even though all of the models mentioned above were found to have certain attractive features from the point of view of their capacities to explain some of the observed physical properties of water, most of them suffer from a too highly idealized and overly rigid arrangement of the water molecules. This becomes particularly obvious if we compare these models with the results of the recent molecular dynamics model of Stillinger and Rahman (1972) mentioned above. Although the results of their computations support the 'mixture model' for water in which water molecules engage in a varying number of hydrogen bonds which locally tend to be tetrahedrally oriented, some bending away from bond linearity was also found to occur, especially at warmer temperatures. Furthermore, no clusters of molecules arranged in the manner of ice-I_h or in any other ice-like or clathrate structures were found for temperatures down to $-8\,°C$, and no obvious separation of water molecules into 'lattice'-molecules and 'interstitial'-molecules was detected. On the other hand, water molecules were frequently found to be arranged in polygons of 4 to 7 sides. Finally, a number of molecules exhibited 'dangling' O—H bonds which were not included in H-bond formation and persisted over times longer than the vibrational period of a water molecule.

Experiments show that the structure of water is altered when water-soluble salts, in part dissociated into ions, are dissolved in water. The aqueous solution resulting from dissolving a salt in water would be an *ideal solution* if the dissolved salt molecules or ions in no way affected the water molecules. In any real aqueous solution this is not the case. For example, some of the salt molecules or ions do not fit into the water 'structure' and, therefore, distort it, causing a *size effect*. Second, solute ions are prone to interact with the water-dipoles which, depending on the size and electric charge of the ion, become grouped

---

* A clathrate is a complex in which molecules of one substance are completely enclosed by molecules of another substance.

around the ion. This effect is called *hydration*. Since large ions have weaker local electric fields than small ions, the hydration effect is greater for small ions. In addition, hydration is more pronounced for positive ions than for negative ones, since a positive ion tends to interact with both lone electron pairs, which blocks the formation of two H-bonds. On the other hand, a negative ion tends to interact with just one H—O group of a water molecule, which blocks the formation of only one hydrogen bond.

Both the size and hydration effects cause hydrogen bonds in the vicinity of an ion to be broken. Such structure breaking and lessening of the four-coordination among the water molecules in water as a result of dissolved salts have been inferred from X-ray, nuclear magnetic resonance, and infrared and Raman spectra studies, as well as from studies on the dielectric properties, the viscosity, thermal conductivity, and heat capacity of aqueous solutions, and from studies on the diffusion of water and ions in aqueous solutions (Kavanau, 1964; Robinson and Stokes, 1970; Horne, 1972; Franks, 1973; Ben-Naim, 1974). According to these investigations, it is useful to visualize the arrangement of water molecules around an ion in the form of three regions: (1) a region close to the ion where the water molecules are immobilized as a result of their electrical interaction with it, (2) a transition region further out in which the water is less ordered than ordinary water because of the structural disruption caused by the size and charge of the ion, and (3) the outermost region consisting of ordinary water.

CHAPTER 4

EQUILIBRIUM BETWEEN WATER VAPOR, WATER, AQUEOUS SOLUTIONS, AND ICE IN BULK

In this chapter we shall discuss the equilibrium thermodynamics of and between the bulk phases of water and aqueous solutions. In addition to providing useful information on the behavior of water substance, this material, with surface effects included, will also serve as a basis for our later discussion on the phase changes which lead to cloud particle formation.

For background on the material covered in this chapter, the reader may wish to refer to texts on chemical thermodynamics and physical chemistry such as Kortüm (1972), Robinson and Stokes (1970), Prigonine and Defay (1967), Reiss (1965), Kirkwood and Oppenheim (1961), Lewis and Randall (1961), and Glasstone (1959), and the review articles by Harrison (1965a, b) and Goff (1949).

4.1 Useful Thermodynamic Relations

Consider an open, homogeneous (single phase) thermodynamic system which may exchange heat, pressure work, and mass with the environment. For small reversible changes the second law of thermodynamics tells us that the heat added may be expressed as $T\,dS$, where T and S are respectively the temperature and entropy of the system; the incremental pressure work done on the system is $-p\,dV$, where p and V are the pressure and volume of the system; and the incremental mass added is measured by dn_k, $k = 1, 2, \ldots, c$, where n_k is the number of moles of chemical component k of the c components comprising the system. According to the first and second laws of thermodynamics, the incremental change in the internal energy $U = U(S, V, n_1, n_2, \ldots, n_c)$ of the system for reversible processes is

$$dU = T\,dS - p\,dV + \sum_{k=1}^{c} \mu_k\,dn_k, \qquad (4\text{-}1)$$

where

$$\mu_k \equiv \left(\frac{\partial U}{\partial n_k}\right)_{S,V,n_{j\neq k}} \qquad (4\text{-}2)$$

is called the chemical potential of component k.

Note that U and the independent state variables S, V, and n_k are extensive (proportional to n_k), in contrast to the intensive variables T, p, and μ_k. Let us denote the extensive and intensive variables by x_i and $y_i = \partial U/\partial x_i$, respectively.

71

Then for constant λ we have $U(\lambda x_1, \lambda x_2, \ldots) = \lambda U(x_1, x_2, \ldots)$, so that

$$U(x_1, x_2, \ldots) = \frac{d}{d\lambda} U(\lambda x_1, \lambda x_2, \ldots) = \sum_i \frac{\partial U(\lambda x_1, \lambda x_2, \ldots)}{\partial(\lambda x_i)} x_i = \sum_i y_i x_i,$$

(4-3a)

or

$$U(S, V, n_1, n_2, \ldots, n_c) = TS - pV + \sum_{k=1}^{c} \mu_k n_k,$$

(4-3b)

which is called Euler's equation.

If we subtract (4-1) from the differential of (4-3b), we obtain the *Gibbs-Duhem* relation

$$\sum_{k=1}^{c} n_k \, d\mu_k = -S \, dT + V \, dp$$

(4-4a)

or

$$\sum_{k=1}^{c} x_k \, d\mu_k = -s \, dT + v \, dp,$$

(4-4b)

where

$$x_k = n_k \Big/ \sum_{i=1}^{c} n_i = n_k/n$$

(4-5)

is the mole fraction of component k, and s and v are mean molar quantities. The result (4-4a) proves to be especially useful for exploring the relationships between phases in equilibrium.

The study of some processes is facilitated by introducing other thermodynamic potentials, in addition to the internal energy. We shall have occasion to use three: the enthalpy $H \equiv U + pV$, the Helmholtz free energy $F \equiv U - TS$, and the Gibbs free energy $G \equiv U + pV - TS = H - TS$. From (4-1), we see that

$$dH = T \, dS + V \, dp + \sum_{k=1}^{c} \mu_k \, dn_k,$$

(4-6)

$$dF = -S \, dT - p \, dV + \sum_{k=1}^{c} \mu_k \, dn_k,$$

(4-7)

$$dG = -S \, dT + V \, dp + \sum_{k=1}^{c} \mu_k \, dn_k.$$

(4-8)

Also, from (4-3b) and (4-8) we find that μ_k may be regarded as a partial molar Gibbs free energy, g_k, i.e.,

$$g_k \equiv \left(\frac{\partial G}{\partial n_k}\right)_{T,p,n_j} = \mu_k; \quad G = \sum_{k=1}^{c} n_k \mu_k = \sum_{k=1}^{c} n_k g_k.$$

(4-9)

Since dG is a perfect differential, we further conclude from (4-8) and (4-9) that

$$\left(\frac{\partial \mu_k}{\partial T}\right)_{p,n_{j\neq k}} = -\left(\frac{\partial S}{\partial n_k}\right)_{T,p,n_{j\neq k}} = -s_k$$

(4-10)

and

$$\left(\frac{\partial \mu_k}{\partial p}\right)_{T,n_{j\neq k}} = \left(\frac{\partial V}{\partial n_k}\right)_{T,p,n_{j\neq k}} = v_k,$$

(4-11)

where s_k and v_k are the partial molar entropy and volume of component k, respectively.

From (4-6) we see that if p and $n_1, n_2, \ldots, n_c$ are held constant, then dH measures the change in heat content in a reversible process. Therefore, the enthalpy is called the heat content of the system at constant pressure, and we may write

$$\left(\frac{\partial h}{\partial T}\right)_{p,n_k} = C_p = T\left(\frac{\partial s}{\partial T}\right)_{p,n_k}, \tag{4-12}$$

where C_p is the mean molar heat capacity at constant p, and $h = H/n$, $s = S/n$. Finally, a useful relationship for evaluating the chemical potential from the enthalpy may be easily derived: $\mu_k = g_k = h_k - Ts_k$, so that

$$\left[\frac{\partial(\mu_k/T)}{\partial T}\right]_{p,n_{j\neq k}} = -\frac{1}{T}\left[\frac{\mu_k}{T} - \left(\frac{\partial \mu_k}{\partial T}\right)_{p,n_{j\neq k}}\right] = -\frac{h_k}{T^2}, \tag{4-13}$$

where we have used (4-10).

4.2 General Conditions for Equilibrium

The second law of thermodynamics provides, as a corollary, a quantitative criterion for thermodynamic equilibrium. Consider an isolated system which is not in equilibrium. In such a system irreversible processes evolve spontaneously. According to the second law, the entropy of such a system will increase until eventually it reaches a state where its entropy is a maximum. In such a state all irreversible processes will have stopped and only those processes, if any, will continue which are completely reversible. The system is then in a state of equilibrium. Thus, for an open homogeneous system held at constant U, V, and $n_1, n_2, \ldots, n_c$, the criterion of equilibrium is

$$(\delta S)_{U,V,n_k} \leq 0, \tag{4-14}$$

where δS refers to the virtual variation in entropy with respect to neighboring states. An alternative expression for a system with constant S, V, and $n_1, n_2, \ldots, n_c$ is

$$(\delta U)_{S,V,n_k} \geq 0. \tag{4-15}$$

In addition, the equilibrium is stable if $(\delta^2 S)_{U,V,n_k} < 0$, unstable if $(\delta^2 S)_{U,V,n_k} > 0$, and conditionally stable or metastable if $(\delta^2 S)_{U,V,n_k} = 0$. Here $\delta^2 S$ is the second virtual variation in entropy with respect to neighboring states.

Unstable equilibrium states cannot be realized in nature since natural systems are continuously exposed to environmental perturbations which, even though very small, are always sufficient to prevent the system from remaining in such a state. On the other hand, metastable states frequently occur in nature. Super-cooled water in an environment of moist air saturated with water vapor is an example of such a system: while the supercooled water is in stable equilibrium with the water vapor surrounding it, it is in unstable equilibrium with respect to ice, into which it would immediately transform if it came into contact with it.

Let us extend these equilibrium conditions to a heterogeneous isolated system containing c chemical components characterized by μ_k and n_k. By definition, a heterogeneous system consists of two or more phases which are separated from each other by surfaces of discontinuity in one or more of the intensive variables. Let us assume that all φ phases of a heterogeneous system are originally isolated and each phase is in internal equilibrium. We may now ask what conditions on the intensive variables are necessary and sufficient to insure equilibrium in the system after the restraint of isolation of the phases has been removed. In seeking these conditions we shall assume that no chemical reactions occur and that the heterogeneous system itself remains isolated. After removing the restraint of isolation of the φ phases of the system, each phase (α) constitutes a homogeneous open system for which the condition of equilibrium is given by (4-15). Also, since extensive variables are additive, we may write

$$U = \sum_{\alpha=1}^{\varphi} U^{(\alpha)}; \quad S = \sum_{\alpha=1}^{\varphi} S^{(\alpha)}; \quad V = \sum_{\alpha=1}^{\varphi} V^{(\alpha)}; \quad n_k = \sum_{\alpha=1}^{\varphi} n_k^{(\alpha)}. \tag{4-16}$$

Then from (4-1) the condition of equilibrium may be expressed as

$$\delta U = \sum_{\alpha=1}^{\varphi} \left[T^{(\alpha)} \delta S^{(\alpha)} - p^{(\alpha)} \delta V^{(\alpha)} + \sum_{k=1}^{c} \mu_k^{(\alpha)} \delta n_k^{(\alpha)} \right] \geqslant 0, \tag{4-17}$$

where S, V, and n_k are held constant, according to

$$\delta S = \sum_{\alpha=1}^{\varphi} \delta S^{(\alpha)} = 0; \quad \delta V = \sum_{\alpha=1}^{\varphi} \delta V^{(\alpha)} = 0; \quad \delta n_k = \sum_{\alpha=1}^{\varphi} \delta n_k^{(\alpha)} = 0. \tag{4-18}$$

For simplicity, let us momentarily consider a system of just two phases, which we denote by (') and (''). Then from (4-17) and (4-18) we may express the equilibrium condition as

$$(T'' - T')\delta S'' - (p'' - p')\delta V'' + (\mu_k'' - \mu_k')\delta n_k'' \geqslant 0. \tag{4-19}$$

The constraints of (4-18) have all been incorporated into this equation, so that $\delta S''$, $\delta V''$, and $\delta n_k''$ can be chosen independently. Therefore, the equation can be satisfied only if the coefficients of each of the variations are equal to zero. Since the same analysis could be applied to any pair of phases in a more complex system, we conclude that the conditions for thermodynamic equilibrium of a heterogeneous system in which all interface surfaces are perfectly deformable, heat conducting, and permeable to all components are

$$T' = T'' = \cdots = T^{(\alpha)}, \tag{4-20}$$

$$p' = p'' = \cdots = p^{(\alpha)}, \tag{4-21}$$

$$\mu_k' = \mu_k'' = \cdots = \mu_k^{(\alpha)}, \quad k = 1, 2, \ldots, c. \tag{4-22}$$

These three equations express the conditions of thermal, mechanical, and chemical equilibrium, respectively.

4.3 Phase Rule

The discussion of systems in equilibrium is facilitated by what is known as the *Gibbs phase rule*. This rule enables us to determine the *variance* of a system, i.e., the number of intensive variables which may be freely specified without causing the system to depart from equilibrium. To derive the phase rule, let us consider again the heterogeneous, isolated system of the previous section. As we have seen, in equilibrium the system is characterized by a common T and p and by a number of mole fractions in the various phases. Let us denote the mole fraction of the kth component in the jth phase by $x_k^{(j)}$ $(= n_k^{(j)}/\Sigma_k n_k^{(j)})$. Then for a system of φ phases and c components, there will be φc mole fractions altogether, giving us a total of $2 + \varphi c$ intensive variables at equilibrium. However, not all of these are independent. Thus, for every phase we have the simple mass conservation constraint that $\Sigma_{k=1}^{c} x_k^{(j)} = 1$, for a total of φ constraints. In addition, we have the condition (4-22) on the chemical potentials, which constitute another $\varphi - 1$ constraints for every k, for a total of $c(\varphi - 1)$ constraints. Therefore, at equilibrium the total variance, or number of thermodynamic degrees of freedom, is

$$w = 2 + \varphi c - \varphi - c(\varphi - 1) = 2 + c - \varphi, \qquad (4\text{-}23)$$

which is the Gibbs phase rule for bulk phases. (As we shall see in Section 5.3, the phase rule assumes a substantially different form if phases with curved interfaces are present in a system.)

Let us consider some simple applications of (4-23). For a homogeneous fluid in equilibrium, we have $\varphi = 1$, $c = 1$, and so $w = 2$. This is consistent with the familiar circumstance that the equation of state of such a system provides one connection among three thermodynamic state variables (e.g., ρ, T, p). For a mixture of two gases, $c = 2$, $\varphi = 1$, and so $w = 3$; obviously, this is like the previous example, except that now we can also freely choose the relative concentration of the gases. For water in equilibrium with its vapor, $c = 1$, $\varphi = 2$, and so $w = 1$: the system is monovariant, and the vapor pressure is a function only of temperature. For water in equilibrium with its vapor and ice, $c = 1$, $\varphi = 3$, and so $w = 0$; equilibrium is possible only for a single choice of T and pressure, which defines the *triple point* temperature T_{tr} of the system. If this system is now exposed to the atmosphere, $c = 2$ (water substance and air), $\varphi = 3$, and so $w = 1$. However, if we make the reasonable assumption that the total gas pressure remains constant, the system can have no further variance if it is to remain in equilibrium. The system is now said to be at its *ice point* temperature T_0. Thus, we see that the ice point temperature is a function of pressure. By convention the concept of the ice point is restricted further by specifying that the pressure on the system should be exactly one atmosphere (see Section 4.9).

4.4 Ideal Versus Real Behavior of Dry Air, Water Vapor, and Moist Air

Let us now consider some of the equations of state we will need in order to apply the equilibrium conditions (4-20)–(4-22). If we assume that water vapor behaves as an ideal gas of non-interacting point molecules, its equation of state may be written in the following familiar forms:

$$e v_{v,0} = \mathscr{R}T; \quad e = \rho_v R_v T, \tag{4-24}$$

where e denotes the water vapor pressure, $v_{v,0}$ is the molar volume of pure water vapor, $\mathscr{R}$ the universal gas constant, ρ_v the vapor density, and $R_v = \mathscr{R}/M_w$ the specific gas constant for water vapor, M_w being the molecular weight of water.

A similar equation of state may be written for dry air, if we regard it as a mixture of ideal gases. Then in a fixed volume V we have for the partial pressure p_k of the kth component, $p_k V = n_k \mathscr{R}T = m_k \mathscr{R}T/M_k$, where m_k is the mass of the kth component. Applying Dalton's law, $p = \Sigma_k p_k$, we obtain

$$p_a v_{a,0} = \mathscr{R}T; \quad p_a = \rho_a R_a T, \tag{4-25}$$

where p_a denotes the pressure of dry air, $v_{a,0} = V/\Sigma_k n_k$ is its molar volume, ρ_a its density, and $R_a = \mathscr{R}/M_a$ its specific gas constant, with

$$M_a = \frac{\sum_k m_k}{\sum_k n_k} = \frac{\sum_k m_k}{\sum_k (m_k/M_k)}. \tag{4-26}$$

It is important to assess the extent of deviations from ideality owing to molecular interactions of the sort we discussed in the previous chapter. This problem has been considered in detail by Goff (1942, 1949) and Goff and Gratch (1945, 1946), who found that a virial expansion of the equation of state truncated at the fourth term could be used to represent the behavior of real air and water vapor. Thus, the real gas equations of state can be expressed adequately in the following form:

$$e v_{v,0} = \mathscr{R}T - A_{ww}e - A_{www}e^2, \tag{4-27}$$

$$p_a v_{a,0} = \mathscr{R}T - A_{aa}p_a - A_{aaa}p_a. \tag{4-28}$$

Values for the virial coefficients (A) may be determined experimentally from accurate measurements of the state variables. Alternatively, they may be computed theoretically, at least in principal, by using the methods of statistical mechanics (e.g., Hirschfelder et al., 1954). Of course, in order to do this one must model the intermolecular forces. Both routes have encountered great difficulties for the case of water vapor (e.g., Kennard, 1938; Harrison, 1965a; Kell et al., 1968; Eisenberg and Kauzmann, 1969; Ben-Naim, 1974). The presently accepted values for A_{aa}, A_{aaa}, A_{ww}, and A_{www} are tabulated in Goff (1949), Harrison (1965a) and the Smithsonian Meteorological Tables (SMT).

It is customary to express the deviation from ideal gas behavior in terms of what is known as the compressibility factor, $C \equiv pv/\mathscr{R}T$, values of which may be computed once the virial coefficients are known. A few selected values for C_a and C_v are given in Table 4-1. These show that the ideal gas law for both dry air and water vapor is in error by less than 0.2% throughout the range of meteorological interest. Fortunately, therefore, the simple expressions (4-24) and (4-25) can be used with confidence.

TABLE 4-1

Deviation of dry air, pure water vapor, and moist air from ideality in terms of the compressibility factor. (Based on values from Harrison, 1965a.)

| Temperature | C_a | | C_v | C_v | C_m |
	1100 mb	300 mb	Water vapor saturated with respect to water	Water vapor saturated with respect to ice	100% Relative humidity, 1100 mb
−30 °C	0.9988	0.9997	0.9999	0.9999	–
0 °C	0.9994	0.9998	0.9995	0.9995	0.9993
30 °C	0.9997	0.9999	0.9982	–	0.9995

In view of these results it is perhaps almost obvious that moist air can also be treated as an ideal gas (see Table 4-1). We say almost, because there remains the possibility that the forces of interaction between water molecules and some species of air molecules might be much greater than the water-water or air-air interactions. A partial explanation of why this in fact does not occur may be given by considering the example of the van der Waal's interaction (see Chapter 3): the strength of this force depends on the mean square fluctuation of the electric dipole moment and the molecular polarizability. Neither of these parameters shows an extremely wide range in nature, and so accordingly the van der Waal's interaction is relatively insensitive to the molecular species involved.

The ideal gas law for moist air may be written in analogy to (4-25) as

$$p v_m = \mathscr{R} T; \quad p = \rho R_m T, \tag{4-29}$$

where R_m is determined in the same fashion as R_a. However, it is customary and more convenient to write the equation of state in the form used for dry air, with the moisture correction associated with the temperature. Proceeding in this way, we have for the pressure and density of moist air, $p = p_a + e$ and $\rho = \rho_a + \rho_v$. Then from (4-24) and (4-25) we find $\rho = (M_a p_a + M_w e)/\mathscr{R} T = M_a[p - (1 - \varepsilon)e]/\mathscr{R} T$, with $\varepsilon = M_w/M_a = 0.622$, or

$$p = \rho R_a T_v, \tag{4-30}$$

where T_v, called the *virtual temperature* of moist air, is given by

$$T_v = T[1 - (1 - \varepsilon)e/p]^{-1}. \tag{4-31}$$

Physically, T_v is the temperature which dry air would have to have in order for its density to match that of the actual air. Since $M_w < M_a$, moist air has a lower density than dry air at the same temperature, so that always $T_v > T$. However, the extent of the difference is not large: a reasonable upper bound for e is the saturation value at 30 °C, which is only about 4% of standard sea level pressure; hence $T_v - T \leqslant 5$ °C.

We have now introduced two quantities, the virtual temperature T_v and the density or *absolute humidity* ρ_v, which provide a measure of the water vapor content of air. There are, in addition, several other such 'moisture variables' in common use. Among the most important are: the *mixing ratio* w_v, the *specific humidity* q_v, the *relative humidity* ϕ_v, the *mole fraction of water vapor* x_v, the *saturation ratio* S_v, and the *supersaturation* s_v. These are defined as follows:

$$w_v \equiv \frac{m_v}{m_a} = \frac{\rho_v}{\rho_a} \quad (4\text{-}32) \qquad q_v \equiv \frac{m_v}{m_m} = \frac{\rho_v}{\rho} \quad (4\text{-}33) \qquad \phi_v \equiv w_v/w_{v,sat} \qquad (4\text{-}34)$$

$$x_v \equiv \frac{n_v}{n_a + n_v} \quad (4\text{-}35) \qquad S_v \equiv \frac{e}{e_{sat}} \quad (4\text{-}36) \qquad s_v \equiv (w_v/w_{v,sat}) - 1, \quad (4\text{-}37)$$

where the subscript sat refers to the maximum possible, saturated value. Air for which $\phi_v = 1$ (100%) is saturated; if $\phi_v > 1$ (> 100%), it is said to be super-saturated, corresponding to $s_v > 0$.

The moisture variables are also connected by various relationships, such as

$$w_v = q_v/(1 - q_v); \quad q_v = w_v/(1 + w_v), \tag{4-38}$$

$$x_v = e/p = w_v/(w_v + \varepsilon), \tag{4-39}$$

$$w_v = \varepsilon x_v/(1 - x_v) = \varepsilon e/(p - e) \approx \varepsilon e/p, \tag{4-40}$$

$$T_v = T\left[\frac{1 + (w_v/\varepsilon)}{1 + w_v}\right] = T\left[1 + \left(\frac{1}{\varepsilon} - 1\right)q_v\right]. \tag{4-41}$$

4.5   Chemical Potential of Water Vapor in Humid Air, and of Water in Aqueous Solutions

We are now in a position to derive the chemical potential of water vapor and, through the equilibrium conditions, the chemical potential of water in aqueous solutions. From (4-11) we have, for an ideal gas k in a mixture of ideal gases, $(\partial \mu_k / \partial p_k)_{T,n_{j \neq k}} = v_k = \mathscr{R}T/p_k$, so that, upon integration,

$$\mu_k = \mu_{k,0} + \mathscr{R}T \ln p_k, \tag{4-42}$$

where the integration constant $\mu_{k,0}$ depends only on the temperature. For such a mixture the partial pressure is $p_k = x_k p$, so that also

$$\mu_k = \mu_{k,0} + \mathscr{R}T \ln p + \mathscr{R}T \ln x_k. \tag{4-43}$$

Therefore, if we assume pure water vapor at pressure e is an ideal gas, its chemical potential is

$$\mu_{v,0}(e, T) = \mu_{v,0}^+(T) + \mathscr{R}T \ln e, \tag{4-44}$$

where $\mu_{v,0}^+(T)$ is the chemical potential at a standard state of unit pressure. Similarly, the chemical potential $\mu_v(p, T)$ of water vapor in humid air at total pressure p is

$$\mu_v(p, T, x_v) = \mu_{v,0}^+(T) + \mathscr{R}T \ln p + \mathscr{R}T \ln x_v. \tag{4-45}$$

From (4-44) with e = p we see that this last result may also be expressed as

$$\mu_v(p, T, x_v) = \mu_{v,0}(p, T) + \mathscr{R}T \ln x_v, \tag{4-46}$$

which shows that $\mu_v \leq \mu_{v,0}$ since $x_v \leq 1$.

In contrast to pure gases whose chemical potentials vary logarithmically with pressure, the chemical potential of a pure liquid is proportional to pressure, to an excellent approximation. This is obvious from (4-11), on realizing that liquids are nearly incompressible. Thus for water we have

$$\left(\frac{\partial \mu_{w,0}}{\partial p}\right)_T = v_{w,0} \approx \text{constant,} \tag{4-47}$$

from which the chemical potential is found to be

$$\mu_{w,0}(p, T) \approx \mu_{w,0}(0, T) + v_{w,0}p. \tag{4-48}$$

As we have seen, if a liquid and gas are in equilibrium the chemical potential of a given component will be the same in both phases (Equation (4-22)); consequently, from (4-42) the chemical potential of component k in a liquid solution, which is in equilibrium with its vapor at partial pressure p_k, is

$$\mu_{k,l} = \mu_{k,0} + \mathcal{R}T \ln p_k. \tag{4-49}$$

In addition, experiments show that for so-called 'ideal' solutions, for which there are no interactions between the solvent and solute molecules, the equilibrium vapor pressure of any component is porportional to its mole fraction in the solution. (This is known as Raoult's law, about which more will be said in the following section.) Assuming Raoult's law, then, we have $p_k = x_{k,l}p_{k,0}$, where $x_{k,l}$ is the mole fraction of component k in the solution, and $p_{k,0}$ is the partial pressure of component k in equilibrium with the pure liquid phase of k at the same temperature. Then as a function of $x_{k,l}$, the chemical potential becomes

$$\mu_{k,l} = \mu_{k,l}^{\circ} + \mathcal{R}T \ln x_{k,l}, \tag{4-50}$$

where $\mu_{k,l}^{\circ}$ is a function of both temperature and total pressure, but is independent of the composition of the solution.

In clouds the liquid phase is rarely present in the form of pure water, but rather is generally a dilute aqueous salt solution. Therefore, (4-50) is especially relevant to us, and we may use it to write the chemical potential for water in an ideal aqueous salt solution in the following form:

$$\mu_w(p, T, x_w) = \mu_w^+(p, T) + \mathcal{R}T \ln x_w, \tag{4-51}$$

where $x_w = n_w/(n_w + n_s) = 1 - x_s$ is the mole fraction of water, n_w and n_s being respectively the number of moles of water and salt in the solution. By analogy, one would expect that the chemical potential of the salt component could be expressed in the same way, viz.,

$$\mu_s(p, T, x_s) = \mu_s^+(p, T) + \mathcal{R}T \ln x_s. \tag{4-52}$$

In passing, we may note that for $x_w = 1$, $\mu_w(p, T) = \mu_{w,0}(p, T) = \mu_w^+(p, T)$, the chemical potential of pure water at p, T. There is no analogous simple physical interpretation for the quantity $\mu_s^+(p, T)$.

Experiments show that most dilute solutions of non-electrolytes are in conformity with (4-51) and (4-52). In general, however, real aqueous solutions depart from such ideal behavior. It is customary to account for non-ideal solutions through the replacement of the mole fraction x by the *activity* a ≡ fx, where f is called the *rational activity coefficient*. Thus for real aqueous salt solutions we write

$$\mu_w(p, T, a_w) = \mu_{w,0}(p, T) + \mathscr{R}T \ln a_w; \quad a_w = f_w x_w, \tag{4-53}$$

$$\mu_s(p, T, a_s) = \mu_s^+(p, T) + \mathscr{R}T \ln a_s; \quad a_s = f_s x_s. \tag{4-54}$$

The importance of the activity to us is that it provides a direct measure of the equilibrium water vapor pressure over a real salt solution, or, in other words, the generalization of Raoult's law to real solutions. We now turn to a demonstration of this property.

4.6 Equilibrium Between an Aqueous Salt Solution and Water Vapor

Consider a system consisting of water vapor in equilibrium with an aqueous salt solution, both at temperature T and pressure e (here $e = e_{sat}$, but for brevity we omit the subscript in the development which follows). From (4-22) we have

$$\mu_{v,0}(e, T, a_w) = \mu_w(e, T, a_w). \tag{4-55}$$

On substituting this equilibrium condition into (4-53) for p = e, we obtain

$$\mu_{v,0}(e, T, a_w) = \mu_{w,0}(e, T) + \mathscr{R}T \ln a_w. \tag{4-56}$$

According to the phase rule, the present system of two components and two phases is divariant (w = 2). Let us now fix T and investigate the variation of a_w with e. Then from (4-55) and (4-56) we see that equilibrium can be maintained for variable e only if

$$\left(\frac{\partial \mu_w}{\partial e}\right)_T de = \left(\frac{\partial \mu_{v,0}}{\partial e}\right)_T de = \left(\frac{\partial \mu_{w,0}}{\partial e}\right)_T + \mathscr{R}T \left(\frac{\partial \ln a_w}{\partial e}\right)_T de. \tag{4-57}$$

Now, on substituting (4-11) and noting that $v_{w,0} \ll v_{v,0}$, (4-57) becomes

$$\left(\frac{\partial \ln a_w}{\partial e}\right)_T = \frac{v_{v,0}}{\mathscr{R}T} = \frac{1}{e}, \tag{4-58}$$

which upon integration yields

$$\ln a_w = \ln e_{sat} + g(T) \tag{4-59}$$

where we have again recognized explicitly that $e = e_{sat}$, and where g(T) is an unknown function of T. We may determine g by taking the limit $a_w \to 1$, which corresponds to the case of pure water; i.e., $g(T) = -\ln e_{sat,w}$, where $e_{sat,w}$ is the saturation vapor pressure over pure water at temperature T. Therefore, (4-59) becomes

$$\frac{e_{sat,s}}{e_{sat,w}} = a_w(T), \tag{4-60}$$

where we have now similarly replaced e_{sat} by the more complete notation $e_{sat,s}$, which denotes the equilibrium vapor pressure over an aqueous salt solution at temperature T.

Equation (4-60) is the desired extension of Raoult's law. For an ideal solution, $a_w = x_w$, and we recover the original Raoult's law:

$$\frac{e_{sat,s}}{e_{sat,w}} = x_w; \quad \frac{\Delta e_{sat}}{e_{sat,w}} = x_s, \tag{4-61}$$

where $\Delta e_{sat} = e_{sat,w} - e_{sat,s}$.

Let us now consider briefly the problem of finding values of a_w for use in (4-60). In the literature of cloud physics the most commonly followed practice in expressing deviations from ideality has been to use the *van't Hoff factor*, i, originally introduced by van't Hoff in his classic studies of osmotic pressure to account, in some poorly understood manner, for the degree of ionic dissociation in electrolytes, McDonald (1953a) effectively defined the factor i through the relation

$$a_w = \frac{n_w}{n_w + in_s}. \tag{4-62}$$

This approach has been followed, for example, in the well known cloud physics texts of Fletcher (1962) and Mason (1971), who use this definition of i in their descriptions of the behavior of solution drops.

However, as pointed out by Low (1969a), the use of the van't Hoff factor has the practical disadvantage that relatively few values for it are available. Also, it is no coincidence that this approach is out of the mainstream of modern physical chemistry, which has largely ignored the van't Hoff factor altogether. Therefore, following Low (1969a, c) we shall now briefly introduce those parameters which are regarded by contemporary physical chemists as providing a more fundamental measure of non-ideality, and for which abundant tabulated data exist.

First of all, we introduce the *molality* concentration scale, in place of the mole fraction. The molality $\mathfrak{M}$ is defined as the number of moles of salt dissolved in 1000 g of water, so that

$$x_s = \frac{n_s}{n_s + n_w} = \frac{\mathfrak{M}}{\mathfrak{M} + (1000/M_w)}, \tag{4-63a}$$

or

$$\mathfrak{M} = 1000 \, n_s/n_w M_w = 1000 \, m_s/M_s m_w, \tag{4-63b}$$

where again, M refers to molecular weight and m to mass. It is convenient also to define the quantity

$$\hat{M} \equiv M_w/1000, \tag{4-64}$$

which in combination with (4-63) gives

$$\hat{M}\mathfrak{M} = n_s/n_w = m_s M_w/m_w M_s. \tag{4-65}$$

For example, if the aqueous salt solution is present in the form of a drop of radius a and density ρ_s'', one finds

$$\hat{M}\mathfrak{M} = \frac{m_s M_w}{M_s\left(\dfrac{4\pi a^3}{3}\rho_s'' - m_s\right)}. \tag{4-66}$$

Now, whereas before we associated the rational activity coefficient f with the mole fraction x, we now associate a quantity called the *mean activity coefficient*, and denoted by $\gamma_\pm$, with the molality $\mathfrak{M}$. Then in terms of $\mathfrak{M}$ and $\gamma_\pm$, the water activity of a solution of one salt in concentration $\mathfrak{M}$ turns out to be expressible in the form (e.g., Robinson and Stokes, 1970; Lewis and Randall, 1961; Low, 1969a):

$$\ln a_w = -\nu \hat{M} \mathfrak{M} \left(1 + \frac{1}{\mathfrak{M}} \int_0^{\mathfrak{M}} \mathfrak{M} d \ln \gamma_\pm \right), \tag{4-67}$$

where ν is the total number of ions a salt molecule dissociates into. This is a useful result, because extensive data for $\gamma_\pm(\mathfrak{M})$ exist.

Another quantity which appears often in the physical chemistry literature is called the *molal* or *practical osmotic coefficient*, Φ_s of the salt in solution. This is just the expression in parentheses in (4-67):

$$\Phi_s \equiv 1 + \frac{1}{\mathfrak{M}} \int_0^{\mathfrak{M}} \mathfrak{M} d \ln \gamma_\pm, \tag{4-68}$$

and so also

$$a_w = \exp \left(-\nu \mathfrak{M} \hat{M} \Phi_s \right). \tag{4-69}$$

For aqueous solutions which contain several salts, and this is generally the case for cloud drops, the practical osmotic coefficient for the mixture is obtained by taking a weighted average over the molality of each component in the solution (Hänel, 1976; Thudium, 1978):

$$\Phi_{s,mix}(a_w) = \sum_k \nu_k \mathfrak{M}_k \Phi_{s,k}(a_w) \Big/ \sum_k \nu_k \mathfrak{M}_k \tag{4-70}$$

This result holds on the assumption that interactions between the salts in solution may be disregarded.

The parameter Φ_s was apparently first brought to the attention of cloud physicists by Byers (1965). As for sources of these various measures of non-ideality, we note that Robinson and Stokes (1970) have tabulated values for Φ_s and $\gamma_\pm$ as a function of $\mathfrak{M}$ for a large number of salts. Values for a_w and i have been computed and tabulated by Low (1969a, b) for some typical salts present in the atmosphere. (Incidentally, we should perhaps emphasize that the parameter i in Byers' description of solution drops is not the van't Hoff factor, but rather is $i = \nu \Phi_s$.) In Table 4-2 we have provided values of Φ_s, a_w, and i for a few salts and concentrations. Note that for a solution to behave ideally, $\Phi_s = 1$, $i = \nu$, and $a_w = x_w$.

Significant departures from ideality are evident in Table 4-2. This behavior is exhibited also in Figure 4-1. It is seen that the interactions of salt ions with

TABLE 4-2

Deviation of aqueous solutions from ideality at 25 °C in terms of the activity of water in solution, the osmotic coefficient, and the van't Hoff factor. For an ideal solution $a_w = 1$, $\Phi_s = 1$, $i = \nu$. (Based on values from Low, 1969a, b.)

m	NaCl $\nu = 2$			NaNO$_3$ $\nu = 2$			(NH$_4$)$_2$SO$_4$ $\nu = 3$		
	a_w	Φ_s	i	a_w	Φ_s	i	a_w	Φ_s	i
0.1	0.99665	0.9324	2.65931	0.99669	0.921	1.84506	0.99586	0.767	2.30577
0.5	0.98355	0.9209	2.03730	0.98440	0.873	1.75980	0.98187	0.677	2.04969
1.0	0.96684	0.9355	1.90392	0.96980	0.851	1.72836	0.96600	0.640	1.95359
2.0	0.93162	0.9833	1.85737	0.94222	0.826	1.70215	0.93488	0.623	1.93337
5.0	0.80675	1.1916	1.86713	0.86766	0.788	1.69335	0.83080	0.672	2.26110

molecules results in a larger reduction of vapor pressure than is predicted by Raoult's law.

Figure 4-1 also illustrates the fairly strong dependence of vapor pressure reduction on the type of salt. This behavior may be used as a measure for the *hygroscopic* nature of salt. It has been customary to express the hygroscopicity of salts in terms of the relative humidity at which a dry salt changes ('deliquesces') into a saturated salt solution (see Table 4-3). Of course, this is also the relative humidity at which the saturated salt solution is in equilibrium with the environmental water vapor. Low (1969a) has proposed an alternative definition in which hygroscopicity is expressed in terms of the amount of salt required per 100 g of pure water to achieve a specified degree of vapor pressure

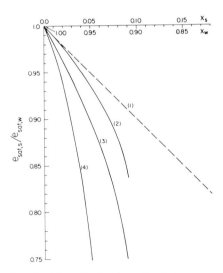

Fig. 4-1. Variation of ($e_{sat,s}/e_{sat,w}$) with mole fraction of salt in aqueous solution and with mole fraction of water in solution at 25 °C. (Based on data of Low, 1969b.) (1) Equation (4-61), (2) sucrose solution, (3) NaCl solution, (4) CaCl$_2$ solution.

TABLE 4-3

Critical values for the water activity of an aqueous salt solution, for the relative humidity of moist air in equilibrium with a saturated aqueous salt solution, and for the relative humidity at which dry salt transforms into a saturated aqueous solution.

Salt	a_w of saturated salt solution (Stokes and Robinson, 1949; Robinson and Stokes, 1970) (25 °C)	Relative humidity (%) for equilibrium between vapor and a saturated salt solution (Stokes and Robinson, 1949; Robinson and Stokes, 1970) (25 °C)	Relative humidity (%) for equilibrium between vapor and a saturated salt solution (Lagford, 1961; Roussel, 1968; Admirat and Grenier, 1975) (0 °C)	(−20 °C)	Relative humidity (%) for the transformation of 2 to 20 μm diameter salt particles to aqueous solution drops (Twomey, 1953, 1954) (20 °C)
K_2SO_4	0.9740	97.40	–	–	97
KNO_3	0.9248	92.48	96	99	–
KCl	0.8426	84.26	88.5	96	86
$(NH_4)_2SO_4$	0.7997	79.97	–	–	80
NH_4Cl	0.7710	77.10	81	90	80.5
NaCl	0.7528	75.28	76.3	80	75–77
$NaNO_3$	0.7379	73.79	–	–	–
NH_4NO_3	0.6183	61.83	78	88	–
$Mg(NO_3)_2$	0.5286	52.86	–	–	56
$Ca(NO_3)_2$	0.4997	49.97	–	–	–
$MgCl_2 \cdot 6H_2O$	0.3300	33.00	35	40	–
$CaCl_2 \cdot 6H_2O$	0.2450	24.50	40	50	–

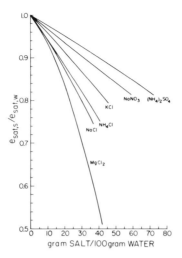

Fig. 4-2.   Variation of $(e_{sat,s}/e_{sat,w})$ with concentration of salt in aqueous solution at 25 °C. (Based on data of Low, 1969b.)

lowering, i.e., a specified activity of water in solution. Low felt this definition to be somewhat more directly relevant to applications in weather modification experiments, where one generally wishes to obtain the maximum possible vapor pressure reduction for a given mass of hygroscopic salt. A comparison between the values given in Table 4-3 and Figure 4-2 shows that each of the two definitions leads to a different ranking for the hygroscopic salts.

Note also that, because of the temperature dependence of the saturation ratio through a_w (Equation (4-60)), the relative humidity at which salts deliquesce changes noticeably with temperature. For example, experiments by O'Brien (1948), Lagford (1961), Hedlin and Trofimenkoff (1965), Roussel (1968), and Admirat and Grenier (1975) show that the relative humidity at which most salts in the atmosphere transform into a saturated salt solution increases by 2 to 30% as the temperature varies from +20 to −20 °C. This reflects the experimental fact that the solubility of the salts studied decreases with decreasing temperature.

4.7   Bulk Density of Ice, Water, and Aqueous Solutions

We shall now present some approximate equations of state for the solid and liquid phases of water.

The density ρ_i of ice can be inferred directly from experiments such as those of Ginnings and Corruccini (1947), or indirectly from a knowledge of the lattice constants of ice which determine the volume of the unit cell, V_{uc}. From V_{uc}, ρ_i is obtained from the relation $\rho_i = 4 M_w/N_A V_{uc}$, where N_A is Avogadro's number. On the basis of the experimental value for ρ_i at 0 °C and a pressure of 1 atm. found by Ginnings and Corruccini, and from the dimensions of the unit cell of ice derived from X-ray studies by La Placa and Post (1960), we have obtained the following empirical curve fit to observed values of $\rho_i(T)$, for $-150\,°C \leqslant T \leqslant 0\,°C$:

$$\rho_i = A_0 + A_1T + A_2T^2, \tag{4-71}$$

with $A_0 = 0.9167$, $A_1 = -1.75 \times 10^{-4}$, $A_2 = -5.0 \times 10^{-7}$, with T in °C, and with ρ in g cm^{-3}. Values computed from (4-71) fit the experimental data with an accuracy better than $\pm5 \times 10^{-4}$ g cm^{-3}. The increase of ρ_i with decreasing T is a reflection of the unit cell temperature dependence shown in Figure 3-5.

As ice melts the bulk density of water substance abruptly increases by about 9.1%. Contrary to what might be expected, X-ray measurements show that during melting the intermolecular distance between first nearest neighbors in water does not decrease, but rather increases over that found in ice by about 3% at 0 °C. Consequently, the density increase must be attributed to a 'filling-in' of space by water molecules which leave regular lattice positions to move into what were cavities in the ice lattice. The X-ray findings are in accordance with this view, and show that the number of first nearest neighbors in water increases from 4.0 in ice to 4.4 at 1.5 °C, reaching 4.9 at 83 °C. But despite this 'filling-in,' water has a very open structure and a density lower than that of an ideal liquid with a close-packed arrangement of molecules. This can be readily seen if we consider that the observed density of water is $\rho_w = 1.0$ g cm^{-3}, and that therefore in water the average volume of a water molecule, $\dot{v}_w = M_w/\rho_wN_A$, is 30×10^{-24} cm^3. On the other hand, if we were to regard a water molecule as a rigid sphere of radius equal to $1/2$ the closest approach distance of two nearest neighbor water molecules, which is 1.38×10^{-8} cm (see Section 3.3), the volume of a water molecule would be 11×10^{-24} cm^3, which is 2.7 times less than that observed. If these spherical molecules were arranged in water in a hexagonal close-packed arrangement (in which case a fraction 0.74 of space is filled with mass), the density of water would have a value of $\rho_w = 0.74 M_w/\dot{v}_wN_A = 2$ g cm^{-3}, or twice the value observed.

According to Kell (1972b), at p = 1 atm. the best experimental values for the density of water can be fitted to

$$10^3\rho_w = \frac{A_0 + A_1T + A_2T^2 + A_3T^3 + A_4T^4 + A_5T^5}{1 + BT}, \tag{4-72}$$

with ρ_w in g cm^{-3}, with $A_0 = 999.8396$, $A_1 = 18.224944$, $A_2 = -7.922210 \times 10^{-3}$, $A_3 = -55.44846 \times 10^{-6}$, $A_4 = 149.7562 \times 10^{-9}$, $A_5 = -393.2952 \times 10^{-12}$, $B = 18.159725 \times 10^{-3}$, and T in °C. Equation (4-72) is applicable to the temperature interval $0\,°C \leqslant T \leqslant 100\,°C$, over which it fits the data to within $\pm1 \times 10^{-6}$ g cm^{-3}. For $T < 0\,°C$, we have fitted the following function to the experimental data of Dorsch and Boyd (1951):

$$\rho_w = A_0 + A_1T + A_2T^2, \tag{4-73}$$

with ρ_w in g cm^{-3}, with T in °C, and $A_0 = 0.99984$, $A_1 = 0.860 \times 10^{-4}$, $A_2 = -0.108 \times 10^{-4}$. This relation is applicable to the temperature interval $-50\,°C \leqslant T \leqslant 0\,°C$, with an accuracy of better than $\pm1 \times 10^{-5}$ g cm^{-3}.

A plot of ρ_w as a function of temperature shows that the density of water undergoes a maximum at about 4 °C (more exactly, at 3.984 °C, where $\rho_{w,max} =$

$0.999972 \, \text{g cm}^{-3}$). As we have discussed, the density decrease for lower temperatures is the result of an increasing number of 4-coordinated water molecules and hydrogen bonds, creating an increasingly 'open' structure. The decrease in density for temperatures larger than 4°C must be attributed to an increase in amplitude of the molecular vibrations, which causes a general expansion of the water volume.

The above equations of state are incomplete, of course, since no dependence on pressure has been included. Fortunately, however, this influence is negligible for the range of pressures encountered in the troposphere. This may be illustrated by the example of the small temperature difference which separates the triple point and the ice point. As we shall discuss more fully in the next section, a relatively extreme change in total pressure (from 6 to 10^3 mb) lowers the melting temperature of ice by only about 0.01 °C.

If an aqueous solution behaved ideally, its density could be computed from the simple relation $\rho'' = (m_s + m_w)/V_{s,0}$, where $V_{s,0} = n_w v_{w,0} + n_s v_{s,0}$, and the subscript 0 refers to the pure substance in isolation. A suitable modification for real solutions can be obtained in the following manner. First, one now writes

$$V_s = n_w v_w + n_s v_s, \qquad (4\text{-}74)$$

where v_w and v_s are the partial molar volumes of water and salt in solution. Next, all deviations from ideality are incorporated into a quantity v_a called the *apparent molal volume* (of salt), and defined by

$$v_a \equiv (V_s - n_w v_{w,0})/n_s. \qquad (4\text{-}75)$$

Combining (4-74) and (4-75) and noting that $v_{w,0} = M_w/\rho_w$, $v_w = M_w/\rho''_s$, and $v_s = M_s/\rho''_s$, we finally obtain

$$\rho''_s = \left(\frac{1000}{\mathfrak{M}} + M_s\right)\Big/\left(v_a + \frac{1000}{\mathfrak{M}\rho_w}\right), \qquad (4\text{-}76)$$

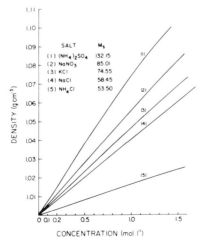

Fig. 4-3. Variation of the density of aqueous salt solutions with concentration of salt, at $T = 20\,°\text{C}$.

where we have used (4-63). Millero (1972) has summarized various methods for determining v_a, and has presented tables from which v_a can be computed for a large number of salts in solution, and for temperatures ranging from 0 to 100 °C. As an example, the variation of ρ''_s with concentration has been plotted in Figure 4-3 for solutions of a few salts typically found in the atmosphere. As expected from the ability of salt ions to break hydrogen bonds (see Section 3.4), ρ''_s increases with increasing salt concentration and with increasing molecular weight of the salt.

4.8   Latent Heat of Phase Change and its Temperature Variation

It is well known that whenever a new phase appears, a certain amount of heat, the latent heat of phase change, is released or consumed. This latent heat can be defined in terms of the difference between the heat content (enthalpy) of the two phases involved in the phase change. Let us assume that inside a closed system consisting of two phases, a unit mass of water substance is reversibly transferred from phase (') of n' moles to phase (") of n" moles, during which time p and T of the system remain constant. The total enthalpy change must then be

$$dH = \left(\frac{\partial H}{\partial n'}\right)_{T,p,n''} dn' + \left(\frac{\partial H}{\partial n''}\right)_{T,p,n'} dn''. \tag{4-77}$$

But since the system is closed, $dn' = -dn''$. Therefore, denoting the partial molar enthalpies by $(\partial H/\partial n')_{p,T,n''} = h'$, $(\partial H/\partial n'')_{p,T,n'} = h''$, we find

$$\left(\frac{dH}{dn''}\right)_{p,T} = h'' - h' \equiv \mathscr{L}'^{'''}, \tag{4-78}$$

which defines the latent heat of phase change per mole in passing from phase (') to phase ("). Also, since for a closed system at constant T and p we have $dH = T\,ds$, we may also write

$$\mathscr{L}'^{'''} = T(s'' - s'). \tag{4-79}$$

Let us denote the molar latent heats of evaporation, sublimation, and melting for pure water substance by $\mathscr{L}_{e,0}$, $\mathscr{L}_s$ and $\mathscr{L}_{m,0}$, respectively. Then simple conservation of energy (the first law of thermodynamics) applied to the triple point state where ice, vapor, and water are in equilibrium tells us that

$$\mathscr{L}_s = \mathscr{L}_{m,0} + \mathscr{L}_{e,0}, \qquad (T = 273.16 \,°K). \tag{4-80}$$

To find the temperature dependence of the latent heat, we may substitute (4-10) into (4-79) to obtain

$$\mathscr{L}'^{'''} = T\left[\left(\frac{\partial \mu'}{\partial T}\right)_p - \left(\frac{\partial \mu''}{\partial T}\right)_p\right]$$

$$= \left[\mu'' - T\left(\frac{\partial \mu''}{\partial T}\right)_p\right] - \left[\mu' - T\left(\frac{\partial \mu'}{\partial T}\right)_p\right], \tag{4-81}$$

using (4-22). If we now take the total differential of this equation, and apply

(4-10) through (4-12), we find

$$\frac{d\mathcal{L}'''}{dT} = (C_p'' - C_p') + \frac{dp}{dT}\left\{(v'' - v') - T\left[\left(\frac{\partial v''}{\partial T}\right)_p - \left(\frac{\partial v'}{\partial T}\right)_p\right]\right\}. \tag{4-82}$$

To a first approximation we may ignore the second term and obtain Kirchoff's equations:

$$\frac{d\mathcal{L}_{e,0}}{dT} = C_{p,v} - C_w; \quad \frac{d\mathcal{L}_s}{dT} = C_{p,v} - C_i; \quad \frac{d\mathcal{L}_{m,0}}{dT} = C_w - C_i. \tag{4-83}$$

These turn out to be excellent approximations for most purposes.

Below we shall give empirical fits to values given in the Smithsonian Meteorological Tables (SMT) for the specific heats of ice, $c_i(= C_i/M_w)$, and of water $c_w(= C_w/M_w)$. The specific heats are expressed in units of IT cal g^{-1} $°C^{-1}$ and are given as a function of T in °C. Thus:

$$c_i = 0.503 + 0.00175\,T, \tag{4-84a}$$

which agrees with the tabulated data to within ± 0.0005 IT cal g^{-1} $°C^{-1}$ for $0 \leqslant T \leqslant 50\,°C$;

$$c_w = 1.0074 + 8.29 \times 10^{-5}\,T^2, \tag{4-84b}$$

which agrees with the tabulated data to within ± 0.005 IT cal g^{-1} $°C^{-1}$ for $-40 \leqslant T \leqslant 0\,°C$; and

$$c_w = 0.9979 + 3.1 \times 10^{-6}(T - 35)^2 + 3.8 \times 10^{-9}(T - 35)^4, \tag{4-84c}$$

which agrees with the tabulated data to within ± 0.0001 IT cal g^{-1} $°C^{-1}$ for $0 \leqslant T \leqslant 35\,°C$. Also, we may note that the specific heat of water vapor, $c_{pv} = C_{pv}/M_w$, varies between 0.44 and 0.46 IT cal g^{-1} $°C^{-1}$ for all meteorological conditions of interest.

Empirical fits to the values given by the SMT for the specific latent heats of evaporation, $L_{e,0}(= \mathcal{L}_{e,0}/M_w)$, and of melting, $L_{e,0}(= \mathcal{L}_{m,0}/M_w)$, are as follows:

$$L_{e,0} = 597.3\left(\frac{273.15}{T}\right)^\gamma, \quad \gamma = 0.167 + 3.67 \times 10^{-4}\,T, \tag{4-85a}$$

with T in °K and $L_{e,0}$ in IT cal g^{-1}; and

$$L_{m,0} = 79.7 + 0.485\,T - 2.5 \times 10^{-3}\,T^2. \tag{4-85b}$$

Equation (4-85a) agrees with the tabulated data to within ± 0.3 IT cal g^{-1} for the temperature range -40 to $40\,°C$, while (4-85b) agrees with the tabulated data to within ± 0.05 IT cal g^{-1} for 0 to $-50\,°C$.

Since ions generally have a structure-breaking effect on water, it is not surprising to find from experiments that salts dissolved in water lower its heat capacity. However, this lowering is small at high dilution and only becomes significant if the salt concentration is larger than 0.1 moles liter^{-1}. Since salts affect the specific heat of water, it is quite reasonable to expect that the latent heats are affected also. A quantitative assessment of the effect of salts is easily made by noting that the partial molar enthalpy of water in an aqueous solution is

$h_w = h_{w,0} + h_m$, where $h_{w,0}$ is the enthalpy for pure water and h_m is the enthalpy of mixing of water in an aqueous salt solution. Then from (4-78) we have for the molar latent heats of evaporation and melting for solutions,

$$\mathcal{L}_e = h_v - h_m = \mathcal{L}_{e,0} - h_m, \tag{4-86}$$

and

$$\mathcal{L}_m = h_m - h_i = \mathcal{L}_{m,0} + h_m. \tag{4-87}$$

Inserting numerical values into these equations one finds that even for concentrations as large as 5 moles of NaCl per liter, h_m affects the magnitude of $\mathcal{L}_e$ by less than 0.2%, and that of $\mathcal{L}_m$ by less than about 2%. Thus, for most purposes we may set $\mathcal{L}_m \approx \mathcal{L}_{m,0}$ and $\mathcal{L}_e \approx \mathcal{L}_{e,0}$.

4.9   Clausius-Clapeyron Equation

The conditions for equilibrium derived in Section 4.2 find a useful application in what is known as the *Clausius-Clapeyron equation*. In order to derive this equation consider a system of one component and two phases, (') and ("). From the phase rule we know there is one degree of freedom, so that, for instance, the pressure is a function only of temperature for those states corresponding to equilibrium between the two phases. The Clausius-Clapeyron equation provides an expression for the slope of this phase boundary curve in the p-T plane. This may be obtained by noting that for small displacements along the curve, $d\mu' = d\mu''$, since $\mu' = \mu''$ along the curve (or, more accurately, on either side of the curve). Then from (4-4b) we find

$$v' \, dp - s' \, dT = v'' \, dp - s'' \, dT,$$

or

$$\frac{dp}{dT} = \frac{s'' - s'}{v'' - v'} = \frac{\mathcal{L}'^{|''}}{T(v'' - v')}, \tag{4-88}$$

using (4-79). This is the Clausius-Clapeyron equation. As one direct application of its use, note that it provides the expression for dp/dT which is needed to integrate (4-82).

Considering the bulk phases to be water and water vapor, we thus find the saturation vapor pressure $e_{sat,w}$ is determined from the equation

$$\frac{de_{sat,w}}{dT} = \frac{\mathcal{L}_{e,0}}{T(v_{v,0} - v_{w,0})} \approx \frac{\mathcal{L}_{e,0}}{T v_{v,0}}, \tag{4-89}$$

since $v_{w,0} \ll v_{v,0}$. Analogously, if the bulk phases are ice and water vapor, or ice and water, we have

$$\frac{de_{sat,i}}{dT} = \frac{\mathcal{L}_s}{T(v_{v,0} - v_i)} \approx \frac{\mathcal{L}_s}{T v_{v,0}}, \tag{4-90}$$

and

$$\frac{dp_m}{dT} = \frac{\mathcal{L}_{m,0}}{T(v_{w,0} - v_i)}. \tag{4-91}$$

If we further assume the ideal gas law (Equation (4-24)), we obtain

$$\frac{d \ln e_{sat,w}}{dT} \approx \frac{\mathscr{L}_{e,0}}{\mathscr{R}T^2} = \frac{\varepsilon L_{e,0}}{R_a T^2},$$ (4-92)

and

$$\frac{d \ln e_{sat,i}}{dT} \approx \frac{\mathscr{L}_s}{\mathscr{R}T^2}.$$ (4-93)

If one includes the approximate temperature dependence given by (4-83), then (4-92) and (4-93) determine $e_{sat,w}$ and $e_{sat,i}$ to an accuracy quite sufficient for applications in cloud microphysics. In addition, as a practical alternative, Lowe and Ficke (1974) have provided expressions which they feel are more convenient for typical modern numerical simulations of cloud physical processes. Their expressions are given in the appendix to this chapter. They are curve fits based on the Goff (1942, 1949) integrations of the Clausius-Clapeyron equation, wherein the virial equation of state for water vapor was used. Goff's accurate values are also tabulated in the Smithsonian Meteorological Tables. Unfortunately, these suffer from the 1954 revision of the temperature scale (Stille, 1961). However, comparison shows that both the SMT values and the values subsequently revised by Goff (1957, 1965) agree to within 0.035% over the whole temperature range of meteorological interest (Murray, 1967, 1970). According to Goff (1965), $e_{sat,w}(T_{tr} \equiv 0.01\,°C) = 6.1112$ mb, and $e_{sat,w}(T_0 \equiv 0\,°C,\ p = 1$ atm.) $= 6.1067$ mb. The former result is in excellent agreement with the most recent experimental value obtained at the U.S. National Bureau of Standards by Guilder et al. (1975) who determined $e_{sat,w}(T_{tr}) = 6.11657$ mb (± 0.00010 mb, at the 90% confidence level).

The temperature variation of the saturation vapor pressures is shown in Figure 4-4. Note that $e_{sat,w} > e_{sat,i}$ for $T < 0\,°C$. This is obvious also on comparison of

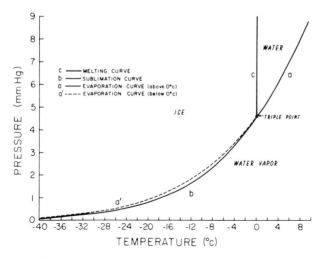

Fig. 4-4. p-T phase diagram for bulk water substance. (Based on data of the Smithsonian Meteorological Tables.)

(4-89) and (4-90), since $\mathscr{L}_s > \mathscr{L}_{e,0}$. A closer inspection of Figure 4-4 reveals a single maximum for the difference $e_{sat,w} - e_{sat,i}$, which we can calculate by noting that the slopes of the curves are equal where the difference is a maximum. Therefore, on setting (4-92) and (4-93) equal for the temperature T_{max} which yields a maximum difference, we find

$$\left(\frac{e_{sat,w}}{e_{sat,i}}\right)_{T=T_{max}} = \left(\frac{\mathscr{L}_s}{\mathscr{L}_{e,0}}\right)_{T=T_{max}}. \tag{4-94}$$

Now, if we integrate the difference between (4-89) and (4-90) from T_{max} to $T_0 = 273.15\,°K$, holding $\mathscr{L}_{m,0}$ constant and taking into account that $e_{sat,i} = e_{sat,w}$ at $T = T_0$, we find that

$$\ln\left(\frac{e_{sat,w}}{e_{sat,i}}\right)_{T=T_{max}} = \frac{\mathscr{L}_{m,0}}{\mathscr{R}}\left(\frac{T_0 - T_{max}}{T_0 T_{max}}\right). \tag{4-95}$$

Combining (4-94) and (4-95), we obtain the following expression for T_{max}:

$$T_{max} = T_0 - \frac{T_0 T_{max}\mathscr{R}}{\mathscr{L}_{m,0}}\ln\left(\frac{\mathscr{L}_s}{\mathscr{L}_{e,0}}\right)_{T=T_{max}}. \tag{4-96}$$

Solving this equation by iteration gives $T_{max} = 261.37\,°K$, or about $-11.8\,°C$. This agrees to within $0.1\,°C$ with the value found from the Goff expressions for e_{sat}, which indicates again the accuracy of the ideal gas law approximation. Finally, the variation of $e_{sat,w} - e_{sat,i}$ is illustrated in Figure 4-5, and the corresponding plots of $\rho_{v,sat,w}$, $\rho_{v,sat,i}$, and $\rho_{v,sat,w} - \rho_{v,sat,i}$ are given in Figure 4-6.

Of course, the interesting point to emphasize here is that air saturated with respect to ice is always subsaturated with respect to water, with the consequence that supercooled water drops and ice crystals cannot coexist in equilibrium. As we mentioned in Chapter 1, this important fact was first realized by Wegener in 1911, and is the basis of the Wegener-Bergeron-Findeisen precipitation mechanism.

It is also worth emphasizing that at sufficiently low temperatures, air may be ice-supersaturated but water-subsaturated. This is illustrated in Figure 4-7. Similarly, Figure 4-8 shows the ice-supersaturations which are required for water-supersaturation to occur also.

Let us now consider the equilibrium between bulk ice and water which is described by (4-91). No simplification can be made in this equation since

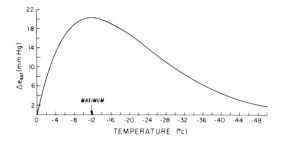

Fig. 4-5.   Variation of $\Delta e_{sat} = (e_{sat,w} - e_{sat,i})$ with temperature.

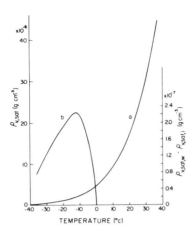

Fig. 4-6. Variation with temperature of the density of saturated water vapor, $\rho_{v,sat}$ and of $(\rho_{v,sat,w} - \rho_{v,sat,i})$. (Curve (a) pertains to the ordinate on the left; curve (b) pertains to the ordinate on the right.)

$v_{w,o} \approx v_i$. Since at all temperatures below $0\,°C$, $v_i > v_{w,0}$, we find that $(dp_m/dT) < 0$. In fact, experiments show that $(dp_m/dT) = -132.2$ atm. $(°C)^{-1}$, which means that, at any given temperature, ice can be melted by applying sufficiently high pressures. The temperature variation of the melting pressure of ice is given in Figure 4-9. As an example, we note that at $-10\,°C$ a very large pressure of 1100 atm. is required to melt ice. As we mentioned briefly earlier, it is a consequence of this melting pressure effect that the triple point of water substance is slightly higher than the ice point. Experiments have shown that by opening a vessel in which ice, water, and water vapor are originally in equilibrium, and exposing it to air of 1 atm. the equilibrium temperature is reduced by $0.0098 \pm 0.0003\,°C$, the pressure effect contributing $0.0075\,°C$, while an additional $0.0023\,°C$ is due to the dissolved air. By international agreement, the total temperature difference between the triple point and ice point has been set equal to $0.0100\,°C$, and the temperature of the

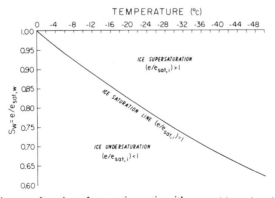

Fig. 4-7. Ice saturation as a function of saturation ratio with respect to water at temperatures below $0\,°C$.

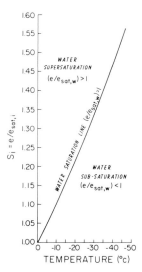

Fig. 4-8. Water saturation as a function of saturation ratio with respect to ice at temperatures below 0 °C.

triple point itself has been set equal to $T_{tr} = 273.16$ °K (Stille, 1961). Therefore, the temperature of the ice point is $T_0 = (T_{tr} - 0.01)$ °K $= 273.15$ °K $\equiv 0.0$ °C.

In addition to affecting the melting temperature of ice, pressure also affects its crystal structure. The phenomenon that a single chemical substance may appear in different crystallographic modifications is called *polymorphy*. At present, 11 polymorphic forms of ice have been found. Hobbs (1974) and Fletcher (1970a) have discussed in detail these crystallographic forms and the thermodynamic conditions for which they are stable and in equilibrium with each other. They are

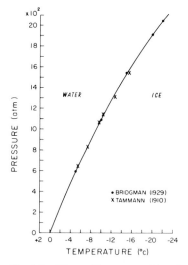

Fig. 4-9. The melting curve of ice-I_h.

of little concern to us since they are not stable at typical atmospheric temperatures and pressures.

Let us consider Figure 4-4 once more and note an interesting observation which can be made regarding the transformation of one phase into another. Suppose, for instance, that water vapor is cooled at constant pressure above the triple point. We see that eventually a temperature is reached at which the vapor is saturated (curve a). Upon further cooling, the evaporation curve is crossed and conditions are reached at which water is the stable phase. On cooling still further, the melting curve is crossed and conditions are reached at which ice is the stable phase. However, observations show that neither water nor ice appear at the temperatures predicted by the equilibrium phase diagram for bulk water substance. Unless suitable impurities are present in the vapor or on the walls enclosing the system, the water vapor *supersaturates* and water *supercools*. (By definition, the supersaturation of water vapor is described by (4-37), and the supercooling of water is defined by the quantity $\Delta T \equiv T_0 - T$, where $T_0 = 273.15 \,°K$.) The reason for this behavior (discussed at length in Chapters 7 and 9) rests in the fact that, during a phase change, the new phase always appears in the form of a small particle with a highly curved surface. The equilibrium conditions for such highly curved phases are not described by the Clausius-Clapeyron equation.

Another, similarly incorrect prediction is made by Figure 4-4. If water vapor is cooled isobarically below the triple point, the sublimation curve is crossed before the evaporation curve. This means that, upon cooling, those conditions are reached at which ice is the stable phase before water is; i.e., the phase diagram for bulk water predicts that ice will appear first at temperatures below the triple point. Observations, however, show that unless suitable impurities are present, the metastable phase, i.e., supercooled water, always appears before ice. The reason for this behavior again lies in the fact that phase change proceeds via the formation of new phase particles with highly curved surfaces. We shall show in Chapters 7 and 9 that, unless suitable impurities are present in the system, the formation of water drops is energetically favored at all temperatures over the formation of ice crystals directly from the vapor.

Thus far we have displayed the equilibrium behavior of bulk water substance in the form of a p-T phase diagram. Further information on the equilibrium behavior may be obtained from a p-V phase diagram (Figure 4-10). Note that during isothermal compression a state (e.g., state A) is reached at which water vapor is saturated with respect to water. If the walls enclosing the vapor are ideally rough, further compression results in the condensation of vapor to liquid. Along the line AB, $e = e_{sat,w}$, which remains constant as the specific volume decreases from that of pure vapor to that of pure liquid water. The small compressibility of liquid water is revealed by the steep excursion of the isotherm to the left of B.

At the top of the phase boundary curve is the *critical point*, where the distinction between liquid and gas vanishes. At this extraordinary point the surface tension or surface energy of the interface separating the phases becomes zero. Atmospheric water always lies far below the critical point, which occurs at

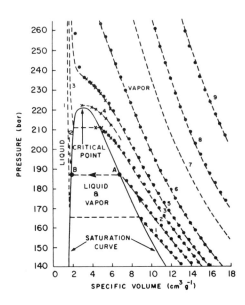

Fig. 4-10. p-V phase diagram for bulk water substance. Dashed lines represent isotherms: (1) 350 °K, (2) 360 °K, (3) 370 °K, (4) 374.4 °K, (5) 380 °K, (6) 400 °K, (7) 450 °K, (8) 500 °K, (9) 600 °K; based on experiments of Bain (1964), and of Nowak *et al.* (1961). ● based on Keyes' (1949) empirical formula. × based on Nowak and Grosh's (1961) empirical formula. (From *The Structure and Properties of Water* by D. Eisenberg and W. Kauzmann, copyrighted by Oxford Univ. Press, 1969.)

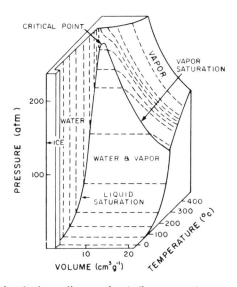

Fig. 4-11. Three-dimensional phase diagram for bulk water substance (dashed lines represent isotherms). (From *The Structure and Properties of Water* by D. Eisenberg and W. Kauzmann, copyrighted by Oxford Univ. Press, 1969.)

$p_{cr} = 221$ bars (1 bar = 10^6 dynes cm^{-2} = 0.9869 atm.)   and $T_{cr} = 374\,°C$.   The concept of the critical point illuminates the distinction between water vapor and the other, permanent, atmospheric gases; the latter have critical temperatures far below atmospheric temperatures, and thus never change phase. Note that this is not a consequence of an insufficiently massive atmosphere – no amount of pressure can liquify them as long as they remain above the critical temperature.

For completeness, in Figure 4-11 the *thermodynamic surface* of water in p-V-T space is shown. The projection of this surface on the p-T plane and the p-V plane yields the phase diagrams shown in Figures 4-4 and 4-10, respectively.

4.10 Equilibrium Between an Aqueous Salt Solution and Ice

Experiments have shown that the equilibrium temperature between ice and an aqueous salt solution is lower than that between ice and pure water. This is a direct consequence of the lowering of vapor pressure over a salt solution, as illustrated in Figure 4-12. In order to derive an expression for this temperature lowering effect, consider a system open to the atmosphere, and consisting of air and an aqueous salt solution in equilibrium with ice (assumed to be free of salt) of chemical potential μ_i. The condition of chemical equilibrium between the pure ice and water in aqueous solution, assuming p and T to be uniform throughout the system, is just $\mu_i(p, T) = \mu_w(p, T)$, from (4-22). Therefore, on substitution from (4-53), we obtain

$$\mu_i(p, T, a_w) = \mu_{w,0}(p, T) + \mathscr{R}T \ln a_w. \tag{4-97}$$

From the phase rule, (4-23), we have $\varphi = 3$ (water, ice, air), $c = 3$ (salt, water, air), and thus $w = 2$. Let us assume the atmospheric air pressure is fixed at 1 atm., so that $w = 1$. Then the equilibrium temperature becomes a function of the salt concentration, and we must therefore investigate the variation of a_w with T. Directly from (4-97), we see that equilibrium can be maintained for variable T only if

$$\left[\frac{\partial\left(\frac{\mu_i}{T}\right)}{\partial T}\right]_p = \left[\frac{\partial\left(\frac{\mu_{w,0}}{T}\right)}{\partial T}\right]_p + \mathscr{R}\left(\frac{\partial \ln a_w}{\partial T}\right)_p. \tag{4-98}$$

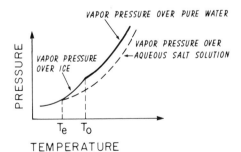

Fig. 4-12.   Schematic to illustrate temperature equilibrium between ice and aqueous salt solution, and between ice and pure water.

On substituting (4-13) and (4-78), this becomes

$$\left(\frac{\partial \ln a_w}{\partial T}\right)_p = \frac{h_{w,0} - h_i}{\mathscr{R}T^2} = \frac{\mathscr{L}_{m,0}}{\mathscr{R}T^2}, \tag{4-99}$$

which upon integration yields, assuming $\mathscr{L}_{m,0}$ is independent of T,

$$\ln a_w = -\frac{\mathscr{L}_{m,0}}{\mathscr{R}T} + f(p). \tag{4-100}$$

The unknown function f(p) may be determined by noting that $a_w = 1$ for $T = T_0$; thus the *equilibrium freezing temperature*, T_e, may be determined from the relation

$$\frac{\mathscr{L}_{m,0}}{\mathscr{R}}\left(\frac{T_0 - T_e}{T_0 T_e}\right) = -\ln a_w. \tag{4-101}$$

Since $a_w \leqslant 1$, we have $T_e \leqslant T_0$, as expected. The extent of the temperature lowering effect is generally measured by the *equilibrium freezing point depression*, defined as

$$(\Delta T)_{e,s} \equiv T_0 - T_e = -\frac{\mathscr{R}T_0 T_e}{\mathscr{L}_{m,0}} \ln a_w \tag{4-102}$$

$$= \frac{\mathscr{R}T_0 T_e M_w}{1000\, \mathscr{L}_{m,0}} \nu \Phi_s \mathfrak{M}, \tag{4-103}$$

using (4-69). Let us now suppose the solution is very dilute, so that $T_0 T_e \approx T_0^2$ and $a_w \approx x_w = 1 - x_s$. Then $\ln a_w \approx -x_x \approx -n_s/n_w$, and (4-102) becomes, using (4-65),

$$(\Delta T)_{e,s} \approx \frac{\mathscr{R}T_0^2 n_s}{\mathscr{L}_{m,0} n_w} = K_f \mathfrak{M}, \tag{4-104}$$

where

$$K_f = \frac{\mathscr{R}T_0^2 M_w}{1000\, \mathscr{L}_{m,0}}, \tag{4-105}$$

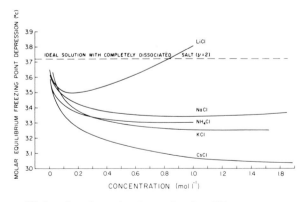

Fig. 4-13. Molar equilibrium freezing point depression for different aqueous salt solutions as a function of salt concentration.

is the *molal equilibrium freezing point depression*. For water at 0 °C, we find $K_f = 1.859$ °C mol^{-1}. For non-ideal or more concentrated solutions, (4-104) is inadequate. For these more realistic cases, $(\Delta T)_{e,s}$ depends noticeably on the nature of the dissolved salt, i.e., its degree of dissociation and the capability of its ions in solution for interacting with each other and the water molecules. The deviation of the molar equilibrium freezing point depression from ideality is illustrated in Figure 4-13 which shows experimental values for $(\Delta T)_{e,s}/\mathfrak{M}$. (In Figure 4-13 the small difference between the molal and molar units has been neglected.)

CHAPTER 5

SURFACE PROPERTIES OF WATER SUBSTANCE

In clouds the liquid and solid phases of water are highly dispersed, with a large surface-to-volume ratio. As might be expected, this necessitates going beyond the bulk phase descriptions of the previous chapter, even for the most rudimentary understanding of the formation and growth of cloud particles. Therefore, in this chapter we shall consider briefly the essential distinctive surface properties of ice and water, and explore some of their more immediate consequences.

Additional relevant material may be found in Hobbs (1974), Samorjai (1972), Bikerman (1970), Flood (1967), Reiss (1965), Osipow (1962), Defay *et al.* (1966), Davies and Rideal (1961), Adamson (1960), Ono and Kondo (1960), and Landau and Lifschitz (1958).

5.1 Surface Tension

Phases in contact are separated by a thin transitional region, generally only a few molecules thick; consequently, a useful abstraction is to regard such an interface as a geometrical surface. This permits a relatively simple and generally adequate description of surface effects via the usual straightforward machinery of macroscopic thermodynamics. Of course, this ceases to be a reasonable procedure when the bulk phases themselves have a similar microscopic thickness. In this chapter we shall not consider such difficult circumstances. Such problems do arise, however, in the theory of homogeneous nucleation (Chapter 7); there we shall see how macroscopic thermodynamics must be supplanted, at least in part, by a detailed statistical mechanics approach in order that a satisfactory understanding of nucleation phenomena can be achieved.

The extension of our thermodynamic systems to include surface effects is conceptually simple: in complete analogy to the contribution $-p\,dV$ of pressure-volume work to the internal energy, we now introduce a contribution $\sigma\,d\Omega$, where Ω denotes the area of the surface of separation, and σ is the *surface tension*. The quantity σ is an intensive thermodynamic variable, and is seen to have the dimensions of energy per unit area, or force per unit length. The physical basis of this formulation is probably familiar to the reader from the example of a liquid drop: on the average, molecules in the drop interior find themselves in a symmetrical, attractive force field, while molecules in the surface layer do not, and in fact experience a net attractive force toward the interior. As a consequence of this inward pull, the surface is in a state of tension, and it requires work to extend the surface further. On the molecular level, this work is seen to be that required to bring molecules from the interior to the surface, against the attractive forces.

The thermodynamic properties of the surface are so distinctive that it is conceptually useful to regard it as a separate phase, (σ), having its own entropy, $S^{(\sigma)}$, adsorbed number of moles of chemical component k, $n_k^{(\sigma)}$, and so forth. Then, in accordance with the discussion above, the change in internal energy $U^{(\sigma)}(S^{(\sigma)}, \Omega, n_k^{(\sigma)})$ of the surface phase for reversible processes is (cf. (4-1)):

$$dU^{(\sigma)} = T^{(\sigma)} \, dS^{(\sigma)} + \sigma \, d\Omega + \sum_{k=1}^{c} \mu_k^{(\sigma)} \, dn_k^{(\sigma)}, \qquad (5\text{-}1)$$

where the $\mu_k^{(\sigma)}$ are the surface chemical potentials.

5.2 Equilibrium Conditions

Let us now generalize our discussion in Section 4.2 of the equilibrium between two bulk phases $(')$ and $('')$, by taking into account also the surface phase (σ) which separates them. Proceeding as before, we imagine that each bulk phase is originally isolated and in internal equilibrium. After removing the constraint of isolation, we seek the conditions on the intensive variables which are necessary and sufficient to insure equilibrium throughout the system, which remains isolated as a whole. The independent extensive variables for the entire system are $V = V' + V''$, $S = S' + S'' + S^{(\sigma)}$, $n_k = n_k' + n_k'' + n_k^{(\sigma)}$, and Ω. Similarly, the total internal energy is $U = U' + U'' + U^{(\sigma)}$, and the generalized equilibrium condition which replaces (4-15) is

$$(\delta U)_{S,V,\Omega,n_k} \geq 0. \qquad (5\text{-}2)$$

Then from (4-1), (5-1), and (5-2), the expanded form of the equilibrium condition is

$$\delta U = T' \delta S' + T'' \delta S'' + T^{(\sigma)} \delta S^{(\sigma)} - p' \delta V' - p'' \delta V'' + \sigma \delta \Omega$$
$$+ \sum_{k=1}^{c} \mu_k' \delta n_k' + \sum_{k=1}^{c} \mu_k'' \delta n_k'' + \sum_{k=1}^{c} \mu_k^{(\sigma)} \delta n_k^{(\sigma)} \geq 0, \qquad (5\text{-}3)$$

where, from the constraint of isolation of the system as a whole, we have the additional conditions

$$\delta S = \delta S' + \delta S'' + \delta S^{(\sigma)} = 0; \quad \delta n_k = \delta n_k' + \delta n_k'' + \delta n_k^{(\sigma)} = 0;$$
$$\delta V = \delta V' + \delta V'' = 0. \qquad (5\text{-}4)$$

None of these conditions is violated if we conceive a set of infinitesimal variations for which $\delta S' = 0$ and $\delta n_k' = 0$. Let us also suppose the bulk phase $('')$ is a sphere of radius a, so that $d\Omega = 2 \, dV''/a$. With these specializations and (5-4), (5-3) becomes

$$(T'' - T^{(\sigma)}) \delta S'' + \left(p' - p'' + \frac{2\sigma}{a} \right) \delta V'' + (\mu_k'' - \mu_k^{(\sigma)}) \delta n_k'' \geq 0. \qquad (5\text{-}5)$$

Since $\delta S''$, $\delta V''$, and $\delta n_k''$ represent independent and arbitrary variations, each coefficient must vanish. A similar result would have been obtained had we originally chosen $\delta S'' = \delta n''_k = 0$. Therefore, the condition of thermodynamic equilibrium leads finally to

$$T' = T'' = T^{(\sigma)}, \tag{5-6}$$

$$p'' - p' = \frac{2\sigma}{a}, \tag{5-7}$$

$$\mu'_k = \mu''_k = \mu_k^{(\sigma)}, \quad k = 1, 2, \ldots, c. \tag{5-8}$$

Comparison with (4-20) to (4-22) shows that only the condition of mechanical equilibrium has a new form; this is expressed by (5-7) and is called the *Laplace formula*. It may also be obtained by more elementary means, e.g., by equating the surface tension force along the circumference of a great circle of the sphere with the pressure difference acting across the great circle area.

The extension of these results to a more general system comprised of φ phases and χ spherically curved interfaces is obvious: for thermodynamic equilibrium to exist, we must have uniform temperatures (thermal equilibrium) and uniform chemical potentials (chemical equilibrium) throughout the entire system. Additionally, a relation of the form of (5-7) (mechanical equilibrium) must hold for every pair of bulk phases, the greater pressure occurring on the concave side of the interface whose radius of curvature replaces a in (5-7).

5.3   Phase Rule for Systems with Curved Interfaces

Let us now consider the generalization of the phase rule (Section 4.3) to a system of φ bulk phases, c components, and χ curved surface phases. As before, we want to determine the number of intensive variables which may be altered independently without causing the system to depart from equilibrium. In this connection, therefore, we must consider what to use for the intensive variable corresponding to the quantity $n_k^{(\sigma)}$, the number of moles of component k adsorbed into the interface (σ). An obvious natural choice is the *adsorption*, defined by

$$\Gamma_k^{(\sigma)} \equiv \frac{n_k^{(\sigma)}}{\Omega^{(\sigma)}}, \quad k = 1, 2, \ldots, c; \quad \sigma = 1, 2, \ldots, \chi, \tag{5-9}$$

where $\Omega^{(\sigma)}$ is the area of the interface (σ).

The number of intensive variables required to specify the state of the system in equilibrium must therefore include: (1) the common temperature T; (2) the $c\chi$ adsorptions $\Gamma_k^{(\sigma)}$ in the surface phases; (3) the $c\varphi$ mole fractions $x_k^{(\alpha)}$ in the bulk phases $(\alpha = 1, 2, \ldots, \varphi)$; (4) the φ pressures $p^{(\alpha)}$ of the bulk phases; and (5) the χ mean radii of curvature $a^{(\sigma)}$. This constitutes a total of $1 + (\varphi + \chi)(c + 1)$ intensive variables. Constraints among them include the following: (1) the mole fractions $x_k^{(\alpha)}$ must sum to unity for each bulk phase, leading to φ constraints; (2) the chemical potentials must be equal for all the phases for every k, leading to $c(\chi + \varphi - 1)$ constraints; (3) each interface gives rise to a condition of mechanical equilibrium like (5-7), leading to χ constraints. This gives a total of $(\varphi + \chi)(c + 1)$

$-c$ constraints, and so we find for the variance of the system,

$$w = c + 1. \tag{5-10}$$

An interesting feature of this result is that the variance is independent of the number of bulk and surface phases.

Note that in the case that one of the components of the system is not present in one of the phases (e.g., humid air surrounding aqueous NaCl solution), one of the equations relating the chemical potentials disappears. On the other hand, for the phase in question, one must write $\Sigma_{k=1}^{c-1} x_k^{(\alpha)} = 1$, and add the relation $x_{i\neq k}^{(\alpha)} = 0$. This expresses the fact that component i is not present in the phase (α). Thus, the total number of relations remains the same and (5-10) does not change.

Let us now consider four simple examples which will illustrate the use of (5-10). First, consider a system of uniform temperature T in which a pure water drop of radius a is surrounded by pure water vapor of pressure e_a. From (5-10), $w = 2$ since $c = 1$. Thus we may choose, for example, to hold T constant and study the dependence of e_a on a. Second, consider a system of uniform temperature T in which a drop of pure water of radius a is surrounded by humid air of total pressure p. From (5-10), $w = 3$; to study the dependence of e_a on a we must hold both T and p constant. Third, consider a system of uniform temperature T in which an aqueous solution drop of radius a is surrounded by humid air of pressure p. Since $c = 3$ (water, salt, air), we find $w = 4$. However, since the total mass of salt in the drop does not change, even though the drop may change its radius by acquiring or losing water as a result of water vapor diffusion to or from the drop, the mole fraction of the water in solution becomes a function of the drop radius. This constitutes an additional relation not considered in (5-10). Thus $w = 3$ as in the case of a pure water drop in humid air, and we may again choose to hold T and the total gas pressure p constant and study the dependence of e_a on a. Fourth, consider a system of uniform temperature T in which a pure water drop of radius a_w and a spherical ice crystal of radius a_i are surrounded by humid air of total gas pressure p. From (5-10), $w = 3$. We may choose to hold p constant and thus dispose of one of the intensive variables, so that the system at equilibrium is di-variant. This is in contrast to a system of 3 bulk phases which, at equilibrium, is non-variant once p is fixed (Section 4.3). Thus, we may independently vary a_w and a_i and investigate the effect of their variation on the equilibrium temperature of the system. It is clear that if a_w and a_i are given, the system has no further variance at equilibrium.

5.4   Water-Vapor Interface

The difference between the pressure p_w inside a water drop of radius a and the pressure $e_{sat,w}$ of vapor with which it is in equilibrium is given by (5-7):

$$p_w - e_{sat,w} = \frac{2\sigma_{w/v}}{a}, \tag{5-11}$$

where now we have introduced the subscript w/v for the surface tension to emphasize that it is the water-vapor interface which is involved. Given that $\sigma_{w/v} \approx 76$ dyne cm^{-1} at 0 °C, we see that the pressure difference is about 1.5 atm. for $a = 1 \mu$m; smaller drops have correspondingly larger internal pressures.

For practical purposes one may replace $\sigma_{w/v}$ in (5-11) by $\sigma_{w/a}$, the surface tension for a water-humid air interface. Experiments by Richards and Carver (1921) and Adam (1941) have shown that $\sigma_{w/a}$ increases by less than 0.05% if air at 1 atm. is replaced by pure water vapor at saturation pressure (at the same temperature).

5.4.1   EFFECT OF TEMPERATURE ON SURFACE TENSION

As would be expected on consideration of the effects of thermal agitation, the surface tension of water decreases with increasing temperature. This behavior has been investigated experimentally by Dorsch and Hacker (1951) and Gittens (1969). Their results are shown in Figure 5-1. Although the observations of Dorsch and Hacker extend into the regime of supercooled water, making their values preferable from a cloud physics point of view, the measurements of Gittens have been carried out by a considerably more refined experimental technique, and are therefore probably more accurate for $T > 0$ °C. For the interval 0 to 40 °C, his data can be reproduced to within ± 0.02 dyne cm^{-1} by the following expression:

$$\sigma_{w/a} = 76.10 - 0.155\,T, \tag{5-12}$$

where $\sigma_{w/a}$ is in dyn cm^{-1}, i.e., erg cm^{-2}, and T is in °C. From Figure 5-1 it appears that (5-12) also represents a reasonable extrapolation for the temperature interval 0 to -40 °C.

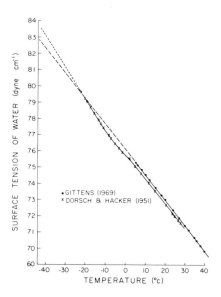

Fig. 5-1.   Variation of the surface tension of water (against air) with temperature.

5.4.2   SURFACE TENSION OF A SALT SOLUTION

Let us now consider the effect of dissolved salts on the surface tension of water. Given that the liquid surface is in a state of strain owing to a residual force field, we expect that the adsorption onto the surface layer of some chemical component to concentrations higher than that which appears in the bulk phase will occur if such behavior will serve to lower the state of strain, and vice versa. Therefore, we expect that if a solute can lower the surface tension, it will appear in a greater relative concentration at the interface than in the bulk solution. Materials which cause this to happen to a marked degree are called *surface active*. Conversely, solutes which can increase the surface tension should appear in a relatively weaker concentration at the interface; this behavior is known as *negative adsorption*.

An expression relating surface tension to adsorption may be derived most easily by integrating (5-1) in the same manner that led to (4-3). The result is

$$U^{(\sigma)} = T^{(\sigma)}S^{(\sigma)} + \sigma\Omega + \sum_{k=1}^{c} \mu_k^{(\sigma)} n_k^{(\sigma)}. \tag{5-13}$$

Now, on subtracting (5-1) from the differential of this equation, we obtain the surface phase form of the Gibbs-Duhem equation (cf. (4-4)):

$$S^{(\sigma)}\, dT + \Omega\, d\sigma + \sum_{k=1}^{c} n_k^{(\sigma)}\, d\mu_k^{(\sigma)} = 0, \tag{5-14}$$

or

$$d\sigma = -s^{(\sigma)}\, dT - \sum_{k=1}^{c} \Gamma_k^{(\sigma)}\, d\mu_k^{(\sigma)}, \tag{5-15}$$

where we have used (5-9), and $s^{(\sigma)} = S^{(\sigma)}/\Omega$ is the surface specific entropy. For constant T, (5-15) reduces to the *Gibbs adsorption isotherm* equation:

$$\Gamma_k^{(\sigma)} = -\left(\frac{\partial\sigma}{\partial\mu_k^{(\sigma)}}\right)_{T,\mu_j^{(\sigma)}}, \quad j \neq k. \tag{5-16}$$

Let us apply this result to the case of a binary solution of a salt in water. Denote the chemical potential of the salt in the solution-vapor interface by $\mu_s^{(s/v)}$. For equilibrium changes we have $d\mu_s^{(s/v)} = d\mu_s$ from (5-8), where μ_s is the chemical potential of the salt in the bulk phase. For the latter, however, we have the form (4-54); therefore, at constant T and p, the adsorption of the salt is described by

$$\Gamma_s^{(s/v)} = -\left(\frac{\partial\sigma_{s/v}}{\partial\mu_s}\right)_{T,p} = -\frac{1}{\mathscr{R}T}\left(\frac{\partial\sigma_{s/v}}{\partial\ln a_s}\right)_{p,T}. \tag{5-17}$$

Also, from (4-4a) we have $n_s\, d\mu_s = -n_w\, d\mu_w$; therefore, on substituting (4-53) and (4-54) for μ_w and μ_s into this expression, and introducing (4-69), we obtain another form for $\Gamma_s^{(s/v)}$:

$$\Gamma_s^{(s/v)} = -\frac{\mathfrak{M}}{\mathscr{R}T\nu\Phi_s}\left(\frac{\partial\sigma_{s/v}}{\partial\mathfrak{M}}\right)_{p,T}. \tag{5-18}$$

In order to interpret this result, we must consider more closely the meaning of the quantity $n_k^{(\sigma)}$ in (5-9). Implicit in our abstraction of the transition zone between bulk phases to a geometric surface of separation has been the assumption of the homogeneity of these adjacent phases up to their contact with the surface. Thus, in the expression $n_k^{(\sigma)} = n_k - n_k' - n_k''$, where n_k is the total number of moles of component k in the real system, the quantities n_k', n_k'' are the corresponding number of moles in the homogeneous phases (') and ("), assumed to retain their bulk properties up to the geometric interface. Referring now to the present example of a solution drop, we see therefore that if $n_s^{(s/v)} > 0$ (i.e., $\Gamma_s^{(s/v)} > 0$), there must be a higher concentration of salt in the transition region than in the interior of the drop, and vice versa. Consequently, the result (5-18) does support our qualitative expectations: for positive (negative) adsorption, the surface tension decreases (increases) with increasing concentration.

For most salts which are present in clouds, it has been found experimentally that if $\mathfrak{M} \leqslant 10^{-3}$ mol $(1000 \text{ g})^{-1}$, then $(\partial\sigma_{s/v}/\partial\mathfrak{M}) < 0$, i.e., $\Gamma_s^{(s/v)} > 0$; at larger concentrations, $(\partial\sigma_{s/v}/\partial\mathfrak{M}) > 0$, meaning $\Gamma_s^{(s/v)} < 0$ (Jones and Ray, 1937). If we evaluate (5-18) for a 1 molal NaCl solution at 0 °C ($\nu = 2$, $\Phi = 0.9355$, $\mathfrak{M} = 1$), we obtain $\Gamma_s^{(s/v)} = -3.82 \times 10^{-11}$ mol cm$^{-2} = -23 \times 10^{12}$ salt molecules cm^{-2}. This implies that the surface of such a solution lacks 23×10^{12} salt molecules per cm^2 to make the surface phase homogeneous with the bulk. Let us now consider a spherical drop of aqueous NaCl solution and seek the drop size below which the error due to omitting the surface salt deficiency is less than 1%. This condition can be expressed by the inequality $\mathfrak{M}(4\pi/3)a^3\rho_s'' \geqslant 100(4\pi a^2)\Gamma_s^{(s/v)}$. For $\rho_s'' \approx \rho_w$ and $|\Gamma_s^{(s/v)}| = 3.82 \times 10^{-11}$ mol cm^{-2}, we find $a \geqslant 1.15 \times 10^{-5}$ cm. Thus, a drop consisting of one-molal sodium chloride solution can be considered a homogeneous salt solution if its radius is larger than about 0.1 μm. We shall see that this criterion is fulfilled by most cloud drops drops during the condensation process (see Chapter 13).

This estimate is also in accord with a detailed study by Tsuji (1950) of the effect of salt adsorption at the surface of a solution drop, formed by condensation of a salt particle, on the equilibrium vapor pressure over the drop. The results of his computations show that the inhomogeneity due to adsorption at the drop surface is negligible for salt masses (NaCl) larger than 10^{-17} g, but becomes increasingly significant for smaller salt masses. As we shall see in Chapters 6 and 8, only salt particles of masses larger than 10^{-17} g contribute importantly to the formation of cloud drops by condensation. The effect of solution inhomogeneity may, therefore, be neglected in studying the condensation process.

Just as one may replace $\sigma_{w/v}$ by $\sigma_{w/a}$, a negligible error results on substituting $\sigma_{s/a}$, the surface tension of an aqueous salt solution exposed to humid air, for the quantity $\sigma_{s/v}$. Experimental values for $\sigma_{s/a}$ for many salts are tabulated in the Handbook of Physics and Chemistry. A few examples of the variation of $\sigma_{s/a}$ with concentration are presented in Figure 5-2. Note that for NaCl and $(NH_4)_2SO_4$, $\sigma_{s/a}(\mathfrak{M})$ is approximately linear. Since the slopes of these curves turn out to have a negligible dependence on temperature, over the range of meteorological interest, Hänel (1970) suggested the following empirical relation for $\sigma_{s/a}(\mathfrak{M}, T)$:

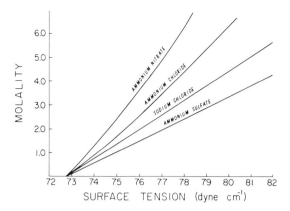

Fig. 5-2. Variation of the surface tension of electrolytic solutions (against air) as a function of salt concentration in solution at 20 °C. (From Low, 1969b; by courtesy of *J. de Rech. Atm.*, and the author.)

$$\sigma_{s/a}(\mathfrak{M}, T) = \sigma_{w/a}(T) + B\mathfrak{M} = \sigma_{w/a}(T) + B'(m_s/m_w), \qquad (5\text{-}19)$$

where $B_{NaCl}(B_{(NH_4)_2SO_4}) = 1.62(2.17)\,\mathrm{dyn\,cm^{-1}}$ per mole salt in 1000 g water, $B'_{NaCl}(B'_{(NH_4)_2SO_4}) = 27(16.4)\,\mathrm{dyn\,cm^{-1}\,g^{-1}}$, and where (4-63) was used to obtain the second form of (5-19).

5.4.3 RADIUS DEPENDENCE OF SURFACE TENSION

A last relevant consideration for this section is the possible dependence of $\sigma_{w/v}$ (or $\sigma_{w/a}$) on the curvature of the water phase. Given that the surface tension arises from attractive forces between molecules near the surface, we might expect that only an alteration of the average geometrical configuration of these molecules on a size scale comparable to the effective range of the attractive forces would significantly affect the surface tension. Thus we would expect a dependence of $\sigma_{w/v}$ on size only for extremely small drops, consisting of merely a few tens or hundreds of water molecules.

Several investigations of this problem have been carried out. Tolman (1949a, b) and Koenig (1950) suggested finding the answer on the basis of a quasi-thermodynamic approach. Kirkwood and Buff (1949) and Buff (1951, 1955) utilized statistical mechanics methods. Benson and Shuttleworth (1951) based their study on molecular interactions. Although all three approaches qualitatively predict that the surface tension of water decreases with decreasing radius of curvature of the water surface, there is little quantitative agreement among them. Tolman estimated that $\sigma_{w/v}$ for a drop which consists of 13 water molecules (equivalent to a drop radius of $4.6 \times 10^{-8}\,\mathrm{cm}$, based on $\dot{v}_w = 30 \times 10^{-24}\,\mathrm{cm^3}$ and $\rho_w = 1.0\,\mathrm{g\,cm^{-3}}$) is 40% smaller than that for a plane water surface. In contrast, Benson and Shuttleworth computed the surface tension of a small group of water molecules by counting the number of bonds which had to be broken in order to cut off the group of molecules from the bulk water structure. In order to estimate the interaction energy between water molecules in water, Benson and Shuttleworth assumed that only the first and second nearest neighbors had to be considered. In this manner they predicted that the surface tension for a drop of 13 water molecules is only about 15% smaller than that for a plane water surface. Since the quasi-thermodynamic approach is not rigorous for such small water drops, one would tend to prefer the result of Benson and Shuttleworth. However, the success of the molecular interaction method obviously depends on the accuracy with which the structure of water can be described. In this context, our discussion in Chapter 3 suggests that a hexagonal, close-packed structure such as that used by Benson and Shuttleworth can hardly describe the actual water structure accurately. Therefore, both the thermodynamic and molecular methods must be treated with caution.

No trustworthy experimental determination of $\sigma = \sigma(a)$ is available for any liquid except for the measurements of Sambles $et\ al.$ (1970), who experimentally tested the Kelvin law (see Chapter 6) for evaporating lead droplets. They concluded that the surface tension of these droplets did not deviate from the values over a flat surface even if the drops were as small as 10^{-7} cm.

We shall now present a simple, approximate, quasi-thermodynamic derivation of the radius dependence of σ, following Defay $et\ al.$ (1966). From (5-11) we have, for a displacement at equilibrium,

$$\frac{2\,d\sigma_{w/v}}{a} + 2\,\sigma_{w/v}d(1/a) = dp_w - de_{sat,w}. \tag{5-20}$$

Also, from (4-4b) we find, at constant T, $d\mu_v = v_v\,de_{sat,w}$, and $d\mu_w = v_w\,dp_w$. But here $d\mu_v = d\mu_w$; consequently, we may express the right side of (5-20) in the form $(v_v - v_w)\,dp_w/v_v$. Then on substituting from (5-15), $d\sigma_{w/v} = -\Gamma_w^{(w/v)}\,d\mu_w = -v_w\Gamma_w^{(w/v)}\,dp_w$, (5-20) becomes

$$\frac{d\sigma_{w/v}}{\sigma_{w/v}} = -\frac{2\,\Gamma_w^{(w/v)}}{(2\,\Gamma_w^{(w/v)}/a) + \rho_w - \rho_v}\,d(1/a). \tag{5-21}$$

Assuming $\Gamma_w^{(w/v)}/(\rho_w - \rho_v)$ is independent of a, we may integrate (5-21) to obtain

$$\sigma_{w/v} = \frac{(\sigma_{w/v})_\infty}{1 + (2/a)[\Gamma_w^{(w/v)}/(\rho_w - \rho_v)]}, \tag{5-22}$$

where $(\sigma_{w/v})_\infty$ is the surface tension of a plane water surface.

The adsorption $\Gamma_w^{(w/v)}$ may be estimated in the following manner. Our thermodynamic formalism requires for complete consistency that we choose the surface of separation between phases to be the same as the $surface\ of\ tension$ in which the net surface forces appear to lie. Only for this choice can we be sure that the bulk volumes defined by the position of the dividing surface are identical with those appearing in the equation for mechanical work on which (5-1) and (5-3) are based. Of course, in practice it is essentially impossible to know exactly where the surface of tension is, and fortunately, for most purposes it turns out not to be necessary. Nonetheless, in the present instance there is some predictive value in realizing that the surface of tension, which is our reference surface for measuring adsorption, must lie slightly below the free surface of a mass of

TABLE 5-1
Variation of surface tension of water for drops of various radii.

a(cm)	$\sigma_{w/v}/(\sigma_{w/v})_\infty$	$[(\sigma_{w/v})_\infty - (\sigma_{w/v})]/(\sigma_{w/v})_\infty$ (%)
∞	1.0	0
10^{-4}	0.9997	0
10^{-5}	0.9969	0.3
10^{-6}	0.9697	3.0
10^{-7} (140 molecules)	0.7622	23.8
4.6×10^{-8} (13 molecules)	0.5959	40.4

water molecules comprising a drop. Since the forces of attraction between the molecules in the first layer act nearly along the lines connecting their centers, we may suppose that the surface of tension is about half a molecular thickness below the free surface. Therefore, the amount of water adsorbed on the dividing surface may be estimated as half the mass of the first molecular layer. Then, taking 9.6 (Å)^2 to be the area occupied by each water molecule at the surface (Defay *et al.*, 1966), their surface density is 1.7×10^{-9} mol cm^{-2}, and we find that $\Gamma^{(w/v)}$ is approximately 0.87×10^{-9} mol cm^{-2}. This is in fair agreement with estimates made by Tolman (1949a, b).

Table 5-1 lists results for $\sigma_{w/v}(a)$ computed from (5-22), using the above value for $\Gamma_w^{(w/v)}$, and the approximation $\rho_v \ll \rho_w = 1$ g cm^{-3}. It is seen that the radius dependence becomes important for $a \lesssim 10^{-6}$ cm, as expected. It is clear from the derivation and discussion that these values are not likely to be very accurate for such small sizes; nevertheless, they should be adequate for our purposes. Unfortunately, more rigorous values are not presently available.

5.5 Angle of Contact

So far we have considered water or solution drops which are surrounded by vapor or moist air only. Let us now consider a drop of water which is bounded by two phases: moist air and a solid phase on which the drop is resting (Figure 5-3). If the water only partially wets the solid, it will form a 'cap' which makes contact with the underlying surface at an angle θ, the *contact angle* for water on this surface. If the water wets the solid completely, $\theta = 0$. A surface which is readily wetted by water is called *hydrophilic*; a surface which is not is called

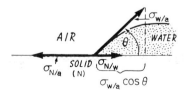

Fig. 5-3.   Mechanical equilibrium conditions for a drop on a horizontal solid surface. Tension $\sigma_{N/a}$ balances the sum of tensions $\sigma_{N/w}$ and $\sigma_{w/a} \cos \theta$.

TABLE 5-2
Contact angle of water on selected solid substances. (Based on data of Head
(1961), Letey *et al.* (1962), Koutsky *et al.* (1965), Zettlemoyer (1968), Miyamoto
and Latey (1971), and Isaka (1972).)

Substrate	Contact angle θ (angular degrees)	$m = \cos\theta$
Polyvinylformal (Formvar)	50	0.64
Polyethylene terphtalate (Therpane)	70	0.34
Polymethylmetacrylate (Plexiglass)	80	0.17
Polyethylene (Tedlar)	94	−0.07
Teflone	100 to 117	−0.17 to −0.45
Platinum (metal)	40	0.77
Gold (metal)	65.5	0.41
Silver (metal)	79.5	0.18
Cadmium iodide	0	1.0
Silver iodide	9 to 17	0.956 to 0.988
Silver chloride	50 to 55	0.57 to 0.64
Lead iodide	64 to 80	0.17 to 0.64
Surface soil	65.2 to 68.9	0.36 to 0.42
Quartz, beach sand	43 to 52	0.62 to 0.73

hydrophobic. The contact angle for water on various solid surfaces is given in
Table 5-2.

The contact angle is determined by the condition of mechanical equilibrium:
there must be no net force component along the solid surface. From Figure 5-3,
this condition, known as *Young's relation*, is easily seen to be given by

$$\sigma_{w/a}\cos\theta = \sigma_{N/a} - \sigma_{N/w}. \tag{5-23}$$

This relation, though quite useful, does rest on some idealizations which, of
course, are not found in practice. Some difficulties which complicate its use
include: (1) the roughness of the substrate (Osipow, 1962); (2) the presence or

TABLE 5-3
Surface energy for a few selected solids, against air and against
water ($\sigma_{w/a} = 72.30 \, \text{erg cm}^{-2}$).

Solid	Surface energy against air (erg cm^{-2}) (20 °C)	Surface energy against water (erg cm^{-2}) (20 °C)
Teflon	18	30.3 to 50.5
PbI$_2$	130	83.7 to 117.7
AgI	128	56.6 to 58.9
AgCl	190	143.7 to 147.8
soil, sand	21 to 43	0 to 17

absence of hydrophilic sites embedded in the surface (Zettlemoyer *et al.*, 1961); (3) the saturation state of the surrounding vapor (Corrin, 1975; also, see Section 5.6); (4) the dependence on whether the cap is advancing or receding ('contact angle hysteresis') (Osipow, 1962). A few values for $\sigma_{N/v}$ and $\sigma_{N/w}$ are presented in Table 5-3.

It is customary in the cloud physics literature to speak of the 'wetting coefficient', or 'compatibility parameter'; this is just the quantity $m_{w/a} \equiv \cos\theta$, apparently introduced by Fletcher (1958). In analogy to the case of water on a solid substrate, Fletcher also defined compatibility parameters for ice on a solid substrate. These definitions are as follows:

$$m_{i/a} \equiv \frac{\sigma_{N/a} - \sigma_{N/i}}{\sigma_{i/a}} \quad (5\text{-}24a) \qquad m_{i/w} \equiv \frac{\sigma_{N/w} - \sigma_{N/i}}{\sigma_{i/w}}. \quad (5\text{-}24b)$$

Of course, formally identical defining equations can be set up for the case of an environment of pure water vapor, and in fact one finds $m_{i/a} \approx m_{i/v}$, and $m_{w/a} \approx m_{w/v}$.

5.6 Adsorption of Gases onto Solid Surfaces

Most solids, especially in highly dispersed form, adsorb water vapor onto their surfaces. This reflects the tendency toward spontaneous reduction of surface energy, in the same way as was discussed in Section 5.4 in the context of the adsorption of dissolved salts onto the surface of tension.

Two main types of forces attract molecules to a solid surface: physical forces (*physical adsorption*) and chemical forces (*chemical adsorption* or *chemisorption*). The former are due to dispersion forces (attractive), forces caused by the presence of permanent dipoles (attractive), and short range repulsion forces. The latter are due to a transfer or electrons between the solid surface and the adsorbed water molecules and thus involve valency forces.

If the binding force between the molecules in the first adsorbed layer and the newly arriving molecules is larger than the binding force between the molecules in the first adsorbed layer and the surface of the solid, a higher vapor pressure is required for formation of the first layer than for any subsequent layer. On such walls, called *hydrophobic walls*, a critical supersaturation is required to form the first adsorbed layer, and then subsequent layers are formed spontaneously. On the other hand, if the water molecules in the first layer are more strongly bonded to the solid surface than to the newly arriving molecules, the wall becomes covered with molecules at relative humidities below 100%. However, for the completion of the first and all subsequent adsorbed layers, the relative humidity in the environment must continuously be raised. On such walls, termed *hydrophilic walls*, the thickness of the adsorbed layer of water molecules increases as the relative humidity of the environment increases, and may be several molecular layers thick before a relative humidity of 100% is reached.

The adsorption behavior of a solid surface is generally characterized by a plot of the amount of gas adsorbed as a function of the gas pressure at constant temperature. The contour which describes such a functional variation is called an *adsorption isotherm*. For physical adsorption, Brunauer *et al.* (1967) distinguish 5 main types of adsorption isotherms (see Figure 5-4). Type I represents monolayered adsorption; types II and III represent monolayered adsorption at low pressures, followed by the adsorption of further layers with increasing pressure; types IV and V represent mono- and multilayered adsorption which occurs in the presence of condensation, at subsaturation pressures, in the capillary pores of the solid surface.

Various theories have been advanced to describe the processes of adsorption of gases and vapors onto solid surfaces. Since the physics of adsorption is a large and quite complicated subject in its own right, we must refrain from treating it here in great detail. However, since studies of the adsorption of water vapor on solid surfaces have frequently and very successfully been used to characterize the nucleating properties of these surfaces, it is important that we at least become a little familiar with the basic features of the most widely used models for the adsorption phenomenon. For further information the reader may refer to sources such as Bowers (1953), Meyer (1958), Pierce (1960), Osipow (1962), Flood (1967), Dunning (1967), Clark (1970), and Samorjai (1972).

The three most widely used theoretical adsorption isotherms are those of Langmuir (1918) (*L-equation*); Brunauer, Emmett, and Teller (1967) (*BET-equation*); and Frenkel (1946), Halsey (1948), and Hill (1946, 1947, 1949, 1952) (*FHH-equation*). Because of its simplicity and because it serves as a prototype for the others, we shall now sketch a derivation of the Langmuir isotherm. Langmuir (1918) was the first to realize that adsorbed films are often just molecular monolayers, owing to the very short range of intermolecular forces. Accordingly, he treated adsorption in terms of a dynamic balance between molecules entering and leaving a unimolecular layer. Proceeding in this way, let $w^{\downarrow}$ denote the magnitude of the gas particle flux, i.e., the number of molecules striking the surface per unit area and time. Suppose a fraction α of these adhere,

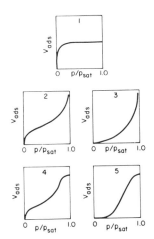

Fig. 5-4.   The five principal isotherms for physical adsorption. (After Brunauer *et al.*, 1967.)

and that the fraction of the surface covered by molecules is f. Then the rate of evaporation of gas molecules per unit area can be expressed as βf, where β is a constant, while the rate of deposition of molecules per unit area is $w^{\downarrow}\alpha(1-f)$. For a condition of dynamic equilibrium these rates are equal, and so $f = w^{\downarrow}\alpha/(\beta + w^{\downarrow}\alpha)$. But $w^{\downarrow}$ is proportional to the gas pressure (see (5-49)), and for a monolayer, $f = V/V_m$, where V is the volume of gas adsorbed at the equilibrium gas pressure p, and V_m is the gas volume necessary to form a complete monolayer. Therefore, for T = constant the balance may be expressed in the form

$$V = V_m bp/(1 + bp),\qquad\qquad(5\text{-}25)$$

where b is a constant for the given adsorbing material. This is the L-equation.

At low pressures, (5-25) predicts that adsorption is proportional to gas pressure; this is known as Henry's law. The type-I isotherm in Figure 5-4 is of the Langmuir form. An experimental example of this type is the adsorption of O_2 or CO onto silica at $0\,°C$.

The BET theory extends the Langmuir theory to include the adsorption of two or more molecular layers. The BET-equation can be written as

$$\frac{p}{V(p_{sat}-p)} = \frac{1}{V_m c} + \frac{c-1}{V_m c}(p/p_{sat}),\qquad\qquad(5\text{-}26)$$

or

$$\frac{V}{V_m} = \frac{n_{ad}}{n_m} = \frac{c(p/p_{sat})}{[1-(p/p_{sat})][1+(c-1)(p/p_{sat})]},\qquad\qquad(5\text{-}26a)$$

where n_{ad} is the number of vapor molecules adsorbed on the surface, n_m is the number of adsorption sites available on the solid surface, c is a constant for a given solid, T = const., and p_{sat} is the saturation vapor pressure for the vapor being adsorbed. The BET theory assumes: (1) that all adsorption sites on the adsorbing surface are equivalent; (2) that each molecule adsorbed in a particular layer is a possible site for adsorption of a molecule in the next layer; (3) that no horizontal interaction between adsorbed vapor molecules takes place; (4) that the heat of adsorption is the same for all molecules in any given adsorbed layer; and (5) that the heat of adsorption is equal to the latent heat of evaporation for the condensed gas in bulk for all adsorbed layers except the first. Of course, assumptions (1), (3), and (4) also apply to the L theory.

If V/V_m is plotted against p/p_{sat} a type-II isotherm is obtained if $c > 2$. A type-III isotherm results if $0 < c < 2$. If $p/V(p_{sat}-p)$ is plotted against p/p_{sat}, (5-26) yields a straight line with a slope of $(c-1)/cV_m$ and an intercept of $1/V_m c$, from which c and V_m can be determined. The total surface area of the adsorbing solid can be computed from a known value of V_m and of the area occupied by one molecule adsorbed on the surface. From this one can compute ω, the specific surface area per unit mass of the adsorbing solid. Nitrogen, argon, and krypton turn out to be the gases best suited for such surface area determinations.

At pressures close to saturation, the adsorbate consists of multilayers and has properties similar to the condensate in bulk. For such conditions, the adsorption mechanism is probably best described by the FHH theory, which can be expressed by the relation

$$\ln (p_{sat}/p) = \frac{A}{(V/V_m)^B},$$ (5-27)

where A and B are constants for any particular adsorbing solid for T = const. Thus, a plot of V/V_m vs. $\ln (p_{sat}/p)$ on a doubly logarithmic scale exhibits a linear variation from which A and B can be determined.

Of course, the Gibbs adsorption isotherm (5-16) may also be used to study the adsorption behavior of a solid surface. As an example of its use, we shall now determine the dependence of the contact angle on the environmental vapor pressure, following Corrin (1975). The reasoning is straightforward: the higher the vapor pressure e, the greater the adsorption $\Gamma_w^{(N/v)}$ of water on a given solid surface; from (5-15), a change in surface tension must result if T = constant. This, in turn, alters the contact angle. Proceeding in this manner, we have $d\mu_w^{(N/v)} = d\mu_v = \mathcal{R}T \, d \ln e$ for equilibrium changes, and assuming ideal gas behavior. Then from (5-15) we have at constant T,

$$d\sigma_{w/v} = -\mathcal{R}T\Gamma_w^{(N/v)} \, d \ln e,$$ (5-28)

which upon integration yields

$$\Pi(e) \equiv \sigma_{N/v}(0) - \sigma_{N/v}(e) = \mathcal{R}T \int_0^e \Gamma_w^{(N/v)} \, d \ln e.$$ (5-29)

The quantity $\Pi(e)$ is known as the *spreading pressure*.

If in a particular experiment the mass $m_{v,ad}$ of adsorbed water vapor is measured as a function of e in an environment of pure water vapor, then $\Gamma_w^{(N/v)} = m_{v,ad}/M_w \omega m_N$, where m_N is the mass of the adsorbing solid (N), and where ω is its specific surface area (determined by a separate experiment); with this information, $\Pi(e)$ may be found from (5-29).

For two different pressures e_1 and $e_2 > e_1$, we have

$$\Pi(e_2) - \Pi(e_1) = \sigma_{N/v}(e_1) - \sigma_{N/v}(e_2) = \mathcal{R}T \int_{e_1}^{e_2} \Gamma_w^{(N/v)} \, d \ln e > 0.$$ (5-30)

Consequently, the corresponding contact angles determined from (5-23) will also be different, and in fact

$$\cos \theta_2 = \cos \theta_1 - [\Pi(e_2) - \Pi(e_1)]/\sigma_{w/v}.$$ (5-31)

This shows that the contact angle for water on a solid substrate increases with increasing vapor pressure. For example, Barchet and Corrin (1972) studied the adsorption of water vapor onto pure silver iodide (AgI) at T = −10 °C, and at an ice supersaturation of $e/e_{sat,i} = 1.025$; this led to $\Pi(e) - \Pi(e_{sat,i}) = 0.84$ erg cm^{-2}. From (5-31) one finds that $\theta(e) - \theta(e_{sat,i}) = 3°$, from which for $\theta(e_{sat,i}) \approx 11°$, we find $\theta(e) \approx 14°$. This trend is as expected; adsorption lowers the surface energy of the solid substrate, enabling the drop on the solid surface to pull itself together further.

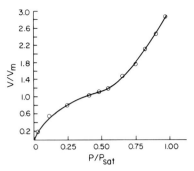

Fig. 5-5a. Adsorption isotherm for nitrogen on silver iodide at $-196\,°C$. (From Birstein, 1954; by courtesy of the Air Force Geophys. Laboratory, Hanscom Air Force Base, Mass.)

Because of its ice nucleating properties, silver iodide has been the object of numerous adsorption studies in the recent past. It is seen in Figure 5-5a, b that the adsorption of nitrogen on a sample of powdered AgI is characterized by a type-II adsorption isotherm which can be fitted to give a straight BET adsorption curve. From such a curve one may determine V_m, the number of adsorption sites available to N_2 molecules, and the total surface area of the absorbing AgI sample. Knowing the total surface area of the sample makes it possible to determine the amount of water adsorbed per unit surface area from the adsorption characteristics of the same sample for water vapor.

The adsorption characteristics of AgI for water vapor have been studied by Coulter and Candela (1952), Birstein (1955, 1956), Zettlemoyer *et al.* (1961, 1963), Tcheurekdjian *et al.* (1964), Corrin *et al.* (1964, 1967), Barchet and Corrin (1972), and Gravenhorst and Corrin (1972) using AgI of various purity. Type-II as well as type-III isotherms were observed, depending mainly on the method of preparing the AgI. Silver iodide samples, strongly contaminated with water-soluble impurities such as $AgNO_3$ and KI salts, characteristically gave type-III isotherms. Figure 5-6a illustrates the adsorption behavior of water vapor onto 'pure' AgI, and Figure 5-6b shows the adsorption behavior of water vapor on AgI of various purity. We notice that the amount of water vapor adsorbed

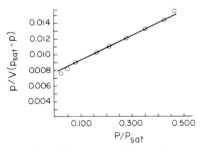

Fig. 5-5b. BET curve for adsorption of nitrogen on silver iodide at $-196\,°C$. (From Birstein, 1954; by courtesy of the Air Force Geophys. Laboratory, Hanscom Air Force Base, Mass.)

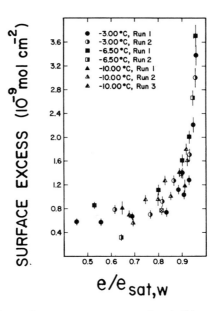

Fig. 5-6a. Adsorption isotherms for water vapor on pure silver iodide at −10 °C, −6.5 °C, and −3 °C. Isotherms at −10 °C and −6.5 °C were terminated as a result of nucleation on the sample at $e/e_{sat,w} = 0.930$ and 0.967, respectively. Vertical bars indicate one standard deviation error estimate. (From Barchet and Corrin, 1972; reprinted with permission from *J. Phys. Chem.*, copyright by the American Chemical Society.)

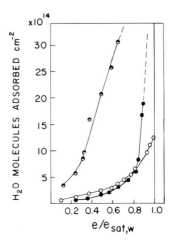

Fig. 5-6b. Adsorption isotherms for water vapor at 0 °C on doped and undoped AgI: ○ pure AgI, ● AgI doped with 0.1% KNO_3, ◐ AgI doped with 1% KNO_3. (From Gravenhorst and Corrin, 1972; by courtesy of *J. de Rech. Atm.*, and the authors.)

increases with increasing vapor pressure, rising particularly strongly as saturation is approached, and that the presence of impurity ions such as K^+ and NO_3^- in the AgI lattice enhances the adsorption of water vapor.

Some AgI samples were found to give adsorption isotherms for water vapor which could be fitted to a linear BET curve from which the number of adsorption sites available to H_2O molecules could be determined. Other samples did not behave in this manner, requiring instead alternative methods to estimate the number of water adsorbing sites (Tcheurekdjian et al., 1964; Corrin and Nelson, 1968). Comparison between the adsorption properties of AgI for water vapor and those for nitrogen demonstrated that the number of AgI surface sites available to water molecules is significantly less than the total number of sites present. This suggested that an AgI surface basically behaves like a *hydrophobic* surface with a few *water receptive*, i.e., *hydrophilic*, sites. Zettlemoyer and co-workers suggested that chemical impurity ions built into the AgI lattice may serve as such hydrophilic sites.

Since water molecules are rather weakly bonded to the AgI surface surrounding a site, they diffuse relatively easily towards the site to form a 3-dimensional (3-D) water cluster. This results in the build-up of adsorbed multilayers before the completion of an adsorbed monolayer. The concept of water clusters was strongly advocated also by Corrin and co-workers. They too interpreted the adsorption behavior on impure AgI surfaces in terms of 3-D water clusters, which they found even at low relative humidities over highly localized surface impurity sites.

However, 'pure' AgI essentially free of impurity ions was found to behave differently. The studies of Corrin and co-workers suggested that on the surface of 'pure' AgI no 3-D water clusters build up at low relative humidities. Instead, the adsorption behavior suggests the formation of two-dimensional water patches, in which the water molecules are distributed over a relatively wide area, exhibiting strongly cooperative, lateral interaction. Multilayers begin to build up only at high relative humidities.

Zettlemoyer (1968) found that, similarly to impure AgI, silica compounds doped with salt ions adsorbed considerably more water than undoped silica characterized by a fully hydroxylized surface. The larger adsorption was attributed to the doped ions acting as hydrophilic sites over which water clusters are built up.

Federer (1968) studied the adsorption behavior of water vapor on surfaces of silicon doped with boron and phosphorous. He noted a pronounced correlation between the amount of water adsorbed and the specific electric resistance of the adsorbens (see Figure 5-7a). Federer found that the samples of higher specific resistance (lower concentration of doping atoms) had a larger total density of adsorption sites, but that the sites on samples of lower specific resistance (higher concentration of doping atoms) were more active, and could adsorb more water. Through a study of the electrical surface potential of doped silicon, he concluded that the amount of charge exchanged between physically adsorbed water molecules and the substrate increases with an increase in doping. Since in giving up charge to the substrate such molecules are chemisorbed, and since

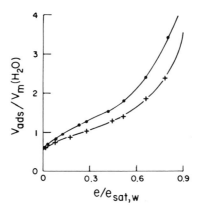

Fig. 5-7a.   Adsorption isotherms at $-10\,°C$ for water vapor on p-type silicon doped with various amounts of boron. Specific electrical resistance of sample: ● $10^{-2}\,\Omega$ cm, $+7.5 \times 10^{3}\,\Omega$ cm. (From Federer, 1968; by courtesy of *Z. Angew. Math. Phys.*, and the author.)

chemisorbed water molecules are preferred sites for subsequent further adsorption (Wanlass and Eyring, 1961), Federer interpreted the positive correlation noted between doping and adsorption in terms of a positive correlation between doping and the creation of active chemisorption sites.

A type of cluster-forming active site quite different from those mentioned above was photographically studied by Pruppacher and Pflaum (1975) on single crystals of $BaTiO_3$. Their studies showed that the tendency for water cluster formation was strongly correlated with the location of the ferroelectric domains in $BaTiO_3$, and was particularly favored on regions where the electric dipole in the surface was oriented horizontally, and on the boundaries separating ferroelectric domains.

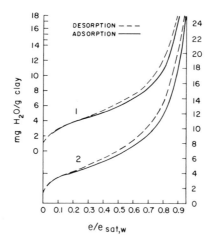

Fig. 5-7b.   Adsorption isotherms for (1) potassium kaolinite and (2) lithium kaolinite at $25.5\,°C$. Upper curve pertains to ordinate on left; lower curve pertains to ordinate on right. Both ordinates have same units. (From Martin, 1959, with changes.)

Experiments indicate that clays (a significant component of the atmospheric aerosol (see Chapter 8)) are uniformly hydrophilic and strongly adsorb water molecules over their entire surface. Nuclear magnetic resonance (NMR) studies by Wu (1964) at temperatures down to $-10\,^\circ$C verified that water molecules are very tightly bound to clay surfaces, where they are arranged close to the surface in a structure significantly different from that of ice. His observations showed that these strongly adsorbed water molecules experience a considerable loss of translational and rotational degrees of freedom. A similar observation was made by Morariu and Mills (1972), who found that at a coverage of one statistical monolayer, the diffusivity of water molecules on clay surfaces was almost one order of magnitude smaller than water diffusivity in bulk. Extrapolation of their data to higher coverage showed that the bulk value of the water diffusivity was approached only after the formation of about 15 monolayers. Expressed in another way, in the temperature interval between -22 and $+15\,^\circ$C the activation energy for self diffusion of water molecules on a clay surface was found to be higher by about 1.6 kcal mol^{-1} than the corresponding value for bulk water.

In accordance with NMR and diffusivity studies, measurements by Palmer (1952) showed that the static dielectric constant for water adsorbed on clays varies between 4 and 3.14. This is considerably below the value of about 81 for water in bulk, and is indicative of a reduction of the freedom of movement of the adsorbed water molecules, so that the degree of dipole alignment in an applied electric field is lessened.

Figure 5-7b illustrates the adsorption behavior of two typical samples of clay. One notices that in the BET classification both depicted adsorption curves are type-II isotherms. Winkler (1970) attempted to fit adsorption isotherms for clays

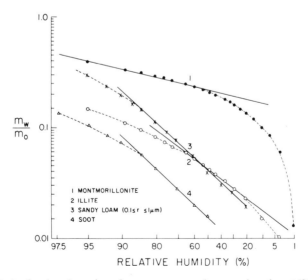

Fig. 5-7c.   FHH plot for the adsorption of water vapor on clays, sand, and soot (25 °C). m_w, mass of water adsorbed; m_0, mass of dry clay. Dashed lines connect actual measured points; straight lines represent approximate fits to data over various humidity ranges. (Based on results of Winkler, 1970.)

to an FHH equation. As illustrated in Figure 5-7c such a fit is also reasonably successful, particularly at higher relative humidities.

5.7   Ice-Vapor Interface

5.7.1   SURFACE ENERGY OF ICE

By the surface energy of ice we mean, in strict analogy to the previous cases considered, the energy required to form a unit area of new surface. For ideal crystalline ice this energy may be identified with one half the energy per unit area, W_c, which is needed to split an infinite crystal parallel to a particular crystallographic plane and separate the two parts by an infinite distance. (The factor 1/2 accounts for the fact that by cleaving the crystal, two new surfaces are created.) It is then natural to take $W_c/2$ as the surface tension or interfacial energy, $\sigma_{i/v}$, between the particular ice crystal face and water vapor or air, assuming the presence of such gases does not significantly affect the surface energy. Owing to the structure of the crystal lattice, it is clear that $\sigma_{i/v}$ will generally be different for different surface orientations with respect to the crystallographic axes, in contrast to the behavior of liquids.

To obtain W_c one must determine the binding energy, or work of cohesion, for a molecule on the crystal surface. Now for a cut perpendicular (parallel) to the crystallographic c-axis, a water molecule which is hydrogen bonded across the cleavage plane loses one nearest neighbor molecule and three (four) next nearest neighbor molecules. A water molecule which is not hydrogen bonded across the cleavage plane loses three (two) next nearest neighbor molecules. Therefore, a water molecule may be regarded as losing one nearest and six next nearest neighbors by cutting along a basal or prism plane. If we disregard the forces of interaction due to third and higher order nearest neighbors, the energy per molecule required for cleavage of an ice crystal can therefore be expressed as

$$E_m = U_1 + 6U_2, \tag{5-32}$$

where U_1 and U_2 are the average interaction potentials between the molecules in the first and second interaction zones. The interaction potentials U_1 and U_2 for intermolecular spacings of 2.76×10^{-8} cm and 4.51×10^{-8} cm, respectively, were computed by Reuck (1957) on the basis of Rowlinson's (1951) force constants, which take account of the multipole electrostatic interaction forces, induction forces, and repulsion forces. As a result, Reuck found $N_A E_m = 6.08$ kcal mol^{-1}, or $E_m = 4.22 \times 10^{-13}$ erg per molecule, for an ice crystal in vacuum at 0 °K.

All that remains is to count the density of molecules on the basal and prism planes of the unit cell of ice. For a basal plane the area is $\sqrt{3}\, a_0^2/2$ (see Figures 3-3, 3-4, 5-9). Since this area is occupied by $2(1/3 + 1/6) = 1$ molecule, the density of molecules in the basal planes is $2/\sqrt{3}\, a_0^2$. At -20 °C, $a_0 \approx 4.52 \times 10^{-8}$ cm, and so the density is 5.65×10^{14} molecules cm^{-2} = 9.38×10^{-10} mol cm^{-2}. Similarly, for a prism plane the area is $a_0 c_0$, and the occupancy is $(4 \times 1/4) + (2 + 1/2) = 2$ molecules. Since $c_0 \approx 7.36 \times 10^{-8}$ cm at $-20\,°C$, we obtain a density of 6.00×10^{14} molecules cm^{-2} = 1.00×10^{-9} mol cm^{-2} for the prism planes.

Now by multiplying these densities by E_m we finally arrive at the estimates $W_c^{(B)} = 238$ erg cm^{-2} for the basal faces of ice and $W_c^{(P)} = 253$ erg cm^{-2} for the prism faces; the corresponding values for the surface energies are $\sigma_{i/v}^{(B)} = 119$ erg cm^{-2} and $\sigma_{i/v}^{(P)} = 126$ erg cm^{-2}. Similar estimates were made by Mason (1952, 1954a) and McDonald (1953b).

A simpler approach leading to very similar results is to suppose that $N_A E_m = \mathscr{L}_s/2$, i.e., for a molecule to get from the interior to the surface requires breaking roughly half the bonds which must be severed for a complete escape. The mean value of $\mathscr{L}_s$ over the range $-40\,°C$ to $0\,°C$ is about 3.05 kcal mol^{-1}, which leads to $\sigma_{i/v}^{(B)} = 120$ erg cm^{-2} and $\sigma_{i/v}^{(P)} = 127$ erg cm^{-2}, in remarkably good agreement with the previous estimates.

McDonald (1953b) has pointed out, however, that the surface energies thus computed pertain to a 'fresh' surface. Since molecules in a freshly cleaved surface will not remain in their original position but will relax into new equilibrium positions, the surface energy of an 'aged' surface is somewhat less than that of a 'fresh' surface. The significance of such relaxation can be appreciated if we calculate $\sigma_{w/a}$ for water at $0\,°C$ in the same way as was just done for ice, and compare results with the known values of surface tension. Proceeding in this manner, we suppose that $N_A E_{m,water} \approx \mathscr{L}_e/2 = 5.38$ kcal mol^{-1} at $0\,°C$. Assuming there are about 10^{15} water molecules cm^{-2} of surface, we estimate $\sigma_{w/a}(0\,°C) \approx 188$ erg cm^{-2}; the corresponding estimate at $-40\,°C$ is $\sigma_{w/a}(-40\,°C) \approx 195$ erg cm^{-2}. However, from experiments we know that $\sigma_{w/a}(0\,°C) \approx 76$ erg cm^{-2} and $\sigma_{w/a}(-40\,°C) \approx 83$ erg cm^{-2}. Thus, for water at least, the real, 'relaxed' surface has less than half the surface energy predicted for a hypothetical 'freshly cut' surface.

In order to arrive at a suitable correction for the ice surface energy calculation, McDonald noted that the analogous calculation for water led to an overestimate by the factor $\mathscr{L}_e/2\,\sigma_{w/a}$; realizing that this may overcorrect because of the difference between $\mathscr{L}_e$ and $\mathscr{L}_s$, McDonald suggested reducing the above factor by $\mathscr{L}_e/\mathscr{L}_s$, giving a total correction factor of $\mathscr{L}_e^2/(2\,\mathscr{L}_s \sigma_{w/a}) \approx 1.20(-40 \leqslant T \leqslant 0\,°C)$. With this correction the surface free energy for a relaxed basal and prism face of ice becomes $\sigma_{i/v}^{(B)} \approx \sigma_{i/a}^{(B)} \approx 100$ erg cm^{-2} and $\sigma_{i/v}^{(P)} \approx \sigma_{i/a}^{(P)} \approx 106$ erg cm^{-2}. These values turn out to be in good agreement with the independent experimental estimate of Ketcham and Hobbs (1969), who deduced $\sigma_{i/a} \approx (109 \pm 3)$ erg cm^{-2} from observing the solid-vapor grain boundary angles of polycrystalline ice.

5.7.2 WULFF'S THEOREM

Wulff's theorem (Wulff, 1901) provides a description of the equilibrium shape of a crystal from a knowledge of the variation of surface tension with crystal face orientation. It should be emphasized that this equilibrium shape is not often observed, since actual crystal geometries are strongly influenced by thermal and diffusion gradients, and other kinetic effects associated with active growth (see Section 13.3).

Because work is required to form new surface, the equilibrium shape must be the one which minimizes the total surface energy for a given volume. For a concise treatment of the problem along these lines, the reader is referred to Landau and Lifschitz (1958). Here we shall outline a simpler, more heuristic derivation following Dufour and Defay (1963).

Consider a crystal which has a volume V'' and is bounded by (σ) faces, each of which has a surface area $\Omega^{(\sigma)}$. Let the crystal be surrounded by its own vapor (or melt) of volume V' and pressure p', and suppose the whole system is contained in a cylinder whose volume $V = V' + V''$ can be varied by a piston. By moving this piston the work $đW = -p'\,dV$ may be done on the system; this may also be written as

$$đW = -p'\,dV' - p''\,dV'' + \alpha\,dV'',\qquad(5\text{-}33)$$

where $\alpha = p'' - p'$, and where we have assumed a uniform pressure p'' within the crystal. But $dV'' = \Sigma_\sigma\,\Omega^{(\sigma)}\,dh^{(\sigma)}$, for normal outward displacement of the faces $\Omega^{(\sigma)}$ by $dh^{(\sigma)}$ (see Figure 5-8); furthermore, the volume of the crystal is $V'' = (\Sigma_\sigma\,h^{(\sigma)}\Omega^{(\sigma)})/3$, and on differentiation this yields

$$3\,dV'' = \Sigma_\sigma\,h^{(\sigma)}\,d\Omega^{(\sigma)} + \Sigma_\sigma\,\Omega^{(\sigma)}\,dh^{(\sigma)},$$

so that

$$dV'' = \frac{1}{2}\sum_\sigma h^{(\sigma)}\,d\Omega^{(\sigma)}.\qquad(5\text{-}34)$$

Therefore, the work done on the system may also be expressed as

$$(đW)_{\text{cryst.}} = -p'\,dV' - p''\,dV'' + \frac{\alpha}{2}\sum_\sigma h^{(\sigma)}\,d\Omega^{(\sigma)}.\qquad(5\text{-}35)$$

Now consider a system just like the above except that the crystal is replaced by a drop of the same volume. The work done may again be expressed as in (5-33). However, on invoking the condition of mechanical equilibrium, (5-7), we can write the last term in (5-33) in the form $2\,\sigma\,dV''/a = \sigma\,d\Omega$, where Ω is the area of the drop; therefore,

$$(đW)_{\text{drop}} = -p'\,dV' - p''\,dV'' + \sigma\,d\Omega.\qquad(5\text{-}36)$$

Comparison of (5-35) and (5-36) shows that a complete analogy between the two systems can be maintained by setting

$$\sigma^{(\sigma)} = \frac{\alpha}{2}h^{(\sigma)},\qquad(5\text{-}37)$$

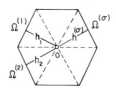

Fig. 5-8.   Schematic of a six-sided crystal to illustrate Wulff's relations.

so that also

$$(\text{d}W)_{\text{cryst.}} = -p' \, dV' - p'' \, dV'' + \sum_\sigma \sigma^{(\sigma)} \, d\Omega^{(\sigma)}. \tag{5-38}$$

From (5-37) we have

$$\frac{\sigma^{(1)}}{h^{(1)}} = \frac{\sigma^{(2)}}{h^{(2)}} = \cdots = \frac{p'' - p'}{2} = \text{constant}. \tag{5-39}$$

This constitutes Wulff's theorem: In equilibrium, the distance of any crystal face from the center of the crystal is proportional to the surface tension of that face.

Let us use Wulff's theorem to estimate the equilibrium crystal shape for ice. From (5-39) we have

$$\frac{h^{(B)}}{h^{(P)}} = \frac{\sigma_{i/v}^{(B)}}{\sigma_{i/v}^{(P)}} \approx 0.94, \tag{5-40}$$

inserting our previously determined values for the surface tensions. Referring to Figure 5-9, we see that $h^{(P)} = a'$ is the radius of the circle inscribed in the basal plane, and that $2 h^{(B)} = H$ is the height of the ice prism. Also, $a' = \sqrt{3} \, a/2$ is the radius of the circle circumscribed in the basal plane, and so

$$\frac{H}{2 \, a} \approx 0.81. \tag{5-41}$$

The equilibrium form is thus predicted to be a hexagonal prism with a ratio of axial length to hexagonal diameter (the observed 'diameter' of an ice crystal is usually given in terms of the diameter of the circle circumscribed within the basal plane of the crystal) of about 0.8 (Krastanow, 1943; Higuchi, 1961).

This result is in excellent agreement with the observations of Kobayashi (1961), who found that at very low excess vapor pressures $(H/2 \, a) \to 0.8$ for ice crystals grown at $-22\,°C \leqslant T \leqslant -10\,°C$. However, at warmer temperatures no trend to a limiting habit could be observed, while at $T < -22\,°C$, $(H/2 \, a) \to 1.4$. These latter results could be explained if it is assumed that Kobayashi's experimental arrangement could not reproduce equilibrium conditions outside the first temperature interval. Another possibility is that $\sigma_{i/v}$ is temperature dependent in a way which is different for the basal and prism faces. This could possibly arise from temperature dependent behavior of surface defects, for example, but at present no information on this point is available.

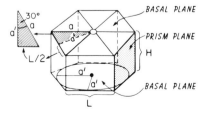

Fig. 5-9. Schematic of an ice crystal to illustrate the determination of the shape factor.

5.7.3 REAL ICE SURFACES

Surface energies characterize the average conditions on a surface. For understanding the detailed behavior of an ice crystal during growth and evaporation by vapor diffusion, we must consider in addition the microscopic, topographic surface features which are typically present.

Experimental studies show that crystalline solids have rough surfaces, i.e., they contain molecular, microscopic, and even macroscopic *steps*. Such steps often are the result of crystallographic dislocations induced in the crystal by mechanical stresses, thermal stresses, and/or accidental assimilation of foreign solid particles during the crystal's growth. These may cause lattice layers to slip along definite boundaries called *dislocations*. There are two main types of dislocation: *edge dislocations* and *screw dislocations*. In a crystal with the former, the boundary between slipped and unslipped regions extends perpendicular to the slip direction. An edge dislocation may thus be thought of as being caused by inserting an extra plane of atoms into the crystal. In a crystal with a screw dislocation, the boundary between the slipped and unslipped regions extends parallel to the slip direction, and so a screw dislocation may be thought of as being caused by cutting part way through a crystal with a knife, then shearing it parallel to the plane of cutting by one atomic spacing. Steps from screw dislocations transform successive atom planes into a helical or screw-type surface, hence the name (Figure 5-10).

Several studies have shown that molecular steps resulting from dislocations can be made visible at the ice surface by the method of *thermal etching* (see, e.g., Hobbs, 1974). During this process the ice surface is subjected to slow evaporation. Since a surface molecule at a topographic imperfection is surrounded by fewer molecules than a molecule elsewhere in the surface, it is less strongly bonded to the surface. Topographic surface imperfections are therefore the location of preferred evaporation. As water molecules are removed preferentially from such locations, topographic imperfections are made visible in the form of *etch-pits* (Figure 5-11). Thermal etching is thus capable of revealing the location of dislocations in the crystal. During thermal etching screw dislocations 'unwind' in a screw-type manner, causing the formation of etch-pits with spirally stepped walls. Step heights have been found to vary between 0.01 and 0.2 μm. It has been suggested that microscopically visible spiral step heights of as much as

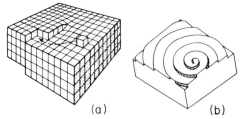

(a)   (b)

Fig. 5-10. Schematic representation of an emerging screw dislocation: (a) schematic of formation, (b) schematic of a well developed screw dislocation 'staircase.' (From Lamb and Scott, 1974, with changes.)

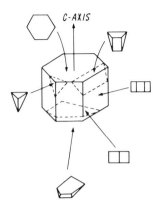

Fig. 5-11. Various types of etch-pits on the surface of ice. (From Higuchi, 1958a; by courtesy of Pergamon Press Ltd.)

several tenths of a micron do not represent the original step height of an emerging screw dislocation, but rather result from a *bunching* of monomolecular layers (Frank, 1958). The number density of etch-pits has been found to range between 10^4 and $10^6 \, \text{cm}^{-2}$, as expected from the number density of dislocations estimated on the basis of other methods.

In analogy to the etch-pits which appear at the locations of emergent crystallographic dislocations during very slow evaporation of an ice surface, small raised surface features termed *hillocks* have been observed to appear on a nascent, slowly growing ice surface at the site of emergent dislocations. This demonstrates that sites of emergent crystal dislocations are not only sites of preferred evaporation but also sites of preferred ice crystal growth. The reason for such a site serving as a preferred growth center lies in the fact that considerably less energy is involved in the propagation of an ice crystal face by the addition of water molecules to steps and ledges already present on the ice surface than by nucleation of new growth layers on a perfectly smooth ice surface.

At conditions where an ice crystal freely grows or evaporates in air, the submicroscopic surface roughness is found to manifest itself in the form of *facetted* surfaces. Such crystals are said to have a *hopper structure* (Figure 5-12). For surface ridges to be visible the ice crystal diameter needs to exceed a few hundred microns. The ridges are considered to result from the bunching of much thinner growth layers.

The bunching mechanism of growth layers at the surface of ice crystals which grow in a vapor environment was studied by Mason *et al.* (1963). According to these authors, the bunching of monolayers on a growing ice surface is the result of interference between propagating steps. At the low supersaturations typical for atmospheric clouds, new layers on an ice surface originate at topographical surface imperfections and at the edges and corners of ice crystals. There the vapor concentration gradient which constitutes the driving force for diffusional growth, as well as the temperature gradient which controls the dissipation of the

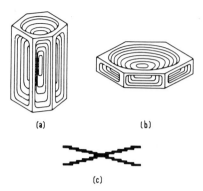

Fig. 5-12. Hopper structure at the surface of ice crystals growing freely in a supersaturated environment; (a) hollow, hexagonal prismatic column, (b) dish-shaped, hexagonal plate, (c) vertical section through center of (b). (From Mason *et al.*, 1963; by courtesy of *Phil. Mag.*, and the authors.)

latent heat released during growth, are high relative to the center of an ice crystal face. Once a step is formed it advances chiefly by the addition of water molecules brought to the step by surface diffusion. In isolation, each layer moves across the ice crystal surface with the same speed v_0. However, if two layers have originated sufficiently close to each other, competition for adsorbed water molecules tends to slow down both steps. If an additional layer originates some distance behind the pair of interfering steps, it will initially travel at the speed of an isolated step until it catches up with the interfering step pair, when it too will slow down, and so on. Eventually, the pile-up or bunching of monolayers will produce a microscopically visible step.

It is not difficult to estimate the time required for the bunching of monolayers. Let d_S denote the average migration distance which an adsorbed molecule travels by surface diffusion before re-evaporating. Then, assuming that direct arrival of molecules to the step front from the vapor is negligible in comparison to the surface diffusion flux, we see that steps grow by collecting molecules from a diffusion zone of width $2\,d_S$. Furthermore, two such fronts can be expected to experience considerable interference when their separation becomes less than this amount. We therefore expect the time needed for the two fronts to merge to be proportional to $2\,d_S$, and inversely proportional to the characteristic step speed v_0. In this manner, or simply by strict dimensional analysis, given that d_S and v_0 are the only relevant parameters, we estimate the time t required for the formation of a step of N unit heights to be within an order of magnitude of the quantity $2\,d_S N/v_0$. A detailed calculation by Mason *et al.* provides a more quantitative estimate, viz.:

$$t = 4 \ln 2\, d_S N/v_0. \tag{5-42}$$

This expression agrees well with their observations, giving sufficient evidence that surface diffusion is, in fact, the dominant process behind the bunching mechanism.

Since, as mentioned, the formation and propagation of layers is favored at the edges and corners of the ice crystal, as compared to its face center where the growing layers slow down, freely growing ice crystals preferentially thicken at the crystal's periphery, leading to the observed hopper structure depicted in Figure 5-12.

In studying the adhesive properties of ice, Faraday (1860) conjectured that a 'quasi-liquid' layer exists at the interface between ice and air, and that this layer solidifies only when sandwiched between two ice surfaces. Although this possibility obviously has considerable bearing on the feasibility of the collection growth of ice crystals in clouds, it was not until much later that the idea was pursued more quantitatively. In support of Faraday's quasi-liquid' film hypothesis, Nakaya and Matsumoto (1954) and Hosler et al. (1957), who measured the force required to separate two ice spheres brought into contact while hanging side by side, noted that the adhesive force was relatively large close to 0 °C, but decreased with decreasing temperature and humidity of the surrounding air. In an ice saturated atmosphere, Hosler et al. found that the adhesive force decreased to zero if the ambient temperature decreased below −25 °C. This result was taken to mean that quasi-liquid films on ice may be stable down to this temperature.

Further indirect evidence for the presence of a quasi-liquid layer at the ice-air interface has been provided by several studies, including Adamson et al. (1967) and Orem and Adamson (1969), who found a discontinuity in the adsorption characteristics of N_2 on ice at temperatures warmer than −35 °C, whereupon a resemblance to the adsorption characteristics of N_2 on a liquid water surface was noted; by Bullemer and Riehl (1966), Jaccard (1967), Ruepp and Käss (1969), Maidique et al. (1971), and Maeno (1973), who showed that the surface electrical conductivity of ice increased significantly at temperatures warmer than −10 °C, and particularly at temperatures warmer than −4 °C; and by Kvlividze et al. (1970), who detected a quasi-liquid layer on ice warmer than −10 °C using a nuclear magnetic resonance method.

Direct evidence for the presence of highly mobile water molecules at the surface of ice has been presented by Bryant et al. (1959), Hallett (1961), and Mason et al. (1963) in the experiment referred to above; these groups all found that water molecules at the surface of ice migrate for considerable distances before they become part of the ice crystal lattice.

An early attempt at a physical explanation for the existence of a disordered, quasi-liquid layer at the ice-air interface was provided by Weyl (1951), who suggested that it can be explained in terms of the tendency of any system to minimize its surface energy. A rearrangement of water molecules to provide such a minimization of surface energy is easily possible in liquid water where the molecules are highly mobile. On the other hand, in ice the long range order of water molecules prevents an easy rearrangement of the molecules. However, according to the views of Weyl, breakdown of this long range order may still occur near 0 °C inside a thin layer at the ice-air interface. Weyl's conjectures were followed up more quantitatively by Fletcher (1962b, 1963, 1968, 1973). By taking into account the structure of a real ice crystal lattice and considering all the defects which occur in such a lattice (see Chapter 3), Fletcher (1973)

developed a molecular thermodynamics model, from which it was possible to
compute the lowering of the surface energy of ice when a quasi-liquid film is
present, and the temperature range for which such a film is stable. The thickness
h of the quasi-liquid film could not be expressed rigorously in a simple manner
as a function of temperature T, but the following approximate relation was
derived graphically from the computed variation of h vs. T:

$$h(10^{-8}\,\text{cm}) \approx (20 \text{ to } 50) - 25 \log (T_0 - T), \tag{5-43}$$

with T in °K and $T_0 = 273$ °K. This result is in good agreement with the recent
experiments of Mazzega *et al.* (1976) who, for the sake of comparison with the
work of Fletcher, fitted their experimental results to an equation of the same form
as (5-43). They found

$$h(10^{-8}\,\text{cm}) \approx 37 - 25 \log (T_0 - T). \tag{5-44}$$

Equation (5-44) suggests that the quasi-liquid layer has decreased to a thickness
of 1 molecular diameter ($\approx 3 \times 10^{-8}$ cm) at a temperature of about -23 °C, in fair
agreement with the earlier experiments of Hosler *et al.* (1957) in an ice saturated
atmosphere.

The work of Kuczynski (1949), Kingery (1960a, b), Kuroiwa (1961, 1962),
Hobbs and Mason (1964), Hobbs and Radke (1967), Itagaki (1967), and Kikuchi
(1972) showed that the quasi-liquid layer mechanism is not the only one which
can explain the sticking together of two ice surfaces. The formation of a 'neck'
joining two ice surfaces in contact may proceed in four additional ways: (1) by
viscous and plastic flow of water substance under surface tension forces, (2) by
evaporation of water substance from the convex surface portion of the ice
system, its transfer through the environment and subsequent condensation onto
the strongly concave neck joining the two ice surfaces, (3) by volume diffusion
of water substance resulting from a local excess of ice lattice vacancies which
arise from the deficit in pressure produced by the surface tension forces in the
neck region, and (4) by surface diffusion of water substance arising from the
difference in concentration of adsorbed molecules existing in the neck and the
rest of the ice system, again set up by the surface tension forces. Theoretical
expressions for the growth rate of the neck by each of these four mechanisms
have been derived by Kuczynski (1949). His theoretical considerations, as well
as the experiments carried out by the authors mentioned above, demonstrated
that the growth rate of the neck joining the two surfaces is of the general form

$$\left(\frac{\Delta}{a}\right)^n = \frac{A(T)}{a^m}\, t, \tag{5-45}$$

where $A(T)$ is a function of temperature and the type of neck-forming ('sintering') mechanism, 2Δ is the width of the neck after time t, and a is the radius of
curvature of the two surfaces in contact. For the case of spherical particles of
radius a in contact, $n = 2$, $m = 1$ for process (1); $n = 3$, $m = 2$ for process (2);
$n = 5$, $m = 3$ for process (3); and $n = 7$, $m = 4$ for process (4).

In an experimental study and re-analysis of earlier work, Hobbs and Mason
(1964) concluded that the adhesion of spherical ice particles is mainly the result

of the evaporation-condensation mechanism (2). Later, however, Hobbs and Radke (1967) and Kikuchi (1972b) showed that volume diffusion of water molecules (mechanism 3), caused by the existence of a large concentration of molecular vacancies in ice just beneath the concave surfaces of the neck, contributes almost equally to its growth. While these two mechanisms may jointly determine the rate of growth of the neck, the initial 'bridging' between the two ice particles in contact is, according to Hobbs (1974), most likely the result of the pseudo-liquid layer at the ice-air interface.

5.8 Ice-Water and Ice-Aqueous Solution Interfaces

Let us now consider a system consisting of an ice crystal surrounded by supercooled water. As we shall see in the next chapter, for sufficiently small particles of ice such a system can be in stable equilibrium, so that it is again possible to speak of the surface tension or interface energy of the boundary separating the phases.

Intuition tells us that the surface tension between ice and supercooled water, $\sigma_{i/w}$, must be considerably less than $\sigma_{i/v}$ simply because the forces between water molecules and the spatial arrangement of molecules in supercooled water are not too different from those in ice (Chapter 3). Unfortunately, it is very difficult to determine $\sigma_{i/w}$ by experimental techniques, and the results of numerous attempts show considerable spread (Figure 5-13). Nevertheless, all measurements agree

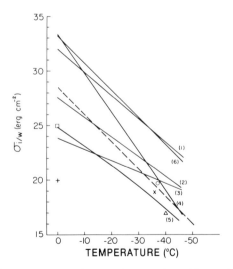

Fig. 5-13. Variation with temperature of the surface tension of ice against supercooled water. (1) Wood and Walton (1970) for spherical ice crystal, (2) Wood and Walton (1970) for Wulff crystal, (3) Dufour and Defay (1963) based on data of Jacobi (1955), (4) Ketcham and Hobbs (1969), (5) Pruppacher (unpub.) based on data of Kubelka and Prokscha (1944), (b) Pruppacher (unpub.) based on data of Turnbull (1950, 1965). × Hoffer (1961), ○ Kuhns and Mason (1968), △ Mason (1957a), + Kotler and Tarshis (1968), □ Coriel *et al.* (1971). The dashed line is the variation presently proposed.

on the fact that $\sigma_{i/w}$ decreases with decreasing temperature. This is to be expected since the structure of water becomes increasingly ice-like as the temperature decreases below $0\,°C$ (Chapter 3).

In addition to the experimental results plotted in Figure 5-13, two other indirect methods for estimating $\sigma_{i/w}$ deserve attention. The first method is based on *Antonoff's rule* (Antonoff, 1907), which states that the interfacial tension between two mutually saturated liquids (l_1, l_2) is given by the absolute difference between their respective surface tensions against the gas surrounding them, i.e.,

$$\sigma_{l_1/l_2} = |\sigma_{l_1/g} - \sigma_{l_2/g}|,\tag{5-46}$$

where g indicates the gas in contact with the liquid. This can be interpreted as a limiting form of Young's relation, (5-23). Antonoff's rule has successfully been applied to the three phases of water substance, assuming they are mutually saturated. The analogous equation for ice is

$$\sigma_{i/w} = \sigma_{i/v} - \sigma_{w/v}.\tag{5-47}$$

It should be emphasized that this relation is largely just a plausible conjecture for the behavior of $\sigma_{i/w}$, with some additional indirect empirical support coming from the success of Antonoff's rule in other applications. For $0\,°C$ with $\sigma_{i/v}^{(B)} = 100\ \mathrm{erg\ cm}^{-2}$, $\sigma_{i/v}^{(P)} = 106\ \mathrm{erg\ cm}^{-2}$, and $\sigma_{w/v} = 76\ \mathrm{erg\ cm}^{-2}$, (5-47) predicts $\sigma_{i/w}^{(B)} = 24\ \mathrm{erg\ cm}^{-2}$ and $\sigma_{i/w}^{(P)} = 30\ \mathrm{erg\ cm}^{-2}$; for $-40\,°C$, one similarly finds $\sigma_{i/w}^{(B)} = 17\ \mathrm{erg\ cm}^{-2}$ and $\sigma_{i/w}^{(P)} = 23\ \mathrm{erg\ cm}^{-2}$.

A second method for estimating $\sigma_{i/w}$ is based on a knowledge of the latent heat of fusion, $\mathscr{L}_m$. To make this estimate, let us imagine that we 'cut' both a body of ice and a body of water, each surrounded by water vapor, into two halves. In so doing the energy $W_{c,i} \approx \mathscr{L}_s/2$ is necessary to cut N_A bonds in the surface of the ice, and the energy $W_{c,w} \approx \mathscr{L}_e/2$ is expended to 'cut' N_A bonds in the surface of the water. On joining one water-half to one ice-half, we may roughly assume that the ice half gains back the energy $\mathscr{L}_e/2$. The net energy expended in cutting the ice body is therefore approximately $\mathscr{L}_s/2 - \mathscr{L}_e/2 \approx \mathscr{L}_m/2$, which amounts to $5.0 \times 10^{-14}\ \mathrm{erg}$ molecule^{-1} at $0\,°C$, and $3.5 \times 10^{-14}\ \mathrm{erg}$ molecule^{-1} at $-40\,°C$. Therefore, on taking into account the molecular coverages on the basal and prism planes, we obtain $\sigma_{i/w}^{(B)} \approx 28(20)\ \mathrm{erg\ cm}^{-2}$ and $\sigma_{i/w}^{(P)} \approx 30(21)\ \mathrm{erg\ cm}^{-2}$ at $0\,°C$ $(-40\,°C)$.

Comparison of results from these two semi-empirical methods and experiment reveals an overall consistency. Unfortunately, however, the theoretical estimates are insufficiently accurate for discriminating among the wide scatter of experimental values. They do guide us, however, in making the following educated guess as to a reasonable relation for the variation of $\sigma_{i/w}$ with temperature:

$$\sigma_{i/w} = 28.5 + 0.25\ T,\tag{5-48}$$

where σ is in $\mathrm{erg\ cm}^{-2}$ and T is in $°C$. Note that (5-48) does not discriminate between the prism and basal planes of the ice lattice.

As with ice crystals growing from the vapor, ice crystals growing from supercooled water do not have a molecularly smooth surface. Direct photographic evidence of the presence of steps at a growing ice-water interface has been provided by Ketcham and Hobbs (1968) and Hobbs and Ketcham (1969). Some of the observed steps had spiral forms appearing at a number concentration of about $10^2 cm^{-2}$, with step heights between 0.1 and 4 μm, and spacings between the steps of 5 to 20 μm. These observed step heights are enormous considering that they probably originated on screw dislocations. No explanation for this observation is currently available except to say that, as in the case of step formation at the ice-air interface, some sort of bunching mechanism may be operating.

Bryant and Mason (1960) have investigated the etched surfaces of ice grown from supercooled water. Shallow etch-pits of about 100 μm diameter and having a number density of $10^3 cm^{-2}$ were observed on all crystal faces. Inside a large, shallow pit small pyramidal etch-pits of 5 to 10 μm diameter were found in number concentrations up to $10^5 cm^{-2}$. Pyramid heights and base diameters were about equal. The sides of the pyramidal pits were facetted in the form of concentric steps, each of which had a height of a few tenths of a micron. In some larger pyramidal pits the concentric spiral steps reached heights of up to 1 μm.

A similar etch pattern had been observed earlier by Truby (1955a, b). He noted pyramidal etch-pits which were 0.2 to 0.5 μm in depth and 0.5 to 20 μm in width. At the pit walls, concentric steps of up to 0.05 μm height were observed. Often the pits had cores up to 20 μm in depth.

These observers have suggested that the etch pattern at the surface of an ice crystal grown from supercooled water is the result of dislocations introduced into the ice during the freezing process by mechanical or thermal stresses. Gentile and Drost-Hansen (1956), elaborating on this mechanism, have suggested that an ice crystal represents a 'socially unhappy arrangement' of water molecules. The 'unhappiness' is caused by the necessity for 'opening up' the bond angle of a water molecule in order to conform to the tetrahedral lattice structure of ice. Such a forced, bond angle opening introduces into the ice lattice a strain which is not uniformly distributed over the entire crystal volume. Rather, the strain energy tends to be concentrated near lines parallel to the c-axis, thereby causing strain cores. According to Gentile and Drost-Hansen, the strain energy may be relieved through the incorporation of suitably-sized, foreign salt ions in optimal concentrations. This may explain the observations of Truby (1955a, b), who noted that ice crystals grown from 10^{-3} molar fluoride solutions did not exhibit any microstructure.

Since cloud drops consist of weak aqueous salt solutions, it is worthwhile to describe briefly some of the processes which take place at the ice-solution interface. Experimental observations such as those by Jaccard and Levi (1961), de Micheli and Iribarne (1963), and Kvajic and Brajovic (1971) demonstrate that at the ice-aqueous solution interface a segregation process takes place which allows a small percentage of salt to enter the ice lattice, while the rest remains dissolved in solution. Salt ions do not enter the ice in stoichiometric proportions. Rather, the experiments show that the interface behaves as a semipermeable membrane, allowing certain types of salt ions to pass through and enter the ice lattice more readily than others. This phenomenon has been found to be strongly dependent on the concentration of the salt in solution and the rate at which the ice-solution interface advances. Earlier work on this subject has been reviewed by Drost-Hansen (1967) and Gross (1968). Recent work has been reported by Pruppacher *et al.* (1968), Cobb and Gross (1969), Gross (1971), Shewchuk and Iribarne (1971), and Iribarne (1972).

These studies lead to the following conclusions: (1) The salt concentration at which the ion selection process operates most efficiently ranges between 10^{-5} and 10^{-3} moles liter^{-1}. (2) The ion separation process is most pronounced at freezing rates between 1 and 50 μm sec^{-1}, i.e., when the ice-solution interface advances slowly. However, ion segregation still is significant at freezing rates of several centimeters per second, which are typical during the spontaneous growth of ice crystals in supercooled aqueous solutions. (3) Generally, negatively charged ions (anions) are more acceptable within the ice lattice than are positively charged ions (cations). (4) Small ions are more readily built into the ice lattice than are large ions or ions of complicated structure. (5) The two ions most readily accepted by the ice lattice are F^- and NH_4^+. F^- is preferred because of its electronegativity, and because of its ionic radius, which is similar to that of the oxygen atom in a water molecule. Although NH_4^+ does not conform well with points (3) and (4), it is nevertheless preferred because of its tetrahedral molecular structure, which is analogous to the tetrahedral structural units in ice, and because of its ionic radius, which is similar to that of O^{2-}, thus making NH_4^+ isomorphous with H_3O^+. (6) With regard to salt ions which are present in cloud and raindrops, NH_4^+ is most readily accepted into the ice lattice, while Cl^- and CO_3^{2-} are preferred against K^+, Na^+, Mg^{2+}; these cations in turn are slightly preferred over SO_2^{2-} and NO_3^-.

Finally, an interesting electrical effect occurs, known as the 'freezing potential.' Owing to the low conductivity of ice, the incorporated ions, distributed throughout the ice volume, behave as a 'frozen-in' space charge. The charge of opposite sign which remains in solution is distributed as a surface charge at the ice-solution interface. This arrangement of charge results in the development of an electrical potential between the ice and the aqueous solution. Typical freezing potentials range from several volts to several tens of volts; the highest observed range up to even a few hundred volts.

There has been considerable interest in the freezing potential phenomenon as a possible mechanism for cloud electrification. However, it now appears unlikely that the mixed aqueous solutions present in clouds are conducive to the development of large freezing potentials. A further difficulty is the apparent lack of any efficient means by which the oppositely charged water and ice might be mechanically separated, so that large scale charge separations could ensue. Therefore, the freezing potential phenomenon appears to be of only limited significance to cloud physical processes, and so will not be pursued further here.

Those who are interested in freezing potentials may wish to read, in addition to the previously mentioned references, the theoretical explanations presented in Gross (1954), Le Febre (1967), Chernov and Melnikova (1971), and Jindal and Tiller (1972).

5.9 Condensation, Deposition, and Thermal Accommodation Coefficients

There are some important gas kinetic relations pertaining to surfaces which we shall need in our discussions of nucleation and diffusion growth. Since they are of an elementary nature, for the most part we wish only to record them here, without derivations, for our future use.

Let $\dot{c}$ denote the concentration of molecules and $\bar{v}$ the mean molecular speed in a Maxwell-Boltzmann gas. Then the number of molecules crossing per unit time to either side of an arbitrarily oriented planar unit area in this gas is

$$w^{\downarrow} = \dot{c}\,\bar{v}/4. \tag{5-49}$$

Also for such a gas, the relation between $\bar{v}$ and temperature is

$$\bar{v} = \left(\frac{8\,kT}{\pi\,\dot{m}}\right)^{1/2}, \tag{5-50}$$

where $\dot{m}$ is the molecular mass. And, if we may assume the gas is ideal, the gas pressure is

$$p = \dot{c}kT. \tag{5-51}$$

Let us now consider the water-vapor interface. On combining (5-49)–(5-51) we find that the molecular flux of water vapor to the surface can be expressed as

$$w^{\downarrow} = e/(2\,\pi\,\dot{m}_w kT_g)^{1/2}, \tag{5-52}$$

where $p = e$ is the vapor pressure, and T_g is the vapor temperature. For example, for saturated conditions at 20 °C, $w^{\downarrow} = 8.5 \times 10^{21}\,cm^{-2}\,sec^{-1}$. Of these impinging molecules only a fraction α_c, called the *condensation coefficient*, actually is retained by the water surface. Under equilibrium conditions then, the rate $w^{\uparrow}$ at which molecules leave the surface must satisfy the relation

$$w^{\uparrow} = \alpha_c w^{\downarrow}, \quad \text{at equilibrium.} \tag{5-53}$$

Experimental values for α_c are listed in Table 5-4 in two categories: (a) those derived from observations on a quiescent or quasi-quiescent water surface, and (b) those derived from a rapidly renewing surface. Since the latter conditions are not likely to be realized in clouds, we recommend the values for α_c given in Table 5-4a for cloud physics computations. These values range approximately from 0.01 to 0.07, with an average of $\bar{\alpha}_c \approx 0.035$.

The mean residence time τ_S of a molecule in the water surface is given by

$$\tau_S = n/w^{\uparrow}, \tag{5-54}$$

where $n \approx 10^{15}\,cm^{-2}$ is the equilibrium number of water molecules present in one cm^2 of water surface. Assuming $\alpha_c = 0.04$, we then find $w^{\uparrow} = 3.4 \times 10^{20}\,cm^{-2}\,sec^{-1}$ and $\tau_S \approx 3\,\mu sec$ at 20 °C. This very short lifetime of a water molecule before it evaporates from the water surface implies an extremely violent agitation; however, because of the strong cohesion in the liquid surface this agitation is confined to a layer of only a few molecular thicknesses. Note that this rapid exchange of molecules applies only at equilibrium. Of course, $w^{\uparrow}$ should not be

TABLE 5-4
Condensation coefficient for water.

(a)
Evaporation From A Quasi-Quiescent Water Surface

OBSERVER	TEMPERATURE (°C)	α_c
Alty (1931)	18 to 60	0.006 to 0.016
Alty and Nicole (1931)	18 to 60	0.01 to 0.02
Alty (1933)	−8 to +4	0.04
Alty and Mackay (1935)	15	0.036
Baramaev (1939)	–	0.033
Pruger (1940)	100	0.02
Yamamoto and Miura (1949)	–	0.023
Hammeke and Kappler (1953)	20	0.045
Delaney et al. (1964)	0 to 43	0.0415
Kiriukhin and Plaude (1965)	7	0.019
Chodes et al. (1974)	20	0.033
Rogers and Squires (1974)	–	0.065
Narusawa and Springer (1975)	18 to 27	0.038
Sinarwalla et al. (1975)	22.5 to 25.7	0.026

(b)
Evaporation From A Rapidly Renewing Water Surface

OBSERVER	TEMPERATURE (°C)	α_c
Hickman (1954)	0	0.42
Berman (1961)	–	1.0
Nabavian and Bromley (1963)	10 to 50	0.35 to 1.0
Jamieson (1965)	0 to 70	0.35
Mills and Seban (1967)	7 to 10	0.45 to 1.0
Tamir and Hasson (1971)	50	0.20
Narusawa and Springer (1975)	18 to 27	0.18

interpreted as a net evaporation rate. Drop evaporation rates are discussed in Chapter 13.

We shall assume now that the flux of molecules leaving the water is independent of the flux entering it, and that it is equal to the flux of molecules which would enter the water if it were in equilibrium with the vapor phase for which $e = e_{sat,w}$; i.e., $w^{\uparrow} = \alpha_c e_{sat,w}/(2 \pi \dot{m}_w k T_a)^{1/2}$, where T_a is the temperature of the water surface. Assuming $T_g \approx T_a = T$, we find for the net flux of molecules into the surface,

$$w_{net,w} = \alpha_c(e - e_{sat,w})/(2 \pi \dot{m}_w k T)^{1/2}. \tag{5-55}$$

Analogously, for the net flux of water molecules to an ice surface, we write

$$w_{net,i} = \alpha_d(e - e_{sat,i})/(2 \pi \dot{m}_w k T)^{1/2}, \tag{5-56}$$

where α_d is called the *deposition coefficient*. Equations (5-55) and (5-56) are different forms of what is known as the Hertz-Knudsen equation.

More recently, Schrage (1953) and Patton and Springer (1969) have modified this equation to account for the effect of net bulk vapor motion of the molecular velocity distribution. According to these authors, a better representation for w_{net} is

$$w_{net,w} = \frac{\alpha_c}{1 - B\alpha_c} (e - e_{sat,w})/(2 \pi \dot{m}_w kT)^{1/2}, \qquad (5\text{-}57)$$

where $B = 0.5$, according to Schrage.

Experimentally determined values for α_d are listed in Table 5-5. We see from this table that the deposition coefficient for water molecules on ice exhibits a trend from values near unity at very low ice surface temperatures to values comparable to α_c for surface temperatures near $0\,°C$. This result may be interpreted as evidence for the pseudo-liquid film which has been postulated to exist on an ice surface.

Conceptually, one expects that water molecules striking a water or ice surface suffer inhibited accommodation with respect to heat as well as mass flow. In order to take this effect into account, one introduces a *thermal accommodation coefficient*. In the context of interest to us, this coefficient is defined as the ratio of water vapor molecules which on collision with a (macroscopic) water drop or ice particle achieve thermal equilibrium with it, to the total number of water vapor molecules striking the surface. This definition may also be expressed as

$$\alpha_T \equiv \frac{T' - T_2'}{T' - T_2}, \qquad (5\text{-}58)$$

where T' and T_2' are the kinetic temperatures of the gas molecules incident on, and reflected by, the surface of the body with which the environment attempts thermal equilibrium, and T_2 is the surface temperature of the body. From our verbal definition of α_T it is clear that (5-58) is simply a statement of the balance of thermal energy into and out of the surface. Experiments by Alty and Mackay (1935), which unfortunately are the only ones available, show that for a water surface, $\alpha_T \approx 0.96$. This indicates that most gas molecules thermally equilibrate with a water surface during their residence time on that surface. No measurements for an ice surface are available.

TABLE 5-5
Deposition coefficient for ice.

Observer	Temperature (°C)	α_d
Delaney et al. (1964)	-2 to -13	0.014
Vulfson and Levin (1965)	-6 to -7	0.04
Vulfson and Levin (1965)	-10 to -11	0.7
Fukuta and Armstrong (1974)	-25	0.12
Davy and Somorjai (1971)	-45	0.36
Kramers and Stemerding (1951)	-40 to -60	0.93
Tschudin (1945)	-60 to -85	0.94
Davy and Somorjai (1971)	-85	1.0
Koros et al. (1966)	-115 to -140	0.83

CHAPTER 6

EQUILIBRIUM BEHAVIOR OF CLOUD DROPS AND ICE PARTICLES

Having established some background material for use in studying the bulk and surface properties of water and aqueous solutions, it is appropriate now to take a closer look at the equilibrium behavior of typical and/or idealized cloud particles of ice and water. In particular, we shall study the equilibrium of (1) a pure water or aqueous solution drop surrounded by water vapor or humid air, (2) an ice crystal in humid air, (3) an ice crystal and a separate solution drop in humid air, and (4) an ice crystal immersed in a solution drop in humid air. We shall see later that the relationships provided by these case studies are needed in order to formulate the conditions for which cloud drops and ice crystals are nucleated in the atmosphere (Chapters 7 and 9).

6.1 General Equilibrium Relation for Two Phases Separated by a Curved Interface

In this section we shall return to the system first discussed in Section 5.2, in which a spherical bulk phase $''$ of radius a is imbedded in another bulk phase $'$. We suppose each phase contains component k and other components constituting a non-ideal mixture. Also, we allow mass transfers to occur between phases, but exclude chemical reactions. We further assume thermal equilibrium, and let T denote the common temperature.

Our goal is to obtain a single equation relating the differentials of T, p', σ, a, and the activities a'_k and a''_k for component k. There are, of course, several possible starting points for accomplishing this; here we shall follow a particularly efficient procedure suggested by Dufour and Defay (1963). We begin with the chemical potential of component k in either of the bulk phases (cf. (4–53)):

$$\mu_k(p, T, a_k) = \mu_{k,0}(p, T) + \mathscr{R}T \ln a_k. \tag{6-1}$$

On dividing this expression by T, forming the total differential, and using (4-11) and (4-13), we find that for equilibrium changes

$$d\left(\frac{\mu_k}{T}\right) = -\frac{h_{k,0}}{T^2} dT + \frac{v_{k,0}}{T} dp + \mathscr{R}\, d \ln a_k. \tag{6-2}$$

Now in equilibrium we have $\mu'_k = \mu''_k$ also; consequently, we may write

$$d\left(\frac{\mu'_k}{T}\right) = d\left(\frac{\mu''_k}{T}\right). \tag{6-3}$$

Therefore, on combining this equality with (6-2) as applied to both bulk phases, we may eliminate direct reference to the chemical potentials and obtain

$$-\frac{(h'_{k,0} - h''_{k,0})}{T^2} \, dT + \frac{v'_{k,0}}{T} \, dp' - \frac{v''_{k,0}}{T} \, dp'' + \mathcal{R} \, d \ln (a'_k/a''_k) = 0. \tag{6-4}$$

Finally, we invoke the condition of mechanical equilibrium, (5-7), and introduce the latent heat $\mathscr{L}'^{''}$, from (4-78); the desired form is thereby obtained from (6-4):

$$-\frac{\mathscr{L}'^{''}_{k,0}}{T^2} \, dT + \frac{(v'_{k,0} - v''_{k,0})}{T} \, dp' - \frac{2 v''_{k,0}}{T} \, d\left(\frac{\sigma}{a}\right) + \mathcal{R} \, d \ln (a'_k/a''_k) = 0. \tag{6-5}$$

This result may well be regarded as the 'master equation' for this chapter, because special cases of it describe nearly every situation we discuss. It is obvious that (6-5) contains the Clausius-Clapeyron equation as a special case which is readily obtained by letting $a'_k = 1$, $a''_k = 1$, and $(1/a) = 0$.

6.2 Effect of Curvature on Latent Heat of Phase Change

Perhaps the reader is disturbed by a bit of sleight-of-hand we used in arriving at (6-5): In substituting $\mathscr{L}'^{''}_{k,0}$ for the enthalpy difference in (6-4), we glossed over the fact that the pressures are not equal in phases ' and ''; thus (4-78) does not strictly apply. We shall now estimate the error incurred by ignoring this pressure difference.

Evidently, the error is measured by

$$\Delta h''_{k,0} \equiv h''_{k,0}(p'', T) - h''_{k,0}(p', T) = \int_{p'}^{p''} \left(\frac{\partial h''_{k,0}}{\partial p''}\right)_T dp''. \tag{6-6}$$

But from (4-6) we see that $(\partial h/\partial p)_T = v + T(\partial s/\partial p)_T$, where s is the molar entropy, while from (4-10) and (4-11) we find $(\partial s/\partial p)_T = -(\partial v/\partial T)_p$; consequently,

$$\left(\frac{\partial h''_{k,0}}{\partial p''}\right)_T = v''_{k,0} - T\left(\frac{\partial v''_{k,0}}{\partial T}\right)_{p''}. \tag{6-7}$$

For either water or ice, the second term in (6-7) is negligible in comparison with the first; furthermore, since the compressibilities of water and ice are very small (see Section 6.4), we may regard $v''_{k,0}$ as constant when inserted into (6-6), and so obtain

$$\Delta h''_{k,0} \approx v''_{k,0}(p'' - p') = 2 v''_{k,0}\sigma/a. \tag{6-8}$$

Now, if we denote the latent heat of pure substance k in passing from phase ' to spherical phase '' of radius a by $(\mathscr{L}'^{''}_{k,0})_a$, we see that

$$(\mathscr{L}'^{''}_{k,0})_a = (\mathscr{L}'^{''}_{k,0})_\infty + \Delta h''_{k,0}. \tag{6-9}$$

Consequently, for the case of a pure water drop in equilibrium with water vapor, we can write, in the simpler notation introduced in Section 4.8,

$$\frac{(\mathscr{L}_{e,0})_a}{(\mathscr{L}_{e,0})_\infty} = 1 - \frac{\Delta h_{w,0}}{(\mathscr{L}_{e,0})_\infty} \approx 1 - \frac{2\,\sigma_{w/v}M_w}{\rho_w(\mathscr{L}_{e,0})_\infty a}. \tag{6-10}$$

This demonstrates that the latent heat of evaporation decreases with decreasing radius of curvature of the water surface. At 0 °C the second term on the right side of (6-10) has the value $6.1 \times 10^{-9}/a$, for a in cm. Thus, the error in setting $(\mathscr{L}_{e,0})_a = (\mathscr{L}_{e,0})_\infty$ becomes less than 1%, as long as $a \geqslant 6 \times 10^{-7}$ cm.

Analogously, we find for the latent heat of sublimation,

$$\frac{(\mathscr{L}_s)_a}{(\mathscr{L}_s)_\infty} \approx 1 - \frac{2\,\sigma_{i/v}M_w}{\rho_i(\mathscr{L}_s)_\infty a}. \tag{6-11}$$

At 0 °C the second term on the right side of this equation has the value $7.7 \times 10^{-9}/a$, so that the error in setting $(\mathscr{L}_s)_a = (\mathscr{L}_s)_\infty$ is less than 1%, as long as $a \geqslant 8 \times 10^{-7}$ cm. Considering (4-80), we find that it is also justified to set $(\mathscr{L}_{m,0})_a = (\mathscr{L}_{m,0})_\infty$ with an error of less than 1%, as long as $a \geqslant 2 \times 10^{-7}$ cm.

6.3 Generalized Clausius-Clapeyron Equation

We noted in Section 6.1 that (6-5) contains, as a special case, the Clausius-Clapeyron equation for a pure substance in bulk phases of negligible curvature. We may now very easily derive its extended form for the case where the curvature matters. For this purpose, consider again a pure water drop in equilibrium with vapor at pressure $p' = e_{a,w}$. We have $a'_k = a''_k = 1$, and so we may rearrange (6-5) to read

$$\frac{de_{a,w}}{dT} = \frac{(\mathscr{L}_{e,0})_a}{T(v_{v,0} - v_{w,0})} + \frac{2\,v_{w,0}}{(v_{v,0} - v_{w,0})}\frac{d(\sigma_{w/v}/a)}{dT}. \tag{6-12}$$

According to the phase rule for curved phases, (5-10), the present system has two degrees of freedom. Let us hold the radius a constant and study the variation of $e_{a,w}$ with T. Then substitution of (6-10) and (4-89) into (6-12) leads to the desired extension of the Clausius-Clapeyron equation:

$$\frac{de_{a,w}}{dT} = \frac{de_{sat,w}}{dT} - \frac{2[\sigma_{w/v} - T(d\sigma_{w/v}/dT)_a]}{aT[(v_{v,0}/v_{w,0}) - 1]}, \tag{6-13}$$

where we should recall that the equilibrium vapor pressure $e_{a,w}$ over a curved water surface actually has the physical meaning $(e_{sat,w})_a \equiv e_a$, and the equilibrium vapor pressure $e_{sat,w}$ over a plane water surface has the meaning $(e_{sat,w})_\infty \equiv e_{sat,w}$. Since $d\sigma_{w/v}/dT < 0$ for all T, curvature is seen to decrease the temperature variation of the saturation vapor pressure. On evaluating the terms in (6-13) at 20 °C, we have

$$\left(\frac{de_{a,w}}{dT}\right)_a \bigg/ \left(\frac{de_{sat,w}}{dT}\right) \approx 1 - \frac{10^{-8}}{a}, \tag{6-14}$$

for a in cm. Thus the quantitative effect is quite small (less than 1% difference for $a \gtrsim 10^{-6}$ cm), as we might expect from our previous studies of the effects of curvature.

In the remainder of this chapter we shall ignore the small influence of curvature on latent heat.

6.4 Equilibrium Between a Pure Water Drop and Pure Water Vapor or Humid Air

We may suppose a system comprised of a pure water drop in an environment of humid air contains just two components, air and water, since there is negligible selective adsorption of the gaseous constituents of air. Therefore, the system has a variance of three, according to the phase rule. We shall fix T and the total gas pressure $p' = p$, and study the variation of water vapor pressure $e_{a,w}$ with radius a.

For these conditions (6-5) reduces to

$$-\frac{2\,v_{w,0}}{T}\,d\left(\frac{\sigma_{w/a}}{a}\right) + \mathscr{R}\,d\ln\left(\frac{a_v}{a_w}\right) = 0. \tag{6-15}$$

Assuming ideal gas behavior, we have $a_v = x_v = e_{a,w}/p$; and for the pure water drop, $a_w = 1$. Making these substitutions and disregarding for the moment the compressibility of water, we may immediately integrate (6-15) between ∞ and a to obtain

$$\frac{e_{a,w}}{e_{sat,w}} = \exp\left(\frac{2\,v_{w,0}\sigma_{w/a}}{\mathscr{R}Ta}\right) = \exp\left(\frac{2\,M_w\sigma_{w/a}}{\mathscr{R}T\rho_w a}\right). \tag{6-16}$$

This is the *Kelvin equation*, first derived by W. Thomson (later Lord Kelvin, 1870). It demonstrates that at any given temperature the saturation vapor pressure over the surface of a water drop is larger than that over a flat surface, and increasingly so with decreasing radius. Accordingly, in the atmosphere large drops must grow by vapor diffusion at the expense of the smaller ones.

Since the Kelvin equation assumes equilibrium between the drop and its environment, we of course have $e_{a,w} = e$, the partial pressure of vapor in the environment. Hence we can also say that equilibrium requires an environmental supersaturation of $s_{v,w} = (e_{a,w}/e_{sat,w}) - 1 > 0$; also, the Kelvin equation may be expressed in terms of the saturation ratio $S_{v,w} = e_{a,w}/e_{sat,w}$ in the form

$$\ln S_{v,w} = \frac{2\,v_{w,0}\sigma_{w/a}}{\mathscr{R}Ta} = \frac{2\,M_w\sigma_{w/a}}{\mathscr{R}T\rho_w a}. \tag{6-17}$$

A numerical evaluation of (6-17) is plotted in Figure 6-1 for 20 °C and −20 °C. Note that the effect of curvature becomes important only for $a \lesssim 10^{-5}$ cm (0.1 μm), and that the temperature dependence is relatively weak.

Let us assume now that air is absent from the system and that the water drop is surrounded instead by pure water vapor. From the phase rule it follows that this system has two independent intensive variables. Of these we shall keep T constant and again determine the variation of $e_{a,w}$ with a. Under these conditions the first and last terms of (6-5) are zero ($a'_k = a_v = x_v = 1$, assuming ideal gas

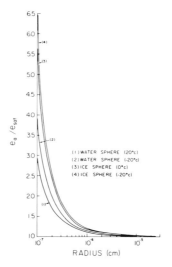

Fig. 6-1. Variation of the equilibrium vapor pressure over a water sphere and over an ice sphere with sphere size, for different temperatures. For (1) and (2), ordinate represents $e_{a,w}/e_{sat,w}$. For (3) and (4), ordinate represents $e_{a,i}/e_{sat,i}$.

behavior), and we obtain

$$\frac{v_{v,0}}{T} de_{a,w} - \frac{v_{w,0}}{T} de_{a,w} - \frac{2 v_{w,0}}{T} d\left(\frac{\sigma_{w/a}}{a}\right) = 0. \tag{6-18}$$

On substituting the ideal gas law for $v_{v,0}$ and integrating between ∞ and a, (6-18) yields

$$\ln (S_{v,w})_{vapor} = \ln \left(\frac{e_{a,w}}{e_{sat,w}}\right)_{vapor} = \frac{2 M_w \sigma_{w/v}}{\mathscr{R}T\rho_w a} + \frac{M_w}{\rho_w \mathscr{R}T}(e_{a,w} - e_{sat,w}), \tag{6-19}$$

which demonstrates that $(S_{v,w})_{vapor} > (S_{v,w})_{air}$. However, the difference is small: Comparison of (6-17) and (6-18) shows that

$$\frac{\ln (S_{v,w})_{vapor}}{\ln (S_{v,w})_{air}} - 1 \approx \frac{e_{a,w} - e_{sat,w}}{(2 \sigma_{w/v}/a)} \ll 1, \tag{6-20}$$

for all cases of interest.

Notice that in our derivation of the Kelvin equation we assumed nowhere that the surface tension is independent of the curvature of the drop. However, we did assume that the compressibility of water is negligible. This latter assumption has been investigated by Dufour and Defay (1963), who found that the inclusion of compressibility leads to the following modification of (6-17):

$$\ln S_{v,w} = \frac{2 \sigma_{w/a} v_{w,0}}{\mathscr{R}Ta} \left(1 - \frac{\kappa_{w,0}\sigma_{w/a}}{a}\right), \tag{6-21}$$

where $v_{w,0}$ is the molar volume of pure water in bulk, in contrast to $v_{w,0}$ in (6-17) which is actually the molar volume evaluated at the internal pressure of the drop, and $\kappa_{w,0} = -(1/v_{w,0})(\partial v_{w,0}/\partial p_w)_T$ is the compressibility. Since in the temperature range $+30$ to $-30\,°C$ $\kappa_{w,0}$ varies between 45×10^{-12} and about $70 \times 10^{-12}\,cm^2\,dyn^{-1}$ (Handbook of Chemistry and Physics), we readily find that $(\kappa_{w,0}\sigma_{w/a}/a) \ll 1$ for all drop radii encountered in clouds. We may therefore represent $v_{w,0}$ by its bulk value with negligible error.

6.5 Equilibrium Between an Aqueous Solution Drop and Humid Air

Let us now investigate the more interesting and realistic case of the equilibrium between a drop of an aqueous salt solution and an environment of humid air. We assume that the dissolved substance has no vapor pressure of its own, and that its mass in the drop remains constant. The first of these assumptions holds for all salts typically found in the atmosphere. The second assumption holds for at least the early stages of cloud drop formation; during the later stages of growth solutes may be added to the drop by means of various scavenging mechanisms.

According to (5-10) this system has three components, and so $w = 4$. However, the required constancy of salt mass in the drop introduces an additional relation between the drop radius and the mole fraction of water in solution, which makes $w = 3$. The volume of the solution drop is given by $4\,\pi a^3/3 = n_w v_w + n_s v_s$. Since $x_w = n_w/(n_w + n_s)$, where n_s, v_w, v_s are constants, the additional relation is

$$\frac{1}{x_w} = 1 + \frac{n_s}{n_w} = 1 + \left[n_s v_w \Big/ \left(\frac{4\,\pi a^3}{3} - n_s v_s \right) \right]. \tag{6-22}$$

Let us now determine the dependence of the saturation vapor pressure e_a on radius, subject to the conditions of constant T, n_s, and total air pressure p. This time only the last two terms of (6-5) survive, and since we now have $a'_k = a_v = e_a/p$ and $a''_k = a_w$, the following result is obtained:

$$d \ln e_a = \frac{2\,v_{w,0}}{\mathscr{R}T} d \left(\frac{2\,\sigma_{s/a}}{a} \right) + d \ln a_w. \tag{6-23}$$

On integration from a, e_a, a_w to $a \to \infty$, $e_a = e_{sat,w}$, $a_w = 1$, this equation yields

$$\frac{e_a}{e_{sat,w}} = a_w \exp \left(\frac{2\,M_w \sigma_{s/a}}{\mathscr{R}T\rho_w a} \right). \tag{6-24}$$

For $a_w = 1$, (6-24) reduces to the Kelvin law, while for a flat water surface, the generalized Raoult's law, (4-60), is recovered.

Unfortunately, no information is available on the curvature dependence of the activity coefficient of water in an aqueous solution. However, in view of our previous discussions of the effects of curvature, it seems very reasonable to regard it as negligible, and we shall do so here. Then for a_w in (6-24) we may use the expressions (4-69) and (4-66) to find

$$\ln a_w = -\frac{\nu \Phi_s m_s M_w/M_s}{(4\,\pi a^3/3)\rho''_s - m_s}. \tag{6-25}$$

Therefore, (6-24) may also be written in the form

$$\frac{e_a}{e_{sat,w}} = \exp \left[\frac{2\,M_w \sigma_{s/a}}{\mathscr{R}T\rho_w a} - \frac{\nu \Phi_s m_s M_w/M_s}{(4\,\pi a^3 \rho''_s/3) - m_s} \right], \tag{6-26a}$$

or

$$\ln \frac{e_a}{e_{sat,w}} = \ln S_{v,w} = \frac{2\, M_w \sigma_{s/a}}{\mathcal{R} T \rho_w a} - \frac{\nu \Phi_s m_s M_w / M_s}{(4\,\pi a^3 \rho_s''/3) - m_s}. \tag{6-26b}$$

Now for a sufficiently dilute solution such that $m_s \ll m_w$, $\sigma_{s/a} \approx \sigma_{w/a}$, $\Phi_s \approx 1$, and $\rho_s'' \approx \rho_w$, (6-26b) reduces to the more convenient form

$$\ln \frac{e_a}{e_{sat,w}} = \frac{A}{a} - \frac{B}{a^3}, \tag{6-27}$$

where

$$A = \frac{2\, M_w \sigma_{w/a}}{\mathcal{R} T \rho_w} \approx \frac{3.3 \times 10^{-5}}{T}; \qquad B = \frac{3\, \nu m_s M_w}{4\,\pi M_s \rho_w} \approx \frac{4.3\, \nu m_s}{M_s}, \tag{6-28}$$

in cgs units and with T in °K. Finally, if $e_a/e_{sat,w} \approx 1$, (6-27) reduces further to

$$\frac{e_a}{e_{sat,w}} = 1 + \frac{A}{a} - \frac{B}{a^3}. \tag{6-29}$$

Equations (6-26) to (6-29) are different forms of the *Köhler equations* (Köhler, 1921a, b, 1922, 1927, 1936).

As we have seen, the vapor pressure over a pure water drop always obeys the inequality $e_{a,w} \geqslant e_{sat,w}$. In contrast, the vapor pressure over an aqueous solution drop may be larger or smaller than $e_{sat,w}$, depending on whether the solute term (the second term on the right side of (6-26) and (6-27)) is smaller or larger than the curvature term. This, in turn, implies that an aqueous solution drop may be in equilibrium with a subsaturated environment. Specifically, if $a^2 A < B$, then $e_a < e_{sat,w}$.

The Köhler equations are plotted in Figure 6-2 for solution drops of two representative salts. The maxima in the curves are found from (6-29) to occur at the *critical radius* $a_c = (3\, B/A)^{1/2}$, corresponding to $[(e_a/e_{sat,w}) - 1]_c = (4\, A^3/27\, B)^{1/2}$. (In the remainder of this chapter, we shall use the subscript c to denote

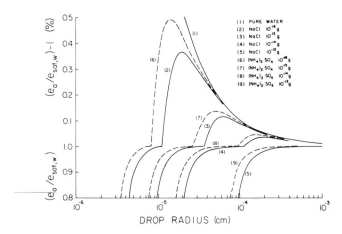

Fig. 6-2. Variation of the equilibrium vapor pressure over an aqueous solution drop with drop size, for various amounts of NaCl (solid lines) and $(NH_4)_2SO_4$ (dashed lines) in solution, and for 20 °C.

conditions at the critical radius.) It is interesting to note that for $a > a_c$ the solution drop is in unstable equilibrium with its environment, just as a pure water drop is in unstable equilibrium at all sizes. For $a < a_c$ on the other hand, the solution drop is in stable equilibrium. This behavior can be understood on realizing the environment provides effectively an infinitely large reservoir of water vapor at constant pressure. For example, suppose the equilibrium state is given by a point on the descending branch ($a > a_c$) of one of the equilibrium growth curves in Figure 6-2. Assume now that a small perturbation causes a few molecules of water to be added to the drop. At the slightly larger new radius, the equilibrium vapor pressure is lower; hence vapor will continue to flow to the drop, and it will grow ever larger. Conversely, a small evaporation excursion will produce a slightly smaller radius for which the equilibrium vapor pressure is higher than that provided by the environment, and the drop will therefore continue to evaporate. If the drop is pure water, it will evaporate completely. On the other hand, if it is a solution drop it will diminish in radius until it has reached a size which corresponds to an equilibrium state on the ascending branch ($a < a_c$) of the given equilibrium curve. Now it will be in stable equilibrium. For $da > 0$ (<0), we see that $de_a > 0$ (<0); i.e., the environmental vapor pressure is insufficient (excessive) for equilibrium at the new radius, and evaporation (condensation) will ensue to oppose the initial radius perturbation.

If the environment has reached a supersaturation equal to or larger than $[(e_a/e_{sat,w}) - 1]_c$, it is said to have reached the supersaturation needed to *activate* the drop. We note also that it is customary to call the radius of an aqueous solution drop which is in equilibrium with an environment of $S_{v,w} = 1.0$, the *potential radius* a_p; i.e., $a = a_p$ for $(e_a/e_{sat,w}) = S_{v,w} = 1.0$, from which $B = Aa_p^2$. Thus, we may write (6-27) in terms of a_p as

$$\ln \frac{e_a}{e_{sat,w}} = A\left(\frac{1}{a} - \frac{a_p^2}{a^3}\right). \tag{6-30}$$

This expression implies that aqueous solution drops which have the same potential radius exhibit the same equilibrium variation of e_a with drop size.

Equation (6-26) shows how the equilibrium behavior of an aqueous solution drop depends on the total mass m_s of salt in the drop, as well as on the type of salt (Φ_s, M_s, ν). This dependence is also illustrated in Figure 6-3, which shows that the smaller the mass of the salt in the drop, the higher the maximum, and the steeper the pre-maximum branch of the equilibrium curve for that drop. The effect of the type of salt in solution is also illustrated in Figure 6-2 for drops containing NaCl and $(NH_4)_2SO_4$. By comparing the figure with (6-27) and (6-28), we see that the dominant influence is the molecular weight of the salt, which is much larger for $(NH_4)_2SO_4$ than for NaCl, so that the equilibrium curves for a solution drop of $(NH_4)_2SO_4$ lie above the corresponding ones for NaCl.

Temperature has only a small effect on the equilibrium conditions for solution drops, as Table 6-1 shows. The trend, such as it is, indicates that the supersaturation necessary to hold a given solution drop in equilibrium increases with decreasing temperature.

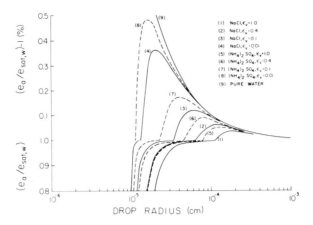

Fig. 6-3. Variation of the equilibrium vapor pressure over an aqueous solution drop formed from a mixed particle of radius $r_N = 0.1\ \mu$m containing various volume fractions ε_v NaCl and $(\dot{N}H_4)_2SO_4$ and a water insoluble substance, for 20 °C. (From Hänel, 1976; by courtesy of *Advances of Geophys.*, and the author.)

Table 6-2 lists the amount of water which is acquired and the salt dilution which is experienced during equilibrium growth of an aqueous solution drop of given salt content at different environmental equilibrium humidities. It is seen that the dilution of the salt solution is generally small, as long as the drop is in equilibrium with an environment of $S_{v,w} \leqslant 0.9$. However, the dilution increases quite rapidly as $S_{v,w}$ increases beyond 1.0 and approaches $(S_{v,w})_c$.

TABLE 6-1

Critical equilibrium size a_c and critical equilibrium supersaturation $(s_{v,w})_c$ for aqueous solution drops containing different amounts of NaCl. $\rho_{NaCl}(20\ ^\circ C) = 2.16\ \mathrm{g\ cm^{-3}}$, $\rho_{NaCl}(-10\ ^\circ C) = 2.17\ \mathrm{g\ cm^{-3}}$. (Based on data of Low, 1969d, and of Hänel, 1976.)

	$m_N(g)$	10^{-16}	10^{-15}	10^{-14}	10^{-13}	10^{-12}
r_N	20 °C	2.227×10^{-6}	4.797×10^{-6}	1.033×10^{-5}	2.227×10^{-5}	4.797×10^{-5}
(cm)	$-10\ ^\circ C$	2.224×10^{-6}	4.792×10^{-6}	1.032×10^{-5}	2.224×10^{-5}	4.792×10^{-5}
$(s_{v,w})_c$	20 °C	3.660×10^{-1}	1.143×10^{-1}	3.578×10^{-2}	1.126×10^{-2}	3.558×10^{-3}
(%)	$-10\ ^\circ C$	4.738×10^{-1}	1.480×10^{-1}	4.628×10^{-2}	1.455×10^{-2}	4.595×10^{-3}
a_c	20 °C	1.947×10^{-5}	6.212×10^{-5}	1.988×10^{-4}	6.358×10^{-4}	2.012×10^{-3}
(cm)	$-10\ ^\circ C$	1.783×10^{-5}	5.679×10^{-5}	1.819×10^{-4}	5.826×10^{-4}	1.844×10^{-3}

TABLE 6-2

Amount of water gained during equilibrium growth of NaCl and $(NH_4)_2SO_4$ solution drops; 20 °C. (Based on data of Low, 1969d, and of Hänel, 1976.)

		NaCl			$(NH_4)_2SO_4$		
mass m_N(g) of salt in solution		1.0×10^{-16}	1.0×10^{-14}	1.0×10^{-12}	1.0×10^{-16}	1.0×10^{-14}	1.0×10^{-12}
mass m_w(g) of water in solution at	$(S_{v,w}) = 0.80$	3.1×10^{-16}	3.2×10^{-14}	3.3×10^{-12}	1.05×10^{-16}	1.1×10^{-14}	1.1×10^{-12}
	$(S_{v,w}) = 0.90$	5.3×10^{-16}	5.8×10^{-14}	6.1×10^{-12}	2.1×10^{-16}	2.4×10^{-14}	2.4×10^{-12}
	$(S_{v,w}) = 1.00$	6.2×10^{-15}	6.8×10^{-12}	7.1×10^{-9}	2.5×10^{-15}	3.7×10^{-12}	3.8×10^{-9}
	$(S_{v,w}) = (S_{v,w})_c$	3.09×10^{-14}	3.29×10^{-11}	3.41×10^{-8}	1.20×10^{-14}	1.61×10^{-11}	1.80×10^{-8}
dilution m_w/m_N at	$(S_{v,w}) = 0.80$	3.1	3.2	3.3	1.05	1.1	1.1
	$(S_{v,w}) = 0.90$	5.3	5.8	6.1	2.1	2.4	2.4
	$(S_{v,w}) = 1.00$	62	680	7 100	25	370	3 800
	$(S_{v,w}) = (S_{v,w})_c$	309	3290	34 100	120	1 610	18 000
Salt concentration in drop at $(S_{v,w})_c$	(g cm^{-3})	3.24×10^{-3}	3.04×10^{-4}	2.93×10^{-5}	8.34×10^{-3}	6.21×10^{-4}	5.55×10^{-5}
	(mol ℓ^{-1})	5.54×10^{-2}	5.20×10^{-3}	5.02×10^{-4}	6.30×10^{-2}	4.69×10^{-3}	4.19×10^{-4}
drop size (cm) at which salt concentration in drop is 1×10^{-3} mol ℓ^{-1}		7.42×10^{-5}	3.44×10^{-4}	1.59×10^{-3}	5.63×10^{-5}	2.62×10^{-4}	1.21×10^{-3}
a_c(cm)		1.95×10^{-5}	1.99×10^{-4}	2.01×10^{-3}	1.42×10^{-5}	1.57×10^{-4}	1.63×10^{-3}

6.6 Equilibrium Between Humid Air and an Aqueous Solution Drop Containing A Solid Insoluble Substance

Most atmospheric aerosol particles are mixed, i.e., they are composed of water soluble and insoluble substances (see Chapter 8). The purpose of this section is to study the effect of a solid insoluble substance within an aqueous solution drop on the equilibrium conditions for that drop. In this study we shall assume the insoluble particle does not take up any water by itself and does not adsorb salt ions, and that it is completely submerged.

Since the molality of the solution drop is unaffected by the addition of the insoluble particle, we may take (6-24) as our starting point. For a_w we invoke (4-65) and (4-69) to obtain

$$\ln a_w = -\nu \Phi_s m_s M_w / m_w M_s. \tag{6-31}$$

The volume of the liquid portion of the solution drop is $V_d^{(\ell)} = n_w v_w + n_s v_s = (m_w/M_w)v_w + (m_s/M_s)v_s$; if we assume the water-insoluble particle is a sphere of radius r_u, then also $V_d^{(\ell)} = (4\pi/3)(a^3 - r_u^3)$. Eliminating $V_d^{(\ell)}$, we find $m_w = (4\pi/3)(a^3 - r_u^3)(M_w/v_w) - m_s v_s M_w/M_s v_w$. Substituting this expression for m_w into (6-31), and the result into (6-24), we obtain finally

$$\frac{e_a}{e_{sat,w}} = \exp\left[\frac{2 M_w \sigma_{s/a}}{\mathscr{R} T \rho_w a} - \frac{\nu \Phi_s m_s M_w/M_s \rho_w}{(4\pi/3)(a^3 - r_u^3) - m_s v_s/M_s}\right]. \tag{6-32}$$

For a dilute solution we may simplify as before and obtain equations analogous to (6-27) and (6-29), with the same values of A and B:

$$\ln \frac{e_a}{e_{sat,w}} = \frac{A}{a} - \frac{B}{a^3 - r_u^3}, \tag{6-33}$$

and for $e_a/e_{sat,w} \approx 1$,

$$\frac{e_a}{e_{sat,w}} = 1 + \frac{A}{a} - \frac{B}{a^3 - r_u^3}. \tag{6-34}$$

Since the size of the water-insoluble portion of an aerosol particle is difficult to determine, it is necessary to have alternative forms of these equations involving either the mass fraction ε_m or volume fraction ε_v of soluble material in a mixed aerosol particle of radius r_N. Assuming that the total drop volume can be approximated by $(m_w/\rho_w) + (m_N/\rho_N)$ from which $m_N/m_w \approx (\rho_N/\rho_w)[(a/r_N)^3 - 1]^{-1}$, we find from (6-31) and (6-24) with $\varepsilon_m \equiv m_s/m_N$:

$$\frac{e_a}{e_{sat,w}} = \exp\left[\frac{2 M_w \sigma_{s/a}}{\mathscr{R} T \rho_w a} - \frac{\nu \Phi_s \varepsilon_m M_w \rho_N r_N^3}{M_s \rho_w (a^3 - r_N^3)}\right], \tag{6-35}$$

and, with $\varepsilon_v = \varepsilon_m(\rho_N/\rho_s)$,

$$\frac{e_a}{e_{sat,w}} = \exp\left[\frac{2 M_w \sigma_{s/a}}{\mathscr{R} T \rho_w a} - \frac{\nu \Phi_s \varepsilon_v M_w \rho_s r_N^3}{M_s \rho_w (a^3 - r_N^3)}\right], \tag{6-36}$$

where $r_N^3 = m_N/(4\pi/3)\rho_N$. Also, the densities ρ_N, ρ_s, and ρ_u of the aerosol particle and its soluble and insoluble fractions are related according to

$$\rho_N = \rho_s \varepsilon_v + \rho_u (1 - \varepsilon_v), \tag{6-37a}$$

and

$$\rho_N = \frac{\rho_u}{1 - \varepsilon_m (\rho_u/\rho_s)}. \tag{6-37b}$$

Equations (6-32) to (6-36) are different forms of a relationship first derived by Junge (1950) and later refined by Junge and McLaren (1971).

Equation (6-36) has been solved by Hänel (1976) for aerosol particles of $r_N = 0.1 \ \mu m$ containing a water-insoluble substance and $(NH_4)_2SO_4$ or NaCl in the volume proportions $\varepsilon_v = 1.0, 0.4, 0.1,$ and 0.01. The results of these computations are summarized in Figures 6-3 and 6-4. We see that the smaller the water-soluble fraction, the higher the supersaturation needed to activate the drop, and the

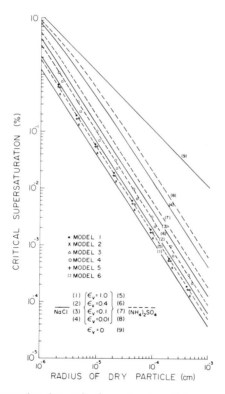

Fig. 6-4. Critical supersaturation for activating aerosol particles composed of various volume proportions of NaCl, $(NH_4)_2SO_4$ and a water-insoluble substance as a function of particle size. (Curve (9) pertains to a water-insoluble but completely wettable particle.) Comparison with computations for natural aerosol particles are made (see Section 6.9). Model 1: average aerosol, summer, 1966, Mainz; Model 2: sea spray aerosol (Aarons and Keith, 1954; Winkler and Junge, 1971); Model 3: maritime aerosol over Atlantic, April, 1969; Model 4: maritime aerosol with Sahara dust over Atlantic, April, 1969; Model 5: urban aerosol, March, 1970, Mainz; Model 6: continental background aerosol, summer, 1970, Hohenpeissenberg (1000 m above sea level); 20 °C. (Based on data of Hänel, 1976; by courtesy of the author.)

TABLE 6-3

Dilution $(m_w/m_N) = (\rho_w/\rho_N)[(a_c/r_N)^3 - 1]$ at $(S_{v,w})_c$, and $(s_{v,w})_c$, for aqueous solution drops which form on an aerosol particle of mass m_N composed of NaCl and SiO_2 (water-insoluble) in various proportions. (a) for 20 °C, (b) for −10 °C; $\rho_s = 2.16$ g cm^{-3} (20 °C), $\rho_s = 2.17$ g cm^{-3} (−10 °C), $\rho_u = 2.65$ g cm^{-3} (20 °C), $\rho_u = 2.66$ g cm^{-3} (−10 °C). (Based on data of Hänel, 1976.)

ϵ_v	ϵ_m	ρ_N(g cm^{-3})	m_w/m_N at $(S_{v,w})_c$ — m_N(g) 1×10^{-16}	m_N(g) 1×10^{-14}	1×10^{-12}	$(s_{v,w})_c$ (%) — 1×10^{-16}	m_N(g) 1×10^{-14}	1×10^{-12}	salt concentration (mol ℓ$^{-1}$) in drop at $(s_{v,w})_c$ — 1×10^{-16}	m_N(g) 1×10^{-14}	1×10^{-12}
(a)											
1.0	1.0	2.163	309	3290	34 100	3.660×10^{-1}	3.578×10^{-2}	3.558×10^{-3}	5.5×10^{-2}	5.2×10^{-3}	5.0×10^{-4}
0.4	0.35	2.455	53.5	560	5 880	6.597×10^{-1}	6.450×10^{-2}	6.389×10^{-3}	1.1×10^{-1}	1.1×10^{-2}	1.0×10^{-3}
0.1	0.0083	2.601	6.5	58.3	620	1.375	1.373×10^{-1}	1.355×10^{-2}	2.2×10^{-1}	2.4×10^{-2}	2.3×10^{-3}
0.01	0.0082	2.601	0.87	2.7	19	3.214	4.130×10^{-1}	4.347×10^{-2}	1.6×10^{-1}	5.2×10^{-2}	7.4×10^{-3}
(b)											
1.0	1.0	2.170	237	2520	26 240	4.738×10^{-1}	4.628×10^{-2}	4.595×10^{-3}	7.2×10^{-2}	6.8×10^{-3}	6.5×10^{-4}
0.4	0.35	2.464	41.6	428	4 534	8.537×10^{-1}	8.351×10^{-2}	8.255×10^{-3}	1.4×10^{-1}	1.4×10^{-2}	1.3×10^{-3}
0.1	0.083	2.611	5.1	44.9	476	1.770	1.778×10^{-1}	1.750×10^{-2}	2.7×10^{-1}	3.2×10^{-2}	3.0×10^{-3}
0.01	0.0082	2.611	0.80	2.3	15	3.978	5.241×10^{-1}	5.599×10^{-2}	1.7×10^{-1}	6.1×10^{-2}	9.3×10^{-3}

steeper the pre-maximum portion of the equilibrium growth curve. Considering that $\rho_s = 2.163 \, \text{g cm}^{-3}$ for NaCl, $\rho_s = 1.769 \, \text{g cm}^{-3}$ for $(NH_4)_2SO_4$, and $\rho_u = 2.65 \, \text{g cm}^{-3}$ for SiO_2 (quartz), the range of r_N shown in Figure 6-4 covers a range of m_N from $10^{-16} \, \text{g}$ to $10^{-9} \, \text{g}$. Table 6-3 lists the corresponding dilutions, i.e., the water uptake of the solution drop at the supersaturation necessary for drop activation under equilibrium conditions. As expected from the equilibrium growth curves (Figure 6-3), the water uptake increases with an increasing water-soluble fraction of the aerosol particle. For a given ε_v or ε_m the water uptake also increases with increasing total mass of the aerosol particle on which the drop forms.

6.7   Equilibrium Conditions for Ice Particles

We now consider three equilibrium situations involving the ice phase: (a) an ice particle in humid air, (b) an ice particle and a separate supercooled solution drop in humid air, and (c) an ice particle in a supercooled solution drop in humid air. These three cases are illustrated in Figure 6-5.

If we may assume a spherical ice particle, then the analysis of case (a) proceeds in strict analogy to the derivation of the Kelvin equation, (6-16), and we find the saturation vapor pressure over ice, $e_{a,i}$, varies with radius according to

$$S_{v,i} \equiv \frac{e_{a,i}}{e_{\text{sat,i}}} = \exp\left(\frac{2\,v_i\sigma_{i/a}}{\mathscr{R}Ta}\right) = \exp\left(\frac{2\,M_w\sigma_{i/a}}{\mathscr{R}T\rho_i a}\right). \tag{6-38}$$

This result is plotted as curves 3 and 4 of Figure 6-1 (for 0 °C and −20 °C). The behavior is seen to be very similar to that for a pure water drop, except that at small radii, $S_{v,i} \gg S_{v,w}$. This is primarily a consequence of the inequality $\sigma_{i/a} > \sigma_{w/a}$.

If we abandon the requirement that the ice particle be spherical, and instead make it a hexagonal prism which follows Wulff's relations (Section 5.7.2), then in place of (6-38) we have

$$\frac{\mathscr{R}T}{2\,v_i} \ln \frac{e_{a,i}}{e_{\text{sat,i}}} = \frac{\sigma_{i/a}^{(P)}}{h^{(P)}} = \frac{\sigma_{i/a}^{(B)}}{h^{(B)}}, \tag{6-39}$$

where $h^{(P)}$ and $h^{(B)}$ are the perpendicular distances from the crystal center to the prism and basal planes, respectively. Thus, the conditions at equilibrium for the hexagonal prism which are compatible with Wulff's relations are formally similar

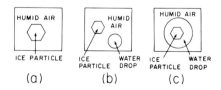

Fig. 6-5.   The three basic equilibrium types involving ice crystals.

to those which apply to the cases of a water drop and an amorphous sphere of ice.

We now turn to case (b) to study the equilibrium behavior of a system comprised of a spherical ice particle of radius a_i and a separate aqueous solution drop of radius a_d, both surrounded by humid air. As in Section 6.5, we have three components, and we assume the constraint n_s = constant, so that w = 3. Let us hold the environmental pressure p constant and determine the independent variations of a_i and a_d with the equilibrium temperature.

On proceeding to specialize (6-5) in the appropriate, and by now familiar manner, we obtain for the solution drop

$$-\frac{\mathscr{L}_{e,0}}{T^2} dT - \frac{2\, v_{w,0}}{T} d\left(\frac{\sigma_{s/a}}{a_d}\right) + \mathscr{R} d \ln e_{a,w} - \mathscr{R} d \ln a_w = 0, \qquad (6\text{-}40)$$

and for the ice particle

$$-\frac{\mathscr{L}_s}{T^2} dT - \frac{2\, v_i}{T} d\left(\frac{\sigma_{i/a}}{a_i}\right) + \mathscr{R} d \ln e_{a,i} = 0. \qquad (6\text{-}41)$$

These equations express the separate equilibrium balances between the drop and its environment, and the ice particle and its environment. To put the drop and ice crystal in equilibrium with each other as well requires that $e_{a,w} = e_{a,i}$. Imposing this condition, we may eliminate the vapor pressure term between (6-40) and (6-41) to obtain

$$\mathscr{L}_{m,0} \frac{dT}{T} - \frac{2\, M_w}{\rho_w} d\left(\frac{\sigma_{s/a}}{a_d}\right) + \frac{2\, M_w}{\rho_i} d\left(\frac{\sigma_{i/a}}{a_i}\right) - \mathscr{R} T\, d \ln a_w = 0. \qquad (6\text{-}42)$$

An approximate integral of (6-42) between $T = T_0$, a_d, $a_i \to \infty$, $a_w = 1$ and $T = T_e$, a_d, a_i, a_w is

$$\bar{\mathscr{L}}_{m,0} \ln \frac{T_e}{T_0} = \frac{2\, M_w \sigma_{s/a}}{\bar{\rho} a_d} - \frac{2\, M_w \sigma_{i/a}}{\bar{\rho}_i a_i} - \frac{\mathscr{R} \bar{T} \nu \Phi_s m_s M_w / M_s}{(4\, \pi a_d^3 \rho_s''/3) - m_s}, \qquad (6\text{-}43)$$

where we have used (6-25) for a_w and employed an overbar to denote mean values over the temperature interval (T_0, T_e). For a pure water drop, this reduces to

$$\ln \frac{T_e}{T_0} = \frac{2\, M_w}{\bar{\mathscr{L}}_{m,0}} \left(\frac{\sigma_{w/a}}{\bar{\rho}_w a_d} - \frac{\sigma_{i/a}}{\bar{\rho}_i a_i}\right). \qquad (6\text{-}44)$$

Inspection of these equations reveals that the presence of salts lowers the equilibrium temperature, as expected from the behavior of water in bulk. This effect is evident in Figure 6-6. However, we see that for NaCl the concentration has to be larger than about 10^{-3} mol ℓ^{-1} to cause a noticeable effect. The figure also indicates that while for a pure drop of given size the equilibrium temperature T_e decreases with decreasing ice particle size, T_e increases with decreasing drop size for a given ice particle size. This opposing behavior derives from the different dependencies of temperature with saturation vapor pressure over ice and supercooled water.

In Figure 6-7 we have plotted the separate solutions to (6-40) (for $a_w = 1$) and

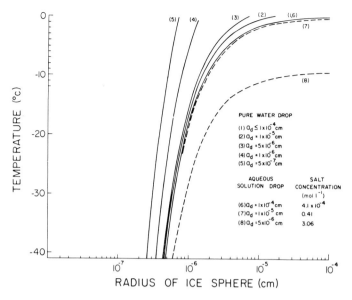

Fig. 6-6. Variation of the equilibrium temperature with drop size, ice particle size, and concentration of salt in solution, for a system which consists of a pure water or aqueous solution drop and an ice particle both surrounded by humid air. $m_s = 10^{-16}$ g, NaCl.

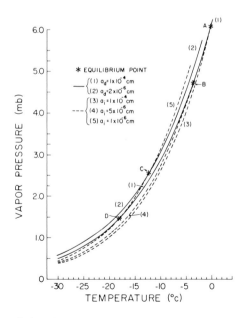

Fig. 6-7. Temperature variation of the vapor pressure over water drops and ice spheres of different sizes, and resulting equilibrium temperatures.

(6-41); the curve intersections therefore constitute states satisfying (6-42). This figure shows that for all drops and ice particles with $(a_d, a_i) > 10^{-4}$ cm, the equilibrium temperature is essentially equal to the triple point for bulk phases (point A). The figure also reveals the contrary temperature dependence referred to above: for a drop of $a_d = 10^{-4}$ cm and an ice particle of $a_i = 5 \times 10^{-6}$ cm, $T_e = -3.8\,°C$ (point B), while for $a_d = 10^{-4}$ cm and $a_i = 10^{-6}$ cm, $T_e = -18\,°C$ (point D). On the other hand, for an ice particle of $a_i = 10^{-6}$ cm and a drop of $a_i = 10^{-6}$ cm, $T_e = -12.5\,°C$ (point C), a warmer temperature than for point D. Finally, we note that for $a_i = a_w$, $e_{a,i} < e_{a,w}$ for $T < 0\,°C$, the same behavior as for ice and water in bulk $(a_i, a_d) \to \infty$.

Let us now consider case (c) and determine the equilibrium temperature for a spherical ice particle inside a supercooled aqueous solution drop which is itself in equilibrium with the environmental humid air. As in case (b), such a system has three independent variables. Of these we shall keep the total gas pressure p constant and study the variation of the equilibrium temperature with the radii a_d and a_i.

For this case (6-5) becomes

$$-\frac{\mathscr{L}_{m,0}}{T^2}\, dT + (v_{w,0} - v_i)\, dp_w - 2\, v_i\, d\!\left(\frac{\sigma_{i/s}}{a_i}\right) + \mathscr{R}T d \ln a_w = 0, \tag{6-45}$$

or, on substituting the condition of mechanical equilibrium, (5-7),

$$-\frac{\mathscr{L}_{m,0}}{T^2}\, dT + 2\!\left(\frac{M_w}{\rho_w} - \frac{M_w}{\rho_i}\right) d\!\left(\frac{\sigma_{s/a}}{a_d}\right) - \frac{2\, M_w}{\rho_i}\, d\!\left(\frac{\sigma_{i/s}}{a_i}\right) + \mathscr{R}T d \ln a_w = 0. \tag{6-46}$$

Integrating as in case (b), we find

$$\bar{\mathscr{L}}_{m,0}\ln\frac{T_0}{T_e} = \frac{2\, M_w\sigma_{s/a}}{a_d}\left(\frac{1}{\bar\rho_i} - \frac{1}{\bar\rho_w}\right) + \frac{2\, M_w\sigma_{i/s}}{\bar\rho_i a_i} + \frac{\mathscr{R}\bar{T}\nu\Phi_s m_s M_w/M_s}{(4\,\pi/3)(a_d^3 - a_i^3)\rho_s'' - m_s}, \tag{6-47}$$

which for a pure water drop becomes

$$\ln\frac{T_0}{T_e} = \frac{2\, M_w\sigma_{w/a}}{\bar{\mathscr{L}}_{m,0} a_d}\left(\frac{1}{\bar\rho_i} - \frac{1}{\bar\rho_w}\right) + \frac{2\, M_w\sigma_{i/w}}{\bar{\mathscr{L}}_{m,0}\bar\rho_i a_i}. \tag{6-48}$$

If the drop is much larger than the ice crystal, this last equation reduces to

$$\ln\frac{T_0}{T_e} = \frac{2\, M_w\sigma_{i/w}}{\bar{\mathscr{L}}_{m,0}\bar\rho_i a_i} \tag{6-49}$$

Finally, if T_e is close to T_0, we may write

$$(\varDelta T)_{e,a} \equiv T_0 - T_e = \frac{2\, M_w T_0\sigma_{i/w}}{\bar{\mathscr{L}}_{m,0}\bar\rho_i a_i}. \tag{6-50}$$

Equations (6-48)–(6-50) are different forms of a relation first derived by J.J. Thomson (1888).

For a solution drop we may obtain a similar simplification of (6-47) for the case $a_d \gg a_i$ and assuming $\sigma_{i/s} = \sigma_{i/w}$:

$$(\varDelta T)_e = \frac{2\, M_w T_0\sigma_{i/w}}{\bar{\mathscr{L}}_{m,0}\bar\rho_i a_i} + \frac{\mathscr{R}T_0\bar{T}\nu\Phi_s m_s M_w/M_s}{\bar{\mathscr{L}}_{m,0}[(4\,\pi a_d^3\rho_s''/3) - m_s]}. \tag{6-51}$$

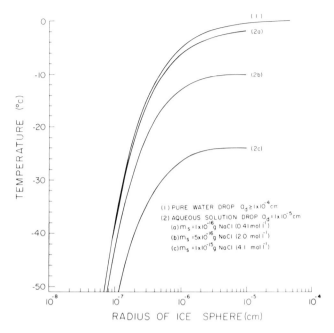

Fig. 6-8.   Variation of the equilibrium temperature with drop size, ice particle size and concentration of salt in solution for a system which consists of an ice particle contained in a pure water or aqueous solution drop surrounded by humid air.

The second term may be recognized as a combination of (4-103) and (4-66), and thus is the equilibrium freezing point depression, $(\Delta T)_{e,s}$, due to the presence of salts in bulk solution. Therefore, to the precision indicated in the derivations of (6-50) and (6-51) we may write

$$(\Delta T)_e = (\Delta T)_{e,a} + (\Delta T)_{e,s}; \qquad\qquad (6-52)$$

i.e., the total equilibrium freezing point depression is simply the sum of contributions from the separate curvature and solute effects.

A numerical solution of (6-46) is displayed in Figure 6-8. In accordance with (6-50), the equilibrium freezing temperature is seen to decrease with decreasing size of the ice particle. This decrease becomes particularly pronounced for $a < 10^{-6}$ cm, and is further enhanced by the solute effect if the salt concentration is larger than about 0.1 mol ℓ^{-1}.

6.8   Experimental Verification

Several experimental difficulties have been encountered in attempting to verify the various equilibrium relationships discussed in the previous sections. The major difficulty in verifying the Thomson equation, (6-50), has arisen from a lack of accurate values for $\sigma_{i/w}$. Generally, therefore, the approach has been to assume the equation is correct and deduce values for $\sigma_{i/w}$; these in turn can be compared with other independently determined values for this quantity. Pawlow

(1910), Meissner (1920), Tammann (1920), and Kubelka (1932) were among the first to verify experimentally that the melting temperature of a pure solid substance is dependent on whether the substance is present in bulk or in the form of small particles. Experiments for the ice-water system were first carried out by Kubelka and Prokscha (1944), Skapski *et al.* (1957), and Skapski (1959). These experiments involved measuring the melting temperature of ice contained in pores and capillaries, or of ice in the form of thin wedges. Unfortunately, this method is subject to a considerable number of errors, and the interface energies obtained only agreed to within 20% with values derived from independent measurements. In subsequent years, more accurate techniques involving electron microscopy and electron diffraction have been developed and perfected (Wronski, 1967; Pocza *et al.*, 1969; Sambles, 1971). These more recent tests of the Thomson equation for tin, indium, lead, and gold particles of radii between 10^{-5} and 3×10^{-7} cm have yielded excellent agreement between experiment and theory.

Uncertainties concerning the Kelvin equation, (6-16), have been due mainly to the fact that the liquids tested were held in capillaries (see Skinner and Sambles, 1972). The first successful attempt to test the Kelvin equation for freely falling drops was carried out by Gudris and Kulikova (1924), who established its validity to within 10%. Subsequently, La Mer and Gruen (1952), experimenting with freely falling droplets in mixtures of dioctylphtalate and toluene, and of oleic acid and chloroform, verified the Kelvin equation to within 5% for drops larger than 0.1 μm radius. Recently, Sambles *et al.* (1970) and Sambles (1971), through electron microscope studies of the evaporation rates of small drops of lead, silver, and gold, have established the correctness of the Kelvin law to within 5% for drops of sizes between 0.1 and 0.003 μm.

Quantitative experimental studies to determine the variation of the size of water-soluble and mixed particles in an environment of increasing and decreasing relative humidity were carried out first by Junge (1936) and by Orr *et al.*

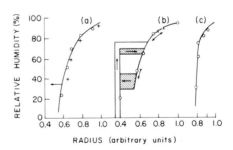

Fig. 6-9. Experimental values for the equilibrium radius of aqueous salt solution drops (o, x, +) and the corresponding theoretical curves. (a) $CaCl_2$ solution drop becoming supersatured at relative humidities less than 35%, O mean of 15 measurements of Junge (1952a) with drops on spider threads; + experimental results of Dalal (1947). (b) NaCl solution drops, O mean of 11 measurements of Junge (1952a) with drops on spider threads; x measurements of Woodcock. (c) Drops of $CaCl_2$ containing water-insoluble $CaSO_4$, O mean of 15 measurements of Junge (1952a) with drops on spider threads. (From Junge, 1952a; by courtesy of *Ann. d. Meteorol.*)

(1958) for Aitken sized particles, and by Junge (1952a) for large and giant particles. In most respects, the results of these tests agreed well with the growth predicted by the Köhler equation, (6-27), for pure aqueous solution drops, and with that predicted by the Junge equation, (6-38), for aqueous solution drops containing water-insoluble material. This is illustrated in Figure 6-9.

On the other hand, two shortcomings of the Köhler and Junge equations were noted: (a) They apply only to solution drops in which all the salt has already dissolved, and consequently do not describe the deliquescence of a dry salt or mixed aerosol particle. (b) They do not predict the *hysteresis effect* which often is observed during decreasing relative humidity. This effect is a consequence of the fact that evaporating solution drops readily supersaturate. Figure 6-9b displays an example of hysteresis behavior; a further discussion of this effect is given in the following section.

6.9   Equilibrium Growth of Atmospheric Aerosol Particles

In order to predict the equilibrium growth of atmospheric aerosol particles, one of three main avenues of approach may be followed. (a) Equations (6-35) and (6-36) may be used in conjunction with a complete chemical and physical analysis of individual aerosol particles occurring in the relevant portion of the spectrum. Considering the large variety and variability of chemical compounds comprising the atmospheric aerosol, such an attempt seems to be impractical. (b) Equations (6-35) and (6-36) may be used in conjunction with an idealization of the chemical composition of the atmospheric aerosol particles. Such an attempt may be justified since often NaCl and $(NH_4)_2SO_4$ are the two most prevalent substances in atmospheric aerosol particles (see Chapter 8). On the other hand, we must not forget that aerosol particles consist of numerous other water-soluble substances, plus a large variety of water-insoluble substances occurring in a wide range of mass or volume proportions. Thus, it becomes apparent that, although the idealized approach can be useful at times, the results may fall significantly short of describing the equilibrium growth behavior of real aerosol particles.

(c) A third semi-empirical approach has been suggested by Hänel (1966, 1968, 1969, 1970, 1972a, b, 1976). His approach has been to study the behavior of aggregate samples of aerosol particles (aerosol deposits) and to infer from such behavior the equilibrium growth curves of representative particles comprising the sample. What makes this approach especially attractive is that the resulting 'typical' particle growth equation is described in terms of the readily measurable amount of water taken up by the aerosol deposit as a whole.

In order to understand his formulation, recall first that for an aqueous solution containing a single salt the water activity, and hence the relative humidity, are related to the salt mass in solution according to (6-31). Now for a mixture of salts, we must replace $n_s = m_s/M_s$ by $\Sigma_s n_s = \Sigma_s m_s/M_s$, ν by $\bar{\nu} = \Sigma_s \nu_s n_s/\Sigma_s n_s$, and Φ_s by $\Phi_s \equiv \Phi_{s,mix}$ given according to (4-70). Of course, the sums are taken over all the component salts. Thus we may express the water activity of a mixture of salts as

$$\ln a_w = -\gamma \bar{\nu} \Phi_{s,mix} M_w m_{s,t} / \bar{M}_s m_w, \qquad (6\text{-}53)$$

where $\bar{M}_s = \Sigma_s\, m_s / \Sigma_s\,(m_s/M_s)$ is the mean molecular weight of the salt mixture, defined as in (4-26), $m_{s,t} = \Sigma_s\, m_s$ is the total water-soluble mass of the aerosol deposit, and where γ is a measure for all the adsorption effects at the solid-solution interface.

Now consider a mixed aerosol deposit of total mass $m_0 = m_{s,t} + m_u$, where m_u denotes the insoluble portion. The soluble portion $m_{s,t}$ can be determined as $m_0 - m_u$ by washing and subsequently drying the deposit. Further, the mass of water $m_w = m - m_0$ taken up by the deposit at each humidity can be found by measuring its equilibrium mass m as a function of the environmental humidity.

Since $\bar{\nu}$, $\Phi_{s,mix}$ and $\bar{M}_s$ are unknown in general, Hänel (1976) absorbed them into the definition of the *mass increase coefficient* β_0:

$$\beta_0 \equiv \gamma \bar{\nu} \Phi_{s,mix} M_w m_{s,t} / \bar{M}_s m_0. \qquad (6\text{-}54)$$

In terms of β_0, (6-35) becomes

$$\ln a_w = -\beta_0 m_0 / m_w. \qquad (6\text{-}55)$$

However, in view of the generalized Raoult's law, (4-60), we may also write this as

$$\beta_0 = -\frac{m_w}{m_0} \ln \left(\frac{e}{e_{sat,w}}\right) \approx -\frac{m_w}{m_0} \ln \phi_v, \qquad (6\text{-}56)$$

using (4-34) and (4-40). All the quantities in the final version of this equation are obtainable from experiment.

Now, for an individual particle (N) of mass m_N, we may write, in analogy to (6-55),

$$\ln a_{w,N} = -\beta_N m_N / m_{w,N}, \qquad (6\text{-}57)$$

where $m_{w,N}$ is the amount of water the particle has taken up corresponding to an activity $a_{w,N}$ of water in solution. Assuming that the volume of the solution drop formed under the influence of the particle is given by the sum of the volumes of the dry particle and the acquired water, (6-57) becomes

$$\ln a_w = -\frac{\beta_N \rho_n}{\rho_w[(a/r_N)^3 - 1]}, \qquad (6\text{-}58)$$

where a is the drop radius, and ρ_N, r_N are the particle density and radius. Finally, on substituting this into (6-24), we obtain the desired equilibrium growth law for the 'typical' aerosol particle:

$$\ln \frac{e_a}{e_{sat,w}} = \frac{2\, M_w \sigma_{s/a}}{\mathcal{R} T \rho_w a} - \frac{\beta_N (\rho_N / \rho_w) r_N^3}{a^3 - r_N^3}. \qquad (6\text{-}59)$$

Assuming the particles in the deposit are approximately uniform in size and are mixed particles of similar composition, we may set $\beta_N \approx \beta_0$. According to Hänel (1972a) we may further set $\rho_N \approx \rho_0$, where ρ_0 is the density of the dry deposit, which may be determined by the pycnometer method (Fischer and Hänel, 1972; Hänel, 1976; Thudium, 1976).

As it is somewhat difficult to measure β_0 at relative humidities above 95%, owing to condensation problems on the measuring apparatus, Hänel (1976) has suggested a simple extrapolation procedure for extending values of $\beta_0(a_w)$ up to $a_w = 1$. Defining

$$\beta_s \equiv \frac{\bar{\nu}\Phi_{s,mix}M_w}{\bar{M}_s} = \frac{\beta_0 m_0}{\gamma m_{s,t}}, \tag{6-60}$$

we can write

$$\beta_0(a_{w,2}) = \beta_s(a_{w,2})\beta_0(a_{w,1})/\beta_s(a_{w,1}), \tag{6-61}$$

where we have assumed that $\gamma(a_{w,2})/\gamma(a_{w,2}) \approx 1$, and considered that neither $m_{s,t}$ nor m_0 is a function of a_w. Suppose $a_{w,2} > a_{w,1}$ and that $a_{w,1}$ is the largest water activity for which a measurement of β_0 is available; then $\beta_s(a_{w,1})$ at this water activity can be determined from (6-60) since $m_{s,t} = m_0 - m_u$, which can be found by the procedure mentioned earlier. Furthermore, the type of salt most prevalent in the deposit can usually be found from a plot of β_0 vs. relative humidity (Hänel, 1976; Schreiber, 1977). Knowing the prevalent salt, $\Phi_s(a_{w,2})$, $\bar{M}_s$, $\bar{\nu}$, and thus $\beta_s(a_{w,2})$ can be estimated with sufficient accuracy from (6-60), which in turn allows estimating $\beta_0(a_{w,2})$ from (6-61).

Using these procedures, Hänel (1976) has computed equilibrium growth curves for typical aerosol particles belonging to deposits whose equilibrium growth behavior has been experimentally determined. These growth curves are displayed in Figure 6-10 for various atmospheric aerosols. We notice that for $r_N > 0.1\ \mu m$ particles of typical urban and maritime aerosols require a threshold supersaturation for activation of less than 0.06%. Even particles of continental background aerosols or maritime aerosols with a significant portion of water-insoluble material require a supersaturation of less than 0.1% for activation.

The peak supersaturations needed to activate these natural aerosol particles are plotted in Figure 6-4, where they are compared to those required for activating particles of given composition. This comparison allows the conclusion that typical atmospheric aerosol particles behave, with regard to their equilibrium growth, as either (1) particles which contain more than 40% (by volume) salt, if the salt is $(NH_4)_2SO_4$ and the insoluble substance is assumed to be a silicate, or (2) as particles which contain more than about 25% salt, if the salt is NaCl. We shall see in Chapter 8 that this result agrees well with the composition of actual aerosol particles.

Let us now take a closer look at the observed behavior of aerosol deposits. This behavior has been studied by means of a microbalance method by Winkler (1967, 1968, 1970, 1973), and by Winkler and Junge (1971, 1972). Examples of the equilibrium uptake of water as a function of environmental humidity are given in Figures 6-11 and 6-12a, b. Figure 6-12 shows the experimental results for pure salt deposits, which are in good agreement with theory, (6-31), when the relative humidity is larger than the value necessary for the deliquescence of the salt into a saturated solution (see Table 4-3). On the other hand, as noted earlier, the

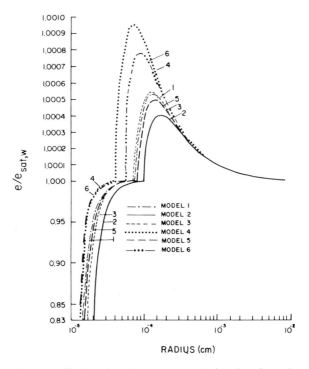

Fig. 6-10. Variation of the equilibrium size of an aqueous solution drop formed on various natural aerosol particles (computed from measured water uptake of natural aerosol deposits). For definition of models 1 to 6 see legend to Figure 6-4. $r_N = 0.1\ \mu$m, 20 °C. (From Hänel, 1976; by courtesy of *Advances of Geophys.*, and the author.)

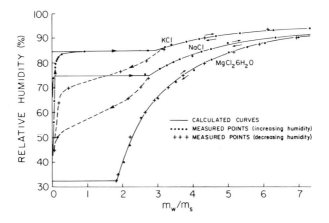

Fig. 6-11. Equilibrium growth curves of pure salt deposits (25 °C). (From Winkler, 1967; by courtesy of the author.)

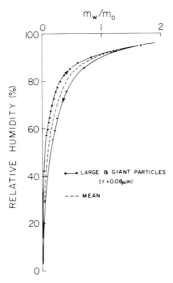

Fig. 6-12a   Equilibrium growth curve of natural aerosol deposit. Urban aerosol, Mainz, June, 1966, northeast wind, continental air trajectory. (From Winkler, 1967; by courtesy of the author.)

equilibrium theory does not account for the deliquescence and hysteresis growth characteristics which are also observed.

Both of these growth characteristics of pure salt deposits are also found in natural aerosol deposits, as can be seen in Figures 6-12a, b. The deposits begin to take up water at relative humidities as low as 20 to 30%, although the actual amount of water taken up is small until relative humidities near 60% are reached. Above this value the water uptake becomes progressively larger, in particular above relative humidities of 70 to 80%. Note that the equilibrium growth curves for deposits of aerosols with a continental trajectory are smooth (Figure 6-12a),

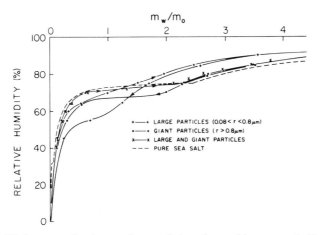

Fig. 6-12b.   Equilibrium growth of natural aerosol deposit; maritime aerosol. Helgoland Island, September, 1966, west wind. (From Winkler, 1967; by courtesy of the author.)

while those for deposits with a maritime trajectory exhibit a characteristic near-discontinuity near 75% relative humidity (Figure 6-12b). This 'sea-salt-discontinuity' is due to the presence of NaCl in the deposit. We further note that at any given relative humidity maritime aerosols take up considerably more water than continental aerosols. This is due to the larger portion of water-soluble, hygroscopic substances in the former deposits, and due to the stronger water uptake of NaCl as compared to $(NH_4)_2SO_4$.

Three factors help explain why aerosol deposits take up water at a value of $e/e_{sat,w}$ less than that necessary for equilibrium with a salt saturated solution. Firstly, the water solubility of any substance is a function of particle size. Solubility is enhanced especially if the particle size decreases below about $0.1 \mu m$. This solubility enhancement was predicted qualitatively by Ostwald (1900) and Freundlich (1926) through a Kelvin law type analysis, and has been quantitatively established by the experiments of Dundon and Mack (1923), May and Kolthoff (1948), and Orr et al. (1958). The experiments of Orr et al. showed that NaCl particles of $0.020 \mu m$ radius go into solution at a relative humidity of 69 to 70%, while NaCl particles of $1 \mu m$ radius need a relative humidity of 74%. Particles of $(NH_4)_2SO_4$ with radii of $0.015 \mu m$ were found to form a saturated solution drop at 68% relative humidity; KCl particles of $0.03 \mu m$ radius went into solution at 78% relative humidity. These values are considerably below the relative humidities necessary for salt in bulk, which are 75.3% for NaCl, 80% for $(NH_4)_2SO_4$, and 84.3% for KCl.

Secondly, aerosol particles contain air capillaries in which condensation of water vapor proceeds at a relatively low saturation ratio. This can be explained if we consider that the meniscus of water in a capillary with water wettable walls is *concave*, in contrast to the convex surface of a water drop. Therefore, instead of (6-16), we now have

$$\frac{e_{a,w}}{e_{sat,w}} = \exp\left(-\frac{2 M_w \sigma_{w/a}}{\mathscr{R} T \rho_w a}\right), \tag{6-62}$$

which means that the smaller the radius of curvature a of the water surface in the capillary, the lower the equilibrium vapor pressure over it. For only partially wettable capillary walls characterized by a contact angle θ, the argument of the exponential must be multiplied by $\cos \theta$, but this does not change the qualitiative effect of capillary spaces in the particle surface.

The third reason for the occurrence of deliquescence is simply that all solids show some affinity for water vapor and thus adsorb it onto their surfaces, as discussed in the previous chapter. The amount of adsorbed water vapor may be considerable, even at low relative humidities.

The development of a hysteresis loop in an equilibrium growth curve has three main causes. Firstly, evaporating salt solutions tend to supersaturate with respect to the salt in solution, as we stated earlier. This is due to the fact that the crystallization of salt requires surmounting an energy barrier unless suitable solid particles are present in the solution to serve as centers for crystallization. Thus, as the relative humidity decreases, the equilibrium growth curve is determined by the water vapor pressure over the supersaturated salt solution,

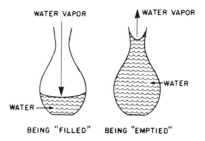

Fig. 6-13.   Behavior of a cylindrical bottle-type air capillary during condensation and evaporation.

until, at some undetermined relative humidity, the salt crystallizes. This behavior is in contrast to the growth during increasing relative humidity, which is assisted by the adsorption and deliquescence behavior of the substances in the aerosol deposit.

Secondly, air capillaries in aerosol particles behave differently during increasing relative humidity when they become filled, than during decreasing relative humidity when they are being emptied. Figure 6-13 demonstrates this hysteresis effect in a cylindrical capillary with an opening narrower than the body of the capillary. From (6-62) it follows that such a capillary begins to fill at a higher relative humidity than that at which it begins to empty.

A third reason for the hysteresis loop is that water-insoluble substances such as clays behave differently during water adsorption than desorption, due to the presence of alkali ions in the silicate lattice. During increasing relative humidity (adsorption), the tendency of these ions to be bonded to the clay surface dominates their tendency to hydrate (to become surrounded by and weakly bonded to water molecules), particularly at low humidities. During the desorption which occurs with decreasing humidities, however, the silicate ions are already largely hydrated, and so their tendency to remain hydrated dominates the tendency to become bonded again to the clay surface.

CHAPTER 7

HOMOGENEOUS NUCLEATION

Our outline in the previous three chapters of the equilibrium thermodynamics of the phases of water is insufficient for an understanding of cloud particle formation, since we did not come to grips with the crucial question of how a new phase is initiated. Consider, for example, that on the basis of the Kelvin equation alone, the formation of a water drop from homogeneous water vapor would be precluded because the vapor pressure required to hold a microscopic quantity of newly formed phase in equilibrium would be quite enormous. This expectation is in disagreement with experimental observations which show that a large but finite supersaturation exists above which homogeneous phase change does take, place. The reason for this behavior is that the formation of a new phase at the expense of a metastable original phase ('mother phase') does not begin in a continuous manner, but rather takes place spontaneously as a result of temperature and density fluctuations ('heterophase fluctuations') in the original phase, provided that a critical supersaturation is exceeded. This spontaneous process is called *nucleation*.

From our studies of adsorption we might expect that nucleation could be greatly assisted if suitable solid surfaces were present. In fact, as we know from Chapter 1, such *heterogeneous nucleation* has long been recognized as being generally responsible for cloud formation. However, in order to clarify the physical principles involved in the nucleation process, we shall assume for now that all foreign substances are absent, and study the *homogeneous nucleation* which occurs when only water substance is present.

Some useful references for the material in this chapter include the texts by Abraham (1974b), Zettlemoyer (1969), Defay *et al.* (1966), Hirth and Pound (1963), Dufour and Defay (1963), Frenkel (1946), and Volmer (1939), and the review articles by Chalmers (1964), McDonald (1962, 1963a), Turnbull (1956), Dunning (1955), and Hollomon and Turnbull (1953).

7.1 Equilibrium Population of Embryos

Within the metastable bulk phase of water vapor are small molecular clusters of liquid water which result from the chance agglomeration of water molecules; these are generally referred to as *embryos* if the vapor pressure is below the critical value required for nucleation. Such embryos have small binding energies and are easily disrupted by thermal agitation. However, at the critical vapor pressure some embryos will reach a critical (*germ*) size, at which point they will be in unstable equilibrium with the mother phase. A germ will proceed to grow spontaneously and thereby produce a macroscopic phase change if, as a result of fluctuations in the mother phase, its size increases by even an infinitesimal amount.

Therefore, in order to understand the nucleation phenomenon one must first learn something about the prenucleation embryos. For the sake of simplicity and on considering the relative populations of *i-mers* (embryos consisting of *i* molecules and denoted by A_i), it is generally assumed that these grow by the capture of single molecules (monomers). A further convention is to assume a state of dynamic equilibrium for the *i*-mers, which we may express in the form

$$A_{i-1} + A_1 \rightleftarrows A_i, \quad i = 1, 2, \ldots \tag{7-1}$$

(the forward and reverse rates are assumed equal). On adding up a series of such equations, we have also

$$i A_1 \rightleftarrows A_i. \tag{7-2}$$

As this represents an equilibrium situation, the corresponding statement in terms of chemical potentials is

$$i\mu_1 = \mu_i, \tag{7-3}$$

where μ_i is the chemical potential of an *i*-mer.

7.1.1 FORMAL STATISTICAL MECHANICS DESCRIPTION

Let us now proceed to determine the number N_i of *i*-mers in a volume V of vapor held at temperature T. As we are dealing in principle with a microscopic fluctuation phenomenon, it is appropriate to apply, insofar as possible, the machinery of statistical mechanics. For this purpose we make the standard assumption that the vapor system consists of a mixture of noninteracting ideal gases; i.e., each collection of *i*-mers is considered to be an ideal gas of indistinguishable particles. Then in view of the fact that we may expect small fluctuations in N_i, it is computationally convenient to determine N_i via the grand partition function of the grand canonical ensemble (see Appendix A-7.1). An alternative description in terms of the canonical ensemble has been given, for example, by Dunning (1969).

The grand canonical partition function for the component gases of *i*-mers is (A.7-15):

$$\mathcal{Q}_i = \sum_{N_i} [\exp (\mu_i N_i / kT)] Q(N_i), \tag{7-4}$$

where $Q(N_i)$ is the canonical partition function for the gas of *i*-mers:

$$Q(N_i) = \frac{q_i^{N_i}}{N_i!}, \tag{7-5}$$

from (A.7-6). In this expression, q_i is the partition function for a single *i*-mer.

Then, from (A7-12) and (A.7-14), the size distribution N_i is given as

$$N_i = kT \frac{\partial \ln \mathcal{Q}_i}{\partial \mu_i}. \tag{7-6}$$

The convenience of the grand partition function is that it can easily be rearranged in a manner which greatly simplifies the indicated calculation in (7-6). Thus, on substituting (7-5) into (7-4) we have

$$\mathcal{Q}_i = \sum_{N_i} \frac{[q_i \exp(\mu_i/kT)]^{N_i}}{N_i!} = \exp[q_i \exp(\mu_i/kT)]. \tag{7-7}$$

Consequently, we find immediately from (7-6) that

$$N_i = q_i \exp(\mu_i/kT). \tag{7-8}$$

Let us now eliminate direct reference to the chemical potential. Letting $i = 1$ in (7-8) and using (7-3), we have

$$N_i = (N_1/q_1)^i q_i, \tag{7-9}$$

which is known as the 'mass action law.' This result may also be expressed in a form containing a Boltzmann factor, viz.,

$$N_i = N_{sat,w} \exp[-\Delta\phi_i/kT], \tag{7-10}$$

where

$$\Delta\phi_i = kT[i \ln(q_1/N_{sat,w}) - \ln(q_i/N_{sat,w}) - i \ln S_{v,w}]. \tag{7-11}$$

Here $N_{sat,w}$ is the number of water molecules in V for conditions at saturation with respect to a flat water surface, and $S_{v,w} = N_1/N_{sat,w} = e/e_{sat,w}$ is the saturation ratio of the system.

The determination of N_i has now been reduced to the problem of evaluating the partition function q_i for the i-mer. Unfortunately, however, no one has yet found an accurate *ab initio* way to do this. This is hardly surprising, since q_i depends on the complex structure of the i-mers which is largely unknown, and on a realistic intermolecular interaction potential suitable for an arbitrary poly-molecular aggregate, which is not available.

Consequently, at this point a much more heuristic approach is necessary. Probably the most successful such procedure is that of Plummer and Hale (1972) and Hale and Plummer (1974a, b), who postulated certain allowed structures for the i-mers, and proceeded to work out the corresponding q_i's. We shall describe their semi-empirical approach in more detail in Section 7.1.4. However, some useful perspective and insight may be gained by first turning briefly to two earlier methods for the determination of N_i, which we shall classify as the 'classical' and 'modified classical' approaches.

7.1.2 CLASSICAL DESCRIPTION

Since we cannot proceed rigorously beyond (7-9) or (7-10) and (7-11), there arises the possibility that an earlier resort to intuition and approximate physical modeling might, in some respects, be more appropriate. Not surprisingly, this is the historical route of development of the subject (e.g., Volmer and Weber, 1926; Farkas, 1927; Becker and Döring, 1935; and Zeldovich, 1942). These workers arrived relatively quickly at a complete description for N_i via two major assumptions: (1) the prenucleation embryos may be regarded as water spheres, characterized by the usual macroscopic densities and surface tensions, and (2) they are distributed according to the Boltzmann law.

The assumption of a Boltzmann distribution like (7-10) is quite reasonable: the i-mers are in thermal equilibrium, and the probability that they have a certain energy $(\Delta\phi)_i$ is just the probability for their existence, if we interpret $(\Delta\phi)_i$ as the energy of formation of the i-mer.

Let us therefore consider the energy of formation of a drop of radius a. We may assume the required phase change occurs at constant temperature. However, it is not a constant pressure process, according to (5-11). On the other hand, we may assume the total volume V of the system considered (the mother phase plus the condensed phase) remains constant. Therefore, the Helmholtz free energy F is the proper thermodynamic potential to use in our description. (Elaborations of the point that F, rather than the traditionally used Gibbs function G, is the proper potential may be found in Abraham (1968) and Dufour and Defay (1963). For practical purposes, the resulting differences turn out to be negligible.)

Suppose the system to be comprised, after the phase change, of n_w moles of water within the drop, and n_v moles of water vapor. We neglect any adsorption of water onto the interface (σ); i.e., we assume $n_w^{(\sigma)} = 0$. The total system Helmholtz free energy at this time is thus given, considering (4-3) and (5-13) and the definition $F \equiv U - TS$, by

$$F_2 = n_v\mu_{v,2} + n_w\mu_w - e_2(V - V_w) - p_wV_w + \sigma_{w/v}\Omega, \qquad (7\text{-}12)$$

where we have used the subscript 2 to denote post-phase change conditions. Similarly, before the phase change we have

$$F_1 = (n_v + n_w)\mu_{v,1} - e_1V. \qquad (7\text{-}13)$$

Further, we may assume that the small amount of new phase negligibly affects the vapor, so that $\mu_{v,1} = \mu_{v,2} = \mu_v$, and $e_1 = e_2 = e$; therefore, by subtraction we find the energy of formation is

$$(\Delta F)_{T,V} \equiv F_2 - F_1 = n_w(\mu_w - \mu_v) - V_w(p_w - e) + \sigma_{w/v}\Omega. \qquad (7\text{-}14)$$

On introducing the mechanical equilibrium condition, (5-11), and noting that $V_w = a\Omega/3$, we may also express this as

$$(\Delta F)_{T,V} = n_w(\mu_w - \mu_v) + \sigma_{w/v}\Omega/3. \qquad (7\text{-}15)$$

Other forms of (7-14) are also of interest. For example, we may express $(\Delta F)_{T,V}$ in terms of bulk volume and surface changes in F. Recall that for $T = $ const., $d\mu_w = v_w\,dp_w$; hence on integration, and ignoring the compressibility of water, we find

$$\mu_w(p_w, T) - \mu_w(e, T) = v_w(p_w - e). \qquad (7\text{-}16)$$

Inserting this into (7-14) and using $V_w = n_w v_w$, we obtain

$$(\Delta F)_{T,V} = n_w[\mu_w(e, T) - \mu_v(e, T)] + \sigma_{w/v}\Omega \tag{7-17}$$

$$= \Delta F_{vol} + \Delta F_{surf}, \tag{7-18}$$

where the first term is the volume or bulk free energy change, and the second is the surface energy change.

Let us now express these results on a molecular scale, assuming that we may still employ macroscopic densities and surface tensions. Then from (7-17), the work of formation of an i-mer is

$$\Delta F_i = i[\mu_w(e, T) - \mu_v(e, T)] + \sigma_{w/v}\Omega_i. \tag{7-19}$$

This form has not made use of the assumption of spherical geometry; consequently, it will hold for complex i-mer shapes. (However, it may be necessary to generalize the surface term. See Section 5.7.) At equilibrium we also have $\mu_w(p_w, T) = \mu_v(e, T)$, where p_w refers to the pressure in the water germ. Thus the bracketed term in (7-19) becomes, on using (7-16),

$$\mu_w(e, T) - \mu_v(e, T) = -\dot{v}_w(p_w - e) = -\frac{2\,\dot{v}_w\sigma_{w/v}}{a_g}. \tag{7-20}$$

In the last step we have invoked mechanical equilibrium again. Going one step further, we may introduce the saturation ratio through the Kelvin law, (6-17), and arrive at the result

$$\mu_w(e, T) - \mu_w(p_w, T) = -kT \ln S_{v,w}. \tag{7-21}$$

Consequently, another form of (7-19) is

$$\Delta F_i = \sigma_{w/v}\Omega_i - ikT \ln S_{v,w}. \tag{7-22}$$

In terms of the i-mer radius, this is

$$\Delta F_i = 4\,\pi a_i^2\sigma_{w/v} - \frac{4\,\pi a_i^3}{3\,\dot{v}_w}kT \ln S_{v,w}. \tag{7-23}$$

We have now reached our goal of describing the distribution of prenucleation embryos via the classical approach: the energy of formation ΔF_i has been determined as a function of embryo size, and so we now merely identify $\Delta F_i = \Delta\phi_i$ in (7-10) to obtain

$$N_i = N_{sat} \exp[-\Delta F_i/kT]. \tag{7-24}$$

(For a highly detailed and quite different derivation of the same result, see Dufour and Defay, 1963.)

The behavior of ΔF_i as given in (7-23) is shown in Figure 7-1. We see that for vapor just saturated with respect to bulk water, the energy of i-mer formation rapidly increases with size (as a_i^2, from (7-23)). However, the behavior is seen to be quite different from supersaturated vapor. In this case the curves each have a single maximum at some radius $a_{i,max}$, so that i-mers of radius $a_i > a_{i,max}$ require a work of formation which decreases with increasing size. Thus $\Delta F_{i,max}$ for $a_i = a_{i,max}$ evidently represents the *energy barrier* to nucleation.

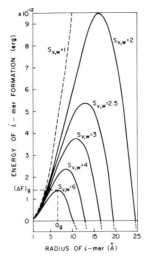

Fig. 7-1. Energy of i-mer formation as a function of i-mer size, for 0 °C. (From *Elements of Cloud Physics* by H. R. Byers, copyrighted by University of Chicago Press, 1965.)

Consequently, through the classical approach we have available not only N_i, but also a description of the germ radius a_g as a function of supersaturation. This is given by $a_g = a_{i,\max}$; by differentiating (7-23) and setting the result equal to zero to obtain the maximum, we find

$$a_g = a_{i,\max} = \frac{2\,\dot{v}_w\sigma_{w/v}}{kT \ln S_{v,w}} = \frac{2\,M_w\sigma_{w/v}}{\mathscr{R}T\rho_w \ln S_{v,w}}, \qquad (7\text{-}25)$$

i.e., the Kelvin law is obtained, as is necessary for consistency since the germ is in (unstable) equilibrium with the vapor. Furthermore, by inserting this result into (7-23), the energy barrier to nucleation, or the free energy of formation of the germ, is found to be

$$\Delta F_g = \Delta F_{i,\max} = \frac{16\,\pi M_w^2\sigma_{w/v}^3}{3[\mathscr{R}T\rho_w(\ln S_{v,w})]^2} = \frac{\sigma_{w/v}\Omega_g}{3}, \qquad (7\text{-}26)$$

where $\Omega_g = 4\,\pi a_g^2$ is the surface area of the germ. This is the classical estimate of the amount of energy which must be supplied by fluctuations in the metastable mother phase in order for nucleation to occur.

It is worthwhile at this point to record also the analogous expressions for homogeneous nucleation of ice in vapor or in supercooled water. Obviously, (7-25) and (7-26) will also hold for a spherical ice germ in unstable equilibrium with water vapor, if we merely replace $\dot{v}_w$ by $\dot{v}_i$, ρ_w by ρ_i, $\sigma_{v/w}$ by $\sigma_{v/i}$ and $S_{v,w}$ by $S_{v,i}$. For the case of a spherical ice germ in supercooled water, the radius is given by (6-49), viz.,

$$a_g = \frac{2\,M_w\sigma_{i/w}}{\bar{\mathscr{L}}_{m,0}\bar{\rho}_i \ln (T_0/T_e)}, \qquad (7\text{-}27)$$

and the corresponding energy of formation is

$$\Delta F_g = \frac{16 \pi M_w^2 \sigma_{i/w}^3}{3[\bar{\mathscr{L}}_{m,0}\bar{\rho}_i \ln (T_0/T_e)]^2}. \tag{7-28}$$

Of course, it is possible to write down similar expressions for an ice germ in the equilibrium Wulff crystal form as well. However, there is little incentive to do so since studies of i-mer shapes (see Sections 7.1.4 and 9.1.3) indicate the simpler spherical geometry is a better approximation to reality.

As we shall see, these results of the classical theory provide a simple basis for predicting nucleation rates which are very similar to those actually observed. This is a much better outcome than we might have expected, in view of the first rather dubious assumption, referred to at the beginning of this section, which serves as half of the foundation for the classical description. Thus, it does not seem very likely that small clusters of molecules should exhibit macroscopic properties. And even if one could assume the macroscopic description is correct in principle, there would still arise conceptual difficulties in its application. In particular, it is not easy to decide where to locate the surface of separation between the phases, since the actual phase transition region may have a thickness comparable to the germ radius (see, for example, Ono and Kondo, 1960). Also, some size correction for the macroscopic surface tension would appear to be in order (recall Section 5.4.3 and Table 5-1).

We have already made some remarks in defense of the assumption of a Boltzmann distribution for the prenucleation embryos. However, here again conceptual difficulties arise in implementing the assumption. The difficulty this time is in describing accurately the contributions to the free energy of formation of the embryo. The classical account of ΔF_i assumes the embryos are at rest in the mother phase. This is obviously incorrect, but the error which results thereby is not obvious. We shall now turn briefly to a discussion of the efforts which have been made to clarify and resolve this problem.

7.1.3 MODIFIED CLASSICAL DESCRIPTION

As the new-phase embryos constitute a polymolecular gas in thermal equilibrium with the mother phase, it is apparent that they must have translational and rotational energies which are overlooked in the classical approach. On the other hand, the bulk free energy term $(\Delta F)_{vol}$ which contributes to ΔF_i (see (7-18)) includes the translational and rotational energy of the bulk water comprising the stationary drop.

According to Lothe and Pound (1962, 1966, 1968, 1969), the free energy of formation ΔF_i appearing in (7-24) should be corrected so that the new value, $\Delta F_i'$, includes the following contributions:

$$\Delta F_i' = \Delta F_i + F_{i,T} + F_{i,R} - F_{i,rep}, \tag{7-29}$$

where $F_{i,T}$ and $F_{i,R}$ result from the translational and rotational degrees of freedom of the embryo. The term $F_{i,rep}$ called replacement term by Lothe and Pound

(1962) has been regarded as resulting from the deactivation of the vibrational degrees of freedom which the cluster would have as bulk water (Lothe and Pound, 1962). In a later paper (Lothe and Pound, 1966), $F_{i,\text{rep}}$ is approximated as due to translational and torsional vibrations about the center of the embryo water mass considered again as a spherical portion of bulk water not yet 'injected' into the vapor. On replacing ΔF_i in (7-24) by $\Delta F'_i$ from (7-29), and by expressing the result in terms of partition functions (see (A.7-7)), the size distribution according to Lothe and Pound becomes

$$N'_i = \left(\frac{q_{i,T} \cdot q_{i,R}}{q_{i,\text{rep}}}\right) N_i \equiv \Phi_{\text{LP}} N_i, \qquad (7\text{-}30)$$

where $q_{i,T}$ and $q_{i,R}$ are the translational and rotational partition functions of the i-mer, described by (A.7-9) and (A.7-10); $q_{i,\text{rep}}$ is the i-mer replacement partition function ($\approx \exp(\dot{s}/k)$, where $\dot{s}$ is the entropy per water molecule in the bulk state (Lothe and Pound, 1962)); and N_i is the uncorrected classical size distribution of (7-24). Unfortunately, the correction factor Φ_{LP} turns out to be of the order of 10^{17}, annihilating the earlier found good agreement between the classical theory and experiment (see Section 7.3).

Not surprisingly, this 'refinement' to the classical approach has been questioned (Reiss and Katz, 1967; Reiss et al., 1968; Kikuchi, 1969, 1971; Reiss, 1970; and Katz and Blander, 1972). Although these authors also found that the classical size distribution lacks an additional factor, they concluded that it is not a consequence of ordinary rotational and translational partition functions. In fact, they found no contribution at all from rotation. The factor was found, instead, to be related to a modified translational partition function referring to the freedom with which the center of mass of a cluster may fluctuate over a portion of the cluster volume.

These studies are altogether too esoteric and controversial to be dealt with here in any detail, and so we are content merely to quote the best estimate to come out of this challenge to Lothe and Pound. This is due to Kikuchi (1971), who extended and refined the work of Reiss (1970) to obtain the following correction factor which they both propose should supplant Φ_{LP} in (7-30):

$$\Phi_{\text{RK}} = p_w \dot{v}_g / kT, \qquad (7\text{-}31)$$

where $\dot{v}_g$ is the volume per molecule in the vapor phase. At 30 °C, $\Phi_{\text{RK}} \approx 3 \times 10^4$; although this is more than 12 orders of magnitude smaller than Φ_{LP}, it still constitutes a significant correction to the classical result.

Unfortunately, the studies of Lothe and Pound, and of Reiss, Kikuchi, and Reiss et al. suffer from a fairly obvious shortcoming: either the work of formation of the germ, or the correction factor itself, or both, have to be evaluated, for lack of information, in terms of macroscopic values of parameters like $\sigma_{w/v}$ and p_w. This tends to seriously compromise the microscopic point of view which is taken when formulating modifications to the classical model. Also, as the reader who searches the literature will soon realize, there are thorny and still partially unresolved conceptual problems which arise in the attempt to merge the classical stationary-drop and statistical mechanics points of view.

At present the most feasible way of circumventing such difficulties is the molecular model approach of Plummer and Hale (1972), which we shall now discuss.

7.1.4 MOLECULAR MODEL METHOD

As we mentioned earlier, the procedure followed by Plummer and Hale is to assume certain structures for the embryo i-mers and to determine the corresponding q_i's. The size distribution is then available directly from (7-10) and (7-11). This formulation has the merit of automatically including the terms which are missing from the classical energy of formation, and of precluding the need for finding 'replacement' energy terms. On the other hand, one has to evaluate q_i, which is not easy.

A direct determination of the partition function q_i would require a realistic description of the interaction potential for the cluster-vapor system. Unfortunately, a realistic intermolecular potential which can be applied to an arbitrary number of water molecules in large clusters is not available. Therefore, Plummer and Hale assumed the following form for q_i:

$$q_i = q_{i,T} q_{i,R} q_{i,V} q_{i,B} q_{i,C}, \qquad (7-32)$$

where $q_{i,T}$ and $q_{i,R}$ have the same meanings as in (7-30), where $q_{i,V} = q_{i,intra} q_{i,lib} q_{i,inter}$ is the vibrational partition function given in terms of the intramolecular, intermolecular, and librational contributions, $q_{i,B} = \exp(-E_{i,B}/kT)$ is the contribution of the intermolecular binding energy $E_{i,B}$ to the partition function, and $q_{i,C}$ is the configurational contribution to the partition function.

Each of these quantities was evaluated semi-empirically in a manner described by Plummer and Hale (1972) and Hale and Plummer (1974a). In these evaluations it was assumed: (1) that i-mers have a well-defined structure, (2) that each structure has a lifetime sufficiently long to characterize its internal vibrational spectrum, and (3) that the internal structure of a water molecule is negligibly affected by cluster formation.

The assumed structure for water clusters in supersaturated vapor is that of closed or partially closed clathrates composed of five-membered rings (Pauling, 1962). An example of such a structure for a 20-mer is shown in Figure 7-2a. These cluster forms fulfil the imposed criteria that the molecules associate by hydrogen bonding with bond angles which are roughly tetrahedral, that the

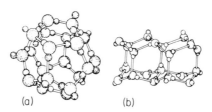

(a)   (b)

Fig. 7-2.   Cluster models used in the theory of Hale and Plummer (1974a, b). (a) cluster of 20 water molecules arranged in a clathrate structure forming a pentagonal dodecahedron. (b) cluster of 20 water molecules arranged in an ice-I_h structure. (Based on Hale and Plummer, 1974a, b.)

number of bonds be maximized, and that the forms possess near spherical symmetry. This choice of geometry is supported by the studies of Lin (1973) and Searcey and Fenn (1974) (see Section 3.2).

However, these perfectly ordered clathrate structures cannot be used to represent arbitrarily large i-mers, since it becomes difficult to maintain the closed 'cages' without grossly distorting the bond angles and lengths. This results in the occurrence of considerable bond strain for $i \geqslant 80$, the effect of which has been studied by Hagen (1973).

For the study of prenucleation embryos of ice in vapor, Hale and Plummer (1974a) assume an ice-I_h structure composed of rings containing six water molecules each (see Section 3.3). A typical structure with 20 molecules is shown in Figure 7-2b.

The model calculations of q_i result in corresponding $\Delta F_i = \Delta \phi_i$ from (7-11). The values for the case of water embryos are shown in Figure 7-3. The effects of bond strain for the larger clusters is also indicated; this causes only a small increase in the formation energy. The general good agreement with the classical model is most impressive and surprising, considering the theoretical deficiencies of the latter.

Before closing this section we should also mention another fairly recent approach to the homogeneous nucleation problem, which is to use Monte-Carlo techniques to evaluate ΔF_i. The method is based on a stochastic process which generates a Boltzmann-weighted set of configurations for a given closed system containing a fixed number of molecules. For details see Abraham (1974b); also, a brief general discussion of the Monte-Carlo method is given in Chapter 14. Though this method holds great promise for the future, it imposes a heavy burden on present generation computers, and it has so far been possible to simulate the growth of only small clusters. Consequently, we shall not consider it further here.

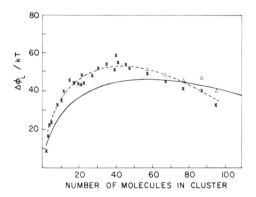

Fig. 7-3. Energy of embryo formation vs. embryo size (in number of molecules) at $S_{v,w} = 5$ and $T = 273\,°K$ for (1) the classical liquid drop model, $(\Delta F)_i/kT = 9.3\,i^{2/3} - 1.6\,i$ (solid line). (2) the molecular model, $(\Delta \phi)_i/kT$ (crosses), (3) the molecular model with strain (triangles), and (4) the least squares fit to $(\Delta \phi)_i/kT\,(= 12.8\,i^{2/3} - 2.40\,i$ (dashed line)). (From Hale and Plummer, 1974b; by courtesy of Amer. Meteor. Soc., and the authors.)

7.2 Nucleation Rate J

When homogeneous nucleation of water from the vapor occurs, what is observed is the (rather sudden) formation of a cloud of small drops. Thus the experimental quantity of interest is the rate at which drops appear in the system as a function of the prevailing saturation ratio $S_{v,w}$. Let us denote this rate by J, measured as the number of drops appearing per unit volume and per unit time. We shall make the traditional assumption that J corresponds completely to the rate of germ formation; i.e., it is the *nucleation rate*.

7.2.1 EQUILIBRIUM APPROXIMATION FOR J

A simple and direct way to estimate J has probably occurred to the reader: assume (7-24) holds for $i = g$, and determine J as the rate at which the $c_g = N_g/V$ germs per unit volume collect single molecules from the vapor; i.e., set

$$J = c_g w^{\downarrow} \Omega_g, \tag{7-33}$$

where $w^{\downarrow}$, given by (5-49), specifies the flux of water molecules to the germs. The estimate (7-33) was first made by Volmer and Weber (1926). It is a quite reasonable first approximation, in that we are interested only in describing the onset of nucleation; for this purpose one might well expect the equilibrium distribution (7-24) for $i = g$ to provide an adequate basis for estimating J.

Let us reflect now on how we might improve our estimate. Two factors which we have so far neglected naturally suggest themselves for consideration. Firstly, germ evaporation as well as condensation should be accounted for. Secondly, we should recognize that as embryos flow through the phase-change 'bottleneck' created by the energy which is required to form a germ-sized embryo, their distribution will not, in principle, be the equilibrium distribution of (7-24), since it assumes no mass flux up the size spectrum.

An improvement which recognizes these features was first carried out by Farkas (1927) and Becker and Döring (1935); other refinements and extensions have been made by Volmer (1939), Zeldovich (1942), Turnbull and Fischer (1949), Farley (1952), and Frenkel (1946). The key simplifying assumption introduced by these authors is that the new size distribution may be regarded as being in a steady state. As we shall see, J can then be found without difficulty.

7.2.2 STEADY STATE RATE APPROXIMATION FOR J

Let us first estimate the time required to reach a steady state once a given vapor supersaturation is achieved. If we denote the time dependent concentration of embryos of size i by f'_i, then the generalization of (7-33) which includes the effect of evaporation is

$$J_i = f'_{i-1} w^{\downarrow} \Omega_{i-1} - f'_i w^{\uparrow} \Omega_i, \tag{7-34}$$

where $w^{\uparrow}$ is the flux of water molecules leaving the embryo surface. Here J_i is the number of embryos $vol^{-1} sec^{-1}$ entering the size category i. A special case of this equation is the equilibrium situation for which $J_i = 0$; for this case we have

$$0 = c_{i-1}w^{\downarrow}\Omega_{i-1} - c_i w^{\uparrow}\Omega_i. \tag{7-35}$$

We may combine these two equations to eliminate $w^{\uparrow}\Omega_i$:

$$J_i = c_{i-1}w^{\downarrow}\Omega_{i-1}\left(\frac{f'_{i-1}}{c_{i-1}} - \frac{f'_i}{c_i}\right). \tag{7-36}$$

From the definition of J_i, the first time variation of f'_i is given by

$$\frac{\partial f'_i}{\partial t} = J_i - J_{i+1}. \tag{7-37}$$

Let us now pass over to an approximately equivalent continuous description. Then in place of (7-36) we have $J_i \approx -c_i w^{\downarrow}\Omega_{i-1}\,\partial(f'_i/c_i)/\partial i$, and (7-37) becomes

$$\frac{\partial f'_i}{\partial t} \approx \frac{\partial}{\partial i}\left[c_i w^{\downarrow}\Omega_{i-1}\frac{\partial}{\partial i}(f'_i/c_i)\right]. \tag{7-38}$$

Assuming further that $c_i w^{\downarrow}\Omega_{i-1}$ is roughly constant, this equation reduces to a diffusion equation in f'_i/c_i with diffusion coefficient $w^{\downarrow}\Omega_{i-1}$:

$$\frac{\partial(f'_i/c_i)}{\partial t} \approx w^{\downarrow}\Omega_{i-1}\frac{\partial^2}{\partial i^2}(f'_i/c_i). \tag{7-39}$$

Thus, the characteristic time to achieve the quasi-steady state germ concentration is just $\tau \approx (g^2/w^{\downarrow}\Omega_{g-1})^{1/2}$ (Farley, 1952). For water germs in typical expansion chambers, $\tau \sim 10^{-6}$ to 10^{-5} sec, which is about 10^{-3} of the time during which the supersaturation remains essentially constant. Thus, the steady state assumption is consistent with usual experimental conditions.

Let us now proceed to find $J_i = J = $ constant, assuming a steady state concentration which we shall denote by f_i. Surprisingly, it turns out we do not have to determine f_i to find J. However, we do need to use boundary conditions on f_i, which we choose as follows: (1) $f_1/c_1 = 1$; this is reasonable since the monomer population is relatively enormous and need not deviate significantly from the equilibrium concentration in order to produce a substantial nucleation rate. (2) $f_G = 0$ for some $G \gg g$; the results are extremely insensitive to the choice of G, which makes this a reasonable working assumption.

From (7-36) we may now immediately obtain J by summing over i as follows:

$$\sum_{i=1}^{G-1} J/c_i w^{\downarrow}\Omega_i = \sum_{i=1}^{G-1}\left(\frac{f_i}{c_i} - \frac{f_{i+1}}{c_{i+1}}\right) = \frac{f_1}{c_1} = 1, \tag{7-40a}$$

or

$$J = \left(\sum_{i=1}^{G-1}\frac{1}{c_i w^{\downarrow}\Omega_i}\right)^{-1}, \tag{7-40b}$$

a result in which f_i does not appear. Now c_i will have a minimum near $i = g$ (recall (7-24) and Figure 7-1), and so the dominant contributions to J will come from terms in that neighborhood. Hence we may approximate (7-40) as

$$J = Zc_g w^{\downarrow}\Omega_g, \tag{7-41}$$

where Z^{-1} effectively counts the number of contributing terms; i.e., it measures the width of the minimum in the curve for c_i. By comparison with (7-33), we see that (7-41) differs from the equilibrium approximation result only by the factor Z, called the 'Zeldovitch factor' (Zeldovitch, 1942).

The Zeldovitch factor is obtained by expanding c_i about the minimum in a Taylor series through terms of the second order in i; this produces a Gaussian approximation to the curve in that neighborhood, and Z^{-1} is identified with the width (i.e., the standard deviation) of the Gaussian curve. Proceeding in this way, we write

$$c_i = c_{sat,w} \exp \left[-\Delta F_i / kT \right],$$ (7-42a)

where

$$\Delta F_i \approx (\Delta F)_g + \frac{1}{2} \left[\frac{d^2}{di^2} (\Delta F_i) \right]_{i=g} (i - g)^2.$$ (7-42b)

Consequently, on abbreviating $B_g \equiv [(d^2/di^2)(\Delta F_i)]_{i=g} (<0)$, we find

$$Z^{-1} = \int_{-\infty}^{+\infty} \exp (B_g x^2 / kT) \, dx = \left(-\frac{2 \pi kT}{B_g} \right)^{1/2}.$$ (7-43)

Using (7-22) and recognizing that $\Omega_i \propto i^{2/3}$, this leads to

$$Z = \left[\frac{\Delta F_g}{3 \pi kT g^2} \right]^{1/2}.$$ (7-44)

This expression is suitable for arbitrary geometries. Two additional forms which hold for spherical water germs are:

$$Z = \left(\frac{\ln S_{v,w}}{6 \pi g} \right)^{1/2} = \frac{2 \dot{v}_w}{\Omega_g} \left(\frac{\sigma_{w/v}}{kT} \right)^{1/2},$$ (7-45)

where $g = (4 \pi /3) a_g / \dot{v}_w$. The mathematical approximations involved in passing from (7-40) to (7-41) and (7-44) produce an error of about 1% (Cohen, 1970), which is insignificant in comparison with the uncertainties in ΔF_i.

Numerical evaluation of (7-45) shows that Z is typically $O(10^{-1})$. This result is in qualitative accord with our expectations: A finite rate of germ production should deplete the embryo population to something below the equilibrium level.

Collecting results, the nucleation rate of water germs from the vapor may be expressed as

$$J = c_{sat,w} w^{\downarrow} \Omega_g Z \exp \left[-\Delta F_g / kT \right].$$ (7-46)

This holds even for non-spherical germs, if (7-44) is used to describe Z. Other versions for spherical germs include

$$J = 2 c_{sat,w} w^{\downarrow} \dot{v}_w (\sigma_{w/v} / kT)^{1/2} \exp \left[-\Delta F_g / kT \right],$$ (7-47)

using the second form of (7-45), and

$$J = \frac{\alpha_c}{\rho_w} \left(\frac{2 N_A^3 M_w \sigma_{w/v}}{\pi} \right)^{1/2} \left(\frac{e_{sat,w}}{\mathcal{R}T} \right)^2 S_{v,w} \exp \left[-\Delta F_g / kT \right],$$ (7-48)

using (5-49) for $w^{\downarrow}$, and the first form of (7-45). Of course, equations exactly analogous to (7-47) and (7-48) hold for the nucleation rate of spherical ice germs from the vapor (with $c_{sat,w} \to c_{sat,i}$, $\dot{v}_w \to \dot{v}_i$, $\alpha_c \to \alpha_d$, $S_{v,w} \to S_{v,i}$, $\rho_w \to \rho_i$, $\sigma_{w/v} \to \sigma_{i/v}$, and $e_{sat,w} \to e_{sat,i}$).

Inspection of (7-46)–(7-48) and (7-26) shows that J is extremely sensitive to $S_{v,w}$, since the term in the exponent varies as $S_{v,w}^{-2}$. This is indicated further in Table 7-1, in which a numerical evaluation of (7-48) and its counterpart for ice are presented. We see, for example, that for water germs J increases by 5 orders of magnitude as $S_{v,w}$ increases from 5 to 6. This behavior enables one to define, from an experimental point of view, a critical saturation ratio, $(S_{v,w})_{crit}$, at which drops suddenly appear in the vapor; by convention, $(S_{v,w})_{crit}$ has been taken to correspond to $J = 1$ germ cm^{-3} sec^{-1}.

In Figure 7-4, (7-48) together with (7-26) is presented as a plot of $S_{v,w}$ vs. T for various values of J and the condensation coefficient α_c. The figure shows that $(S_{v,w})_{crit}$ increases from about 3 to 8 as T decreases from 30 °C to −30 °C. Note that lower values of α_c raise $S_{v,w}$ for given J and T.

Numerical evaluation of (7-48) shows that for $\alpha_c = 1.0$ the prefactor to the exponential term varies between 9.0×10^{21} cm^{-3} sec^{-1} at −30 °C and 3.7×10^{25} cm^{-3} sec^{-1} at 30 °C. However, this variation only marginally affects $(S_{v,w})_{crit}$; e.g., at 0 °C, $(S_{v,w})_{crit}$ decreases by only 7% although the prefactor increases by 10^3.

Values for $(S_{v,w})_{crit}$ evaluated by Hale and Plummer from their molecular model are also plotted as a function of temperature in Figure 7-4. These theoretical points were determined by insertion of their values of ΔF_g into the standard nucleation rate equation, (7-46). Thus, the good agreement with the classical theory reflects the proximity of the curve maxima in Figure 7-3; i.e., the classical and molecular cluster determinations of ΔF_g yield quite similar results.

We should note further that values for $(S_{v,w})_{crit}$ predicted by the statistical theories of Reiss et al. (1968), Reiss (1970), and Kikuchi (1971) would lie between the curves $J = 10^{-3}$, and 10^{-5} cm^{-3} sec^{-1} in Figure 7-4, while the predictions of Lothe and Pound (1969) would lie near the curve for $J = 10^{17}$ cm^{-3} sec^{-1}, i.e., outside the range shown in the figure. This illustrates again the substantial disagreement between the statistical modifications of the classical theory and the molecular cluster theory.

TABLE 7-1

Variation of J, g, and a_g as a function of saturation ratio for homogeneous nucleation of water drops and ice crystals in vapor at −12 °C and for $\alpha_c = \alpha_d = 1.0$. (Based on data of Dufour and Defay, 1963.)

$S_{v,w}$	2	3	4	5	6
$S_{v,i}$	2.249	3.374	4.499	5.623	6.748
J (drops cm^{-3} sec^{-1})	1.9×10^{-112}	7.0×10^{-31}	1.1×10^{-10}	7.1×10^{-2}	6.0×10^3
J (ice crystals cm^{-3} sec^{-1})	9.2×10^{-394}	2.7×10^{-163}	1.2×10^{-98}	4.4×10^{-69}	3.4×10^{-52}
g (molecules per water germ)	899	226	122	72	52
g (molecules per ice germ)	2384	706	373	247	182
a_g (water) (cm)	1.9×10^{-7}	1.2×10^{-7}	9.3×10^{-8}	8.0×10^{-8}	7.2×10^{-8}
a_g (ice) (cm)	2.2×10^{-7}	1.5×10^{-7}	1.2×10^{-7}	1.1×10^{-7}	9.5×10^{-8}

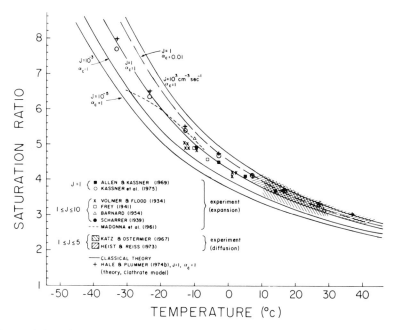

Fig. 7-4. Variation of the saturation ratio with temperature for homogeneous nucleation of water drops in supersaturated water vapor occurring at a given rate J. Comparison of theory with experiment.

Table 7-1 indicates that the nucleation rate of ice germs from the vapor remains near zero for all realizable supersaturations. This follows from the fact that $\Delta F_g \propto \sigma^3$; then since $\sigma_{i/v} > \sigma_{w/v}$, we have $\Delta F_{g,ice} > \Delta F_{g,water}$ and hence $J_{ice} \ll J_{water}$ for a given $S_{v,w}$. This behavior apparently holds for T down to at least $-100\,°C$; i.e., for such T ice particles will not appear directly from the vapor, but rather by way of the freezing of supercooled drops. This is in contrast to predictions based on the phase diagram for bulk water, which merely reinforces again the notion that surface effects dominate in nucleation phenomena. On the other hand, the nucleation prediction is in agreement with *Ostwald's rule of stages* (*Ostwald's Stufenregel*) (Ostwald, 1902) which states that a supersaturated phase (water vapor) does not directly transform into the most stable state (ice), but rather into the next most stable or metastable state (supercooled water). Although Krastanow (1940) proposed a reversal of this rule for water substance below about $-65\,°C$, Dufour and Defay (1963) have shown Krastanow's result to be erroneous since it was based on incorrect values for $\sigma_{i/v}$ and v_i, and neglected the variation of $\sigma_{w/a}$ with temperature. A correct evaluation of the nucleation rate equations demonstrates that at temperatures warmer than $-100\,°C$ no reversal of Ostwald's rule takes place. It must be stressed that this result is only applicable to *homogeneous* nucleation. If nucleation is *heterogeneous*, Ostwald's rule does, indeed, reverse under certain conditions (see Chapter 9).

Figure 7-5 shows that there are larger differences between the classical and molecular cluster models for the case of ice nucleation from the vapor. Nevertheless, both models uphold the Ostwald rule of stages and predict that in homogeneous vapor, ice always forms via the freezing of supercooled drops.

In order to apply the nucleation rate equation to the case of ice-germ formation in supercooled water, we must realize that the main difference between nucleation of ice embryos from supersaturated vapor and from supercooled water lies in the growth mechanism of the embryos. As we have seen, in the former case the growth of an embryo is controlled by the monomer flux from the vapor. In the latter case, where water molecules are essentially already in contact with the ice-embryo, growth is a matter of molecular reorientation involving the breaking of water-to-water bonds and the formation of water-to-ice bonds. During this process a water molecule must pass from its average equilibrium position of minimum potential energy in water to a new equilibrium position in ice, the two positions being separated by an energy barrier ΔF^+. Expressed in another way, ΔF^+ is the energy of activation for diffusion across the phase boundary. Turnbull and Fischer (1949) have shown, by use of the so-called 'absolute reaction rate theory' (e.g., Glasstone *et al.*, 1941; Eyring and Jhon, 1968), that this energy barrier leads to the following expression for the diffusive flux density of water molecules to the ice surface:

$$w_{diff}^{\downarrow} = \frac{N_c kT}{h} \exp\left(-\Delta F^+/kT\right), \tag{7-49}$$

where h is Planck's constant and N_c is the number of monomers of water in contact with unit area of the ice surface. (The energy of activation (kcal mol^{-1})

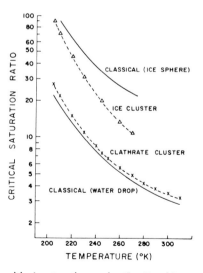

Fig. 7-5. Variation of the critical saturation ratio (J = 1) with temperature for homogeneous nucleation of water drops and ice crystals in supersaturated water vapor. Comparison of the classical theory (for α_c = 1.0) with the cluster theory of Hale and Plummer (1974a, b). (Based on Hale and Plummer, 1974a, with changes.)

for diffusion of water molecules across the water-ice boundary, $\Delta F^{\ddagger}$, can be approximated by the energy of activation for the displacement of water molecules in bulk water, ΔF_{act}, which is given by the following fit to recent experimental studies of the self diffusion coefficient for water (Miles, 1971, 1973; Gillen et al., 1972; Pruppacher, 1972) and of the viscosity of water (Hallett, 1963a; Stokes and Miles, 1965; Kell, 1972b):

$$\Delta F_{act} = 5.5 \exp\left(-1.330 \times 10^{-2}\,T + 2.74 \times 10^{-4}\,T^2 + 1.085 \times 10^{-6}\,T^3\right)$$

$$(7\text{-}50a)$$

for $-30 \leqslant T \leqslant 0\,°C$, and

$$\Delta F_{act} = 5.5 \exp\left(-1.16 \times 10^{-2}\,T + 1.064 \times 10^{-4}\,T^2\right) \qquad (7\text{-}50b)$$

for $0 \leqslant T \leqslant 30\,°C$, with T in $°C$. Both expressions fit the data to within ± 0.1 kcal mol^{-1} over their respective ranges.

Using (7-49) for $w_{diff}^{\downarrow}$ in place of the prior $w^{\downarrow}$, we can immediately write down the homogeneous nucleation rate of ice germs in water in a form analogous to (7-47):

$$J = 2\,c_w w_{diff}\dot{v}_i \left(\frac{\sigma_{i/w}}{kT}\right)^{1/2} \exp\left[-\frac{\Delta F^{\ddagger} + \Delta F_g}{kT}\right], \qquad (7\text{-}51)$$

or, in a more expanded version,

$$J = 2\,N_c \left(\frac{\rho_w kT}{\rho_i h}\right)\left(\frac{\sigma_{i/w}}{kT}\right)^{1/2} \exp\left[-\frac{\Delta F^{\ddagger} + \Delta F_g}{kT}\right]. \qquad (7\text{-}52)$$

In both of these expressions (7-28) is to be used for ΔF_g.

Table 7-2 gives some results of a numerical evaluation of (7-52). As would by now be expected, an extremely strong dependence of J on the supercooling is exhibited, corresponding to observations that the onset of ice nucleation in water occurs almost discontinuously at some stage during cooling. Numerical evaluation of (7-52) also shows that the prefactor to the exponential term varies between 2.7×10^{35} at $0\,°C$ and 1.9×10^{35} cm^{-3} sec^{-1} at $-40\,°C$. These relatively small variations in the prefactor have only a minor effect on the critical supercooling.

Unfortunately, no molecular theory is presently available for homogeneous nucleation of ice in supercooled water. Although some attempts towards solving the problem have been made by Eadie (1971), who proceeded in a manner

TABLE 7-2

Variation of J, g, and a_g with supercooling for homogeneous ice nucleation in supercooled water for $N_c = 5.85 \times 10^{14}$.

$\Delta T = (T - T_0)$ $(°C)$	30	35	40	45	50
J (cm^{-3} sec^{-1})	4.6×10^{-11}	5.1	5.8×10^{6}	4.4×10^{9}	1.5×10^{11}
a_g (cm)	1.4×10^{-7}	1.2×10^{-7}	9.6×10^{-8}	8.2×10^{-8}	7.1×10^{-8}
g (molecules/ice germ)	348	195	116	71	46

similar to that of Hale and Plummer, a complete solution must await a more realistic description of the structure of water (see Section 3.4).

Let us note now that the values given in Table 7-2 were computed for the case of ice germ formation in $1 \, cm^3$ of supercooled water, which is equivalent to a water drop of 6.2 mm radius. Since cloud drops are considerably smaller than this size, it is necessary to formulate ice nucleation theory for smaller volumes of water. For this purpose we shall consider a population of isolated water drops all having the same temperature T and the same volume V_d. We shall assume further that a nucleation event in any one drop is independent of that in any other drop. Given these conditions, we may express the number of ice-germs produced during the time dt in the volume $N_u V_d$ of unfrozen water as

$$N_{i,g}(t + dt) - N_{i,g}(t) = N_u V_d J(T) \, dt, \qquad (7\text{-}53)$$

where N_u is the number of unfrozen drops and $J(T)$ is the rate of ice-germ formation.

We shall also assume that ice formation is the result of only one nucleation event per drop. This assumption is reasonable since the growth velocity of ice is very large at the supercoolings where homogeneous ice-nucleation takes place. This makes it very likely that the first germ formed grows quickly enough to convert the drop into ice before any other germ is formed. The first germ receives additional protection from the fact that during its growth latent heat is released which immediately raises the temperature of the water in the drop. This reduces the nucleation rate of other germs to a negligible value.

For these conditions the increase in the number of ice-germs is given by the increase in the number of frozen drops N_f. Also, since the total number N_0 of drops is constant, we have $dN_u = -dN_f$. Hence we arrive at the simple differential equation

$$dN_u = -N_u V_d J(T) \, dt. \qquad (7\text{-}54)$$

Upon integration, assuming constant T, we find

$$\ln \frac{N_0}{N_u} = V_d J(T) t, \qquad (7\text{-}55)$$

which indicates how, at constant temperature, the number of unfrozen drops decreases with increasing time. On the other hand, for a constant rate of cooling, $\gamma_c = -dT/dt$, we obtain the result

$$\ln \frac{N_0}{N_u} = \frac{V_d}{\gamma_c} \int_T^{T_0} J(T) \, dT. \qquad (7\text{-}56)$$

This indicates that the smaller the rate of cooling or the larger the drop volume, the smaller the number of unfrozen drops at any one temperature T. This implies, in turn, that the smaller the rate at which a drop is cooled or the larger the volume of a drop, the less it can be supercooled before it freezes.

Frequently in the literature, the freezing temperature of a population of drops is characterized by the median freezing temperature T_m, i.e., the temperature

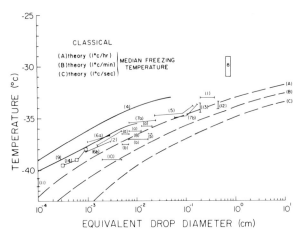

Fig. 7-6. Variation with size of the median freezing temperature of a population of uniformly sized water drops. Comparison of theory with experiment. (1) Meyer and Pfaff (1935), lowest, (2) Bigg (1953a), mean, (3) Wylie (1953), lowest, (4) Jacobi (1955a, b), mode, 0.5 °C min⁻¹, (5) Mossop (1955), lowest, 0.5 °C min⁻¹, (6) Carte (1956), median, a) 0.5 °C min⁻¹, b) 20 °C min⁻¹, (7) Langham and Mason (1958), a) median, b) lowest, 1 °C min⁻¹, (8) Hoffer (1961), a) median, b) lowest, 1 °C min⁻¹, (9) Kuhns and Mason (1968), median, 6 °C min⁻¹, (10) Pound *et al.* (1953, 1955), median, 4 °C min⁻¹, (11) Mossop (1955), cloud chamber, (12) Bayardelle (1955), median-lowest, 3 °C hr⁻¹, (13) Pruppacher (1963), median-lowest, 3 °C hr⁻¹, (14) Day (1958), lowest.

where $N_u = N_0/2$. With this, (7-56) becomes

$$\int_{T_m}^{T_0} J(T) \, dT = 0.693 \, \gamma_c/V_d.$$ (7-57)

This relationship is shown in Figure 7-6 as a plot of T_m vs. drop diameter for a population of equal-sized drops, for $\gamma_c = 1$ °C hr⁻¹, 1 °C min⁻¹, and 1 °C sec⁻¹. It is evident from the figure that the smaller the cooling rate, the warmer T_m of a population of uniformly sized drops. In addition, we note that the larger the drops in the population, the warmer T_m for a given γ_c. These findings are physically reasonable if we consider that the larger the volume of a drop, the larger is the probability for a density fluctuation in the drop and thus the larger is the probability that an ice germ will be produced. It is also reasonable that in a given volume of supercooled water, the probability for a density fluctuation, and thus the probability for ice-formation, increases with increasing time during which the water is exposed to a certain change in temperature.

7.3 Experimental Verification

The various shortcomings inherent in the theories of homogeneous nucleation have made extensive comparison with experiment especially important. Following Wilson (1899) many experimenters have employed the expansion chamber technique to determine the onset of homogeneous water drop formation in supersaturated vapor. A review of many such studies has been given by Mason (1957a). More recently, Katz and Ostermier (1967), and Heist and Reiss (1973) have used the diffusion chamber technique to study homogeneous water drop formation. Both experimental techniques have several major shortcomings and basic difficulties which are hard to overcome. Problems inherent in the diffusion cloud chamber technique have been discussed by Fitzgerald (1970, 1972), while difficulties of the expansion chamber technique have been discussed by Barnard (1953), Mason (1957a), Allard and Kassner (1965), Carstens et al. (1966), Carstens and Kassner (1968), Kassner et al. (1968a, b), and Allen and Kassner (1969).

In addition to the experimental problems of design and technique, Allen and Kassner (1969) gave evidence of the fact that the molecules of any carrier gas with which a cloud chamber is purged may act as nucleation centers to form clathrates. Under these conditions nucleation is not truly homogeneous. Still another uncertainty arises from the fact that the condensation coefficient α_c is not accurately known (see Table 5-4).

In spite of these reservations, Figure 7-4 indicates a reasonably good agreement between experiment and both the classical theory and the molecular clathrate cluster theory of Hale and Plummer (1974b). This is particularly true for the most recent expansion chamber studies of Kassner et al. (1971; 1975, personal communication), and the diffusion chamber studies of Heist and Reiss (1973). On the other hand, the statistical modification of the classical theory is not supported by the experiments. From the discussion of Section 7.1.3, we must interpret the agreement between the classical theory and experiment as being partly fortuitous and due to compensating errors in the theory (see also Hale and Plummer, 1974a, and Lee et al., 1973).

Some of the earlier experimental ice nucleation studies by Sander and Dam-köhler (1943), Cwilong (1947), and Pound et al. (1955) were interpreted to mean that, below a certain temperature, ice forms directly from the vapor. However, Fournier d'Albe (1949), Mason (1952a), Mossop (1955), Kachurin et al. (1956), and Maybank and Mason (1959) demonstrated conclusively that in supersaturated homogeneous vapor and at temperatures between 0 and −70 °C, ice is always the result of the freezing of supercooled drops. Thus, present experiments support both the molecular ice cluster theory of Hale and Plummer (1974a, b) and the classical theory in their prediction that Ostwald's rule of stages indeed applies to the phase change of water substance.

Great experimental difficulties have also been encountered by those who have attempted to verify homogeneous nucleation theory for ice in supercooled water. One of the chief difficulties is the fact that water cannot easily be purified to such an extent that it consists almost entirely of water molecules, although some workers have gone far in devising techniques to reach such a desired state of purification (Mossop, 1955; Haller and Duecker, 1960; Pruppacher, 1963a). Additional uncertainties in providing homogeneous conditions arise from the need to support the water samples. In view of these difficulties it is hardly surprising that the observed median freezing temperatures of populations of drops lie somewhat above the theoretical predictions. On the other hand, the lowest observed freezing temperatures seem to be in fair agreement with results derived from the classical theory. As in the case of drop nucleation from the vapor, this agreement should be regarded as partly fortuitous, and cannot be accepted as proof of the classical theory. Molecular cluster theories or the equivalent in theoretical rigor should eventually supplant the classical theory as a basis for understanding the mechanisms of homogeneous ice nucleation in supercooled water.

CHAPTER 8

THE ATMOSPHERIC AEROSOL

An understanding of the cloud forming processes in the atmosphere requires knowledge of the physical and chemical characteristics of the atmospheric aerosol. In discussing this gaseous suspension of solid and liquid particles, it is customary to include all gases except water vapor, and all solid and liquid particles except hydrometeors, i.e., cloud and raindrops, and ice particles. In the present chapter we shall present a brief discussion of the characteristics of the gaseous constituents followed by a more detailed description of the main physical and chemical characteristics of the aerosol particles. For background on the subjects covered, the reader is referred to the texts of Rasool (1973), Butcher and Charlson (1972), Hidy (1972), McCormac (1971), Matthews *et al.* (1971), Hidy and Brock (1970), Singer (1970), Stern (1968), Davies (1966), Fuchs (1964), Green and Lane (1964), Junge (1963a), Cadle (1961, 1966); to the survey reports SMIC (1971), SCEP (1970), and Robinson and Robbins (1968, 1969, 1971); and to the survey articles by Junge (1969a, 1971, 1972a, b, c, 1974). Most of the data on which this chapter is based are derived from these sources.

8.1 Gaseous Constituents of the Atmosphere

Up to an altitude of about 85 km the composition of the atmosphere is essentially uniform. In this layer, called the *homosphere*, the quasi-constant gaseous constituents are present in constant proportions (Table 8-1). Above about 85 km the composition of the atmosphere begins to vary markedly due to gravitational separation of the chemical constituents, and to solar radiation which dissociates some of the constituents and stimulates the formation of new chemical species. This outer portion of the atmosphere is called the *heterosphere*. For a discussion

TABLE 8-1
Composition of clean, dry air. (Based on U.S. Standard Atmosphere, 1962.)

Constituent gas and formula	Content (% by vol.)	Constituent gas and formula	Content (% by vol.)
Nitrogen, N_2	78.084	Helium, He	0.000524
Oxygen, O_2	20,9476	Krypton, Kr	0.000114
Argon, A	0.934	Hydrogen, H_2	0.00004
Neon, Ne	0.001818	Xenon, Xe	0.0000087

TABLE 8-2a
Sources and sinks of major atmospheric gases, other than N_2

Gas	Sources	Sinks
O_2	Exchange with biosphere, exchange with ocean	
He	He_2^3: cosmic rays He_2^4 radioactive decay in earth crust	Gravitational escape
H_2	Decomposition of organic matter, photolysis of H_2O in upper atmosphere, oxidation of CH_4, influx of protons from sun, volcanoes	Photo-oxidation, utilization by bacteria, gravitational escape
CO	Combustion, forest fires, biological activity in soil and in oceans, volcanoes	Oxidation in upper atmosphere, biological uptake by bacteria, uptake by soil-surface
CO_2	Combustion, cement production, volcanoes, burning of carbon compounds by animals and bacteria (respiration), forest fires	Photosynthesis by biosphere, absorption by oceans
O_3	Dissociation of O_2 by solar radiation and subsequent collision of O with O_2, dissociation of NO_2 by solar radiation and subsequent collision of O with O_2	Photochemical dissociation in upper atmosphere, collision of O_3 with O
CH_4 and other hydro-carbons	Escape from oil wells and mines, bacterial decomposition of organic matter, forest fires, exhalations from biosphere; combustion, solvent use	Photochemical reaction with NO, NO_2, and O_3 in upper atmosphere, dissociation by solar radiation at high altitudes
N_2O	Bacterial decomposition in soil and in oceans, photochemical reactions in upper atmosphere	Photodissociation in upper atmosphere, decomposition by biological activity in soil
NO_x	Bacterial activity in soil, combustion, volcanoes	Chemical reactions (with hydrocarbons, oxidation to nitrates, etc.)
NH_3	Bacterial activity in soil, combustion, volcanoes	Chemical reactions (with SO_2, oxidation to nitrates, etc.)
H_2S	Decomposition of organic matter in soil and in stagnating water, volcanoes, combustion	Chemical reactions (oxidation to sulfate, etc.), uptake by soil and oceans
SO_2	Combustion, volcanoes	Chemical reactions (oxidation to sulfate, etc.), uptake by soil and oceans

Total concentration, production rate, and residence time of major atmospheric gases, other than N_2.

	Mean mixing ratio in troposphere (Northern Hemisphere)	Production rate of gas	Residence time of gas in troposphere
O_2	20.946% (Vol)		2000 to 9700 yr
He	He_2^3: $2.4 \times 10^7\ m^3$ (STP) total amount in He_2^4: $2.0 \times 10^{13}\ m^3$ (STP) atmosphere	He_2^3: $4 \times 10^2\ m^3$ (STP) yr^{-1} He_2^4: $7 \times 10^6\ m^3$ (STP) yr^{-1}	6×10^4 yr 3×10^6 yr
H_2	0.575 ppm (Vol) = 51.3 $\mu g\ m^{-3}$ (STP)	Anthropogenic: $0.30 \times 10^{11}\ kg\ yr^{-1}$ Natural: $0.11 \times 10^{11}\ kg\ yr^{-1}$	4 to 13 yr
CO	0.15 ppm (Vol) = 188 $\mu g\ m^{-3}$ (STP)	Anthropogenic: $6 \times 10^{11}\ kg\ yr^{-1}$ Natural: 16 to $41 \times 10^{11}\ kg\ yr^{-1}$	0.2 to 1 yr
CO_2	1970: 320 ppm (Vol) = 0.627 $g\ m^{-3}$ (STP) Increasing during the past 100 years	Biosphere uptake: $2.1 \times 10^{14}\ kg\ yr^{-1}$ Uptake by oceans: $3.7 \times 10^{14}\ kg\ yr^{-1}$ Release by combustion: $1.5 \times 10^{13}\ kg\ yr^{-1}$ Soil and plant respiration: $2.1 \times 10^{14}\ kg\ yr^{-1}$ Release by oceans: $3.6 \times 10^{14}\ kg\ yr^{-1}$	~7 yr (ocean-atmosphere-biosphere cycle) ~13 to 15 yr (atmosphere-biosphere cycle)
O_3	Earth surface: 0.015 to 0.03 ppm (Vol) = 23 to 64 $\mu g\ m^{-3}$ (STP) Level of concentrating maximum (lower stratosphere): 3 to 6 ppm (Vol) = 320 to 640 $\mu g\ m^{-3}$ (ambient)	Injection rate into troposphere: 4×10^{11} to $1.8 \times 10^{12}\ kg\ yr^{-1}$	1 to 4 months (troposphere) 1 to 2 yr (lower stratosphere)
CH_4	1.4 ppm (Vol) = $10^3\ \mu g\ m^{-3}$ (STP)	Anthropogenic: 0.3 to $0.9 \times 10^{11}\ kg\ yr^{-1}$ Natural: 10 to $14 \times 10^{11}\ kg\ yr^{-1}$	2 to 12 yr
N_2O	0.25 ppm (Vol) = 530 $\mu g\ m^{-3}$ (STP)	Natural: $2.5 \times 10^{11}\ kg\ yr^{-1}$	4 to 16 yr
NO_x	1 to 10 ppb (Vol) = 2 to 20 $\mu g\ m^{-3}$ (STP), as NO_2	Anthropogenic: $0.5 \times 10^{11}\ kg\ yr^{-1}$ Natural: $4.5 \times 10^{11}\ kg\ yr^{-1}$	few days
NH_3	1.3 to 13.2 ppb (Vol) = 1 to 10 $\mu g\ m^{-3}$ (STP)	Anthropogenic: $4 \times 10^9\ kg\ yr^{-1}$ Natural: $1200 \times 10^9\ kg\ yr^{-1}$	few days
H_2S	0.67 to 6.7 ppb (Vol) = 1 to 10 $\mu g\ m^{-3}$ (STP)	Anthropogenic: $2.7 \times 10^9\ kg\ yr^{-1}$ Natural: 0.8 to $1.3 \times 10^{11}\ kg\ yr^{-1}$	few days
SO_2	0.34 to 1.70 ppb (Vol) = 1 to 5 $\mu g\ m^{-3}$ (STP)	Anthropogenic: 0.6 to $0.8 \times 10^{11}\ kg\ yr^{-1}$ Natural: (volcanoes) $2 \times 10^9\ kg\ yr^{-1}$	few days

of the chemical characteristics of the higher atmosphere, the reader is referred
to the texts of Ratcliffe (1960), Whitten and Poppoff (1971), Webb (1966),
and Craig (1965).

One may classify the gaseous constituents of the atmosphere according to
their residence times. For *quasi-constant* constituents this is of the order of
thousands of years or more; *slowly varying* constituents have residence times of
a few months to a few years; and *fast varying* constituents have residence times
of a few days or less.

The quasi-constant gaseous constituents of the atmosphere are N_2, O_2, A, He,
Ne, Kr, and Xe. Slowly varying gaseous constituents include CH_4, O_3, N_2O, CO,
CO_2, and H_2. Fast varying gaseous constituents include SO_2, H_2S, NO, NO_2, and
NH_3. The only gases which can be considered 'permanent' are the nobel gases,
Ne, A, Kr, Xe, since they have negligible sources or sinks in the atmosphere.
All other gases, including O_2, N_2, He, and H_2, have sources and sinks and
therefore a finite residence time. Table 8-2 summarizes the sources and sinks of
these gases (except for N_2), their rate of production or destruction, their average
total concentration for background conditions in the atmosphere, and their
residence time. The numbers given are very rough estimates and can only serve as a
general guide. They are based on the literature mentioned above and on articles by
Junge (1962), Fabian and Junge (1970), Schütz *et al.* (1970), Seiler and Schmidt
(1973), Seiler (1975), Georgii and Müller (1974), Georgii *et al.* (1974), and Georgii
(1975). The values given for the residence time of gases in the atmosphere have
been estimated (1) from the ratio between the amount of a particular gas in the
atmosphere and its rate of removal or injection (Junge, 1963a; Seiler, 1975), and (2)
on the basis of the spatial and temporal variability of the concentration of the
particular gas (Junge, 1971, 1974). The data given are typical for air over areas which
are relatively unaffected by urban complexes. In contrast to these data, typical
concentrations of gases in air over large cities, as quoted by Cadle and Allen (1970)
and Ludwig *et al.* (1971), are usually higher than 'background' concentrations (see
Table 8-3).

TABLE 8-3

Typical concentration of trace gases in photochemical smog. (From Cadle and Allen,
1970; by courtesy of the authors; copyright by the American Association for the
Advancement of Science.)

Constituent	Concentration ppm (vol.)	Constituent	Concentration ppm (vol.)
Oxides of nitrogen	0.20	CH_4	2.50
NH_3	0.02	Higher paraffins	0.25
H_2	0.50	C_2H_4	0.50
H_2O	2×10^4	Higher olefins	0.25
CO	40	C_2H_2	0.25
CO_2	400	C_6H_6	0.10
O_3	0.50	Aldehydes	0.60
		SO_2	0.20

8.2 Atmospheric Aerosol Particles (AP)

Aerosol particles (AP) in the atmosphere have sizes which range from that of clusters of a few molecules to $100\,\mu m$ and larger. It is useful to follow the suggestion of Junge (1955, 1963a) and divide the AP into 3 size categories. Particles with dry radii $r < 0.1\,\mu m$ are called *Aitken particles*, particles with dry radii $0.1 \le r \le 1.0\,\mu m$ are called *large particles*, and particles with dry radii $r > 1.0\,\mu m$ are called *giant particles*.

Particles are injected into the atmosphere from natural and anthropogenic or man-made sources. Most come from the Earth's surface, but some arise from the interior through volcanic action, while others enter the atmosphere from space. The concentration of AP varies greatly with time and location, and depends strongly on the proximity of sources, on the rate of emission, on the strength of convective and turbulent diffusive transfer rates, on the efficiency of the various removal mechanisms (see Chapter 12), and on the meteorological parameters which affect the vertical and horizontal distributions as well as the removal mechanisms. Observations confirm that the concentration of AP decreases with increasing distance from the Earth's surface. This is expected from the atmospheric density profile, and also because the surface constitutes the major source of AP, while removal mechanisms operate continuously throughout the atmosphere. In fact, it is estimated that 80% of the total aerosol particle mass is contained below the lowest kilometer of the troposphere. The AP concentration also decreases with increasing horizontal distance from the seashore towards the open ocean, because the land is a more efficient source of particles than the ocean. Thus, it is estimated that 61% of the total AP is introduced in the Northern Hemisphere, as compared to the Southern Hemisphere which is covered with a smaller land mass. Within the Northern Hemisphere, most of the aerosol particle mass enters the atmosphere between 30 °N and 60 °N, since this latitude belt contains about 88% of all anthropogenic sources for particulates.

The removal rate of aerosol particles by self-coagulation is proportional to the square of the particle concentration, while the removal by interaction with cloud drops and raindrops is proportional to the first power of the particle concentration (see Chapter 12). Consequently, the removal rate of AP may become very small if their concentration is sufficiently small. Indeed, Junge (1957b, 1963a) and Junge and Abel (1965) have demonstrated the existence of a rather stable background AP population of a few hundred particles cm^{-3}. This fairly uniform atmospheric background aerosol exists over land at heights above about 5 km, and over the oceans far from shore above about 3 km (Junge, 1969a).

Aerosol particles of terrestrial origin are formed by two major mechanisms: gas-to-particle conversion, and mechanical and chemical disintegration of the solid and liquid Earth surface. Let us now discuss these two modes of particle formation.

8.2.1 FORMATION OF AP BY GAS-TO-PARTICLE CONVERSION (GPC)

Plant exhalations and combustion products generally include vapors which have low boiling point temperatures. These vapors readily condense to drops or directly to solid particles relatively close to their source. Some substances typically involved in this mechanism are soot, tars, resins, oils, sulfuric acid, sulfates, carbonates, and others. Most of these substances are the result of industrial operations and man-made or natural fires. AP formed in this manner cover a wide range of sizes, but the majority lay within the Aitken particle size range. Detailed discussions of this mode of AP formation are given by Dunham (1966), Sutugin and Fuchs (1968, 1970), and Sutugin *et al.* (1971).

GPC may also be the result of chemical reactions between various gaseous substances. Many of these reactions are catalyzed by the ultraviolet portion of the Sun's radiation. These effects have been discussed by Briccard *et al.* (1968, 1971), Mohnen and Lodge (1969), Vohra *et al.* (1969), Cox and Penkett (1970), Vohra and Nair (1970), Vohra *et al.* (1970), and Mohnen (1970, 1971).

Several reactions which lead to a GPC are of interest to cloud physics, since they result in the formation of particles which later may become involved in the formation of cloud drops. Robbins *et al.* (1959) and Vohra *et al.* (1970) suggested that nitric acid and, subsequently, sodium nitrate are formed by the reactions

$$2NO + O_2 \rightarrow 2NO_2, \tag{8-1}$$

$$2NO_2 + H_2O \rightarrow HNO_2 + HNO_3, \tag{8-2}$$

$$HNO_3 + NaCl \rightarrow NaNO_3 + HCl^\uparrow, \tag{8-3}$$

if NaCl is present in the form of sea salt particles.

Cadle and Powers (1966) suggested that, in polluted air, sulfur trioxide (SO_3) may be produced by three body collisions of the type

$$SO_2 + O + M \rightarrow SO_3 + M, \tag{8-4}$$

where M is a chemically neutral gas molecule (e.g. N_2) which must be present for the reaction to proceed. Alternately, SO_3 may be formed in polluted air containing ozone by the reaction (Junge, 1963a)

$$SO_2 + O_3 \rightarrow SO_3 + O_2. \tag{8-5}$$

Gerhard and Johnstone (1955) proposed that SO_3 may also be the result of the following chain of reactions proceeding at the Earth's surface in bright sunlight:

$$SO_2 \xrightarrow{h\nu} SO_2^*, \tag{8-6}$$

$$SO_2^* + M \rightarrow SO_2 + M, \quad SO_2^* + O_2 \rightarrow SO_4, \tag{8-7}$$

$$SO_4 + SO_2 \rightarrow 2SO_3, \quad SO_4 + O_2 \rightarrow SO_3 + O_3, \tag{8-8}$$

where SO_2^* is an activated state of SO_2. Once formed, SO_3 quickly hydrates to sulfuric acid according to the reaction

$$SO_3 + H_2O \rightarrow H_2SO_4. \tag{8-9}$$

Then, if sea salt particles are present, sodium sulfate is formed according to

$$H_2SO_4 + 2NaCl \rightarrow Na_2SO_4 + HCl^\uparrow, \tag{8-10}$$

as suggested by Erikson (1959).

There is evidence that a significant portion of the atmospheric sulfate particle load is formed by chemical reactions which take place inside cloud drops. Radke and Hobbs (1969), Saxena *et al.* (1970), Dinger *et al.* (1970), Radke (1970), and Hobbs (1971) have observed that the concentration of AP which participate in the formation of cloud drops is considerably larger in the immediate vicinity of evaporating clouds than in air farther away. This increased concentration was attributed to the formation of sulfates inside cloud drops. Chemical reactions leading to the formation of sulfates in cloud drops have been studied by Junge and Ryan (1958), van den Heuvel and Mason (1963), Schmidtkunz (1963), Scott and Hobbs (1967), McKay (1971), Easter and Hobbs (1974), Barrie (1975), and Barrie and Georgii (1976). In a system where SO_2 is in equilibrium with distilled water the following reactions occur:

$$SO_2 + H_2O \rightleftarrows SO_2 \cdot H_2O, \tag{8-11}$$

$$SO_2 \cdot H_2O \rightleftarrows H^+ + HSO_3^-, \tag{8-12}$$

$$HSO_3^- \rightleftarrows H^+ + SO_3^{2-}, \tag{8-13}$$

where the dot between SO_2 and H_2O indicates that SO_2 is in the liquid phase of water. The concentration of the sulfite ion (SO_3^{2-}) is controlled by the acidity (i.e. the pH) of the solution. If, simultaneously, heavy metal ions (Me^{n+}) such as Cu^{2+}, Fe^{2+} or Mn^{2+} are present in the solution, the rate of oxidation of sulfide (SO_3^{2-}) to sulfate (SO_4^{2-}) is determined by the concentration of SO_3^{2-} in solution, by the type of metal ions present, and by the presence of molecularly dissolved oxygen, as shown by the studies of Junge and Ryan (1958), Schmidtkunz (1963), and Barrie and Georgii (1976). The formation of sulfate occurs then by a series of reactions, involving O_2, Me^{n+} and the radical O_2H, which may be summarized by the equation

$$SO_3^{2-} \xrightarrow{\quad O_2, O_2H, Me^{n+} \quad} SO_4^{2-}. \tag{8-14}$$

The studies of Junge and Ryan (1958), Scott and Hobbs (1967), McKay (1971), and Easter and Hobbs (1974) suggest that ammonia (NH_3) also acts as a catalyst speeding up the transformation of SO_2 to SO_4^{2-} in water. Scott and Hobbs showed that in the presence of CO_2 the following reactions must be considered in addition to (8-11) through (8-13)

$$H_2O \rightleftarrows H^+ + OH^-, \tag{8-15}$$

$$NH_3 + H_2O \rightleftarrows NH_3 \cdot H_2O \rightleftarrows NH_4^+ + OH^- \tag{8-16}$$

$$CO_2 + H_2O \rightleftarrows CO_2 \cdot H_2O \rightleftarrows H^+ + HCO_3^- \rightleftarrows 2H^+ + CO_3^{2-}. \tag{8-17}$$

$$SO_3^{2-} \xrightarrow{\quad O_2 \quad} SO_4^{2-}. \tag{8-18}$$

According to Scott and Hobbs, the rate of conversion of SO_2 to sulfate in solution, containing a sufficient amount of molecularly dissolved oxygen, is determined by the concentration of SO_3^{2-} and ammonia.

In the presence of NH_4^+ and SO_4^{2-} ammonium sulfate, $(NH_4)_2SO_4$, forms in aqueous solution according to

$$2NH_4^+ + SO_4^{2-} \rightleftarrows (NH_4)_2SO_4. \tag{8-19}$$

Some of the reactions which form particulates by a GPC lead to what is commonly termed 'smog.' *Coal-burning smog* is mainly caused by particulates formed by condensation of low boiling point vapors, and by ashes and similar substances which are the result of heat disintegration of organic substances. Natural fires emit similar particles. In contrast, *photochemical smog* is caused by particles which are the result of a chain of reactions, often initiated by the formation of ozone according to

$$NO_2 \xrightarrow{h\nu} NO + O, \tag{8-20}$$

$$O + O_2 + M \rightarrow O_3 + M, \tag{8-21}$$

which proceeds in bright sunlight near the Earth's surface. Once ozone is present, complicated chemical reactions evolve whereby O and O_3 oxidize the hydrocarbons emitted by the various combustion mechanisms which are typical for urban areas (see Leighton, 1961; Haagen-Smit and Wayne, 1968; Cadle, 1973).

GPC is also largely responsible for the formation of AP in the stratosphere. Over the past few years several chemical mechanisms have been suggested to account for the formation of sulfate particles in the stratosphere. Thus, (8-5) and (8-6) to (8-8) may be possible mechanisms. However, experiments show that these reactions are too slow to account for the formation of sulfur particles in the stratosphere, despite the relatively large amounts of ozone and ultraviolet radiation present at that level. Cadle (1965), Cadle and Powers (1966), and Friend *et al.* (1973) suggested the first step in the formation of sulfate particles is the trimolecular process (8-4), in the oxygen atoms having been provided by photodissociation of oxygen molecules. According to Friend *et al.* (1973), these reactions are followed by a bimolecular collision of the kind given in (8-9). Through the presence of water vapor embryonic nuclei of sulfuric acid $H_2SO_4 \cdot nH_2O$ are then formed through a step by step addition of individual water molecules. In the presence of small amounts of ammonia (NH_3) the acid embryos become neutralized to form salt embryos of composition $NH_4^+HSO_4^- \cdot nH_2O$ and $(NH_4)_2SO_4^{2-} \cdot nH_2O$. It is thought that within these embryonic salt solution nuclei rapid catalytic oxidation of SO_2 to H_2SO_4 takes place, with NH_4^+ acting as the catalyst. The reaction taking place in the salt solution embryo was summarized by Friend *et al.* as

$$2SO_2 + 2H_2O + O_2 \xrightarrow[\text{salt embryo}]{NH_4^+} 2H_2SO_4 \text{ (solution)}, \tag{8-22}$$

where the individual steps in this reaction are the same as those outlined by Scott and Hobbs (1967) (see Equations (8-15) to (8-17) and (8-19)).

Another group of reactions leading to sulfate particles was suggested by Scott *et al.* (1969), Hartley and Matteson (1975), and Vance and Peters (1976). Based on experiments they proposed the formation of two intermediate compounds according to

$$NH_3 + SO_2 \rightleftarrows NH_3 \cdot SO_2, \quad NH_3 \cdot SO_2 + NH_3 \rightleftarrows (NH_3)_2 \cdot SO_2, \qquad (8\text{-}23)$$

which in the presence of traces of water vapor and molecular oxygen may transform into $(NH_4)_2SO_4$.

Davis (1973) has proposed that the following trimolecular reaction may also be important for sulfate formation in the stratosphere:

$$HO + SO_2 + M \rightarrow HSO_3 + M. \qquad (8\text{-}24)$$

Alternatively, Davis considered the bimolecular reactions

$$HO_2 + SO_2 \rightarrow HO + SO_3, \quad NO_3 + SO_2 \rightarrow NO_2 + SO_3. \qquad (8\text{-}25)$$

In reactions (8-25) it is assumed that the resulting sulfur trioxide quickly hydrates to form sulfuric acid aerosol particles, some of which become neutralized by ambient ammonia. The significance of reactions (8-23) to (8-25) as a sulfate forming mechanism was stressed by Harrison and Larson (1974) and Harker (1975), who showed by means of a one-dimensional diffusion model that reactions involving HO, HO_2, and SO_2 may indeed be sufficiently fast and efficient to explain the formation of a sulfate layer above the tropopause.

8.2.2 FORMATION OF AP BY MECHANICAL AND CHEMICAL DISINTEGRATION
      AND DISPERSAL AT THE SOLID EARTH SURFACE

It is well known that plants release various types of organic particulates, such as pollen, seeds, waxes, and spores which are distributed by air motions throughout the atmosphere (Gregory, 1961). A detailed study of the seasonal variation of pollen and spores was made by Grosse and Stix (1968), Stix (1969), and Stix and Grosse (1970). Diameters of these particles were found to range typically between 3 and 150 μm. It has also been shown that the atmosphere harbors large collections of microbial bodies, both living and dead (Gregory, 1967; Valencia, 1967; Parker, 1968). Lodge and Pate (1966) reported that substantial amounts of organic materials rise from tropical forest floors into the atmosphere. They concluded that these particles were produced by aerobic bacterial decay of tree leaf litter. In related work, Parkin *et al.* (1972) found humus from vegetation, dark plant debris, and fungus debris on aerosol particles which were carried in the westerlies over the Atlantic from the eastern United States. Rasmussen and Went (1965) and Went *et al.* (1967) captured organic particles which originated from decaying mid-latitude forest litter. Went's studies showed that the number concentration of organic particles was largest during periods of rapid plant litter decay. A variety of organic compounds were also identified in snow and rain by Shutt (1907), Fonselius (1954), and Munzah (1960).

An extensive field study of the organic constituents of the atmospheric aerosol was conducted by Kesteridis *et al.* (1976). Aerosol samples were collected in air over three regions of the European continent with different degrees of air pollution, and in three air mass types over the Atlantic ocean. The ether soluble fraction of the sampled aerosol particles was separated into five groups of organic compounds: organic acids, organic bases, aliphatic hydrocarbons, aromatic hydrocarbons, and neutral compounds. In all analyzed samples the organic fraction of the atmospheric aerosol had a very complex composition. However, the relative composition with respect to the above mentioned main groups of organic compounds remained fairly constant. In addition, the concentration of organic matter in the background aerosol was rather constant with an average magnitude of about 1 μg m^{-3} (STP). Kesteridis et al. suggested that their observations may be interpreted in either of two ways: (1) Most of the organic atmospheric aerosol matter is formed over highly polluted continental areas and remains essentially unaltered as it becomes dispersed in the atmosphere, implying that there is no substantial formation of organic particulates in background air. (2) Organic aerosol matter is formed not only over continental areas but also in background air outside these areas, e.g., by GPC involving organic trace gases, implying that the general composition of the organic aerosol material does not differ significantly from that found in polluted areas over the continents. No preference to either of these two interpretations was given.

A significant portion of the Earth's surface is covered by rocks or soil devoid of vegetation. The exposed silicate compounds are chemically and mechanically disintegrated by the combined action of wind, water, temperature variations, and gases such as oxygen, carbon dioxide, and others. This weathering forms particles which have diameters mostly larger than 0.1 μm. The loose silicate material, usually with considerable amounts of organic material attached, is then transported upward by air motions. Clays, which are layer silicates consisting mostly of SiO_2, Al_2O_3, Fe_2O_3, and MgO, are most easily disrupted by weathering. Frequently they are present as kaolinite, montmorillonite, illite, attapulgite, halloysite, and vermiculite (for details on the chemical composition and special properties of clays, see, e.g., Grim, 1953).

Airborne silicates emitted by the Sahara desert have been identified by Prospero and Bonatt (1969), Prospero et al. (1970), Chester and Johnson (1971), Jaenicke et al. (1971), and Parkin et al. (1972) in air over the Atlantic; by Delany et al. (1967) and Prospero (1968) in air over the Isles of Barbado; by Abel et al. (1969) in air over the Island of Teneriffa; and by Rex and Goldberg (1958) and Ferguson et al. (1970) in air over the Pacific and Indian Oceans. The Saharan dust was found to occur preferentially at atmospheric levels between 1.5 and 3.7 km (Prospero and Carlson, 1972). The size of the silicate particles ranged typically between 0.3 and 20 μm radius, with a mode (most probable radius) near 2 to 5 μm. Deserts and semi-arid regions in North China and Mongolia are another significant source of silicates. Isono et al. (1959) identified clay particles in air over Japan which originated in North China or Mongolia, where they became airborne during large dust storms. Isono et al. (1970) have provided evidence that some of these clay minerals are transported by the upper level westerlies across the Pacific and deposited over the northwestern coast of the U.S.

The mechanism by which clays and other soil or sand particles become airborne has been investigated by Chepil (1951, 1957, 1965), Chepil and Woodruff (1957, 1963), Owen (1964), and Gillette *et al.* (1974). It is well known that adjacent to a smooth surface and at wind speeds below some critical value a laminar boundary layer exists, even if the air flow is otherwise turbulent. On the other hand, if the surface is rough due to the presence of irregular soil and sand particles, turbulent motion may prevail right down to the surface. Such turbulent flow can cause a rough soil or sand surface to be eroded either by direct aerodynamic pick-up of the particles, or as a result of the bombardment by particles performing a jumping motion called *saltation*. Conditions for saltation derived by Owen (1964) are summarized in Figure 8-1. During saltation individual grains follow distinctive trajectories determined by the air resistivity and gravity. The air in the saltation layer is strongly sheared. The lift force responsible for the particles' saltation ensues from the combined action of the particles' momenta and the environmental vorticity.

According to observations by Chepil (1951), the minimum threshold wind velocity for direct aerodynamic pick-up of soil particles is a strong function of particle size. If particles were entrained by aerodynamic pick-up into the air layers adjoining the surface, one would expect greatly differing particle size distributions with height, owing to gravitational sorting and the effects of vertical wind shear. Such a dependence was not found by Gillette *et al.* (1974), who experimentally studied the size distribution of particles from soil surfaces eroded by wind. Therefore, they concluded that sandblasting (saltation erosion) of the soil surface is the dominant mechanism by which particles become airborne. This conclusion is supported also by the observations of Owen (1964) and Bagnold (1965), who found that at wind speeds above about 50 cm sec^{-1} even very large particles become airborne. The recent observations of Schütz and Jaenicke (1974) in the Sahara desert provide further support for these views.

After becoming airborne, silicate particles readily coagulate with other AP, thus becoming of mixed composition. It is therefore not surprising that the bulk density of continental aerosol material varies over a range which may differ considerably from the bulk density of SiO_2 (quartz), which is $\rho(SiO_2) = 2.65$ to 2.66 g cm^{-3}. Thus, Hänel and Thudium (1977) found, for $0.05 \leq r \leq 5$ μm, $\rho = 2.6$ to 2.7 g cm^{-3} (desert, Israel), $\rho = 1.8$ to 3.5 g cm^{-3} (urban, Mainz), $\rho = 2.9$ g cm^{-3}

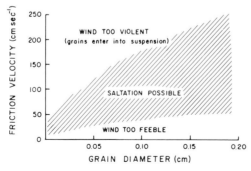

Fig. 8-1. Range of wind speed and grain size in which saltation of quartz grains can occur in the atmosphere. (From Owen, 1964; by permission of Cambridge University Press.)

(Jungfraujoch, Switzerland), and $\rho = 1.8$ to $3.3\,\mathrm{g\,cm^{-3}}$ (rural, Deuselbach, Germany).

Particles emitted by volcanoes also are often the result of a combination of both mechanical disintegration and gas-to-particle conversion. For example, during the eruptions of Krakatoa in 1883 in the East Indies and Gunung Agung in Bali in 1963, some of the emitted particles consisted simply of silicates from the crater walls, others consisted of finely divided solidified lava, while still others consisted of sulfates, halides, and sulfuric acid. The particles, which ranged from submicron size to greater than $100\,\mu\mathrm{m}$, were injected into both the troposphere and stratosphere.

Particles injected into the atmosphere by industrial processes are also often the result of mechanical disintegration and gas-to-particle conversion. Hobbs and Radke (1970) found, downstream of paper mill exhausts, particles of Na_2SO_4, NH_4HSO_3, $Ca(HSO_3)_2$, NaOH, Na_2SO_3, and H_2SO_4 which ranged from submicron size to several hundred microns, with a mode from 1 to $10\,\mu\mathrm{m}$ diameter. Serpolay (1958, 1959) and Soulage (1961) found a large number of metal and metal oxide particles downstream of steel foundries and electric steel mills.

8.2.3   FORMATION OF AP BY MECHANICAL DISINTEGRATION AND DISPERSAL AT THE SURFACE OF OCEANS

Wind blowing across an ocean surface causes the formation of waves which produce spray drops at their crests. The finer of these drops remain airborne and eventually evaporate to give solid AP. A more important source of AP results from the bursting of bubbles produced by the entrainment of air at the wave crests.

Experimental studies of Woodcock *et al.* (1953), Kientzler *et al.* (1954), Knelman *et al.* (1954), Mason (1954b), Moore and Mason (1954), and Blanchard (1954) show that each air bubble which reaches the ocean surface develops a spherical cap which strains, thins, and then bursts. After the bubble cap has burst, fragments of the cap-film are thrown upward by the air which escapes from the bubble orifice. Now deprived of its cap, the bubble fills with water rushing down the sides of the cavity, which subsequently emerges from the center as a narrow jet. As the jet rises it becomes unstable and eventually disintegrates into a few large and several small drops. Bubbles of 2 mm diameter project drops up to heights of nearly 18 cm above the ocean surface; drops from both larger and smaller bubbles generally reach lower heights (Blanchard, 1963). Depending on the relative humidity and turbulence of the air, some of the drops formed by the collapsing jet and by the shattered bubble cap fall back to the ocean surface. The remainder evaporate while airborne, leaving a small sea salt particle light enough to be carried aloft by air motions. The different stages in the production of sea salt particles by this *bubble-burst mechanism* are described schematically in Figure 8-2.

A sea salt particle is mainly composed of NaCl, the most abundant salt in ocean water. In addition, it often contains small amounts of CO_3^{2-}, SO_4^{2-}, K^+,

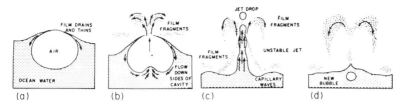

Fig. 8-2. Four stages in the production of sea salt particles by the bubble-burst mechanism. (a) Film cap protrudes from the ocean surface and begins to thin. (b) Flow down the sides of the cavity thins the film which eventually ruptures into many small fragments. (c) Unstable jet breaks into few drops. (d) Tiny salt particles remain, as drops evaporate; new bubble is formed. (From Day, 1965, with changes.)

Mg^{2+}, and Ca^{2+}. However, the chemical composition of sea salt particles deviates considerably from the composition of ocean water. Junge (1972b) suggested that this is partly a result of ion fractionation occurring during the bubble burst mechanism. Since organic materials often reside in the ocean surface they, too, may become airborne and become part of the sea salt particles (Blanchard, 1964, 1968; Garret, 1965, 1969). Subsequent to its formation, a sea salt particle may change its composition further as a result of both chemical reactions with atmospheric trace gases and coagulation with other AP in the atmosphere. Thus, in air over the Atlantic near the Sahara desert, sea salt particles may have silicates admixed. It is therefore not surprising that the bulk density of maritime aerosol material varies over a range which may differ considerably from the bulk density of the salt that crystallizes from evaporating ocean water. Such salt has a bulk density of $2.25 \, g \, cm^{-3}$. In contrast, Hänel and Thudium (1977) observed $\rho = 1.93 \, g \, cm^{-3}$ (W. coast, Ireland), and Fisher and Hänel (1972) observed $\rho = 2.45$ to $2.64 \, g \, cm^{-3}$ (N. Atlantic, near Sahara). Note that $\rho(NaCl) = 2.165 \, g \, cm^{-3}$.

During the bursting of a bubble, the jet-break up produces 1 to 5 larger drops with diameters of about 1/10 the diameter of the parent bubble. Assuming that the density of a sea salt particle is $2.25 \, g \, cm^{-3}$, that the density of ocean water is $1.03 \, g \, cm^{-3}$, and that its salinity is $35 \, g \, kg^{-1}$, we find a bubble of 2 mm diameter produces a salt particle of $1.5 \times 10^{-7} \, g$, which is equivalent to a dry radius of about $25 \, \mu m$. Similarly, bubbles of $100 \, \mu m$ and $20 \, \mu m$ diameter produce salt particles of $1.9 \times 10^{-11} \, g$ and $1.5 \times 10^{-13} \, g$, which are equivalent to dry radii of about $1.3 \, \mu m$ and $0.3 \, \mu m$, respectively. In addition to these large drops, jet-break up bubbles of diameters larger than 2 mm also eject a group of smaller drops at low angles to the horizontal. These drops have diameters of 5 to $30 \, \mu m$ and produce sea salt particles of masses $2.4 \times 10^{-12} \, g$ to $5.1 \times 10^{-10} \, g$, corresponding to dry radii of 0.6 to $3.8 \, \mu m$. There has been some speculation that salt particles of even smaller size are produced by the splintering of the drying remnants of solution drops (Mason, 1971).

The bursting of the bubble cap also produces a large number of small salt particles. Mason (1954b, 1957b) found that most of the salt particles from the bubble cap have masses less than $2 \times 10^{-14} \, g$, equivalent to a dry radius of less than about $0.1 \, \mu m$. The largest particles had masses up to $2 \times 10^{-13} \, g$, equivalent to a dry radius of about $0.3 \, \mu m$, while the smallest particles had masses as low as $10^{-15} \, g$, equivalent to a dry radius of 0.07 μm. These results were essentially confirmed by Twomey (1960). Blanchard (1963) suggested, and Day (1964) confirmed, that the number of bubble cap droplets decreases with decreasing bubble size, and that bubbles of diameter smaller than 100 μm produce no cap drops (see Figure 8-3). Thus bubbles smaller than 100 μm in diameter produce sea salt particles only as a result of jet breakup. However, there also appears to be a lower size limit to air bubble production. According to Woodcock (1972) and Woodcock and Duce (1972), air bubbles smaller than 20 μm in diameter are unlikely to exist near the surface, since bubbles of such sizes rise extremely slowly in ocean water, providing sufficient time for them to be dissolved. Consequently, sea salt particles produced by jet droplets necessarily have masses larger than about 10^{-13} g.

According to Blanchard and Woodcock (1957), the number of bubbles of radii R to R + dR bursting per cm^2 per sec in a foam patch on the ocean surface is given approximately by 3×10^{-6} (dR/R^4). Assuming that each bubble produces one jet drop which remains airborne, the rate of jet particle production in cm^{-2} sec^{-1} by bubbles of radii larger than R$_0$ is then $3 \times 10^{-6} \int_{R_0}^{\infty} dR/R^4 = 1 \times 10^{-6}$ (1/R$_0^3$). One may assume this relation is applicable to bubble radii larger than 50 μm. Day (1964) determined that for bubbles of radii larger than 50 μm the number of cap drops per bubble varies with bubble size as 10^4 R^2. Mason (1971) deduced from this information that the rate of cap particle production (cm^{-2} sec^{-1}) by bubbles of radii larger than R$_0$ can be described by $3 \times 10^{-2} \int_{R_0}^{\infty} dR/R^2 = 3 \times 10^{-2}$ (1/R$_0$), which is applicable to bubbles larger than 50 μm radius. Note that these results only apply to drop production by single bubbles. Experiments by Mason (1957b) and Twomey (1960) suggest that the rate of cap drop production is considerably larger if bubbles break in clusters.

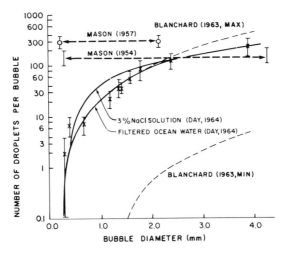

Fig. 8-3. Mean number of droplets resulting from the disintegration of an air bubble cap as a function of equivalent diameter of the air bubble. (From Day, 1964; by courtesy of *Quart. J. Roy. Meteor. Soc.*)

Observations by Moore and Mason (1954) at a height of 10 m over the Atlantic showed that sea salt particles of mass $m_s > 10^{-13}$ g are formed at a rate of $40 \text{ cm}^{-2} \sec^{-1}$, particles of $m_s > 2 \times 10^{-14}$ g at a rate of $100 \text{ cm}^{-3} \sec^{-1}$, and particles of $m_s > 10^{-15}$ g at a rate of $1000 \text{ cm}^{-2} \sec^{-1}$, over areas where bubbles are bursting. Blanchard (1969) found, off the Hawaiian coast, the somewhat larger total particle number of $3000 \text{ cm}^{-2} \sec^{-1}$. Using photographic observations of the state of the ocean surface at different times during a whole year, Blanchard (1963) estimated that, on a global average, 3.4% of the Earth's oceans are covered with breaking bubbles. Although Monahan (1968, 1969, 1971) and Williams (1970) argued that this figure was too large, Blanchard (1971) upheld his original estimate by pointing out that the bubble-burst mechanism operates not only in the areas covered by white-caps, but also in the areas immediately adjacent to them. Thus, Blanchard (1969) suggested that the average sea salt particle production is $0.034 \times 3000 \text{ cm}^{-2} \sec^{-1} \approx 100 \text{ cm}^{-2} \sec^{-1}$, corresponding to a global production rate of about 10^{28} particles yr^{-1}. Assuming that on the average an airborne sea salt particle has a mass of 3×10^{-14} g, one estimates $3 \times 10^{11} \text{ kg yr}^{-1}$ for the global production rate of sea salt. This is in agreement with estimates by Erikson (1959), and by Peterson and Junge (1971) as revised in SMIC (1971).

Observations show that the ocean surface is not only a source of inorganic salts, but of organic materials as well. Wilson (1959), Blanchard (1964, 1968), Goetz (1965), and Garret (1967, 1970) found organic substances present in the ocean surface which become airborne through the bubble-burst mechanism. Goetz (1965) and Blanchard (1969) captured airborne, submicron-sized, organic particles which originated at the ocean surface. Zoebell and Matthews (1936) and Stevenson and Collier (1962) showed that air above oceans contained numerous micro-organisms indigenous to marine water. Studies of organic compounds in snow and rain also suggest a marine source for the materials (Wilson, 1959; Newmann *et al.*, 1959).

8.2.4   AP FROM EXTRATERRESTRIAL SOURCES

Extraterrestrial particles continuously enter the Earth's atmosphere at speeds great enough to produce strong frictional heating. The resulting light phenomenon is called a *meteor*, and the particle itself is called a *meteoroid*. The vast majority of meteoroids are believed to be permanent members of the solar system. Meteoroids produce meteors at an average height of about 95 km, and nearly all of them completely disintegrate and their meteors disappear by the time they have reached an altitude of about 80 km. The small fraction which do survive the fall to Earth are termed *meteorites*. Some of these are quite large, one of the largest on display in a single piece having a mass of 14 tons. This contrasts with an average meteoroid mass of a few grams or less.

Most meteorites are very small, having been nearly consumed before slowing sufficiently to reach temperatures below the burning point; such particles are called *micrometeorites*. Most of these are derived from a large pool of inter-planetary dust which is concentrated in a lens-shaped volume located about the plane of the ecliptic. These particles move around the Sun in orbits similar to that of the Earth, and enter the Earth's atmosphere with small geocentric velocities. Their diameters range between about $1 \mu m$ and $1000 \mu m$. Particles which result from the condensation of evaporated meteoroids are considerably smaller, ranging between $1 \mu m$ and a few ångströms (Rosinski and Snow, 1961).

Chemically, meteorites are divided into four main groups: irons (siderites), stony irons (siderolites), stones with small spheroidal aggregates (chondrites), and stones without such aggregates (achondrites). Common minerals in meteorites include kamacite, taenite, troilite, olivin, orthopyroxene, pigeonite, diopside, and plagioclase.

8.2.5 RATE OF EMISSION OF PARTICULATE MATTER INTO THE ATMOSPHERE

Attempts to estimate the rate at which particulate matter is injected from the Earth's surface or from space into the atmosphere have encountered considerable observational difficulties. Similar difficulties are involved in estimating the rate at which aerosol particles are produced inside the atmosphere. The estimates of Robinson and Robbins (1971) and of Peterson and Junge (1971) are summarized in the form of a table in SMIC (1971), which is reproduced here as Table 8-4. Although the values given in this table are only rough estimates, they

TABLE 8-4

Rate ($10^9 \, kg \, yr^{-1}$) at which aerosol particles of radius less than $20 \mu m$ are produced in, or emitted into, the atmosphere. (Reprinted from *Inadvertent Climate Modification* (SMIC Report), 1971, by permission of MIT Press, Cambridge, Mass.)

NATURAL PARTICLES	
Soil and rock debris	100–500
Forest fires and slash-burning debris	3–150
Sea salt	300
Volcanic debris	25–150
Gas to particle conversion	
sulfate from H_2S	130–200
ammonium salts from NH_3	80–270
nitrate from NO_x	60–430
hydrocarbons from plant exhalations	75–200
Subtotal	773–2200
ANTHROPOGENIC PARTICLES	
Particles by direct emission	10–90
Gas to particle conversion	
sulfate from SO_2	130–200
nitrate from NO_x	30–35
hydrocarbons	15–90
Subtotal	185–415
Total	958–2615
Extraterrestrial	0.5–50

testify to the rather large amounts of material which enter the atmosphere. The data also suggest that the man-made contribution to the total particulate load in the atmosphere ranges between 7 and 43%.

8.2.6 RESIDENCE TIME (τ_{AP}) OF AP

It is considerably more difficult to estimate residence times for aerosol particles than for gaseous constituents of the atmosphere. The reason for the additional difficulty lies mainly in the fact that aerosol particles, particularly those at the smaller end of the size spectrum, undergo a continuous change in size and composition, and thus lose their identity, as a result of coagulating with other particles. Consequently, the term 'residence time' does not apply to individual aerosol particles, but rather to the suspended particulate matter in its entirety. It may also apply to one specific substance suspended in air as particulates, such as silicate particles or sea salt particles.

The results of estimates made prior to 1970 have been summarized in SCEP (1970). More recent measurements have been obtained by using Pb^{210}, Po^{210}, Bi^{210}, Sr^{90}, Cs^{137}, and HTO isotopes (Martell, 1970, 1971; Poet et al., 1972; Moore et al., 1973; Ehhalt, 1973; Weickmann and Pueschel, 1973; Martell and Moore, 1974; Tsunogai and Fukuda, 1974). Estimates for τ_{AP} made before and after 1970 are compared in Table 8-5, where it can be seen that recent data tend toward smaller values than the earlier estimates. Generally, however, the residence time of AP can be seen to increase with height in the atmosphere. It ranges from a very few days in the lower troposphere to a few weeks in the upper troposphere, to several weeks, months, and even years at increasingly higher levels above the tropopause.

The residence time of sea salt in the atmosphere is of special cloud physical interest, since these particles play an important role in the formation of cloud drops. Junge (1972b) showed that unrealistic values for $\tau_{sea\,salt}$ are obtained if sedimentation is the only removal mechanism considered. He suggested that sea salt particles have their major sink at the cloud level. Using his own estimates

TABLE 8-5

Residence time of atmospheric aerosol particles at various levels in the atmosphere.

Level in the Atmosphere	τ_{AP}	
	Based on evidence prior to 1970	Based on evidence after 1970
Below about 1.5 km	–	0.5 to 2 days
Lower Troposphere	6 days to 2 weeks	2 days to 1 week
Middle and upper Troposphere	2 weeks to 1 month	1 to 2 weeks
Tropopause level	–	3 weeks to 1 month
Lower Stratosphere	6 months to 2 years	1 to 2 months
Upper Stratosphere	2 years to 5 years	1 to 2 years
Lower Mesosphere	5 to 10 years	4 to 20 years

and those of Erikson (1959), Junge suggested $\tau_{sea\,salt} \approx 1$ to 3 days for $m_s > 10^{-9}$ g. We may combine this result with the earlier indicated estimate for the emission rate (j) of sea salt particles into the atmosphere of about 50 to 100 particles cm^{-2} sec^{-1}. Assuming that all sea salt particles are contained in an atmospheric layer of thickness $h = 3$ km, we find their average concentration is roughly $j\tau/h \approx 10$ to 99 cm^{-3}. We shall see that this range agrees well with observations (Section 8.2.9).

8.2.7 WATER-SOLUBLE FRACTION OF AP

Atmospheric AP have a wide range of water-solubilities. Compounds such as NaCl, $NaNO_3$, $(NH_4)_2SO_4$, NH_4NO_3, Na_2SO_4, and other salts typically found in the atmosphere are highly water-soluble, while substances such as silicates and metal oxides are practically water-insoluble. Substances such as $CaCO_3$ and $CaSO_4$ have a measurable but rather low solubility. In contrast to solid particulates, most gases exhibit measurable water-solubility. While SO_2, NO_2, NH_3, and CO_2 are highly soluble in water, where they dissociate into ions, other gases such as CO, N_2, and O_2 have a moderate solubility and are molecularly dissolved.

The water-solubility of any compound is temperature dependent. Most salts suspended as particulates in the atmosphere dissolve by means of an endothermic (heat-consuming) process, so that their solubility increases with increasing temperature. On the other hand, gases such as N_2, O_2, CO_2, H_2S, CO, NH_3, NO, and SO_2 dissolve by means of an exothermic (heat-releasing) process, causing their solubility to increase with decreasing temperature. It is worth noting also that the solubility of a gas increases as its partial pressure increases.

The solubility of a salt (given in terms of the maximum mass of salt which can be dissolved in a given mass of water) is not directly related to its hygroscopicity or vapor pressure reducing power (see Chapter 4), but is a function of the interaction energy between the water molecules and the salt ions in water, and of the lattice energy of the salt crystal. Therefore, the solubility of a salt does not predict the relative humidity at which a particle of that salt changes into a solution drop; nor, once dissolved, does it predict the effect which the salt has on the equilibrium growth behavior of that drop.

The increasing water-solubility of gases with decreasing temperature has an interesting cloud physical implication. Since the solubility of gases in ice is negligible, this implies that, upon freezing, the amount of gas released and trapped in the ice as gas bubbles will increase with decreasing temperature. This behavior, in turn, affects the bubble structure of frozen drops, rimed ice crystals, graupel particles, and hailstones in clouds.

Chemical analysis of AP shows that most individual AP are of a mixed chemical nature, and contain both water-soluble and water-insoluble substances. This fact was first pointed out by Junge (1950), who termed these AP *mixed particles* or *Mischkerne*. A mixed particle may be formed: (1) by condensation or adsorption of foreign gases onto the surface of AP; (2) by coagulation of AP with other AP; (3) by solution of gases in cloud and raindrops followed by chemical reaction with other dissolved substances or insoluble particles in the water, and subsequent evaporation of the drop; (4) by coagulation of cloud drops with other cloud drops containing substances of different chemical nature in dissolved form, and subsequent evaporation of the drops; and (5) as a result of the simultaneous condensation of vapors emitted during combustion processes. Most particles emitted into the atmosphere are also of a mixed type, such as sea salt particles, organic plant material, and soil derived particles.

According to Junge (1972c), one generally may assume that AP at the Earth's surface consist typically of about 50% water-soluble inorganic material, about 30% water-insoluble inorganic material, and about 20% organic material. A detailed analysis of the composition of an urban (Mainz, Germany) aerosol and a rural (Deuselbach, Germany) aerosol was performed by Winkler (1970). Winkler's results are summarized in Figure 8-4. Figure 8-4a shows that in the urban aerosol the water-soluble fraction of the dry aerosol deposit (of total mass m_0) was always between 30 and 90% of m_0. In half the cases the water-soluble fraction was more than 57% of m_0. Corresponding figures for the rural aerosol are 40%, 90%, and 75% of m_0, which shows that the rural aerosol contained considerably more water-soluble material than the urban aerosol. Figure 8-4 shows that the rural aerosol contained less soluble and less volatile material than the urban aerosol.

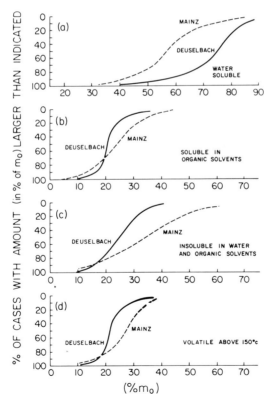

Fig. 8-4.   Composition of a rural and an urban aerosol, (a) water-soluble portion, (b) portion soluble in organic solvents, (c) insoluble portion, (d) volatile portion. (Based on the results of Winkler, 1970.)

Laktinov (1972) observed somewhat lower fractions of soluble material in AP. He collected AP of $0.15 \leqslant r \leqslant 0.3 \, \mu$m in air at 1500 to 4000 m altitude over the U.S.S.R., and found that soluble substances accounted for 20% of the AP mass if $r = 0.15 \, \mu$m, and for 10% of the mass if $r = 0.3 \, \mu$m. Similarly, Meszaros (1968), who collected AP near Budapest, found they contained only about 17% (by mass) soluble substances for $r \leqslant 0.14 \, \mu$m, about 7 to 8% if $r > 0.14 \, \mu$m, and only about 1 to 2% if $r > 0.3 \, \mu$m. On the other hand, Twomey (1965, 1972) concluded that AP collected at Chesapeake Bay (Maryland) and at Robertson (N.S.W., Australia) consisted almost entirely of water-soluble material, mostly $(NH_4)_2SO_4$.

8.2.8   TOTAL CONCENTRATION AND VERTICAL VARIATION OF AP OVER LAND

Table 8-6 summarizes typical values for the number concentration of AP of all sizes, irrespective of their chemical type. The total particle concentration in air over land generally ranges from 10^3 to $10^5 \, cm^{-3}$. In air over cities the concentration may even be as large as $10^6 \, cm^{-3}$, while in air over rural areas, near seashores, and at mountain stations the concentration is usually only a few thousand cm^{-3}. These trends are also evident in Table 8-7, which gives representative values determined by Ludwig *et al.* (1971) for the total mass concentration of AP along with the major chemical constituents of the particles. Values consistent with these have been obtained also by Robinson and Robbins (1971), who observed mass concentrations of 10 to 30 μg m^{-3} in air over rural areas, and by Cadle (1973) who measured mass concentrations which varied between 80 and 228 μg m^{-3} in air over some major U.S. cities.

Landsberg (1938) (see Table 8-6) found an average minimum AP concentration of 1050 cm^{-3} over rural areas, while Auer (1966) reported 1000 cm^{-3} at the remote Yellowstone Park (Wyoming). Lower values were found by Landsberg at mountain stations where the average minimum decreased to a few hundred cm^{-3}. Similarly, Junge *et al.* (1969) found at Crater Lake (Oregon, 2200 m) a total AP concentration of 700 cm^{-3} during subsidence conditions. However, under normal

TABLE 8-6

Number concentration (cm^{-3}) of atmospheric aerosol particles at different locations at the Earth surface. (Based on data of Landsberg, 1938.)

Locality	No. of places	No. of observations	Average concentrations	Average maximum	Average minimum
City	28	2500	147 000	379 000	49 100
Town	15	4 700	34 300	114 000	5 900
Country inland	25	3500	9 500	66 500	1 050
Country seashore	21	2700	9 500	33 400	1 560
Mountain:					
500–1000 m	13	870	6 000	36 000	1 390
1000–2000 m	16	1000	2 130	9 830	450
2000 m	25	190	950	5 300	−160
Islands	7	480	9 200	43 600	460

TABLE 8-7

Average mass concentration of atmospheric aerosol particles at different locations at the Earth's surface. (From Ludwig *et al.*, 1971; in *Man's Impact on Climate*, 1971, reprinted by permission of MIT Press, Cambridge, Mass.)

	Urban (217 Stations)		Nearby (5)		Nonurban Intermed. (15)		Remote (10)	
	$\mu g\ m^{-3}$	%	$\mu g\ m^{-3}$	%	$\mu g\ m^{-3}$	%	$\mu g\ m^{-3}$	%
Suspended particulates	102.0		45.0		40.0		21.0	
Benzene-soluble organics	6.7	6.6	2.5	5.6	2.2	5.4	1.1	5.1
Ammonium ion	0.9	0.9	1.22	2.7	0.28	0.7	0.15	0.7
Nitrate ion	2.4	2.4	1.40	3.1	0.85	2.1	0.46	2.2
Sulfate ion	10.1	9.9	10.0	22.2	5.29	13.1	2.51	11.8
Copper	0.16	0.15	0.16	0.36	0.078	0.19	0.060	0.28
Iron	1.43	1.38	0.56	1.24	0.27	0.67	0.15	0.71
Manganese	0.073	0.07	0.026	0.06	0.012	0.03	0.005	0.02
Nickel	0.017	0.02	0.008	0.02	0.004	0.01	0.002	0.01
Lead	1.11	1.07	0.21	0.47	0.096	0.24	0.022	0.10

conditions the concentration varied typically between 1000 and 2000 cm^{-3}. Bullrich *et al.* (1966) found a value of 600 cm^{-3} in air above the trade wind inversion over Hawaii on Mt. Haleakala (3050 m).

Detailed studies by Selezneva (1966) of the vertical variation of the total AP concentration (approximated by the predominant Aitken particle concentration) showed an exponential decrease for the lowest 6 km over various stations in the U.S.S.R. (Figure 8-5). A similar variation with height was found by Weickmann (1957b), Junge (1961, 1963b), and Junge *et al.* (1961a) in air over Germany, the eastern U.S., and India. These latter results are summarized in Figure 8-6. Note that, following the exponential decrease of the particle concentration in the lowest 5 km of the troposphere, the concentration in the upper

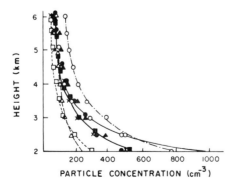

Fig. 8-5.   Variation of the number concentration of Aitken particles with height over various cities of the U.S.S.R. (From Selezneva, 1966; by courtesy of *Tellus.*)

troposphere appears to remain nearly constant at a value varying between 60 and 600 cm^{-3}, with most frequent values near 300 cm^{-3}. Thus, over land, tropospheric background particle concentrations seem to be present at heights above about 5 km.

The observations of Junge (1961, 1963b) and of Junge *et al.* (1961a) show that above the local tropopause the concentration of Aitken particles rapidly decreases to between 1 and 10 particles cm^{-3} (ambient), and remains near this value at least up to 28 km (Figure 8-6). Recent studies of the Aitken particle load of the stratosphere by Rosen (1974), Podzimek *et al.* (1974, 1975), Käselau (1975), and Cadle and Langer (1975) have shown that immediately above the tropopause the concentration of Aitken particles may be subject to orders-of-magnitude variations. Podzimek *et al.* and Cadle and Langer suggested that these fluctuations in concentration may be the result of gravity waves lifting particle-rich air masses from below.

Quite in contrast to the vertical variation in concentration of Aitken nuclei, the concentration of 'large' particles decreases in the troposphere to a minimum ranging between 10 and 20 ℓ^{-1} (ambient) (Penndorf, 1954). In the lower stratosphere the concentration of 'large' particles increases with height, reaching a maximum of 50 to 200 ℓ^{-1} between 15 and 25 km (Junge *et al.*, 1961a, b; Chagnon and Junge, 1961; Junge and Manson, 1961), as shown in Figure 8-7. This *Junge aerosol layer* is a world-wide phenomenon (Lazrus and Gandrud, 1974). The concentration maximum exhibits a definite trend with geographic latitude. It is located near 24 km over the equator and near 17 km

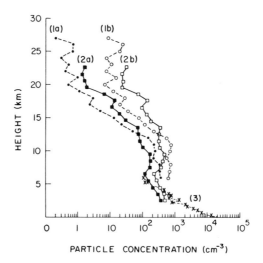

Fig. 8-6. Variation of the average number concentration of Aitken particles with height over various locations. (1a) Seven flights over Sioux Falls (S. Dakota, U.S., 44° N, June, 1959; July, 1960). Concentration in number cm^{-3} (ambient); data of Junge. (1b) Same as (1a) but concentration in number cm^{-3} (STP); data of Junge. (2a) Flights over Hyderabad (India, 17° N, March to April, 1961). Concentration in number cm^{-3} (ambient); data of Junge. (2b) Same as (2a) but concentration in number cm^{-3} (STP); data of Junge. (3) Data of Weickman (1957a, b) based on flights over Germany. (From Junge, 1963b; by courtesy of *J. de Rech. Atmos.*, and the author.)

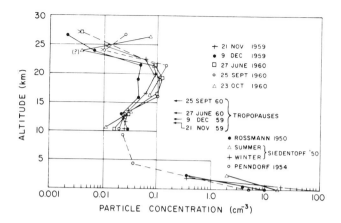

Fig. 8-7. Variation of the average number concentration (ambient) of "large" particles (mean diameter 0.3 μm) over Sioux Falls (S. Dakota, U.S., 44° N). Data obtained by Junge over the U.S. are compared with data of Rossmann (1950), Siedentopf (1950), and Penndorf (1954) taken over Germany. (From Chagnon and Junge, 1961; by courtesy of Amer. Meteor. Soc., and the authors.)

over the poles, in response to local tropopause heights (Rosen *et al.*, 1975). The layer also exhibits a seasonal variation as noticed by Rosen *et al.* (1975) from soundings over Laramie (Wyoming). According to their studies, highest concentrations seem to be present during the winter months, while lowest concentrations occur during the summer months. The maximum concentration in the Junge layer is typically 0.1 particles cm^{-3} (ambient), i.e., 100 particles ℓ^{-1} (ambient). Mossop (1963a) found, at 20 km, 40 to 90 ℓ^{-1} (ambient) or 0.7 to 1.6 cm^{-3} (STP), and at another time (1965), 17 to 42 ℓ^{-1} (ambient) or 0.3 to 0.7 cm^{-3} (STP). Friend (1966) found, at 12 to 18 km, a number concentration of 14 to 69 ℓ^{-1} (ambient), equivalent to a mass concentration of 5×10^{-2} to 2.6×10^{-3} μm m^{-3} (ambient).

The transport of 'large' particles from the troposphere to the stratosphere by turbulent diffusion through the tropopause, or by large scale air exchange with the troposphere, has to be ruled out as a cause of the Junge layer, since such an explanation would not be consistent with the sharp decrease in the number of Aitken nuclei above the tropopause. Chemical analysis of particles in the Junge layer by Junge and Manson (1961), Junge *et al.* (1961a), Chagnon and Junge (1961), Friend (1966), Schedlovsky and Paisley (1966), Lazrus *et al.* (1971), and Cadle (1973) has demonstrated that sulfur, present as SO_4^{2-}, is the predominant compound of these particles, with H$^+$ and NH$_4^+$ as the major cations. Since a large portion of these particles deliquesce to solution drops at a relative humidity of 72 to 80%, Junge and Manson (1961) and Cadle (1972) believe that $(NH_4)_2SO_4$ is the major constituent (see Table 4-3). Further quantitative evidence for the presence of $(NH_4)_2SO_4$ in stratospheric particles has recently been given by Bigg (1975) who, during 6 balloon flights over Wyoming, analyzed stratospheric particles forced to impact on specially treated electron microscope screens. Bigg found that the submicron particles consisted predominantly of ammonium sulfate near the tropopause, and of sulfuric acid at higher altitudes. The acid particles were often, but not always, in a frozen state.

Analysis of individual AP in and immediately above the Junge layer reveals that numerous particles contain one or more water-insoluble, dense inclusions (Mossop, 1963a, 1965). Photographs with an electron microscope show that some of the AP, sampled above the Junge layer at 20 to 40 km, have the shape of compact spherules, while others look like fluffy, highly branched chains, and still others have crystalline shapes or appear as irregular lumps (Bigg et al., 1970, 1971, 1972). The observations of Bigg et al. could be interpreted to mean that the sulfate formation mechanism in the stratosphere is heterogeneous, in that insoluble particles, some possibly of extraterrestrial origin, act as nuclei which surround themselves with sulfate or sulfuric acid.

To close this section we shall briefly touch upon the concentrations, size ranges, and vertical distributions of some AP of specific chemical or physical type. Schütz and Jaenicke (1974) found near the ground over the Libyan desert silicate particle concentrations of $260\ \ell^{-1}$ ($0.2\ mg\ m^{-3}$), $9600\ \ell^{-1}$ ($9.3\ mg\ m^{-3}$), and $8800\ \ell^{-1}$ ($1.2\ mg\ m^{-3}$) at mean wind speeds of $7.6\ m\ sec^{-1}$, $8\ m\ sec^{-1}$, and $8.7\ m\ sec^{-1}$, respectively.

Grosse and Stix (1968), Stix (1969), and Stix and Grosse (1970) found that in air over Darmstadt (Germany) the daily average concentration of pollen and spores ranged from a few hundred m^{-3} (maximum values of $1000\ m^{-3}$) during the months of December to April, to a few thousand m^{-3} (maximum values of a few ten thousand m^{-3}) during the months of May to October. During the months of October and November intermediate values were found.

Particles containing sulfates, nitrates, chlorides, and calcium and ammonium compounds are found to be present in preferred size ranges. Junge (1953, 1954) found, from an analysis of air over Frankfurt (Germany), over Round Hill near Boston (U.S.), at the Taunus Observatory (800 m, Germany), and at the Zugspitze Observatory (3000 m, Germany), that particles in the size range $0.08\ \mu m \leqslant r \leqslant 0.8\ \mu m$ seemed to consist mainly of ammonium sulfate. These results were supported by Georgii et al. (1971), who found that in air over West Germany more than 95% of the sulfate mass of AP was contained in particles with $r < 1\ \mu m$. Similarly, in air over Crater Lake (2200 m Oregon), 86% of the sulfur was contained in particles less than $1\ \mu m$ radius (Junge et al., 1969). Junge (1953, 1954) also noticed that much of the NaCl was contained in particles of $0.8 \leqslant r \leqslant 8\ \mu m$. In agreement with this, Junge et al. (1969) observed at Cape Blanco (Oregon) that 92% of the chloride was contained in particles larger than $1\ \mu m$ radius. According to Meszaros (1969), AP of purely continental origin contain most of the nitrate in the 'large' AP range, while the water-soluble calcium is found primarily in AP of the Aitken size range. In contrast to this, Junge (1954) found that in air near the east coast of the U.S., nitrate was predominant in AP of the 'giant' size range.

Mass concentrations of AP separated according to the major size ranges and constituents have been determined by Junge (1956) for air over various continental and coastal stations (Figures 8-8a, b). Note that with increasing proximity to oceans, the concentration of 'large' particles containing SO_4^{2-}, NH_4^+,

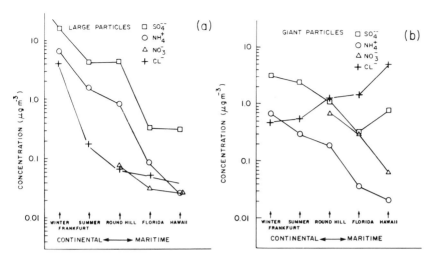

Fig. 8-8. Variation of the mass concentration of SO_4^{2-}, NH_4^+, NO_3^-, and Cl^- in (a) 'large' particles, (b) 'giant' particles, captured at Frankfurt (Germany), at Round Hill near Boston, Mass., in Florida, and in Hawaii. (From Junge, 1956; by courtesy of *Tellus*.)

NO_3^-, and Cl^- decreases. The same holds true for the concentration of 'giant' particles containing SO_4^{2-}, NH_4^+, and NO_3^-, while particles containing Cl^- in this size range become more frequent with increasing proximity to oceans. This trend has also been found by Rossknecht *et al.* (1973) in Oregon, Lodge (1955) in Puerto Rico, and Twomey (1955) in Australia. The results of some of these observations are given in Figure 8-9.

The vertical variation of chloride particle concentrations has been measured by Twomey (1955) over Australia, by Byers *et al.* (1957) over various locations between central Illinois and the Gulf of Mexico, and by Podzimek and Cernoch (1961) over northern Bohemia (Czechoslovakia). All three investigators agreed that chloride particles are very efficiently removed at cloud level.

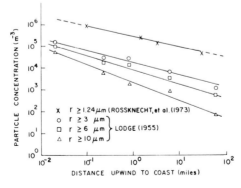

Fig. 8-9. Variation of the concentration of chloride particles of various sizes at ground level with increasing distance from the sea shore. (Based on data of Lodge, 1955, in Puerto Rico, and of Rossknecht *et al.*, 1973, in Oregon.)

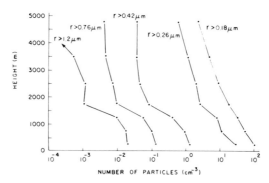

Fig. 8-10. Variation of the average concentration of sulfate particles with height over various locations in West Germany. (From Georgii *et al.*, 1971; by courtesy of the authors.)

Georgii *et al.* (1971) made a detailed study of the vertical variation of sulfate particles over various locations in Germany. Although the concentration varied strongly from day to day, average concentrations decreased with increasing height in a manner illustrated in Figure 8-10. Two additional observations of cloud physical significance were made by Georgii *et al.*: (1) Local maxima in the concentration of sulfate particles were found at cloud level and most often just below cloud base. This observation was interpreted in terms of a sulfate-forming process taking place inside cloud drops. Further evidence for this effect was given by Radke and Hobbs (1969), and by Easter and Hobbs (1974) (see Section 8.2.1). (2) The sulfate content of AP of given size was found to increase with height, while at the same time the size of particles containing a given percentage of sulfate was found to decrease with height (see Figure 8-11).

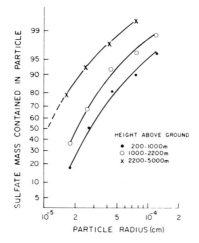

Fig. 8-11. Percentage of sulfate mass contained in atmospheric aerosol particles smaller than given size at different heights. (From Georgii *et al.*, 1971; by courtesy of the authors.)

8.2.9 TOTAL CONCENTRATION AND VERTICAL VARIATION
       OF AP OVER OCEANS

We have already mentioned that the total number concentration of AP is significantly lower over the oceans than over the continents, owing to the difference in the number and efficiency of sources. Table 8-8 illustrates that at locations over oceans, relatively remote from anthropogenic sources, the total concentration of AP usually ranges between 300 and 600 cm^{-3}, with minimum values close to 100 cm^{-3}.

A very small and variable portion of the AP over oceans is made up of water-insoluble materials, some of which are silicates originating in the Sahara desert. According to Prospero and Bonatt (1969), Parkin *et al.* (1970, 1972), Jaenicke *et al.* (1971), Junge and Jaenicke (1971), Junge (1972b), and Prospero and Carlson (1972), the mass concentration of silicate particles in air over the Atlantic may range up to 100 μg m^{-3} near the African coast within the Sahara dust layer typically located at about 3 km over the Atlantic. Outside this particle plume, the concentration is significantly lower, usually 0.05 μg m^{-3} or less. The size of these silicate particles generally ranges between 0.3 and 20 μm radius.

One would expect that in the air layers close to the ocean surface, sea salt particles would dominate the AP concentrations. However, this does not

TABLE 8-8
Total concentration of aerosol particles over oceans.

Observer	Location	Total concentration of aerosol particles (number cm^{-3})
Shiratori (1934)	Atlantic, Pacific	<400 (68 of cases); < 600 (79 of cases)
Ohta (1951)	Pacific	290 (mean); 690 (max.); 70 (min.)
Parkinson (1952)	West Atlantic	676 (mean;) <300 (7%); 300 to 600 (40%) 600–1000 (40%), >1000 (3%)
Moore (1952)	North Atlantic	445 (mean, 1952); 703 (mean, 1951) 77 (min.)
Day (1955)	East Atlantic	200 (mean)
Hogan (1968)	Atlantic (30° to 40° N)	520 (mean)
Jaenicke *et al.* (1971)	West Atlantic (10° S to 60° N)	510 (mean), 120 (min.) 705 (with Sahara dust)
Flyger (1973)	North Atlantic	600 (over North Atlantic) 300 (over Greenland ice cap)
Meszaros and Vissy (1974)	Atlantic and Indian Ocean	300 to 450
Hogan (1975)	Antarktika	50 to 150

happen, and at average wind conditions the concentration of such particles is found to be less than 10 cm^{-3}. Only in air over very agitated seas during storms does the concentration increase to a few tens cm^{-3}. During storm conditions off the coast of Denmark, Schmidt (1972) found sea salt particles with $m_s > 10^{-15}$ g in concentrations of 24 to 43 cm^{-3}. In air over the southern hemispheric Atlantic and Indian Oceans, Meszaros and Vissy (1974) found that the maximum concentration of sea salt particles varied between 4 and 23 cm^{-3}, accounting only for from 5 to 49% of the total number of particles, which ranged from 12 to 82 cm^{-3} for r(dry) $> 0.03 \mu$m.

As is the case of the number concentration, the total mass concentration of sea salt particles over oceans is also found to be strongly dependent on the wind speed at ocean level. Figure 8-12 shows that this concentration varies from about $1 \mu\text{g m}^{-3}$ at very low speeds to as much as $400 \mu\text{m m}^{-3}$ at gale force winds.

The vertical variation of sea salt was studied by Woodcock (1953, 1957) and Lodge (1955) over subtropical oceans, by Junge et al. (1969) over Cape Blanco (Pacific Coast, Oregon), and by Hobbs (1971) over the Pacific off the northwest coast of the U.S. Although on individual days a vertical profile may look rather irregular (Lodge, 1955), in the mean an exponential decrease is observed above 0.5 km, suggesting that sea salt is essentially confined to the lowest 2 to 3 km of the troposphere. This behavior is illustrated in Figure 8-13.

Somewhat unexpectedly, various observations point to the fact that, even in air layers close to the ocean, water-soluble sulfates such as $(NH_4)_2SO_4$ are more abundant than NaCl. In air over the southern hemispheric oceans, Meszaros and Vissy (1974) found that maximum values for the concentration of $(NH_4)_2SO_4$ particles, which varied between 17 and 61 cm^{-3}, accounted for 36 to 74% of the

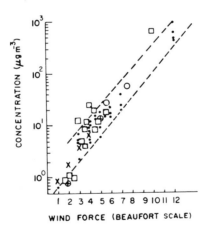

Fig. 8-12. Variation of the total sea salt mass concentration in air over the ocean as a function of wind force. Beaufort 1, 3, 5, 7, and 9 correspond to: 0.3–1.5, 3.4–5.4, 8.0–10.7, 13.9–17.1, and 20.8–24.4 m sec^{-1}. (Based on data compiled by Junge, 1963a, from: Woodcock, 1953 (●). Pacific near Hawaii and Atlantic near Florida; Moore, 1952 (⊕), N. Atlantic; Fournier d'albe, cited in Moore, 1952 (○), Bay of Monaco; Junge, 1954 (×), Atlantic E. Coast of U.S. Based on data of Wilknis and Bressan, 1972 (□), Atlantic, Pacific, Caribbean, Greenland Sea. Dashed line represents upper and lower bounds given by Woodcock, 1953.)

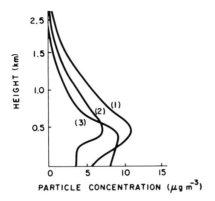

Fig. 8-13. Vertical variation of mass concentration of sea salt in air (1) over the Pacific near Hawaii, (2) over the Atlantic near Florida, and (3) over the Caribbean. (From Erikson, 1959, based on data of Woodcock; by courtesy of *Tellus*.)

total number of observed particles with $r_{dry} > 0.03$ μm. This fraction is considerably larger than that found for sea salt particles over the same region. Also, the observations show that sea salt particles, sulfate particles, and long-shaped unidentified particles account for 75 to 95% of the AP load over oceans.

8.2.10 SIZE DISTRIBUTION OF AP

The size distribution of aerosol particles may be expressed in various ways. If $n(r)$ dr denotes the number of AP cm^{-3} with radii between r and r + dr, then the total concentration of AP of radii greater than r is

$$N(r) = \int_r^\infty n(r)\, dr. \qquad (8\text{-}26)$$

Accordingly, we also have

$$-\frac{dN}{dr} = n(r). \qquad (8\text{-}27)$$

Because of the wide range of particle sizes, it is often more convenient to express the size distribution in logarithmic form by defining $n^*(r)$ d log r as the number of AP cm^{-3} in the interval r, r + d log r. Then the relation between N and n^* is

$$-\frac{dN}{d\log r} = n^*(r). \qquad (8\text{-}28)$$

It is customary to use logarithms to base 10. Adopting this convention and using the usual notation ln for the natural logarithm, we see that

$$n^*(r) = (\ln 10)r n(r). \qquad (8\text{-}29)$$

On the basis of his own observations and those of others, Junge (1952b, 1953, 1955) found that for r > 0.1 μm the concentration of AP decreases with increasing size such that n*(r) can be expressed approximately as a power law function of r, so that

$$n^*(r) = \frac{C}{r^\alpha},\qquad\qquad(8\text{-}30)$$

where C and α are constants. The other forms of this law become

$$n(r) = \frac{C}{(\ln 10)r^{\alpha+1}}\qquad\qquad(8\text{-}31)$$

and

$$N(r) = \frac{C}{(\alpha \ln 10)r^\alpha}.\qquad\qquad(8\text{-}32)$$

The corresponding log radius volume (V) distribution is

$$-\frac{dV}{d\log r} \equiv -\frac{4\pi r^3}{3}\frac{dN}{d\log r} = \frac{4\pi}{3}Cr^{(3-\alpha)}.\qquad\qquad(8\text{-}33)$$

Note that in the special case of $\alpha = 3$, dV/d log r = constant.

Examples of the size distribution of AP over continents are given in Figures 8-14 and 8-15. These figures demonstrate that for r > 0.1 μm the concentration of AP decreases with increasing particle size, the decrease roughly following a power law with $3 \leq \alpha \leq 4$, on the average. For $0.01 \leq r \leq 0.1\ \mu$m, the AP concentration reaches a maximum. This might have been expected on the basis of the

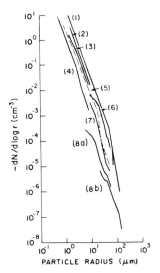

Fig. 8-14.   Size distribution of aerosol particles (1 μm $\leq$ r $\leq$ 100 μm) in air over various locations in Central Europe, Japan, and the U.S. (1) Data of Noll and Pilat (1971), Seattle, State of Washington. (2) to (8) Data of Jaenicke and Junge (1967): (2) Frankfurt (Germany), 1952; (3) Taunus (Germany), 800 m, 1952; (4) Zugspitze (Germany), 3000 m, 1952; (5) Yokomansbetsu (Japan) 1050 m (Okita, 1955); (6) Mainz (Germany), 1962; (7) Taunus, 1962; (8a, b) Jungfraujoch (Switzerland), 3570 m, 1963.

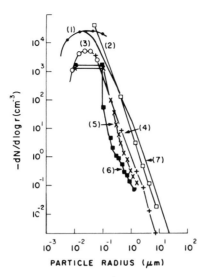

Fig. 8-15. Size distribution of aerosol particles ($10^{-2} \mu m \leq r \leq 10 \mu m$) in air over various locations in Central Europe and the U.S. (Data of Junge (1955): (1) Frankfurt, ion counter; (2) Frankfurt, nuclei counter and impactor; (3) Zugspitze, 3000 m, ion counter; (4) Zugspitze, 3000 m, nuclei counter and impactor. Data of Junge et al., 1969, with Royco counter and nuclei counter: (5) Crater Lake, 2200 m without subsidence; (6) Crater Lake, 2200 m with subsidence. Data of Noll and Pilat (1971) with impactor: (7) Seattle, State of Washington.)

rapid attachment of particles smaller than $0.1 \mu m$ to larger ones by thermal (Brownian) coagulation, in a way which reduces their number concentration progressively with decreasing size (see Section 12.5). However, Junge (1972b) speculated that, following a local minimum caused by the coagulation effect, the AP concentration rises once more as a result of continuous gas-to-particle conversion for particles $r < 0.01 \mu m$. A local minimum and a second maximum was indeed observed by Junge (1972b) over the North Atlantic. It is also implied by the measurements of Meszaros and Vissy (1974) over the oceans of the Southern Hemisphere, since the total concentration of Aitken particles they observed was larger by one order of magnitude than the concentration of AP with $r > 0.03 \mu m$.

Figures 8-16 to 8-18 display some recent measurements of AP size distributions over the northern and southern hemispheric oceans. It is evident that, for $r > 0.1 \mu m$, the size distributions over oceans may be represented adequately by power laws with $3 \lesssim \alpha \lesssim 5$. As expected, the distributions shift towards lower concentrations with increasing height (compare Figures 8-16 and 8-17). Note that Figure 8-17 describes the background aerosol. Figures 8-17 and 8-18 further indicate that the size distribution becomes steeper for $r > 20 \mu m$, a trend also indicated in the continental distribution. Note that no clear upper limit in particle size can be seen. Although particles of $r > 20 \mu m$ are produced by the bubble-burst mechanism at the ocean surface, Toba's (1965a, b) theoretical estimates suggest that gravity prevents such particles from penetrating the turbulent boundary layer over the oceans to reach higher layers. As an explanation of their

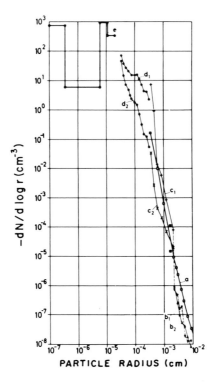

Fig. 8-16.   Size distribution of aerosol particles in air over Atlantic taken during voyage of ship 'RV Meteor' along 30° W from 10° S to 60° N: (a) sea salt distribution, (b_1, c_1) data from impactors on days with Sahara dust, (b_2, c_2) data from impactors on days without Sahara dust, (d_1) data from Royco counter on days with Sahara dust, (d_2) data from Royco counter on days without Sahara dust, (e) data from a combination of a photographic nucleus counter, diffusion boxes, and electric denuders. (From Junge, 1972b, by courtesy of the author; copyrighted by American Geophysical Union.)

presence, Toba (1965a, b) and Junge (1963, 1972b) have suggested that these particles represent the residues of evaporated drops.

The size distributions obtained by Junge (1972b) and Meszaros and Vissy (1974) for AP over oceans show an 'inflection' near $r \approx 0.3 \mu$m. Such an inflection is also present in the size distribution of sea salt particles. Meszaros and Vissy suggested that this is due to a change in the particle production mechanisms. Thus, sea salt particles of $m_s > 2 \times 10^{-13}$ g ($r_{dry} = 0.28 \mu$m) are due almost exclusively to jet drops formed during the bubble-burst mechanism, while particles of $10^{-15} \leqslant m_s \leqslant 2 \times 10^{-13}$ g ($0.07 \leqslant r_{dry} \leqslant 0.28 \mu$m) result from the disintegration of the bubble cap. Woodcock and Duce (1972) and Woodcock (1972) independently drew the same conclusions from their observations of sea salt particles over the Pacific near Hawaii and over the Gulf of Alaska. Woodcock's findings are reproduced in Figure 8-19. Curve (b) exhibits an inflection in the mass distribution of sea salt particles near $m_s = 10^{-13}$ g, in agreement with the findings of Meszaros and Vissy. It was suggested by Woodcock (1972) that the reason for not finding an inflection in the size distribution of sea salt particles

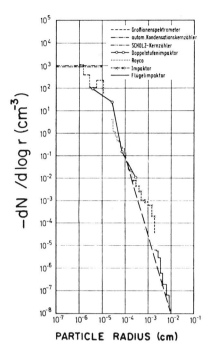

Fig. 8-17.   Size distribution of aerosol particles in air over the North Atlantic at Izana Obs., Teneriffa (28°17′56″N, 16°29′32″W). Observations taken at a height of 2370 m. The straight dashed line represents the results of the theoretical model discussed in Section 12.8.5. (From Abel *et al.*, 1969; by courtesy of *Meteor, Rundschau*.)

over the Gulf of Alaska, which is rich in organic materials, lies in the fact that the bubble cap disintegration mechanism only works when no compressed organic films are present on the ocean surface (Blanchard, 1963; Patterson and Spillane, 1969).

A power law also describes approximately the distribution of particles of a specific chemical type. For silicate particles in air over the Libyan desert, Schütz and Jaenicke (1974) found $\alpha \approx 2$ for $1 < r < 10 \ \mu m$, and $\alpha \approx 3$ for $r > 10 \ \mu m$, on the average. The observations of Gillette *et al.* (1972) on silicate particles over rural Nebraska, and of Gillette *et al.* (1974) near Big Spring, Texas, indicate a similar trend. Also, a power law with $4.2 \leqslant \alpha \leqslant 4.8$ was found to apply to sulfate particles sampled by Georgii and Gravenhorst (1972) over Germany. In these studies it was assumed that the particle radius is that of a spherical particle of mass equal to the sulfate mass in the actual mixed AP.

Although the power law description (8-30) of AP size distributions has the appeal of great simplicity, and often provides a surprisingly good fit to observational data, for some purposes it may be more appropriate to turn to more complex representations which can provide greater accuracy over a larger size range. A candidate of this kind is the modified gamma distribution,

$$n(r) = Ar^{\beta} \exp(-Br^{\gamma}), \tag{8-34}$$

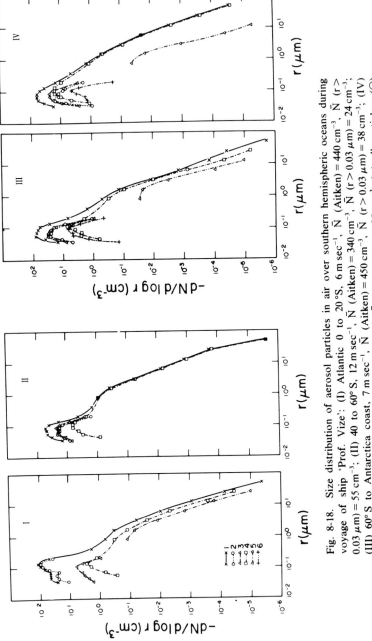

Fig. 8-18. Size distribution of aerosol particles in air over southern hemispheric oceans during voyage of ship 'Prof. Vize': (I) Atlantic 0 to 20 °S, 6 m sec^{-1}, $\bar{N}$ (Aitken) = 440 cm^{-3}, $\bar{N}$ (r > 0.03 μm) = 55 cm^{-3}; (II) 40 to 60 °S, 12 m sec^{-1}, $\bar{N}$ (Aitken) = 340 cm^{-3}, $\bar{N}$ (r > 0.03 μm) = 24 cm^{-3}; (III) 60 °S to Antarctica coast, 7 m sec^{-1}, $\bar{N}$ (Aitken) = 450 cm^{-3}, $\bar{N}$ (r > 0.03 μm) = 38 cm^{-3}; (IV) Indian Ocean, 9 m sec^{-1}, $\bar{N}$ (Aitken) = 310 cm^{-3}, $\bar{N}$ (r > 0.03 μm) = 35 cm^{-3}. ($\times$) all particles; ($\bigcirc$) (NH$_4$)$_2$SO$_4$; (--$\triangle$-) long shaped particles; ($\square$) NaCl; (···$\triangle$··-) non cubic, crystalline particles; (+) H$_2$SO$_4$. (From Meszaros and Vissy, 1974; by courtesy of Pergamon Press Ltd.)

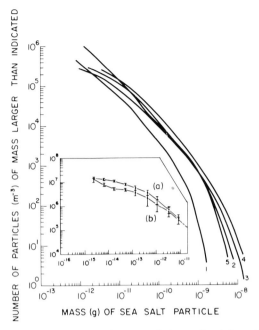

Fig. 8-19. Mass distribution of sea salt particles at wind speeds of 3 to 11 m sec^{-1}. (1) Pacific, Hawaii, 1951, 561 m; (2) Pacific, Hawaii, 1952, 518 m; (3) Pacific, Oahu, 1952, 732 m; (4) Atlantic, east of Florida, 1951, 790 m; (5) Pacific, south of Australia, 1952, 640 m; (a) Gulf of Alaska, 1969, 300 m; (b) Pacific, Hawaii, 1969, 300 m. (Based on data of Woodcock, 1953, 1972.)

where A, B, β, and γ are positive parameters. This distribution has been discussed, for example, by Deirmendjian (1969). The maximum of the distribution occurs at r_m, an observable quantity which relates the parameters of the distribution according to

$$\beta = B\gamma r_m^\gamma. \tag{8-35}$$

Another relationship is provided in terms of the concentration at r_m, i.e.,

$$n(r_m) = Ar_m^\beta \exp(-\beta/\gamma). \tag{8-36}$$

A third constraint is available in terms of the total concentration N (the zeroth moment), viz.,

$$N = \frac{AB^{-(\beta+1)/\gamma}}{\gamma} \Gamma\left(\frac{\beta+1}{\gamma}\right), \tag{8-37}$$

where Γ is the gamma function. On a double log plot the slope of the distribution (8-34) is given by

$$\frac{d[\log n(r)]}{d \log r} = \beta\left[1 - \left(\frac{r}{r_m}\right)^\gamma\right]. \tag{8-38}$$

(The corresponding slope for the power law distribution (8-31) is $-(\alpha + 1)$.) If, for example, an estimate of β or γ can be obtained by comparing (8-38) with the shape of the experimental curve, then from the other constraints (8-35)–(8-37) the distribution (8-34) may be specified completely.

An obvious computational advantage of (8-34) is that the applicable domain, $r_{min} \leqslant r \leqslant r_{max}$, may be chosen as large as one wishes, an option which is not available with the strict power law form. It has been shown recently by Mallow (1975) that the distribution (8-34) can be applied successfully to hazes and fogs.

To conclude this section we shall now briefly mention some of the work conducted on the size distribution of AP in the lower stratosphere. Junge et al. (1961a, b) and Junge and Manson (1961) concluded from an examination of AP collected by balloon-borne impactors that the radii of such particles were distributed approximately according to a power law with $\alpha \approx 2$ for the range $0.1 \, \mu\text{m} \leqslant r \leqslant 1.0 \, \mu\text{m}$. In contrast to these observations, later measurements of Mossop (1965) and Friend (1966) showed that the size distribution of 'large' stratospheric AP peaks near $0.35 \, \mu\text{m}$ and typically has a shape as given in Figure 8.20. From this figure one can infer that the total number of AP in the stratosphere, which according to Junge et al. (1961a, b) and Junge (1961) is of the order of 1 to 10 cm^{-3}, at least up to 33 km, cannot be accounted for by the number of 'large' particles alone. If the distribution function for 'large' particles as given in Figure 8-20 is correct, then in the region of radii less than $0.1 \, \mu\text{m}$ the number concentration must increase again so that the total number concentration will agree with the Aitken nuclei concentration. Although there is no direct evidence available concerning the average size of Aitken nuclei in the stratosphere, Junge et al. (1961a, b) have suggested that a radius of $0.04 \, \mu\text{m}$ would be consistent with some theoretical estimates of mixing by turbulent diffusion from a tropospheric source. Thus, Friend (1966) conjectured that the actual distribution of particle sizes in the stratosphere is bimodal with peaks near $0.35 \, \mu\text{m}$ and $0.04 \, \mu\text{m}$ radius, as shown schematically in Figure 8-20.

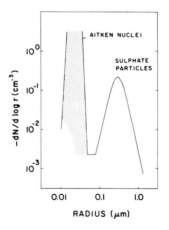

Fig. 8-20. Schematic size distribution of stratospheric aerosol particles: 'large' particles according to observations of Mossop (1965) and Friend (1966), Aitken particles conjectural. (From Friend, 1966; by courtesy of *Tellus*.)

8.3   Aerosol Substances in Cloud and Precipitation Water

To a certain extent, cloud water, rain water, and water from melted ice particles reflect the chemical composition of AP, since they become incorporated in cloud drops and ice crystals as a result of serving as nuclei for drop and ice crystal formation. In addition, AP may be captured as a result of Brownian diffusion, thermo- and diffusiophoresis, turbulent motions, electrical forces, and inertial impaction (see Chapter 12). Gaseous substances also become incorporated in cloud and raindrops due to their solubility in water. As a result of such nucleation and capture processes, the following ions are typically found in cloud and rain water: NH_4^+, K^+, Na^+, Ca^{2+}, Mg^{2+}, H^+, SO_4^{2-}, Cl^-, NO_3^-, SO_3^{2-}, HSO_3^-, HCO_3^-, and CO_3^{2-}.

The chemical composition of fog, cloud, and rain water has been the subject of several investigations, including those of Mrose (1966) (fogs in Germany), Houghton (1953) (fogs and clouds in eastern U.S.), Miyake (1948) (mountain fogs in Japan), Georgii (1965) (rain in Germany), and Petrenchuk and Drozdova (1966) and Petrenchuk and Selezneva (1970) (clouds and rain at various locations in the Soviet Union). These studies confirm the expectation that the chemical compounds detected in cloud and rain water correspond to the major water-soluble substances present as salts in the atmospheric aerosol (Table 8-9). The data also show that the amount of water-soluble substance available for pick-up by clouds and precipitation decreases with increasing height. Petrenchuk and Drozdova (1966) and Petrenchuk and Selezneva (1970) show further that strati-form clouds located over major cities and below inversions are particularly likely to contain large amounts of soluble materials in their cloud water (Table 8-10a), and that the concentration of salts in cloud water is a function of cloud type (Table 8-10a, b). The salt concentration was found to be larger in non-precipitating clouds, where drops are smaller and subject to more evaporation, than in precipitating clouds where drops grow rapidly by diffusion and collection. Turner (1955) and Georgii and Wötzel (1970) have also found that smaller drops generally contain higher salt concentrations than larger ones.

TABLE 8-9

Variation of concentration (mg ℓ^{-1}) of salt ions in rain water as a function of height above sea level. (Based on data of Georgii, 1965.)

	NH_4^+	NO_3^-	SO_4^{2-}	Cl^-	Na^+	Ca^{2+}	Total
Frankfurt a. M., city (100 m) 1960–1961	3.2	2.8	16.3	3.9	1.1	1.9	29.2
Langen (Hessen), rural, (~120 m) 1960–1961	3.9	2.6	15.3	6.7	1.1	1.4	31.0
Taunus observatory on Kl. Feldberg (800 m) 1957, 1960	1.5	2.3	4.9	1.8	1.1	1.6	13.2
Zugspitze observatory (2966 m) 1957, 1960–1962	1.1	0.8	2.1	1.6	1.1	1.6	8.3

TABLE 8-10a

Variation of salt ion concentrations (mg ℓ^{-1}) in cloud water as a function of cloud type over the Soviet Union. (Based on data of Petrenchuk and Drozdova, 1966, and Petrenchuk and Selezneva, 1970.)

Cloud or Precipitation type	SO_4^{2-}	Cl^-	NO_3^-	HCO_3^-	NH_4^+	Na^+	K^+	Mg^{2+}	Ca^{2+}	Total
St, and Sc cloud over Kiev	91.5	7.76	12.2	–	24.6	1.8	4.8	–	10.7	153.36
Subinversion cloud	45.3	3.4	1.3	2.4	6.4	3.3	1.2	2.7	7.4	73.4
Nonprecipitation cloud	12.3	1.8	0.6	2.4	1.7	1.1	0.5	1.2	2.3	23.9
Precipitation from clouds with frontal rain	3.1	0.8	0.1	1.5	0.5	0.3	0.2	0.4	0.7	7.6
Average in precipitation water	9.2	2.1	1.3	5.6	0.9	1.5	0.7	1.5	2.0	24.8

Junge and Werbey (1958) and Whitehead and Feth (1964) studied the geographic distribution of the concentration of various salts in rain water collected by an extensive network of surface observation stations in the U.S. Granat (1972) reported on similar measurements carried out by means of a large network of observation stations in Europe. Both sets of observations reflect the effect of the oceans on the concentration of Cl^- in rain water. This is exemplified in Figure 8-21, which demonstrates a strong decrease of the Cl^- concentration in rain water with increasing distance inland. Similar maps were drawn by Junge and Werbey (1958) and by Junge (1958) for the concentration of SO_4^{2-}, Na^+, K^+, Ca^{2+}, NO_3^-, and NH_4 in rain water. Observations in both the U.S. and in Europe show that Cl^- and SO_4^{2-} are the ions most prevalent in rain water. Similar results were obtained by Handa (1969) for rain water collected at Calcutta (India).

Table 8-11 shows that in water from melted ice particles, Cl^- and SO_4^{2-} are the ions most prevalent. The fact that columnar ice crystals contain less foreign material than dendritic crystals or graupel particles must be attributed to the lower efficiency with which they scavenge AP and cloud drops. This behavior will be discussed in Chapters 12 and 14.

Analysis of the residues of evaporated cloud particles provides further information as to the incorporated materials. Table 8-12 shows that in Japan sea

TABLE 8-10b

Salt ion concentrations (mg ℓ^{-1}) in precipitating and non-precipitating cloud water. (Based on data on Petrenchuk and Drozdova, 1966, and Petrenchuk and Selezneva, 1970.)

Cloud type	SO_4^{2-}	Cl^-	NO_3^-	HCO_3^-	NH_4^+	Na^+	K^+	Mg^{2+}	Ca^{2+}	Total
St, Sc	11.4	2.2	1.0	0.01	2.4	0.7	0.7	0.5	0.9	19.8
Ns	5.3	2.0	0.4	1.1	0.9	0.7	0.5	0.4	1.0	12.2

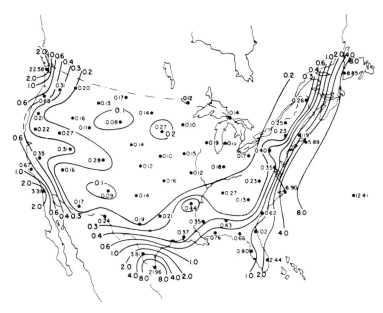

Fig. 8-21. Average Cl⁻ concentration (mg ℓ^{-1}) in rain water collected at the ground in the U.S. (1955–56). (From Junge and Werbey, 1958; by courtesy Amer. Meteor. Soc., and the authors.)

TABLE 8-11

Concentration (mg ℓ^{-1}) of salt ions in water from melted columnar and dendritic ice crystals, and in water from melted graupel particles. (Based on data of Takahashi, 1963.)

Water from melted:	Cl⁻	SO_4^{2-}	Na^+	NH_4^+	Mg^{2+}	Ca^{2+}	NO_3^-	$Fe^{2+,3+}$	Total
columnar ice crystals	1.1	1.6	0.42	0.05	0.15	0.037	0.0008	0.026	3.38
dendritic ice crystals	4.1	4.4	1.5	0.27	0.34	0.27	0.001	0.042	10.92
graupel particles	4.3	1.3	1.9	0.03	0.28	0.045	0.001	0.03	7.89

TABLE 8-12

Composition of residues of cloud and fog drops.

	Kuroiwa (1951, 1953)		Yamamoto and Ohtake (1953, 1955)		Isono (1957, 1959a)	
	number	%	number	%	number	%
sea salt	5	13	11	16	37	54
combustion products	20	51	25	36	15	22
soil material	11	28	16	23	10	14
unknown	3	8	17	25	7	10
Total	39	100	69	100	69	100

salt particles and particles derived from combustion account together for more than 50% of the residue left by evaporated cloud and fog drops. Again, sea salt is found to be more abundant near or over the ocean, while combustion products are more abundant inland. Similar results have been obtained by Ogiwara and Okita (1952). In addition, Naruse and Maryama (1971) have found that about 95% of all cloud and fog drops with diameters between 5 and 50 μm contain residue masses between 8×10^{-14} g and 9×10^{-13} g. The largest residue observed had a mass of 831×10^{-13} g.

The residues of ice crystals have been studied by Kumai (1951, 1957, 1961, 1976), Isono (1955), Kumai and Francis (1962b), Rucklidge (1965), and Isono et al. (1959, 1966) through electron microscopy or electron diffraction techniques. Quite frequently, just one solid particle was found in the central portion of a snow crystal. These central particles had diameters between 0.1 and 15 μm, with a mode between 0.4 and 1 μm. Their composition is given in Table 8-13 for snow crystals collected at various locations. These studies showed that, most frequently, central particles consist of clay minerals such as illite, kaolinite, halloysite, and vermiculite, the last being the most abundant (Kumai, 1976), although some central particles may also be composed of hygroscopic materials, combustion products, and micro-organisms. Soulage (1955, 1957) dissected the residue of snow crystals and found that the larger particles consisted of a mixture of soluble and insoluble materials. Kumai (1966b, 1969b) found that about 2% of the central particles of his sample of ice crystals consisted of spherical particles (spherules) of diameter between 0.6 and 6 μm, some of which were identified as extraterrestrial material. In addition to a larger central particle, numerous smaller particles of diameters between 0.05 and 0.15 μm were found in the outer portions of the crystals. This led to the supposition that the central particle was instrumental in the nucleation of the snow crystal, while all other particles were captured subsequently.

Ishizaka (1972, 1973) has made a detailed study of the amount and type of solid, water-insoluble material contained in rain and snow water collected in Japan. Water from melted snow which fell during the NW monsoon contained 4.6 mg of solid material per liter of snow water. Of this material, 70% by weight consisted of α-quartz, feldspar, illite, chlorite, kaolinite, halloysite, montmorillonite, and talc. Rain water from a storm which originated in central China contained 1.3 mg ℓ^{-1} of solid material, of which about 20% consisted of α-quartz, feldspar, illite, chlorite, kaolinite, and vermiculite. Rain water collected during Typhoon 7002 was relatively clean and contained small amounts of α-quartz, α-crystobalite, and pyroxene, while rain water from a storm which had a relatively long trajectory over Japan contained large amounts of amorphous carbon and other organic material.

Rosinski (1966, 1967a), Rosinski and Kerrigan (1969), and Rosinski et al. (1970) studied the number concentration and size of water-insoluble particles in rain water, in individual raindrops, and in ice particles. Most frequently, the particles had diameters much less than 40 μm, but rain water collected from severe storms contained appreciable numbers of particles with even larger diameters. The particle concentration in rain water depended strongly on the collection time

TABLE 8-13

Composition of aerosol particle in central portion of snow crystal.

Composition of center particle	Hokkaido (Japan) Number	%	Honshu (Japan) Number	%	Michigan (U.S.A.) Number	%	Missouri (U.S.A.) Number	%	Thule (Greenland) Number	%	Amundsen-Scott (South Pole) Number	%
Clay mineral	176	57	46	88	235	87	70	28	302	84	55	59.1
Hygroscopic particle	57	19	0	0	2	1	5	2	2	1	19	20.4
Combustion product	26	8	2	4	6	2	7	3	0	0	–	–
Micro-organism	3	1	0	0	0	0	3	1	0	0	–	–
Unidentified material	30	10	4	8	25	9	100	40	39	11	5	5.4
Not observed	15	5	0	0	3	1	65	26	13	4	14	15.1
Total	307	100	52	100	271	100	250	100	356	100	93	100
Reference	Kumai (1961)				Rucklidge (1965)				Kumai and Francis (1962b)		Kumai (1976)	

during the life cycle of the storm. Highest particle concentrations were always found at the onset of precipitation. Typical values are given in Table 8-14. Unfortunately, no counts were taken of particles with diameters less than 1 μm. Using a power law of the type $N \sim r^{-3}$, Vali (1968a) estimated by extrapolation that particles with diameters of 0.01 μm may be present in rain water in concentrations of 10^{10} to 10^{11} cm^{-3}. Also, Rosinski et al. (1970) observed that even at the cirrus cloud level the number of water-insoluble particles in water from melted crystals is apppreciable, ranging in their samples from 7.4×10^3 cm^{-3} to 8.4×10^6 cm^{-3} for particles with diameters larger than 2 μm. Particles designated as magnetic spherules were also present in water from melted cirrus ice crystals, and appeared in concentrations of 5 to 7670 cm^{-3}.

TABLE 8-14

Concentration (number of particles cm^{-3}) of water-insoluble particles in rain water and in water of melted graupel and hail, as a function of particle size. (Based on data of Rosinski, 1966, 1967a; and of Rosinski and Kerrigan, 1969.)

Diameter of particle (μm)	1.5 to 3	3 to 15	15 to 50	50 to 100	100 to 200
	2.6×10^4	7.4×10^3	2.1×10^2	9	2
Concentration (number cm^{-3})	to	to	to	to	to
	2.7×10^6	2.5×10^5	1.2×10^4	1100	520

CHAPTER 9

HETEROGENEOUS NUCLEATION

Observations summarized and discussed in Chapter 2 show that supersaturations as high as several hundred percent, which would be necessary for drop formation in homogeneous water vapor (see Chapter 7), do not occur in the atmosphere, but that typically supersaturations remain below 10% and most often even below 1%. This indicates that drop formation in the atmosphere occurs via heterogeneous nucleation involving aerosol particles (AP). AP which are capable of initiating drop formation at the observed low supersaturations are called *cloud condensation nuclei* (CCN).

Observations summarized in Chapter 2 show also that cloud glaciation generally begins at temperatures much too warm for homogeneous freezing of water. For example, on one occasion Mossop *et al.* (1968) observed ice crystals in a long lived cumulus cloud whose top was probably never colder than $-4\,°C$, and which was not seeded with ice particles from clouds at higher altitudes. Such behavior indicates some fraction of the local AP also can serve as *ice forming nuclei* (IN).

In this chapter we shall discuss the atmospheric CCN and IN, including their modes of action, sources, concentrations, and other characteristic features.

9.1 Cloud Condensation Nuclei (CCN)

9.1.1 NUMBER CONCENTRATION OF CCN

The results of a comprehensive study by Twomey and Wojciechowski (1969) of CCN concentrations over various parts of the world are summarized in Figures

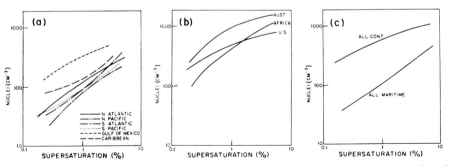

Fig. 9-1. Median world-wide concentration of CCN as a function of supersaturation required for activation; (a) in air over oceans, (b) in air over continents, (c) all observations. (From Twomey and Wojciechowski, 1969; by courtesy of Amer. Meteor. Soc., and the authors.)

9-1a, b, c. The results reveal no systematic latitudinal variation in concentration. The observations also confirm previous conclusions that continental air masses are generally richer in CCN than are maritime air masses. Within a particular air mass at flight level the variation of the median CCN concentration was surprisingly small. At supersaturations between 0.1 and 10%, the median concentration of CCN was found to range from a few tens to a few hundred cm^{-3} in air over oceans, and from a few hundred to a few thousand cm^{-3} in air over the continents.

Figure 9-1 also shows that the concentration N_{CCN} of CCN increases with increasing supersaturation $s_{v,w}$, as would be expected. This behavior can often be expressed adequately by a relation of the form

$$N_{CCN} = Cs_{v,w}^{k}, \tag{9-1}$$

where k and C are approximately constant. Twomey and Wojciechowski found $k \approx 0.7$, $C \approx 100$ cm^{-3} for maritime air, and $k \approx 0.5$, $C \approx 600$ cm^{-3} for continental air. In Australia Twomey (1959a) found $k \approx 0.4$, $C \approx 2000$ cm^{-3} for continental air, and $k \approx 0.3$, $C \approx 125$ cm^{-3} for maritime air. Kocmond (1965) observed $k \approx 0.9$, $C \approx 3500$ cm^{-3} at Buffalo, N.Y., and Jiusto (1967) found $k \approx 0.46$, $C \approx 53$ cm^{-3} at Hilo (Hawaii), and $k \approx 0.63$, $C \approx 105$ cm^{-3} at an island station in Hawaii.

At a given location the CCN concentration is found to vary with time over several orders of magnitude, depending on the proximity of sources and on meteorological factors such as wind direction, air mass type, precipitation, and decreasing or increasing cloudiness (Twomey, 1959a; Jiusto, 1966; Radke and Hobbs, 1969). Figure 9-2 illustrates a typical time variation of the CCN concentration. Notice the effect of air mass changes, wind speed, and wind direction on the CCN concentration. Twomey and Davidson (1970, 1971) showed that at a given location, repeatable patterns can be detected in the diurnal variation of the

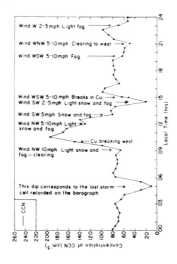

Fig. 9-2. Variation with time of the CCN concentration activated at 1% supersaturation during March 29, 1968, in air at observation station (2025 m) in Olympic Mts., Washington State. (From Radke and Hobbs, 1969; by courtesy of Amer. Meteor. Soc., and the authors.)

CCN concentration. Thus, a noon maximum and a late evening maximum were observed consistently during a one year observation period at Robertson (N.S.W., Australia).

Generally, CCN concentrations in maritime and modified maritime air masses which have been over land less than 2 days rarely exceed $100 \, cm^{-3}$, while concentrations in excess of $10^3 \, cm^{-3}$ are found in air which has been over land for several days (Twomey, 1959a; Jiusto, 1966; Jiusto and Kocmond, 1968; Radke and Hobbs, 1969). This behavior is illustrated in Figure 9-3.

Although CCN concentrations often tend to be particularly large in air over urban areas, a large total concentration of AP does not always necessitate a large number of CCN, as shown in Table 9-1. Similarly, measurements of Terliuc and Gagin (1971) near Jerusalem indicated little, if any, relationship between the Aitken particle and CCN concentrations. On the other hand, Kocmond and Mack (1972) measured concentrations of both CCN (at 0.3% saturation) and Aitken particles which were significantly larger in air downwind of pollution sources in Buffalo (N.Y.) than in the upwind background. Similar observations were made at supersaturations of 0.1 to 1% by Fitzgerald and Spyers-Duran (1973) downstream of pollution sources in St. Louis (Missouri), and by Alkezweeny and Lockhart (1972) in Los Angeles smog.

Table 9-1 implies that in air over land only a very small fraction of the total number of AP is capable of serving as CCN at supersaturations of 1% or less. This fraction may be as large as 1/10, but is typically 1/100 or less. Measurements by Twomey (1963) in the U.S. and Australia support this view. On the other hand, in air over oceans the fraction of total AP to CCN concentration is considerably larger, ranging from 1/5 to 1/10 for a typical AP concentration of $600 \, cm^{-3}$.

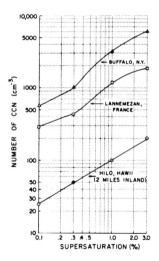

Fig. 9-3. Variation of the CCN concentration required for activation at various locations, as a function of supersaturation. (From Jiusto and Kocmond, 1968; by courtesy of *J. de Rech. Atmos.*, and the authors.)

TABLE 9-1

Comparison between total concentration of aerosol particles and concentration of cloud condensation nuclei activated at 1% supersaturation at various locations.

Location \ Type of Nuclei	Number of Aitken particles (cm^{-3})	Number of CCN (cm^{-3})
Washington D.C.	78 000	2000
	68 000	2000
	57 000	5000
(Allee, 1970)	50 000	7000
Long Island (N.Y.)	51 000	220
	18 000	110
	6 500	150
(Twomey and Severynse, 1964)	5 700	30
Yellowstone National Park (Wyoming) (Auer, 1966)	1 000	15

In relatively pure air with close to 'background' AP concentrations at locations distant from sources, the concentration of CCN is very small. For example, at Yellowstone Park (Wyoming), Auer (1966) measured CCN concentrations at 1% supersaturation which ranged from zero to 78 cm^{-3}, with an average of 15 cm^{-3}. In 1968 Kikuchi (1971) observed at an Antarctic Station (69° S) CCN concentrations as low as zero, but generally near 100 cm^{-3} at 1% supersaturation. In flights over the North Atlantic, Iceland, and Greenland, Flyger *et al.* (1973) found at 1% supersaturation a CCN concentration which in 45% of the cases was less than 10 cm^{-3}, in 80% of the cases was less than 50 cm^{-3}, and in 92% of the cases was less than 100 cm^{-3}.

Observations by Squires and Twomey (1966) in continental air over Colorado and in maritime air over the Caribbean, and by Hoppel *et al.* (1973) in continental air over Arizona and Florida, and in maritime air over the Caribbean, the Pacific, and Alaska have shown that the CCN concentration generally decreases with increasing height over continents in the troposphere, whereas in oceanic environments and polar regions the CCN concentration may remain fairly constant with height and sometimes may even increase above the marine inversion, with the net effect being that at higher altitudes no systematic difference may be found between oceanic and continental air masses. This behavior is illustrated in Figure 9-4.

From our discussions in Chapter 6 we expect that those AP which consist of water-soluble, hygroscopic substances are most suitable for initiating the formation of water drops from water vapor, and therefore will most likely act as CCN. It would seem reasonable, therefore, to assume that the oceans are the most significant source for CCN. However, we recall from Chapter 8 that even close to the ocean surface, the concentration of sea salt particles is too small by a factor of 10 to account for the total of CCN. This conclusion is based on direct measurements of the sea salt particle concentration, as well as on indirect estimates from their rate of production and their residence time. For example, Twomey (1968, 1969, 1971) and Dinger et al. (1970) found that most soluble AP over both oceans and land were volatile when subjected to temperatures above 300 °C, and thus behaved analogously to $(NH_4)_2SO_4$ or possibly NH_4Cl. On the other hand, NaCl aerosols withstood temperatures up to 500 °C. Similarly, Dinger et al. (1970) measured the concentration of CCN activated at 0.75% supersaturation in air over the North Atlantic and over the east coast of Barbados (West Indies), and found that out of a typical population of 100 cm^{-3} near the ocean surface as many as 50 to 90 cm^{-3} were volatile at 300 °C and thus did not consist of NaCl. With increasing height, the percentage of volatile AP increased, reaching almost 100% above about 3 km. On the basis of these measurements, Twomey and Dinger et al. suggested that over the ocean only a small percentage of CCN consist of NaCl, while most are very likely composed of $(NH_4)_2SO_4$. This suggestion is in good agreement with the recent findings of Meszaros and Vissy (1974) discussed in Chapter 8. It is also quite likely that over the continents $(NH_4)_2SO_4$ is the most abundant inorganic component of CCN.

Little is known about the abundance of organic substances in the atmosphere, and their capability of serving as CCN. However, the work of Winkler (1970) suggests that a considerable fraction of AP consists of volatile organic material. Further, Table 8-4 demonstrates that cloud and fog drop residues often contain combustion products in addition to salt, suggesting that the former also serve as CCN. This notion is supported by the observations of Twomey (1960), Warner and Twomey (1967), Warner (1968b), and Woodcock and Jones (1970), who observed a significant local increase in the concentration of CCN as a result of the burning of sugar cane leaves in Hawaii and Australia, and by Hobbs and Radke (1969), who observed a similar CCN increase as a result of forest fires.

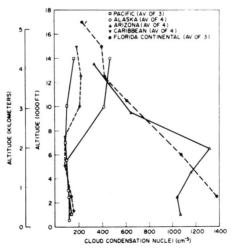

Fig. 9-4. Vertical variation of the CCN concentration activated at 0.7% supersaturation over various locations. (From Hoppel et al., 1973; by courtesy of Amer. Meteor. Soc., and the authors.)

Considering the definition of CCN, it is reasonable to assume that their concentration in a given air volume is to a large extent indicative of the drop concentration in a cloud which forms in that air volume. In fact, the concentrations would indeed be equivalent if the cloud updraft reached the particular supersaturation at which the CCN concentration was determined. The actual supersaturation reached in a given cloud will depend on the size distribution of AP present in the rising air, their chemical nature, the moisture content of the atmosphere, and its thermodynamic state which largely determines the updrafts that can develop (see Chapter 13). However, we may assume roughly that cloud supersaturations rarely exceed a few percent. At these values the concentrations of CCN over continents typically range from 100 to 1000 cm^{-3}, while over oceans they range from a few tens to a few hundred cm^{-3} (Twomey and Wojciechowski, 1969). These values agree well with the drop concentrations found in continental and maritime clouds (see Chapter 2).

9.1.2   MODE OF ACTION OF WATER-SOLUBLE AND MIXED CCN

The nucleation of water drops by water-soluble or mixed AP is controlled by the mass and chemistry of the water-soluble component. As we have seen, at a specific relative humidity, which for most soluble compounds in the atmosphere is well below 100% (Table 4-3), the soluble components of the AP deliquesce into aqueous solution drops. As the relative humidity in the environment rises, such a drop will undergo a slow equilibrium growth by diffusion of water vapor until, if it

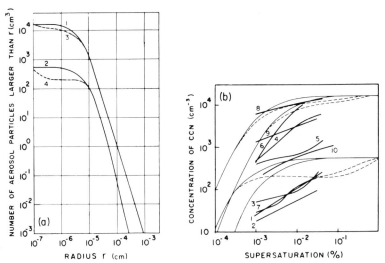

Fig. 9-5. Theoretical computation of CCN concentration. (a) Model cumulative size distributions of atmospheric aerosol particles used in computation. Curves 1, 3 continental type size distributions, curves 2, 4 maritime type size distributions. (b) Comparison between computed and observed CCN concentrations. Upper theoretical curve pair ($\varepsilon_v = 1.0, 0.1$) for continental type size distributions. Lower theoretical curve pair for maritime type size distributions ($\varepsilon_v = 1.0, 0.1$). (1) (2) Jiusto (1967), Hawaii; (3) (5) (8) Twomey (1963), Australia; (9) Twomey (1963), U.S.A.; (4) Kocmond (1965), Buffalo, N.Y.; (6) Wieland (1956), Switzerland; (10) Twomey and Wojciechowski (1969), continental.
(From Junge and McLaren, 1971; by courtesy of Amer. Meteor. Soc., and the authors.)

reaches a critical supersaturation, it will become *activated* and henceforth grow freely and comparatively swiftly, again by vapor diffusion, into a macroscopic cloud drop.

In Chapter 6 we showed that the critical equilibrium supersaturation for activation can be computed from (6-35) or (6-36) for any given composition of the mixed AP. Junge and McLaren (1971) used an equation similar to (6-36) to determine the critical equilibrium supersaturation as a function of AP size, and as a function of the volume fraction of the water-soluble portion of the AP. Assuming two typical size distributions for atmospheric aerosol particles (see Figure 9-5a), their computations permitted a determination of the number of AP which become activated to drops (which is the number of CCN) at any specified supersaturation. A comparison of the computed CCN concentrations with observed CCN spectra is given in Figure 9-5b. It is seen that the chemical composition of AP has little effect on the shape of the CCN spectrum, but strongly affects the CCN concentration at any given supersaturation. On the other hand, the shape of the CCN spectrum was found to depend strongly on the AP size distribution. Both facts can be considered as reasons for the poor agreement between the observations and theoretical predictions of the CCN concentration, evident from Figure 9-5b, since neither the chemical composition nor the AP size distribution was determined at the time the CCN counts given in Figure 9-5b were made. An additional reason for the poor agreement hinges on the difficulty of measuring precisely the supersaturation in the diffusion cloud chambers in which the CCN concentrations are determined. Similar results were obtained by Fitzgerald (1973, 1974).

9.1.3 NUCLEATION ON WATER-INSOLUBLE, PARTIALLY WETTABLE CCN

Let us now consider AP which are wettable by water but completely water-insoluble. For the sake of simplicity it is useful to assume that a water embryo or

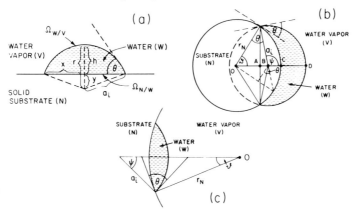

Fig. 9-6. Illustration for determining the volume and surface area of a spherical cap embryo of water: (a) on a planar, water-insoluble, partially wettable substrate, (b) on a spherically convex, water-insoluble, partially wettable substrate, (c) on a spherically concave, water-insoluble, partially wettable substrate.

germ nucleated in supersaturated vapor on a water-insoluble, partially wettable surface assumes the shape of a cap. Although no direct observations of the shape of water germs are available, some experimental justification for the spherical cap assumption has been provided by Gretz (1966a), who used an electron microscope to study the nature and appearance of silver embryos deposited from supersaturated silver vapor on a tungsten substrate.

9.1.3.1 Nucleation on a Planar Substrate

Our present goal is to determine the rate at which water drops are nucleated on an insoluble AP surface. Let us first consider the simplest possible model for this process, and assume the surface on which the nucleation occurs is planar and energetically homogeneous.

We shall adopt the classical theory of nucleation (Chapter 7) for this problem. This procedure is amply justified in view of the success of the theory, its simplicity, and the ease with which it may be extended to new situations. Also, we may note that the 'replacement term' difficulties associated with the attempt to refine the classical approach do not arise here, since the nucleating substrate may be taken to be at rest. In this and the next section, we follow most closely the works of Fletcher (1958, 1959a, b, 1962a), Hirth and Pound (1963), and Pound *et al.* (1954).

Let $c_{1,S}$ be the concentration of single water molecules adsorbed on a surface, and assume they are in metastable equilibrium with the new phase embryos residing there also. By invoking the classical theory of nucleation, we can express the *i*-mer embryo concentration on the surface in a form analogous to (7-24), viz.,

$$c_{i,S} = c_{1,S} \exp\left[-\Delta F_{i,S}/kT\right], \tag{9-2}$$

where $\Delta F_{i,S}$ is the energy of *i*-mer formation on the surface. Then the rate J_s at which germs are formed per unit time and per unit surface area may be written, in analogy to (7-41), as

$$J_S = Z_S c_{g,S} w^{\downarrow} \Omega_{g,S}, \tag{9-3}$$

where $Z_S = [\Delta F_{g,S}/3\, G^2 \pi kT]^{1/2}$ is the Zeldovich factor for surface nucleation (see (7-44)).

In this expression $w^{\downarrow} \Omega_{g,S}$ may be taken to have the same meaning as in (7-41), if we assume the germs grow by direct addition of water molecules from the vapor. For this case we have from (5-49), assuming $\alpha_c = 1$, the result

$$w^{\downarrow} \Omega_{g,S} = \frac{\pi a_g^2 e}{(2\, \pi \dot{m}_w kT)^{1/2}}. \tag{9-4}$$

where we have approximated the cap surface area by πa_g^2.

On the other hand, if we assume that growth occurs primarily by surface diffusion of adsorbed water molecules to the germs, then the quantity $w^{\downarrow} \Omega_{g,S}$ must be interpreted as the product of the number of adsorbed water molecules in a position to join the germ, and the frequency $\nu_S \exp\left[-\Delta F_{sd}/kT\right]$ with which such an adsorbed molecule will jump to join the germ; i.e., in this case we have

$$w^{\downarrow} \Omega_{g,S} = 2\, \pi \bar{\delta} c_{1,S} \nu_S a_g \sin\theta \exp\left[-\Delta F_{sd}/kT\right], \tag{9-5}$$

where $2 \pi a_g \sin \theta$ is the circumference of the cap germ (see Figure 9-6a), ΔF_{sd} is the activation energy for surface diffusion of a water molecule on the substrate, ν_S is the frequency of vibration of an adsorbed molecule normal to the surface ($\approx 10^{13}$ sec^{-1}), and $\bar{\delta}$ is the average distance a molecule moves in a diffusion step or jump.

The assumed steady concentration $c_{1,S}$ may be determined, in principle at least, by equating the flux density of water molecules to the surface with the outward flux of desorbed molecules; the latter is given by

$$w^{\uparrow} = c_{1,S} \nu_S \exp [-\Delta F_{des}/kT], \tag{9-6}$$

where ΔF_{des} is the energy of desorption per molecule. On setting this result equal to (5-49), we obtain

$$c_{1,S} = \frac{e}{(2 \pi \dot{m}_w kT)^{1/2} \nu_S} \exp [\Delta F_{des}/kT]. \tag{9-7}$$

Therefore, on combining (9-3), (9-4), and (9-7), the rate of surface nucleation for the case of germ growth by direct vapor deposition is

$$J_S = \frac{\pi Z_S e a_g^2}{(2 \pi \dot{m}_w kT)^{1/2}} c_{1,S} \exp [-\Delta F_{g,S}/kT] \tag{9-8a}$$

$$= \frac{Z_S e^2 a_g^2}{2 \dot{m}_w kT \nu_S} \exp [(\Delta F_{des} - \Delta F_{g,S})/kT]. \tag{9-8b}$$

Similarly, combining (9-3), (9-5), and (9-7) gives the corresponding rate for the case of growth by surface diffusion of adsorbed molecules:

$$J_S = \frac{Z_S e^2 \bar{\delta} a_g \sin \theta}{\dot{m}_w kT \nu_S} \exp [(2 \Delta F_{des} - \Delta F_{sd} - \Delta F_{g,S})/kT]. \tag{9-9}$$

If we assume $\bar{\delta} \sin \theta \approx a_g$ and compare (9-8) with (9-9), we find the surface diffusion nucleation rate is faster by the factor $\exp [(\Delta F_{des} - \Delta F_{sd})/kT]$; experimental determination of the energy terms is difficult, but generally the qualitative outcome is $\Delta F_{des} > \Delta F_{sd}$.

Let us now turn to the determination of the energy of germ formation, $\Delta F_{g,S}$. As is always done in the classical approach, we shall assume macroscopic values of all relevant parameters. Then on writing $\Delta F_{i,S}$ in terms of bulk volume and surface contributions (recall (7-18) and (7-19)), we have only to cope with a simple geometry problem, namely that of finding the surface and volume of the cap embryo shown in Figure 9-6a. Proceeding in this manner, we write the energy of i-mer formation on the surface in the form

$$\Delta F_{i,S} = V_i \Delta f_{vol} + \sigma_{w,i} \Omega_{w,i} \tag{9-10}$$

where V_i is the i-mer volume, $\Delta f_{vol} = (\mu_w(e, T - \mu_v(e, T))/v_w$ is the bulk energy change per unit volume $[= -\rho_w \mathcal{R} T (\ln S_{v,w})/M_w$, from (7-21)], and the last term in (9-10) represents the total surface energy of the i-mer:

$$\sigma_{w,i} \Omega_{w,i} = \sigma_{w/v} \Omega_{w/v} + (\sigma_{N/w} - \sigma_{N/v}) \Omega_{N/w}. \tag{9-11}$$

In terms of the lengths x, y, and h defined by Figure 9-6a, the i-mer volume is

$$V_i = \frac{a_i}{3}(2\,\pi a_i h) - \frac{\pi x^2 y}{3} = \frac{\pi a_i^3}{3}(2 + m_{w/v})(1 - m_{w/v})^2, \qquad (9\text{-}12)$$

where $m_{w/v} = \cos\theta$. Similarly, $\Omega_{w/v} = 2\,\pi a_i h = 2\,\pi a_i^2(1 - m_{w/v})$, and $\Omega_{N/w} = \pi x^2 = \pi a_i^2(1 - m_{w/v}^2)$, so that

$$\sigma_{w,i}\Omega_{w,i} = 2\,\pi a_i^2 \sigma_{w/v}(1 - m_{w/v}) + \pi a_i^2(\sigma_{N/w} - \sigma_{N/v})(1 - m_{w/v}^2)$$
$$= \pi a_i^2 \sigma_{w/v}(2 + m_{w/v})(1 - m_{w/v})^2, \qquad (9\text{-}13)$$

making use of Young's relation (5-23). Therefore, (9-10) becomes

$$\Delta F_{i,S} = \left(\frac{\pi a_i^3}{3}\Delta f_{vol} + \pi a_i^2 \sigma_{w/v}\right)(2 + m_{w/v})(1 - m_{w/v})^2. \qquad (9\text{-}14)$$

As in the case of homogeneous nucleation, we identify the energy of germ formation as the maximum in the curve of $\Delta F_{i,s}$ vs. a_i; i.e., we set $\partial(\Delta F_{i,S})/\partial a_i = 0$ for fixed $m_{w/v}$ to find the germ radius, which is

$$a_g = -\frac{2\,\sigma_{w/v}}{\Delta f_{vol}} = \frac{2\,M_w\sigma_{w/v}}{\mathscr{R}T\rho_w \ln S_{w,v}}. \qquad (9\text{-}15)$$

This result, the same as for homogeneous nucleation, could have been written down immediately on realizing that the curved surface of the germ must be in equilibrium with the vapor. Substituting this result into (9-14) for $i = g$, the energy of germ formation is

$$\Delta F_{g,S} = \frac{4\,\pi a_g^2 \sigma_{w/v}}{3}[(2 + m_{w/v})(1 - m_{w/v})^2/4] \qquad (9\text{-}16)$$

$$= \frac{16\,\pi M_w^2 \sigma_{w/v}^3}{3[\mathscr{R}T\rho_w \ln S_{v,w}]^2}f(m_{w/v}), \qquad (9\text{-}17)$$

where

$$f(m) = (2 + m)(1 - m^2)/4 \qquad (9\text{-}18)$$

(Volmer, 1939).

If the substrate is completely wettable by water, then $m = 1$, $f = 0$, and there is no energy barrier to nucleation. At the other extreme of a non-wettable surface, $m = -1$, $f = 1$, and $\Delta F_{g,S} = \Delta F_g$, the energy of germ formation for homogeneous nucleation. Of course, this last result is also as expected: The germ drop rests on the non-wettable surface, but does not otherwise interact with it. In general, $0 \leqslant f(m) \leqslant 1$, confirming that the presence of a foreign surface serves to lower the free energy barrier to nucleation.

As we have seen, the rate of embryo growth by surface diffusion of adsorbed molecules dominates, to some extent, the rate of growth by direct deposition of molecules from the vapor. Unfortunately, however, the former process involves contributions which are not accurately known, and so we must be content to proceed with the latter process in order to arrive at a numerical evaluation of the nucleation rate. Thus, Fletcher (1962a) used (9-8a) to estimate J_S. The monomer concentration $c_{1,S}$ was obtained by assuming adsorption yields essentially a

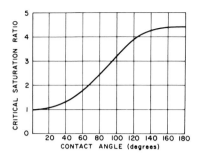

Fig. 9-7. Critical saturation ratio for water drop nucleation ($J_S = 1$ cm^{-2} sec^{-1}) on a planar, water-insoluble, partially wettable substrate as a function of contact angle θ at 0 °C. (From *The Physics of Rain Clouds* by N. H. Fletcher, copyrighted by Cambridge University Press, 1962a.)

monolayer at the supersaturations involved. In this manner Fletcher estimated the prefactor to the exponential term in (9-8a) (the 'kinetic coefficient') to be of the order of 10^{24} to 10^{27} cm^{-2} sec^{-1}. Choosing 10^{25} cm^{-2} sec^{-1} as representative, he used (9-8a) and (9-17) to evaluate $(s_{v,w})_{crit}$ as a function of contact angle θ for $J_S = 1$ cm^{-2} sec^{-1} and T = 0 °C; the results are shown in Figure 9-7. It is seen that the critical saturation ratio for nucleation of water on a water-insoluble, partially wettable, planar substrate increases monotonically with contact angle. The increase is small at small and large wetting angles, and is almost linear for $60 \leq \theta \leq 120°$.

9.1.3.2 *Nucleation on a Curved Substrate*

Fletcher (1958, 1959a) has extended the theory presented in the last section to include an account of the effects of the finite size of the nucleating particle. In this extension the substrate is assumed to be a sphere of radius r_N. The determination of the energy of germ formation is carried out just as before; only the geometry has changed somewhat. Thus, from Figure 9-6b, if we let $\overline{AD} \equiv b = a_i(1 - \cos \psi)$ and $\overline{AC} \equiv c = r_N(1 - \cos \vartheta)$, then the volume of the embryo is $V_i = [\pi b^2(3 a_i - b)/3] - [\pi c^2(3 r_N - c)/3]$, and the surfaces bounding the embryo are $\Omega_{N/w} = 2 \pi r_N c$ and $\Omega_{w/v} = 2 \pi a_i b$. With these we find

$$V_i = \frac{\pi a_i^3}{3}(2 - 3 \cos \psi + \cos^3 \psi) - \frac{\pi r_N^3}{3}(2 - 3 \cos \vartheta + \cos^3 \vartheta) \tag{9-19}$$

and

$$\sigma_{w,i}\Omega_{w,i} = \sigma_{w/v}(\Omega_{w/v} - \Omega_{N/w} \cos \theta)$$
$$= 2 \pi \sigma_{w/v}[a_i^2(1 - \cos \psi) - r_N^2 \cos \vartheta(1 - \cos \vartheta)], \tag{9-20}$$

where $\cos \vartheta = (r_N - a_i \cos \theta)/d$,
$\cos \psi = -(a_i - r_N \cos \theta)/d$,
and $d = (r_N^2 + a_i^2 - 2 a_i r_N \cos \theta)^{1/2}$.

By substituting (9-19) and (9-20) into (9-10) we find, as before, that $(\Delta F_{i,S})_{max} = \Delta F_{g,S}$ occurs at $a_{i,max} = a_g$, where the latter is given by (9-15). Accordingly, the energy of germ formation becomes

$$\Delta F_{g,S} = \frac{16 \pi M_w^2 \sigma_{w/v}^3}{3[\mathscr{R} T \rho_w \ln S_{v,w}]^2} f(m_{v/w}, x), \tag{9-21}$$

where

$$2 f(m, x) = 1 + \left(\frac{1 - mx}{\phi}\right)^3 + x^3 \left[2 - 3\left(\frac{x - m}{\phi}\right) + \left(\frac{x - m}{\phi}\right)^3\right]$$
$$+ 3 mx^2\left(\frac{x - m}{\phi} - 1\right)$$

(9-22)

with

$$\phi = (1 - 2 mx + x^2)^{1/2}, \quad x = r_N/a_g.$$

(9-23)

Let us consider the nucleation rate per particle, J_S'; to a first approximation we simply have $J_S' = 4 \pi r_N^2 J_S$, or

$$J_S' = \frac{4 \pi^2 r_N^2 Z_S e}{(2 \pi \ln kT)^{1/2}} a_g^2 c_{1,S} \exp\left[-\Delta F_{g,S}/kT\right],$$

(9-24)

using (9-8a). The kinetic coefficient in this equation has been estimated by Fletcher (1958, 1959a, b) to be about $10^{26} r_N^2$ (this assumes $r_N > a_g$).

Figure 9-8 presents a plot of the relations (9-21)–(9-24), giving the critical saturation ratio for $J_S' = 1$ germ (particle)$^{-1}$ sec^{-1} at 0 °C, as a function of r_N for various contact angles. It is seen that for all values of $m_{w/v}$, $(S_{v,w})_{crit}$ increases rapidly if $r_N < 0.01 \mu$m, and that for a given substrate particle size, $(S_{v,w})_{crit}$ increases monotonically with increasing contact angle. These results clearly indicate that water-insoluble, partially wettable, spherical AP must be large and exhibit low contact angles for water if they are to serve as CCN.

McDonald (1964) has extracted an approximate solution from (9-21)–(9-24) for the most interesting and relevant case of small contact angles and super-saturations; the result is

$$\cos \theta = 1 - \left(\frac{x - 1}{x}\right) [0.662 + 0.022 \ln r_N]^{1/2} \ln (S_{v,w})_{crit},$$

(9-25)

where $(S_{v,w})_{crit}$ corresponds again to $J_S' = 1$ germ (particle)$^{-1}$ sec^{-1} at 0 °C and where again $x = r_N/a_g$. This equation is plotted in Figure 9-9, where comparison is also made with the critical supersaturation required to activate mixed AP containing various volume proportions of NaCl. The figure reveals a strong

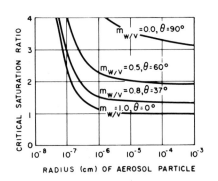

Fig. 9-8. Critical saturation ratio with respect to water for water nucleation ($J_S' = 1$ sec^{-1} particle^{-1}) on a spherical, water-insoluble particle of radius r_N as a function of $m = \cos \theta$, for 0 °C. (From *The Physics of Rain Clouds* by N. H. Fletcher, copyrighted by Cambridge University Press, 1962a.)

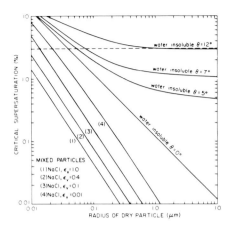

Fig. 9-9.   Critical supersaturation for water nucleation ($J'_S = 1 \sec^{-1}$ particle^{-1}) on a spherical substrate particle of radius r_N, as a function of contact angle θ for water-insoluble particles, and as a function of the volume percentage of NaCl in mixed particles; for 0 °C.

dependence of $(s_{v,w})_{crit}$ on contact angle. For example, if we consider that the supersaturation reached in clouds is typically smaller than 3%, we must require that AP of radii $\geqslant 0.01$ μm have contact angles less than 12° if they are to serve as CCN in this supersaturation range.

Unfortunately, little is known about the contact angle for insoluble AP, except for some measurements of water against certain silicates (see Table 5-2). These values suggest that silicate particles are not likely to serve as CCN. Even if the contact angle on insoluble AP were zero, it would still be very unlikely that many such particles would become involved in the condensation process, because sufficient soluble and mixed particles are generally available to nucleate at significantly lower supersaturations. Although continents release predominantly insoluble AP, many of these soon coagulate with soluble AP to become mixed particles, which are then competitive with CCN from the oceans.

Recently, Mahata and Alofs (1975) modified the heterogeneous nucleation theory of Fletcher to consider nucleation on an insoluble, partially wettable, spherically concave substrate of radius of curvature r_N (see Figure 9-6c). The only change from the previous theory is one of geometry. Thus, in place of (9-19) we now have

$$V_i = \frac{\pi a_i^3}{3} (2 - 3 \cos \psi + \cos^3 \psi) + \frac{\pi r_N^3}{3} (2 - 3 \cos \vartheta + \cos^3 \vartheta), \qquad (9\text{-}26)$$

while the surface energy contribution still has the form of (9-20); however, in contrast to the convex case, we now have $\cos \vartheta = (r_N + a_i \cos \theta)/d$, $\cos \psi = (a_i + r_N \cos \theta)/d$, and $d = (r_N^2 + a_i^2 + 2 r_N a_i \cos \theta)^{1/2}$. The resulting free energy of

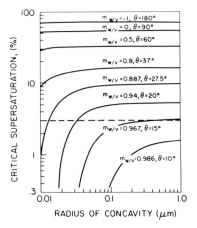

Fig. 9-10. Critical supersaturation for water nucleation ($J_S = 1\ \text{sec}^{-1}\,\text{cm}^{-2}$) on a concave, water-insoluble, partially wettable substrate, as a function of the radius of curvature and contact angle θ, for 23 °C. (From Mahata and Alofs, 1975; by courtesy of Amer. Meteor. Soc., and the authors.)

germ formation is given by (9-21), where instead of (9-22) and (9-23), we have

$$2\,f(m, x) = 1 - \left(\frac{1 + mx}{\phi}\right)^3 - x^3\left[2 - 3\left(\frac{x + m}{\phi}\right) + \left(\frac{x + m}{\phi}\right)^3\right]$$
$$+ 3\,mx^2\left(\frac{x + m}{\phi} - 1\right), \qquad\qquad (9\text{-}27)$$

and

$$\phi = (1 + 2\,mx + x^2)^{1/2}, \quad x = r_N/a_g. \qquad\qquad (9\text{-}28)$$

Mahata's computations, based on Fletcher's value of $10^{25}\,\text{cm}^{-2}\,\text{sec}^{-1}$ for the kinetic coefficient in (9-8a) (and with the minor change of setting $T = 23$ °C instead of 0 °C), are shown in Figure 9-10. The results indicate that concave surface features can significantly enhance the capacity of AP to serve as CCN.

Of course, one should realize that this enhancement results from an 'inverse Kelvin law' effect which must break down at some minimum radius. Mahata assumes, somewhat arbitrarily, that this cutoff should occur near 0.01 μm radius. The figure shows that surface roughness of this resolution has the effect, at 3% supersaturation, of inducing nucleation for contact angles $\theta \lesssim 30°$. This is considerably less stringent than the 12° figure for convex surface features, but is still insufficient to indicate that many insoluble AP could act as CCN.

9.1.4 EXPERIMENTAL VERIFICATION OF HETEROGENEOUS WATER DROP NUCLEATION

Since water-soluble and mixed AP deliquesce and subsequently grow with increasing relative humidity, it is appropriate to test the theory of nucleation on such particles in two ways: (1) by experimentally determining the equilibrium growth of these particles, and (2) by determining the number of particles of given size and composition which become activated at a given supersaturation. The results of such equilibrium growth experiments, discussed in Chapter 6, have been found to be in good accord with theory. Observed CCN concentrations have been cited in Section 9.1.2, where it was pointed out that the agreement with theory is satisfactory as regards the shape of the CCN spectral curve, but that there is some difficulty in predicting correctly the actual concentrations of activated drops.

Attempts to verify experimentally the predictions of water drop nucleation on water-insoluble, partially wettable substrates have been made by Twomey (1959b), Koutsky *et al.* (1965), Isaka (1972), van de Hage (1972), and by Mahata and Alofs (1975). Their experiments were carried out in diffusion chambers where nucleation was forced to take place on *plane* substrates for which the contact angle of water had been measured separately. Unfortunately, the results derived from the experiments show considerable disagreement. While the critical supersaturations necessary for onset of drop nucleation were found by van de Hage and by Isaka to be significantly lower for all wetting angles than those predicted by theory, those measured by Twomey agree well with theory up to $\theta = 100°$.

In a critical review, Mahata and Alofs (1975) pointed out some serious experimental difficulties involved in such measurements, and expressed the opinion that these were the cause of the contradictory results. In an attempt to avoid the errors of previous investigators, they carried out new experiments, the results of which are reproduced in Figure 9-11. It is seen that their measured critical supersaturations as a function of contact angle agree fairly well with

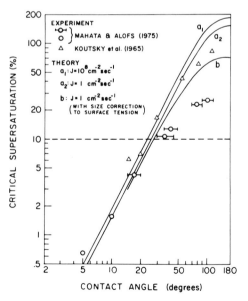

Fig. 9-11. Critical supersaturation for onset of water nucleation on a plane, water-insoluble, partially wettable substrate, as a function of contact angle of water on substrate. Comparison of experiment with theory. (From Mahata and Alofs, 1975, with changes.)

those of Koutsky *et al.* (1965) up to a supersaturation of about 10%, corresponding to a contact angle of about 25°. In this range the experimental results also
agree with the predictions of the classical theory, expecially if a size correction
for $\sigma_{w/v}$ is made, such as (5-22). Since the supersaturation in atmospheric clouds
rarely exceeds 10%, the disagreement between experiment and theory at larger
contact angles is of little significance.

The success of the classical nucleation theory in its simple extension to
heterogeneous nucleation is surprisingly good, as it has several apparent
deficiencies: (1) The theory assumes that a macroscopic contact angle and a
macroscopic interfacial free energy and bulk density applies to submicron-sized
water germs. (2) It treats the surface of a nucleating substrate as energetically
homogeneous. (3) The theory, in Fletcher's version, assumes that a water
embryo on the nucleating substrate grows by the addition of water molecules
directly from the vapor.

Let us now elaborate on some of these criticisms. (1) The concept of a contact
or wetting angle was discussed in Section 5.5. The angle is defined as the limiting
angle of water toward the solid surface on which the drop rests, the system
being in equilibrium with saturated vapor. As pointed out by Corrin (1975),
contact angles are measured under the microscope with a linear resolution
generally no better than 1 μm. This implies that the angle is not truly measured
at the surface of the substrate, but is inferred by extrapolation from measurements of the drop profile well removed, on a molecular scale, from the surface.
The phenomenon of hysteresis adds to the difficulties in measuring contact
angles, making any measurement hard to reproduce. Even if contact angles
could be measured accurately, the values would only apply to macroscopic
systems and would have little meaning in the case of water-'caps' of germ-size.
In addition, one has to recall that the contact angle is not only a function of the
surface properties of the substrate, but is also a function of temperature and
supersaturation.

Criticisms of the assumption of macroscopic interface free energies and bulk densities for small systems such as water germs have already been expressed in Chapter 7 in the context of homogeneous nucleation, and need not be repeated here.

(2) The classical theory assumes that the thermodynamic functions employed
are independent of location on the nucleating substrate. This implies that the
substrate surface is treated as energetically homogeneous. It is well known,
however, that surfaces are energetically heterogeneous. This behavior is well
documented by numerous adsorption studies, some of which were discussed in
Chapter 5. These studies show that the surfaces of many solids do not adsorb
water molecules uniformly, but rather preferentially at certain active sites. It is
at these locations that phase changes are most likely to occur.

One may distinguish among three types of active sites for preferred adsorption
of water molecules from the vapor, and hence preferred water drop formation.
The first type of site is represented by a morphological surface inhomogeneity
such as a step, crack, or cavity at the surface of the nucleating substrate. The
second type is represented by a chemical inhomogeneity in the surface, generally
caused by the presence of a foreign ion, which is hydrophilic relative to the rest
of the solid surface. The third type of site is represented by electrical in-

homogeneities other than ions in the surface of the nucleating substrate. Such sites may consist of sharply defined boundaries between surface regions of different electric field sign, or of locations where the electric field vector in the substrate surface is oriented parallel to the surface.

Morphological surface inhomogeneities are high energy sites where surface forces are available to effectively tie water molecules to the surface. Chemical inhomogeneities attract water molecules to the substrate surface by means of electric forces which develop between the dipole moment of the water molecule and the net dipole or charge on the foreign atom or ion. Growth of a water cluster at such a site is also aided by the relatively higher mobility of molecules on the substrate surface surrounding the hydrophilic site. Electric in-homogeneities other than ions attract water molecules to the substrate surface through interaction between local electric dipoles in the solid substrate and the dipole of a water molecule. Growth of water clusters at such sites can be aided if the diffusivity of water molecules on surface regions with either an inward-directed or an outward-directed electric field is high as compared to the diffusivity over the boundary between regions of electrically different sign, or as compared to the diffusivity over an area where the electric field vector in the substrate is oriented parallel to the substrate, since then both the positive and the negative ends of the water molecule are partially tied down.

(3) The adsorption studies of Corrin *et al.*, Zettlemoyer *et al.*, Federer, and Pruppacher and Pflaum referred to in Chapter 5 indicate that new phase embryos do not grow solely by the addition of water molecules directly from the vapor, but rather to a large extent as a result of surface diffusion of water molecules. (See also the statement immediately following (9-9).)

We recall that for the case of homogeneous nucleation, the deficiencies of the classical approach have been circumvented, in some instances, by the molecular model method of Hale and Plummer. Unfortunately, it is difficult to extend this method to a study of heterogeneous nucleation. This is because little is known of the mode and energy of interactions between clusters of water molecules and substrate molecules. The major problem in determining the interaction energy is that it is specific to each solid and to each crystallographic face of that solid. Furthermore, it is a function of position on each crystallographic face, due to the heterogeneous nature of the surface. In spite of the difficulties, preliminary studies of Plummer and Hale (1973) have produced at least qualitative agreement with observations by predicting that the weaker the interaction between the surface cluster and the solid substrate, the more likely it is that water molecules will arrange in isolated multilayered structures instead of a single layer.

9.2 Ice Forming Nuclei (IN)

9.2.1 NUMBER CONCENTRATION OF IN

Ice forming nuclei exhibit three basic modes of action. In the first, water is adsorbed directly from the vapor phase onto the surface of the IN where, at sufficiently low temperatures, the adsorbed vapor is transformed into ice. In the second mode, the IN initiates the ice phase from inside a supercooled water drop. (Any of the scavenging mechanisms discussed in Chapter 12 may cause AP to enter a cloud or raindrop. Alternatively, IN find their way into cloud drops as a result of becoming involved in the condensation process.) In the third mode of action, the IN initiates the ice phase at the moment of contact with the supercooled drop. Any of the scavenging mechanisms discussed in Chapter 12 may provoke such a contact.

The first mode of action is called the *deposition mode*, and AP which exhibit this behavior are called *deposition IN*; the second is called the *freezing mode*, and the corresponding AP are *freezing nuclei*; the third is called the *contact mode*, participated in by *contact nuclei*. Unfortunately, none of the presently available devices which count the fraction of AP acting as IN is capable of allowing for all the different modes of action, nor can they realistically simulate the time scale over which temperatures and supersaturations vary in at-mospheric clouds. Therefore, the IN concentrations quoted in the literature have to be treated with considerable caution.

In Figures 9-12 and 9-13 a selected number of IN concentration measurements are reproduced as a function of temperature and location. One notices that the mean or median IN concentration exhibits no systematic variation with geo-graphic location. This suggests that, far from sources, the atmospheric aerosol is quite uniform with respect to its ability to initiate the ice phase. Figure 9-12 shows further that at any given temperature the mean concentration of IN varies

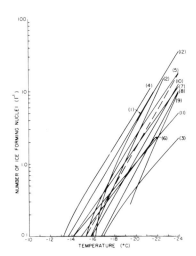

Fig. 9-12. Variation of the mean or median number concentration of IN with temperature and geographic location. (1) Bracknell (England), 51° N, 0° W, (2) Clermont-Ferrand (France), 46° N, 3° E, (3) Corvallis (Oregon, U.S.), 44° N, 123° W, (4) Tokyo (Japan), 36° N, 140° E, (5) Tucson (Arizona, U.S.), 32° N, 111° W, (6) Jerusalem (Israel), 32° N, 35° E, (7) Palmbeach (Florida, U.S.), 27° N, 80° W, (8) Hawaii (U.S.), 20° N, 158° W, (9) Swakopmund (S. Africa), 34° S, 14° E, (10) Sidney (Australia), 34° S, 151° E, (11) Tasmania (Australia), 43° S, 147° E, (12) Antarctica, 78° S, 166° E. The dashed line represents $N_{IN} = 10^{-5} \exp(0.6 \Delta T)$. (Data from Hefferman and Bracewell, 1959; Isono *et al.*, 1959; Carte and Mossop, 1960; Bigg and Hopwood, 1963; Kline, 1963; Soulage, 1964; Bigg, 1965; Droessler and Hefferman, 1965; Gagin, 1965; Stevenson, 1968; and Mossop *et al.*, 1970.)

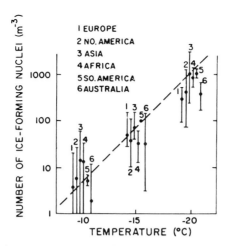

Fig. 9-13. Range of median number concentration of IN as function of temperature for various geographic locations; 44 stations. The dashed line represents $N_{IN} = 10^{-5} \exp(0.6 \, \Delta T)$. (From Bigg and Stevenson, 1970; by courtesy of *J. de Rech. Atmos.*, and the authors.)

by factors up to 10. Figure 9-14 illustrates a day to day variation in the IN concentration observed with one counting technique at one particular site. Note that even at remote locations such as the Antarctic, pronounced variations in the daily IN concentrations do occur. Similar observations were made by Kikuchi (1971), who measured at an Antarctic station IN concentration maxima as high as 30 ℓ^{-1} at -20 °C. Such high concentration counts were found to last up to 2 days.

Short term, positive anomalies in the IN concentration have also been found to occur at other latitudes. Such anomalies, termed *IN-storms*, are characterized by a sudden rapid increase of the IN concentration within a day or less, to values which may be several orders of magnitude larger than the typical average. After a few days this rapid rise is followed by a similar rapid decrease in concentration.

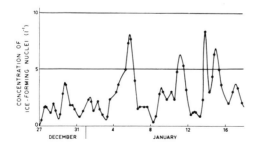

Fig. 9-14. Variation of the daily number concentration of ice forming nuclei at -20 °C measured during December, 1961, to January, 1962, at Antarctica station. (From Bigg and Hopwood, 1963; by courtesy of Amer. Meteor. Soc., and the authors.)

Various explanations for these IN-storms have been offered. Bigg and Miles (1963, 1964), Droessler (1964), and Bigg (1967) attempted to explain this phenomenon on the basis of an extraterrestrial source of IN. They hypothesized that local subsidence of air from a nucleus-rich stratosphere, occurring in the vicinity of jet streams, gives rise to layers in the troposphere which have relatively high concentrations of IN. It was supposed that the particles in these layers are transported to the ground by vertical mixing. In support of this hypothesis, Droessler (1964) found that a pronounced IN-storm which occurred in Dec., 1963, in S.E. Australia was accompanied by a stable jet stream situation, stratospheric subsidence in the vicinity of the IN-storm area, and an associated unstable troposphere. Further support to this concept was given by Telford (1960), Bigg et al. (1961), and Bigg and Miles (1963), who found that the IN concentration was significantly higher in the upper troposphere and lower stratosphere than at the ground, and by Rosinski (1967b) who found the highest IN concentrations in the vicinity of jet streams.

A second explanation for the phenomenon of IN-storms was given by Isono et al. (1959, 1970) and Hobbs et al. (1971b, c), based on air trajectory analysis. They concluded that local IN sources, such as volcanic eruptions, or dust storms in N. China and Mongolia or in the Sahara Desert, inject IN which are advected over thousands of miles by strong tropospheric winds such as jet streams; intermittent vertical mixing and deposition should then follow. As a third alternative, Isono and Tanaka (1966), Georgii and Kleinjung (1968), and Ryan and Scott (1969) attributed IN-storms to the local formation of IN by evaporation of cloud and precipitation particles. A fourth explanation was given by Higuchi and Wushiki (1970) who noted that aerosols, sampled at Barrow (Alaska) at air temperatures between −20 and −40 °C, and on Mt. Fuji (3776 m, Japan) at air temperatures between 0 and −27 °C, contained IN concentrations which were considerably higher when the aerosol samples were kept below 0 °C than when the samples were heated above 0 °C. They concluded, therefore, that terrestrial AP become activated at sufficiently cold temperatures, such as those found in the upper troposphere or in polar air masses.

Figures 9-12 and 9-13 also show that the IN counts increase nearly exponentially with decreasing temperature. A convenient approximate statement of this behavior, due to Fletcher (1962a), is the following:

$$N_{IN} = A \exp(\beta \Delta T), \tag{9-29}$$

where $\beta = 0.6(°C)^{-1}$, $A = 10^{-5} \ell^{-1}$, where N_{IN} is the number concentration of IN active at a temperature warmer than T, and where $\Delta T = T_0 - T$.

If we compare the number concentration of IN with the total AP, it becomes clear that the ice forming process is very selective. Consider, for example, that at a temperature as low as −20 °C, at which the atmosphere typically contains 1 IN ℓ^{-1}, the ratio of the number concentration of IN to that of AP is as small as 10^{-6} for a total AP concentration of only $10^3 \, cm^{-3}$. In polluted areas and at warmer temperatures, this ratio is even smaller.

Experiments show that the three modes of action by which AP initiate the ice phase are strong functions of temperature. All three modes share the characteristic that the number of AP which act as IN increases with decreasing temperature. But while the freezing and contact modes depend exclusively on the air temperature, the deposition mode depends, in addition, on the humidity.

For a number of years it was assumed that water saturation is needed for AP to act as IN in the deposition mode (Weickmann, 1949). Experiments by Bryant *et al.* (1959), Mason and van den Heuvel (1959), Roberts and Hallett (1968), Gagin (1972), Matsubara (1973), and by Huffman (1973a, b) have shown, however, that such ice nucleation may proceed at relative humidities (with respect to water) less than 100%, as long as there is supersaturation with respect to ice. Gagin and Huffman found that, at any given temperature, the number concentration of IN increases with increasing relative humidity (Figure 9-15a), and correlates logarithmically with the supersaturation over ice, independently of temperature (Figure 9-15b), according to the relation

$$N_{IN} = Cs_{v,i}^k, \qquad\qquad (9\text{-}30)$$

where C and k are 'constants.' Huffman found that k assumed a value of 3 in air over rural N.E. Colorado, a value of 4.5 at Laramie (Wyoming), and a value of 8 for air near St. Louis (Missouri). Note that the form of (9-30) is the same as that of (9-1) for the CCN concentration.

The experiments mentioned demonstrated further that, in contrast to the case of homogeneous nucleation, Ostwald's rule of stages (see Section 7.2.2) becomes inverted below a certain critical temperature which is specific to each IN. Below this temperature ice nucleation in the deposition mode may proceed at ice supersaturations less than water saturation, while at warmer temperatures water

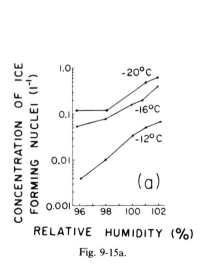

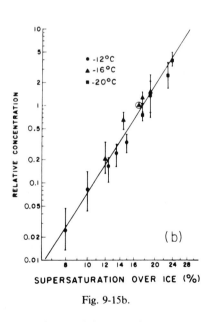

Fig. 9-15a.   Fig. 9-15b.

Figs. 9-15a, b.   Variation of the IN concentration with (a) relative humidity and with (b) super-saturation over ice. (From Huffman, 1973a; by courtesy of Amer. Meteor. Soc., and the author.)

saturation is required (Figure 9-16). (Note the requirement for water saturation does not imply that the IN will be submerged in water, and hence nucleate via the freezing mode, since we recall from Figure 9-9 that water *super*-saturations are needed for drop formation on even completely wettable insoluble surfaces.) Some additional comments on this result will be made in Section 9.2.3.5.

Considering that the AP concentration decreases with increasing height, one might expect the same for the concentration of IN. However, as we have noted already, Telford (1960), Bigg *et al.* (1961), and Bigg and Miles (1963) found that in air over Australia the concentration of IN was significantly higher at 13 to 27 km altitude than at the ground. Analogously, Rosinski (1967b) found over Colorado that on some occasions the IN concentration was highest near the jet stream. In more detailed investigations, Bigg and Miles (1964) and Bigg (1967) found that at ground level in the Southern Hemisphere high IN concentrations occurred in long strips of 100 to 300 km width, as well as inside elongated layers at 4 and 11 km altitude. These banded regions were narrower and had higher IN concentrations aloft than at the ground. Isolated smaller regions with high concentrations were also found. These regions of high concentration were found to be mostly confined to the latitude belt from 23 to 30° S. In addition to these anomalous regions, broad sections at 10 to 12 km altitude were found near the equator where the air was markedly deficient in IN.

Huffman (1973b) observed that the vertical IN concentration profiles often exhibit a pronounced layer-structure even in the lowest few kilometers above ground. For example, a strong concentration maximum was observed during a few days of summertime sampling at about 500 m above ground over St. Louis and over N.E. Colorado. No temperature inversion was present at the time over N.E. Colorado, and the inversion over St. Louis was considerably above the IN concentration maximum; thus the concentration maxima appear not to be correlated with the temperature inversion. It is likely that IN concentration maxima in the atmosphere and IN-storms at the ground are closely related and have the same causes.

If one compares the concentrations of IN and cloud ice particles at nearby locations, a rather unexpected discovery may be made. One finds that in many

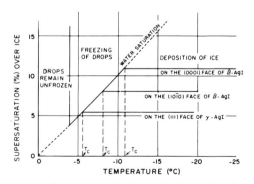

Fig. 9-16. Minimum supersaturation over ice for ice nucleation in the deposition mode on various AgI faces. (From Isono and Ishizaka, 1972; by courtesy of *J. de Rech. Atmos.*, and the authors.)

TABLE 9-2

Fraction in percent of soil particles (N.E. Colorado) nucleating ice in the deposition mode at given conditions. (Based on data of Langer and Rodgers, 1975.)

Temperature (°C)	Supersaturation with respect to water (%)	Particle size interval (μm)				
		10–44	44–74	74–117	117–250	>250
−6	2	0.0	0.0	0.0	2.7	14
−12	2	0.3	0.6	4.0	23	90
−18	1	1.0	0.6	7.3	66	96
−17	9	0.6	1.4	12	88	100

clouds, particularly at relatively warm temperatures, the concentration of ice particles may exceed by many orders of magnitude the concentration of IN determined at the cloud top temperature. Observations to this effect were made by Hobbs (1969) and Hobbs *et al.* (1974b) over the Cascade Mts. (State of Washington), by Auer *et al.* (1969) in stable cap clouds over Wyoming, by Isono (1965), Ono (1972), and Magono and Lee (1973) over Japan, by Mossop (1970, 1971, 1972), Mossop *et al.* (1967, 1968, 1970, 1972), Mossop and Ono (1969) in cumulus clouds over Australia and Tasmania, and by Braham (1964) and Koenig (1963, 1965) in cumulus clouds over Missouri.

These observations show that the maximum *enhancement ratio* or *enhancement factor* R_M, defined as the ratio of the ice crystal concentration to the IN concentration determined at the cloud top temperature, can be as large as 10^4 to 10^5 at temperatures between −5 and −15 °C. Also, R_M tends to decrease with decreasing cloud top temperature, and reaches unity between −25 and −30 °C. Figure 9-17 presents some observations of R_M over the Cascade Mts. during the

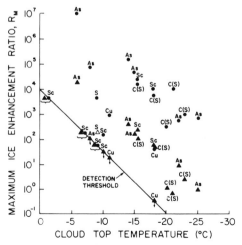

Fig. 9-17. Maximum ice enhancement ratio (R_M) in clouds over the Cascade Mts. (Washington State) during winter, 1971–72. (If value is below the detection threshold, it is shown on the threshold line with an arrow attached.) As–altostratus, S–stratus, Sc–stratocumulus, Cu–cumulus, C(S)–cumulus with stratified tops. (From Hobbs *et al.*, 1974b; by courtesy of the authors.)

winters of 1971 to 1973, which were summarized by the best-fit relation log $R_M = 5.02 + 0.204 \, T_c$, where T_c is the cloud top temperature in °C.

Heymsfield (1972) and Heymsfield and Knollenberg (1972) found an opposite situation prevailing in cirrus clouds. While at cirrus cloud level and $-30°C$ the IN concentration was typically $10^3 \, \ell^{-1}$, the maximum ice crystal concentration was only about 50 ℓ^{-1}. This behavior was attributed to a dilution mechanism operating in cirrus clouds. Ice multiplication, if any, was found to be due to the break-up of rosettes into single bullets, providing at most an enhancement factor of 9 (Kikuchi, 1968). Similarly, little or no ice multiplication was found in mid-level or low-level layer clouds in Australia (Mossop, 1972; Mossop et al., 1972), nor in stratus clouds over Alaska (Jayaweera, 1972a). Negligible ice multiplication was also reported by Gagin (1971) in winter cumuli over Israel, the maximum enhancement ratio varying at most between 1 and 10. A description of the various mechanisms responsible for ice multiplication in clouds is given in Section 16.6.

9.2.2 SOURCES OF IN

Some clues as to possible sources of IN are provided by the chemical identification of AP found at the center of snow crystals. Such studies were carried out by Kumai (1951, 1957, 1961), Isono (1955), Isono et al. (1961), Kumai and Francis (1962b), and Rucklidge (1965) using electron microscope and electron-diffraction techniques. Typically, one solid silicate particle, usually identified as clay, was found in the central portion of a snow crystal (see Table 8-13). The diameters of the particles ranged from 0.1 to 15 μm, with a mode between 0.4 and 1 μm. These findings suggest that desert and arid regions of the Earth surface are a major source of IN.

In support of this possibility, Isono et al. (1959, 1970) and Hobbs et al. (1971b, c) used an air mass trajectory analysis to show that high IN concentrations over Japan and the northwestern U.S. are often the result of local dust storms over arid regions of N. China and Mongolia. During such storms, clay particles become airborne and are transported towards Japan and eastward to the U.S. continent via the jet stream.

The notion that surface soils act as an IN source is also strongly supported by laboratory experiments. The ice-forming capability of silicate particles collected at various parts of the Northern Hemisphere has been tested by Schaefer (1950), Pruppacher and Sänger (1955), Mason and Maybank (1958), Isono et al. (1959), Mason (1960a), Isono and Ikebe (1960), and Roberts and Hallett (1968). When these particles were allowed to function as IN in the deposition mode at water saturation, the threshold ice-forming temperature (which is conventionally taken as the temperature at which 1 particle in 10^4 produces an ice crystal) was found to range typically between -10 and $-20°C$. Clay particles such as kaolinite, anauxite, illite, and metabentonite have a threshold temperature as warm as $-9°C$ (Mason, 1960a).

Surprisingly, clays such as kaolinite often exhibited varying ice nucleating ability, or 'nucleability,' depending on the location at which the clays were sampled. Similarly, soils from different parts of Australia were found to be generally less active than those from the Northern Hemisphere, the threshold temperature of Australian soils ranging between −18 and −22 °C (Paterson and Spillane, 1967). The observations suggest that other substances admixed to soils in small quantities may importantly affect their ice nucleation behavior.

Experiments by Roberts and Hallett (1968) showed that clay particles, having been involved once in ice crystal formation, can exhibit a considerably improved nucleability. AP which behave in such a manner are termed *preactivated*. Roberts and Hallett showed that the IN activity spectrum for each of the tested clays, containing particles of diameters between 0.5 and 3 μm, shifted by more than 10 °C towards warmer temperatures (Figure 9-18), provided that the temperature of the air in which the clay particles were kept never rose above 0 °C, and its relative humidity never fell below 50%.

Clays also seem to exhibit significant differences in their ice nucleating ability according to their mode of action. Isono and Ikebe (1960), Mason (1960a), and Roberts and Hallett (1968) showed that kaolinite acting in the deposition mode has a typical threshold temperature of about −9 °C, and reaches full activity (1 to 1 ice crystal production ratio) near −20 °C, while montmorillonite has a threshold temperature of about −25 °C, and reaches full activity near −30 °C. In contrast, Hoffer (1961) found that the warmest freezing temperature of 50 to 60 μm radius drops containing kaolinite and montmorillonite was −13.5 °C, and that the median freezing temperatures were −24.0 and −32.5 °C, respectively. Gokhale and Spengler (1972), on the other hand, found that drops of 2 to 3 mm radius, freely suspended in the air stream of a wind tunnel, froze at temperatures as warm as −2.5 °C, with full activity between −6 and −8 °C, when contacted by red soil, sand, or clay particles. In agreement with the observations of Hoffer and of Gokhale and Spengler, Pitter and Pruppacher (1973) found from wind tunnel experiments that freely falling drops of 325 μm radius froze at temperatures below −14 °C when freezing was initiated from within a drop by kaolinite or montmorillonite (freezing mode), with full activity being achieved below −28 °C. On the other hand, the same drops froze at temperatures as warm

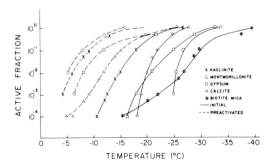

Fig. 9-18. Variation of the ice nucleating ability of clay mineral particles with temperature: Initial ice nucleation ability and ice nucleation ability after preactivation. (From Roberts and Hallett, 1968; by courtesy of *Quart. J. Roy. Meteor. Soc.*)

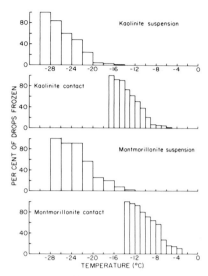

Fig. 9-19. Comparison between the freezing temperature of drops frozen by contact with clay particles and the freezing temperature of drops frozen by clay particles suspended in the drops. (From Pitter and Pruppacher, 1973; by courtesy of *Quart. J. Roy. Meteor. Soc.*)

as −4 °C and with full activity near −14 °C when contacted by dry clay particles (contact mode). These results are illustrated in Figure 9-19.

While monitoring the IN concentration in Japan, investigators detected a second source of IN. Isono and Komabayasi (1954), Isono (1959b), and Isono *et al.* (1959) observed that pronounced IN storms arose following the eruption of volcanoes. As an example of this effect, the time variation of the IN concentration during the eruption of volcano Asama is given in Figure 9-20. In support of the notion that volcanoes act as IN sources, Mason and Maybank (1958),

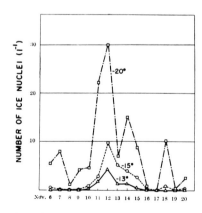

Fig. 9-20. Effect of the eruption of volcano Asama on Nov. 10, 1958, in Japan on the concentration of IN. (From Isono, 1959b; by courtesy of *Nature*, and the author.)

Isono *et al.* (1959), and Isono and Ikebe (1960) found during laboratory experiments that volcanic ash and other volcanic materials were capable of serving as IN with threshold temperatures as warm as $-7.5\,°C$. However, in contrast to these findings, Price and Pales (1963) found no significant increase of the IN concentration during the eruption of volcanoes on the Hawaiian Islands, indicating that not all volcanic material has a high ice nucleability.

From Table 8-13 it is evident that some of the center particles of ice crystals consist of combustion products. This finding suggests natural or anthropogenic combustion sources for IN. In support of this notion, Hobbs and Locatelli (1969) observed a significant increase in the concentration of IN downwind of a forest fire, and Pueschel and Langer (1973) observed increased IN concentrations during sugar cane fires in Hawaii.

Evidence for the effectiveness of other sources of anthropogenic IN has been provided by Soulage (1958, 1964), Telford (1960), Admirat (1962), Langer and Rosinski (1967), and Langer (1968), who showed that certain industries, in particular, steel mills, aluminum works, sulfide works, and some power plants, release considerable amounts of IN into the atmosphere. The particles emitted from electric steel furnaces in France were found to be of particularly high effectiveness as IN, with a threshold temperature of about $-9\,°C$. Wirth (1966), Soulage (1966), and Georgii and Kleinjung (1968) showed that the mean IN concentration is comparatively high throughout heavily industrialized Europe. For example, the mean IN concentration observed during summer, 1964, at 10 different locations in Europe ranged from 2.6 to 53.7 ℓ^{-1} at $-20\,°C$.

These high IN concentrations can be understood in part from the fact that some of the particles emitted during industrial processes consist of metal-oxides, most of which are known to have a high nucleability. Thus, Fukuta (1958), Serpolay (1958, 1959), Mason and van den Heuvel (1959), and Katz (1960) found that the oxides Ag_2O, Cu_2O, NiO, CoO, Al_2O_3, CdO, Mn_3O_4, and MgO exhibit threshold temperatures between -5 and $-12\,°C$, while oxides such as CuO, MnO_2, SnO, ZnO, and Fe_3O_4 have threshold temperatures between -12 and $-20\,°C$. Also, particles of portland cement ($3CaO·SiO_2$, $2CaO·Si_2O$, $3CaO·Al_2O_3$, $4CaO·Al_2O_3·Fe_2O_3$) were found to act as IN at temperatures as warm as $-5\,°C$ (Murty and Murty, 1972).

Although some specific anthropogenic sources may emit IN, the anthropogenic emission from urban complexes as a whole is generally deficient in IN. Thus, studies by Braham (1974) and Braham and Spyers-Duran (1975) carried out in the area of St. Louis show that, on the average, fewer IN were found downwind of the city than upwind. This suggests that anthropogenic combustion products emitted into the air over urban areas are generally poor IN and, in addition, are capable of deactivating existing IN.

Two processes have been suggested by which IN become deactivated. Georgii (1963) and Georgii and Kleinjung (1967) showed that foreign gases such as SO_2, NH_3, and NO_2 severely reduce the ability of atmospheric AP to serve as IN. The higher the concentration of these gases, the stronger is the deactivation. A second mechanism of deactivation was also proposed by Georgii and Kleinjung (1967). They suggested that in urban areas, where the concentration of Aitken particles may reach $10^6 \, \text{cm}^{-3}$ and higher, IN become deactivated as a result of coagulation with these particles, which are generally found to be poor IN. In support of this mechanism, Georgii and Kaller (1970) computed the time necessary for deactivation of IN of various sizes by coagulation with Aitken particles of various sizes and number concentrations. In these computations it was assumed (1) that coagulation is the result of Brownian diffusion, (2) that the effects of forced convection and turbulence can be neglected, (3) that an IN can be considered deactivated when covered with a monolayer of Aitken particles, (4) that the sticking efficiency is unity, (5) that the Aitken particles and the IN are spherical, (6) that the size of an IN remains constant, implying that the small size increase due to the added Aitken particles can be neglected, and (7) that there is no depletion of Aitken particles in the environment of an IN. Since some of these assumptions have opposite effects, it was considered that the over-all model should be fairly realistic.

The deactivation time determined by Georgii and Kaller is

$$t = \frac{2\,\nu^*\rho}{K_B(r_N, r)N(r)}, \tag{9-31}$$

where the coverage number $\nu^* = 2/\{1 - [(r_N^2 + 2r_N r)^{1/2}/(r_N + r)]\}$, $\rho \approx 0.91$ is the packing density of the Aitken particles around the IN, r is the radius of the Aitken particle, r_N is the radius of the IN, N(r) is the number concentration of Aitken particles, and K_B is the coagulation constant (see Chapter 12). The result of a numerical evaluation of (9-31) is given in Figure 9-21. It is seen that the time to deactivate IN decreases with increasing concentration of Aitken particles, with decreasing IN size for a constant Aitken particle size, and with increasing Aitken particle size for a constant IN size. Figure 9-21 implies that IN may become completely deactivated during 0.5 to 3 days, i.e., 12 to 72 hours (which is the residence time of AP in the lowest atmospheric layers) if the air is strongly polluted, i.e., if $N \geq 10^6 \, \text{cm}^{-3}$. If $N < 10^6 \, \text{cm}^{-3}$, partial coverage of the surface of IN may result in partial deactivation.

Recent laboratory and field studies demonstrate that IN also have a biogenic source. Vali (1968a, b) noted during experiments that soils with a relatively high

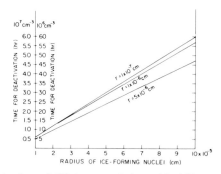

Fig. 9-21. Time for deactivation of IN by coagulation with Aitken particles of radius r and concentration $N = 10^6 \, \text{cm}^{-1}$ and $N = 10^7 \, \text{cm}^{-1}$. (Based on data of Georgii and Kaller, 1970.)

content of organic matter exhibited a higher ice nucleability than pure clays or sand. This observation led him to suggest that decaying plant material contributes to the IN content of the atmosphere. Subsequent detailed studies by Schnell (1972, 1974), Fresch (1973), Vali and Schnell (1973), Schnell and Vali (1976), and Vali et al. (1976) showed that some IN in soils are produced by decomposition of naturally occurring vegetation such as tree leaves. Leaf derived nuclei (LDN) were found to initiate ice by the freezing mode at temperatures typically between −4 and −10 °C (Figure 9-22). The diameter of these particles ranged between 0.1 and 0.005 μm.

The global ubiquity of these IN was established by testing plant litters collected at various geographic locations in different climactic zones (Vali and Schnell, 1973). Highly active IN (active as freezing nuclei at −1.3 °C) were found to be present during the early stages of decay of aspen leaves. Fresch (1973) demonstrated that the ice forming capability of these decaying leaves is closely related to a single strain of aerobic bacteria (pseudomonas syringae), which by themselves act as IN. Such nuclei were termed bacteria derived nuclei (BDN). Whether or not LDN and BDN are to be regarded as acting independently of each other has not, as yet, been established. Schnell (1974) also found that some organic material from the ocean surface, termed ocean derived nuclei (ODN), can be quite effective IN. His experiments showed that LDN, BDN, and ODN may act as IN in both the freezing and deposition modes, being generally more efficient in the freezing mode. Unfortunately, no studies on the contact mode of these particles are available.

Except for a limited number of organic substances concentrated in the ocean surface, the world oceans are not a source of IN. Field observations by Mossop (1956), Georgii and Metnieks (1958), Georgii (1959a), Isono et al. (1959), Carte and Mossop (1960), Murty (1969), and Hobbs and Locatelli (1970) have demonstrated that maritime air masses are consistently deficient in IN. The inverse

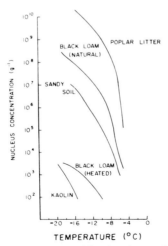

Fig. 9-22. Freezing nucleus spectra (number of ice forming nuclei per gram) for poplar leaf litter, loam, sandy soil, heated sample of loam, and pure kaolinite. (From Schnell, 1974; by courtesy of the author.)

correlation between air masses with maritime trajectories and IN concentration is illustrated in Figure 9-23. Laboratory studies carried out in cloud chambers by Hosler (1951), Birstein and Anderson (1955), Pruppacher and Sänger (1955), Fukuta (1960), and Sano *et al.* (1960) corroborate this notion. Their experiments showed that AP consisting of NaCl, NH_4Cl, NH_4NO_3, $(NH_4)_2SO_4$, and $CaCl_2$, salts typically found in the atmosphere, did not act as IN at temperatures warmer than $-18\,°C$.

Following a suggestion by Bowen (1953, 1956a, b), several researchers have advocated an extraterrestrial source of IN. Accounts of earlier arguments for and against the Bowen hypothesis have been given by Fletcher (1962) and Mossop (1963b). More recent evidence seems to be just as controversial. For example, Bigg (1963) and Bigg and Miles (1964) found a close correlation between the moon phase and the IN concentration, which was interpreted by them in terms of a lunar modulation of the extraterrestrial influx of IN. Bigg and Miles (1963, 1964) and Bigg (1967) suggested that an extraterrestrial influx of IN would explain the increase in the IN concentration with height, and the peculiar layered vertical structure in the concentration of IN which is often found over the Southern Hemisphere. Support for the Bowen hypothesis was also given by Maruyama and Kitagawa (1967), who gave evidence of a positive correlation between the IN concentration and the occurrence of meteorite showers.

However, no such correlation was found by Georgii (1959b), Gagin (1965), or Isono *et al.* (1970). Reinking and Lovill (1971) also took issue with the Bowen hypothesis. They found at a high mountain observatory that a large IN concentration was not accompanied by a high ozone concentration, as would have been expected were the IN of stratospheric, and hence probably extraterrestrial, origin. Evidence against the Bowen hypothesis was also given by Mason and Maybank (1958) and by Qureshi and Maybank (1966), who tested the ice nucleating efficiency of ground meteorite materials. All samples tested by Mason and Maybank acted as IN below $-17\,°C$, while Qureshi and Maybank found three samples to be active between -13 and $-15\,°C$, and all others below $-16\,°C$.

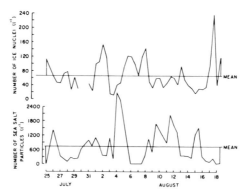

Fig. 9-23.   Correlation between number concentration of ice forming nuclei (0 to $-30\,°C$) and sea salt particle concentration on the island of Valencia during July and August, 1958. (From Georgii, 1959a; by courtesy of *Ber. Beutsch Wett. Dienst.*, U.S. Zone.)

Bigg and Giutronich (1967), on the other hand, criticized the early laboratory experiments as being unrealistic and showed that freshly condensed particles formed from artificially evaporated meteorite material acted as an abundant source of IN at −10 °C and water saturation. In contrast to these experiments, however, Gokhale and Goold (1969) found that particles of extraterrestrial material which were sampled at an altitude of 80 km did not act as IN at −20 °C.

However, as was pointed out somewhat earlier by Junge (1957a), even if clear evidence were available that certain types of meteorite materials possess ice forming capabilities at temperatures warmer than −15 °C, it would remain physically and meteorologically quite unrealistic to assume that meteorite shower particles, which typically have a broad size distribution, could reach the troposphere as a sharply defined cloud, and then become responsible for world-wide rainfall anomalies after a time lag of 30 days between the meteor shower and the rainfall occurrence, as envisioned by the Bowen hypothesis. It is obvious from this controversy that more evidence is needed to settle the question of an extraterrestrial source of IN. Our present knowledge of the behavior of terrestrial AP seems, however, to suggest that no extraterrestrial source of significant strength needs to be invoked in order to explain the observed characteristics of atmospheric IN.

9.2.3 CHARACTERISTIC FEATURES AND MODE OF ACTION OF IN

Observations show that ice forming nuclei have to meet a number of specific characteristics. The most important of these shall now be briefly discussed.

9.2.3.1 Insolubility Requirement

In general, IN are highly water-insoluble. The negative correlation which is observed between the concentrations of IN and sea salt particles gives some evidence of this fact (Figure 9-23). The obvious disadvantage of a soluble substrate is that its tendency to disintegrate under the action of water prevents it from providing the structural order needed for ice germ formation. In addition, the presence of salt ions causes a lowering of the effective freezing temperature (recall Section 6.7; also, see Section 9.2.5).

9.2.3.2 Size Requirement

Field studies have shown that AP of the Aitken size range are considerably less efficient IN than 'large' AP. For example, it can be seen from Figure 9-24a that the concentration of IN active at temperatures warmer than −20 °C is positively correlated with the concentration of large particles, but is uncorrelated with the number of Aitken particles (Figure 9-24b). Although it is tempting to interpret such observations in terms of an IN size effect alone, they may partly reflect a dependence of AP chemistry on size (e.g., silicate particles, which are known to be good IN, are mostly confined to the 'large' size range).

However, there are other clear indications that IN size is important. For example, IN must have a size comparable to or larger than that of the critical ice embryo or germ. Estimates of the size cutoffs imposed by this requirement can

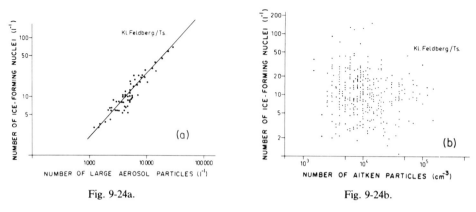

Fig. 9-24a.   Fig. 9-24b.

Figs. 9-24a, b.   Relation between the concentration of IN and aerosol particles on Mt. Kleiner Feldberg/Taunus (Germany): (a) large aerosol particles, (b) Aitken particles. (From Georgii and Kleinjung, 1967; by courtesy of *J. de Rech. Atmos.*, and the authors.)

be obtained from inspection of Figures 6-7 and 6-9. From the former figure we see that in vapor at water saturation the radius of an ice germ is $0.035\,\mu$m at $-5\,°$C and $0.0092\,\mu$m at $-20\,°$C, while from the latter figure we find that inside a water drop at $-5\,°$C and $-20\,°$C the ice germ radius is $0.010\,\mu$m and $0.0024\,\mu$m, respectively. For AP to act as IN at the temperatures indicated, they evidently must have radii larger than these. Observations show that, indeed, AP found in the center portions of ice crystals are considerably larger than the germ sizes indicated.

Such trends have been confirmed by several experiments. For example, Hosler and Spalding (1957), Sano *et al.* (1960), Edwards and Evans (1960), Edwards *et al.* (1962), and Gerber (1972) showed that below about $0.1\,\mu$m radius the effectiveness of IN progressively decreases with decreasing size, becoming increasingly temperature dependent and negligibly small if the particle radius is less than about $0.01\,\mu$m (100 Å). However, the exact value of this cutoff seems to be somewhat dependent on the chemical composition of the IN and on its mode of action. For IN acting in the deposition mode, the cutoff seems to depend on the vapor supersaturation of the environment. IN acting in the freezing mode tend to be less affected by size. Unfortunately, no studies have been reported on the size effect of IN acting in the contact mode.

A size dependence quite different from that just described was found by Vali (1968a), who studied the freezing temperature of aqueous suspensions of surface soils which contained large quantities of organic material. By filtering these suspensions Vali showed that organic particles attached to surface soil may initiate water freezing at temperatures as warm as $-8\,°$C, even though their diameters may be as small as $0.01\,\mu$m. These results were confirmed by Schnell (1972, 1974) who, as mentioned earlier, chemically isolated some organic, ice-nucleation active materials. He found that organic particles such as those derived from tree leaf litter could have diameters as small as $0.025\,\mu$m to $0.05\,\mu$m and still be capable of initiating freezing at temperatures between -5 and $-8\,°$C.

Since the solubility of a substance increases with decreasing particle size (see Section 6.9), we might have expected, on this basis alone, to find that IN occur predominantly in the larger size ranges of AP. The existence of very small organic IN is consistent with the fact that such particles are known to be highly water-insoluble.

9.2.3.3 Chemical Bond Requirement

Numerous experimental studies show that the chemical nature of an IN, expressed in terms of the type and strength of the chemical bonds exhibited at its surface, also affects its nucleation behavior. Considering the fact that an ice crystal lattice is held together by hydrogen bonds (O—H---O) of specific strength and polarity, it is quite reasonable to assume that an IN must have similar hydrogen bonds available at its surface in order to exhibit good ice nucleability. Fukuta (1966) found that, in addition to having similar bond strength and polarity, a hydrogen-bonding molecule at the IN surface should also possess rotational symmetry. While asymmetric molecules tend to point their active H-bonding groups inward to achieve minimum free energy at the solid surface, molecules with rotational symmetry cannot avoid exposing their active H-bonding groups, thus allowing maximum interaction with an oncoming water molecule.

In view of this bond requirement, it is not surprising that certain organic compounds have been found to behave as excellent IN. Head (1961a, b, 1962a, b) was one of the first to demonstrate experimentally that hydrogen-bonding groups such as —OH, —NH$_2$, =O, and their geometric arrangement at the surface of an organic substrate, are of importance to ice nucleation. Other experiments have shown that in the deposition mode metaldehyde (CH$_3$CHO$_4$) has an ice nucleation threshold temperature as warm as -0.4 °C (Fukuta, 1963), cholesterol (C$_{27}$H$_{46}$O·H$_2$O) has a threshold of -1 to -2 °C (Fukuta, 1963; Fukuta and Mason, 1963), and phloroglucinol (C$_6$H$_3$(OH)$_3$·2H$_2$O) has a threshold temperature of -2 to -4 °C (Langer et al., 1963; Fukuta, 1966). Many other organic compounds have also been found to be ice nucleation active at temperatures warmer than -10 °C (Komabayasi and Ikebe, 1961; Fukuta and Mason, 1963; Langer et al., 1963; Barthakur and Maybank, 1963, 1966; Garten and Head, 1964; Parungo and Lodge, 1965, 1967a; Evans, 1966; Fukuta, 1966).

9.2.3.4 Crystallographic Requirement

Experiments have also shown that the geometrical arrangement of bonds at the substrate surface is often of equal or greater importance than their chemical nature. Since ice nucleation on a foreign substrate may be regarded as an oriented (or *epitaxial*) overgrowth of ice on this substrate, it is quite reasonable to assume that this overgrowth is facilitated by having the atoms, ions, or molecules which make up the crystallographic lattice of the substrate exhibit, in any exposed crystallographic face, a geometric arrangement which is as close as possible to that of the water molecules in some low index plane of ice. In this manner, atomic matching across the interface between ice and the substrate particle may be achieved.

If there are but small crystallographic differences between ice and the substrate, either or both the ice lattice and the substrate may elastically deform so that they may join coherently. Thus, strain considerations suggest that the solid substrate should have an elastic shear modulus which is as low as possible in order to minimize the elastic strain energy. If there are large crystallographic differences between ice and the substrate, dislocations at the ice-substrate interface will result, leaving some molecules unbonded across the interface and causing the ice germ to be incoherently joined to the substrate. The interface may then be pictured as being made up of local regions of good fit bounded by line dislocations. These dislocations at the interface will raise the interface free energy. In addition, any elastic strain within the ice embryo will raise its bulk free energy. Both effects will reduce the ability of the AP to serve as an IN.

The apparent crystallographic misfit or *disregistry* is usually defined by

$$\delta \equiv \frac{n a_{0,N} - m a_{0,i}}{m a_{0,i}}, \tag{9-32}$$

where $a_{0,N}$ is the crystallographic lattice parameter of a particular face of the nucleus, $a_{0,i}$ is the corresponding constant in the ice lattice, and n, m are integers chosen such that δ is minimal. An example of the crystallographic matching between the (0001) planes of ice and CuS is shown in Figure 9-25. In order to determine the actual crystallographic misfit, it is necessary to allow for strain. Assuming the embryo can be strained by the amount $\varepsilon = (a'_{0,i} - a_{0,i})/a_{0,i}$, where $a_{0,i}$ and $a'_{0,i}$ are the lattice parameters of the ice in the strain-free and the strained conditions, respectively, and assuming the substrate strain is negligible, the actual crystallographic misfit between the embryo and a particular face of the IN is given by $\delta - \varepsilon$. If $\delta = \varepsilon$, the embryo fits the surface element of the IN coherently. If the ice embryo cannot be strained by the full amount, then $\delta - \varepsilon > 0$ and the ice embryo joins the IN incoherently.

The effect of the crystallographic properties of IN on ice nucleation has been studied by a large number of investigators (see Mason, 1971). Some selected

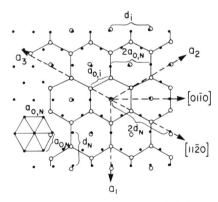

Fig. 9-25. Relative atomic position in the (0001) plane of CuS and ice. ● Cu atoms, ○ H_2O molecules; $d_N = a_{0,N}/\sqrt{3}$, $d_i = a_{0,i}/\sqrt{3}$, where $a_{0,N}$ and $a_{0,i}$ are the lattice constants in the (0001) plane for CuS and ice, respectively. (After Bryant *et al.*, 1959, with changes.)

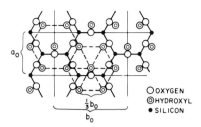

Fig. 9-26. Crystallographic structure of kaolinite. Projection on (001). (From *Clay Mineralogy* by R. E. Grim, copyrighed by McGraw-Hill Book Co., 1953.)

results of these studies are summarized in Table 9-3, where the threshold temperature for ice nucleation in the deposition mode is listed for various chemical compounds as a function of their crystallographic characteristics. Note in particular the relatively good ice nucleability of kaolinite. From Figure 9-26 this appears to be due to the pseudohexagonal arrangement of the hydroxyl (—OH) groups at the surface of the lattice. Other substances listed in Table 9-3 also seem to derive their warm ice nucleation threshold from a close fit of their crystal lattice to that of ice. An unequivocal proof of the necessity, although not of the sufficiency, for a substrate to meet certain crystallographic requirements for good ice nucleability was given by Evans (1965). From a study of the effectiveness with which AgI nucleates ice-I_h and the high pressure modification ice-*III*, he was able to show that even under conditions where ice-*III* is the stable phase, ice-I_h, which has a closer crystallographic fit to AgI than ice-*III*, was consistently nucleated by AgI. However, despite the obvious importance of the crystallographic properties of a substrate to its ice nucleability, Table 9-3 demonstrates clearly that no unique correlation can be established between ice nucleation threshold and any of the crystallographic characteristics such as symmetry or misfit. The main reason for this irregular behavior is the role played by active sites, which we shall discuss next.

9.2.3.5 Active-Site Requirement

Experiments have established that ice nucleation, like heterogeneous water nucleation, is a very localized phenomenon in that it proceeds at distinct active sites on a substrate surface. Not surprisingly, it happens that sites which are capable of initiating ice nucleation are also generally active with respect to water vapor adsorption and water nucleation. Therefore, our previous descriptions of the adsorption properties of active sites in Section 5.6, and of the character and types of water active sites given at the end of Section 9.1.4, both serve to describe the behavior of IN active sites as well. Consequently, for the most part in the material which follows we will attempt merely to supplement our previous discriptions.

Experimental evidence for the effectiveness of topographic surface features to initiate the ice phase from the vapor has been given by Bryant *et al.* (1959), Fukuta *et al.* (1959), Fletcher (1960), Hallett (1961), and Kobayashi (1965), who studied the ice nucleating properties of CuS, AgI, and PbI_2 in the depositional

TABLE 9-3a

Crystallographic misfit between various chemical compounds and ice, and its relation to the ice nucleability of these compounds. (Crystallographic data from Wyckoff, 1963; B. L. Davis et al., 1975; and Burley 1963, 1964; data on nucleation threshold from references cited in text.)

Chemical Compound	Crystallographic Symmetry	Lattice constants (Å)			Substrate plane	Ice Matching plane	Misfit between substrate lattice in directional ice (%)	Approx. ice nucleation threshold (°C)
		a_0	b_0	c_0				
Ice-I_h	D^4_{6h} hex.	4.52(0 °C)		7.36(0 °C)		prism	$[0001] + 8.1(1:1)$; $[1\bar{2}10] + 1.5(1:1)$	
β-A_gI	C^4_{6v} hex.	4.592		7.510	(0001)	basal	$[10\bar{1}0] + 1.5(1:1)$; $[0001] + 2.0(1:1)$; $[1\bar{2}10] + 1.5(1:1)$	-3 to -6
γ-A_gI	T^2_d cubic	6.496			(111)	prism	$[1\bar{2}10] + 1.5(1:1)$; $[0001] + 8.1(1:1)$; $[1\bar{2}10] + 1.5(1:1)$	-3 to -6
MgTe	C^4_{6v} hex.	4.52		7.33	(0001)	basal	$[10\bar{1}0]\ 0.0(1:1)$; $[0001] - 0.41(1:1)$; $[1\bar{2}10]0.0(1:1)$	-4 to -5
PbI$_2$	D^3_{3d} trig.	4.54		6.86	(0001)	basal	$[10\bar{1}0] + 0.4(1:1)$; $[0001] - 6.8(1:1)$; $[1\bar{2}10]0.4(1:1)$	-4 to -7
HgTe	T^2_d cubic	6.36			(111)	basal	$[1\bar{2}10] - 0.4(1:1)$	-4 to -7
CdTe	T^2_d cubic	6.41			(111)	basal	$[1\bar{2}10] + 0.2(1:1)$	-4 to -6
CuI	T^2_d cubic	6.04			(111)	basal	$[1\bar{2}10] - 5.5(1:1)$	-4 to -7
Cu$_2$O	T^2_h cubic	4.25			(100)	basal	$[1\bar{2}10] - 6.0(1:1)$; $[10\bar{1}0] + 8.7(2:1)$; $[0001] + 15.5(2:1)$	-4 to -6
Mg$_3$Sb$_2$	D^3_{3d} trig.	4.57		7.23	(0001)	basal	$[10\bar{1}0] + 1.1(1:1)$; $[0001] - 1.8(1:1)$; $[1\bar{2}10]1.1(1:1)$	-4 to -5
CuS	D^4_{6h} hex.	3.80		16.43	(0001)	basal	$[10\bar{1}0] - 2.8(2:1)$; $[0001] + 11.6(1:2)$; $[1\bar{2}10] - 2.8(2:3)$	-4 to -8
Ag$_2$S	C^5_{2h} monocl.	4.20	6.93	9.50	(010)	basal	$[1\bar{2}10] - 7.1(1:1)$; $[10\bar{1}0] - 0.3(1:1)$; $[0001] - 5.9(1:1)$	-5 to -8
HgI$_2$	C^{12}_{2v} orthorh.	4.67	13.76	7.32	(010)	basal	$[1\bar{2}10] + 3.3(1:1)$; $[10\bar{1}0] - 6.4(1:1)$; $[0001] - 6.5(1:2)$	-5 to -8
Ag$_2$O	T^2_h cubic	4.72			(001)	prism	$[1\bar{2}10] + 4.4(1:1)$; $[10\bar{1}0] - 9.5(3:2)$; $[0001] - 3.8(3:2)$	-5 to -11
CdSe	C^4_{6v} hex.	4.30		7.02	(0001)	basal	$[10\bar{1}0] - 4.9(1:1)$; $[0001] - 4.6(1:1)$; $[1\bar{2}10] - 4.9(1:1)$	-5 to -7
ZnSe	T^2_d cubic	5.67			(111)	basal	$[1\bar{2}10] - 11.5(1:1)$	-5 to -12

TABLE 9-3b

Chemical Compound	Crystallographic Symmetry	Lattice constants (Å) a_0	b_0	c_0	Substrate plane	Ice Matching plane	Misfit between substrate lattice in directional ice (%)	Approx. ice nucleation threshold (°C)
NiO	O_h^1 cubic	4.17(0 °C)			(100)	basal	$[1\bar210]-7.7(1:1)$; $[10\bar10]+6.6(2:1)$; $[0001]+13.3(2:1)$	−6 to −11
CuO	C_{2h}^6 monoc.	4.65	3.41	5.11	(001)	prism	$[1\bar210]+2.9(1:1)$; $[0001]-7.4(2:1)$	−7 to −12
					(010)	basal	$[1\bar210]+2.9(1:1)$; $[10\bar10]-2.0(3:2)$	
SnO$_2$	D_{4h}^{14} tetrag.	4.72		3.16	(010)	basal	$[1\bar210]+4.4(1:1)$; $[10\bar10]-14.2(2:1)$; $[0001]-3.8(3:2)$	−8 to −14
CdO	O_h^1 cubic	4.69			(100)	basal	$[1\bar210]+3.8(1:1)$; $[10\bar10]-10.0(3:2)$; $[0001]-4.4(3:2)$	−9 to −10
V$_2$O$_5$	D_{2h}^{13} orthorh.	11.48	4.36	3.55	(100)	basal	$[1\bar210]-3.5(1:1)$; $[10\bar10]-2.1(2:3)$; $[0001]-3.5(2:1)$	
					(001	prism	$[1\bar210]-3.5(1:1)$; $[10\bar10]-2.1(2:3)$; $[0001]-3.5(2:1)$	
I$_2$	D_{2h}^{18} orthorh.	4.78	7.25	9.77	(001)	basal	$[1\bar210]+5.8(1:1)$; $[10\bar10]-7.3(1:1)$; $[0001]-11.3(2:3)$	−11 to −15
MnO$_2$	D_{4h}^{14} tetrag.	4.44		2.89	(010)	prism	$[1210]-1.8(1:1)$; $[10\bar10]+10.9(3:1)$; $[0001]-9.5(3:2)$	−12 to −14
CdI$_2$	D_{3d}^3 trig.	4.24		6.84	(0001)	basal	$[10\bar10]-6.2(1:1)$; $[0001]-7.1(1:1)$; $[12\bar10]-6.2(1:1)$	−12 to −14
Kaolinite (OH)$_8$Si$_4$Al$_4$O$_{10}$	C_s^4 triclin.	5.16	8.94	7.38	(001)	basal	$[1\bar210]-1.1(3:2)$; $[10\bar10]-2.9(1:1)$	−12 to −15
	(pseudohex. OH groups)	2.98		7.15				−9 to −11
Metaldehyde (CH$_3$CHO)$_4$	tetragonal	10.40		4.11	(110)	(0001)	c-axis 0.2	−1
					(1$\bar1$0)	(01$\bar1$0)	$[01\bar10]$ 6.1	
					(001)	($\bar2$110)	a-axis 9.1	
α-Phenacin C$_6$H$_4$NC$_6$H$_4$N	monoclinic	13.22	5.06	7.09	(001)	(1$\bar1$00)	a-axis 2.6	
							c-axis 3.0	
Phloroglucinol-dihydrate C$_6$H$_3$(OH)$_3$2H$_2$O	orthorhombic	6.73	13.58	8.09	(011)	(0001)	c-axis 8.7	−3 to −4
					(010)	(112$\bar0$)	a-axis 0.1	
					(001)	(10$\bar1$0)	$[10\bar10]$ 3.3	

mode. Photographs taken during the course of these studies revealed that ice crystals appear preferentially at cleavage and growth steps, at cracks, and in cavities, and at the edges of the substrate surface (see Plate 8). These results are supported by the study of Hallett and Srivastava (1972), who showed that the ice nucleability of an AgI single crystal surface could be improved by etching or scratching. However, such behavior has not always been observed. For example, Federer (1968), working with single crystals of Si, GaAs, and GaSe, found that the area density of ice crystals formed by vapor deposition on these substrates was not closely correlated to the area density of steps caused by emerging dislocations.

Adsorption studies by Gravenhorst and Corrin (1972) have established that particles from AgI samples containing impurity ions have a considerably higher ice nucleation efficiency than 'pure' AgI. This behavior is shown in Figure 9-27. The lower nucleation efficiency of 'pure' AgI was attributed by Barchett and Corrin (1972) to the presence of relatively inactive physical adsorption sites at its surface, causing water molecules initially to be adsorbed in the form of extended water 'patches' within which the water molecules exhibit strong lateral interaction. Only on approaching water saturation do multilayers develop with an adsorbate-vapor interface which assumes the energetic properties of a liquid-like surface prior to nucleation.

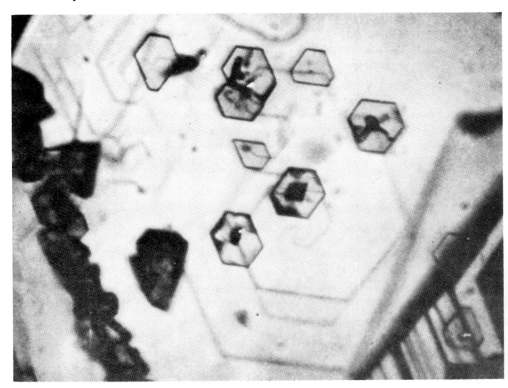

Plate 8.   An epitaxial deposit of ice crystals growing at the steps of a hexagonal growth spiral on cadmium iodide. (From Briant *et al.*, 1959; by courtesy of Pergamon Press Ltd.)

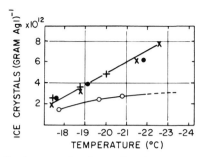

Fig. 9-27. Ice nucleation effectiveness of AgI as a function of doping; ○ pure AgI, ● AgI doped with 0.05% KNO₃, + AgI doped with 0.1% KNO₃ × AgI doped with 1% KNO₃. (From Gravenhorst and Corrin, 1972; by courtesy of *J. de Rech. Atmos.*, and the authors.)

We recall from Section 5.6 that the adsorption efficiency of active sites on doped silicon was found by Federer (1968) to be negatively correlated with its specific electrical resistance. As expected, a similar dependence for the nucleation efficiency of these sites occurs also, as shown in Figure 9-28.

Experiments with AgI and numerous other substances show that the region surrounding active sites should have a hydrophobic character for the surface to exhibit a high ice nucleation efficiency. One reason for this is that it is easier for water molecules to join a disordered water cluster on a low energy (hydrophobic) surface than to enter an oriented array of water molecules on a polar (hydrophylic) surface. Similarly, it is easier for a water cluster than an oriented film to achieve an ice-like structure. Also, as we have indicated previously, the growth of an ice embryo is facilitated by surface diffusion of weakly adsorbed molecules near the active site.

The advantage of having a low energy, non-polar substrate was pointed out by Fletcher (1959b). As we know from Chapter 3, water dipoles in ice may assume a large number of orientations. Hence, Fletcher was able to demonstrate

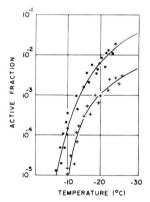

Fig. 9-28. Ice nucleation effectiveness of p-type silicon as a function of doping with boron. Specific electric resistance of sample: ● $10^{-2}\,\Omega\,cm$, + $7.5 \times 10^{3}\,\Omega\,cm$. (From Federer, 1968; by courtesy of *Z. Angew. Math. Phys.*, and the author.)

theoretically that a polar substrate can reduce the configurational entropy, and therefore raise the free energy of an ice embryo growing on it. In consequence, Fletcher predicted that AgI surfaces exposing either Ag^+ or I^- exclusively should be poorer ice nucleation substrates than those where both ions are present.

Partial qualitative confirmation of this prediction was provided by Isono and Ishizaka (1972), who determined that the critical supersaturation for the onset of ice nucleation in the deposition mode is lower, and the critical temperature warmer, on a hexagonal AgI face of type $(10\bar{1}0)$ where both Ag^+ and I^- are present than on a hexagonal face of type (0001) where only one or the other is present. However, an arrangement of Ag^+ or I^- identical to that on a hexagonal AgI face of type (0001) is also present on a cubic AgI face of type (111), but for this surface Isono and Ishizaka found exactly contrary behavior (Figure 9-16). Thus, the experiment of Isono and Ishizaka is somewhat inconclusive. Their results may have been affected by a difference in the abundance and efficiency of active sites on the different faces.

Many experiments have shown that the ice-nucleating active fraction of a sample of AP increases with decreasing temperature (Serpolay, 1959; Katz, 1960, 1961, 1962; Edwards et al., 1962; Roberts and Hallett, 1968; Edwards and Evans, 1968; Mossop and Jayaweera, 1969). Some representative results are shown in Table 9-2 (see p. 247), and Figures 9-18, 9-28, and 9-29. Such behavior is observed even for AP of a given size and chemical composition.

A reasonable interpretation of these results is that a given solid substance has a characteristic area density of ice nucleation active sites of varying quality. This implies another size effect in addition to that discussed in Section 9.2.3.2. It also

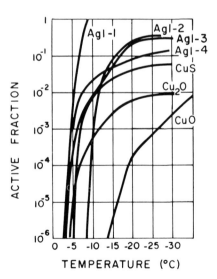

Fig. 9-29. Temperature dependence of active fractions of particles active as IN for various substances. AgI—1: $r_N = 2\,\mu$m, freezing mode; AgI—2: $r_N = 0.015$ to $0.2\,\mu$m (median, $0.045\,\mu$m), freezing mode; AgI—3: $r_N = 0.2$ to $4\,\mu$m (mode, $1\,\mu$m), deposition mode at water saturation; AgI—4: $r_N = 2\,\mu$m, deposition mode at water saturation. (Based on data of Katz, 1961; Edwards and Evans, 1968; Mossop and Jayaweera, 1969.)

implies that experiments on nucleation thresholds should be interpreted with caution, since the outcome may well depend on the area of the substrate under study. For example, the experiments by Mason and van den Heuvel (1959), Isono and Ishizaka (1972), and Bryant et al. (1959) typically involved fields of view under the microscope of about 5×10^{-3} cm², and led to observations of ice nucleation at water saturation (ice supersaturations of from 7 to 12%) for temperatures between -7.5 and $-12\,°C$, and at ice supersaturations of about 12% for temperatures between -12 and $-20\,°C$. In contrast to these results, Anderson and Hallett (1976), who observed considerably larger substrate areas, found that ice could be nucleated at ice supersaturations as low as 3% for temperatures between -7.5 and $-20\,°C$. These more recent results are also consistent with the observations of Barchett (1971), who found from adsorption studies that the onset of ice nucleation in the deposition mode (1 ice germ per AgI sample of about 0.8 g, with a specific area of as large as 4900 cm² g⁻¹) at temperatures between -4 and $-10\,°C$ required an ice supersaturation, independent of temperature, of only about 3% (equivalent to a water-subsaturation of about 7% at $-10\,°C$ and of about 3.3% at $-6.5\,°C$).

Two additional effects which demonstrate the importance of active sites for nucleation in the deposition mode are worth mentioning. In order to discuss the first effect, let us recall from Section 9.2.3 that under certain conditions IN may become activated (or 'trained'), and in this state exhibit a considerably improved ice nucleability (memory effect). To behave in this manner the IN either must have been previously involved in an ice nucleation process and formed a macroscopic ice crystal, or they must have been exposed to temperatures below $-40\,°C$. Earlier explanations of this phenomenon conjectured the retention of ice remnants in small cavities or capillaries where they could survive relative humidities considerably below ice saturation due to the negative curvative effect, as long as the temperature of the environment remained below $0\,°C$. However, the work of Roberts and Hallett (1968) and of Edwards and Evans (1971) has shown that the observations are better explained in terms of the retention of patches of ordered, ice-like layers of water molecules at the surface of a substrate, where each patch can be considered to represent the remnant of a macroscopic ice crystal which developed over an active site.

The second effect involves the previously mentioned observation that foreign gases or vapors such as NO_2, SO_2, NH_3 strongly reduce the nucleability of IN (Georgii, 1963; Georgii and Kleinjung, 1967). Since the adsorption studies we discussed in Chapter 5 have demonstrated that such foreign gases are adsorbed at active sites, we see that the observed suppression of nucleation is caused by the occupancy of active sites by molecules other than H_2O.

So far we have discussed the significance of active sites to ice nucleation in the deposition mode, and it remains for us to consider their significance in the freezing and contact modes. Unfortunately, no quantitative experiments are available to help elucidate this problem. One can merely speculate that the ice nucleation process proceeds somewhat differently in the freezing mode than it does in the deposition mode. Once a water-insoluble AP becomes submerged in water, it is surrounded by an abundance of water molecules. At any one moment, a large number of these molecules are linked together into small structural units in which some of them tend to be tetrahedrally bonded, while other bonds seem to 'dangle' (Stillinger and Rahman, 1972; see also Section 3.4). Suppose now that the surface of the submerged solid particle is generally hydrophobic, but contains hydrophilic sites where water molecules are preferentially adsorbed. The molecules most likely to be adsorbed are those which exhibit 'dangling' bonds. In this manner the already existing structural units become 'anchored' to the solid surface, causing them to be less vulnerable to destruction by the heat motion in the water. As the temperature of the water is lowered, more and more 'dangling' bonds become 'hooked' to the particle's surface, thus allowing individual structural units to be 'joined' together to form clusters in which individual water molecules have considerable freedom to move their dipoles into an orientation most favorable for a tetrahedral, ice-like arrangement. Eventually, the anchored 3-D cluster may reach germ size.

If, on the other hand, we consider a particle with a surface which has a strong, uniform affinity to water due to the presence of an array of strongly hydrating ions, polar groups, hydroxyl (OH^-), or oxygen (O^-) atoms in the solid surface, water molecules will become adsorbed in a close array with most of the dipoles of individual molecules oriented more or less alike. Such an arrangement is not conducive to ice nucleation, due to the structural entropy penalty imposed on such an adsorbed layer. In this case, a second or even more adsorbed layers may be required before the freedom of orientation among the water molecules in the outermost adsorbed layer is sufficiently large for some of them to assume ice-like orientations while others remain anchored. Mechanisms similar to these have been proposed by Evans (1967a, b) and Edwards et al. (1970), based on their studies of heterogeneous ice nucleation in supercooled water.

There is also little known quantitatively about the importance of active sites to ice formation in the contact mode. Observations have shown that dry particles of many compounds such as clays, sand, soil, CuS, organic compounds, etc., are considerably better IN when acting in the contact mode rather than in the freezing or deposition modes (Rau, 1950; Levkov, 1971; Fletcher, 1972; Gokhale and Spengler, 1972; Pitter and Pruppacher, 1973). In an attempt to explain this effect, Fletcher (1970b) and Guenadiev (1970) pointed out that the observed differences between nucleation in the contact and freezing modes may be caused by the partial solubility of any solid, especially when in the form of small particles. Thus, it is reasonable to assume that active sites at the surface of a particle are especially vulnerable to erosion by dissolution after a particle has become immersed in water. In support of this notion, Hallett and Shrivastava (1972) found that prolonged immersion in water reduced the nucleation efficiency of AgI single crystals. However, although the erosion effect may account for some differences between the contact and freezing modes, it cannot explain the significant difference between the ice nucleability of some clays in the contact and deposition modes. In addition, the erosion effect is unable to account for the fact that in all three modes, AgI exhibits practically the same threshold temperature of $-4\,°C$ for particles of the same size.

A different explanation for contact nucleation was given by Evans (1970). He suggested that only those compounds initiate ice formation in the contact mode which exhibit a strong affinity for water, thus adsorbing water molecules from the liquid or vapor in a close array. In such a case, as mentioned previously, ice nucleation is hindered due to a structural entropy penalty imposed on the adsorbed layer. However, during the initial brief moments of contact between a particle and a supercooled water drop, adsorption is incomplete and disordered despite the strong affinity. Thus, during this period the energy barrier to the formation of a more ordered, ice-like arrangement may be considerably lower, and thus nucleation may be much more likely, than in an adsorbed and firmly attached oriented array. Although this explanation is attractive, it hinges on the assumption that the time required for building up an oriented water film is much longer than the time needed to form an ice germ in the disordered-adsorbed layer. Unfortunately, this assumption has not yet been justified.

A third explanation for contact nucleation has been given by Guenadiev (1972) and Cooper (1974). They conjectured that an IN acting in the contact mode must build up a critical ice embryo which is in equilibrium with the water of the supercooled drop, rather than with the surrounding water vapor. Since at any given temperature the former requirement is less stringent, an IN may nucleate on contact with a supercooled drop even though the ice particle is of sub-germ size with respect to ice formation from the vapor. Although this mechanism can account for clay particles being better IN in the contact mode than in the deposition mode, it cannot explain, for example, why AgI exhibits the same nucleation threshold in both modes, nor can it explain why clay particles exhibit a better ice nucleability in the contact mode than in the freezing mode. In addition, computations by Grover (1978) show that the water vapor density decreases very rapidly with distance from the surface of a drop falling in sub-saturated air and reaches water saturation only at the drop surface itself. Thus, an AP which approaches a supercooled drop on a collision trajectory in a water subsaturated environment will not encounter a water saturated air layer, as was assumed by Cooper (1974), although it may be exposed to a thin layer of air which is ice supersaturated. The thickness of this layer depends strongly on the drop size and its fall speed, and on the relative humidity of the environmental air. One would therefore expect that contact nucleation is a function of drop size and of the relative humidity of the air. However, no dependence on either of these parameters was detected in the nucleation studies by Rau (1950), Gokhale and Goold (1968), Levkov (1971), Gokhale and Spengler (1972), and Pitter and Pruppacher (1973). Also, the question is left open as to whether the short time spent by a particle on its collision trajectory inside the region which is ice supersaturated is sufficient to build up ice embryos of germ size on its surface.

Another interesting explanation of the contact nucleation mechanism has recently been given by Fukuta (1975a, b), who studied the freezing behavior of supercooled drops frozen by contact with various organic substances. As other workers had found for inorganic substances, Fukuta observed that the drop freezing temperature via contact nucleation was significantly warmer than that of drops frozen by the same particles submerged within them. He interpreted his results to indicate that the reduction in ice-forming ability which the particles suffered when immersed in drops was not due to the dissolution of nucleation active sites. Rather, Fukuta suggested on the basis of experiments carried out by Pruppacher (1963) that the difference in freezing temperature was associated with the movement of the water-air interface relative to the solid substrate surface during contact. He suggested that the mechanically forced rapid spreading of water along the hydrophobic solid surface forces its local wetting, and thereby temporarily creates local high interface-energy zones which can increase the likelihood of ice nucleation. Although the explanation of the creation of such zones seems somewhat incomplete, it is a simple matter to show that, should they exist, the consequence would be an enhanced nucleation efficiency. Thus, on adapting (9-9) and (9-10) to the case of ice nucleation inside the drop on the solid surface it contacts, we find the energy of embryo formation, assuming a planar substrate, is

$$\Delta F_{i,S} = V_i \Delta f_{vol} + \sigma_{i/w} \Omega_{i/w} + \Omega_{N/i} (\sigma_{N/i} - \sigma_{N/w}).$$ (9-33)

The creation of a transient high energy zone would correspond to an increase in $\sigma_{N/w}$, which (9-33) shows would lower the formation energy of ice embryos.

9.2.4   THEORY OF HETEROGENEOUS ICE NUCLEATION

9.2.4.1 *Classical Model*

Let us now turn to the classical theoretical model for heterogeneous ice nucleation. The treatment closely parallels that given previously in Section 9.1.3 for heterogeneous water nucleation. In view of the complex nature of IN, and our previous discussions of the limitations of the classical approach, we cannot expect too much of the following description. Nevertheless, it is the most comprehensive theory available and is capable of correctly predicting at least some of the observed features of ice nucleation.

We shall follow Fletcher (1958, 1959b, 1962a) and assume that an ice embryo on a curved solid substrate can be described by the spherical cap model. Then the work of ice-germ formation from the vapor may, through arguments strictly analogous to those presented in Section 9.1.3.1, be expressed as

$$\Delta F_{g,S} = \frac{16 \pi M_w^2 \sigma_{i/v}^3}{3[\mathcal{R} T \rho_i \ln S_{v,i}]^2} f(m_{i/v}, x),$$ (9-34)

where f is given by (9-22) and $m_{i/v}$ by (5-24a). If the embryos are assumed to grow by direct vapor deposition, then the nucleation rate per particle, J_s', is given by combining (9-24) and (9-34).

For an environment which is saturated with respect to water, one may write $s_{v,i} = e_{sat,w}/e_{sat,i}$, and therefore from (4-95), $\ln S_{v,i} \approx (\mathcal{L}_{m,0}/\mathcal{R} T_0^2) \Delta T$, with $\Delta T = T_0 - T$. For these conditions, and assuming $\sigma_{i/v} \approx 100$ erg cm^{-2}, $J_S' = 1$ germ cm^{-2} sec^{-1}, and, further, that the kinetic coefficient has a value of about $10^{26} r_N^2$, Fletcher has determined the variation of temperature with particle radius r_N and compatibility parameter $m_{i/v}$; the results are shown in Figure 9-30. It is seen that there is little dependence on radii $r_N \gtrsim 0.1 \ \mu m$ for a given $m_{i/v}$, but that below this size the ice

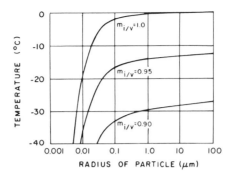

Fig. 9-30. Temperature at which a spherical aerosol particle will nucleate an ice germ in 1 second by the deposition mode at water saturation as a function of its radius and compatibility. (From *The Physics of Rain Clouds* by N. H. Fletcher, copyrighted by Cambridge University Press, 1962a.)

nucleation efficiency decreases rather sharply. The nucleability also decreases rapidly with decreasing $m_{i/v}$, i.e., with increasing interface free energy between ice and the substrate particle, for a given particle size.

Let us now turn to the case of heterogeneous ice nucleation in supercooled water. The rate J_S of germ formation per unit area per unit time may be obtained by combining (9-2), (9-3), and (7-49):

$$J_S = \frac{kT}{h} Z_S N_c \Omega_{g,S} c_{1,S} \exp[-(\Delta F^{\ddagger} + \Delta F_{g,S})/kT]. \qquad (9\text{-}35)$$

This equation may be simplified somewhat, since it happens that the factor $Z_S N_c \Omega_{g,S}$ is approximately unity under typical conditions. (One might have expected this to be so, since we have noted in Section 7.2.2 that $Z_S \approx 10^{-1}$, while $N_c \Omega_{g,S}$ counts the number of water molecules contacting the ice germ; an estimate that this count is $0(10)$ is quite reasonable.) Therefore, an approximate description of the nucleation rate per particle is

$$J'_S \approx \frac{kT}{h} 4 \pi r_N^2 c_{1,S} \cdot \exp[-(\Delta F^{\ddagger} + \Delta F_{g,S})/kT]. \qquad (9\text{-}36)$$

Assuming a spherical cap germ as before, and following familiar arguments, the quantity $\Delta F_{g,S}$ in (9-35) and (9-36) may be expressed as one third the surface energy of the germ, multiplied by the geometric factor f of (9-22); using (6-49) for the germ radius, we have

$$\Delta F_{g,S} = \frac{16 \pi M_w^2 \sigma_{i/w}^3 f(m_{i/w}, x)}{3[\mathscr{L}_{m,0} \bar{\rho}_i \ln (T_0/T_e)]^2}, \qquad (9\text{-}37)$$

where $m_{i/w}$ is given by (5-24b). Computations of (9-36) and (9-37) have been carried out by Fletcher (e.g., 1959b, 1969), but, unfortunately, these contain small errors and do not include the temperature dependences of $\sigma_{i,w}$, $\mathscr{L}_{m,0}$, $\bar{\rho}_i$, and $\Delta F^{\ddagger}$. However, a suitably corrected computation by Pruppacher and Hall (1974, unpubl.), supports the results of Fletcher (1969). Their results are shown in

Figure 9-31, where it has been assumed that $c_{1,S} = 6 \times 10^{28}$ cm^{-2} and $J'_S = 1$ germ (particle)$^{-1}$ sec^{-1}. It is seen from the figure that for $m_{i/w} \geqslant 0.3$ and $r_N \geqslant 0.03 \mu$m, the ice nucleation efficiency is relatively insensitive to r_N; however, it decreases rapidly for smaller sizes. For a fixed r_N, the required supercooling increases rapidly with decreasing $m_{i/w}$. Comparison with Figure 9-30 shows that the nucleability is less sensitive to the size and compatibility parameter of an IN in the freezing mode than in the deposition mode at water saturation.

Recalling now the experimental observations discussed in the previous section, we find that the classical theory is successful in qualitatively predicting the observed decrease in ice nucleation ability with decreasing AP size. Quantitatively, however, observations show a considerably stronger size dependence, for small sizes in particular. The classical theory also qualitatively predicts the observed decrease in nucleability with decreasing m-factor, i.e., with increasing interface free energy between ice and the substrate. On the other hand, in practice it cannot discriminate among various aerosol particles, since $m_{i/v}$ and $m_{i/w}$ (i.e., $\sigma_{i/N}$) have not been determined for any substance. Even if these quantities were accessible to experimental determination, the derived values would pertain to an average, macroscopic behavior of the particular substrate face studied. Such values would clearly be unsatisfactory since heterogeneous ice nucleation occurs preferentially at the location of microscopic active sites, and is controlled by the nature of these sites rather than by the average behavior of the surface.

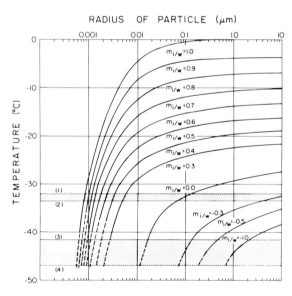

Fig. 9-31. Temperature at which a spherical aerosol particle will nucleate on an ice germ in supercooled water in 1 second by the freezing mode as a function of its radius and compatibility factor. Shaded areas: homogeneous freezing temperature of various size drops at various cooling rates: (1) $a = 4$ mm, $\gamma_c = 1$ °C hr^{-1}, (2) $a = 4$ mm, $\gamma_c = 1$ °C sec^{-1}, (3) $a = 1 \mu$m, $\gamma_c = 1$ °C hr^{-1}, (4) $a = 1 \mu$m, $\gamma_c = 1$ °C sec^{-1}.

In view of these shortcomings, it is not surprising, for example, that the classical theory makes the seriously erroneous prediction that all AP of a given size and chemical composition will have the same ice nucleation efficiency. This prediction is in complete disagreement with observations which show that at any given temperature the nucleability varies from particle to particle, and that the ice nucleating active fraction of such a homogeneous set of particles increases with decreasing temperature. In addition, experiments do not substantiate the theoretical prediction that the nucleation efficiency is less dependent on the compatibility between ice and the substrate in the freezing mode than in the deposition mode.

9.2.4.2 Extensions of the Classical Model

Thus far, three attempts have been made to improve the classical theory. First of all, efforts have been made to include the effects of the elastic strain ε produced within an ice germ due to the misfit δ between the ice and substrate lattices. This gives rise to modifications in the interface energy, and so the m-factor, as well as the bulk free energy contribution Δf_{vol} of the ice germ. According to Turnbull and Vonnegut (1952) the concentration of dislocations at the interface depends linearly on $(\delta - \varepsilon)$, while to Δf_{vol} the term $C\varepsilon^2$ has to be added. In the case of coherent nucleation, i.e., $\varepsilon = \delta$, the concentration of dislocations at the interface is zero, and the m-factor requires no correction. In most cases where $\delta \ll 1$, this is probably a reasonable approximation to make.

For the case of ice germ formation from the vapor, the only formal change required by these modifications is that $\mathscr{R}T\rho_i \ln S_{v,i}$ in (9-34) must be replaced by $-\mathscr{R}T\rho_i \ln S_{v,i} + C\varepsilon^2$, where C is a constant whose value depends on the elastic properties of ice. Similarly, for the case of ice germ formation in supercooled water, $-\mathscr{L}_{m,0}\bar{\rho}_i \ln (T_0/T_e) + C\varepsilon^2$ replaces $\mathscr{L}_{m,0}\bar{\rho}_i \ln (T_0/T_e)$ in (9-37). The effect is thus easy to interpret as a shift in $\ln S_{v,i}$ or $\ln (T_0/T_e)$. As an example, the additional required depression of the ice nucleation threshold temperature as a function of ε is plotted in Figure 9-32, where the value for C is taken as 1.7×10^{11} erg cm^{-3}, following Turnbull and Vonnegut (1952). Granted that this refinement to the classical theory is highly idealized, it is nevertheless encouraging to note the qualitative agreement of the figure with observation,

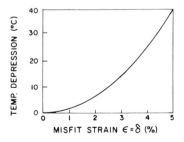

Fig. 9-32. Additional depression of the ice nucleation temperature as a function of misfit strain for coherent nucleation, $\varepsilon = \delta$. (From Turnbull and Vonnegut, 1952; copyright by American Chemical Society.)

namely that the temperature for the onset of ice nucleation is caused to decrease with an increase in the crystallographic misfit between ice and the nucleating substrate, in particular for misfits larger than 1%.

Fletcher (1969) has provided a second extension of the classical theory by constructing a simple model for the effect of active sites on the nucleating substrate. He supposed that, for the case of ice nucleation in the freezing mode on a particle of radius r_N, an active site could be represented by a patch of area αr_N^2 on which $m_{i/w} = 1$. The remainder of the surface was assumed to be characterized by $m_{i/w} < 1$.

The formulation is quite straightforward. Thus, without the active site, the total surface energy of the ice embryo $\sigma_{i,i}\Omega_{i,i}$ would be, in strict analogy to (9-11) $\sigma_{i/w}\Omega_{i/w} + (\sigma_{N/i} - \sigma_{N/w})\Omega_{N/i} = \sigma_{i/w}\Omega_{i/w} - m_{i/w}\sigma_{i/w}\Omega_{N/i}$, using (5-24b). We may now take the presence of the active site into account by replacing the last term above with $-[m_{i/w}\sigma_{i/w}(\Omega_{N/i} - \alpha r_N^2) + \sigma_{i/w}\alpha r_N^2]$. In this manner the expression analogous to (9-10) and (9-11) for the energy of i-mer formation is

$$\Delta F_{i,S} = V_i \Delta f_{vol} + \sigma_{i/w}(\Omega_{i/w} - m_{i/w}\Omega_{N,i}) - \alpha r_N^2 \sigma_{i/w}(1 - m_{i/w}). \tag{9-38}$$

The last term on the right side of this expression represents the correction due to the presence of the active site. Since it does not involve the geometry of i-mer, it is also not involved in the process of maximizing $\Delta F_{i,S}$ to determine $\Delta F_{g,S}$. Therefore, we have immediately the result

$$\Delta F_{g,S} = \frac{4\pi a_g^2}{3} \sigma_{i/w} f(m_{i/w}, x) - \alpha r_N^2 \sigma_{i/w}(1 - m_{i/w}), \tag{9-39}$$

where a_g is given by (6-49) (it is assumed tha germ still possesses the standard spherical cap geometry), and f is given by (9-22).

The solution of (9-36) together with (9-39) for $J_S' = 1$ germ (particle)$^{-1}$ sec^{-1} is shown in Figure 9-33. It can be seen that, as expected, the larger α is for a given r_N, the warmer is the temperature at which the ice nucleation in the freezing mode is initiated. Thus, while a particle of $r_N = 1 \, \mu$m bearing no active site

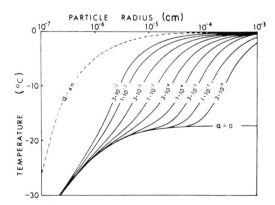

Fig. 9-33. Temperature at which a spherical aerosol particle of compatibility factor $m_{i/w} = 0.5$ will nucleate an ice germ in 1 second as a function of its radius r_N and active site area αr_N^2. (From Fletcher, 1969; by courtesy of Amer. Meteor. Soc., and the author.)

($\alpha = 0$) would initiate ice nucleation at $-17\,°C$, a particle bearing an active site of $\alpha = 2 \times 10^{-4}$, i.e., $\alpha r_N^2 = 2 \times 10^{-12}\,cm^2$, is capable of initiating ice formation at $-4\,°C$, a truly potent effect. If $r_N = 1\,\mu m$ and $\alpha = 2 \times 10^{-5}$, or $\alpha r_N^2 = 2 \times 10^{-13}\,cm$, the particle is capable of initiating ice nucleation at a temperature near $-12\,°C$.

Of course, a major drawback to this extension of the classical theory is that neither the area occupied by an ice nucleation active site, nor the m-factor which characterizes the site is known *a priori*.

A third modification of the classical theory recognizes the possibility that the triple interface boundary between ice, substrate, and saturated water vapor may have some distinctive properties. In particular, it is assumed that any energy excess associated with its existence may be characterized by a *line tension*, $\lambda (\geq 0)$. For the case of a spherical cap embryo on a planar substrate, this evidently leads to the additional contribution $2\pi\lambda a_i \sin\theta$ to the $\Delta F_{i,S}$ of (9-14), assuming the necessary trivial changes in subscripts are made in the shift in application from a water to an ice embryo.

However, before proceeding with the prior development we must first extend Young's relation, (5-23), to include line tension. For this purpose we follow Gretz (1966b, c) and consider the total surface contributions to $\Delta F_{i,S}$, viz.,

$$(\Delta F_{i,S})_{surf} = 2\pi a_i^2 \sigma_{w/v}(1 - \cos\theta) + \pi a_i^2(1 - \cos^2\theta)(\sigma_{N/w} - \sigma_{N/v})$$
$$+ 2\pi\lambda a_i \sin\theta. \tag{9-40}$$

Gretz used the principle that $(\Delta F_{i,S})_{surf}$ should achieve a minimum value at equilibrium, subject to the constraint of constant volume; i.e., he set $[\partial(\Delta F_{i,S})/\partial\theta]\,d\theta + [\partial(\Delta F_{i,S})/\partial a_i]\,da_i = 0$, where $da_i/d\theta$ is determined from the condition that $V_i = (4\pi/3)a_i^3 f(\cos\theta) = $ constant, with f given by (9-18). In this way the resulting generalization of Young's relation is found to be

$$\sigma_{w/v} \cos\theta = (\sigma_{N/v} - \sigma_{N/w}) - \lambda/a_i \sin\theta. \tag{9-41}$$

Now on substituting (9-41) into (9-40), and proceeding to maximize $\Delta F_{i,S}$ for fixed θ, we again obtain the form (9-15) for a_g. Finally, the free energy of ice germ formation becomes (cf. (9-17)):

$$\Delta F_{g,S} = \frac{16\pi M_w^2 \sigma_{i/w}^3 f(m_{i/w})}{3[\mathcal{R}T\rho_i \ln S_{i/v}]^2} + \frac{2\pi\sigma_{i/v}M_w\lambda \sin\theta}{\mathcal{R}T\rho_i \ln S_{i,v}}. \tag{9-42}$$

It is seen that line tension always increases $\Delta F_{g,S}$.

Evans and Lane (1973) used (9-42) in conjunction with (9-8a) to describe ice nucleation on a planar substrate. They assumed water saturation, so that $\ln S_{i/v} \approx (\mathcal{L}_{m,0}/RT_0^2)(T_0 - T)$, from (4-95); also assumed were the values $\sigma_{i/v} = 100\,erg\,cm^{-2}$, a kinetic coefficient of $10^{25}\,cm^{-2}\,sec^{-1}$, and, as usual, $J_S = 1\,germ\,cm^{-2}\,sec^{-1}$. Their results are shown in Figure 9-34. It is seen that line tension can cause a significant lowering of the ice nucleation temperature. For example, the nucleation temperature is lowered from $-5\,°C$ ($\lambda = 0$) to $-10\,°C$, if what Evans and Lane regard as a conservative value for line tension, $\lambda = 5 \times 10^{-7}\,erg\,cm^{-1}$, is assumed. As might be expected, however, accurate values for the line tension of ice on a substrate are not known.

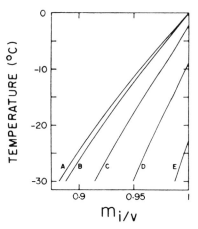

Fig. 9-34. Temperature at which a plane substrate nucleates an ice germ per second and per cm^2 as a function of compatibility factor $m_{i/v}$ and line tension λ (erg cm^{-1}). (A) $\lambda_{i/v} = 0$, (B) $\lambda_{i/v} = 10^{-7}$, (C) $\lambda_{i/v} = 5 \times 10^{-7}$, (D) $\lambda_{i/v} = 10^{-6}$, (E) $\lambda_{i/v} = 1.5 \times 10^{-6}$. (From Evans and Lane, 1973; by courtesy of Amer. Meteor. Soc., and the authors.)

The above discussions suggest that the improved versions of the classical heterogeneous ice nucleation theory do not constitute major steps toward an overarching quantitative description. At best, they can be considered as relations which help to probe qualitatively the importance of some neglected quantities. It is hoped that significant progress toward a quantitative solution of the heterogeneous nucleation problem will eventually be made by applying *ab initio* approaches of the type considered by Hale and Plummer (see Chapter 7) and by Hale and Kiefer (1975).

9.2.5 Heterogeneous Freezing of Supercooled Water Drops

Experiments with water drops containing various impurities have revealed that their freezing temperature (usually expressed in terms of the median freezing temperature of a population of drops) is a function of the drop volume. Such a dependence was first suggested by Heverly (1949), and subsequently verified in a quantitative manner by Dorsch and Hacker (1950), Levine (1950), Bigg (1953b, 1955), Mossop (1955), Carte (1956, 1959), Kiryukhin and Pevzner (1956), Langham and Mason (1958), Barklie and Gokhale (1959), Stansbury and Vali (1965), and Vali and Stansbury (1965, 1966). The experimentally derived volume relationship can be expressed as

$$T_s = A - B \ln V_d, \tag{9-43}$$

where $T_s = T_0 - T$ is the temperature in °C below 0 °C, and A and B are constants for a particular water sample. Figure 9-35 demonstrates this behavior for various water samples. In order to understand the physical basis of such a relationship, we shall consider two different points of view.

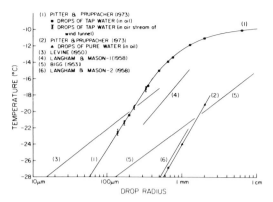

Fig. 9-35. Variation of the median freezing temperature of water drops as a function of their size. (From Pitter and Pruppacher, 1973; by courtesy of *Quart. J. Roy. Meteor. Soc.*)

The 'classical' point of view was adopted by Bigg (1953a, b, 1955), Carte (1956, 1959), and Dufour and Defay (1963), who attempted to explain the freezing behavior of a population of water drops by assuming that at a given temperature all equal-sized ice embryos formed in a population of equal-sized supercooled water drops have an equal probability of reaching the size of a critical embryo or germ as a result of random fluctuations among the water molecules. Although these fluctuations were envisioned to form ice germs more efficiently in the presence of foreign particles, the effect of such particles in the water was considered nonspecific, i.e., they were assumed to enhance the efficiency of the random nucleation process but not disturb its stochastic nature (*stochastic hypothesis*). Under these conditions the theory becomes equivalent to that given for homogeneous freezing in Section 7.2.2. In particular we may use (7-54) to express the relative change in the number of unfrozen drops during the time interval dt as

$$-\frac{dN_u}{N_u} = V_d J(T_s)\, dt. \tag{9-44}$$

From his experiments, Bigg (1953b) also deduced that

$$J(T_s) = B(e^{aT_s} - 1), \tag{9-45}$$

where a is a constant of the order of unity. Therefore, for constant $T_s \geqslant 5\,°C$, we have from (9-44) and (9-45) the approximate result that

$$-\frac{1}{N_u}\left(\frac{dN_u}{dt}\right)_{T_s} = \frac{1}{N_u}\left(\frac{dN_f}{dt}\right)_{T_s} = BV_d \exp\,(aT_s) = \text{constant}, \tag{9-46}$$

since $dN_u = -dN_f$. Thus, the stochastic hypothesis leads to the prediction that at constant supercooling the fraction of drops of volume V_d frozen per unit time interval is constant, and is larger, the larger the supercooling.

On the other hand, if the drop population is cooled at a constant cooling rate $\gamma_c = -(dT/dt) = dT_s/dt$, then instead of (9-46) we find

$$-\frac{1}{N_u}\left(\frac{dN_u}{dT_s}\right)_{\gamma_c} = \frac{1}{N_u}\left(\frac{dN_f}{dT_s}\right)_{\gamma_c} = \frac{BV_d}{\gamma_c}\exp(aT_s). \tag{9-47}$$

In words, the fraction of drops of volume V_d frozen per unit temperature interval, while being cooled at a constant rate, increases exponentially with increasing supercooling, the fraction being smaller, the larger the cooling rate at any one supercooling.

The integral of (9-46) is

$$\ln\frac{N_0}{N_u(t)} = BV_d[\exp(aT_s)]t, \tag{9-48}$$

where $N_0 = N_u(t = 0)$. Thus, the number of unfrozen drops decays exponentially with time at a given supercooling. In terms of the time t_m needed to freeze one half the drop population, i.e., $N_u(t_m) = N_0/2$, (9-48) states that

$$\ln t_m = A - aT_s, \tag{9-49}$$

where $A = \ln(\ln 2/BV_d) = $ constant. We see that the median freezing time t_m decreases exponentially with supercooling.

Similarly, integration of (9-47) at constant γ_c yields

$$\ln\frac{N_0}{N_u(T_s)} = \frac{BV_d}{a\gamma_c}\exp(aT_s), \tag{9-50}$$

so that for any given rate of cooling the number of unfrozen drops decreases with increasing supercooling, and for any given supercooling the number of unfrozen drops decreases with decreasing rate of cooling. In terms of the median freezing temperature T_{sm}, where $N_u(T_{sm}) = N_0/2$, (9-50) becomes

$$T_{sm} = T_0 - T_m = C + \frac{\ln\gamma_c}{a}, \tag{9-51}$$

where $C = a^{-1}\ln(a\ln 2/BV_d) = $ constant. Thus, the median freezing temperature lowers logarithmically with increasing cooling rate. Given the form of C it is also clear that for fixed γ_c this relationship is consistent with (9-43), expressing the volume dependence of the drop freezing temperature.

A second point of view was developed by Levine (1950) and Langham and Mason (1958), who attributed heterogeneous drop freezing entirely to the singular freezing characteristics of AP which have become incorporated in drops (*singular hypothesis*). They assumed that every particle contained inside a drop has one characteristic temperature at which freezing will be initiated in the drop. According to the singular hypothesis, then, the freezing temperature of a drop is determined by that particle in the drop which has the warmest characteristic temperature. It also implies that the number of ice germs formed in a drop volume V_d during time t at a supercooling T_s is given by the number of particles $n_P(T_s)$ inside the drop which become active as IN between 0 °C and T_s; i.e.,

$n_P(T_s) = \int_0^t J(T_s) \, dt$. Therefore, on integrating (9-44) we have

$$\ln \frac{N_0}{N_u} = V_d n_P(T_s). \tag{9-52}$$

For the distribution n_P, Langham and Mason assumed the empirical form of (9-29):

$$n_P = n_{P_0} \exp(bT_s), \tag{9-53}$$

so that also

$$\ln \frac{N_0}{N_u(T_s)} = n_{P_0} V_d \exp(bT_s). \tag{9-54}$$

In contrast to the classical stochastic model, (9-54) predicts the number of unfrozen drops is independent of the cooling rate. The equation also predicts the number of frozen or unfrozen drops will not change with time for a given supercooling. In terms of the median freezing temperature, (9-54) states that

$$T_{sm} = T_0 - T_m = D - \frac{1}{b} \ln V_d, \tag{9-55}$$

where $D = b^{-1} \ln(\ln 2/n_{P_0}) = $ constant. Thus we see further that the experimentally observed volume dependence is predicted both by the stochastic and the singular models, and so cannot be used to choose between them.

However, there are other corresponding predictions of the two models, such as (9-50) and (9-54), which may be checked against experimental data. Such comparisons have been made by Barklie and Gokhale (1959), Stansbury and Vali (1965), and Vali and Stansbury (1965, 1966). For drops of distilled water containing a large population of relatively small particles, it was found that, at a given cooling rate, the fraction of drops frozen per unit temperature interval increased exponentially with increasing supercooling (Figure 9-36), in accord with the dependence predicted by the classical stochastic model, and in direct

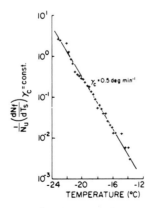

Fig. 9-36. Fraction of drops frozen per unit temperature interval as a function of temperature for distilled water; $V_d = 0.01 \text{ cm}^3$. (From Vali and Stansbury, 1966; reproduced by permission of the National Research Council of Canada.)

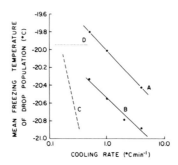

Fig. 9-37. Variation of the mean freezing temperature of a population of drops with cooling rate; $V_d = 0.01$ cm³; lines A, B are for two sets of experiments; lines C and D represent the prediction of the stochastic and singular hypothesis, respectively; distilled water. (From Vali and Stansbury, 1966; reproduced by permission of the National Research Council of Canada.)

conflict with the singular model, which predicts the fraction should be in-dependent of γ_c. However, the variation with γ_c was found to be less than that expected on the basis of the stochastic model (Figure 9-37). Also, at a constant temperature the fraction of drops frozen per unit time interval was found to decrease exponentially with time as $\exp(-t/T_s)$ (Figure 9-38), in contrast to both the stochastic model which predicts the fraction remains constant, and the singular hypothesis which predicts that the number of new freezing events is zero.

These studies suggest that the actual drop freezing mechanism is better represented by some combination of the stochastic and singular mechanisms than by either one acting alone. This observation is also more consistent with the known characteristic features of IN acting in the freezing mode, as discussed in Section 9.2.3. Thus, the singular characteristics of drop freezing can be attri-buted to the nature of the active sites on particles contained in the drop, and to the average chemical and crystallographic properties of the particles. On the other hand, the stochastic aspect of the drop freezing process can be attributed to the random manner in which, at any given supercooling, the water molecules

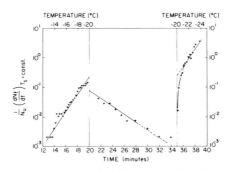

Fig. 9-38. Fraction of drops frozen per unit time interval as a function of time for distilled water; $V_d = 0.01$ cm³, $t = 0$ at 0 °C; at -20 °C temperature was held constant for 15 minutes. (From Vali and Stansbury, 1966; reproduced by permission of the National Research Council of Canada.)

in water join the clusters adsorbed at the active sites to eventually form an ice-like germ.

However, it should be recognized that the linear relationship between $\ln V_d$ and T_s in the singular model derives from a particular (exponential) size distribution for n_P. There is no reason to believe that such a functional dependence holds even approximately for all temperatures. Indeed, curve (1) in Figure 9-35 suggests that, for the particular water sample studied, the variation of the median freezing temperature of a population of water drops is not linear with $\ln V_d$ at temperatures warmer than $-15\,°C$. However, the behavior observed is physically reasonable since, on the basis of the singular hypothesis, one must expect that at temperatures approaching $0\,°C$ the number of particles serving as IN in the freezing mode will rapidly decrease. This implies that above a certain temperature the volume of a water sample which contains a particle that can nucleate ice will be progressively larger than that predicted by the $\ln V_d - T_s$ law.

In addition, we must recall that all the results summarized in Figures 9-36 to 9-38, which provided evidence for the above arguments, were derived from experiments with distilled water. Such water contains solid particles which, although numerous, are relatively small and relatively uniform in composition. Consequently, they undergo only weak physicochemical interactions, and so their size distributions remain relatively simple. This is in contrast to the case of cloud water, which contains a strongly heterogeneous mixture of AP which have entered the cloud and precipitation particles through the condensation or freezing process, and through various scavenging mechanisms. Also, the dissimilar sizes and chemical composition of the particles lead to their further modification through coagulation and chemical reactions. Accordingly, one finds that the AP in cloud water have a relatively complex size distribution. It is therefore not surprising that the freezing spectra of drops made from cloud and rain water (Fig. 9-39b) is quite irregular in comparison to that from distilled water (Figure 9-39).

Such complicated freezing spectra transcend the simple analytical formulations for the singular hypothesis given above. However, the manner in which the

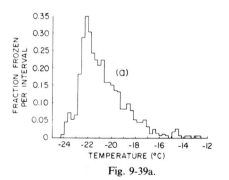

Fig. 9-39a.

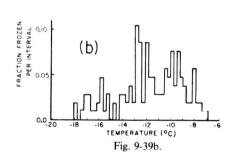

Fig. 9-39b.

Figs. 9-39a, b. Freezing spectrum (fraction of drops frozen per unit temperature interval, $(1/N_0)(dN_f/dT_s)$): (a) Distilled water at constant cooling rate, 288 drops. (b) Rain water at constant cooling rate, 144 drops. (From Vali and Stansbury, 1966; reproduced by permission of the National Research Council of Canada; and from Vali, 1971; by courtesy of Amer. Meteor. Soc., and the author.)

freezing of drops of such water depends on time and temperature does indicate that – as for distilled water – the freezing mechanism is of a mixed singular and stochastic type, and is therefore not completely dominated by the IN in the water, as one might have expected.

9.2.6 HETEROGENEOUS FREEZING OF SUPERCOOLED AQUEOUS SOLUTION DROPS

We have pointed out in Chapter 8 that the major portion of aerosol particles consists of water-soluble substances. Hence, through various nucleation and scavenging processes cloud water acquires both soluble and insoluble substances. The question then arises as to the effect of dissolved salts on the nucleating behavior of solid particles.

This problem was studied by Pruppacher (1963a) and Pruppacher and Neiburger (1963), who also critically summarized relevant earlier work. Through experiments with drops of filtered solutions of various salts, they found that alkali- and earth alkali-halides, as well as other salts commonly found in the atmospheric aerosol, have a negligible effect on the freezing temperature of water drops if their concentrations are less than 10^{-3} mol ℓ^{-1}. For larger concentrations, the salts studied invariably depressed the drop freezing temperatures. Their results can be summarized by the empirical relation

$$(\Delta T)_s = (\Delta T)_w + (\Delta T)_{e,s} + \delta T, \tag{9-56}$$

where $(\Delta T)_s$ is the median supercooling of a population of aqueous solution drops containing particles of radius less than $0.01\,\mu$m, $(\Delta T)_w$ is the median supercooling of a population of drops of the same size and same water from which the solution was made, $(\Delta T)_{e,s}$ is the equilibrium freezing point depression of the particular salt in solution (see Section 4.10), and δT is a small temperature departure which varied between 0 and 2.5 °C, being largest for solutions with salt ions which have a large tendency to disrupt the water structure, and hence inhibit the formation of ice embryos.

Kuhns (1968), Parungo and Lodge (1967b), Pena et al. (1969), and Pena and Pena (1970) studied the effect of gases, in solution equilibrium with pure water drops, on the freezing temperature of these drops. Kuhns found that the freezing temperature was affected by less than 1 °C if the drops were in solution equilibrium with He, H_2, O_2, and air. Pena and Pena's observations supported Kuhns' results, but in addition they found that some organic gases such as CH_4, C_2H_6, and CH_4Cl slightly depressed the freezing temperature, while SO_2 depressed it strongly. Pena and Pena concluded that the clathrate structures which are induced by these gases at low temperatures are not conducive to ice formation. A similar conclusion was reached by Pruppacher (1962), and Parungo and Wood (1968), who studied the effect of organic macromolecules in water. Parungo and Wood showed that the freezing temperature of water drops which contained macromolecules of dissolved substances such as agar, gelatin, citrus pectin, ovalbumin, bovin albumin, ribonucleic acid (RNA), and deoxyribonucleic acid (DNA) deviated by less than 1 °C from that of pure water if the solute concentrations ranged between 0.01 and 1%. These results imply that water molecules, although immobilized by their interaction with macromolecules, are not arranged in an ice-like manner which would promote ice-germ formation, despite the fact that some macromolecules such as DNA have atomic spacings which closely agree with those present in ice (Jacobson, 1953; Watson and Crick, 1953).

Hoffer (1961) and Pruppacher and Neiburger (1963) suggested that the surfaces of insoluble particles of radii >0.01 μm in aqueous solution drops may be rendered ice nucleation active by physical adsorption of ions or by chemical reactions between the salt ions and the particle. However, Hoffer showed that the freezing temperature of drops which consisted of solutions of typical atmospheric salts such as $MgCl_2$ and Na_2SO_4, and containing typical aerosol insolubles, such as illite, kaolinite, montmorillonite, or halloysite, was not affected by the salt if its concentration was less than 10^{-3} mol ℓ^{-1}, and was progressively lowered at larger concentrations. These results were essentially confirmed by Reischel (1972) and Reischel and Vali (1975), who found that salts present in the atmosphere affected the freezing temperature of water drops, which contained leaf derived IN, by less than 1.5 °C. A similar result was found for drops containing clay particles, except when $(NH_4)_2SO_4$ or NH_4Cl were present. In the latter case, the freezing temperature was shifted to warmer temperatures by up to 4 °C at a salt concentration of about 10^{-2} mol ℓ^{-1}, but by less than 1 °C if the concentration was less than 10^{-3} mol ℓ^{-1}. No explanation for the effects at larger concentrations was offered.

Junge (1952c) suggested that the freezing temperature of drops formed by condensation on mixed AP is affected by the presence of soluble material, particularly during the early stages of condensation, i.e., prior to activation of the drop. This fact is evident from Table 6-3, where the concentration of salt at the point of activation is listed for AP composed of NaCl and SiO_2 in various proportions. By comparing the results given in this table with the concentration requirement for a negligible effect on the freezing temperature of a solution drop, which is 10^{-3} mol ℓ^{-1}, we learn that mixed particles have to have masses larger than 10^{-13} g, and have to contain more than 35% (by mass) salt in order for the drop to grow large enough so that the concentration of salt in solution is less than 10^{-3} mol ℓ^{-1} at the point of activation of the drop. This requirement is not fulfilled for most atmospheric AP. Thus, prior to and at their point of activation, most atmospheric solution drops will consist of salt solutions too concentrated for ice nucleation. This implies, as Junge (1952) suggested, that most mixed AP must form drops which have sizes beyond activation before the salt concentration is sufficiently low for a freezing or contact nucleus to initiate freezing.

CHAPTER 10

HYDRODYNAMICS OF SINGLE CLOUD AND
PRECIPITATION PARTICLES

Once formed, cloud particles immediately begin to move under the action of gravity and frictional forces, the latter arising from their motion relative to the air. Some fraction of these particles will undergo complex hydrodynamic interactions causing some to collide. The particles will experience growth if the collision results in a permanent union. However, most of the time most cloud particles will simply fall with negligible interaction. It is this basic mode of isolated motion that we address ourselves to in this chapter. Furthermore, for simplicity we shall defer to Chapter 17 the consideration of the complicating influence of electrical forces.

As we shall see, the smallest of the cloud drops and ice crystals fall slowly, with speeds typically less than 1 cm sec^{-1}, so that very gentle updrafts suffice to keep them suspended. On the other hand, large raindrops and hailstones have fall speeds of 5 m sec^{-1} and more, and generally cannot be supported by the prevailing updrafts.

In applying hydrodynamic theory to the motion of isolated cloud and precipitation particles, we shall first restrict our attention to droplets small enough to be regarded as rigid impermeable spheres. Later we shall consider the phenomena of drop deformation, internal circulation, vibration, and breakup. The complicated shapes of ice particles makes a quantitative description of their hydrodynamic behavior extremely difficult. However, it turns out that the motions of simple plate-like ice crystals may be understood reasonably well through the expedient of studying flows past disks and thin oblate spheroids. Similarly, we shall use circular cylinders and prolate spheroids as idealizations of simple columnar ice crystals. The present chapter will conclude with a discussion of the motion of more complex ice particles.

10.1 Basic Governing Equations

The principle of mass conservation for a fluid in motion is shown in Appendix A-10.1 to lead to the *continuity equation*:

$$\frac{1}{\rho}\frac{d\rho}{dt} = -\nabla \cdot \breve{u}, \tag{10-1}$$

where ρ and $\breve{u}$ are the fluid density and velocity, respectively. It is well known that flow past an object may be regarded as incompressible ($d\rho/dt = 0$) whenever the following conditions are met (see, for example, Chapter 1 of Rosenhead (1963)): (1) the characteristic flow speed U satisfies $U \ll c$, where c is the speed of sound in the fluid ($c \approx 340$ msec^{-1} in air at 15 °C and 1 atm.); (2) the dominant flow oscillation frequency f satisfies $f \ll c/L$, where L is the characteristic length scale for changes in $\vec{u}$ (in our context, L is the order of the size of the falling object); (3) $L \ll g/c^2$, where g is the magnitude of the gravitational acceleration (this is equivalent to the condition that the static pressure difference between two points separated in the vertical by length L must be very small compared to the absolute pressure; i.e., L is a small fraction of the atmospheric scale height H (≈ 10 km)); (4) the fractional temperature difference between obstacle and stream is small, i.e., $|T - T_\infty| \ll T_\infty$, where T and T_∞ are the characteristic temperatures of the obstacle and streaming fluid, respectively. As these inequalities hold for all cloud particle motions, we shall henceforth assume the flows under consideration are incompressible, so that from (10-1)

$$\nabla \cdot \vec{u} = 0. \tag{10-2}$$

For ordinary (Newtonian) fluids, of which air and water are examples, the momentum equation for incompressible flow in the presence of gravity acquires the form of (A.10-13):

$$\frac{\partial \vec{u}}{\partial t} + \vec{u} \cdot \nabla \vec{u} = -\frac{\nabla p}{\rho} + \nu \nabla^2 \vec{u} + \vec{g}, \tag{10-3}$$

where $\vec{g}$ is the local gravitational acceleration, p is the fluid pressure, and ν is the local kinematic viscosity, which like ρ is assumed constant over distances large compared to L. Equation (10-3) is known as the *Navier-Stokes equation*.

This system of equations must be supplemented with suitable boundary conditions. The most important of these recognizes that real fluids adhere to any material surface; this is known as the 'no-slip' boundary condition. Thus, at any solid boundary surface (S) the fluid velocity must satisfy the condition

$$\vec{u}|_S = \vec{v}_S, \tag{10-4}$$

where $\vec{v}_S$ is the local surface velocity. If the boundary is a surface of separation between two immiscible fluids, then in addition to (10-4) we must require that the stresses the fluids exert on each other at the boundary are equal and opposite; at a free surface, for example, the stress must be zero.

If the fluid is at rest, (10-3) reduces to the equation for the static pressure, p_s:

$$\nabla p_s = \rho \vec{g} = -\nabla(\rho g z), \tag{10-5}$$

assuming constant ρ and $|\vec{g}| = g$, and letting z denote height above the Earth's surface. The total static pressure force on a particle of volume V and surface S may therefore be expressed as

$$-\int_S \hat{n} p_s \, dS = -\int_V \nabla p_s \, dV = \int_V \nabla(\rho g z) \, dV = \rho V g \hat{e}_z, \tag{10-6}$$

where $\hat{n}$ is the unit outward normal to dS and $\hat{e}_z$ is the unit vector in the z-direction. This result is just Archimedes' principle, which states that an object immersed in a fluid experiences a buoyancy force equal to the weight of fluid it displaces.

Hence we have an opportunity for another small simplification: We may hereafter ignore the gravity term in (10-3), as its only effect on the motion of the particle in the flow is to provide the simple buoyancy force given by (10-6). This may more conveniently be introduced separately later when considering the equation of motion of the falling cloud particle. (In any case, for practical purposes the buoyancy force is negligible in comparison to the gravitational force on the particle, since $\rho_a/\rho_P \approx 10^{-3}$, where ρ_a is the density of air and ρ_P is the bulk density of the particle.) Formally, this simplification amounts to rewriting $-\nabla p/\rho + \vec{g}$ in (10-3) as $-\nabla p'/\rho$, where $p' = p - p_s$, and then dropping the prime for brevity. The pressure profile obtained in this manner is called the dynamic pressure, since the static distribution due to gravity has been subtracted out.

Another simplification is possible for most of the applications considered in the present chapter. Suppose a particle is released from rest in the atmosphere. If it is not too large or irregular, the fluid forces resisting its motion will eventually equilibrate with gravity, and a steady state fall at some terminal velocity will result. We may analyze such motion from the point of view of the particle, past which the flow streams. Therefore, assuming the conditions are such that the flow is not intrinsically unsteady (e.g., due to oscillations of some kind), we may drop the $\partial \vec{u}/\partial t$ from (10-3) for times long compared to the transient period of velocity buildup. Of course, for this to be a useful simplification the transient time should be short; i.e., most of the particle's fall time should be at terminal velocity. Observations show this to be true generally. Also, some simple theoretical estimates which support this conclusion are offered in Section 10.3.5. Consequently, we shall be concerned here almost exclusively with the steady state Navier-Stokes equation, viz.,

$$\vec{u} \cdot \nabla \vec{u} = -\frac{\nabla p}{\rho} + \nu \nabla^2 \vec{u}. \tag{10-7}$$

10.2   Flow Past a Rigid Sphere

10.2.1   CLASSIFICATION OF FLOWS ACCORDING TO REYNOLDS NUMBER

A glance at (10-7) warns of great difficulties, as the *convective acceleration term* $\vec{u} \cdot \nabla \vec{u}$ (also called the *inertia term*) is nonlinear. As a matter of fact, complete solutions to (10-2) and (10-7) have been found for only a very few special situations, among which the case of flow past a sphere, of great importance to cloud physics, is unfortunately not included. Nevertheless, useful approximate analytical and numerical solutions are available for a wide range of conditions.

For the problem of a falling sphere, the relative importance of $\vec{u} \cdot \nabla \vec{u}$ and the linear viscous acceleration term $\nu \nabla^2 \vec{u}$ may be assessed by simple dimensional arguments. Physically, the flow is characterized by the size of the sphere, for which we may take either the radius a or diameter d as a natural measure, and by the streaming velocity (or terminal velocity) U_∞. Therefore, we might expect $\bar{u} \cdot \nabla \bar{u}$ to be of order U_∞^2/a, and $\nu \nabla^2 \bar{u}$ to be of order $\nu U_\infty/a^2$. This leads to the estimate

$$\frac{|\bar{u} \cdot \nabla \bar{u}|}{|\nu \nabla^2 \bar{u}|} \approx \frac{U_\infty a}{\nu} \equiv R, \tag{10-8}$$

where R is the *Reynolds number*. (An alternative definition,

$$N_{Re} \equiv \frac{U_\infty d}{\nu} = 2R, \tag{10-9}$$

also appears often in the literature and in this book.) Equation (10-8) implies that $\bar{u} \cdot \nabla \bar{u}$ may be omitted from (10-7) if $R \ll 1$.

A more precise way to come to almost the same conclusion is to introduce the dimensionless variables $\bar{r}' \equiv \bar{r}/a$ and $\bar{u}' \equiv \bar{u}/U_\infty$. Then we have

$$\frac{|\bar{u} \cdot \nabla \bar{u}|}{|\nu \nabla^2 \bar{u}|} = R \frac{|\bar{u}' \cdot \nabla' \bar{u}'|}{|\nabla'^2 \bar{u}'|}, \tag{10-10}$$

where $\nabla' \equiv a\nabla$ is the dimensionless gradient operator. The factor multiplying R in (10-10) is a function only of $\bar{r}'$, and must be of order unity if (10-8) is to be consistent with (10-10). As we shall see below in Section 10.2.4, this is in fact not always the case. Nevertheless, it is generally correct to say that the inertia term becomes less important with decreasing R.

Many cloud particles indeed have Reynolds numbers much smaller than unity. For example, a cloud drop of $a = 10\ \mu m$ has a fall velocity of about 1.2 cm sec^{-1} at $T = 20\ °C$ and $p = 1000$ mb. Since under these conditions the kinematic viscosity for air is $\nu \approx 0.15$ cm^2 sec^{-1}, we find $R \approx 0.01$. Similarly, for a drop of $a = 30\ \mu m$, we have $R \approx 0.2$. To a good approximation the flow field, drag, and terminal velocities for such drops are described by (10-7) without the $\bar{u} \cdot \nabla \bar{u}$ term. The resulting equation governs what is known as *Stokes flow*.

For larger drops, things are not as simple. Thus, a drop of $a = 50\ \mu m$ has $R \approx 1$, while for $a = 150\ \mu m$, $R \approx 10$. For a raindrop a few mm in radius, $R > 10^3$. As R increases, the flow becomes more complicated, reflecting the greater contribution of the nonlinear term. An additional complication is the change of shape with size for the larger drops, which by itself influences the flow pattern. Also, the larger the drop the greater the tendency for development of a complex internal circulation, which contributes to its overall behavior.

In order not to unduly complicate matters at the outset, we shall restrict ourselves in this section to a discussion of the flows past drops small enough to be regarded as rigid spheres. As we shall see in Section 10.3, this is a good assumption for radii less than about 500 μm, corresponding to $R \leqslant 130$ ($N_{Re} \leqslant 260$).

Before presenting any detailed results for the flow past a sphere, it is worthwhile to consider briefly the qualitative features to be expected at various Reynolds numbers. If we use a coordinate system in which the sphere is at rest, the no-slip boundary condition requires that the fluid velocity must decrease to zero at the surface. This causes the surface to act effectively as a source of fluid shearing motion and angular momentum. The latter may be measured by the *vorticity* $\nabla \times \bar{u}$ (it is easy to show that $\nabla \times \bar{u} = 2 \bar{\omega}$, where $\bar{\omega}$ is the local fluid angular velocity). When $R \ll 1$ and fluid inertia is negligible, the flow is characterized by the diffusion of vorticity away from the sphere. This is easily seen by taking the curl of (10-7) without the $\bar{u} \cdot \nabla \bar{u}$ term, and recalling that $\rho = $ constant; the result is $\nabla^2(\nabla \times \bar{u}) = 0$, which is the steady state diffusion equation for vorticity. Such flow is relatively simple and has 'fore-aft' symmetry, i.e., symmetry with respect to the plane separating the sphere into upstream and downstream hemispheres.

As R increases to order unity (corresponding to increasing U_∞ for a given sphere), there is a tendency for part of the vorticity generated at the sphere surface to be convected downstream. This leads to an asymmetry in the flow, with most of the vorticity confined to the rear of the sphere in a roughly paraboidal region, known as the wake, with its vertex in the sphere and with symmetry about the axis of motion. If R increases to O(10), the wake becomes narrower and the vorticity within it more intense. Incipient wake instability occurs near $R = 65$ ($N_{Re} = 130$); for larger R the flow thus becomes intrinsically unsteady. At about $R = 10$ ($N_{Re} = 20$) a region of circulating fluid forms behind the sphere. This 'standing eddy' grows in size and strength with further increase in R. For $R \geq 200$ ($N_{Re} \geq 400$) the eddy oscillates while lumps of circulating fluid are torn away from it and travel downstream.

Outside the wake there is little vorticity caused by the sphere, and since we assumed none upstream to begin with (the flow is assumed unbounded and undisturbed except for the sphere), the flow outside the wake tends to be *irrotational* ($\nabla \times \bar{u} = 0$), and increasingly so with increasing R. Viscosity has no effect where the flow is irrotational and nondivergent, since then $\nabla^2 \bar{u} = \nabla(\nabla \cdot \bar{u}) - \nabla \times (\nabla \times \bar{u}) = 0$. Therefore, such flow behaves like frictionless or *inviscid* irrotational flow, which is called *potential* flow. This type of flow is especially simple since, as the name implies, its velocity field may be expressed in terms of the gradient of a potential: $\nabla \times \bar{u} = 0$ implies $\bar{u} = \nabla \Phi$, where $\nabla^2 \Phi = 0$ since $\nabla \cdot \bar{u} = 0$.

Potential flow prevails almost everywhere when $R \gg 1$, and reflects the dominance of the fluid inertia. It cannot exist close to the sphere however, since there the no-slip condition creates large shears and viscous forces of the same magnitude or larger than the inertia forces. Neither can it exist in the vorticity-carrying wake. The thickness of the fluid layer adjacent to the sphere over which the transition from viscosity-dominated to inertia-dominated flow takes place, called the *boundary layer*, can be shown to vary as $R^{-1/2}$ (see Section 10.2.5). Outside the boundary layer and wake, the essentially potential flow possesses streamlines very much like those of low Reynolds number flow.

With this qualitative picture of the flow regimes as a background, we shall now look in more detail at the problem of flow past a sphere.

10.2.2 STREAM FUNCTION

The problem of a sphere falling in the $+z$-direction is the same as that of flow streaming in the $-z$ direction past a fixed sphere. The obvious advantage of the latter point of view, which we shall adopt here, is that the flow is steady relative to the sphere for moderate Reynolds numbers, as we have discussed above.

Such flow has axial asymmetry about the z-axis, meaning there is no azimuthal component of velocity, and that the motion is the same in every meridian plane (i.e., every plane containing the z-axis and defined by ϕ = constant, where ϕ is the azimuthal angle). As shown in Appendix A-10.2.2, the constraint of incompressibility on the two components of the velocity field makes it possible to describe the flow in terms of the derivatives of a single scalar function, ψ, called the *stream function*.

In spherical coordinates, with the polar angle θ measured from the $+z$ direction, the velocity field $\vec{u} = (u_r, u_\theta, 0)$ is given in terms of ψ (A.10-17):

$$u_r = -\frac{1}{r^2 \sin\theta}\frac{\partial\psi}{\partial\theta}, \quad (10\text{-}11a) \qquad u_\theta = \frac{1}{r\sin\theta}\frac{\partial\psi}{\partial r}. \qquad (10\text{-}11b)$$

Similarly, according to (A.10-22) and (A.10-26), the *vorticity* may be expressed as

$$\nabla \times \vec{u} = \hat{e}_\phi \zeta = \frac{\hat{e}_\phi E^2\psi}{r\sin\theta}, \qquad (10\text{-}12)$$

where $\zeta \equiv |\nabla \times \vec{u}|$ and

$$E^2 = \frac{\partial^2}{\partial r^2} + \frac{\sin\theta}{r^2}\frac{\partial}{\partial\theta}\left(\frac{1}{\sin\theta}\frac{\partial}{\partial\theta}\right). \qquad (10\text{-}13)$$

Finally, the steady state Navier-Stokes equation, (10-7), is given in terms of the stream function by (A.10-25):

$$\sin\theta\left[\frac{\partial\psi}{\partial r}\frac{\partial}{\partial\theta} - \frac{\partial\psi}{\partial\theta}\frac{\partial}{\partial r}\right]\left(\frac{E^2\psi}{r^2\sin^2\theta}\right) = \nu E^4\psi, \qquad (10\text{-}14)$$

where $E^4 \equiv E^2(E^2)$. The boundary condition that the velocity must vanish on the sphere surface leads from (10-11) to $\partial\psi/\partial r|_a = \partial\psi/\partial\theta|_a = 0$. If we choose to label the center streamline in the flow by $\psi = 0$, the surface boundary conditions are thus

$$\psi|_a = 0, \quad \left.\frac{\partial\psi}{\partial r}\right|_a = 0. \qquad (10\text{-}15)$$

Far from the sphere there is a uniform flow given by $-U_\infty\hat{e}_z$; hence for $r \gg a$ we have

$$-U_\infty\cos\theta = -\frac{1}{r^2\sin\theta}\frac{\partial\psi}{\partial\theta}, \quad U_\infty\sin\theta = \frac{1}{r\sin\theta}\frac{\partial\psi}{\partial r}. \qquad (10\text{-}16)$$

Since we want $\psi = 0$ for $\theta = 0$, an equivalent form for the boundary condition 'at infinity' is

$$\psi \to \frac{U_\infty r^2 \sin^2\theta}{2} \quad \text{as} \quad r \to \infty. \qquad (10\text{-}17)$$

10.2.3 THE DRAG PROBLEM

The components F_i of the hydrodynamic force on the sphere are given by

$$F_i = \int_S T_{ij} n_j \, dS, \tag{10-18}$$

where the quantities T_{ij} denote the components of the stress tensor $\bar{T}$ (see Appendix A.10-1), and n_j is the jth component of the unit outward normal vector to the surface element dS. By symmetry there is only a z-component, F_z, which is generally referred to as the *drag* D.

For the calculation of D it is appropriate to express $\bar{T}$ in spherical coordinates. By the symmetry of the problem only T_{rr} and $T_{r\theta}$ are involved, as follows:

$$D = 2 \pi a^2 \int_0^\pi (T_{rr} \cos \theta - T_{r\theta} \sin \theta)_{r=a} \sin \theta \, d\theta. \tag{10-19}$$

Unfortunately, the development of the components of $\bar{T}$ in various coordinate systems is quite tedious, and so we will merely list the needed expressions for T_{rr} and $T_{r\theta}$:

$$T_{rr} = -p + 2 \eta \frac{\partial u_r}{\partial r}, \quad (10\text{-}20a) \qquad T_{r\theta} = \eta \left(\frac{1}{r} \frac{\partial u_r}{\partial \theta} + \frac{\partial u_\theta}{\partial r} - \frac{u_\theta}{r} \right). \tag{10-20b}$$

In these expressions $\eta = \rho \nu$ is the *dynamic viscosity*. (For the interested reader, we mention what is probably the most elementary, though somewhat impractical, way to obtain such results, which is to make direct use of the fact that the (r, θ, ϕ) system is locally orthogonal. Consequently, we use the fundamental rule for transforming $\bar{T}$ between two orthogonal systems S and S', i.e., $T'_{ij} = \gamma_{ik} \gamma_{jl} T_{kl}$, where γ_{ik} is the direction cosine between the ith axis of S' and the kth axis of S, and we have used the convention of summation over a double index. Thus, for example, $T_{rr} = \gamma_{ri} \gamma_{rj} T_{ij}$, which may be reduced directly to (10-20a) on substituting the Cartesian form (A.10-10) for T_{ij}. A good source for more elegant and powerful methods is Aris (1962).)

On substituting (10-20) into (10-19), and simplifying through the use of the conditions $\nabla \cdot \tilde{u} = 0$ and $\tilde{u}|_a = 0$, we obtain

$$D = -2 \pi a^2 \int_0^\pi \left[p \cos \theta + \eta \left(\frac{\partial u_\theta}{\partial r} \right) \sin \theta \right]_{r=a} \sin \theta \, d\theta \tag{10-21}$$

$$= D_p + D_f, \tag{10-22}$$

where D_p, denoting the first term on the right side of (10-21), is called the *form* or *pressure* drag, and the second term, D_f, is called the *skin-friction* drag.

The dependence of D on U_∞, a, η, and ρ may be elucidated by writing the equations of motion in dimensionless form. Thus, on introducing $r' \equiv r/a$, $p' \equiv p/\rho U_\infty^2$, $\tilde{u}' \equiv \tilde{u}/U_\infty$, and $\nabla' \equiv a\nabla$ into (10-2) and (10-7), we obtain

$$\tilde{u}' \cdot \nabla' \tilde{u}' = -\nabla' p' + \frac{1}{R} \nabla'^2 \tilde{u}', \quad (10\text{-}23a) \qquad \nabla' \cdot \tilde{u}' = 0. \tag{10-23b}$$

The corresponding boundary conditions in dimensionless form are

$$\vec{u}'|_{r'=1} = 0, \quad (10\text{-}24a) \qquad \lim_{r'\to\infty} \vec{u}' = -\hat{e}_z. \qquad (10\text{-}24b)$$

It is apparent that the solutions to (10-23) and (10-24) must be functions only of $\vec{r}'$ and R. The same is therefore true of the dimensionless stress $T'_{ij} = T_{ij}/\rho U_\infty^2$, given by

$$T'_{ij} = -p'\delta_{ij} + \frac{1}{R}\left(\frac{\partial u'_i}{\partial x'_j} + \frac{\partial u'_j}{\partial x'_i}\right). \qquad (10\text{-}25)$$

In consequence, we see the drag must be of the form

$$D = a^2 \rho U_\infty^2 h(R), \qquad (10\text{-}26)$$

where h is a function of R (or N_{Re}) alone.

This characteristic dependence of the drag on the Reynolds number is traditionally expressed in terms of the *drag coefficient* C_D, defined as

$$C_D \equiv \frac{D}{(\rho U_\infty^2/2)A_c}, \qquad (10\text{-}27)$$

where A_c is the cross-sectional area exhibited by the body normal to the flow. Thus, C_D is a function of R only, and the drag problem for any ρ, U_∞, and A is solved once $C_D = C_D(R)$ is determined.

For a sphere we may combine (10-21) and (10-22) with (10-27) to obtain expressions for the form drag coefficient, $C_{D,p}$, and the skin friction drag coefficient, $C_{D,f}$. For this purpose it is convenient to introduce another dimensionless pressure parameter, namely,

$$k \equiv (p - p_\infty)/(\rho U_\infty^2/2), \qquad (10\text{-}28)$$

and also the dimensionless vorticity magnitude,

$$\zeta' \equiv \zeta a/U_\infty. \qquad (10\text{-}29)$$

The constant p_∞ in (10-28) is the pressure far from the sphere. In terms of these quantities we find $C_D = C_{D,p} + C_{D,f}$, where

$$C_{D,p} = 2 \int_0^\pi [k(\theta)]_{r'=1} \cos\theta \sin\theta \, d\theta, \qquad (10\text{-}30)$$

and

$$C_{D,f} = \frac{8}{N_{Re}} \int_0^\pi [\zeta'(\theta)]_{r'=1} \sin^2\theta \, d\theta. \qquad (10\text{-}31)$$

Equations (10-27), (10-30), and (10-31) demonstrate that the drag on a sphere can be found from a knowledge of the pressure and vorticity distributions on its surface. Furthermore, the surface pressure may be expressed in terms of the vorticity through the straightforward but lengthy process of integrating ∇p in

(10-7) along the center streamline, recognizing that $\nabla^2 \vec{u} = -\nabla \times (\zeta \hat{e}_\phi)$. The result is

$$[k(\theta)]_{r'=1} = k_0 + \frac{4}{N_{Re}} \int_0^\theta \left(\frac{\partial \zeta'}{\partial r'} + \zeta' \right)_{r'=1} d\theta, \tag{10-32}$$

where

$$k_0 = 1 + \frac{8}{N_{Re}} \int_{r'=1}^\infty \left(\frac{\partial \zeta'}{\partial \theta} \right)_{\substack{r'=1 \\ \theta=0}} \frac{dr'}{r'}. \tag{10-33}$$

In this manner the drag coefficients can be determined solely from the surface vorticity distributions.

10.2.4 ANALYTICAL SOLUTIONS FOR THE SPHERE

1. *Stokes Flow.* As we have said, Stokes flow is governed by (10-2) and (10-7) without the inertia term. In the stream function formulation, this term appears on the left side of (10-14) (note it is nonlinear in ψ), and so the governing form of the Stokes stream function ψ_S is

$$E^4 \psi_S = 0. \tag{10-34}$$

The boundary condition (10-17) motivates a trial solution of the form $\psi_S = f(r) \sin^2 \theta$. This proves to be successful, and reduces (10-34) to an ordinary linear differential equation in r (for a detailed treatment of this and many other low Reynolds number problems, see Happel and Brenner, 1965). This can be integrated easily, and the constants of integration can be determined from (10-15) and (10-17). The result is

$$\psi_S = \frac{U_\infty a \sin^2 \theta}{4} \left(\frac{2 r^2}{a^2} - \frac{3 r}{a} + \frac{a}{r} \right), \tag{10-35}$$

which upon substitution into (10-12) yields for the vorticity

$$\zeta_S = \frac{3 U_\infty a \sin \theta}{r^2}. \tag{10-36}$$

Similarly, the pressure may be recovered from (A.10-27):

$$p_S = p_\infty + \frac{3 \eta a U_\infty \cos \theta}{r^2}. \tag{10-37}$$

Therefore, from (10-30) and (10-31) the drag coefficients are

$$(C_{D,P})_S = \frac{8}{N_{Re}}, \qquad (C_{D,f})_S = \frac{16}{N_{Re}}, \qquad C_{D,S} = \frac{24}{N_{Re}}, \tag{10-38}$$

and hence from (10-27) the drag on a sphere in Stokes flow is

$$D_S = 6 \pi a \eta U_\infty, \tag{10-39}$$

(Stokes, 1851).

2. *Oseen Flow.* The Stokes approximation assumes $\tilde{u} \cdot \nabla \tilde{u}$ is negligible every-where. One way to test this assumption is to form the ratio given in (10-10), using the Stokes velocity field for $\tilde{u}$. The result is

$$\left(\frac{|\tilde{u} \cdot \nabla \tilde{u}|}{|\nu \nabla^2 \tilde{u}|}\right)_S \sim \frac{Rr}{a}. \tag{10-40}$$

This states that at sufficiently large distances $(r > a/R)$ the assumption of negligible inertia breaks down, no matter how small R is. Therefore, the Stokes flow field is inaccurate at large distances. This ensures the failure of iteration attempts to improve upon Stokes flow past objects by using the Stokes solution to approximate previously neglected inertia terms in the equation of motion. (This predicament was puzzling to Whitehead (1889) and others who first tried such a procedure, and so became referred to as 'Whitehead's Paradox'.)

A way around this difficulty was proposed by Oseen (1910, 1927), who pointed out that a good approximation to $\tilde{u} \cdot \nabla \tilde{u}$ at large distances is $\tilde{U}_\infty \cdot \nabla \tilde{u}$, where $\tilde{U}_\infty$ is the free stream velocity. He therefore suggested the following linear governing equations for the far field velocity distribution:

$$U_\infty \cdot \nabla \tilde{u} = -\frac{\nabla p}{\rho} + \nu \nabla^2 \tilde{u}, \quad (10\text{-}41a) \qquad \nabla \cdot \tilde{u} = 0. \tag{10-41b}$$

The solution to (10-41) for the case of flow past a sphere subject to the conditions $\tilde{u}|_{surface} = 0$ and $\tilde{u} \rightarrow \tilde{U}_\infty$ as $r \rightarrow \infty$, expressed in terms of potential functions for p and $\tilde{u}$, is summarized briefly in Appendix A-14.4.4. Here we merely wish to note that Oseen obtained an approximate solution, and from it a new expression for the drag (D_0), including a term contributed by the fluid inertia:

$$D_0 = 6 \pi a \eta \, U_\infty \left(1 + \frac{3}{16} N_{Re}\right). \tag{10-42}$$

The corresponding drag coefficient is

$$C_{D,0} = \frac{24}{N_{Re}} \left(1 + \frac{3}{16} N_{Re}\right) = C_{D,S} + 4.5. \tag{10-43}$$

It is not obvious these results constitute any real improvement over (10-38) and (10-39), since $\tilde{U}_\infty \cdot \nabla \tilde{u}$ misrepresents the convective acceleration close to the sphere. However, it has been proved that in fact the Oseen equations, (10-41a, b), do give the correct drag on bodies of arbitrary shape, to first order in N_{Re} (Brenner and Cox, 1962). The reason is that the first order contribution of fluid inertia to the drag depends on the inertia forces far from the body, where the Oseen equations provide a valid representation.

A more complete solution of the Oseen equations was obtained by Goldstein (1929) in terms of a series expansion. The drag coefficient obtained by him is

$$C_{D,G} = \frac{24}{N_{Re}} \left(1 + \frac{3}{16} N_{Re} - \frac{9}{1280} N_{Re}^2 + \frac{71}{20\,480} N_{Re}^3\right.$$
$$\left. - \frac{30\,179}{34\,406\,400} N_{Re}^4 + \frac{122\,519}{550\,502\,400} N_{Re}^5 \cdots \right. \tag{10-44}$$

The last term is that corrected by Shanks (1955). Unfortunately, there are no theoretical reasons for regarding the extra terms supplied by Goldstein as providing an improved physical description. As we shall see in Section 10.2.6, however, comparison with experiment indicates (10-44) is slightly better than (10-43).

The stream function ψ_0 for Oseen flow past a sphere, valid to $O(N_{Re})$, is

$$\psi_0 = \frac{U_\infty a^2 \sin^2 \theta}{2} \left(\frac{r}{a} - 1\right)^2 \left[\left(1 + \frac{3}{16} N_{Re}\right)\left(1 + \frac{a}{2r}\right) \right. \\ \left. + \frac{3}{16} N_{Re} \left(1 + \frac{a}{r}\right)^2 \cos \theta \right]. \qquad (10\text{-}45)$$

It can be seen that $\psi_0 \to \psi_S$ as $N_{Re} \to 0$. A plot of streamlines, $\psi_0 = $ constant, from (10-45) reveals the presence of a wake for $N_{Re} \neq 0$. This is to be expected, since Oseen flow is characterized by both diffusive and convective transport of vorticity, as is evident from (10-41).

A simple argument shows that the wake has a paraboidal shape, i.e., its width ℓ varies as $z^{1/2}$, where z is the distance downstream of the sphere (the argument holds for any finite obstacle shape). Thus, an element of fluid containing vorticity generated at the sphere surface is convected downstream in the wake a distance $\sim U_\infty \delta t$ in time δt. If we imagine moving along with the element at the speed U_∞, its vorticity ζ will be seen to undergo a local transverse spreading by viscous diffusion; from our point of view this will be characterized by the equation $\partial \zeta / \partial t \sim \nu \nabla^2 \zeta$. The boundary of this 'diffusive wave' must correspond approximately to the local boundary of the wake. Therefore, also in time δt the vorticity will change significantly over the distance ℓ, where $1/\delta t \sim \nu/\ell^2$. Hence we find $\ell \sim (\nu z/U_\infty)^{1/2}$.

It is interesting to note also from (10-45) that $\psi_0 = 0$ on the sphere, along the center streamline where $\theta = 0, \pi,$ *and* along the curve

$$\cos \theta = - \frac{16 \left(1 + \frac{3}{16} N_{Re}\right)\left(1 + \frac{a}{2r}\right)}{3 N_{Re} \left(1 + \frac{a}{r}\right)^2}, \qquad (10\text{-}46)$$

which may be interpreted as describing the boundary of a standing eddy. According to (10-46) the eddy first appears at the rear of the sphere ($\theta = \pi$, $r = a$) for $N_{Re} = 3.2$, which correlates somewhat with observations giving $N_{Re} \approx 20$. Even this limited success appears rather fortuitous, since one would expect (10-45) to be capable of meaningful predictions only for Reynolds numbers less than unity.

More recent research has confirmed the existence of Oseen flow eddies behind spheres. Thus, a numerical solution of the Oseen equations by Bourot (1969) for $N_{Re} \leq 30$ has shown that a standing eddy develops at $N_{Re} = 7.6$, and grows steadily with increasing N_{Re}. (This is in contrast to an earlier and less accurate numerical solution by Pearcey and McHugh (1955), who found no evidence of an eddy for $N_{Re} \leq 10$.)

3. *Carrier's Modification.* Carrier (1953) proposed a simple semi-empirical modification of Oseen flow past obstacles. He argued that since the Stokes theory neglects inertia altogether, while the Oseen theory overestimates it, at least close to the body, perhaps a better representation might be found by some sort of compromise between the two approaches. Carrier suggested the inertial term in the Navier-Stokes equation be replaced by $cU_\infty \cdot \nabla \tilde{u}$, where c is a number between 0 and 1. Thus the Stokes and Oseen approximations are given by $c = 0$ and $c = 1$, respectively. According to the idea that either of the classical approximations mày be interpreted as replacing the factor $\tilde{u}$ in $\tilde{u} \cdot \nabla \tilde{u}$ by a weighted average, Carrier conjectured a better weighting might be found. From an analytical study of flow past a flat plate, he proposed $c = 0.43$. (According to Murray (1967), this value may be understood as a consequence of forcing the integral of the difference between the exact and approximate forms of the convective terms to vanish over the whole field of flow; i.e., for a plate in the plane $y = 0$, parallel to oncoming flow $U\hat{e}_x$, it happens that $\int_0^\infty (\partial u/\partial x) \times (u - 0.43\, U_\infty)\, dy \approx 0$, where $0.43\, U\ \partial u/\partial x = v\, \partial^2 u/\partial^2 y$.) Carrier then found that this same value, if used to describe the drag on spheres and cylinders, produced fair agreement with experimental data for $N_{Re} \leqslant 40$. This he took to imply that the theory describes general properties of the flow and is not strongly dependent on the geometry of the obstacles.

It is a simple matter to show that the Carrier drag coefficient is related to Oseen's by $C_{D,C} = cD_{D,0}(cN_{Re})$. For a sphere we therefore have

$$C_{D,C} = \frac{24}{N_{Re}} \left(1 + \frac{3\,c}{16} N_{Re}\right). \tag{10-47}$$

4. *Matched Asymptotic Expansions.* As we have seen, the problem of obtaining an expansion of the flow at small N_{Re} is complicated by the existence of two flow regimes: an inner regime where viscosity dominates, and an outer one where inertia forces are comparable to or larger than viscous forces. This difficulty has been surmounted, at least in principle, by the development of the method of matched asymptotic expansions (see, for example, Proudman and Pearson, 1957, and Van Dyke, 1964). The method employs two expansions in N_{Re}, one suitable for each regime. The no-slip boundary condition is used with the inner 'Stokes-type' expansion, and the uniform stream condition with the outer 'Oseen-type' expansion. Since the two expansions represent different forms of the same solution function, it is possible to complete the solution by matching the inner and outer expansions, term by term, in an intermediate region of common validity.

The first two terms of the inner expansion produce the following stream function in the vicinity of the sphere, valid to $O(N_{Re})$ (Proudman and Pearson, 1957):

$$\psi_{PP,1} = \frac{U_\infty a^2}{4} \left(\frac{r}{a} - 1\right)^2 \sin^2 \theta \left[\left(1 + \frac{3}{16} N_{Re}\right)\left(2 + \frac{a}{r}\right)\right.$$
$$\left. + \frac{3}{16} N_{Re} \left(2 + \frac{a}{r} + \frac{a^2}{r^2}\right) \cos \theta\right]. \tag{10-48}$$

Like ψ_0, this stream function predicts a standing eddy for sufficiently large N_{Re}. In the present case the boundary of the eddy is described by

$$\cos \theta = -\frac{16\left(1 + \frac{3}{16}N_{Re}\right)\left(2 + \frac{a}{r}\right)}{3\,N_{Re}\left(2 + \frac{a}{r} + \frac{a^2}{r^2}\right)}, \tag{10-49}$$

so that an eddy is predicted to form when $N_{Re} = 16$, in remarkably good agreement with the observed value of $N_{Re} \approx 20$.

The next approximation to the stream function includes a term proportional to $N_{Re}^2 \log N_{Re}$:

$$\psi_{PP,2} = \frac{U_\infty a^2 \sin^2 \theta}{4}\left(\frac{r}{a} - 1\right)^2\left[\left(1 + \frac{3}{16}N_{Re} + \frac{9}{160}N_{Re}^2 \ln \frac{N_{Re}}{2}\right)\left(2 + \frac{a}{r}\right)\right.$$
$$\left. + \frac{3}{16}N_{Re}\left(2 + \frac{a}{r} + \frac{a^2}{r^2}\right)\cos \theta\right] + O(N_{Re}^3). \tag{10-50}$$

Unlike (10-48), this more accurate stream function (at least for $N_{Re} \ll 1$) does not predict a standing eddy. It thus remains unclear why the prediction made by (10-48) should agree so well with experiment. It may simply be a fortuitous result, or the result of an effective cancellation of higher order terms in N_{Re} for $N_{Re} \gg 1$.

The stream function $\psi_{PP,1}$ reproduces the first order Oseen drag, while from $\psi_{PP,2}$ Proudman and Pearson obtained the new result

$$C_{D,PP} = \frac{24}{N_{Re}}\left[1 + \frac{3}{16}N_{Re} + \frac{9}{160}N_{Re}^2 \ln (N_{Re}/2)\right]. \tag{10-51}$$

Finally, a further extension of the Proudman and Pearson analysis by Chester and Breach (1969) led to the inclusion of two more terms:

$$C_{D,CB} = \frac{24}{N_{Re}}\left[1 + \frac{3}{16}N_{Re} + \frac{9}{160}N_{Re}^2 \ln (N_{Re}/2)\right.$$
$$\left. + \frac{9}{160}N_{Re}^2\left(\gamma + \frac{5}{3}\ln 2 - \frac{323}{360}\right) + \frac{27}{640}N_{Re}^3 \ln (N_{Re}/2)\right], \tag{10-52}$$

where $\gamma = 0.57722\ldots$ is Euler's constant. The fourth term in (10-52) has been verified by Ockendon and Evans (1972), who used the method of matched asymptotic expansions in conjunction with Fourier transforms of the solution expansions.

5. *Potential Flow and Boundary Layer Theory.* There are no known analytical solutions capable of an accurate overall description of the flow past a sphere for intermediate or large Reynolds numbers, i.e., for $N_{Re} \gtrsim 10$. However, as we have pointed out already in Section 10.2.1, for $N_{Re} \gg 1$ the flow is such that viscous effects and vorticity are noticeable only within a thin boundary layer near the surface and in a downstream wake, and that elsewhere the flow is essentially potential. Fortunately, it turns out the flow in the boundary layer is amenable to

analysis, and so it is possible to obtain an approximate, though somewhat incomplete, account of high Reynolds number flow by piecing together the properties of potential and boundary layer flow.

Let us first consider potential flow. This is described either in terms of a potential Φ, viz.,

$$\vec{u}_P = \nabla\Phi, \qquad \nabla^2\Phi = 0, \tag{10-53}$$

or in terms of a stream function ψ_P by

$$\vec{u}_P = \nabla\phi \times \nabla\psi_P, \qquad E^2\psi_P = 0, \tag{10-54}$$

from (A.10-18) and (A.10-22). To solve for potential flow past a sphere, we must also take into account its frictionless character and hence abandon the no-slip boundary condition (10-4), replacing it instead by the weaker condition that at the surface (S) the flow must not penetrate the surface:

$$\vec{u}_P \cdot \hat{n}|_S = 0. \tag{10-55}$$

As in the case of Stokes flow, the condition (10-17) of streaming flow at infinity suggests a trial solution of the form $\psi_P = f(r) \sin^2\theta$. Along with (10-54) and (10-55), this leads directly to the solution

$$\psi_P = \frac{U_\infty r^2}{2} \sin^2\theta \left(1 - \frac{a^3}{r^3}\right), \tag{10-56}$$

and

$$\vec{u}_p = U_\infty \sin\theta \left(1 + \frac{a^3}{2\,r^3}\right)\hat{e}_\theta - U_\infty \cos\theta \left(1 - \frac{a^3}{r^3}\right)\hat{e}_r. \tag{10-57}$$

Since the convective acceleration term for potential flow can be expressed as $\vec{u}_P \cdot \nabla\vec{u}_P = \nabla(u_P^2/2) - \vec{u}_P \times (\nabla \times \vec{u}_P) = \nabla(u_P^2/2)$, the Navier-Stokes equation (10-7) reduces to $\nabla(p/\rho + u_P^2/2) = 0$, or

$$\frac{p}{\rho} + \frac{u_P^2}{2} = \text{constant}, \tag{10-58}$$

which is one version of *Bernoulli's law*. From (10-57) and (10-58) we find that the pressure distribution around a sphere in potential flow may be expressed, in terms of the dimensionless form of (10-28), as

$$k(\theta) = 1 - \frac{u_\theta^2(a, \theta)}{U_\infty^2} = 1 - \frac{9}{4}\sin^2\theta. \tag{10-59}$$

Now let us turn to an elementary discussion of the properties of boundary layer flow, of which the basic theory is due to Prandtl (1904). Consider a small region of the flow near the sphere surface where the boundary layer thickness δ_u is assumed to be well defined and much smaller than the sphere radius a. Let x denote distance along the drop surface in the direction of the local flow, and y denote distance normal to the surface. Then, using simple scaling arguments as in Section 10.2.1, we expect that in the region considered $\vec{u}$ will experience changes of order U_∞ over a length a in the x direction, and over a length δ_u in the y direction. Hence, from the condition $\nabla \cdot \vec{u} = 0$ we conclude immediately that in the boundary layer,

$$u_y/u_x \sim \delta_u/a \ll 1. \tag{10-60}$$

By the same logic it is also obvious that $|\bar{u} \cdot \nabla \bar{u}| \sim U_\infty^2/a$ and $|\nu\nabla^2\bar{u}| \sim \nu U_\infty/\delta_u^2$ in the boundary layer. Moreover, since in this region the viscous and inertial forces are of comparable magnitude, we may set these two estimates equal to obtain $\delta_u^2 \sim a\nu/U_\infty$, or

$$\delta_u \sim \frac{a}{R^{1/2}}; \tag{10-61}$$

i.e., the boundary layer thickness decreases with increasing Reynolds number as $R^{-1/2}$. From (10-60) and (10-61) we find also

$$u_y \sim \frac{u_x}{R^{1/2}}. \tag{10-62}$$

Comparison of the x and y components of the equations of motion in the boundary layer shows that the respective terms involving $\bar{u}$, and hence generally the pressure terms as well, differ by a factor of order u_y/u_x, so that

$$\frac{\partial p}{\partial y} \Big/ \frac{\partial p}{\partial x} \sim \frac{u_y}{u_x} \sim \frac{1}{R^{1/2}} \ll 1. \tag{10-63}$$

Basically, this says that the pressure gradient normal to the boundary layer may be neglected because of its relatively small width, and that therefore the pressure in the layer is approximated well by the potential pressure profile just outside the layer. Then from (10-58) we may also write $\partial p/\partial x \approx -\rho u_P \, du_P/dx$, i.e., the pressure gradient in the layer may be approximated directly in terms of the potential velocity profile just outside the layer.

We are now in a position to write down the equations of motion for the assumed steady and laminar (i.e., non-turbulent) boundary layer flow. In view of our scale analysis, we see that we need be concerned only with the x-component of (10-7), and that in this equation only the $\partial^2 u_x/\partial x^2$ term in the Laplacian may be neglected. Hence, if we assume for the moment that our x and y coordinates are strictly Cartesian, we obtain the following governing equations:

$$u_x \frac{\partial u_x}{\partial x} + u_y \frac{\partial u_x}{\partial y} - \nu \frac{\partial^2 u_x}{\partial y^2} = -\frac{1}{\rho}\frac{\partial p}{\partial x} = u_P \frac{\partial u_P}{\partial x}, \tag{10-64}$$

and

$$\frac{\partial u_x}{\partial x} + \frac{\partial u_y}{\partial y} = 0. \tag{10-65}$$

These equations are strictly valid only for rectilinear two dimensional flow, but they turn out also to be very accurate for describing transverse flow past an infinite cylinder, so long as the boundary layer thickness is very small compared to the cylinder radius. However, it happens that the geometry of axisymmetric flow, of interest to us here, is a little more complicated: By taking into account the curvature of our choice of x–y coordinates for the sphere, one can show (e.g., Pai, 1956) that (10-64) is unaltered but that the continuity equation takes on the new form

$$\frac{\partial(\bar{\omega}u_x)}{\partial x} + \frac{\partial(\bar{\omega}u_y)}{\partial y} = 0, \tag{10-66}$$

where $\bar{\omega}$ is the distance from the flow symmetry axis to the sphere surface.

Let us now introduce the dimensionless variables $x' = x/a$, $y' = y/\delta_u$, $u_x' = u_x/U_\infty$, $u_y' = R^{1/2}u_y/U_\infty$, $\bar{\omega}' = \bar{\omega}/a$, and $u_P' = u_P/U_\infty$. Then the boundary layer equations for a sphere acquire the form

$$u_x' \frac{\partial u_x'}{\partial x'} + u_y' \frac{\partial u_x'}{\partial y'} = \frac{\partial^2 u_x'}{(\partial y')^2} + u_P' \frac{du_P'}{dx'}, \tag{10-67}$$

and

$$\frac{\partial(\bar{\omega}'u_x')}{\partial x'} + \frac{\partial(\bar{\omega}'u_y')}{\partial y'} = 0. \tag{10-68}$$

Since neither these equations nor the boundary conditions which must be used with them ($u_x' = u_y' = 0$ at $x' = 0$; $u_x' = 1$ as $y' \to \infty$) involve the Reynolds number, we see that flows for different R are related by a simple similarity transformation; i.e., when R changes, the flow pattern changes only by having distances and velocities in the direction normal to the surface vary as $R^{-1/2}$.

From our previous discussion of the standing eddies and wakes which exist behind spheres for even moderate Reynolds numbers, the question arises whether or not a well defined boundary layer can exist over the entire sphere surface. In fact it cannot; rather, at some location from the front of the sphere the boundary layer detaches itself from the surface and flows into the main stream, carrying its load of vorticity with it. The occurrence of this phenomenon of *separation* can be understood qualitatively in terms of the potential pressure profile which is impressed on the boundary layer. From (10-59) we see that the potential pressure achieves a maximum at the front and rear of the sphere, and a minimum at the equator. Hence over the back hemisphere there is a pressure force which acts to retard the flow. Of course this adverse pressure gradient cannot reverse the free stream, but it is sufficient to reverse the relatively weak flow in the boundary layer, and thus cause separation.

When separation occurs, u_y is evidently no longer small compared to u_x, as is assumed in the equations for the boundary layer. Nevertheless, one can use these equations to estimate the location of separation, which will be at the first angle θ (or ring) from the front of the sphere at which the flow parallel to the surface stops even for $y \neq 0$; i.e., separation occurs where $\partial u_x/\partial y = 0$, or where the shear stress falls to zero. According to (10-67) and (10-68), the separation position should also be independent of the Reynolds number, since there are no scale changes in the x-direction with changing R.

As might be expected, the flow and pressure profiles near the surface downstream of the ring of separation are relatively complicated, and the model of potential plus boundary layer flow breaks down. And even where their use is justified, the boundary layer equations, though vastly simpler that the original governing set, are still non-linear and hence quite formidable.

Nevertheless, some useful approximate analytical solutions for boundary layer flow past a sphere have been obtained. The better results take into account the fact that experimental observations show a marked difference between the actual pressure distribution at the surface, and the theoretical potential distribution (Figure 10-1). Thus, an improvement in the description is possible by adopting the measured pressure profile as that which is impressed on the boundary layer. In one such study by Tomotika (1935), the pressure measurements of Flachsbart (1927) were used to obtain an approximate series solution for the velocity distribution in the boundary layer. Tomotika predicted the ring of separation should occur at $\theta = 81°$, which is in good agreement with Fage's (1934, 1937a, b) measured value of 83°. For the boundary layer thickness at the ring of separation, Tomotika obtained

$$\delta_{u,\text{sep}} = \frac{6.8 \, \nu}{u_{\text{sep}}} \, N_{\text{Re}}^{1/2}, \tag{10-69}$$

where u_{sep} denotes the tangential velocity just outside the boundary layer at the separation ring. Measurements show also that for $10^3 \lesssim N_{\text{Re}} \lesssim 10^5$, the surface pressure achieves a minimum near $\theta = 75°$ (see Figure 10-1); for this point Tomotika found

$$\delta_{u,75°} = \frac{3.8 \, \nu}{u_{75°}} \, N_{\text{Re}}^{1/2}. \tag{10-70}$$

These last two expressions illustrate the fact that the boundary layer thickness increases with increasing θ. Qualitatively, this trend can be understood by the same kind of argument we used earlier to explain the parabolic shape of the downstream wake.

Several numerical solutions to (10-67) and (10-68) have also been obtained (e.g., Smith and Clutter, 1963; Wang, 1970; and Blottner and Ellis, 1973). Wang's study provides a comparison between the numerically computed values for the

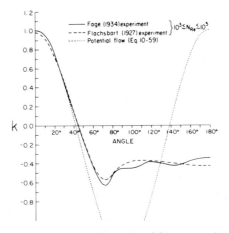

Fig. 10-1.   Variation of the pressure at the surface of a rigid sphere with polar angle and Reynolds number for $N_{\text{Re}} > 500$.

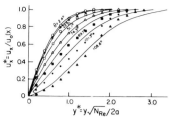

Fig. 10-2. Variation of the tangential velocity u_x^* with polar angle θ and distance y* from the surface of a rigid sphere, based on boundary layer theory. (From Wang, 1970; by permission of Cambridge University Press.)

tangential velocity u_x as a function of angle θ and distance y from the sphere surface, and the analytical series solution of Holstein and Bohlen (1940) (reproduced in Schlichting, 1968). This comparison is given in Figure 10-2. As expected, the series solution becomes poorer with increasing θ, i.e., with increasing boundary layer thickness.

10.2.5 NUMERICAL APPROACH TO THE NAVIER-STOKES EQUATION

The preceding discussions should serve to underscore the fact that the non-linear Navier-Stokes equation is tractable by analytical methods only for very special circumstances, and that often a great deal of effort is required for rather meagre results. Fortunately, however, the development of digital computers with increasingly large memories and fast execution times has provided an alternative approach of direct numerical solution which has become progressively more attractive and fruitful.

The basic idea of the numerical approach is to represent the continuous flow field at only a finite number of points by means of a grid or lattice, and to approximate the solution of the governing differential equation by satisfying a finite difference version of it which relates function values at neighboring points. Since finite difference representations of second order derivatives are simpler than those for fourth order derivatives, it turns out to be advantageous to write the fourth order governing equation (10-14) as two coupled second order equations. This is easily accomplished by using the vorticity as the second dependent variable. Thus on substituting (10-12) into (10-14), and introducing for convenience the dimensionless variables $r' \equiv r/a$, $\psi' \equiv \psi/U_\infty a^2$, and $\zeta' \equiv \zeta a/U_\infty$, the two coupled equations suitable for numerical treatment may be expressed as

$$\sin \theta \left[\frac{\partial \psi'}{\partial r'} \frac{\partial}{\partial \theta} - \frac{\partial \psi'}{\partial \theta} \frac{\partial}{\partial r'} \right] \left(\frac{\zeta'}{r' \sin \theta} \right) = \frac{2}{N_{Re}} E'^2(\zeta' r' \sin \theta), \tag{10-71a}$$

and

$$\zeta' = \frac{E'^2 \psi'}{r' \sin \theta}, \tag{10-71b}$$

where $E'^2 \equiv a^2 E^2$. The boundary conditions to be used in conjunction with (10-71) include: (1) at the sphere surface ($r' = 1$), $\partial \psi'/\partial r' = \psi' = \partial \psi'/\partial \theta = 0$, $\zeta' = E'^2 \psi'/\sin \theta$; (2) along the axis of symmetry ($\theta = 0, \pi$), $\psi' = \zeta' = 0$; (3) far from the sphere surface ($r' = r'_\infty$), $\psi' = (1/2) r'^2_\infty \sin^2 \theta$, $\zeta' = 0$.

After the stream function and vorticity fields have been determined for a given Reynolds number, the surface pressure distribution can be found from (10-32) and (10-33), and hence the form and skin friction drag coefficients from (10-30) and (10-31). Numerical solutions following this formulation have been provided by Jenson (1959), Hamielec *et al.* (1967), Le Clair (1970), Le Clair *et al.* (1970), and Pruppacher *et al.* (1970), among others.

10.2.6 COMPARISON OF ANALYTICAL AND NUMERICAL SOLUTIONS OF THE NAVIER-STOKES EQUATION WITH EXPERIMENTAL RESULTS

Let us now make a comparison between the characteristics of viscous, steady state, incompressible flow past a sphere as determined by analytical and numerical solutions of the Navier-Stokes equation, and the flow characteristics determined experimentally. First consider the stream function and vorticity distribution determined numerically by Le Clair (1970) and Le Clair *et al.* (1970). Le Clair found that the streamlines at $N_{Re} = 0.01$ show fore-aft symmetry. However, the vorticity contours clearly reveal that asymmetry in the flow exists even at this low Reynolds number, which is in pronounced disagreement with Stokes flow. At $N_{Re} = 10$ (Figures 10-3a, b) the flow is already strongly asymmetric, as shown by both the streamlines and the vorticity contours. At $N_{Re} = 30$ (Figures 10-3c, d) a standing eddy is present at the downstream end of the sphere

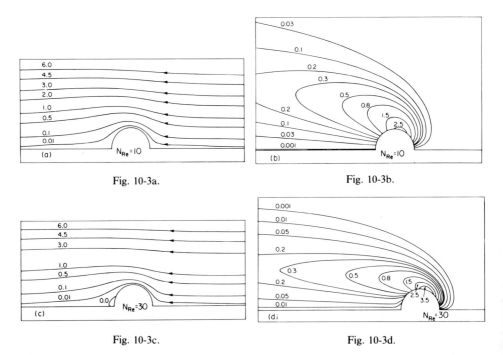

Fig. 10-3a.   Fig. 10-3b.

Fig. 10-3c.   Fig. 10-3d.

Figs. 10-3a, b, c, d. Numerically computed stream function (a, c) and vorticity (b, d) distribution around a rigid sphere at $N_{Re} = 10$(a, b) and $N_{Re} = 30$(c, d). (From Le Clair, 1970; by courtesy of the author.)

and the flow asymmetry has increased further. At $N_{Re} = 300$ the standing eddy extends over more than a sphere diameter downstream.

In Figure 10-4 comparison is made between the theoretically predicted and experimentally measured eddy lengths. It is seen that the length increases almost linearly with N_{Re} if $N_{Re} > 50$. Extrapolation suggests that the eddy begins to develop at $N_{Re} \approx 20$. Note that the experimental results are in good agreement with the numerical predictions except at low Reynolds numbers where flow visualization, which suffers from the finite fall velocity of the tracer particles, slightly underestimates the eddy length. The simple analytical result (10-49) of Proudman and Pearson is again surprisingly successful, and only slightly overestimates the eddy length. The numerical solution of the Oseen equation by Bourot (1969), not shown in the figure, considerably overestimates the eddy length. And, of course, the Stokes equations predict no eddy at all.

The pressure distribution computed by Le Clair *et al.* for the sphere surface is exhibited in Figure 10-5. Note the large discrepancy between these results and

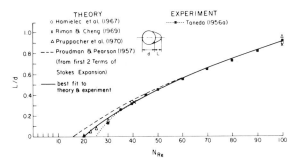

Fig. 10-4. Variation with Reynolds number of the length of the standing eddy at the downstream end of a rigid sphere. (From Pruppacher *et al.*, 1970; by permission of Cambridge University Press.)

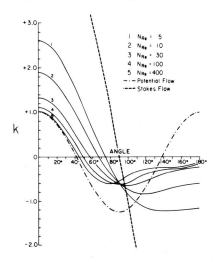

Fig. 10-5. Variation of the nondimensionalized pressure at the surface of a rigid sphere with polar angle and Reynolds number for $N_{Re} < 500$. (Based on data of Le Clair *et al.*, 1970.)

the predictions of Stokes flow. Note, too, that on the front half of the sphere the potential flow solution becomes acceptable for $N_{Re} \geqslant 400$. Comparison between the numerical results of Le Clair *et al.* (Figure 10-5) and the experimental measurements of Flachsbart (1927) and Fage (1937a, b) (Figure 10-1) shows that for $500 \leqslant N_{Re} \leqslant 10^5$ the normalized pressure distribution around a sphere remains essentially constant for $\theta \geqslant 80°$.

Values for the drag force coefficients computed by Le Clair *et al.* are reproduced in Table 10-1. It is evident that there is a crossover in dominance from the skin friction to the pressure or form drag coefficient for Reynolds numbers between 100 and 300. The numerical solutions for C_D are in excellent agreement with experiment, as shown in Figure 10-6. On the other hand, the analytical solutions of Stokes (Equation (10-38)) and Oseen (Equation (10-43)) are seen to significantly underestimate and overestimate C_D, respectively, for $N_{Re} \geqslant 1$. Curve 3 of Figure 10-6 indicates that more complete solutions of the Oseen equation are somewhat more physically realistic than the solution to O(R) found by Oseen; this was not anticipated theoretically.

TABLE 10-1

Comparison between the drag coefficients for rigid and fluid, circulating spheres at various Reynolds numbers. (Based on data of Le Clair *et al.*, 1972.)

N_{Re}	$C_{D,f}$		$C_{D,p}$		C_D	
	rigid	liquid	rigid	liquid	rigid	liquid
10	2.77	2.71	1.52	1.51	4.29	4.23
30	1.30	1.29	0.81	0.81	2.11	2.10
57	0.88	0.88	0.63	0.63	1.51	1.51
100	0.59	0.59	0.51	0.49	1.10	1.08
300	0.28	0.29	0.35	0.34	0.63	0.63

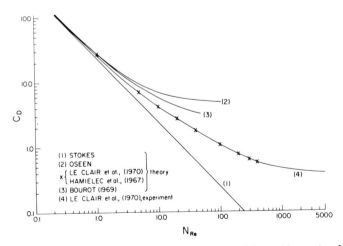

Fig. 10-6. Variation of the drag force coefficient as a function of Reynolds number for flow past a rigid sphere.

Further comparisons are given in Figure 10-7, where the fractional deviation from the Stokes drag is plotted against the Reynolds number. The experimental data in this plot confirm the proof, mentioned earlier, that the Oseen drag is correct to O(R); i.e., the experimental curves approach zero for $N_{Re} \rightarrow 0$ via the Oseen theory, rather than the Stokes theory.

It is interesting to note also that for small N_{Re} the solution of Chester and Breach (1969) is inferior to that of Proudman and Pearson (1957), even though in principle the former is correct to a higher order in N_{Re} than the latter. Proudman (1969) attempted to remedy this problem of apparent poor convergence of successive solutions obtained through the method of matched asymptotic expansions. He suggested the poor convergence is due, at least in part, to the inappropriateness of the choice of the function D for expansion in terms of N_{Re}. Through semi-empirical arguments he recast the results of Chester and Breach in a new form involving a free parameter m. In Figure 10-7 his results are plotted as curve 6 for m = 5, which choice gives the best fit. The outcome is a fit to the experimental and numerical results which is roughly as good as that provided by Carrier (1953) (curve 5) (however, the Carrier theory has the advantage of being relatively simple). The other analytical results shown are not as good.

To summarize, we can say that the numerical solutions agree excellently with the experimental drag determinations for $0.01 \leqslant N_{Re} \leqslant 400$. (An exception to this observation is provided by the results of Jenson (1959), whose flow fields suffer

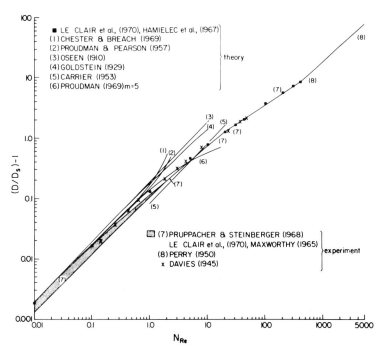

Fig. 10-7. Variation with Reynolds number of the nondimensionalized drag (D/D_S) for rigid spheres. D_S is the Stokes drag. (From Le Clair *et al.*, 1970; by courtesy of Amer. Meteor. Soc., and the authors.)

from step size and wall effect errors for $N_{Re} < 20$.) The analytical solutions do not fare as well, and are generally limited in their applicability to low Reynolds numbers. Probably the best, and one of the simplest, of these for $N_{Re} < 20$ is (10-47) with $c = 0.43$, due to Carrier. For the indicated range this expression overestimates the drag on a sphere by a maximum of 13% (at $N_{Re} = 20$). Finally, we should not lose sight of the fact that the simplest and most familiar analytical result, the Stokes drag given by (10-38) or (10-39), is quite adequate for many applications of interest in cloud physics. For example, it is accurate to within about 10% for $N_{Re} \lesssim 2$, which, as we shall see below in Section 10.3.5, corresponds to drop radii $\lesssim 50 \ \mu$m.

Close inspection of Figure 10-7 reveals another interesting feature, namely that there are three drag regimes. These are characterized by an almost constant slope over the Reynolds number intervals $0.01 \lesssim N_{Re} \lesssim 20$, $20 \lesssim N_{Re} \lesssim 400$, and $400 \lesssim N_{Re} \lesssim 5000$. It is interesting to note that the change of drag regime at $N_{Re} \approx 20$ coincides with the Reynolds number at which a standing eddy begins to form at the downstream end of the sphere, while the change of regime at $N_{Re} \approx 400$ coincides with the Reynolds number at which vortex shedding begins (see below). This suggests that flow regimes make themselves felt as drag regimes, which is certainly a physically plausible relationship.

As we indicated earlier in our discussion of the classification of flows past spheres according to the Reynolds number, intrinsic unsteadiness sets in for $N_{Re} \approx 130$. Close to this Reynolds number, the experiments of Möller (1938), Taneda (1956a), Goldburg and Florsheim (1966), Toulcova and Podzimek (1968), and Zikmundova (1970) have indicated the onset of faint, periodic, pulsative motions downstream of the standing eddy. These pulsations become increasingly pronounced and move toward the sphere as N_{Re} increases until finally, for some N_{Re} on the range $300 \lesssim N_{Re} \lesssim 450$, the standing eddies begin periodically to shed lumps of rotating fluid. In response to this behavior a sphere freely falling under gravity exhibits an increasingly pronounced spiral motion. This behavior implies that numerical solutions for axially symmetric, steady state flow will have decreasing relevance for increasing N_{Re} beyond some maximum value. The comparisons given in Figure 10-7 imply that the instability which sets in near $N_{Re} = 130$ causes a negligible departure from the strictly steady state drag. However, the observed change of drag regime at $N_{Re} \approx 400$ suggests this might be a reasonable cutoff point for steady state numerical approach. Thus, for $N_{Re} \gtrsim 400$ the drag on a sphere should best be determined by experiment.

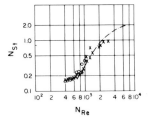

Fig. 10-8.   Variation of the Strouhal number with Reynolds number for flow past a rigid sphere.
(From Achenbach, 1974; by permission of Cambridge University Press.)

The frequency f of vortex shedding from the downstream end of a sphere has been measured by Möller (1938) and Achenbach (1974). Their results are shown in Figure 10-8 as a plot of the *Strouhal number*, $N_{St} \equiv 2 \, fa/U_\infty$, versus N_{Re}. One finds, for example, that for water drops (considered to be rigid spheres) falling in air with $\nu_a = 0.15 \, cm^2 \, sec^{-1}$, the shedding frequencies are $f \approx 5.1 \times 10^2 \, sec^{-1}$ for $N_{Re} = 400$ ($a = 662 \, \mu m$), $f \approx 1.9 \times 10^3 \, sec^{-1}$ for $N_{Re} = 1000$ ($a = 1.12 \, mm$), and $f \approx 1.9 \times 10^3 \, sec^{-1}$ for $N_{Re} = 2000$ ($a = 1.79 \, mm$).

10.3 Hydrodynamic Behavior of Water Drops in Air

We must now consider the extent to which cloud and raindrops depart from the idealization of rigid spheres. Considerable information on this point is provided by the comparison shown in Figure 10-9 of the drag on rigid spheres and on water drops falling in air. It is seen that good agreement exists for Reynolds numbers corresponding to drop radii less than about 500 μm. At larger sizes the drag on drops progressively increases above that for rigid spheres. In order to explain this behavior we must consider three observed characteristics of falling drops: their internal circulation, their distortion from a spherical shape, and their oscillation.

10.3.1 INTERNAL CIRCULATION IN DROPS

The existence of an internal circulation inside drops falling through another immiscible liquid, or inside gas bubbles rising through a liquid, has been established by many experimenters (e.g., Spells, 1952; Savic, 1953; Garner *et al.*, 1954). For water drops falling at terminal velocity in air, qualitative evidence for the presence of an internal circulation has been given by Blanchard (1949) and Garner and Lane (1959), and quantitatively by Pruppacher and Beard (1970) and

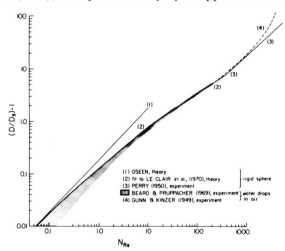

Fig. 10-9. Comparison between the Reynolds number dependence of the nondimensionalized drag D/D_s for rigid spheres and liquid circulating water drops.

Le Clair *et al.* (1972). Pruppacher and Beard and Le Clair *et al.* used a wind tunnel to determine the flow pattern and the average speed of the circulating water inside drops of 100 μm to 2 mm radius. Over the whole size range the maximum internal velocity near the drop surface was found to be close to $u_i \approx (1/25)U_\infty$, where U_∞ is the terminal velocity of the drop. A schematic representation of the observed flow pattern inside and outside a water sphere at $N_{Re} = 140$ and $N_{Re} < 1$ is given in Figure 10-10. For the case $N_{Re} = 140$, the reverse circulation in the standing eddy at the downstream side of the drop is seen to cause a neighboring region of fluid within the drop to be stagnant, or to exhibit a weak reverse circulation.

This observed behavior may be confirmed theoretically through a straightforward extension of our previous formulation of flow past a rigid sphere to the case of flow past and within a fluid sphere. Thus, if we use the vorticity-stream function approach, the exterior flow (ψ'_0, ζ'_0) will be governed by (10-71), as before, while the interior flow (ψ'_i, ζ'_i) will obey the same equation set, with the understanding that the relevant Reynolds number now depends on ν_i, the kinematic viscosity of the interior fluid. The boundary conditions which must supplement these equations are as follows: (1) the normal velocity components must vanish at the interface, so that $\psi'_0 = \psi'_i = 0$, $\zeta'_0 = E^2\psi'_0/\sin\theta$, and $\zeta'_i = E^2\psi'_i/\sin\theta$ at $r' = 1$; (2) the tangential velocities must be equal at the interface, so that $\partial\psi'_i/\partial r' = \partial\psi'_0/\partial r'$ at $r' = 1$; (3) the shear stress must be continuous at the interface, so that $(T_{r\theta})_0 = (T_{r\theta})_i$ at $r' = 1$; (4) along the symmetry axis $(\theta = 0, \pi)$, $\psi'_0 = \psi'_i = \zeta'_0 = \zeta'_i = 0$; (5) far from the sphere surface $(r' = r'_\infty)$, the flow streams freely, so that $\psi'_0 = (1/2)r'^2_\infty\sin\theta$ and $\zeta'_0 = 0$.

After the stream function and vorticity fields have been determined, the pressure distribution at the surface of the fluid sphere can be found, as before, from (10-32) and (10-33), except that now the centrifugal term $-u^2_\theta$ must be added to the right side of (10-32). Finally, given the surface pressure and vorticity distributions, the drag force coefficients can be computed from (10-30) and (10-31).

An accurate numerical solution to this problem has been obtained by Le Clair

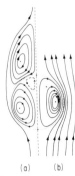

(a)   (b)

Fig. 10-10.   Observed flow pattern inside and outside a water drop in air. Inside flow observed by Pruppacher and Beard (1970); outside flow observed by Taneda (1956a); for $N_{Re} = 140$. (a) Observation, (b) Stokes flow. (From Pruppacher and Beard, 1970; by courtesy of *Quart. J. Roy. Meteor. Soc.*)

(1970) for exterior flow Reynolds numbers $N_{Re} \le 400$. The stream function distributions inside and outside a circulating fluid sphere are given in Figure 10-11 for a few selected Reynolds numbers, assuming the values $\eta_0 = \eta_{air} = 1.82 \times 10^{-4}$ poise and $\eta_i = \eta_{water} = 1.00 \times 10^{-2}$ poise, at 20 °C. We see from the figures that, in its main features, the flow of air around a circulating water sphere strongly resembles the flow past a rigid sphere, a result which we might have expected from the fact that the viscosity of water is about 55 times larger than that of air. We further note that, as in the case of a rigid sphere, a standing eddy develops at the downstream end of the sphere at $N_{Re} \approx 20$. However, while the length of this standing eddy differs little from that for a rigid sphere, its angular extent is significantly less. In particular, close to the fluid surface the eddy stream line $\psi = 0$ is shifted towards the rear stagnation point due to the effect of internal circulation. The general similarity between the gross hydrodynamic behavior of a liquid circulating sphere and that of a rigid sphere is further documented by the very small difference in values for the drag force coefficient. Comparison of the values listed in Table 10-1 shows that for $10 \le N_{Re} \le 300$, C_D (fluid) differs from C_D (rigid) by less than $\sim 1\%$.

Le Clair's numerical solution for the flow within the liquid sphere is in good qualitative agreement with the flow patterns observed in wind tunnel experiments. The computations show that a water sphere falling in air has a vigorous internal circulation with a stagnation ring slightly upstream of the equator. Also, a reverse circulation toward the rear of the sphere develops for $100 < N_{Re} < 300$, in agreement with observations. Le Clair *et al.* (1972) computed tangential

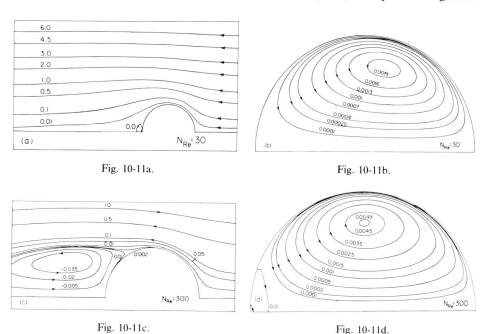

Fig. 10-11a.   Fig. 10-11b.

Fig. 10-11c.   Fig. 10-11d.

Figs. 10-11a, b, c, d. Numerically computed distribution of stream function and vorticity inside and outside a circulating spherical water drop in air at $N_{Re} = 30$(a, b) and at $N_{Re} = 300$(c, d).

velocities at the sphere surface using the values of terminal velocities observed by Gunn and Kinzer (1949), and Beard and Pruppacher (1969). The results for $\theta = 75°$, shown as curve 3 in Figure 10-12, agree well with observations (curve 4) for $a < 400 \, \mu$m. The extrapolation of the numerical results beyond $a = 630 \, \mu$m or $N_{Re} = 400$ (the dashed portion of curve 3) was based on the computed trend that $u_i(a, 75°) \to 0.042 \, U_\infty$ as $N_{Re} \to 400$.

Analytical models for drop internal circulations exist for very small and very large Reynolds numbers. For the case $N_{Re} \ll 1$, a simple analytical solution is available through the assumption of Stokes flow. The same form of trial solution as was invoked in Section 10.2.4 for the case of Stokes flow past a rigid sphere is successful also for both the interior and exterior flows for a fluid sphere, and the following solutions may be obtained without difficulty:

$$\psi_0 = \frac{U_\infty \sin^2 \theta}{4} \left[-3 \, ar \left(\frac{1 + \frac{2}{3}\gamma}{1 + \gamma} \right) + 2 \, r^2 + \frac{a^3}{r} \left(\frac{1}{1 + \gamma} \right) \right], \tag{10-72}$$

and

$$\psi_i = \frac{U_\infty \sin^2 \theta}{4} \left(\frac{\gamma}{1 + \gamma} \right) r^2 \left(\frac{r^2}{a^2} - 1 \right), \tag{10-73}$$

where $\gamma = \eta_0/\eta_i$ is the ratio of the outside and inside dynamic viscosities (Hadamard, 1911; Rybczinski, 1911). From (10-73) one finds a stagnation ring (where $u_i = 0$) at $\theta = \pm \pi/2$ and $r = a/2^{1/2}$. The pattern of streamlines is shown in Figure 10-10b.

The Stokes tangential surface velocity for a falling water drop is

$$u_{i\theta}(a, \theta) = \frac{U_\infty \sin \theta}{2} \left(\frac{\gamma}{1 + \gamma} \right) \approx \frac{\eta_a U_\infty \sin \theta}{2 \, \eta_w} \approx 0.009 \, U_\infty \sin \theta, \tag{10-74}$$

using the previously quoted values for η_0 and η_i. This expression predicts that the internal circulation of a small falling drop is only a small fraction of its terminal velocity, and is largest at $\theta = 90°$. Equation (10-74) for $\theta = 75°$ is plotted as curve 1 in Figure 10-12. The description is seen to be accurate to within about 10% for drops smaller than $100 \, \mu$m in radius; for larger sizes, (10-74) progressively underestimates the internal circulation.

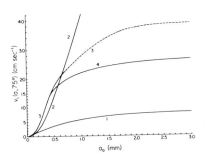

Fig. 10-12. Variation of the internal circulation velocity at the surface of a water drop in air ($\theta = 75°$). (1) Stokes flow model, (2) boundary layer model, (3) numerical, (4) observation. (From Le Clair *et al.*, 1972; by courtesy of Amer. Meteor. Soc., and the authors.)

The drag according to the Hadamard-Rybczinski (HR) theory is

$$D_{HR} = 6 \, \pi a \eta \, U_\infty \left(\frac{1 + \frac{2}{3} \gamma}{1 - \gamma} \right). \tag{10-75}$$

For a falling drop in air at 20 °C, $(1 + \frac{2}{3} \gamma)/(1 + \gamma) = 0.994$; thus $D_{HR} \approx D_S$ and the drop behaves like a rigid sphere as far as the drag is concerned. For a gas bubble in a liquid, $D_{HR} \approx 2 \, D_S/3$. Note there is no dependence of D_{HR} on surface tension. This is reasonable since surface tension only serves to alter the internal pressure by a constant amount, and thus should have no dynamic effect. These results for the drag may become invalid if certain surface active impurities are adsorbed onto the drop interface (Levich, 1962).

It is interesting to note from (10-71b) that the vorticity corresponding to (10-73) has the form $\zeta_i = Ar \sin \theta$, where A is constant. Therefore, from (10-71a) it can be seen that this same vorticity would enable the equation of motion to be satisfied also for the case of negligible viscosity ($N_{Re} \to \infty$). In fact, the stream function (10-73) is an exact solution of the Navier-Stokes equation, and provides a description of a flow pattern which is possible for both the Stokes and inviscid limits. In the latter application, it is known as 'Hill's spherical vortex' (Hill, 1894).

In terms of the constant A, the stream function for Hill's spherical vortex is $\psi_i = -[A(a^2 - r^2)r^2 \sin^2 \theta]/10$. If this is to describe inviscid flow, it is natural to fix A by assuming potential flow past the vortex. Therefore, for the exterior flow we let $\psi_0 = \psi_P$ as given by (10-56). If we now impose the condition of velocity continuity at $r = a$ (in this application conditions on shear stress do not apply), we obtain

$$\psi_i = -\frac{3 \, U_\infty}{4 \, a^2} (a^2 - r^2)r^2 \sin^2 \theta. \tag{10-76}$$

The corresponding interior tangential velocity is

$$u_{i\theta}(r, \theta) = \frac{3 \, U_\infty}{2 \, a^2} (2 \, r^2 - a^2) \sin \theta. \tag{10-77}$$

At the surface this is $u_{i\theta}(a, \theta) = (3/2)U_\infty \sin \theta$, which matches the outer potential flow.

It is tempting to regard (10-76) and (10-77) as a model for the internal circulation at high Reynolds numbers. However, the predicted surface velocities are too large; e.g., $u_{i\theta}(a, 75°) = 1.46 \, U_\infty$, which lies far above the experimental curve 4 in Figure 10-12, assuming experimental values for the terminal velocity U_∞. Of course the failure of the model is due to the complete omission of viscous effects. As for the case of flow past a rigid object, one expects that for $N_{Re} \gg 1$ there will be a viscous boundary larger near the drop surface which will play a dominant role in the adjustment of the internal circulation to the exterior flow.

McDonald (1954) and Le Clair et al. (1972) have used boundary layer theory to estimate the shear stress which drives the internal circulation in a falling spherical drop. The results of Le Clair et al. are shown as curve 2 in Figure

10-12. It can be seen that their boundary layer method, which is a refinement of McDonald's, drastically overestimates the true strength of the internal circulation for large drops. On the other hand, fair agreement with experiment and the numerical computations is found for smaller drops with radii $\leqslant 500$ μm ($N_{Re} \leqslant 260$). This outcome seems puzzling, since the boundary layer approach should work best at large Reynolds numbers.

The failure of the boundary layer model of McDonald and Le Clair et al. has a simple explanation (Klett, 1977). The boundary layer thickness expression used by these authors was borrowed from the theory of flow past a rigid sphere, and thus varies as $N_{Re}^{-1/2}$. Consequently, the shear stress at the drop surface, and hence the internal circulation within it, are predicted to increase strongly with increasing N_{Re}. It is clear that this description must eventually lead to large overestimates of both quantities with increasingly larger N_{Re}, since in reality the drop surface can relieve stress by moving. This motion will, in turn, tend to limit the thinning of the boundary layer.

From the observation, referred to earlier, that the strength of the internal circulation becomes proportional to U_∞ for $N_{Re} \geqslant 400$, an alternative boundary layer model suggests itself in which the boundary layer thickness approaches an asymptotic limit at $N_{Re} = 400$. The results of the theory of Tomotika (1935) for a rigid sphere can be used to estimate the limiting thickness $\delta_{u,\infty}$. Thus, from (10-70) we find $\delta_{u,\infty} \approx 0.33 \, a$, assuming the value $u_{75°} = 1.33 \, U_\infty$ from the numerical solution of Le Clair et al., and letting $N_{Re} = 400$. Assuming further that the internal circulation has approximately the pattern of Hill's spherical vortex, the internal tangential velocity has the form $u_{i,\theta} = BU_\infty (2 \, r^2/a^2 - 1) \sin \theta$ (cf. (10-77)). The constant B may be determined through the condition of continuity of shear stress at the interface. Thus, from (10-20b) the shear stress on the interior side of the interface is $(T_{r\theta})_i = 3 \, BU_\infty \eta_i \sin \theta/a$. If this is equated, at $\theta = 75°$, to the exterior shear stress, which on insertion of Tomotika's boundary layer velocity profile corrected for the drop surface motion is $2 \, \eta_0(u_{75°} - BU_\infty \sin 75°)/\delta_{u,\infty}$, the constant B is found to be B = 0.044. Support for this model comes from the fact that the resulting predicted surface velocities for $0 \leqslant \theta \leqslant \theta_{\text{separation}}$ agree closely with the computed values tabulated by Le Clair et al. for $N_{Re} = 300$. (For $\theta = 75°$, $u_i(0, 75°) = 0.043 \, U_\infty$, so that in the context of Figure 10-12 the results of this boundary layer model lie along the dashed portion of curve 3.)

From Figure 10-12 it can be seen that neither the numerical solution nor the modified boundary layer approach described above agree well with the experimental results for the strength of the internal circulation if the equivalent radius $a_0 \geqslant 500$ μm. Two principal reasons for this failure may be cited. Firstly, as we shall discuss in some detail in the next section, drops with $a_0 \geqslant 500$ μm are deformed into shapes resembling oblate spheroids; the theoretical descriptions have neglected this feature. Secondly, experiments show drop oscillations tend to occur when $a_0 \geqslant 500$ μm. These oscillations disrupt the internal circulation, so that the observed circulation strengths tend to be somewhat less than the true steady state values.

10.3.2 SHAPE OF WATER DROPS

Our discussion in the previous section suggests that internal circulation, although a marked characteristic of a water drop falling in air, contributes only negligibly to the drag on the drop. In the present section we shall show that the observed drag increase exhibited by drops of $N_{Re} \geqslant 200$ is primarily the result of a progressive change of their shape. Earlier experimental studies (Lenard, 1904; Flower, 1928; Laws, 1941; Best, 1947; Blanchard, 1948, 1950, 1955; Kumai and Itagaki, 1954; Magono, 1954a; Jones, 1959; Garner and Lane, 1959) on the variation of drop shape with size showed considerable differences due to the numerous difficulties involved in determining photographically the exact shape of a water drop falling at its terminal velocity in air. Most of these difficulties were circumvented by Pruppacher and Beard (1970) and Pruppacher and Pitter (1971), who freely suspended the drops in the air stream of a vertical wind tunnel. Some selected photographs from their studies on drop shape are given in Plate 1. As we have described previously in Section 2.1.2, drops falling at terminal velocity are nearly perfect spheres if their equivalent radius a_0 is less than $140 \, \mu m$ ($N_{Re} \leqslant 20$). For $140 \leqslant a_0 \leqslant 500 \, \mu m$ ($20 \leqslant N_{Re} \leqslant 260$), the drops are slightly deformed and resemble oblate spheroids. For $a_0 = 500 \, \mu m$, the ratio of the major and minor axes a and b is $b/a \approx 0.98$. Near $a_0 = 0.14 \, cm$ ($N_{Re} = 1.4 \times 10^3$, $b/a = 0.85$), the oblate drops develop flattened bases. The flattening becomes increasingly pronounced until finally near $a_0 = 0.20 \, cm$ ($N_{Re} = 1.9 \times 10^3$, $b/a = 0.78$), a concave base depression begins to develop. With further increase in size the concavity deepens until, at some critical size near $a_0 = 5 \, mm$, the drop becomes hydrodynamically unstable and breaks up, even in the absence of any measurable turbulence in the air stream.

The equilibrium shape of a falling drop can be determined in principle from the condition that local interface forces must be in balance. Early accounts of this force balance were somewhat incomplete. For example, Lenard (1904) assumed the drop shape was controlled by an equilibrium between surface tension and the centrifugal force resulting from internal circulation. Somewhat later, Spilhaus (1948) attributed the flattening of large drops to the combined action of surface tension and aerodynamic pressure, and thus ignored the effects of internal circulation and the hydrostatic pressure gradient within the drop.

To date most purely analytical treatments of the drop shape problem have required a considerable sacrifice of physical realism, or a restricted scope in applications, for the sake of mathematical tractability. Thus, Imai (1950) assumed an unrealistic unseparated potential flow past the drop, while the analysis of Taylor and Acrivos (1964) is restricted to very low Reynolds numbers and very small drop deformations. (However, the recent very simple analytical model of Green (1975), which ignores flow effects altogether, appears to yield results of surprising accuracy – see below.)

We shall now briefly outline the semi-empirical approach to the drop shape problem employed by Pruppacher and Pitter (1971). Their work follows a method proposed originally by Savic (1953), and has the chief merit of providing good agreement with independent wind tunnel observations.

First of all, we must obtain a generalization of the mechanical equilibrium condition (5-7) for a non-spherical drop shape. Consider a point on the surface where the principal radii of curvature are R_1 and R_2. Let the local area element $dS = dl_1 \, dl_2$ undergo an outward displacement (away from the drop center) by a small distance $\delta\xi$, where dl_1 and dl_2 are the surface arc length elements associated with R_1 and R_2. Then if the interior and exterior pressures are p_i and p_e, the pressure work done on dS by the displacement is $(p_e - p_i)\delta\xi \, dS$, neglecting the change $\delta(dS)$ in dS which the displacement brings about. On the other hand, this area change results in a change of surface energy in the amount $\sigma\delta(dS)$. Therefore, for a situation of static equilibrium, in which the net energy change should be zero for an arbitrary small displacement, we find

$$(p_e - p_i)\delta\xi \, dS + \sigma\delta(dS) = 0. \tag{10-78}$$

Now the first order change in the area element is $\delta(dS) = \delta(dl_1) \, dl_2 + dl_1 \, \delta(dl_2)$, and the corresponding change in the arc length elements is $\delta(dl_i) = dl_i(\delta\xi/R_i)$, $i = 1, 2$. Consequently, (10-78) implies

$$(p_i - p_e) = \sigma \left(\frac{1}{R_1} + \frac{1}{R_2}\right), \tag{10-79}$$

which is the desired result (Laplace, 1806). Note that (10-79) reduced to (5-7) for a spherical surface with $R_1 = R_2 = a$.

The further generalization of (10-79) to the case of nonzero exterior and interior flows is achieved simply through the replacement of the static pressures by the full stress tensor components. Thus, the dynamic boundary condition of stress continuity at the drop interface, including the effect of surface tension stresses, becomes

$$(T_{ij}n_j)_{\text{exterior}} - (T_{ij}n_j)_{\text{interior}} = \sigma \left(\frac{1}{R_1} + \frac{1}{R_2}\right) n_i, \tag{10-80}$$

(cf. (10-18)).

In their drop shape study, Pruppacher and Pitter used (10-80) to describe the normal stresses acting on the surface, and ignored the viscous stress contribution. In terms of the notation of Figure 10-13, their final force balance equation reads as follows:

$$\sigma \left(\frac{1}{R_1} + \frac{1}{R_2}\right) = g\rho_w(r_0 + r \cos \theta) - [p_e(\theta) - p_\infty - p_{ic}(\theta)] + [p_i(\pi) - p_\infty]. \tag{10-81}$$

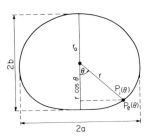

Fig. 10-13. Schematic of a deformed water drop.

The first term on the right side represents the stress contribution due to hydrostatic pressure within the drop (ρ_w is the density of water, and the buoyancy of the air has been neglected). The second term, in which $p_{i,c}(\theta)$ denotes the pressure due to internal circulation, Pruppacher and Pitter approximated by $(\tilde{p}_e(\theta) - p_\infty)_{\text{rigid sphere}}$, and then used the pressure measurements of Fage (1937) for this quantity.

The equation of the deformed drop surface of revolution may be expressed as follows:

$$r = a_0 \left(1 + \sum_{n=0}^{\infty} c_n \cos n\theta\right), \tag{10-82}$$

where the c_n are coefficients representing the deformation. If the deformation is small, i.e., $r = a_0 + \xi$ with $\xi \ll a_0$, we may express the curvature terms in (10-81) in terms of the c_n by means of (A.10-33), and obtain

$$\sigma \left[\frac{1}{R_1} + \frac{1}{R_2}\right] = \frac{\sigma}{a_0} \left[2 + \sum_{n=0}^{\infty} (n^2 - 2)c_n \cos n\theta + \sum_{n=0}^{\infty} n c_n \sum_{m=1}^{m=n} \cos (n - 2 m)\theta\right]. \tag{10-83}$$

Finally, in order to exploit the orthogonality properties of the expansions (10-82) and (10-83), the (known) pressure field is also expressed as a cosine series, viz.,

$$[p_e(\theta) - p_\infty]_{\text{rigid sphere}} = \frac{\rho_a U_\infty}{2} \sum_{n=0}^{\infty} q_n \cos n\theta, \tag{10-84}$$

where ρ_a is the density of air.

By substituting (10-82)–(10-84) into (10-81), the c_n may be determined in terms of the q_n and the physical parameters of the problem. To avoid excessive computation Pruppacher and Pitter truncated the infinite series at $n = 9$. They also bypassed the need to evaluate the constant terms in (10-81) by invoking the constraint of constant drop volume, which determines c_0 in terms of the other c_n. That is, on substituting (10-82) into the integral for the drop volume, $(2 \pi/3) \int_0^\pi r^3 \sin \theta \, d\theta$, and assuming small deformations, the condition of volume conservation becomes $\int_0^\pi \sum_{n=0}^{\infty} c_n \cos n\theta \sin \theta \, d\theta = 0$, which leads to

$$c_0 = \sum_{n=1}^{\infty} c_{2n}/(4 n^2 - 1). \tag{10-85}$$

Some computed meridional outlines of the drop shapes are presented in Figure 10-14. The results are based on the values $\sigma = 72.75 \text{ erg cm}^{-2}$, $\rho_a = 1.88 \times 10^{-3} \text{ g cm}^{-3}$, and $\nu_a = 0.153 \text{ cm}^2 \text{ sec}^{-1}$. The progressive flattening of the drop base and the development of a concave depression for $a_0 > 2$ mm is in good agreement with observations. The distortion in terms of a plot of b/a versus a_0 is shown in Figure 10-15. The results can be seen to match the experimental values quite well in predicting little change in drop shape for radii up to about 500 μm, and an almost linear decrease in b/a for larger increasing a_0. An approximate statement of this latter behavior is

$$b/a \approx 1.05 - 0.131 \, a_0, \tag{10-86}$$

for $0.5 \le a_0 \le 3$ mm, with a_0 in mm.

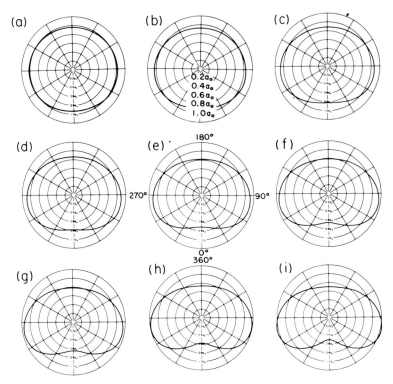

Fig. 10-14. Variation of the shape of water drops in air with their size. (a) $a_0 = 0.11$ cm, (b) $a_0 = 0.14$ cm, (c) $a_0 = 0.18$ cm, (d) $a_0 = 0.20$ cm, (e) $a_0 = 0.25$ cm, (f) $a_0 = 0.29$ cm, (g) $a_0 = 0.30$ cm, (h) $a_0 = 0.35$ cm, (i) $a_0 = 0.40$ cm. (From Pruppacher and Pitter, 1971; by courtesy of Amer. Meteor. Soc., and the authors.)

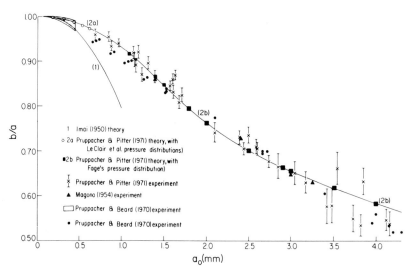

Fig. 10-15. Variation of the deformation of water drops in air with their size. (From Pruppacher and Pitter, 1971; by courtesy of Amer. Meteor. Soc., and the authors.)

Recently, Green (1975) has provided a simplified analysis of the drop shape problem which yields drop aspect ratios and maximum diameters which agree well with the theoretical and experimental results of Pruppacher and Pitter. His model assumes the drop shape is oblate spheroidal for all deformations, and that hydrostatic and surface tension stresses alone are sufficient to determine the equilibrium shape. This simple model appears to be reasonable since the more complete treatment of the problem by Pitter and Pruppacher has shown that the dynamic stresses cause only weak to moderate distortions from the shape of a pure oblate spheroid.

By applying (10-79) at the equator of the spheroid, Green's model yields

$$\sigma\left(\frac{1}{R_1} + \frac{1}{R_2}\right) = \sigma\left(\frac{a}{b^2} + \frac{1}{a}\right) = \frac{2\sigma}{a_0} + \rho_w g b, \tag{10-87}$$

where a and b are the semi-major and semi-minor axes, respectively, and $a_0 = a^{2/3}b^{1/3}$ is the equivalent radius. Equation (10-87) may be solved for a_0 in terms of the axis ratio to obtain

$$a_0 = \left(\frac{\sigma}{g\rho_w}\right)^{1/2} \frac{[(b/a)^{-2} - 2(b/a)^{-1/3} + 1]^{1/2}}{(b/a)^{1/6}}. \tag{10-88}$$

This agrees very well with the theoretical curve of Pruppacher and Pitter shown in Figure 10-15, except for a progressive divergence toward smaller deformations for $a_0 \gtrsim 3$ mm. However, even in this large size range (10-88) is quite consistent with the experimental results shown in the figure.

10.3.3 DROP OSCILLATION

As we discussed in Section 10.2.6, periodic vortex shedding from the downstream end of falling rigid spheres is observed to occur for $N_{Re} \gtrsim 400$. As one might expect, such vortex shedding can initiate internal vibrations in falling water drops. Drop oscillations may also be excited through a resonance with turbulent eddies in a turbulent air stream, or by the collision or near collision of two or more drops.

As one might also expect, the oscillatory motions of a drop can significantly affect its internal circulation. Indeed, Le Clair et al. (1972) observed that the internal circulation in millimeter size drops is only intermittently present – breaking up, reforming, and breaking up again in fast succession as dictated by the drop vibrations. A similar fluctuation is induced in the drag on a vibrating drop, since the drag is in part a function of drop shape.

A simple estimate of the characteristic vibration frequency f is available through the use of dimensional analysis. If we ignore viscous effects and the weak coupling to the exterior medium (air), the only relevant characteristic physical parameters are the drop's surface tension σ, density ρ_w, and equivalent radius a_0. Consequently, the necessity for dimensional consistency immediately tells us that

$$f \sim \left(\frac{\sigma}{\rho_w a_0^3}\right)^{1/2}. \tag{10-89}$$

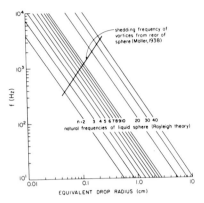

Fig. 10-16.   Variation of the natural oscillation frequency of a water sphere in air, and variation of the shedding frequency of eddies from a rigid sphere with size of sphere. (From Pruppacher and Pitter, 1971; by courtesy of Amer. Meteor. Soc., and the authors.)

A complete analysis of the problem, based on the assumption of potential flow and small amplitudes of vibration (Appendix A.10-33), shows that there is a discrete spectrum of allowed frequencies. From (A.10-38), these are

$$f_n = \left[\frac{n(n-1)(n+2)\sigma}{4\,\pi^2\rho_w a_0^3} \right]^{1/2}. \tag{10-90}$$

The choice $n = 0$ corresponds to radial oscillations, which are prohibited by the condition of incompressibility. The value $n = 1$ corresponds to translatory motion of the drop as a whole. Therefore, the fundamental mode of drop oscillation occurs for $n = 2$, with the frequency

$$f_2 = \left(\frac{2\,\sigma}{\pi^2\rho_w a_0^3} \right)^{1/2}. \tag{10-91}$$

Thus we see that the simple estimate (10-89) for the gravest mode is too large only by a factor of $\pi/\sqrt{2}$. Higher modes have correspondingly higher frequencies.

Equation (10-90) is plotted in Figure 10-16 for $\sigma = 72.75$ erg cm^{-2} and $n = 2$ to 40, together with the shedding frequency of vortices from the downstream end of a rigid sphere found by Möller (1938). Note from this figure that the fundamental mode of oscillation can be excited by the shedding of vortices from the drop if its radius is about 500 μm. This fact, first pointed out by Gunn (1949), is quite well documented by wind tunnel observations.

Measurements of the oscillation frequency of drops in air have been carried out by Blanchard (1948, 1949), Brook and Latham (1968), and Nelson and Gokhale (1972). Their observations are summarized in Figure 10-17. It can be seen that the experimental results agree well with the Rayleigh theory for $n = 2$. According to Nelson and Gokhale, the observed frequencies follow the empirical relation

$$f = 4.22\,a_0^{-1.47}, \tag{10-92}$$

with f in sec^{-1} and a_0 in cm.

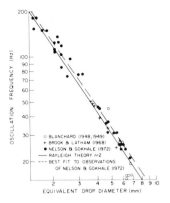

Fig. 10-17. Variation with size of the oscillation frequency of water drops in air. (From Nelson and Gokhale, 1972; by courtesy of the authors; copyrighted by American Geophysical Union.)

From (A.10-37), we see that the drop shape for axisymmetric oscillations may be described by an equation of the form

$$r = a_0 + \sum_n a_n \cos \omega_n t P_n(\cos \theta). \tag{10-93}$$

The normal mode oscillations for $n = 2$ and 3 are shown in Figure 10-18 for the case of an assumed maximum distortion axial ratio of 1.7. While (10-93) is strictly valid only for infinitesimal amplitudes, the results shown in the figure are in close agreement with observations, and with a detailed numerical calculation of finite amplitude drop oscillations by Foote (1971). (Foote simulated the drop dynamics by means of the marker-and-cell (MAC) method, a fast numerical scheme suitable for the study of transient incompressible flow problems, and particularly adapted for the study of flows with free surfaces (see Welch *et al.*, 1966).)

Drop oscillations will decay in time owing to the effects of viscosity. For small ν, the frequencies are not affected, but the amplitudes decay exponentially with

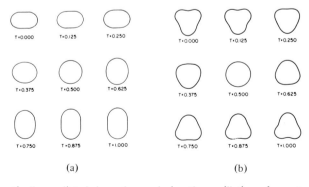

Fig. 10-18. Theoretically predicted drop shapes during the oscillation of a water drop in air; (a) $n = 2$, (b) $n = 3$. $T = \omega t$ is in units of π-radians. (From Foote, 1971; by courtesy of the author.)

a time constant T_n given by

$$T_n = \frac{a_0^2}{(n-1)(2n+1)\nu}$$ (10-94)

(Lamb, 1881; also 1945, p. 640). (Recently, Prosperetti (1976) has shown that (10-94) holds rigorously only for small times; for large times, vorticity generated at the drop surface diffuses to the interior and brings about an effective increase in damping.) According to (10-94), the fundamental mode (n = 2) of a 5 mm diameter water drop will decay in amplitude to only 10% of its original value after about 63 oscillations, each with a period of 33 msec ($f_2 = 30$ sec^{-1}); the same amplitude reduction for a 100 μm diameter drop occurs in 9 oscillations, each of 0.1 msec duration ($f_2 = 10^4$ sec^{-1}). The fact that the higher modes damp out first explains the previously mentioned predominance of the fundamental oblate-prolate oscillation (Figure 10-18a) observed in wind tunnel experiments.

10.3.4 DROP INSTABILITY AND BREAKUP

In this section we shall consider the conditions under which isolated falling drops become dynamically unstable and break up. Experiments carried out by Fournier d'Albe and Hidayetulla (1955) with drops falling through a long column of air at rest, and by Blanchard (1948, 1950) and Beard and Pruppacher (1969) with drops suspended in low turbulence wind tunnels, have demonstrated that, in quiet air, drops may be as large as 4.5 mm in equivalent radius before breaking up. The mechanism of breakup of such drops is closely tied to the development of the previously mentioned concave depression in the base of the falling drop. At a critical drop size, this depression almost explosively deepens and develops rapidly into an expanding bag supported by an annular ring which contains the bulk of the water. As this bag-like drop bursts, the bag portion breaks into a large number of small drops while the annular ring breaks into a smaller number of larger drops. The different stages of breakup are schematically drawn in Figure 10-19. This *bag breakup mechanism* has been studied by Blanchard (1948, 1949, 1950), Magarvey and Taylor (1956), Matthews and Mason (1964), and Koenig (1965a).

While a complete theoretical description of the growth of large amplitude drop

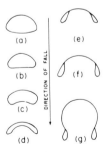

Fig. 10-19.   Stages during the bag-breakup of a water drop in air. (From Matthews and Mason, 1964; by courtesy of *Quart. J. Roy. Meteor. Soc.*)

distortions remains out of reach, the superficial aspects of the breakup process are well known. Probably the simplest order of magnitude estimate for the maximum stable drop diameter d_{max} is obtained through the statement that incipient instability occurs when the drag stress on the drop exceeds the surface tension stress. Thus, d_{max} may be estimated from the relation $C_D \rho_a U_\infty^2/2 \approx 4 \sigma/d_{max}$, or

$$d_{max} \approx \frac{8\sigma}{C_D \rho_a U_\infty^2}. \tag{10-95}$$

An equivalent dimensionless statement of this is

$$\frac{\rho_a d_{max} U_\infty^2}{\sigma} \equiv N_{We,max} = \frac{8}{C_D}, \tag{10-96}$$

where N_{We} is the *Weber number*. The Weber number can be seen to be a measure of the relative strengths of the Bernoulli pressure (which is effective because of flow separation) and the stress due to surface tension. It is clear that if $N_{We} \gg 1$ the surface tension stress, which tends to maintain a spherical shape, is negligible in comparison to the pressure, so that the latter can therefore strongly distort or disrupt the drop.

The value of C_D in (10-95) and (10-96) is unknown, but should lie between the values for a sphere (≈ 0.46) and a disk (≈ 1.17) at high Reynolds numbers. If we choose $C_D = 0.85$, (10-95) and (10-96) accord well with the experiments of Lane (1951), who found $U_\infty^2 d_{max} = 6.1 \times 10^5$ (c.g.s. units). Thus, the critical Weber number for breakup (based on d_{max}) is approximately 10. Estimates equivalent to (10-95) and (10-96) have been obtained, for example, by Dodd (1960), Levich (1962), and Matthews and Mason (1964).

Komabayasi *et al.* (1964) adopted a somewhat different approach to arrive at a similar estimate of the maximum stable drop size. They assumed that at the surface of a freely falling drop capillary and gravity waves are induced which, under certain conditions, become unstable and amplify. Thus, they consider drop breakup as a manifestation of the well known *Rayleigh-Taylor instability* of two superposed fluids of different density in a gravitational field. (An exhaustive treatment of this subject is available in Chapter 10 of Chandrasekhar (1961). A brief description, adequate for our purposes here, is given in Appendix A-10.3.4.)

Komabayasi *et al.* assumed the surface waves are plane parallel. From (A.10-48) we see that the critical wave number for instability is

$$k_c = \left[\frac{g(\rho_w - \rho_d)}{\sigma}\right]^{1/2} = 3.67 \text{ cm}^{-1}, \tag{10-97}$$

assuming $\sigma = 72.8$ erg cm^{-2}. The corresponding minimum stable wave length is $\lambda_{min} = 2\pi/k_c = 1.71$ cm. The critical maximum stable base width may therefore be estimated as $\lambda_{min}/2 = 0.855$ cm, i.e., the base width is just sufficient to accommodate the fundamental standing wave corresponding to λ_{min}.

When expressed in terms of a dependence on the relevant physical parameters, the estimate of Kombayasi *et al.* is

$$d_{max} = \frac{\lambda_{min}}{2} = \frac{\pi}{k_c} = \pi \left[\frac{\sigma}{g(\rho_w - \rho_a)} \right]^{1/2}. \tag{10-98}$$

This can be compared with the previous estimate (10-95) by noting that at terminal velocity the drag force balances the net gravitational force, so that $4 \pi a_0^3(\rho_w - \rho_a)g/3 = C_D(\rho_a U^2/2)\pi(d_{max}/2)^2$. Assuming an oblate spheroidal shape of axial ratio $x \equiv b/a$, where $a = d_{max}/2$, we have $4 \pi a_0^3/3 = 4 \pi a^2 b/3$, or $a_0 = x^{1/3} d_{max}/2$. Consequently, we may write

$$g(\rho_w - \rho_a) = \frac{3 C_D \rho_a U_\infty^2}{4 x d_{max}}. \tag{10-99}$$

Substitution of (10-99) into (10-98) gives

$$d_{max} = \frac{4 \pi^2 x \sigma}{3 C_D \rho_a U_\infty^2}, \tag{10-100}$$

which is the same as (10-95) except for a numerical factor. We may estimate $x = 0.55$ by extrapolation of the measurements of Pruppacher and Pitter (1971) to the case $a_0 = 0.45$ cm (see below); this gives $4 \pi^2 x/3 = 7.2$. Given the approximate and disparate nature of the estimates, we conclude that (10-95) and (10-100) agree surprisingly well.

The above comparisons provide additional physical support for the use of the Rayleigh-Taylor instability mechanism as a means for estimating the maximum stable drop size. A further refinement of this technique is available by employing a model more compatible with the geometry of the drop (Klett, 1971a). The symmetry of a falling drop about a vertical axis through its center suggests it is appropriate to consider the stability of two dimensional circular waves rather than one dimensional plane parallel waves. Also, the curvature of the drop surface near the edge of its base will have a constraining effect on the wave motion. This influence may be modeled in an approximate manner by regarding the effect of the drop's walls on its stability as lying between two extreme cases: (1) There are no walls; i.e., the bottom surface of the drop behaves as if it were part of an infinite interface between two otherwise unbounded fluids. (2) There are rigid vertical walls such that the stability of the drop is the same as that of two superposed fluids contained in a vertical circular tube.

As shown in Appendix A-10.3.4, if we assume inviscid flow the velocity potentials in both cases have a radial dependence of the form $J_n(kr)$, where J_n is the Bessel function of the nth order and k is the wave number. Also in both cases, the instability condition (10-97) holds. Following the standing wave criterion described just before (10-98), we see that for case 1 the radial distance from the z-axis to the lowest point on the drop bottom (half of what we call the base width $d_{c,1}$) must be the smallest of those distances r_{nm} such that $k_c r_{nm}$ is the mth root of $J_n(x) = 0$. The smallest such root is $k_c r_{01} = 2.40$, so that $d_{c,1} = 2 r_{01} = 1.31$ cm. The most unstable mode in case 1 is axisymmetric. In case 2 the velocity potential satisfies (A.10-39), so that the base width $d_{c,2}$ is twice the smallest of those distances r_{np} such that $k_c r_{np}$ is the pth root of $dJ_n(x)/dx = 0$. In this case, the smallest root is $k_c r_{11} = 1.84$, from which $d_{c,2} = 1.00\,\text{cm}$. The most unstable mode in case 2 is seen to be asymmetric.

The preceding analysis thus suggests that the largest stable base width of a water drop falling in quiet air should lie between 1.00 and 1.31 cm, i.e., a value somewhat larger than that predicted by Komabayasi. Wind tunnel measurements by Pruppacher and Pitter (1971) have shown that the largest drops which are stable in quiet air have $a_0 \approx 0.45\,\text{cm}$, and base widths ranging between 1.00 and 1.05 cm. This agreement with the refined but still very approximate Rayleigh-Taylor analysis is perhaps better than could be expected, since flow within and past the drop was ignored, and the effect of drop curvature was neglected except insofar as it determined boundary conditions at the 'edge' of the drop. A final comment in defense of the Rayleigh-Taylor instability model seems worth mentioning. By its nature, the model predicts that instability will occur on the bottom surface of the drop. This is in accord with wind tunnel observations.

Before leaving this section, let us briefly consider the question of whether or not atmospheric turbulence might have the capacity to disrupt drops. This is another nearly intractable problem, but again some estimates can be made on the basis of elementary arguments.

Let us assume the drop is immersed in homogeneous, isotropic turbulence in the inertial subrange, and that its size is larger than the turbulent microscale length (see Section 12.6.2). Kolmogorov (1949) obtained an estimate of the maximum stable drop size under the further assumption that the dynamic pressure difference across the drop surface is due solely to turbulent fluctuations. When this is true, one may proceed as in the argument leading to (10-95), except that now U_∞ is to be replaced by Δu, the characteristic velocity fluctuation over the length d_{max}; from (12-57) this is of order $(\varepsilon d_{max})^{1/3}$, where ε is the turbulent energy dissipation rate. On making the appropriate replacement in (10-95), it immediately follows that

$$d_{max} \approx \left(\frac{8\,\sigma}{C_D \rho_a}\right)^{3/5} \varepsilon^{-2/5}. \tag{10-101}$$

Unfortunately, however, the assumption that the dynamic pressure difference is caused solely by turbulent fluctuations does not hold for drops falling in air. It would be true only if the drop were essentially completely entrained by the turbulent eddies; i.e., (10-101) is applicable only for cases where drop and medium have comparable densities. Levich (1962) has obtained another estimate for the case where the density difference is large. However, his expression is also inapplicable in the context of interest to us, since he disregarded the effect of gravity on the drop motion.

For the case of water drops falling in air, it would seem reasonable to assess the influence of turbulence by comparing the pressure increment due to turbulent fluctuations with the Bernoulli pressure. The strength of the pressure fluctuations increases with eddy size, and the maximum relevant size should be of the order of the distance through which the drop falls during its relaxation time τ_{rel} for velocity fluctuations. Larger eddies are not felt as strongly since the drop is able, to some extent, to move with them.

Since for $N_{Re} \gg 1$ the drag varies as the square of the velocity, the fall speed v of the drop may be estimated from the equation $dv/dt = -Av^2 + g$, where $AU_\infty^2 = g$. Letting $v = U_\infty + v'$, where the fluctuation $v' \ll U_\infty$, we thus have $dv'/dt \approx -2 AU_\infty v'$, so that the relaxation time is

$$\tau_{rel} \approx (2 AU_\infty)^{-1} = U_\infty/2 g. \tag{10-102}$$

Hence the maximum effective eddy size is $\lambda \approx U_\infty^2/2 g$, corresponding to the turbulent velocity fluctuation $v_\lambda \approx (\varepsilon \lambda)^{1/3} \approx (\varepsilon U_\infty^2/2 g)^{1/3}$. Therefore, the ratio of the turbulent and Bernoulli pressures is

$$v_\lambda^2/U_\infty^2 \approx (\varepsilon/2 g U_\infty)^{2/3}. \tag{10-103}$$

For large drops turbulent disruption should be significant only if this ratio is not much smaller than unity. A relatively large value of ε is $10^3 \, cm^2 \, sec^{-3}$; assuming this value and $U_\infty \approx 10^3 \, cm \, sec^{-1}$, the magnitude of (10-103) is $O(10^{-2})$. Hence it appears turbulence, if it is of the type assumed here, cannot be effective in breaking up drops that are not already close to the critical size in the absence of turbulence.

Finally, since raindrops are rarely observed to have diameters exceeding 2 to 3 mm (recall Section 2.1.2), it appears that isolated drop breakup is not the predominant breakup mode in the atmosphere. Rather, the evidence suggests that breakup is generally a consequence of collision and temporary coalescence between pairs of drops. We shall discuss this breakup mode in Section 14.8.2.

10.3.5 TERMINAL VELOCITY OF WATER DROPS IN AIR

In cloud physics studies there is often a need to know the terminal velocities U_∞ of water drops at various levels in the troposphere. Of course, U_∞ is determined simply through the condition of balance between the buoyancy-corrected gravitation force and the drag force acting on the drop. Unfortunately, however, some complications arise in the attempt to describe the drag forces accurately. Some of these problems we have encountered already, while others we have avoided until now.

Consider first the case of small drops in air falling in the Stokes regime of negligible Reynolds numbers. Here matters seem especially simple: from (10-39) we write $6 \pi a \eta_a U_\infty = 4 \pi a^3 g(\rho_w - \rho_a)/3$, so that

$$U_\infty = U_S = \frac{2 a^2 g(\rho_w - \rho_a)}{9 \eta_a}, \tag{10-104}$$

where U_S is the *Stokes terminal velocity*. However, the Stokes drag description assumes continuum flow, and this assumption begins to break down for just those drops which are small enough to have negligible Reynolds numbers. Thus, for great accuracy it is necessary to correct (10-104) for the effects of slip-flow (see also Section 14.4.3). The correction becomes more important with increasing height in the atmosphere, since the molecular mean free path λ_a increases with decreasing air density.

As we have seen, there is a second range of intermediate Reynolds numbers for which drops may be assumed to fall as rigid spheres in a continuum flow for which both inertial and viscous forces are significant. For this range we may invoke the experimental and theoretical drag information which has been presented in Section 10.2.5. Finally, for larger sizes the problem of drop shape, and possibly also the intrinsic flow unsteadiness, must be taken into account.

It is convenient and natural, therefore, to resolve the terminal velocity problem into three drag regimes. Of course, there is some latitude in the proper choice of the regime boundaries. In the sequel we shall follow the classification and description of Beard (1976).

In regime 1: $0.5 \lesssim a \lesssim 10 \ \mu m \ (10^{-6} \lesssim N_{Re} \lesssim 10^{-2})$, the terminal velocity is

$$U_{\infty} = (1 + 1.26 \ \lambda_a/a) U_S, \tag{10-105}$$

where U_S is given by (10-104). The slip-flow 'Cunningham' correction factor multiplying U_S in (10-105) increases the terminal velocity above the Stokes value by about 1% for $a = 10 \ \mu m$ and 17% for $a = 1 \ \mu m$, at standard conditions $(p_0 = 1013.25 \ mb, \ T_0 = 293.15 \ °K, \ \lambda_{a,0} = 6.6 \times 10^{-6} \ cm)$. For other levels in the atmosphere, the mean free path may be obtained from the expression

$$\lambda_a = \lambda_{a,0} \left(\frac{p_0}{p}\right)\left(\frac{T}{T_0}\right). \tag{10-106}$$

Similarly, the dynamic viscosity for other than standard conditions is given with an accuracy of $\pm 0.002 \times 10^{-4}$ poise by

$$\eta_a(\text{poise}) = (1.718 + 0.0049 \ T) \times 10^{-4}, \quad T(°C) \geqslant 0 \ °C, \tag{10-107a}$$

$$\eta_a(\text{poise}) = (1.718 + 0.0049 \ T - 1.2 \times 10^{-5} \ T^2) \times 10^{-4}, \quad T(°C) < 0 \ °C. \tag{10-107b}$$

These expressions are based on data given by the Smithsonian Meteorological Tables. (Atmospheric water vapor has a negligible effect on ρ_a and η_a; see, for example, Kestin and Whitelaw, 1965.)

In regime 2: $10 \lesssim a \lesssim 535 \ \mu m \ (10^{-2} \lesssim N_{Re} \lesssim 3 \times 10^2)$, we may equate the drag and gravitational forces to obtain $6 \ \pi a \eta_a U_{\infty}(C_D N_{Re}/24) = 4 \ \pi a^3 g(\rho_w - \rho_a)/3$. When multiplied by ρ_a/η_a^2, this yields for spherical drops

$$C_D N_{Re}^2 = \frac{32 \ a^3(\rho_w - \rho_a)\rho_a g}{3 \ \eta_a^2}, \tag{10-108}$$

which is sometimes referred to as either the *Davies* or *Best number*. On the other hand, from (10-27) and (10-39) we also have

$$D/D_S = C_D N_{Re}/24, \tag{10-109}$$

where D_S is the Stokes drag. Hence

$$C_D N_{Re}^2 = 24 \ N_{Re}(D/D_S) = f(N_{Re}), \tag{10-110}$$

since D/D_S is known as a function of N_{Re} from the data displayed in Figure 10-9. In fact, Beard has suggested the following empirical fit for the curves shown in Figure 10-9:

$$Y = B_0 + B_1 X + \cdots + B_6 X^6, \tag{10-111}$$

where $B_0 = -0.318657 \times 10^1$, $B_1 = +0.992696$, $B_2 = -0.153193 \times 10^{-2}$, $B_3 = -0.987059 \times 10^{-3}$, $B_4 = -0.578878 \times 10^{-3}$, $B_5 = +0.855176 \times 10^{-4}$, $B_6 = -0.327815 \times 10^{-5}$, and where $X = \ln(C_D N_{Re}^2)$, and $N_{Re} = \exp(Y)$. Thus, for given a and atmospheric conditions ρ_a and η_a, $C_D N_{Re}^2$ is specified through (10-108), after which N_{Re} may be found from (10-111). Once N_{Re} is known, the terminal velocity may be determined from

$$U_\infty = \eta_a N_{Re} / 2 \rho_a a. \tag{10-112}$$

Some results for U_∞ versus a under different atmospheric conditions are shown in Figure 10-20. Notice that U_∞ increases almost linearly with radius for $100 < a < 500 \, \mu m$. Note also the U_S progressively overestimates actual fall velocities; for $a = 30 \, \mu m$ ($N_{Re} = 0.4$) the deviation is about 6%, while for $a = 42 \, \mu m$ ($N_{Re} = 1.0$) it is 13%.

In regime 3: $535 \, \mu m \lesssim a_0 \lesssim 3.5 \, mm$ ($3 \times 10^2 \lesssim N_{Re} \lesssim 4 \times 10^3$), the drops can no longer be considered spherical (see Section 10.3.2). Foote and du Toit (1969) pointed out that the procedure outlined above for computing U_∞ is only justified for spherical drops, for only then are C_D and D/D_S a function of N_{Re} alone. If a drop is deformed, the drag and thus the terminal velocity are functions also of the amount of the drop's deformation. We have seen that the deformation is a function of ρ_a, which varies with height in the atmosphere, and also a function of σ and ρ_w, which vary with temperature and, therefore, also with height. Furthermore, the methods presented in Section 10.3.2 permit a determination of a drop's deformation only if its terminal velocity is known a priori. Therefore, the procedures of that section are inadequate to determine U_∞ at arbitrary heights, since the only measurements of U_∞ for drops have been carried out at pressures prevailing at the Earth's surface (Gunn and Kinzer, 1949; Beard and Pruppacher, 1969).

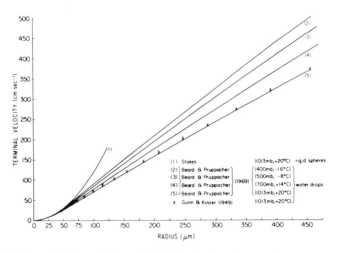

Fig. 10-20. Variation with size of the terminal fall velocity of water drops smaller than 500 μm radius in air. (From Beard and Pruppacher, 1969; by courtesy of Amer. Meteor. Soc., and the authors.)

Methods to get around this problem have been proposed by Wobus *et al.* (1971), Dingle and Lee (1972), and Berry and Pranger (1974). Unfortunately, these attempts are based on the Gunn and Kinzer observations which pertain to the Earth surface, and thus cannot be used to extrapolate C_D, $C_D N_{Re}^2$, and hence U_∞ to other pressures and temperatures. Therefore, these methods do not correctly account for drop shape. Foote and du Toit (1969) recognized the need for including a pressure dependence, and used the data of Davies (as reported by Best, 1950b) and Sutton (1942), derived from measurements of U_∞ at reduced air densities. Unfortunately, however, a careful scrutiny of the data reveals that under the conditions used by Davies the water drops did not reach terminal velocity, and thus could not develop their equilibrium shape.

In order to remedy this situation Beard (1976) suggested following an approach originally proposed by Garner and Lihou (1965). These authors found from studies of drops of various liquids freely suspended in the air stream of a wind tunnel that U_∞ depends on 3 independent dimensionless parameters: the Reynolds number $N_{Re} = 2\, a_0 U_\infty \rho_a / \eta_a$, the *Bond* number $N_{Bo} = g(\rho_w - \rho_a) a_0^2 / \sigma$, and the 'physical property' number $N_P = \sigma^3 \rho_a / \eta_a^4 g (\rho_w - \rho_a)$. The Bond number can be seen to measure the relative strength of gravitational and surface tension force, i.e., the relative strength of the drag and surface tension forces for a drop at terminal velocity. The parameter N_P is formed by eliminating the radius between the Davies and Bond numbers. From their observations Garner and Lihou established that

$$N_{Bo} N_P^{1/6} = f(N_P^{-1/6} N_{Re}). \tag{10-113}$$

Beard specified this functional relationship from a fit to the experimental results of Gunn and Kinzer (1949) for water drops in air. He found for drops of $535\ \mu\text{m} \leqslant a_0 \leqslant 3500\ \mu\text{m}$ that

$$Y = B_0 + B_1 X + \cdots + B_5 X^5, \tag{10-114}$$

with $B_0 = -0.500015 \times 10^1$, $B_1 = +0.523778 \times 10^1$, $B_2 = -0.204914 \times 10^1$, $B_3 = +0.475294$, $B_4 = -0.542819 \times 10^{-1}$, $B_5 = +0.238449 \times 10^{-2}$, and where $X = \ln[(16/3) N_{Bo} N_P^{1/6}]$, and $N_{Re} = N_P^{1/6} \exp(Y)$. If we now specify the temperature and pressure of the atmosphere, ρ_a, η_a, ρ_w, and σ are determined, and thus for a given drop size a_0, N_{Bo} and N_P are specified. Since these determine X, Y can be found via (10-114). From Y the Reynolds number and hence U_∞ may be determined.

Some results of this computational scheme are shown in Figures 10-21 and 10-22, in which the respective plots of $N_{Re} = N_{Re}(a_0)$ and $U_\infty = U_\infty(a_0)$ are shown. As would be expected, in the latter figure the curve for 1013 mb and 20 °C agrees with the experimental data of Gunn and Kinzer. It can be seen that the velocity of large drops increases noticeably with height in the atmosphere from about 9 m sec^{-1} at sea level to a value in excess of 12 m sec^{-1} at 500 mb.

It is interesting to note from Figure 10-22 that U_∞ becomes independent of size for $a_0 \geqslant 2.5$ mm. This indicates that the larger the drop, the more it is flattened, and thus the larger is the cross-section presented to the flow. The consequent increased drag resistance compensates for the increase in gravitational force.

A prediction of a limiting, size-independent terminal velocity follows directly

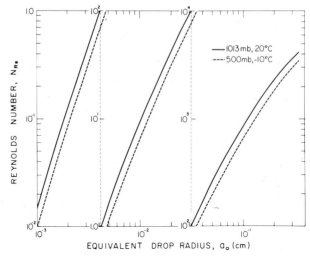

Fig. 10-21.   Reynolds number as a function of equivalent radius for water drops falling in air. (Based on the relations of Beard, 1976.)

from the Rayleigh-Taylor instability analysis of the previous section. Thus, from (10-98) and (10-100) we find immediately that for drops of the maximum stable size,

$$U_\infty \approx \left(\frac{4\,\pi b}{3\,a C_D}\right)^{1/2}\left(\frac{\rho_w g \sigma}{\rho_a^2}\right)^{1/4},\qquad\qquad (10\text{-}115)$$

which is independent of size, except possibly for a weak dependence in the first factor. An equivalent expression was obtained in an entirely different manner by Levich (1962). Equation (10-115) predicts limiting velocity ratios for different levels which agree to within 10% with those displayed in Figure 10-22.

Because of drop vibrations which are excited by the shedding of vortices for $N_{Re} \gtrsim 300$, one might also expect corresponding fluctuations in drop drag and

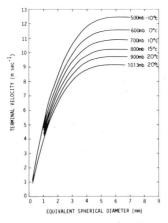

Fig. 10-22.   Variation with size of the terminal fall velocity of water drops larger than 500 μm in air. (From Beard, 1976; by courtesy of Amer. Meteor. Soc., and the author.)

hence terminal velocity. However, as we have seen the velocity response is limited by the relaxation time τ_{rel}, given by (10-102). In view of the results shown in Figure 10-22, we see that $\tau_{rel} \approx 0.5$ sec. This implies the excited vibration frequencies of 400 Hz and more (recall Figure 10-16) are too fast to result in significant fluctuations in the terminal velocity. This expectation is confirmed by experimental studies of drops falling in still air.

On the other hand, in a turbulent atmosphere there occur velocity fluctuations of characteristic periods which are comparable to or larger than τ_{rel}; these can be expected to cause noticeable fluctuations in U_∞. Unfortunately, there is not yet sufficient theoretical or experimental evidence to establish a relationship between the time-averaged terminal velocity and the character and intensity of turbulence. However, Jones (1959) found from photographic evidence that turbulence can induce significant fluctuations in drop shape, from oblate to prolate spheroidal. In the mean, the deformation of a drop of given size was found to be somewhat less, i.e., its axis ratio (b/a) somewhat larger, than that observed by Pruppacher and Pitter (1971) for a drop of the same size falling in still air. From this result one might conjecture that the time-averaged drag on a raindrop is lower in turbulent air than in still air.

Finally, let us estimate the distances and times which are necessary for water drops of various sizes to reach terminal velocity after their release from rest. A completely rigorous treatment of this problem of accelerating motion is complicated by the need to include the $d\bar{u}/dt$ term in the equation of motion for the fluid (recall (10-3)). However, arguments presented in Section 14.4.1 and in Appendix (A.14.5) demonstrate that the effect of local fluid acceleration is negligible since $\rho_a/\rho_w \ll 1$. Hence it is sufficient to use the steady state drag formulas to describe the hydrodynamic resistance experienced by the drops.

In view of the above remarks and (10-109), the equation of motion for a spherical drop of mass m_d and velocity v is

$$m_d \frac{dv}{dt} = m_d g \left(1 - \frac{\rho_a}{\rho_w}\right) - 6 \pi a \eta_a \left(\frac{C_D N_{Re}}{24}\right) v. \qquad (10\text{-}116)$$

In general this equation must be solved numerically since both C_D and N_{Re} are functions of v. However, an approximate analytical solution may be obtained for very small drops falling in the Stokes regime. Then $C_D N_{Re}/24 = 1$ and the resulting linear equation may be integrated immediately to obtain

$$v(t) = U_\infty \left[1 - \exp\left(-\frac{6 \pi a \eta_a}{m_d} t\right)\right]. \qquad (10\text{-}117)$$

Thus the *viscous relaxation time* for a spherical drop falling in Stokes flow is

$$\tau_S = \frac{m_d}{6 \pi a \eta_a} = \frac{2 a^2 \rho_w}{9 \eta_a} \approx 1.3 \times 10^3 a^2, \qquad (10\text{-}118)$$

for $T = 0\,°C$, a in cm and τ_S in sec. Similarly, the time required for the drop to reach a fraction β of its terminal velocity is seen to be $t_\beta = -\tau \ln (1 - \beta)$; e.g., $t_{99\%} = 1.02 a^2 \rho_w/\eta_a$. Integration of (10-117) gives for the distance s_β the drop must fall to reach a fraction β of its terminal velocity the result

$$s_\beta = -\frac{m_d U_\infty}{6 \pi a \eta_a} [\ln(1-\beta) + \beta].$$ (10-119)

For example, $s_{99\%} = 0.70 \, a^2 \rho_w U_\infty / \eta_a$. Thus for a 30 μm radius drop falling in air at 20 °C, we find $t_{99\%} \approx 5$ sec and $s_{99\%} \approx 5$ mm ≈ 170 drop radii.

Numerical solutions of (10-116) have been computed by Wang and Pruppacher (1977a) for conditions under which the assumption of Stokes flow is not valid. The drag was represented by (10-111) or (10-114), depending on the drop size regime, and the atmospheric conditions assumed were p = 700 mb, T = 10 °C. For $a_0 = 100$, 200, 300, 400, 500 μm, 1.0 mm, and 2.0 mm, they found $s_{99\%}$(m) = 0.18, 0.90, 2.1, 3.6, 5.4, 12.6, and 19.8, respectively. Experimental confirmation of the numerical results has been provided by Sartor and Abbott (1975) and Wang and Pruppacher (1977a). It is clear from these results that the often used assumption that isolated cloud particles always fall in quiet air at their terminal velocities is fully justified. On the other hand, in experiments on drop behavior in which drops are released from rest, some care must be taken to ensure a sufficient fall distance for terminal velocities to be achieved at the observation location.

10.4 Hydrodynamic Behavior of Disks, Oblate Spheroids, and Cylinders

We have devoted considerable attention to the hydrodynamic behavior of water drops partly because they are the most abundant cloud particles, and also because their approximately spherical geometry is relatively easy to deal with. We shall now turn to a discussion of the hydrodynamics of ice crystals, based largely on idealized geometric models which approximate their shape. As might be expected, we shall find much behavior which is analogous to that of spherical particles. In such cases therefore, where the differences in the descriptions are due mainly to mathematical details arising from differences in geometry, we shall provide only brief summaries of the essential results, and refer the reader to the references for details.

We remarked in Chapter 2 that to the casual observer ice crystals appear in a variety of different shapes. Actually, however, these only represent variations of two fundamental shapes: that of a columnar hexagonal prism and that of a plate-like hexagonal prism. One is tempted, therefore, to approximate the hydrodynamic behavior of ice crystals by that of finite circular cylinders, and of thin circular disks or oblate spheroids. Indeed, Jayaweera and Cottis (1969) have shown by model experiments that for simple columnar and plate-like ice crystals this analogy does hold (Figure 10-23). We shall therefore briefly discuss the relevant hydrodynamics of circular disks, oblate spheroids, and cylinders.

10.4.1 CIRCULAR DISKS AND OBLATE SPHEROIDS

An analytical solution to the Navier-Stokes equation for the (Stokes) flow past an oblate spheroid at negligible Reynolds numbers, parallel to its axis of revolution, was obtained by Oberbeck (1876) (see also Happel and Brenner, 1965). He found the drag to be given by the expression

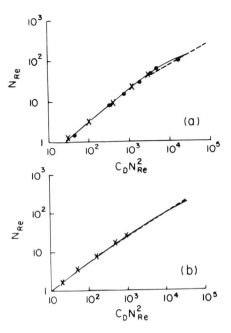

Fig. 10-23. Variation of $C_D N_{Re}^2$ vs. N_{Re} for disks and cylinders. (a) —— Schmiedel (1928), circular disks; ● Willmarth *et al.* (1964), circular disks; – – – Podzimek (1968), hexagonal disks; × Jayaweera and Cottis (1969), hexagonal disks. (b) —— Jayaweera and Mason (1965), circular cylinders; – – – Podzimek (1968), hexagonal cylinders; × Jayaweera and Cottis (1969), hexagonal cylinders. (From Jayaweera and Cottis, 1969; by courtesy of *Quart. J. Roy. Meteor. Soc.*)

$$D_{OB} = 8 \, \pi \eta \, U_\infty \delta / [\lambda_0 - (\lambda_0^2 - 1) \cot^{-1} \lambda_0], \qquad (10\text{-}120)$$

where $\delta = (a^2 - b^2)^{1/2}$ and $\lambda_0 = b/\delta$. Since the cross-sectional area perpendicular to the flow is πa^2, the corresponding drag coefficient is, from (10-27),

$$C_{D,OB} = \frac{32 \, \delta}{a \, N_{Re} [\lambda_0 - (\lambda_0^2 - 1) \cot^{-1} \lambda_0]}, \qquad (10\text{-}121)$$

where the Reynolds number is based on the major axis $2 \, a$. If the oblate spheroid is infinitely thin, i.e., $b \to 0$, $\lambda_0 \to 0$, $\delta \to a$, we obtain the drag expression for a disk:

$$\lim_{b \to 0} D_{OB} = 16 \, \eta a U_\infty, \quad \lim_{b \to 0} C_{D,OB} = \frac{64}{\pi N_{Re}}. \qquad (10\text{-}122)$$

An extension of these results to include first order inertial effects was accomplished by Aoi (1955). From a study of Oseen flow past an oblate spheroid he obtained the result

$$C_{D,A} = \frac{32}{N_{Re} S} \left[1 + \frac{N_{Re}}{S} \right], \qquad (10\text{-}123)$$

where $S = a \delta^{-1} [\lambda_0 - (\lambda_0^2 - 1) \cot^{-1} \lambda_0]$. For the case of a disk this reduces to an expression first obtained by Oseen (1915):

$$C_{D,O} = \frac{64}{\pi N_{Re}} + \frac{32}{\pi^2}. \qquad (10\text{-}124)$$

A further extension to include higher order terms in N_{Re} was carried out by Breach (1961), who used the method of matched asymptotic expansions to obtain the result

$$C_{D,B} = \frac{8\,m}{3\,N_{Re}} \left[1 + \frac{mN_{Re}}{48} + \frac{m^2}{1440} N_{Re}^2 \ln\,(N_{Re}/2) + O(N_{Re}^2) \right], \qquad (10\text{-}125)$$

where $m = 12\,e^3\{e(1-e)^{1/2} + (2\,e^2 - 1)\,\tan^{-1}\,[e/(1-e^2)^{1/2}]\}$ and $e = [1 - (b/a)^2]^{1/2}$. For a thin disk with eccentricity $e \to 1$, (10-125) becomes

$$\lim_{e \to 1} C_{D,B} = \frac{64}{\pi N_{Re}} \left[1 + \frac{N_{Re}}{\pi} + \frac{2\,N_{Re}^2}{5\,\pi^2} \ln\,(N_{Re}/2) + O(N_{Re}^2) \right]. \qquad (10\text{-}126)$$

All of these results are quite analogous to those which hold for spheres, and so it should come as no surprise to learn they are accurate only for small values of N_{Re}. For moderate to large N_{Re} one must again resort to numerical solutions of the Navier-Stokes equation. This can be done following procedures like those we discussed earlier for the case of flow past a sphere, the only additional complication being the need to use oblate spheroidal coordinates (e.g., Rimon and Lugt (1969), Masliyah and Epstein (1970), and Pitter et al. (1973)).

As an example of the behavior of viscous flow past an oblate spheroid, the stream function and vorticity fields (as determined by Pitter et al., 1973) are plotted in Figure 10-24 for the case of an axial ratio $b/a = 0.05$ and for $N_{Re} = 1$ and 20. Figure 10-24b shows that flow asymmetry is noticeable even at $N_{Re} = 1$. A standing eddy develops at $N_{Re} \approx 1.5$ and grows in length with increasing N_{Re}, reaching a length of $2\,a$ at $N_{Re} \approx 30$. The larger the axis ratio of the spheroid, the smaller the extent of the eddy and the higher the Reynolds number at which it begins to form. This is in good agreement with the experimental observations of Masliyah (1972). In Figure 10-25 the length of the standing eddy at the downstream end of the oblate spheroid is given as a function of N_{Re}. In Figure 10-26a comparison is made between the pressure distributions at the surfaces of a sphere and of oblate spheroids of various b/a. It is seen that with decreasing b/a the variation of the pressure with polar angle becomes increasingly pronounced.

The variation of C_D with N_{Re} for an oblate spheroid of $b/a = 0.05$ is plotted in Figure 10-27. Note that Oberbeck's low Reynolds number solution underestimates C_D on a disk progressively with increasing N_{Re}, while Oseen's solution progressively overestimates it. In Figure 10-28 the dimensionless drag $D/D_{OB} = C_D/C_{D,OB}$ is shown as a function of N_{Re}. This figure shows clearly that the analytical solutions are very accurate only for $N_{Re} < 1$, while the numerical solutions agree excellently with experiment for $N_{Re} \lesssim 100$. Also note from Figure 10-28 that, in analogy to flow past a sphere, the variation of D/D_{OB} reveals 'drag regimes' within which $\ln\,(D/D_{OB} - 1)$ varies roughly linearly with $\ln\,(N_{Re})$, and with rather pronounced changes in drag near $N_{Re} = 1.5$ and $N_{Re} = 100$. The experiments of Masliyah (1972) and the numerical results of Pitter et al. (1973) demonstrate that a standing eddy develops behind a thin oblate spheroid or disk at $N_{Re} \approx 1.5$, while the experiments of Willmarth et al. (1964) show that at $N_{Re} \approx 100$ the shedding of eddies begins from the downstream end of the disk. Thus, in further analogy to flow past a sphere, its drag regimes are a close

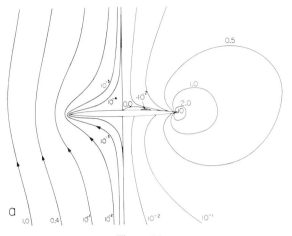

Fig. 10-24a.

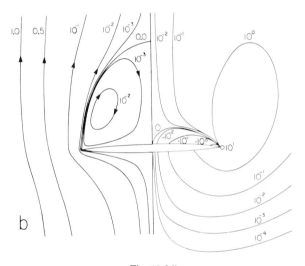

Fig. 10-24b.

Figs. 10-24. Numerically computed stream function (b/a) and vorticity distribution around an oblate spheroid of $b/a = 0.05$ at (a) $N_{Re} = 1.0$, and (b) $N_{Re} = 20$. (From Pitter *et al.*, 1973; by courtesy of Amer. Meteor. Soc., and the authors.)

manifestation of its flow field regimes.

Assuming that the variation of $C_D N_{Re}^2$ with N_{Re} is only negligibly dependent on b/a for $b/a \leqslant 0.2$, and is given essentially by the variation found for $b/a = 0.2$ (this is justifiable from the experiments of Jayaweera and Cottis, 1969), we may express the empirical functional relationship between $C_D N_{Re}^2$ and N_{Re} as

$$\log N_{Re} = B_0 + B_1 X + B_2 X^2, \tag{10-127}$$

where $X = \log (C_D N_{Re}^2)$, and $B_0 = -1.3300$, $B_1 = 1.0217$, $B_2 = -0.049018$. The cor-

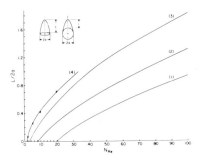

Fig. 10-25. Variation with Reynolds number of the length of the standing eddy at the downstream end of oblate spheroids of different axis ratios: (1) sphere, (2) oblate spheroid, $b/a = 0.5$ (Masliyah and Epstein, 1970), (3) oblate spheroid, $b/a = 0.2$ (Masliyah and Epstein, 1970), (4) oblate spheroid, $b/a = 0.05$. (From Pitter *et al.*, 1973; by courtesy of Amer. Meteor. Soc., and the authors.)

responding expression for an oblate spheroid of $b/a = 0.5$ is

$$\log N_{Re} = B_0 + B_1 X + B_2 X^2 + B_3 X^3, \tag{10-128}$$

where $X = \log (C_D N_{Re}^2)$, and $B_0 = -1.3247$, $B_1 = 1.0396$, $B_2 = -0.047556$, $B_3 = -0.002327$.

These relations may be used to approximate the terminal fall velocity of planar ice crystals. For this purpose we may follow a procedure similar to that outlined in Section 10.3.5, and express $C_D N_{Re}^2$ as a function of the relevant basic atmospheric parameters. Then from (10-127) and (10-128) we have N_{Re} and hence U_∞ for given atmospheric conditions. The desired expression for $C_D N_{Re}^2$ is found by equating the drag and gravitational forces; from (10-27) this leads to

$$C_D = \frac{2 V_c (\rho_c - \rho_a) g}{A_c \rho_a U_\infty^2}, \tag{10-129}$$

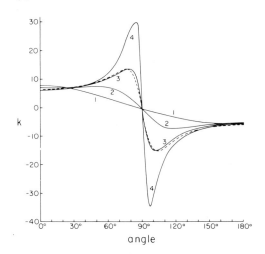

Fig. 10-26. Variation of the nondimensionalized pressure with polar angle at the surface of oblate spheroids at $N_{Re} = 1$; (1) sphere (Le Clair *et al.*, 1970), (2) oblate spheroid, $b/a = 0.5$ (Masliyah and Epstein, 1970), (3) oblate spheroid, $b/a = 0.2$ (Masliyah and Epstein, 1970), (4) oblate spheroid, $b/a = 0.05$. (From Pitter *et al.*, 1973; by courtesy of Amer. Meteor. Soc., and the authors.)

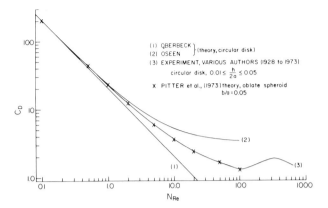

Fig. 10-27. Variation of the drag force coefficient with Reynolds number for thin circular disks and oblate spheroids of $b/a = 0.05$. The experimental results are those of Pitter *et al* (1973), Jayaweera and Cottis (1969), Willmarth *et al.* (1964), and of Schmiedel (1928). (From Pitter *et al.*, 1973; by courtesy of Amer. Meteor. Soc., and the authors.)

where V_c and ρ_c are the volume and density of the ice crystal, and A_c is its equatorial area oriented normal to the flow direction. Thus for a circular disk of thickness h and radius a, it follows from (10-129) that

$$C_D N_{Re}^2 = \frac{8\,a^2 h(\rho_c - \rho_a)\rho_a g}{\eta_a^2} \approx \frac{8\,m_c \rho_a g}{\pi \eta^2}, \tag{10-130}$$

where m_c is the mass of the crystal (the small buoyancy force has been ignored

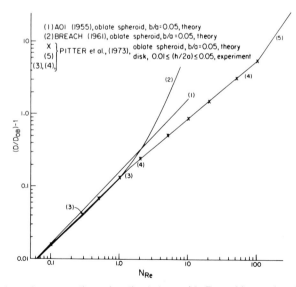

Fig. 10-28. Variation of the nondimensionalized drag with Reynolds number for thin disks and oblate spheroids of $b/a = 0.05$. The experimental results are those of the authors cited in Figure 10-27. (From Pitter *et al.*, 1973; by courtesy of Amer. Meteor. Soc., and the authors.)

in the last form of this equation). Similarly, for an oblate spheroid we have

$$C_D N_{Re}^2 = \frac{32\, a^2 b\, (\rho_c - \rho_a)\rho_a g}{3\, \eta_a^2} \approx \frac{8\, m_c \rho_a g}{\pi \eta_a^2}. \tag{10-131}$$

In view of the close similarity between the motion of circular disks and plate-like hexagonal prisms (recall Figure 10-23), we may also include the case of a hexagonal plate of maximum (circumscribed) radius a and thickness h, for which

$$C_D N_{Re}^2 = \frac{8\, a^2 h (\rho_c - \rho_a)\rho_a g}{\eta_a^2} \approx \frac{16\, m_c \rho_a g}{3\sqrt{3}\; \eta_a^2} \tag{10-132}$$

Depending on the geometry of interest, any of the relations (10-130)–(10-132) may be used in conjunction with either (10-127) or (10-128) to estimate the terminal velocity of planar ice crystals.

   This method is only applicable for $N_{Re} \lesssim 100$, since only for these Reynolds numbers do such objects as thin disks fall stably with their broadest dimension normal to the direction of fall. For $N_{Re} > 100$, oscillations occur which increase in amplitude with increasing N_{Re}, until eventually a glide-tumbling motion ensues. Finally, for still higher $N_{Re}\ (= O(10^3))$ the gliding stops and the objects simply tumble (Willmarth *et al.*, 1964; Stringham *et al.*, 1969).

10.4.2   CIRCULAR CYLINDERS

It is obvious from our discussions in Chapter 2 that the shape of certain columnar ice crystals can be idealized by the shape of a circular cylinder of finite length. Unfortunately, however, even this idealized geometry has proven overwhelmingly complicated, and there are no known solutions for the flow past such objects. Obviously, the difficulties are associated with the 'end' geometry. On the other hand, there is some empirical evidence which indicates that end effects are often not of great importance. For example, a glance at Plates 3 and 4 shows that while plate-like ice crystals rime preferentially at the edges, indicating a significant flow edge effect, columnar ice crystals rime rather uniformly over their entire length. This indicates that as far as the trajectories of drops past such crystals are concerned, end effects imposed on the streamlines are not significant. Thus for our idealization we shall take one further step and assume the cylinder has infinite length.

   At this point the reader might well anticipate a description of the drag per unit length on a solid cylinder immersed in an otherwise unbounded Stokes flow moving at right angles to the cylinder axis. Curiously enough, such an analogy with our previous discussions cannot be provided, for there is no solution to this problem. A simple demonstration of this fact is possible using elementary dimensional arguments. Thus, if the free stream velocity is U_∞ and the cylinder radius is a, the force per unit length, f, can only depend on a, U_∞, and η. However, since the only dimensionless group that can be formed from these variables is $f/\eta U_\infty$, we conclude that $f = A\eta U_\infty$, where A is a dimensionless

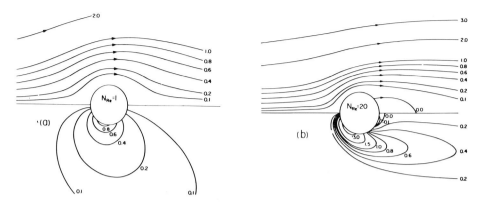

Fig. 10-29. Numerically computed stream function and vorticity distribution around a circular cylinder at (a) $N_{Re} = 1$ and (b) $N_{Re} = 20$. (From Schlamp *et al.*, 1977; by courtesy of *Pure and Appl. Geophys.*, and the authors.)

constant. This is clearly impossible, since the force per unit length must also depend on the radius a; e.g., we should have f = 0 if $a = 0$.

The resolution of this so-called 'Stokes paradox' requires recognition of the fact that the assumption of negligible fluid inertia everywhere cannot hold at infinity (see Section 10.2.4). Thus, one must turn to the Oseen equations for the simplest analytical estimate of the drag per unit length on an infinite cylinder in low Reynolds number flow. This was achieved by Lamb (1911) (also in Lamb, 1945, §343), who found the drag coefficient per unit length to be given by

$$C_D' \equiv \frac{D}{(1/2)\rho U_\infty^2 d} = \frac{8\pi}{N_{Re}[\ln(8/N_{Re}) - \gamma + \frac{1}{2}]}, \qquad (10\text{-}133)$$

where $\gamma = 0.57722\ldots$ is Euler's constant, and where d is the diameter of the cylinder. This expression progressively overestimates the drag for $N_{Re} \gtrsim 0.2$.

Numerical solutions to the problem of flow past an infinite cylinder can be constructed, for example, by following a vorticity-stream function approach similar to that we discussed earlier for the case of a sphere. Numerical solutions have been obtained by Thom (1933), Kawaguti (1953), Hamielec and Raal (1969), Takami and Keller (1969), Griffin (1972), and Schlamp *et al.* (1975, 1977), among others. Some representative results are displayed in Figure 10-29, which shows the streamline and vorticity fields for $N_{Re} = 1$ and 20. Note that, as was the case for disks and thin spheroids, the flow has fore-aft asymmetry even at $N_{Re} = 1$. At $N_{Re} = 20$ a standing eddy of considerable length has developed at the downstream end of the cylinder. In Figure 10-30 a comparison shows excellent agreement between the theoretical and experimental eddy lengths. Note that the eddy begins to develop at $N_{Re} \approx 6$, and subsequently grows linearly with increasing N_{Re}. Comparison with the case of flow past a sphere shows that at a given N_{Re} the standing eddy on a cylinder is considerably larger than that on a sphere. This result implies a longer range wake influence on cloud particles in the vicinity of a columnar ice crystal than would be the case for a spherical drop of comparable radius.

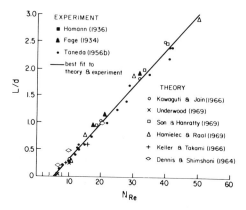

Fig. 10-30. Variation with Reynolds number of the length of the standing eddy at the downstream end of a circular cylinder. (From Pruppacher *et al.*, 1970; by permission of Cambridge University Press.)

The experiments of Homann (1936), Roshko (1954), and Taneda (1956b) have shown that the standing eddy at the downstream end of an infinite cylinder begins to 'shed' at $N_{Re} \approx 40$. Measurements by Relf and Simmons (1924) and Jayaweera and Mason (1965) have demonstrated that the shedding frequency for a given N_{Re} is somewhat larger than for the corresponding case of a sphere.

A comparison of drag coefficients as determined by theory and experiment is given in Figure 10-31. The agreement for the case of 'infinite' cylinders is seen to be excellent. The figure also shows that C_D for a finite cylinder depends relatively strongly on the shape factor d/L, where L is the length and d the diameter of the cylinder. This is in contrast to the dependence of C_D on the axis ratio b/a for thin oblate spheroids and disks. It is obvious from these results that the numerical determinations of C_D for an infinite cylinder cannot provide an accurate description of the behavior of most columnar ice crystals.

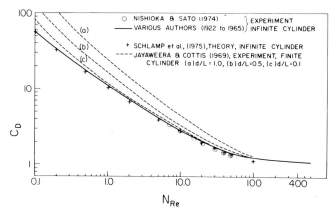

Fig. 10-31. Variation with Reynolds number of the drag force coefficient for circular cylinders of various diameter-to-length ratios. The experimental results for C_D on an infinite cylinder are those of Wieselberger (1922), Finn (1953), Tritton (1959), and Jayaweera and Mason (1965). (From Schlamp *et al.*, 1977; by courtesy of *Pure and Appl. Geophys.*, and the authors.)

The experimental values of Jayaweera and Cottis for C_D shown in Figure 10-31 may be represented adequately by the following empirical expressions:

$$\log N_{Re} = B_0 + B_1 X + B_2 X^2 + B_3 X^3, \qquad (10\text{-}134)$$

where $X = \log (C_D N_{Re}^2)$. For $(d/L) = 1.0$, $B_0 = -1.3100$, $B_1 = 0.98968$, $B_2 = -0.042379$, $B_3 = 0$. For $(d/L) = 0.5$, $B_0 = -1.11812$, $B_1 = 0.97084$, $B_2 = -0.058810$, $B_3 = 0.002159$. For $(d/L) = 0.1$, $B_0 = -0.90629$, $B_1 = 0.90412$, $B_2 = -0.059312$, $B_3 = 0.0029941$. For $(d/L) = 0$ (infinite cylinder), $B_0 = -0.79888$, $B_1 = 0.80817$, $B_2 = -0.030528$, $B_3 = 0$. As we have done previously for spheres, spheroids, and disks, we may use (10-134) as a basis for estimating the terminal velocity of columnar ice crystals. Thus, from an elementary and by now familiar calculation, the Best or Davies number for a circular ice cylinder of length L, radius a, and mass m_c is

$$C_D N_{Re}^2 = \frac{4 \pi a^3 (\rho_c - \rho_a)\rho_a g}{\eta_a^2} \approx \frac{4 \, m_c a \rho_a g}{L \eta_a^2}. \qquad (10\text{-}135)$$

Similarly, for a hexagonal cylinder of length L and radius a of the base circumscribed circle, we find

$$C_D N_{Re}^2 = \frac{6\sqrt{3} \, a^3 (\rho_c - \rho_a)\rho_a g}{\eta_a^2} \approx \frac{4 \, m_c a \rho_a g}{L \eta_a^2}. \qquad (10\text{-}136)$$

Therefore, for a given columnar crystal geometry and atmospheric conditions, N_{Re} ($= 2 U_\infty a \rho_a / \eta_a$) and hence U_∞ may be found by substituting (10-135) or (10-136) into (10-134).

There are stability limitations on the use of this method for determining U_∞, just as there were for oblate spheroidal configurations. Jayaweera and Mason (1965) observed that cylinders with $d/L = 0.4$ fall stably with their broadest extension perpendicular to the fall direction if $N_{Re} \lesssim 1000$; for $d/L = 0.5$ the fall was stable if $N_{Re} \lesssim 200$; for longer cylinders with $d/L = 0.1$ the fall was stable if $N_{Re} \lesssim 100$. At Reynolds numbers larger than those indicated, the cylinders were observed to oscillate during their fall. However, the horizontal excursions during oscillation have been found to be rather small, even at $N_{Re} > 8000$ (Stringham et al., 1969).

10.5   Motion of Ice Crystals, Snowflakes, Graupel, and Hailstones

The fall velocities of basically hexagonal, plate-like, or columnar ice crystals can be estimated from the drag data presented in the previous section. For example, from the thickness h, maximum radius a, and density ρ_c of a planar crystal a value for $C_D N_{Re}^2$ may be determined for given atmospheric conditions from (10-132). Then from (10-127) or (10-128) one may compute N_{Re} for the equivalent circular disk, which is the disk with the same mass and thickness as the crystal. Since the radius of this disk is $a = (m_c / \pi h \rho_c)^{1/2}$, the terminal velocity U_∞ may now be estimated from the relation $U_\infty = \eta_a N_{Re} / 2 \, a \rho_a$.

This approach has been followed by Jayaweera and Cottis (1969), Jayaweera (1972b), Jayaweera and Ryan (1972), and Kajikawa (1971, 1972, 1973); some selected results are summarized in Figures 10-32 to 10-34. Figure 10-32a shows,

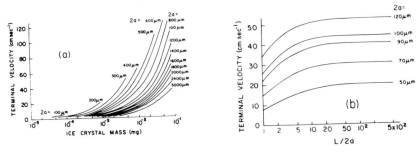

Fig. 10-32. Variation of the terminal fall velocity of ice crystals with crystal size, computed from drag data. (a) Hexagonal plates, −10 °C, 1000 mb, (b) hexagonal columns, −10 °C, 1000 mb. (From Jayaweera and Cottis, 1969; by courtesy of *Quart. J. Roy. Meteor. Soc.*)

as would be expected, that U_∞ of a simple hexagonal ice crystal of given mass decreases with increasing diameter; Figure 10-32b shows the corresponding behavior for a simple hexagonal columnar ice crystal. Note also that U_∞ of columnar ice crystals with $L/2\,a > 50$ remains essentially constant with increasing crystal length. Figure 10-33 indicates that the more pronounced are the dendritic shape features, the lower the fall velocity for a plate-type ice crystal of given maximum radius. This follows from the decrease in mass with an increase in dendritic features, for a given crystal size.

Figure 10-33 also shows very good agreement between the observed fall speeds and those computed on the basis of the equivalent circular disks. This is somewhat in conflict with the experimental drag coefficient determinations for planar ice crystal shapes by List and Schemenauer (1971) and Jayaweera (1972b). For example, Jayaweera's results displayed in Figure 10-34 indicate that a stellar crystal may fall as much as 25% slower than its equivalent ice disk, while no noticeable differences are found between the fall speeds only of simple

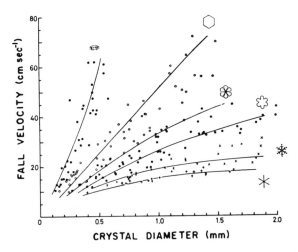

Fig. 10-33. Variation with size of the observed and computed terminal fall velocity of ice crystals of various shapes, −10 °C, 1000 mb. (From Kajikawa, 1972; by courtesy of *J. Meteor. Soc., Japan*.)

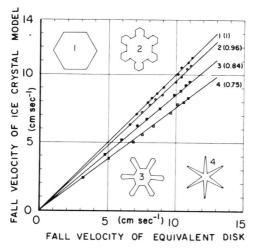

Fig. 10-34. Comparison between the terminal fall velocity of ice crystals models and their equivalent circular disks. (From Jayaweera, 1972b; by courtesy of Amer. Meteor. Soc., and the author.)

hexagonal plates and their equivalent disks. On the other hand, there is general agreement as to the close correspondence between the observed fall speeds of columnar ice crystals and those computed from their equivalent cylinders (Jayaweera and Ryan, 1972; Kajikawa, 1972). Similarly, studies of ice particles in cirrus clouds by Heymsfield (1972) have demonstrated that U_∞ for a bullet-shaped ice crystal may be determined from its equivalent cylinder if one sets $(C_D N_{Re}^2)_{bullet} = 1.67 \ (w/L)^{0.24} \ (C_D N_{Re}^2)_{equivalent \ cylinder}$, where w is the width of the bullet crystal.

Another approach for determining U_∞ of ice crystals is based on observations of terminal velocities for a given crystal type, which then must be corrected to apply to environmental conditions other than those at which the experiments were carried out. For example, Davis (1974) has fitted the terminal velocities observed by Bashkirova and Pershina (1964a, b), Fukuta (1969), Brown (1970), Jayaweera and Ryan (1972), Kajikawa (1972), and Locatelli and Hobbs (1974) to an expression of the form $U_\infty = A d^B$, where d is the ice crystal diameter and A, B are constants for a particular shape. The formulae arrived at by Davis are compiled in Table 10-2. They apply to ice crystals with $0.005 \leqslant N_{Re} \leqslant 300$, and to $-15\,°C$ and 1000 mb. In order to extend these results to other atmospheric conditions, note that if the crystal shape may be idealized by that of a sphere, circular disk, oblate spheroid, or circular cylinder, one can make use of the empirical finding that $N_{Re} \approx (C_D N_{Re}^2)^\alpha$. For all of these shapes we have, approximately, $\alpha = 1$ for $N_{Re} < 0.1$, $\alpha = 0.9$ for $0.1 \leqslant N_{Re} \leqslant 4$, $\alpha = 0.75$ for $4 \leqslant N_{Re} \leqslant 20$, $\alpha = 0.65$ for $20 \leqslant N_{Re} \leqslant 400$, and $\alpha = 0.57$ for $400 \leqslant N_{Re} \leqslant 1000$. Then, from the fact that $C_D N_{Re}^2 \sim \rho_a/\eta_a$ and $U_\infty \sim N_{Re}\eta_a/\rho_a$, we may estimate U_∞ for arbitrary conditions in terms of the known reference values $U_{\infty,0}$, $\rho_{a,0}$, and $\eta_{a,0}$ from the scaling relation.

TABLE 10-2

Empirical formulae for the terminal velocity of snow crystals of various shapes and sizes, L(cm), d(cm). (From C. E. Davis, 1974; by courtesy of the author.)

Symbolic ice crystal shape	Terminal velocities (cm sec^{-1})		Applicable to ice crystal diameter (μm)
Pla	$U_\infty = 2.96 \times 10^2 \, d^{0.824}$		10 to 300
Plb	$U_\infty = 2.96 \times 10^2 \, d^{0.824}$		10 to 40
Plb	$U_\infty = 2.96 \times 10^2 \, d^{0.824}$		41 to 2000
Plc-r	$U_\infty = 1.39 \times 10^2 \, d^{0.748}$		10 to 90
Plc-r	$U_\infty = 1.39 \times 10^2 \, d^{0.748}$		91 to 500
Plc-s	$U_\infty = 1.39 \times 10^2 \, d^{0.748}$		10 to 100
Plc-s	$U_\infty = 1.39 \times 10^2 \, d^{0.748}$		101 to 1000
Pld	$U_\infty = 4.22 \times 10^1 \, d^{0.442}$		500 to 3000
Clg	$U_\infty = 3.48 \times 10^3 \, d^{1.375}$		10 to 1000
Cle	$U_\infty = 7.31 \times 10^3 \, L^{1.415}$	$(L/d) \leqslant 2,$	10 to 1000
Cle	$U_\infty = 2.43 \times 10^3 \, L^{1.309}$	$(L/d) > 2,$	10 to 1000
Clf	$U_\infty = 7.31 \times 10^3 \, L^{1.415}$	$(L/d) \leqslant 2,$	10 to 1000
Clf	$U_\infty = 2.43 \times 10^3 \, L^{1.309}$	$(L/d) > 2,$	10 to 1000
Sl	$U_\infty = 7.92 \times 10^2 \, d^{0.99}$		400 to 1200

$$\frac{U_\infty}{U_{\infty,0}} = \left(\frac{\rho_a}{\rho_{a,0}}\right)^{\alpha-1}\left(\frac{\eta_a}{\eta_{a,0}}\right)^{1-2\alpha}. \qquad (10\text{-}137)$$

The drag force coefficient for smooth conical graupel models was determined by List and Schemenauer (1971), who released the models both in an apex-up position (Figure 10-35) and in an apex-down position. It is seen from Figure 10-35 that for $N_{Re} > 100$, both C_D and $C_D N_{Re}^2$ are larger for the cone models than for a sphere, while for $N_{Re} \lesssim 100$, C_D and $C_D N_{Re}^2$ are approximately the same as those for a sphere. Zikmunda and Vali (1972) determined C_D and $C_D N_{Re}^2$ for

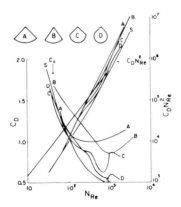

Fig. 10-35. Variation with Reynolds number of drag force coefficient for graupel models released apex-up; S = sphere; A: 90° cone-spherical sector, B: 70° cone-spherical sector, C: 90° cone hemisphere, D: 90° teardrop. (From List and Schemenauer, 1971; by courtesy of Amer. Meteor. Soc. and the authors.)

natural graupel particles, and found that C_D was larger than for the smooth conical models, in particular at $N_{Re} < 500$. This result was attributed to surface roughness of the natural graupel.

List *et al.* (1973) measured C_D for smooth oblate spheroidal hailstone models of various axis ratios and for $4 \times 10^4 \leqslant N_{Re} \leqslant 4 \times 10^5$. Depending on the angle of attack ($\theta = 0°$ to $90°$) of the flow past the spheroid, C_D was found to vary between 0.2 and 0.85. The variation of C_D with N_{Re} for $\theta = 45°$ is illustrated in Figure 10-36, where comparison is made with C_D for a sphere. Note the sharp drop in C_D which occurs for large N_{Re}. Qualitatively, this 'drag crisis' corresponds to a transition from a laminar to a turbulent boundary layer. Turbulence in the boundary layer brings more fluid momentum closer to the boundary surface. This counteracts the effect of the adverse pressure gradient acting on the downstream side of the surface, so that the flow separation phenomenon is suppressed, and hence the pressure drag is reduced (recall the discussion of separation in Section 10.2.4). Figure 10-36 shows that the drag crisis transition point is affected by the axis ratio of the oblate spheroid.

Young and Browning (1967) and List *et al.* (1969) used spherical hailstone models with close packed hemispherical roughness elements on their surfaces to study the effect of surface roughness on C_D of hailstones. Their observations showed that, except for uncommonly high surface roughness, the drag force coefficient varied unsystematically in value and sign, and differed by less than $\pm 10\%$ from that of a smooth sphere at the same Reynolds number. Both studies agreed, however, that with increasing surface roughness the transition to a turbulent boundary layer occurred at progressively lower Reynolds numbers and became less pronounced.

List (1959), Macklin and Ludlam (1961), Bailey and Macklin (1968a), and Landry and Hardy (1970) measured C_D of natural and artificially grown hailstones. Depending on the angle of attack of the air flow, C_D was found to vary between about 0.3 and 0.8 for N_{Re} between 10^3 and 10^5.

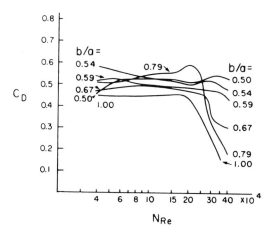

Fig. 10-36.   Variation with Reynolds number of the drag force coefficient of oblate spheroids, for an angle of attack of 45°. (From List *et al.*, 1973; by courtesy of Amer. Meteor. Soc., and the authors.)

Despite the availability of values for C_D (and thus $C_D N_{Re}^2$) vs. N_{Re}, there is little justification for computing the terminal velocity of graupel particles, ice crystal aggregates, or hailstones from their $C_D N_{Re}^2$ values, since neither the volume of these particles nor their cross-sectional area perpendicular to the flow are well-defined. Additional complications arise from their complicated oscillatory-, spinning-, and tumbling motions while falling. Thus, it is necessary to resort to experimental observations to find U_∞ for these ice particles, and to estimate U_∞ at other pressure and temperature levels by the use of (10-137).

Early studies of U_∞ for spatial ice crystals, ice crystal aggregates, and graupel by Nakaya and Tereda (1935), Magono (1951, 1954b), Langleben (1954), Litvinov (1956), Bashkirova and Pershina (1964), and Magono and Nakamura (1965) show inconsistencies and cannot be considered reliable. In addition, some of the earlier data were given in terms of the drop size into which the ice crystal aggregate would melt, while others were given in terms of the size of the aggregate itself. Comparison of the data is therefore difficult, making the results hard to generalize. However, these early data did show that most crystal aggregates fall with speeds of 1.0 to 1.5 cm sec^{-1}, which increase very little with increasing size. Somewhat different behavior is exhibited by the terminal velocities of rimed ice crystals and graupel. A study by Pflaum and Pruppacher (1976) and Pflaum et al. (1978) of such particles, freely suspended in the vertical air stream of a wind tunnel, showed that during the initial stages of 'dry-growth' riming (see Section 16.1), their terminal velocities stayed constant or even decreased with time. The result was considered to imply that during the deposition of rime of sufficiently low density the mass of a growing ice particle may increase less rapidly than its drag-determining cross-sectional area presented to the flow.

Detailed quantitative studies of U_∞ for natural, rimed ice crystals, graupel particles, and ice crystal aggregates have recently been carried out as a function of ice particle type and their maximum dimension by Jiusto and Bosworth (1971), Zikmunda (1972), Zikmunda and Vali (1972), Locatelli and Hobbs (1974), and Kajikawa (1975a, b). Although these authors have provided best fit relations between U_∞ and the maximum dimension of each ice particle type studied, we shall not reproduce them here since the U_∞ of such particles varies greatly from cloud to cloud, and even with the life cycle of any particular cloud, due to differences in the degree of riming and crystal aggregation. Such behavior is excluded from the empirical relations. Nevertheless, these recent observations, exemplified in Figures 10-37 and 10-38 by the data of Locatelli and Hobbs (1974), allow one to conclude that: (1) U_∞ increases as the maximum dimension of the ice particle increases; (2) U_∞ increases as the mass of the ice particle increases; (3) U_∞ of a densely rimed ice particle may be up to twice as large as that of a similar unrimed ice particle of the same maximum dimension; (4) U_∞ of ice crystal aggregates is generally larger than that of their component crystals.

The fall behavior of smooth conical graupel models was studied by List and Schemenauer (1971). Their study showed that all models described in Figure 10-35 began to exhibit sideways oscillatory motions if the Reynolds number exceeded 200 to 800, depending on the model and its release mode. Some models tumbled while

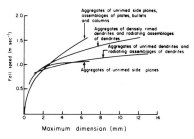

Fig. 10-37. Best fit curves for fall speed versus maximum dimension of various ice crystal aggregates; 750 to 1500 m above sea level, Cascade Mts., State of Washington. (From Locatelli and Hobbs, 1974; by courtesy of the authors; copyrighted by American Geophysical Union.)

falling. Thus, model B released in apex-up position, model C released in apex-up or apex-down position, and model D released in apex-down position, each began to tumble at a Reynolds number larger than 500 to 800.

   Zikmunda and Vali (1972) studied the fall behavior of natural graupel particles. Their observations showed that lump graupel particles generally fall vertically with only small horizontal movements, although some were observed to tumble and spin. Conical graupel were observed to oscillate over an angular range of 0 to 20 °C, and with a frequency of as much as 50 sec^{-1}. In the mean, conical graupel fell with their apex oriented upward. Rimed columnar ice crystals with $L/d < 3$ fell steadily and with their major axis oriented perpendicular to the fall direction. Rimed columnar ice crystals with $L/d > 3$ exhibited rotation in both

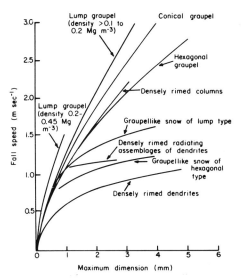

Fig. 10-38. Best fit curves for fall velocity versus maximum dimension of single solid precipitation particles of various types. (From Locatelli and Hobbs, 1974; by courtesy of the authors; copyrighted by American Geophysical Union.)

the vertical and horizontal planes about the minor ice crystal axes. Although most of these ice particles fell with their major axis inclined by less than 15° from an orientation perpendicular to the direction of fall, some exhibited angular deviations of as much as 75°. This behavior is consistent with the experiments of Jayaweera of Mason (1966) and Podzimek (1968, 1969) with loaded cylinders. These experiments showed that an uneven distribution of mass results in a change of orientation of the major axis of a cylinder during its fall. Zikmunda and Vali (1972) observed further that rimed capped columns fell with the columnar axis oriented vertically if $L/d < 1.0$, and that ice particles with $L/d > 1.0$ fell with the columnar axis generally horizontal, but with frequent changes in orientation. Rimed plates exhibited no oscillations or side motions while falling if $N_{Re} < 100$, in agreement with the laboratory studies of Podzimek (1965, 1968, 1969).

Recently, Pflaum and Pruppacher (1976) and Pflaum *et al.* (1978) studied the hydrodynamic behavior of growing graupel particles freely suspended in the vertical air stream of a wind tunnel. Spinning-, helical-, and oscillatory motions, or simple straight fall with a fixed orientation of the ice particle, were observed. Although these fall modes appeared to respond to the instantaneous shape, surface roughness, and mass loading of the ice particle, they appeared to vary during growth of the ice particle in quite an unpredictable manner, so that no quantitative correlations could be drawn except for the fact that rear capture of cloud drops frequently initiated a tight pendulum motion which in turn caused the ice particle to assume conical shape.

The motion of frozen drops was studied by Blanchard (1957), Spengler and Gokhale (1972), and Pitter and Pruppacher (1973) by seeding supercooled drops suspended freely in a vertical wind tunnel. Supercooled drops with radii less than $500\,\mu m$ fell at a steady terminal velocity with no horizontal drift and without rotation. At the onset of ice nucleation, freezing supercooled drops underwent an abrupt decrease of 6 to 7% in terminal velocity, and drifted erratically in circulating paths. After the freezing process was completed, the trajectories of the rigid particles became more regular, as they spun and tumbled while following more or less helical paths. The presence of pronounced surface features such as knobs and spikes reduced the amount of tumbling and stabilized the ice particles in gyro-like spinning motion and helical fall paths of small radii.

We know from Section 10.3.2 that large deformed drops fall with a relatively flat bottom surface. Although the center of gravity of these drops is above the point of action of the resultant pressure force – which seems to constitute an unstable equilibrium – the drop remains in this orientation since in its liquid form it can always adjust its shape to changing conditions. However, once the drop is frozen or has an ice shell on its surface, its shape is fixed and the pressure force 'tips' the rigid drop such that its flat portion turns upwards (Spengler and Gokhale, 1972).

Terminal fall velocities of large graupel and hailstones have been determined by Bilham and Relf (1937), List (1959), Macklin and Ludlam (1961), Williamson and McCready (1968), Auer (1972b), and Roos (1972) from direct observation as well as from observed drag force coefficients. From (10-27) we find for a roughly spherical hailstone falling at terminal velocity,

$$U_\infty(\text{m sec}^{-1}) \approx 0.36 \, (\rho_H d / C_D \rho_a)^{1/2}, \tag{10-138}$$

where ρ_H and ρ_a are the densities of the hailstone and the air, respectively, $d(\text{cm})$ is the average diameter of the hailstone, and we have assumed $g = 981$ cm sec^{-2}. Using appropriate values for ρ_H, C_D, and ρ_a, terminal velocities for hailstones can be computed from (10-138) which are in fair agreement with those directly observed. A summary of observed terminal velocities of hailstones and large graupel has been given by Auer (1972b). At the 800 mb level and 0 °C these values can be fitted to the relation

$$U_\infty(\text{m sec}^{-1}) \approx 9 \, d^{0.8}, \tag{10-139}$$

for the range $0.1 \leqslant d \leqslant 8$ cm with $d(\text{cm})$. Note that giant hailstones may have terminal fall velocities of up to 45 m sec^{-1}. These large terminal velocities imply that comparable updraft velocities must exist inside clouds to permit the growth of such particles. Considering their size, hailstones have Reynolds numbers which range typically between 10^3 and 5×10^5.

Many observational difficulties have so far prevented direct quantitative studies of the fall mode of hailstones falling from hail-bearing clouds, although a number of field experiments have been carried out to simulate such fall. For example, Macklin and Ludlam (1961) and Landry and Hardy (1970) dropped spheres and model hailstone in the free atmosphere, and tracked them with radar. Although these experiments yielded drag force coefficients, they could not resolve the detailed motion of the falling models. Knight and Knight (1970c) employed a skydiver to photograph freely falling oblate spheroids. It was found that these objects rotate preferentially around the major axis, which generally remains oriented horizontally. Unfortunately, little additional quantitative knowledge was gained.

Since large wind tunnels which are capable of freely supporting hailstones in an air stream are not yet available, laboratory studies have been confined to indirect determinations of the fall mode of hailstones, or to model experiments. For example, from water tank experiments List (1959) determined that large, smooth oblate spheroids prefer to fall with the minor axis vertical. Browning (1966), Browning and Beimers (1967), and Knight and Knight (1970c) interpreted the observed structure of hailstones in terms of a constant fall attitude at some times and a random tumbling at others. Bailey and Macklin (1967) deduced from wind tunnel studies on artificially grown hailstones that hailstones indeed tumble while falling. List et al. (1973) used experimentally determined values for the drag force, lift force, and torque acting on a smooth oblate spheroid of $4 \times 10^4 \leqslant N_{Re} \leqslant 4 \times 10^5$ to solve numerically for its motion. From their computations they predicted that subsequent to a small perturbation on a spheroid's steady state fall attitude (major axis horizontal), damping may either restore the initial state, or coupling between rotational and horizontal translational components of the spheroid's motion may give rise to amplification of the initial perturbation. Depending on the axis ratio and size of the spheroid, and on the magnitude of damping, such amplification was found to lead to either a constant amplitude oscillation or continuous tumbling around the horizontal major axis.

Although all these studies point to the fact that hailstones, like graupel particles, oscillate and tumble while falling, there is not yet a clear quantitative picture of the conditions under which one or the other fall mode prevails. In particular, more observations on falling natural hailstones are needed in order to understand their growth rate, structure, and morphological features – all of which depend importantly on the hailstones' fall mode.

CHAPTER 11

COOLING OF MOIST AIR

As mentioned in the historical review of Chapter 1, it has been known for over 150 years that only cooling by expansion of humid air during its ascent can provide clouds with sufficient condensed water to account for the observed amounts of precipitation. It is our main purpose in this chapter to describe the essentials of this thermodynamic cooling process. This will provide us with the minimum necessary mathematical framework for describing the background environmental (cloud) conditions of temperature, pressure, and humidity which control the rate of activation and subsequent diffusional growth of a representative population of cloud particles.

Some useful references for this chapter include the texts of Iribarne and Godson (1973), Turner (1973), Hess (1959), van Mieghem and Dufour (1948), the survey articles of Simpson (1976), Cotton (1975), Turner (1969), and Simpson *et al.* (1965), and the reports of Dufour (1965a, b).

11.1 Water in the Atmosphere

As a prelude to our discussion of the basic thermodynamics of moist air, we shall touch here briefly on some of the more relevant characteristics of the atmospheric water cycle. Hopefully, this will set some of the later material in better perspective.

The amount of water vapor present in the atmosphere is a complicated function of (1) the amount which enters the atmosphere through evaporation and sublimation, (2) its transport by motions of various scales throughout the troposphere and lower stratosphere, and (3) the amount which leaves the atmosphere intermittently and almost exclusively as a flux of rain, hail, and snow. The fact that all three phases of water contribute to this cycle at the prevailing terrestrial temperatures and pressures is most fundamentally a consequence of the molecular structure of water, which permits the strong association of water molecules through hydrogen bonding (see Chapter 3). This is also the principal determinant for the relatively small amount of atmospheric water – only about 1.3×10^{16} kg or about 10^{-5} of the total surface store – in spite of the presence of extensive water surfaces for evaporation and sublimation.

Since the Earth's surface is the primary source of water vapor, we expect a decrease in the water vapor mixing ratio w_v with height. From a slightly different point of view, one may also attribute this expectation to the observed decrease in temperature with height in the troposphere: Since the maximum possible

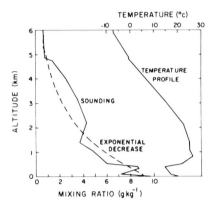

Fig. 11-1. Variation with height of the mixing ratio of water vapor in the lower troposphere. Radiosonde data taken over San Diego, California. (From Wang, 1974; by courtesy of Amer. Meteor. Soc., and the author.)

(saturation) mixing ratio $w_{v,sat}$ decreases with decreasing temperature (recall Section 4.9), on the average water is squeezed out of air parcels during their ascent. Thus w_v is not a conservative quantity in the troposphere, but in fact decreases with height. An example of the observed vertical distribution of mixing ratio of water vapor in the lower troposphere is shown in Figure 11-1, and can be seen to be approximately exponential.

Many measurements have been made of the distribution of water vapor above the tropopause (i.e., above about 10 km) (see Harries, 1976). Some typical results are shown in Figure 11-2, from which we can see that the stratosphere is generally very dry (e.g., $w_{v,sat} = 1.7 \times 10^{-4}$ g g^{-1} at 40 mb and -60 °C, compared to the observed value of $w_v \approx 2 \times 10^{-6}$ g g^{-1}), with occasional relatively moist layers. The distributions may very likely be extrapolated to 70 km, above which level photodissociation of water by ultraviolet radiation becomes important.

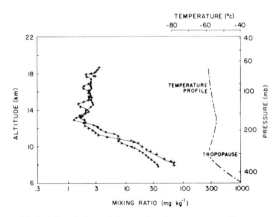

Fig. 11-2. Variation with height of the mixing ratio of water vapor in upper troposphere and lower stratosphere over Alaska on March 25, 1968. (From McKinnon and Morewood, 1970; by courtesy of Amer. Meteor. Soc., and the authors.)

In accordance with its arid state, the stratosphere is generally cloudless. Occasionally, however, 'mother-of-pearl' or 'nacreous' clouds form at altitudes between about 17 and 35 km, as a result of air flows over mountain ranges which induce lifting and subsequent cooling to saturation of substantial amounts of upper tropospheric air.

The mean global annual rate of precipitation and evaporation both correspond to nearly 1000 mm, or 5×10^{17} kg of water. (It is perhaps a little disconcerting to realize that this is only about 100 times larger than the projected world-wide annual demand for water by the year 2000, which will have increased by a factor of 3 from 1965 (Lvovich, 1969, in SMIC Report, 1971, p. 178).) Since the atmospheric store of water vapor is equivalent to a layer of precipitated water approximately 25 mm thick, we see that it is replaced about $1000/25 = 40$ times a year, which is equivalent to a global residence or turn-over time of about 9 days.

We may also obtain a simple estimate of the residence time of liquid water in the atmosphere. If we make the reasonable assumptions of an average cloud depth of 3 km, a cloud water content of $0.5 \, \mathrm{g \, m^{-3}}$, and a fractional global coverage of 0.5, we find the total cloud water corresponds to a world-wide layer 0.8 mm thick. Thus only about 1/30 of the total atmospheric moisture is present as water in clouds. On comparison with the global mean precipitation rate, we find cloud water is replaced about 1200 times a year, yielding a residence time of about 7 hr.

Only a small fraction of this average instantaneous cloud water is capable of reaching the ground. In fact, McDonald (1958) has estimated that only about 1/10 of the cloud cover is in a precipitating state, and evaporation losses lead to a net precipitation efficiency of about 0.5 (Foote and Fankhauser, 1973; Marwitz, 1972, 1974). Therefore, on the average the global cloud cover has to go through a precipitating stage about $20/7 \approx 3$ times per hour. Thus the average duration of precipitation events is estimated to be about 20 min, and the average rain intensity at ground level is estimated at roughly 2 mm/hr. These numbers appear to be reasonably representative of the observed durations and intensities of isolated precipitating cloud 'cells.' Finally, if we assume the average cloud lifetime ranges typically between 30 min and 1 hr, then on comparison with the liquid water residence time of about 7 hr we estimate the number of cloud evaporation-condensation cycles required before a precipitating system is formed to be O(10). Similar estimates have been made by Junge (1963a) and Junge and Abel (1965).

11.2 Isobaric Cooling

Although most cooling of moist air to saturation occurs via the expansion which accompanies lifting, there are also a variety of cooling mechanisms which can occur isobarically (at constant pressure). As these often lead to the formation of ground fogs and stratus clouds, isobaric cooling processes are of sufficient importance to be mentioned here in passing.

Consider a volume of moist air which is cooled isobarically. Assuming we may neglect any exchange of air between the parcel and its environment, its mixing ratio w_v must remain constant during cooling. Therefore, after sufficient cooling $w_v = w_{v,sat}$, and the parcel will be vapor-saturated. Assuming the proper condensation or ice forming nuclei are present, a phase change will thus commence. This occurs at the *dew point* temperature, T_d, if the condensate is in the form of water drops; if ice crystals appear, it is called the *frost point* temperature, T_f.

Dew or frost are often formed by the night-time radiational cooling of calm moist air in ground contact. The same cooling process may also produce ground fogs. Isobaric cooling leading to fog or stratus cloud formation may also occur when a mass of moist air moves horizontally over a colder land or water surface, or over a colder air mass.

An expression giving T_d as a function of the prevailing T and w_v, or relative humidity $\phi_v \equiv w_v / w_{v,sat}(T)$, may be obtained by recognizing that $e = e_{sat,w}(T_d)$ when $T = T_d$. Then from the Clausius-Clapeyron equation, (4-89), and noting that $w_v \approx \varepsilon e/p$ (Equation (4-40)), we find, for $p = $ constant,

$$\frac{dT_d}{T_d^2} = \frac{R_v}{L_e} \frac{de}{e} \approx \frac{R_v}{L_e} \frac{dw_v}{w_v}. \tag{11-1}$$

Integration between T_d, w_v, and T, $w_{v,sat}(T)$ yields

$$T_d \approx T + R_v \frac{TT_d}{L_e} \ln \phi_v \approx T + \frac{R_v T^2}{L_e} \ln \phi_v. \tag{11-2}$$

11.3   Adiabatic Cooling of Unsaturated Air

We shall now consider the expansion cooling of a rising parcel of dry air. Although heat $đQ$ may be added to the parcel through the effects of radiation, frictional dissipation, and mixing with the environment, in many situations the resulting temperature changes are of secondary importance to that arising from the expansion process. Hence it is a reasonable and useful idealization to assume the expansion is strictly adiabatic, with no heat exchanges. We shall further assume it is reversible, so that $đQ = T \, dS = 0$, i.e., the process will be assumed isentropic ($S = $ constant).

We shall also assume the dry air is an ideal gas. Now a property of an ideal gas, which we have not made use of previously, is that its internal energy U is a function of temperature only. (This is true because ideal gases have no intermolecular potential, so that U is just the total molecular kinetic energy, which is a function only of T.) Therefore, its enthalpy $H \equiv U + pV$ is also just a function of T, owing to the form of the ideal gas law. Consequently, from (4-12) the specific enthalpy h_a of a unit mass of dry air satisfies the relation

$$dh_a = c_{pa} \, dT, \tag{11-3}$$

where c_{pa} is the specific heat of dry air at constant pressure ($c_{pa} \approx 0.240 \text{ cal g}^{-1} \text{°C}^{-1}$). Therefore, (4-6) applied in the present context of a reversible adiabatic expansion of a closed parcel becomes

$$c_{pa} \, dT = \alpha_a \, dp, \tag{11-4}$$

where $\alpha_a \equiv 1/\rho_a$ is the specific volume of the dry air. Introducing the gas law (4-25) into (11-4) and integrating, we obtain

$$T = A p_a^{\kappa_a}, \tag{11-5}$$

where A is a constant and $\kappa_a \equiv R_a/c_{pa} \approx 0.286$. Logarithmic differentiation of this equation with respect to height z yields the temperature lapse rate of the parcel:

$$\frac{dT}{dz} = \frac{R_a T}{c_{pa} p_a} \frac{dp_a}{dz}. \tag{11-6}$$

Assuming $p_a = p_a'(z)$, the environmental pressure at level z, and that the latter is in hydrostatic equilibrium, so that $dp_a'/dz = -g\rho_a' = -gp_a'/R_a T'$, (11-6) can be written in the form

$$\Gamma_a \equiv -\left(\frac{dT}{dz}\right)_a = \frac{g}{c_{pa}}\left(\frac{T}{T'}\right) \approx \frac{g}{c_{pa}} \approx 9.76\,°C\ km^{-1}, \tag{11-7}$$

where Γ_a is the *dry adiabatic lapse rate*, and where g is the magnitude of the gravitational acceleration. Thus a parcel of dry air cools by about 1 °C for every 100 m of lift.

For moist unsaturated air the only modification required in the above development is a recognition that the composition of the ideal gas under consideration is now slightly different, which must be reflected in the parameters κ and c_p. For example, the lapse rate of the parcel becomes $\Gamma_m = g/c_{pm}$, where c_{pm} is the specific heat of the moist air. Since air of mixing ratio w_v contains w_v grams of vapor for every gram of dry air, c_{pm} is determined from the balance condition $c_{pm}(1 + w_v)\,dT = (c_{pa} + w_v c_{pv})\,dT$ for arbitrary dT and where c_{pv} is the specific heat of water vapor; i.e., c_{pm} is a mass weighted average of the component specific heats. Since generally $w_v \ll 1$, we find $c_{pm} \approx c_{pa}$ and $\Gamma_m \approx \Gamma_a$, the small differences being negligible for all practical purposes. In a similar fashion we can show that $\kappa_m \approx \kappa_a(1 - 0.23\,w_v) \approx \kappa_a$.

Let us now consider the height to which a parcel of unsaturated air would have to be lifted adiabatically to become saturated. This height is called the *lifting condensation level* (LCL). During the ascent $w_v \approx \varepsilon e/p$ remains constant, so that from (11-5) we may write $T \sim e^{\kappa_a}$. Hence the temperature at the LCL is given in terms of w_v and the initial T and p by

$$T_{LCL} \approx T\left[\frac{e_{sat,w}(T_{LCL})}{e}\right]^{\kappa_a} \approx T\left[\frac{\varepsilon e_{sat,w}(T_{LCL})}{w_v p}\right]^{\kappa_a}. \tag{11-8}$$

This implicit relation for T_{LCL} may be solved iteratively, since $e_{sat,w}(T_{LCL})$ is known as a function of T_{LCL} and T from the integral of (4-92). Once T_{LCL} is determined, the LCL itself can be found from the lift distance, which is $(T - T_{LCL})/\Gamma_a$.

In order to avoid tedious computations, one may use the following approximate, explicit, semi-empirical relations for the LCL (Inman, 1969), and for the *lifting sublimation level* (LSL) (Chappel *et al.*, 1974) at which $e = e_{sat,i}(T_{LSL})$:

$$T_{LCL} = T_d - (0.212 + 0.001571\,T_d - 0.000436\,T)(T - T_d), \tag{11-9}$$

$$T_{LSL} = T_d - (0.182 + 0.00113\,T_d - 0.000358\,T)(T - T_d), \tag{11-10}$$

with T and T_d in °C. The use of T_d as a moisture variable follows standard practice. Finally, graphical solutions for the LCL are readily obtained by use of various thermodynamic diagrams (e.g., see Chapter 5 of Hess, 1959).

11.4 Adiabatic Cooling of Saturated Air

If we lift a parcel of moist air beyond the LCL, latent heat of condensation will be released. (For simplicity we shall consider here only the onset of the liquid phase.) Thus the parcel will cool at less than the dry adiabatic rate, the exact rate depending on whether part or all of the water remains inside the parcel. If it all remains, we continue to have a closed system, and the process can be carried out reversibly. This is called a 'reversible saturated adiabatic process'. If all the water is assumed to fall out immediately upon formation, the process can be neither strictly adiabatic nor reversible; this extreme is called a 'saturated pseudo-adiabatic process.' Fortunately, the heat capacity of the water is negligible relative to that of the parcel air, so that there is no significant difference between the two resulting cooling rates. Since the pseudo-adiabatic process requires less bookkeeping, we shall use it exclusively in the sequel.

Let us calculate the cooling rate of a saturated parcel containing 1 gram of dry air and $w_{v,sat}$ grams of vapor. For a temperature change dT the enthalpy change is $dh = (c_{pa} + w_{v,sat}c_{pv})\,dT = dQ + V\,dp$, while the incremental heat of condensation dQ for the corresponding change $dw_{v,sat}$ is just $-L_e\,dw_{v,sat}$. Substituting the gas law for V, we thus obtain

$$(c_{pa} + w_{v,sat}c_{pv})\,dT + L_e\,dw_{v,sat} - (R_a + w_{v,sat}R_v)T\frac{dp}{p} = 0. \tag{11-11}$$

(Obviously, had we considered the reversible adiabatic process instead, the additional heat change $w_L c_w\,dT$ of the condensate water would have been added to (11-11), where w_L and c_w are the mixing ratio and specific heat capacity of the water, respectively. Since typically $w_L c_w / c_{pa} = O(10^{-2})$, the difference between the approaches is of no practical significance.) Usually, $w_{v,sat} < 2 \times 10^{-2}\,g\,g^{-1}$, so that (11-11) may be simplified further to

$$c_{pa}\frac{dT}{T} + L_e\frac{dw_{v,sat}}{T} - R_a\frac{dp}{p} = 0. \tag{11-12}$$

Division by dz results in the lapse rate expression

$$\Gamma_s \equiv -\left(\frac{dT}{dz}\right)_s = \frac{g}{c_{pa} + L_e(dw_{v,sat}/dT)}, \tag{11-13}$$

where we have assumed the parcel is in hydrostatic equilibrium with the environment. This may be reduced further to an explicit function of T and $w_{v,sat}$

by substitution of (4-40) and (4-92) into the denominator. For 1000 mb and 0 °C, (11-13) gives $\Gamma_s = 5.8$ °C km^{-1}, or about 60% of the dry adiabatic lapse rate.

11.5 Cooling with Entrainment

Let us now abandon the fiction that a rising air parcel is a closed system. We have already discussed in Section 2.1.1 how the entrainment of relatively dry environmental air into convective clouds lowers their liquid water contents substantially below that expected on the basis of a closed saturated ascent (recall Figure 2-15). Observations also show that updrafts in most cumulus clouds are warmer than the environment by only 1 °C or less, rather than the 2 to 3 °C temperature excess which is typical for computations of rising closed air parcels.

The description of the previous section may be easily extended to the case of an entraining parcel by following a procedure originated by Austin and Fleisher (1948). According to their scheme we must add the following additional heat terms to (11-11) multiplied by the mass m of the saturated parcel under consideration: (1) a term $L_e(w_{v,sat} - w'_v)\,dm$ expressing the heat needed to evaporate sufficient water to increase the mixing ratio w'_v of mass dm, entrained during the parcel displacement dz, to the saturation value in the parcel, and (2) a term $c_{pa}(T - T')\,dm$, expressing the heat required to warm the entrained air from its original temperature T' to the temperature T of the parcel. With these additions the generalization of (11-12) to include the effects of entrainment is

$$c_{pa}\frac{dT}{T} + \frac{L_e\,dw_{v,sat}}{T} - R_a\frac{dp}{p} + \left[L_e(w_{v,sat} - w'_v) + c_{pa}(T - T')\right]\frac{dm}{m} = 0.$$

(11-14)

Similarly, in place of (11-13) we obtain the 'cloud lapse rate' Γ_c:

$$\Gamma_c \equiv -\left(\frac{dT}{dz}\right)_c = \frac{g + \dfrac{1}{m}\dfrac{dm}{dz}[L_e(w_{v,sat} - w'_v) + c_{pa}(T - T')]}{c_{pa} + L_e(dw_{v,sat}/dT)}.$$

(11-15)

Since dm/dz > 0, we see that $\Gamma_c > \Gamma_s$. As an example (from Hess, 1959, p. 108), if we suppose that p = 700 mb, T = 0 °C, T' = −1 °C, $w'_v/w_{v,sat} = 0.67$, and d ln m/dz = 0.25 km^{-1}, we find $\Gamma_c = 6.6$ °C km^{-1}, compared to $\Gamma_s = 5.8$ °C km^{-1}.

Note that in (11-15) the *entrainment rate*, μ, has the simple dimensional form

$$\mu \equiv (1/m)\frac{dm}{dz} = 1/l,$$

(11-16)

where l is a length scale characterizing the mixing process. There is an extensive literature on buoyant convective processes, involving a variety of physical models of varying complexity, which yield estimates for l. We shall not discuss these here in any detail, not only because it would lead us too far afield, but also because all of these models somewhat oversimplify the complex growth processes of real convective clouds.

There are three especially prominent models for the entraining convective

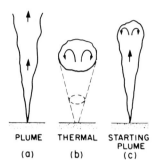

elements, namely, (1) that of a 'steady state plume' or jet (Figure 11-3a), (2) that of a 'thermal' or warm bubble (Figure 11-3b), and (3) that of a 'starting plume' (Figure 11-3c), which is like a jet capped by a thermal. Laboratory studies of dry jets have provided the estimate $l = 5 R_J$ ($\mu_J = 0.2/R_J$), where R_J is the jet radius (e.g., Squires and Turner, 1962). Similar studies of thermals have led to $l \approx 5 R_T/3$ ($\mu_T = 0.6/R_T$), where R_T is the thermal radius (e.g., Turner, 1962, 1963). Generally the thermal cap of a starting plume is regarded as only a small fraction of the 'cloud,' so that the entrainment rate is taken to be the same as that for a steady state jet. One-dimensional numerical cloud models based on the entrainment hypothesis generally work best in predicting cloud top height when the choice $\mu \approx 0.2/R$ is made (Cotton, 1975a).

An unrealistic feature of all cloud models of this type has been pointed out by Warner (1970). By comparing observational data with numerical computations of a representative, entraining, one-dimensional cloud model, he showed that such models cannot simultaneously predict cloud top height and liquid water content. (Further discussion of this point may be found in Warner (1975a), McCarthy (1975), Simpson (1971, 1972), Warner (1972), Cotton (1971), and Weinstein (1971).) In spite of this shortcoming (the predicted liquid water content may be twice as large as in the corresponding real cloud of the same height), we shall use the entraining parcel model in the sequel. It is adequate for our purpose of demonstrating microphysical behavior, and has the virtue of being relatively simple.

11.6   Governing Equations for a One-Dimensional Cloud

In this section we shall formulate the differential equations describing the properties of a vertically rising, entraining parcel of cloudy air. In doing this we must account for the presence of condensed water, even though we have shown it has a negligible effect on the parcel cooling rate.

Suppose the parcel has mass m and density ρ (exclusive of liquid water), vertical velocity W, and liquid water mixing ratio w_L*. Then, if we include body forces arising from buoyancy and the weight of the liquid water, the momentum equation becomes

$$\frac{d(mW)}{dt} = g(m' - m) - mgw_L = gm\left(\frac{\rho' - \rho}{\rho} - w_L\right),$$ (11-17)

where m' is the environmental mass of air displaced by the parcel. We have not included a drag force due to turbulent entrainment of momentum, on the assumption that the environmental momentum is small. However, the air outside the parcel must be accelerated to some extent by the parcel's motion. In order to account for the resulting reaction force on the parcel, we shall follow a suggestion of Turner (1963) and include an 'induced mass' acceleration term, $(m'/2)\, dW/dt$, on the left side of (11-17) (see also A-14.5). With this addition the parcel acceleration may be expressed as

$$\frac{dW}{dt} = \frac{g}{1 + \gamma}\left(\frac{T - T'}{T'} - w_L\right) - \frac{\mu}{1 + \gamma}W^2,$$ (11-18)

where $\gamma \equiv m'/2m \approx 0.5$.

The time rate of change of temperature in the parcel may be written down immediately from (11-15), noting that $d/dt = W\, d/dz$. For our applications in the next chapter, we must take into account the slight supersaturations which develop from the lifting of air. Therefore, it is no longer sufficient to regard the parcel as just saturated, and so we must replace $w_{v,sat}$ in (11-15) by w_v. The result is

$$-\frac{dT}{dt} = \frac{gW}{c_{pa}} + \frac{L_e}{c_{pa}}\frac{dw_v}{dt} + \mu\left[\frac{L_e}{c_{pa}}(w_v - w'_v) + (T - T')\right]W.$$ (11-19)

An equation relating the time rates of change of w_v and the parcel supersaturation $s_{v,w}$ is also needed. From (4.37) and (4-40) we have $w_v = (1 + s_{v,w})(\varepsilon e_{sat,w}/p)$. On differentiating this expression with respect to time, and assuming the environment is in hydrostatic equilibrium so that

$$\frac{dp}{dt} = -\frac{gpW}{R_aT'} \approx -\frac{gpW}{R_aT},$$ (11-20)

we find approximately

$$\frac{ds_{v,w}}{dt} = \frac{p}{\varepsilon e_{sat,w}}\frac{dw_v}{dt} - (1 + s_{v,w})\left[\frac{\varepsilon L_e}{R_aT^2}\frac{dT}{dt} + \frac{gW}{R_aT}\right],$$ (11-21)

where we have used (4-92).

The condensed water is related to w_v through an obvious statement of water conservation, viz., $m(w_v + w_L) + w'_v\, dm = (w_v + dw_v + w_L + dw_L)(m + dm)$. Neglecting products of differentials, this leads to

$$\frac{dw_v}{dt} = -\frac{dw_L}{dt} - \mu(w_v - w'_v + w_L)W.$$ (11-22)

* Note that w_L here has units of g water/g air, while in Chapter 2 the units assigned for w_L were g water/m³ air.

We shall assume constant environmental profiles of T' and w'_v. Therefore, from the point of view of the parcel, the time variations in T' and w'_v are given by

$$\frac{dT'}{dt} = -\Gamma'_T W, \tag{11-23}$$

and

$$\frac{dw'_v}{dt} = \Gamma'_v W, \tag{11-24}$$

where $\Gamma'_T = -dT'/dz$ and $\Gamma'_v = -dw'_v/dz$ are the observed height variations of the environmental T' and w'_v, respectively.

Entrainment will cause the parcel volume to increase with time. If we assume the convective element is a sphere or thermal bubble of radius R_T, then since $m = 4\pi\rho R_T^3/3$ we can relate the growth of the radius to the entrainment rate μ_T according to

$$\frac{d\ln R_T}{dt} = \frac{1}{3}\left[\mu_T W - \frac{d\ln\rho}{dt}\right]. \tag{11-25}$$

For a jet or plume of radius $R_J(z)$ and vertical velocity $W(z)$, entrainment is described in terms of a change in mass flux $F_m \equiv \pi R_J^2 \rho W$ along the vertical plume axis. Thus the change in mass flux over Δz is $\Delta z(dF_m/dz)$, and by continuity this must equal the radial mass influx over Δz, viz., $2\pi R_J \Delta z \rho' U_R$, where $U_R(z)$ is the inward radial velocity. Therefore, for a jet the entrainment rate is

$$\mu_J = \frac{1}{F_m}\frac{dF_m}{dz} = \frac{C}{R_J}, \tag{11-26}$$

where $C = 2\rho' U_r/\rho W$ is the *entrainment parameter*; as we have indicated earlier, laboratory studies show that $C \approx 0.2$. Finally, the equation for the jet radius which is analogous to (11-25) is

$$\frac{d\ln R_J}{dt} = \frac{1}{2}\left[\mu_J W - \frac{d\ln\rho}{dt} - \frac{d\ln W}{dt}\right]. \tag{11-27}$$

To establish a connecting link between the parcel properties and microphysical drop growth processes, note that the liquid water mixing ratio can be expressed as

$$w_L = \frac{4\pi\rho_w}{3\rho}\int_0^\infty a^3 n(a)\,da, \tag{11-28}$$

where $n(a)$ describes the size spectrum of the drops. Hence we have

$$\frac{dw_L}{dt} = \frac{4\pi\rho_w}{3\rho}\int_0^\infty \left[3a^2 n(a)\frac{da}{dt} + a^3\frac{dn}{dt}\right]da. \tag{11-29}$$

(Note there is no change in the quantity n/ρ due to expansion.) In Chapter 13 an expression will be derived for the drop radius growth rate da/dt as a function of

T, $s_{v,w}$, and the assumed spectrum of condensation nuclei. A simple model for dn/dt, adequate for our purposes, may be obtained in terms of an exchange rate between the parcel and the environment. Thus we assume that the fractional depletion of the drop population in a small time interval is equal to the fractional change in parcel mass from entrainment, i.e., $(dn)_- = -(dm/m)n$. At the same time, a similar influx of drops from the environment is assumed to occur, viz., $(dn)_+ = (dm/m)n'$, where n' counts the number per unit volume of (unactivated) drops on the specified mixed CCN at the prevailing (or specified) environmental humidity. Thus we arrive at the equation

$$\frac{dn}{dt} = -\mu W(n - n').$$

(11-30)

We now have a complete set of governing equations. For a specified size distribution of aerosol particles (AP) of given composition, for given Γ'_T, Γ'_v, and entrainment parameter μ, and assuming the gas law (4-25), the Clausius-Clapeyron equation (4-92), and the growth rate equation for da/dt (Section 13.2.1), we can determine the 11 quantities W, T, μ, w_v, p, $s_{v,w}$, w_L, T', w'_v, R, and n from the 11 equations (11-18)–(11-24), (11-25) or (11-27), (11-26) or its equivalent for a thermal, (11-29), and (11-30). To model a closed (non-entraining) parcel, the only changes required are to omit (11-24) and set $\mu = 0$ everywhere. We shall use these equations in Chapter 13 to study the growth of a population of AP into cloud drops.

CHAPTER 12

MECHANICS OF THE ATMOSPHERIC AEROSOL

The main purpose of this chapter is to outline the basic dynamical behavior of aerosols. We shall therefore discuss the phenomena of Brownian motion, diffusion, sedimentation, and coagulation of aerosol particles, including some effects of turbulence and the so-called 'phoretic' forces. (Possible electrical influences will be discussed in Chapter 17.) We shall also extend and/or apply the various formulations to the problems of evaluating mechanisms which remove aerosol particles from the atmosphere, and to the problems of explaining some observed features of the size distributions of atmospheric aerosol particles. This, in turn, will provide much of the mathematical framework which we shall subsequently apply to the study of the individual or collective growth of cloud particles by diffusion (Chapter 13), and by collision and coalescence (Chapters 15 and 16).

Some useful references for this chapter include Rasool (1973), Hidy and Brock (1970, 1971, 1972), Davies (1966), Fuchs (1964), Greene and Lane (1964), Junge (1963a), Levich (1962), and Bird *et al.* (1960).

12.1 Brownian Motion of Aerosol Particles

Let us first review briefly the classical theory of Brownian motion as it applies to the atmospheric aerosol. Brownian motion is the name given the irregular motion ('random walk') of particles due to thermal bombardment with gas molecules. A satisfactory account of it may be deduced from a simple model which ignores the detailed structure of the participating particles and, more importantly, assumes that successive particle displacements are statistically independent. This latter assumption is often described by saying Brownian motion may be idealized as a *Markoff process*, which means a stochastic process in which what happens at a given instant of time t depends only on the state of the system at time t.

A direct consequence of the randomness of Brownian motion is that the mean square distance traversed by a Brownian particle is proportional to the length of time it has experienced such motion. One simple way to prove this and evaluate the constant of proportionality is by means of the form of Newton's second law known as *Langevin's equation* (see, for example, Chandrasekhar (1943), p. 20), which reads as follows:

$$\frac{d\vec{v}}{dt} = -\beta \vec{v} + \vec{A}(t), \tag{12-1}$$

where $\vec{v}$ denotes the velocity of the particle. The first term on the right describes the continuous frictional resistance of the air to the motion of the particle, and the second term denotes the fluctuating acceleration which is characteristic of Brownian motion. That the equation of motion can be broken up into continuous and discontinuous pieces like this is an *ad hoc* assumption which is justifiable in part because it is intuitively appealing, but more importantly because it is successful in predicting behavior.

The frictional term represents the loss of organized energy which contributes to the thermal energy responsible for the fluctuations. If we apply ordinary macroscopic hydrodynamic arguments, then, since the Reynolds number for an aerosol particle is quite small, we may assume the frictional retarding force is described adequately by Stokes law (see Section 10.2.4). Obviously, for a particle with characteristic dimension r comparable to the mean free path λ_a of the air molecules, there will also be a dependence of the frictional force on the *Knudsen number* $N_{Kn} = \lambda_a/r$, but for now we will ignore this refinement. Then for a spherical particle of radius r and mass m we have

$$\beta = 6\,\pi r \eta_a/m. \tag{12-2}$$

If we now take the dot product of (12-1) with the position vector $\vec{r}$, and average the result over many trials or 'realizations' of the motion (denoting the averaging process by angular brackets), we find

$$\left\langle \vec{v} \cdot \frac{d\vec{v}}{dt} \right\rangle = \frac{d}{dt} \langle \vec{r} \cdot \vec{v} \rangle - \langle v^2 \rangle = -\beta \langle \vec{r} \cdot \vec{v} \rangle + \langle \vec{r} \cdot \vec{A} \rangle. \tag{12-3}$$

Assuming the law of equipartition of energy holds, so that $\langle v^2 \rangle = 3\,kT/m$, and considering that the complete directional isotropy of collisions implies $\langle \vec{r} \cdot \vec{A} \rangle = 0$, we find

$$\frac{d}{dt} \langle \vec{r} \cdot \vec{v} \rangle = \frac{3\,kT}{m} - \beta \langle \vec{r} \cdot \vec{v} \rangle, \tag{12-4}$$

which upon integration yields

$$\langle \vec{r} \cdot \vec{v} \rangle = \frac{3\,kT}{\beta m} + c \exp(-\beta t), \tag{12-5}$$

where c is an arbitrary constant. The parameter β^{-1} is just the viscous relaxation time τ_s given by (10-118). Generally, τ_s is quite small for atmospheric aerosols, e.g., for a particle of $r = 1\,\mu m$ and density $\rho_p = 1\,g\,cm^{-3}$ in air of 1 atm and 20°C, $\tau_s = 10^{-5}$ sec. For times $t \gg \tau_s$, any initial velocity disturbance will have decayed so that the stationary mean of $\vec{r} \cdot \vec{v}$, characteristic of Brownian motion, is

$$\langle \vec{r} \cdot \vec{v} \rangle = \frac{3\,kT}{\beta m}. \tag{12-6}$$

Since $\vec{r} \cdot \vec{v} = d(r^2/2)/dt$, one further integration immediately yields

$$\langle r^2 \rangle = \frac{6\,kTt}{\beta m} = \frac{kTt}{\pi \eta_a r}, \tag{12-7}$$

or

$$\langle z^2 \rangle = \langle x^2 \rangle = \langle y^2 \rangle = \frac{\langle r^2 \rangle}{3} = \frac{kTt}{3 \, \pi \eta_a r}. \tag{12-8}$$

This basic result was derived originally by Einstein (1905) in a somewhat different manner. It has been confirmed experimentally in numerous ways, and has been used, for example, to determine the Boltzmann constant, k, and Avogadro's number, $N_A = \mathcal{R}/k$, where $\mathcal{R}$ is the universal gas constant.

12.2   Particle Diffusion

The dependence $\langle z^2 \rangle \propto t$ of (12-8) is characteristic of a diffusion process, and in fact we may describe the motion of a large number of particles undergoing random walks without mutual interference as a process of diffusion. We now turn to a brief discussion of the connection between these two points of view. According to the well known macroscopic theory of diffusion, if $n(\vec{r}, t)$ denotes the concentration of the diffusing aerosol particles at $\vec{r}$ and at time t, then the current density in a stationary medium is given by 'Fick's first law,' namely

$$\vec{j} = -D\nabla n, \tag{12-9}$$

where D is the *diffusion coefficient* or *diffusivity*. Application of the equation of continuity for the diffusing substance,

$$\frac{\partial n}{\partial t} = -\nabla \cdot \vec{j}, \tag{12-10}$$

immediately results in the *diffusion equation* ('Fick's second law'):

$$\frac{\partial n}{\partial t} = D\nabla^2 n, \tag{12-11}$$

where we have assumed D is a constant.

Let us now determine the mean square displacement (second moment) of a distribution of diffusing particles, according to the foregoing equation. Let N identical aerosol particles per unit area be introduced at time $t = 0$ in an infinitesimally thick slab near $z = 0$ (we assume no dependence on x or y). The resulting diffusion corresponds to many simultaneous realizations of a single particle random walk from the origin. Multiplying the diffusion equation by z^2 and integrating over z, we obtain

$$\int_{-\infty}^{+\infty} z^2 \frac{\partial n}{\partial t} \, dz = N \frac{\partial \langle z^2 \rangle}{\partial t} = D \int_{-\infty}^{+\infty} z^2 \frac{\partial^2 n}{\partial z^2} \, dz = 2\,DN, \tag{12-12}$$

where the right side of the equation has been integrated by parts twice. Consequently, the mean square displacement is

$$\langle z^2 \rangle = 2\,Dt, \tag{12-13}$$

which agrees with (12-8) if

$$D = \frac{kT}{\beta m} = \frac{kT}{6 \pi \eta_a r}. \tag{12-14}$$

For particles in the submicron range, the Knudsen number correction referred to earlier becomes important. It has been found experimentally (Cunningham, 1910) that a suitable form of the diffusion coefficient for small particles is

$$D = \frac{kT(1 + \alpha N_{Kn})}{6 \pi \eta_a r}, \tag{12-15}$$

where $(1 + \alpha N_{Kn})$ is termed the Cunningham slip-flow correction. According to Davies (1945),

$$\alpha = A + B \exp(C/N_{Kn}), \tag{12-16}$$

where $A = 1.257$, $B = 0.400$, and $C = -1.10$. The dependence of λ_a, and hence N_{Kn}, on temperature and pressure is described adequately by (10-106), with $\lambda_{a,0} = 6.6 \times 10^{-6}$ cm. Table 12-1 lists a few representative values for D, computed from (12-15) and (12-16), and the corresponding values of $\langle r^2 \rangle^{1/2}$ (from (12-7) multiplied by $(1 + N_{Kn})$) after one minute of diffusion time. On the basis of the small and rapidly decreasing values of D and $\langle r^2 \rangle^{1/2}$ with increasing particle size, one can reasonably anticipate that Brownian diffusion of aerosol particles in the troposphere becomes of secondary importance to other transport processes for $r \geqslant 1\ \mu m$.

12.3 Mobility and Drift Velocity

Let us now generalize the Langevin equation (12-1) by letting an external force, $\vec{F}_{ext}$, act on the diffusing particle:

$$m\frac{d\vec{v}}{dt} = \vec{F}_{ext} - \beta m \vec{v} + m\vec{A}. \tag{12-17}$$

On taking mean values and assuming a steady state (which requires only that the

TABLE 12-1

Diffusion coefficient, Schmidt number, root mean square distance travelled in 1 min, and terminal fall velocity, for Brownian particles of radius r for 15 °C, 1 atm, $\lambda_a = 7.37 \times 10^{-6}$ cm, $\eta_a = 1.78 \times 10^{-4}$ poise, $\nu_a = 0.148$ cm^2 sec^{-1} (NACA Standard Atmosphere).

r (μm)	D (cm^2 sec^{-1})	$N_{Sc} = \nu/D$	$\langle r^2 \rangle^{1/2}$ (μm), after 1 min	V_S (cm sec^{-1})
0.01	1.56×10^{-4}	9.46×10^2	2333	3.1×10^{-5}
0.1	2.35×10^{-6}	6.28×10^4	290	4.3×10^{-4}
1.0	1.29×10^{-7}	1.14×10^6	68	2.7×10^{-2}
10	1.19×10^{-8}	1.24×10^7	21	2.5

characteristic time for a change in $\langle \vec{v} \rangle$ be large compared to β^{-1}) this yields

$$\langle \vec{v} \rangle = \frac{\vec{F}_{ext}}{\beta m} \equiv \vec{v}_{drift}. \tag{12-18}$$

This average response to an impressed force is called the *drift velocity*. It is customary to define the particle *mobility*, B, by the equation

$$\vec{v}_{drift} \equiv B \vec{F}_{ext}. \tag{12-19}$$

Therefore,

$$B = \frac{1}{\beta m} = \frac{1 + \alpha N_{Kn}}{6 \pi \eta_a r}, \tag{12-20}$$

for a spherical particle obeying Stokes law corrected for slip. Further, on combining (12-15) and (12-20) we obtain an intimate connection between the mobility and diffusion coefficients, namely

$$D = BkT, \tag{12-21}$$

which is known as the Einstein relation.

12.4   Sedimentation and the Vertical Distribution of Aerosol Particles

The drift velocity of small particles falling under gravity is obviously just the terminal velocity in slip-corrected Stokes flow, V_s. Thus, for spherical particles of density ρ_p we have

$$\vec{v}_{drift} = V_s = \frac{2(1 + \alpha N_{Kn}) r^2 g (\rho_p - \rho_a)}{9 \eta_a}, \tag{12-22}$$

(cf. Section 10.3.5). This result is plotted in Figure 12-1 for $\rho_p = 2 \, g \, cm^{-3}$, which shows that the fall velocity of a particle increases rapidly with height, particularly above 10 km. Also, the velocities are seen to be quite small for $r \lesssim 1 \, \mu m$; in fact, a time of the order of years is required for particles of $r < 0.1 \, \mu m$ to fall through a layer 1 km thick at altitudes less than 25 km.

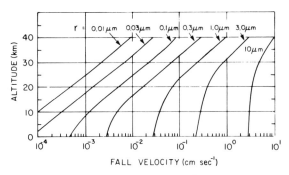

Fig. 12-1.   Fall velocity of spherical particles ($\rho_P = 2 \, g \, cm^{-3}$) as a function of particle radius and altitude in the NACA Standard atmosphere. (From Junge *et al.*, 1961; by courtesy of Amer. Meteor. Soc., and the authors.)

Let us now consider the problem of determining the distribution of particles undergoing simultaneous Brownian diffusion and sedimentation. In this case we write the total particle current density $\vec{j}$ as the sum of the contributions from diffusion and drift in the gravity field:

$$\vec{j} = -D\nabla n + n\vec{v}_{\text{drift}}. \tag{12-23}$$

Application of the continuity equation for the particles, $\partial n/\partial t = -\nabla \cdot \vec{j}$, yields the governing ('Smoluchowski') equation for the concentration $n(r, t)$:

$$\frac{\partial n}{\partial t} = D\nabla^2 n + V_s \frac{\partial n}{\partial z}, \tag{12-24}$$

where we have assumed gravity acts in the $-z$ direction and that $V_s = \text{constant}$. If we suppose the plane $z = 0$ forms the bottom absorbing boundary of a semi-infinite homogeneous aerosol which was initially of uniform concentration n_0 for $z > 0$, then the solution for the vertical distribution $n(z, t)$ is

$$n(z, t) = \frac{n_0}{2}(1 - e^{-V_s t/D}) + \frac{n_0}{2}\left[e^{-V_s t/D}\,\text{erf}\left(\frac{z - V_s t}{\sqrt{4Dt}}\right) + \text{erf}\left(\frac{z + V_s t}{\sqrt{4Dt}}\right)\right], \tag{12-25}$$

where erf denotes the error function (see Appendix A-12.4 for details). The deposition rate of particles on the surface $z = 0$ is therefore given by

$$I(t) = -\vec{j}(0, t) \cdot \hat{e}_z = D\frac{\partial n}{\partial z}\bigg|_{z=0} = n_0\left[\frac{V_s}{2}\left(1 + \text{erf}\sqrt{\frac{V_s^2 t}{4D}}\right) + \sqrt{\frac{D}{\pi t}}\,e^{-V_s^2 t/4D}\right]. \tag{12-26}$$

According to (12-26) the diffusion current is infinite at $t = 0$. This happens because of the artificial specification of an infinite concentration gradient at $z = t = 0$. For $t \ll t_c = 4D/V_s^2$, (12-26) becomes $I(t) = n_0[(D/\pi t)^{1/2} + V_s/2]$, which is equal to the diffusion deposition rate when would occur in the absence of sedimentation, plus half of the pure sedimentation rate. For particles of radius $0.1\,\mu m$ and $1\,\mu m$ in air at $p = 1000$ mb and $T = 20\,°C$, t_c is about 40 sec and 4×10^{-3} sec, respectively, assuming a particle density of $2\,\text{g cm}^{-3}$. For $t \gg t_c$ the deposition rate is $I(t) = n_0 V_s$, so that Brownian motion no longer has any effect.

In reality, of course, the atmosphere is not motionless, nor is the surface of the Earth purely an aerosol sink. There generally will be an upward flux of material, usually by turbulent diffusion, which will tend to equilibrate with sedimentation and other processes over various time scales. If we assume a steady state balance of vertical turbulent diffusion and sedimentation, a crude account of the resulting vertical distribution of aerosol may be obtained by the following plausible modification of the particle current density in (12-23): let D be replaced by D_e, where D_e is an effective eddy diffusivity of aerosol particles describing the transport capability of turbulence within the framework of classical diffusion theory, and let $V_s \rightarrow V_s - W$, where W is the average updraft velocity of the air. Further, that part of the concentration gradient of particles due to the decrease in air density with height will not be effective in the turbulent transport, so to obtain results for heights comparable to the scale height of the atmosphere we should replace n in (12-23) by $\sigma = n/n_a$, where $n_a = n_a(z)$ is the number concentration of air molecules at height z. Assuming the net particle flux is zero at all levels, the equation governing the mixing ratio σ becomes

$$\frac{d\sigma}{dz} = \frac{-(V_s - W)\sigma}{D_e}, \tag{12-27}$$

so that with the ground at $z = 0$ the distribution is

$$\sigma(z) = \sigma(0) \exp(-z/z_e), \tag{12-28}$$

where

$$z_e = \frac{D_e}{V_s - W}, \tag{12-29}$$

assuming constant z_e. If desired, the particle concentration may be recovered by noting that (12-28) may also be expressed in the form $n(z)/n(0) = [n_a(z)/n_a(0)] \exp(-z/z_e)$, and that from the ideal gas law and the assumption of hydrostatic equilibrium we have

$$\frac{n_a(z)}{n_a(0)} = \frac{T(0)}{T(z)} \exp\left[-\frac{g}{R_a}\int_0^z \frac{dz}{T(z)}\right], \tag{12-30}$$

where $T(z)$ is the temperature profile and R_a the gas constant for air.

As we have discussed in Chapter 8, an exponential decrease of aerosol particle concentration with height, in qualitative conformity with (12-28), is often observed in the lower troposphere (recall Figures 8-5 and 8-6). However, the model used above leaves much to be desired. For example, we have neglected the effects of local sources and sinks, meteorological conditions of wind and stability, dilution by mixing with air of different properties, 'washout' or 'scavenging' by cloud and precipitation particles, and various coagulation processes operating in a heterogeneous aerosol. More complete models to describe the vertical distribution of aerosols have been formulated by Junge (1957b), Erickson (1959), Junge et al. (1961a), and Toba (1965a). Models for explaining the horizontal distribution of the atmospheric aerosol have been formulated by Toba (1965a), Tanaka (1966), Toba and Tanaka (1968), and Rossknecht et al. (1973).

12.5 Brownian Coagulation of Aerosol Particles

Particles undergoing relative Brownian diffusion have a finite probability of colliding and sticking to one another; i.e., they may experience *thermal* or *spontaneous coagulation*. The sticking probability or 'efficiency' of aerosol particles is a complicated function of their shape and surface conditions (roughness, absorbed vapors, etc.), the relative humidity of the air, the presence of foreign vapors in the air, and other factors. Although little is known quantitatively about the sticking efficiency of aerosol particles, the fact that the kinetic energy of the colliding particles is very small makes bounce-off unlikely.

We shall therefore assume a sticking efficiency of unity in the sequel. (For a review of the topic of sticking efficiency, see Corn (1966).)

To formulate the coagulation process we first need to find the diffusion coefficient D_{12} which characterizes the diffusion of particles of radius r_2 relative to those of radius r_1. Suppose the particles experience displacements $\Delta \vec{r}_1$ and $\Delta \vec{r}_2$ in time Δt, respectively. Then their mean square relative displacement is

$$\langle |\Delta \vec{r}_1 - \Delta \vec{r}_2|^2 \rangle = \langle \Delta r_1^2 \rangle + \langle \Delta r_2^2 \rangle - 2 \langle \Delta \vec{r}_1 \cdot \Delta \vec{r}_2 \rangle = \langle \Delta r_1^2 \rangle + \langle \Delta r_2^2 \rangle,$$

since the motions are independent. Now from the meaning of D_{12} and the results (12-8) and (12-13) we have $\langle |\Delta \vec{r}_1 - \Delta \vec{r}_2|^2 \rangle = 6 D_{12} \Delta t$; on the other hand, we also have $\langle \Delta r_1^2 \rangle = 6 D_1 \Delta t$ and $\langle \Delta r_2^2 \rangle = 6 D_2 \Delta t$. Therefore we obtain the simple result

$$D_{12} = D_1 + D_2. \tag{12-31}$$

By the preceding argument we may regard the r_1 particle as stationary, and suppose the r_2 particles are diffusing toward it with diffusion coefficient D_{12}. Using the center of the r_1 particle as the origin of coordinates, we have by the isotropy of the process that the concentration n_2 of the r_2-particles is a function only of r, the distance from the origin, and time t. Further, the boundary condition of adhesion on contact may be expressed as

$$n_2(r_{12}, t) = 0, \quad r_{12} = r_1 + r_2. \tag{12-32}$$

So we must solve

$$\frac{\partial n_2}{\partial t} = D_{12} \nabla^2 n_2 = \frac{D_{12}}{r} \frac{\partial^2}{\partial r^2}(r n_2), \tag{12-33}$$

subject to (12-32) and the initial condition

$$n_2(r, 0) = n_2(\infty), \quad r > r_{12}, \tag{12-34}$$

where $n_2(\infty)$ is the ambient concentration of r_2-particles at the beginning of the coagulation process.

The solution to (12-32)–(12-34) is easily obtained, since (12-33) is just a one dimensional diffusion equation in the dependent variable $r n_2$. It may be solved, for example, by application of the method presented in A-12.4, using the independent variable $z = r - r_{12}$. By this or other means the solution is found to be

$$n_2(r, t) = n_2(\infty) \left[1 - \frac{r_{12}}{r} + \frac{r_{12}}{r} \operatorname{erf}\left(\frac{r - r_{12}}{\sqrt{4 D_{12} t}} \right) \right]. \tag{12-35}$$

Therefore, the coagulation rate (loss rate) of the r_2-particles with the r_1-particle is

$$-\int_{S_{12}} \vec{j} \cdot d\vec{S} = D_{12} \int_{S_{12}} \frac{\partial n_2}{\partial r} dS = 4 \pi r_{12} D_{12} n_2(\infty) \left(1 + \frac{r_{12}}{\sqrt{\pi D_{12} t}} \right), \tag{12-36}$$

where the integration is over the surface S_{12} given by $r = r_{12}$. We may generally ignore the time factor in this expression since $r_{12}^2 / \pi D_{12}$ (i.e., the characteristic time for diffusion over distances comparable to aerosol radii) is less than $\sim 10^{-5}$ sec for $r \lesssim 0.1 \,\mu m$ and less than $\sim 10^{-2}$ sec for $r \lesssim 1 \,\mu m$.

Now let us carry out an obvious generalization of these results. Suppose initially we have a homogeneous aerosol of particles of volume v_1 in concentration $n_1(0) = n_0$. Coagulation sets in and soon there appear particles of volume $v_2 = 2 v_1$ in concentration n_2, $v_3 = 3 v_1$ in concentration n_3, etc. Then to an excellent approximation the rate of coagulation of v_i and v_j particles per unit volume of aerosol to form particles of volume $v_{i+j} = (i + j)v_1$ is given by $4 \pi D_{ij} r_{ij} n_i n_j$, with $D_{ij} = D_i + D_j$ and $r_{ij} = r_i + r_j = (3/4 \,\pi)^{1/3}(v_i^{1/3} + v_j^{1/3})$, where n_i and n_j are the ambient concentrations of the v_i and v_j particles at time t.

Strictly speaking of course, this result can apply only if the multiple particles are spherical (i.e., behave as small liquid drops and not as solid angular particles). This is one of those difficulties that is usually passed over in polite silence, in the same category as questions concerning the probability of sticking, and the effects of short range induced forces like Van de Waal's interaction. However, careful experiments by Devir (1966) suggest the Brownian coagulation rate of aerosols is generally in agreement with that indicated by the above formulation.

We can now write down the governing equation for coagulation. From our generalization above the overall rate of formation of v_k particles per unit volume of aerosol is evidently given by $(1/2) \Sigma_{i+j=k} 4 \pi D_{ij} r_{ij} n_i n_j$, where the factor of $1/2$ is included to avoid counting the same interaction event twice $(i + j = j + i)$. Similarly, the loss rate of the v_k particles per unit volume of aerosol is given by $-\Sigma_i 4 \pi D_{ik} r_{ik} n_i n_k$. Therefore, the equation which determines the discrete particle size distribution for binary collisions is

$$\frac{dn_k}{dt} = \frac{1}{2} \sum_{i+j=k} 4 \pi D_{ij} r_{ij} n_i n_j - \sum_i 4 \pi D_{ik} r_{ik} n_i n_k. \qquad (12\text{-}37)$$

This is the discrete form of the *kinetic coagulation equation* for Brownian coagulation.

According to (12-14), the diffusion coefficient is inversely proportional to the radius, so that we may write

$$D_{ij} r_{ij} = Dr \left(\frac{1}{r_i} + \frac{1}{r_j} \right)(r_i + r_j), \qquad (12\text{-}38)$$

where D and r are the diffusion coefficient and radius of the primary particles. Since the Cunningham slip correction has been ignored, this expression applies only for $N_{kn} \lesssim 0.1$. For $0.5 \lesssim r_i/r_j \lesssim 2$, (12-38) may be approximated by $D_{ij} r_{ij} = 4 Dr$, to within a maximum error of 12.5%. Since this approximation assumes $r_i = r_j$, it is said to ignore the aerosol 'polydispersity.'

Smoluchowski (1916, 1917) obtained an approximate solution to (12-37) for the case of an initially homogeneous aerosol, assuming $D_{ij} r_{ij} = 4 Dr$. If we introduce $f_k(\tau) \equiv n_k/n_0$ and $\tau \equiv 8 \pi Drn_0 t$, (12-37) becomes

$$\frac{df_k}{d\tau} = \sum_{i+j=k} f_i f_j - 2 f_k \sum_i f_i. \qquad (12\text{-}39)$$

Summing over k, we obtain

$$\frac{d}{d\tau}\left(\sum f_k\right) = \sum_{i=1}\sum_{j=1} f_i f_j - 2\sum_{k=1}\sum_{i=1} f_i f_k = -\left(\sum_k f_k\right)^2, \qquad (12\text{-}40)$$

which upon integration yields

$$\sum_k f_k = \frac{1}{1+\tau}. \qquad (12\text{-}41)$$

With this result, the solutions for f_k may be obtained successively. Thus the equation for f_1 is

$$\frac{df_1}{d\tau} = -2 f_1 \sum_k f_k = \frac{-2 f_1}{1+\tau}, \qquad (12\text{-}42)$$

which leads to $f_1 = (1+\tau)^{-2}$. By induction one may easily show that for arbitrary k,

$$f_k = \frac{\tau^{k-1}}{(\tau+1)^{k+1}}, \qquad (12\text{-}43)$$

which is Smoluchowski's solution.

The reciprocal dependence of total particle concentration on time predicted by (12-41) has been verified by extensive experimental studies over the last forty years (e.g., Whytlaw-Gray and Patterson (1932), Patterson and Cawood (1932), Whytlaw-Gray (1935), Artemov (1946), Devir (1963)). The total particle concentration is reduced by a factor of one half in a time $t_{1/2}$ given by

$$t_{1/2} = (8\,\pi Drn_0)^{-1} = \frac{3\,n_a}{4\,kTn_0(1+\alpha N_{kn})}, \qquad (12\text{-}44)$$

including the slip correction. As would be expected, the *coagulation time* $t_{1/2}$ decreases with decreasing size, increasing concentration, and increasing temperature of the aerosol. One may use $t_{1/2}$ to estimate the size range of tropospheric aerosols for which Brownian coagulation is important. As an example, we may assume that a typical concentration for Aitken particles centered near 0.01 μm radius is $n_0 = 10^5$ cm^{-3}, which means $t_{1/2} = 2.6 \times 10^3$ sec at 15 °C and 1000 mb, a time short enough to indicate coagulation is the dominant loss mechanism for such particles; for particles of 1 μm radius in concentration $n_0 = 10^3$ cm^{-3}, the coagulation time is greater by a factor of 10^3.

The most serious limitation of Smoluchowski's solution is that it assumes an initially homogeneous aerosol. In order to investigate the effect of Brownian coagulation on realistic aerosol spectra, Junge (1955, 1957b) and Junge and Abel (1965) have carried out numerical solutions of the kinetic coagulation equation, using measured tropospheric aerosol spectra for initial conditions and allowing for the dependence of the coagulation rate on particle size. For this purpose they used a continuous form of the coagulation equation, involving the continuous size distribution function n(r, t), where n(r, t) dr is the number of particles at time t with radius between r and r + dr, per unit aerosol volume.

In order to derive the coagulation equation for $n(r, t)$, first note that (12-37) may be written in the form

$$\frac{dn_k}{dt} = \frac{1}{2} \sum_{i=1}^{k-1} K_{i,k-i} n_i n_{k-i} - n_k \sum_{i=1}^{\infty} K_{ik} n_i, \tag{12-45}$$

where

$$K_{ij} = 4 \pi D_{ij} r_{ij} = K_{ji}. \tag{12-46}$$

The quantity K_{ij} is called the *collection kernel* for Brownian coagulation. An obvious continuous counterpart of this equation is obtained by making the changes

$$n_i(t) \to n(v, t) \, dv \quad \text{(12-47a)} \qquad K_{ij} \to K(v, u) = K(u, v), \tag{12-47b}$$

where $n(v, t) \, dv$ is the number of v-particles at time t with volume between v and $v + dv$, per unit aerosol volume, and $K(u, v) n(u) \, du$ is the coagulation rate of u-particles with a v-particle. The continuous form of (12-45) involving the distribution function $n(v, t)$ is then

$$\frac{\partial n(v, t)}{\partial t} = \frac{1}{2} \int_0^v K(u, v - u) n(u, t) n(v - u, t) \, du - n(v, t) \int_0^{\infty} K(u, v) n(u, t) \, du. \tag{12-48}$$

We may transform this directly into an equation for $n(r, t)$ by noting that $n(r, t) \, dr = n(v, t) \, dv$, which merely reflects the fact that more than one measure may be used to count the same particles. Therefore if r, s, and $r' = (r^3 - s^3)^{1/3}$ are the radii corresponding respectively to v, u, and $v - u$, we have

$$n(v - u, t) \, dv = n(v - u, t) d(v - u) \Big|_{u = \text{const.}}$$

$$= n(r', t) \, dr' \Big|_{s = \text{const.}} = \frac{n(r', t) r^2 \, dr}{(r^3 - s^3)^{2/3}}, \tag{12-49}$$

so that (12-48) becomes

$$\frac{\partial n(r, t)}{\partial t} = \frac{1}{2} \int_0^r K(s, r')(1 - s^3/r^3)^{-2/3} n(s, t) n(r', t) \, ds$$

$$- n(r, t) \int_0^{\infty} K(r, s) n(s, t) \, ds. \tag{12-50}$$

The collection kernel $K_B(r, s)$ for Brownian coagulation is

$$K_B(r, s) = 4 \pi (D_r + D_s)(r + s) = \frac{2 \, kT}{3 \, \eta_a}(r + s)\left[\left(\frac{1}{r} + \frac{1}{s}\right) + \alpha \lambda_a\left(\frac{1}{r^2} + \frac{1}{s^2}\right)\right], \tag{12-51}$$

where we have included the slip correction.

Equations (12-50) and (12-51) were solved numerically by Junge (1957b, 1963a) and Junge and Abel (1965) for the case of an initial spectrum representative of average tropospheric conditions, and assuming that the colliding aerosol particles have a sticking efficiency of unity and behave as droplets. The result of this computation is shown in Figure 12-2 in terms of the concentration $N(r, t)$ of particles with radii $\geq r$ (cf. (8-26)). It is seen that the modification of the size distribution due to Brownian coagulation is confined mainly to particles with radii less than $0.1\ \mu m$, i.e., the Aitken particle size range. With increasing time, particles of radii less than $0.1\ \mu m$ rapidly disappear, while the maximum of the original size distribution shifts to larger sizes, shifting less and less and centering over $r \approx 0.1\ \mu m$ after a few days. From the computed corresponding particle volume changes Junge, and Junge and Abel also concluded that coagulation causes a steady flux of aerosol material from the Aitken size range into the 'large' particle size range, causing the formation of 'mixed' particles in the size range of 'large' particles.

We mentioned in Section 8.2.10 that observations over the ocean suggest that the aerosol particle concentration does not decrease to zero for particles of radii below $0.1\ \mu m$ but, after a minimum for $0.01 \leqslant r \leqslant 0.1\ \mu m$, increases again to a second maximum for particles with $r \approx 0.001\ \mu m$. We indicated that it is reasonable to attribute this second maximum to a continuous production of primary particles by gas-to-particle conversion. In order to take this behavior into account, Walter (1973) solved the coagulation equation (Equations (12-50) and (12-51)) with a source-term included. Walter assumed an initially narrow normal distribution of primary aerosol particle sizes whose mean radius was 1.2×10^{-7} cm, and whose production rate was 10^5 particles cm^{-3} sec^{-1}. Figure 12-3 shows that after only a short time coagulation produces a secondary maximum of the size

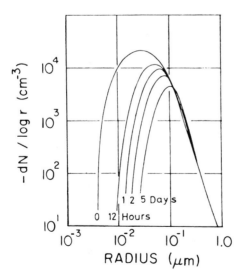

Fig. 12-2. Numerical calculation of the change of an average atmospheric aerosol particle distribution due to Brownian coagulation. (From Junge and Abel, 1965; by courtesy of the authors.)

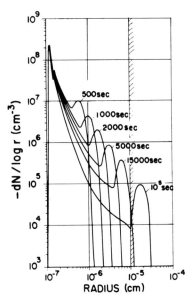

Fig. 12-3. Variation with time of the distribution function $-dN/d\log r$, up to the attainment of a quasi-stationary state in the range $r = 10^{-7}$ to 10^{-5} cm; particle production rate $= 10^5\,\mathrm{cm}^{-3}\,\mathrm{sec}^{-1}$; total volume of aerosol generated $= 7.8 \times 10^{-15}\,\mathrm{cm}^{-3}\,\mathrm{sec}^{-1}$. (From Walter, 1973; by courtesy of Pergamon Press Ltd.)

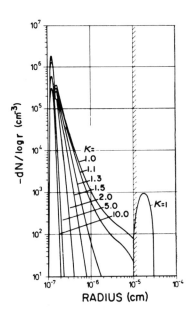

Fig. 12-4. Steady aerosol size distribution for different concentrations of large particles of the same original size. Particle production rate $= 10^2\,\mathrm{cm}^{-3}\,\mathrm{sec}^{-1}$. Concentration of $0.1\,\mu$m radius particles $= \kappa \times 270\,\mathrm{cm}^{-3}$. (From Walter, 1973; by courtesy of Pergamon Press Ltd.)

distribution. With increasing time this maximum becomes more distinct and shifts to larger sizes until, after about 10^5 sec, the size distribution for r between 2×10^{-7} and 10^{-5} cm reaches a quasi-steady state, which means that in any size interval as many particles are formed as are lost by coagulation. The presence of coarse particles of $r = 10^{-5} \mu m$ in concentrations varying from 270 to $\kappa \times 270$ (where $\kappa = 1 \ldots 10$) caused the gap between the first and second maximum in the particle concentration to deepen rapidly (Figure 12-4). Walter concluded from his computations that the two concentration maxima observed in the size distribution of natural aerosol particles are the result of continuous particle production superimposed on Brownian coagulation.

12.6 Laminar Shear, Turbulence, and Gravitational Coagulation

In (12-45), (12-48), and (12-50) we have a framework for evaluating the effects on aerosol spectra of other processes besides Brownian coagulation, since any process which gives rise to a non-zero probability of particle interaction can be represented by some appropriate form of the collection kernel K_{ij} or K. In this section we consider some possible influences of shear flows, turbulence, and relative motion under gravity.

12.6.1 COAGULATION IN LAMINAR SHEAR FLOW

Velocity gradients in the air cause relative motion and may, as a result, induce collisions between aerosol particles. The simplest model for this collection process was worked out by Smoluchowski (1916). His model assumes a uniform shear field, no fluid dynamic interaction between the particles, and no Brownian motion. In order to determine the rate at which v_2-particles are collected by one v_1-particle, we assume flow relative to the origin of v_1 in the $\pm z$-direction with the shear in the y-direction. The velocity field is then given by

$$\vec{u} = \Gamma y \hat{e}_z, \tag{12-52}$$

where Γ is the constant velocity gradient or shear rate. If we again use the general notation of the previous section, we can express the particle flux or current density vector relative to the v_1-particle as

$$\vec{j} = n_2 \Gamma y \hat{e}_z, \quad r > r_{12} \tag{12-53}$$

and the collection rate is given by the positive flux J_{LS} of $\vec{j}$ into a sphere of radius r_{12} concentric with the origin:

$$J_{LS} = 2 \Gamma n_2(\infty) r_{12}^3 \int_0^\pi \sin \phi \, d\phi \int_0^{\pi/2} \sin^2 \theta \cos \theta \, d\theta = \tfrac{4}{3} \Gamma n_2(\infty) r_{12}^3. \tag{12-54}$$

The collection kernel K_{LS} for laminar shear flow may therefore be expressed as $J_{LS}/n_2(\infty)$, or

$$K_{LS}(r, s) = \frac{4 \Gamma}{3}(r + s)^3. \tag{12-55}$$

One would expect that (12-55) might overestimate the actual shear collection rate, since viscous effects should tend to make the particles move around each other, thus lowering the collision cross section below the geometric value. However, at least one experimental study (Manley and Mason, 1952) showed that (12-55) is essentially correct for the case of large glass spheres (60 to 70 μm diameter) in a quite viscous fluid with $\Gamma = 0.4$ to $0.8\ \mathrm{sec}^{-1}$.

The relative strength of laminar shear to Brownian coagulation for a homogeneous aerosol is given by the ratio

$$\frac{K_{LS}(r, r)}{K_B(r, r)} = \frac{2\,\Gamma r^2}{3\,\pi D}. \tag{12-56}$$

From the values of D given in Table 12-1, we find the shear rates required for the two processes to be comparable are rather high for particles smaller than a few microns: $\Gamma \approx 10^5\ \mathrm{sec}^{-1}$ for $r = 0.1\ \mu$m; $\Gamma \approx 60\ \mathrm{sec}^{-1}$ for $r = 1.0\ \mu$m; $\Gamma \sim 6 \times 10^{-2}\ \mathrm{sec}^{-1}$ for $r = 10\ \mu$m. Accordingly, one would expect shear collection to become important only for particles larger than a few microns in radius. However, since in the atmosphere velocity gradients are generally associated with turbulent motions, one should hesitate to draw definite conclusions from a model based on laminar flow. Therefore, we now turn to a consideration of the collection rates in turbulent flow. However, in support of (12-55), it will be seen that the length and time scales over which particles interact are small enough so that the assumption of constant shear during the interaction does not lead to large errors.

12.6.2   COAGULATION IN TURBULENT FLOW

Since a comprehensive treatment of turbulent flow would go far beyond the scope of this book, we shall base our discussion in this section on a rather simple description using dimensional analysis. For more detailed information on the theory of turbulence, the reader is referred to such texts as Tennekes and Lumley (1972), Monin and Yaglom (1971), and Landau and Lifshitz (1959).

Turbulence is characterized in large part by a disordered vorticity field, which may be pictured qualitatively as a distribution of eddies of various sizes. The largest eddies have a size ℓ and characteristic fluctuating velocity Δu of the same order as the length and velocity characterizing the flow as a whole. The corresponding Reynolds number $N_{Re,\ell} = \Delta u \ell / \nu$ is very large, indicating viscosity has a small effect on the dynamics of the large eddies.

The largest eddies contain most of the kinetic energy of the flow, which is passed on with little dissipative loss to eddies of smaller scale by some as yet poorly understood nonlinear break-up process (the 'energy cascade'), until finally it is dissipated into heat by viscosity on the smallest scales of motion. With a little dimensional analysis this picture of turbulence leads to some interesting and useful predictions, which are considered to be correct to within an order of magnitude.

According to the above description the rate of dissipation of kinetic energy per unit mass, ε, is controlled by the rate of break-up of the largest eddies, for which viscosity plays a negligible role. Thus it is reasonable to assume that ε is not a direct function of ν, but instead can only be a function of the physical parameters characterizing the large scale flow, namely Δu, ℓ, and possibly ρ, the density of the fluid. Then dimensional consistency demands that

$$\varepsilon \approx \frac{\Delta u^3}{\ell}. \tag{12-57}$$

There is some evidence that ε is in fact a weak function of $N_{Re,\ell}$, and hence ν. A discussion of this point appears in Saffman (1968).

Note that (12-57) says the characteristic time for kinetic energy loss by the large eddies is $\ell/\Delta u$, i.e., the eddies lose a significant fraction of their energy by break-up in one revolution. On the other hand, the characteristic time for decay of the large eddies by diffusion is ℓ^2/ν. Therefore, the ratio of the diffusion dissipation time scale to the turnover time scale is $\ell \Delta u/\nu = N_{Re,\ell} \gg 1$, which is consistent with the assumption that viscous dissipation of the large eddies is unimportant.

Let us now introduce the concept of an eddy Reynolds number,

$$N_{Re,\lambda} \equiv \frac{v_\lambda \lambda}{\nu}, \tag{12-58}$$

where λ is the size and v_λ the fluctuating velocity which characterizes the eddy. Consider eddies with $N_{Re,\lambda} \gg 1$ but with $\lambda \ll \ell$. Since $N_{Re,\lambda} \gg 1$, it is natural to assume that the flow on this scale must also be independent of ν. However, since $\lambda \ll \ell$, the additional assumption may be made that the flow does not depend on ℓ or Δu (which means that the turbulence on scale $\lambda \ll \ell$ is isotropic), except insofar as ℓ and Δu determine ε. Then the fluctuating velocity v_λ can depend only on λ, ρ, and ε. By dimensional reasoning we then conclude that

$$v_\lambda \approx (\varepsilon\lambda)^{1/3} \approx (\lambda/\ell)^{1/3}\Delta u \tag{12-59}$$

(Kolmogorov, 1941). Combining (12-58) and (12-59), we may also write

$$N_{Re,\lambda} \approx \frac{\varepsilon^{1/3}\lambda^{4/3}}{\nu}. \tag{12-60}$$

Suppose now that λ_0 denotes the eddy size for which viscosity becomes important. A natural statement of this condition is $N_{Re,\lambda_0} \approx 1$, so that λ_0 may be estimated from (12-60) as

$$\lambda_0 \approx \left(\frac{\nu_a^3}{\varepsilon}\right)^{1/4} \equiv \lambda_K, \tag{12-61}$$

where λ_K is called the 'Kolmogorov microscale length'. It may be found a little more directly from dimensional considerations on the assumption that the parameters governing the small scale motion include ρ, ε, and ν. In the same way one may also write down the Kolmogorov microscales of time and velocity:

$$\tau_K \equiv (\nu/\varepsilon)^{1/2} \quad (12-62) \qquad v_K \equiv (\nu\varepsilon)^{1/4} = (\lambda_K/\tau_K). \tag{12-63}$$

The relative sizes of the large and small scales of length, time, and velocity are now easily determined as a function of the Reynolds number of the mean flow:

$$\frac{\lambda_K}{\ell} \approx N_{Re,\ell}^{-3/4} \quad (12\text{-}64a); \qquad \frac{\tau_K}{\ell/\Delta u} \approx N_{Re,\ell}^{-1/2}; \quad (12\text{-}64b) \qquad \frac{v_K}{\Delta u} \approx N_{Re,\ell}^{-1/4}. \qquad (12\text{-}64c)$$

Qualitatively, these relations indicate what is borne out experimentally, namely that the small scale structure of turbulence becomes finer with increasing Reynolds number. They also indicate the assumption of isotropic turbulence for eddies in the 'inertial subrange,' $\lambda_K \ll \lambda \ll \ell$, becomes better with increasing $N_{Re,\ell}$.

Measurements by Ackerman (1967, 1968) under various conditions in cloudy air produced values of ε in the range $3 \leqslant \varepsilon \leqslant 114 \, \text{cm}^2 \, \text{sec}^{-3}$. Values in clear air should generally be close to the low end of this range. Therefore, from (12-61) we expect λ_K to vary typically from 8×10^{-2} to $2 \times 10^{-1} \, \text{cm}$ in the troposphere, which means almost all aerosol particles are much smaller than the microscale length.

12.6.2.1 Turbulent Shear Coagulation

Velocity gradients in turbulent air should cause relative particle motion and possibly collisions in a manner analogous to the laminar shear process treated in the previous section. Since aerosol particles are much smaller than λ_K, it is the shearing motion on length scales $\lambda < \lambda_K$ which is relevant. Since such motion must be strongly affected by viscosity, it is clear that turbulent shear coagulation may be represented by an expression similar to (12-55). To find such an expression we must obtain an estimate for the velocity gradients for $\lambda < \lambda_K$. On this scale velocity causes the flow to vary relatively smoothly, so that we may expand the velocity fluctuation v_λ in powers of λ and retain just the first term. Thus $v_\lambda \approx c\lambda$, where c is a constant. Such treatment should be applicable for $\lambda \leqslant \lambda_K$, so that we may also write $v_K \approx c\lambda_K$, or $v_\lambda/\lambda \approx v_K/\lambda_K = 1/\tau_K = (\varepsilon/v_a)^{1/2}$, which is the relevant characteristic shear rate. By analogy to (12-55) we may then estimate the collection kernel for turbulent shear as

$$K_{TS}(r, s) \approx (r + s)^3 (\varepsilon/\nu_a)^{1/2}. \qquad (12\text{-}65)$$

Note this result also follows from dimensional considerations and the assumption that the only characteristic length of the process is the geometric collision length, $r + s$: The dimensions of K_{TS} are $(\text{length})^3 \, (\text{time})^{-1}$, and since τ_K is the characteristic time scale of the flow for $\lambda \leqslant \lambda_K$, (12-65) then follows. An elaborate model calculation by Saffmann and Turner (1956) provides a numerical coefficient:

$$K_{TS}(r, s) = 1.30 \, (r + s)^3 (\varepsilon/\nu_a)^{1/2}. \qquad (12\text{-}66)$$

Earlier less detailed (but not necessarily less rigorous) calculations by Tunitskii (1946) and Levich (1954a, b) gave coefficients of 0.5 and 3.1, respectively. Ackermann's (1967, 1968) measured values of ε referred to earlier correspond to turbulent shear rates τ_K^{-1} which vary between $4 \, \text{sec}^{-1}$ and $28 \, \text{sec}^{-1}$. This means the crossover in dominance from Brownian to turbulent shear coagulation usually occurs at a particle size of a few microns.

12.6.2.2 *Turbulent Inertial Coagulation*

A second mode of coagulation in turbulent flow is due to local turbulent accelerations, which produce relative particle velocities for particles of unequal mass. The characteristic acceleration, a_K, of eddies of size λ_K is

$$a_K = v_K^2/\lambda_K = \varepsilon^{3/4}/v_a^{1/4}. \tag{12-67}$$

For $\lambda < \lambda_K$ we find from the relation $v_\lambda/\lambda \approx v_K/\lambda_K$ that the acceleration is $v_\lambda^2/\lambda \approx (\lambda/\lambda_K)(v_K^2/\lambda_K) < a_K$. Also, from (12-59) the eddy accelerations are found to decay with size as $\lambda^{-1/3}$ when $\lambda > \lambda_K$. Therefore, (12-67) provides a reasonable estimate of the maximum turbulent acceleration experienced by particles smaller than λ_K.

For particle radii less than $10\ \mu m$, the viscous relaxation time β^{-1} is generally at least two orders of magnitude smaller than τ_K. Accordingly, we may estimate the magnitude v' of the velocity response of a particle of radius r and mass m to local turbulent accelerations by assuming a_K to be constant in time, so that by Stokes law

$$v' \approx \frac{ma_K}{6\,\pi\eta_a r} = \beta_r^{-1}a_K, \tag{12-68}$$

from (12-2). Similarly then, the induced relative velocity of a pair of particles of radii r and s (r > s) in close proximity (separation less than λ_K, so that both particles experience the same acceleration) is $(\beta_r^{-1} - \beta_s^{-1})a_K$. Therefore, if we assume a geometric collision cross section, the collection kernel for this process of turbulent 'inertial' coagulation is

$$K_{TI}(r, s) \sim \pi(r + s)^2(\beta_r^{-1} - \beta_s^{-1})a_K = \pi(r + s)^2(\beta_r^{-1} - \beta_s^{-1})\varepsilon^{3/4}v_a^{-1/4} \tag{12-69}$$

(Levich, 1954b). Essentially the same result has been obtained also by Saffman and Turner (1956), with the slight difference that in their expression π is replaced by the numerical factor 5.7.

The relative strength of the two turbulent coagulation processes is given by the ratio

$$\frac{K_{TI}}{K_{TS}} \approx \frac{\rho_P(r - s)}{2\,\rho_a\lambda_K} \approx \frac{10^3(r - s)}{\lambda_K}. \tag{12-70}$$

The appearance of the factor (r − s) emphasizes that, according to the models presented, collisions between equal sized particles may occur as a result of turbulent shear, but not as a result of turbulent accelerations. For a small particle size ratio we find the two processes are comparable when $r \approx 10^3\ \lambda_K$; e.g., $r \approx 10^{-1}\ \mu m$ for $\varepsilon = 5\ cm^2\ sec^{-3}$, and $r \approx 1\ \mu m$ for $\varepsilon = 10^3\ cm^2\ sec^{-3}$. For larger particles the inertial coagulation process becomes dominant.

According to the model of Saffman and Turner, the proper way to describe the overall rate of turbulent coagulation due to the simultaneous action of turbulent accelerations and shearing motions is by taking the square root of the sum of the squares of the separate rates.

In deriving (12-65) and (12-69) for turbulent coagulation and (12-55) for laminar shear coagulation we assumed that the viscous, hydrodynamic interaction between particles undergoing relative motion in close proximity is negligible. We now turn to a brief discussion of a coagulation mechanism, relative sedimentation under gravity, where it is known that such effects are quite important, and in fact reduce the effective collision cross section below the geometric value by factors smaller than 10^{-1} for particle radii less than $10 \, \mu$m. The implication of these results for the other models of coagulation is considered briefly in turn.

12.6.3   GRAVITATIONAL COAGULATION

To date the most accurate estimates of the effective collision cross section of small particles settling under gravity have been obtained by use of a slip-corrected Stokes flow model in which the mutual interference of the particles and fluid (air) are fully accounted for. This rather complicated approach is discussed in Chapter 14, especially Section 14.4.3, in the context of drop collisions. In this section we shall describe a much simpler and less accurate model for the process, but one which nevertheless is adequate for the purpose of illustrating the effects of the viscous interaction between small particles.

The model is due to Fuchs (1951) and Friedlander (1957). It involves the basic assumptions that: (i) The flow is in the Stokes regime. (ii) The small particles near a large one move as if in a stream caused by the air flow around the large one in isolation ('superposition scheme – see also Section 14.3). More precisely, the velocity of a small particle of radius r_2 relative to a larger one of radius r_1 is taken as $\vec{V}_{S,2} - \vec{u}_{S,1}(\vec{r})$, where $\vec{V}_{S,2}$ is the terminal velocity of the small particle in isolation, and $\vec{u}_{S,1}(\vec{r})$ is the Stokes velocity field which would be induced at the location $\vec{r}$ of the center of the small particle if the large one were falling at its terminal velocity in isolation. The center of the large particle is taken as the origin for $\vec{r}$.

Therefore, in the notation of Section 12.5 the r_2-particle current density relative to the r_1-particle is

$$\vec{j} = (\vec{V}_{S,2} - \vec{u}_{S,1}(\vec{r}))n_2, \tag{12-71}$$

and, with the flow streaming past the r_1-particle in the positive z-direction, the rate of collection by the r_1-particle is given by the positive flux J_G of $\vec{j}$ into the lower hemisphere ($z \leqslant 0$) of radius r_{12} concentric with the origin:

$$J_G = -\int_{S_H} \vec{j} \cdot d\vec{S} = n_2 \int_{S_H} \vec{u}_{S,1}(\vec{r}) \cdot d\vec{S} + n_2 V_{S,2} \int_{S_H} \hat{e}_z \cdot d\vec{S}, \tag{12-72}$$

where S_H denotes the lower hemisphere. The last term on the right side of (12-72) is simply $-\pi r_{12}^2 n_2 V_{S,2}$. The first term on the right side is due to motion following the streamlines, and is found most easily by using the Stokes stream function ψ_S, (10-35), which we write in the form

$$\psi_S = \frac{V_{S,1} r^2 \sin^2 \theta}{2}\left(1 - \frac{3r_1}{2r} + \frac{r_1^3}{2r^3}\right), \tag{12-73}$$

where $V_{S,1}$ is the terminal velocity of the r_1-particle and θ is the polar angle between $\vec{r}$ and the positive z-axis. The streamline which approaches to within a distance r_{12} from the origin when $\theta = \pi/2$ is at a distance δ from the z-axis far upstream of the origin, where from (12-73) we must have $\delta^2 = r_{12}^2(1 - 3\, r_1/2\, r_{12} + r_1^3/2\, r_{12}^3)$. By the same kind of reasoning the incompressibility of the flow requires that $\int_{S_H} \vec{u}_{S,1} \cdot d\vec{S} = \pi V_{S,1}\delta^2$, so that the first term in J_G is $\pi n_2 V_{S,1} r_{12}^2 [1 - 3\, r_1/2\, r_{12} + r_1^3/2\, r_{12}^3]$. This represents the attachment rate by the effect known as 'direct interception.' If we now combine the two contributions to J_G, and note that the corresponding gravitational collection kernel is $K_G = J_G/n_2$, we obtain finally

$$K_G(r, s) = \pi A(1 + p)^2 r^4 \left(1 - \frac{3}{2(1+p)} + \frac{1}{2(1+p)^3} - p^2\right), \quad (12\text{-}74)$$

$$\underset{r > s}{}$$

where $A = (2\, g/9\, \eta_a)(\rho_P - \rho_a)(1 + \alpha N_{Kn})$ (cf. (12-22)), and we have introduced the 'p-ratio' defined by $p \equiv s/r$ (for $s < r$).

It is clear from the assumptions underlying the superposition scheme that (12-74) is strictly valid only for $p \ll 1$; in that limit (12-74) reduces to

$$K_G(r, s) = \frac{\pi}{2} Ar^4 p^2 = \frac{\pi}{2} s^2 V_{S,r}, \quad (12\text{-}75)$$

$$\underset{p \ll 1}{}$$

which says the collection rate vanishes with the radius of the smaller particle. Of course this is not a rigorous result since we have neglected Brownian motion, which becomes increasingly significant for the motion of the smaller particle as $s \to 0$. The function K_G should also vanish as s approaches r, since in that limit the Stokes terminal velocities are equal. In order to obtain an approximate form for K_G which exhibits the correct behavior for p near zero and unity, Friedlander (1965) generalized (12-75) to read as follows:

$$K_G(r, s) = \frac{\pi}{2} Ar^4 p^2 (1 - p^2) = \frac{\pi}{2} s^2 (V_{S,s} - V_{S,r}). \quad (12\text{-}76)$$

It is apparent that if there were no hydrodynamic deflection of approaching particles, the collection rate would be controlled simply by the relative velocity of approach and the geometric collision cross section. It is customary to describe the effect of the hydrodynamic interaction on the collection rate by use of the concept of *collision efficiency*, $E = E(r, s)$, defined to be the ratio of the actual cross section to the geometric cross section for a pair of interacting particles (see also Section 14.2). Then we can also express K_G as follows:

$$K_G(r, s) = \pi(r + s)^2 E(r, s)[V_{S,r} - V_{S,s}]. \quad (12\text{-}77)$$

On comparing this with (12-76), we obtain

$$E(r, s) = \frac{p^2}{2(1+p)^2}, \quad (12\text{-}78)$$

for the collision efficiency according to the model of Fuchs (1951) and Friedlander (1957). Table 12-2 shows that this simple result agrees surprisingly well with the more elaborate computations referred to earlier. (Although in principle E

TABLE 12-2
Collision efficiency for spherical particles of
radii less than 10 μm. (From Equation (12-78)
and from values computed by M. H. Davis,
1972.)

p	E Equation (12-78)	E Davis (1972)
0.2	1.4×10^{-2}	1.3×10^{-2}
0.4	4.1×10^{-2}	2.7×10^{-2}
0.6	7.0×10^{-2}	3.4×10^{-2}
0.8	9.9×10^{-2}	3.4×10^{-2}

depends on the absolute size of the particles as well as the p-ratio – see Section 14.2, in practice the size dependence becomes insignificant for $r \lesssim 10 \mu$m. This is physically reasonable since for sufficiently small particles the motion is controlled by viscous forces, the role of particle inertia being negligible.)

Since Table 12-2 shows that for $r \lesssim 10 \mu$m the effective collision cross section for gravitational coagulation is generally less than three percent of the geometric value, it is appropriate to reconsider briefly the assumption that $E = 1$ for laminar shear, turbulent shear, and turbulent inertial coagulation. Particularly for this last process the model used is close to the gravitational coalescence model, the only obvious difference being that the relative velocity of approach in the turbulent case is due to the terminal velocity difference of particles 'falling' in the constant acceleration field a_K instead of g. However, in the derivation leading to (12-69) a_K was regarded as constant only in the sense that $\beta_r^{-1} \ll \tau_K$. Besides β_r^{-1}, another relevant time scale is the characteristic time of hydrodynamic interaction. If this were much smaller than τ_K also, the analogy between the gravitational and turbulent models would be complete and we should then use values of E such as those in Table 12-2 for the case of turbulent coagulation also.

An estimate of the hydrodynamic interaction time t_i is given by dividing some representative interaction length by the relative velocity of approach, i.e., $t_i \sim n(p)ra_K^{-1}(\beta_r^{-1} - \beta_s^{-1})^{-1}$, where $n(p)$ is a number of order 1 to 10^2 which increases with increasing p (the mutual interference of particles and fluid becomes of longer effective range as the radii become comparable.) Therefore, for small p we have $t_i/\tau_K \sim r\beta_r/\tau_K a_K = r\beta_r/v_K$, where v_K is the microscale of velocity, typically of order 1 cm sec^{-1}. Since $r\beta_r \approx 10^2 \text{ cm sec}^{-1}$ and 1 cm sec^{-1} for $r = 0.1 \mu$m and 10μm, respectively, we see that generally the particle relative velocities and accelerations will not remain constant in direction over the time of hydrodynamic interaction. This means the turbulent inertial coagulation process is similar to one of 'gravitational' coagulation in which gravity is allowed to vary randomly in strength and direction during interaction, at least over a limited range of values, with a characteristic frequency of order τ_K^{-1}. The consequences of such a model have not been worked out. (Sedunov (1960, 1963, 1964) has made a detailed study of the relative motion of small particles in turbulent flow, but this has not resolved the question of what is the effective collision cross section.)

From the above discussion we must conclude that accurate estimates of E for turbulent coagulation processes are unavailable. However, we must also not forget that in the atmosphere turbulent and gravitational coagulation occur simultaneously. And, as we discuss in Section 14.5, the problem of the enhancement of gravitational coagulation by turbulence has been treated recently in a fairly comprehensive manner (de Almeida, 1975, 1976). de Almeida considered the motion of particles (cloud drops) generally much larger than those found in the atmospheric aerosol. However, the trend of his results indicates that particles with $r \lesssim 10 \mu m$ might well experience as much as a two orders of magnitude increase in the gravitational-collision efficiency even for quite weak levels of turbulence (see Figure 14-8). This gives at least some further support to the simple results of this section, although it is clear that much remains to be learned about the coagulation of aerosols in a turbulent medium (see also the criticism of de Almeida's formulation near the end of Section 14.5).

In Figure 12-5 a comparison is made of the collection kernels for Brownian, turbulent shear, turbulent inertial, and gravitational coagulation according to (12-51) (with $\alpha = 0$), (12-66), (12-69), and (12-76), respectively. The dotted portions of the curves for turbulent shear and turbulent inertial coagulation indicate regions where the assumption of a geometric collision cross section is not likely to be accurate. According to the figure, Brownian coagulaton is most important for particles of about one micron radius or smaller, while turbulence and especially gravity control coagulation for particles larger than a few microns, assuming a modest energy dissipation rate of $5 \, \text{cm}^2 \, \text{sec}^{-3}$.

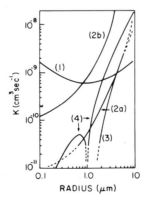

Fig. 12-5. Comparison of various collision mechanisms for a spherical particle of density $1 \, \text{g cm}^{-3}$ and radius $1 \, \mu m$ interacting with a second spherical particle of radius r (dotted lines: regions where the assumption of a geometric collision cross section is not likely to be accurate). (1) Brownian coagulation, (2a) turbulent shear coagulation ($\varepsilon = 5 \, \text{cm}^2 \, \text{sec}^{-3}$), (2b) turbulent shear coagulation ($\varepsilon = 1000 \, \text{cm}^2 \, \text{sec}^{-3}$), (3) turbulent inertial coagulation, ($\varepsilon = 5 \, \text{cm}^2 \, \text{sec}^{-3}$), (4) gravitational coagulation (sedimentation). (Adapted from Klett, 1975 and from Hidy, 1973.)

12.7 Scavenging of Aerosols

The mechanisms discussed in the previous sections of this chapter eventually cause aerosol particles to be lost from the atmosphere. Since none of the mechanisms discussed so far has involved cloud or precipitation particles, they are often collectively referred to as 'dry removal processes'. We shall now turn our attention to various 'wet removal processes,' by which aerosol particles may become attached to, or 'scavenged' by, cloud drops and ice crystals, some fraction of which may subsequently fall to earth as precipitation. In our discussions here we shall consider the attachment mechanisms of (1) convective diffusion, (2) thermophoresis and diffusiophoresis, (3) turbulent shear and turbulent inertial capture, and (4) gravitational or 'inertial' capture. The incorporation of aerosol particles into cloud liquid water or ice phases by nucleation is considered in detail in Chapters 9 and 13. Some electrical influences are discussed in Chapter 17.

From (12-50) we see that the loss rate of aerosol particles per unit volume of air by virtue of scavenging by cloud drops is given by

$$-\frac{\partial n(r, t)}{\partial t} = n(r, t) \int_0^\infty K(r, a) n_d(a, t) \, da, \qquad (12\text{-}79)$$

where a denotes drop radius, and $n_d(a, t) \, da$ is the number of drops per unit volume of air at time t in the size interval a to $a + da$. The fractional depletion rate of the aerosol concentration by scavenging is called the *scavenging coefficient*, Λ. From (12-79) we thus have

$$-\frac{1}{n} \frac{\partial n(r, t)}{\partial t} \equiv \Lambda(r, t) = \int_0^\infty K(r, a) n_d(a, t) \, da. \qquad (12\text{-}80)$$

The scavenging problem is therefore basically one of determining the collection kernel $K(r, a)$ for the various attachment processes of interest. If the drop distribution does not vary with time then Λ becomes constant in time, and we have the simple result that

$$n(r, t) = n(r, 0) \, e^{-\Lambda(r)t}. \qquad (12\text{-}81)$$

12.7.1 SCAVENGING BY CONVECTIVE DIFFUSION

Stationary drops capture aerosol particles by simple Brownian diffusion. The discussion of Section 12.5 is therefore directly applicable to this process. Often however, the fall velocity of drops is large enough so that there is a significant enhancement of the diffusion rate caused by the convection of particles relative to the drop. We shall therefore briefly discuss the mechanism of scavenging by forced convective diffusion.

Suppose particles in concentration n are transported by Brownian motion and by virtue of being suspended in a medium moving with velocity $\vec{v}$. The particle current density is

$$\vec{j} = -D\nabla n + n\vec{v}, \tag{12-82}$$

so that for the case of a steady state and constant diffusivity the condition of particle continuity ($\nabla \cdot \vec{j} = 0$) yields the following governing equation for n:

$$\vec{v} \cdot \nabla n = D\nabla^2 n. \tag{12-83}$$

This is the steady state convective diffusion equation. To obtain (12-83) we have assumed the flow to be incompressible.

Let us now discuss the use of this equation in estimating the rate of attachment of small aerosol particles to a spherical drop of radius a falling with velocity $\vec{U}_\infty$. If we adopt the usual point of view of a reference frame moving with the drop (regarded as a rigid sphere), the boundary conditions for the problem are $\vec{v}|_{drop} = 0$, $\vec{v}|_\infty = -\vec{U}_\infty$, $n|_{drop} = 0$, and $n|_\infty = n_0$, where n_0 is the concentration of particles far from the drop. On using the same simple scaling arguments which demonstrated the significance of the Reynolds number in Section 10.2.1, we find the relative strengths of the particle diffusion and convective transport processes are measured by the dimensionless *Peclet number*, N_{Pe}:

$$\frac{|\vec{v} \cdot \nabla n|}{|D\nabla^2 n|} \propto \frac{V_\infty n_0/d}{Dn_0/d^2} = \frac{U_\infty d}{D} \equiv N_{Pe}, \tag{12-84}$$

where $d = 2a$ is the drop diameter. (Although the radius a would be a more natural choice for the characteristic length in (12-84), we use d in order to conform with the conventional definition of N_{Pe} without having to keep track of factors of two.) Also, the Peclet number may be expressed in terms of the Reynolds number and the particle *Schmidt number*, N_{Sc}:

$$N_{Pe} = N_{Sc}N_{Re}, \quad (12\text{-}85a) \qquad \text{with} \quad N_{Sc} \equiv \frac{\nu}{D}. \tag{12-85b}$$

As can be seen from Table 12-1, the Schmidt number for aerosols is generally quite large. Therefore, the Peclet number for convective diffusion of aerosols will be large for essentially all cloud droplets, even though they may have very small Reynolds numbers. We recall from Section 10.2.4 that for $N_{Re} \gtrsim 1$ there is a momentum boundary layer of characteristic thickness $\delta_u \sim N_{Re}^{-1/2}$. From the analogous definitions of N_{Re} and N_{Pe} (from (12-84) and (12-85) we see N_{Pe} is the 'Reynolds number' for particle diffusion) and the fact that $N_{Sc} \gg 1$, we can anticipate finding a *diffusion boundary layer*, whose characteristic thickness δ_D is a monotonic decreasing function of N_{Pe}. At distances from the drop surface smaller than δ_D, diffusion dominates, while beyond δ_D convection controls the particle concentration.

Let us now use scaling arguments to find the form of δ_D for the case of small drops, with $N_{Re} \ll 1$. The procedure consists simply of exploring the consequences of the fact that both sides of (12-83) have comparable magnitudes at a distance δ_D from the drop surface. Consider a small region of the flow where δ_D is assumed to be well defined, letting x denote distance along the drop surface in the direction of the local flow, and y denote distance normal to the surface. Then we may write $\tilde{u} \cdot \nabla n = u_x \, \partial n / \partial x + u_y \, \partial n / \partial y$. Both terms on the right side are of comparable magnitude, by the condition $\nabla \cdot \tilde{u} = 0$. So it suffices to estimate the term $u_x \, \partial n / \partial x$ at $y = \delta_D$. For small N_{Re} flow, u_x will scale up to the free stream value in a characteristic distance $y \sim d$, so that if $\delta_D \ll d$ we expect $u_x(y = \delta_D) \approx U_\infty(\delta_D/d)$; i.e., the velocity shear is linear for small y. On the other hand, the characteristic length for the gradient in x is just d. So we estimate $u_x \, \partial n / \partial x \sim U_\infty n_0 \delta_D / d^2$ at $y = \delta_D$. For the right side of (12-83) we have the obvious estimate $D \nabla^2 n \approx D \, \partial^2 n / \partial y^2 \sim D n_0 / \delta_D^2$, at $y = \delta_D$. On equating these estimates we find $\delta_D^3 \sim D d^2 / U_\infty$, or

$$\frac{\delta_D}{d} \sim N_{Pe}^{-1/3}, \qquad N_{Re} \ll 1. \tag{12-86}$$

According to (12-86) the concentration gradient in the neighborhood of the droplet, and hence the particle flux to it, will be enhanced by a factor proportional to $N_{Pe}^{1/3}$ over the magnitude due to pure diffusion, for $N_{Pe} \gg 1$. Therefore, an interpolation formula which gives a fairly good approximation to the total convective Brownian diffusion flux $J_{CD} = \int_S \vec{j}_{CD} \cdot d\vec{S}$ for arbitrary N_{Pe} and $N_{Re} \ll 1$ is

$$J_{CD} = J_0(1 + A N_{Pe}^{1/3}), \qquad N_{Re} \ll 1, \tag{12-87}$$

where J_0 is the pure steady state diffusive flux $(= 4 \pi a D n_0$ from (12-36), assuming $r \ll a)$ and A is a dimensionless positive constant. The analyses of convective diffusion to a sphere in Stokes flow by Friedlander (1957), Baird and Hamielec (1962), Levich (1962), and Ruckenstein (1964) have resulted in the estimates $A = 0.45, 0.50, 0.50$, and 0.52, respectively.

For larger droplets with $N_{Re} > 1$ the results are modified somewhat. For this case there exists a momentum boundary layer as well as the diffusion boundary layer. The relevant inequalities are $\delta_u \ll d$ and $\delta_D \ll \delta_u$, the latter one arising from $N_{Sc} \gg 1$. Therefore, in estimating the convective term $u_x \, \partial n / \partial x$, we note that u_x scales up to U_∞ in a distance $y \sim \delta$, rather than $y \sim d$ as before, so that $u_x \, \partial n / \partial x \sim U_\infty n_0 \delta_D / d \delta_u$ at $y = \delta_D$. The diffusion term in (12-83) is still characterized by $D \nabla^2 n = D n_0 / \delta_D^2$, which on being set equal to the convective term leads to $U \delta_D / d \delta \sim D / \delta_D^2$, or

$$\frac{\delta_D}{\delta_u} \sim N_{Sc}^{-1/3} \quad (12\text{-}88a), \qquad \frac{\delta_D}{d} \sim N_{Sc}^{-1/3} N_{Re}^{-1/2}. \tag{12-88b}$$

It is thus seen that for aerosol particle diffusion, δ_D is typically less than one-tenth the thickness of the momentum boundary layer. The corresponding interpolation formula for the total particle flux has the form

$$J_{CD} = 4 \pi a D n_0 (1 + B N_{Re}^{1/2} N_{Sc}^{1/3}), \qquad N_{Re} \gg 1. \tag{12-89}$$

Unfortunately, no experiments are available to verify the low Reynolds number result ((12-87)) and the constants of proportionality determined by Friedlander, Baird and Hamielec, Levich, and Ruckenstein. On the other hand, experiments carried out at high Reynolds numbers by Steinberger and Treybal

(1960), Rowe *et al.* (1965), and Gibert *et al.* (1972) showed that over limited Reynolds number ranges one indeed may express the convection effect in accordance with (12-89) for $0.6 \leqslant N_{Sc} \leqslant 3 \times 10^3$, the value of B ranging between 0.25 and 0.50 due to experimental scatter. Recently, investigators of scavenging and contact nucleation mechanisms (Slinn and Hales (1971), Young (1974)) have used the value B = 0.30, which derives from the correlation of data presented by Ranz and Marshall (1952) (also in Bird *et al.* (1960)). In Young's study aerosol collection rates for $10 \, \mu m$ droplets are presented. This corresponds to the case $N_{Re} \ll 1$, so that in principle it would have been better to use an expression of the form of (12-87) rather than (12-89). However, the quantitative differences are probably not significant.

Numerical computations of convective diffusion to a sphere have been carried out by Woo (1971) for $1 \leqslant N_{Re} \leqslant 300$ and $0.25 \leqslant N_{Sc} \leqslant 5$. Woo found a flux enhancement factor of $J_{CD}/J_0 \approx a N_{Sc}^b$, where $a = 1.552 + 3.41 \times 10^{-2} N_{Re} - 1.17 \times 10^{-4} N_e^2 + 1.83 \times 10^{-7} N_{Re}^3$ and $b = 0.198 N_{Re}^{0.096}$. Unfortunately, Woo did not extend his calculations to higher values of N_{Sc}.

It should be noted that the small values of N_{Sc} considered by Woo correspond to the case where the diffusing particles are of molecular size; then the diffusivity for momentum transfer, namely ν (the kinematic viscosity of the medium), is of the same magnitude as that for mass transfer (both diffusivities being characterized by the mean free path and thermal speed of the molecules), so that $N_{Sc} = O(1)$. This means that Woo's results are especially pertinent to another problem of interest in cloud physics, namely that of determining the evaporation rate of a falling drop. This problem is discussed in detail in Section 13.2.3. As pointed out there, it is customary in the cloud physics literature to describe the enhancement of the water mass flux from a falling drop in terms of the mean ventilation coefficient, $\bar{f}$. In the present context the ventilation coefficient for diffusing aerosol particles is simply the ratio J_{CD}/J_0. Thus by extrapolating the ventilation factors determined through drop evaporation experiments to higher values of N_{Sc}, we have available alternative expressions to the interpolation formulas (12-87) and (12-89). In particular, we note the wind tunnel investigation of drop evaporation by Beard and Pruppacher (1971b), in which close agreement with Woo's theoretical results for $N_{Sc} = 0.71$ was achieved. The values of $\bar{f}$ obtained by Beard and Pruppacher are given in (13-57) and (13-58). According to Beard (1974a), these values agree with the results of Woo to within a few percent for the case of water vapor diffusion in air ($N_{Sc} = 0.71$), and to within $\pm 20\%$ when extrapolated, along with Woo's results, to $5 \leqslant N_{Sc} \leqslant 10^5$.

The collection kernel corresponding to the total particle flux per drop, J_{CD}, is just $K_{CD} = J_{CD}/n_0$, or, by including the effects of ventilation

$$K_{CD} = 4 \pi a D \bar{f} \qquad (12\text{-}90)$$

(cf. (12-46)). The principal dependence on particle size comes from the factor D. The corresponding scavenging coefficient may be determined from (12-80), given the drop spectrum $n_d(a, t)$. As a simple example, we shall obtain an approximate expression for the scavenging coefficient Λ_B for particles of radius $r \ll a$ for the case of negligible convective enhancement of Brownian diffusion, and assuming the Khrgian-Mazin drop spectrum (given by Equations (2-2), (2-6), and (2-7)). The result is

$$\Lambda_B = 4\,\pi D \int_0^\infty a\,n_d(a)\,da \approx \frac{1.35\,w_L D}{\bar{a}^2},\qquad\qquad (12\text{-}91)$$

where w_L is the liquid water content of the cloud in $g\,cm^{-3}$, and $\bar{a}$ is average drop radius in cm (Sax and Goldsmith, 1972). Considering (12-81), we see that the corresponding half-life of the aerosol particles (the time $t_{1/2}$ such that $n(t_{1/2}) = n_0/2$) is

$$t_{1/2} \approx \frac{(\ln 2)\bar{a}^2}{1.35\,w_L D}.\qquad\qquad (12\text{-}92)$$

For example, for $w_L = 10^{-6}\,g\,cm^{-3}$, $\bar{a} = 10\,\mu m$, and $D = 2.3 \times 10^{-4}\,cm^2\,sec^{-1}$ (corresponding to $r = 0.01\,\mu m$, $T = -5°$, and $p = 600\,mb$), we get $t_{1/2} \approx 38\,min$; for $15\,°C$ and $1000\,mb$, $D = 1.6 \times 10^{-4}\,cm^2\,sec^{-1}$ and $t_{1/2} \approx 55\,min$. Thus, if we assume that the total concentration of Aitken particles is typically about $10^4\,cm^{-3}$, a number of this magnitude is predicted to be scavenged by the cloud per cm^3 of cloud air, corresponding to $O(10^{10})$ particles absorbed per cm^3 of cloud water, within the span of one hour. These estimates are consistent with the field observations of Rosinski (1966, 1967a) as extrapolated to small particles by Vali (1968a) (see Section 8.3).

A review of the presently available literature shows that little attention, if any, has been paid to the problem of Brownian capture of aerosol particles by ice crystals. There seems to be little reason for this neglect as this problem can readily be treated in a manner completely analogous to the diffusion of water vapor to ice crystals by means of the electrostatic analogue technique discussed in Section 13.3.1.

12.7.2   SCAVENGING BY THERMOPHORESIS AND DIFFUSIOPHORESIS

The phenomena of thermo- and diffusiophoresis were predicted by Stefan (1873) and subsequently observed by Facy (1955, 1958, 1960). Quantitative studies using various experimental techniques have been carried out by Goldsmith et al. (1963), Goldsmith and May (1966), Waldmann and Schmitt (1966), and Vittori and Prodi (1967). The theoretical aspects of these mechanisms have been discussed in detail by Waldmann and Schmitt (1966), Slinn and Shen (1970), Slinn and Hales (1971), and Derjaguin and Yamalov (1972).

Thermophoresis is the name given to the motion of particles caused by a kind of thermally induced (radiometric) force, which arises from the non-uniform heating of particles due to temperature gradients in the suspending gas. This phenomenon naturally depends strongly on the Knudsen number of the particles. For $N_{Kn} \gg 1$ the mechanism of thermophoresis is relatively simple: The temperature gradients cause the gas molecules to deliver a greater net impulse on the 'warm' side of the particle than on the 'cold' side, thus driving it in the direction of colder gas temperatures. For $N_{Kn} \ll 1$ the organization of this mechanism is somewhat more involved. In this limit one can imagine a portion of the aerosol particle surface layer, large compared to λ, over which a temperature gradient is established. The layer of gas closest to this surface will acquire a temperature gradient which conforms approximately to that of the surface, which means that gas molecules from the hotter direction will impart a greater impulse to the surface locally than those from the colder direction. Thus the entire particle can experience a force against the temperature gradient in the gas.

The mechanism of thermophoresis involves gas motion relative to the particle surface, which means theoretical models for the case $N_{Kn} \ll 1$ must abandon the usual hydrodynamic boundary condition of no slip. What is often done for this case is to assume the flow is described by continuum hydrodynamics with slip-flow boundary conditions. The analysis of this type which appears to yield results of widest application is due to Brock (1962), who obtained the following approximate expression for the thermophoretic force $\vec{F}_{Th}$ on a particle of radius r in air:

$$\vec{F}_{Th} = - \frac{12 \, \pi \eta_a r (k_a + c_t k_p N_{Kn}) k_a \nabla T}{5 \, (1 + 3 \, c_m N_{Kn}) (k_p + 2 \, k_a + 2 \, c_t k_p N_{Kn}) p}, \tag{12-93}$$

where p and ∇T denote gas pressure and temperature gradient, k_a and k_p are the air and particle thermal conductivities, η_a is the dynamic viscosity of air, and c_m and c_t are phenomenological coefficients known as the 'isothermal slip' coefficient and the 'temperature jump' coefficient. These coefficients depend on T as well as air and surface properties. Comparison with other theories and experiment indicates that (12-93), which includes friction slip, is fairly accurate through the entire interval $0 < N_{Kn} < \infty$, except for a high conductivity aerosol. Further details may be found in Waldmann and Schmitt (1966) and Hidy and Brock (1970).

For a quasi-steady state the thermophoretic velocity is easily obtained by setting $\vec{F}_{Th}$ equal to the slip-flow corrected Stokes drag, $6 \, \pi \eta_a r \vec{v}_{th} (1 + \alpha N_{Kn})^{-1}$, acting on the particle. Therefore, the thermophoretic velocity of a particle in air may be written as

$$\vec{v}_{Th} = - \frac{B k_a \nabla T}{p}, \tag{12-94}$$

where

$$B = \frac{0.4 \, (1 + \alpha N_{Kn}) (k_a + 2.5 \, k_p N_{Kn})}{(1 + 3 \, N_{Kn}) (k_p + 2 \, k_a + 5 \, k_p N_{Kn})}, \tag{12-95}$$

and where α is given by (12-16). In (12-95) the values $c_t = 2.5$ and $c_m = 1.0$ have been adopted, following Brock (1962). Note that $\vec{v}_{Th}$ is proportional to the heat flux vector, $\vec{j}_h = -k_a \nabla T$ (see Section 13.1.2).

With these results we may write down the thermophoretic flux of particles in concentration n_0 to a stationary evaporation drop of radius a. Since the thermophoretic particle flux vector is $\vec{j}_{Th} = n_0 \vec{v}_{Th}$, the total particle flux $J_{Th} =$

$\int_S \vec{j}_{Th} \cdot d\vec{S}$ becomes

$$J_{Th} = \frac{4 \pi a^2 n_0 B j_h}{p},$$ (12-96)

where j_h is the magnitude of the heat flux vector ($= k_a(T_\infty - T_a)/a$). For a falling water drop, we may obtain an approximate description of the particle flux by multiplying (12-96) by an appropriate ventilation factor for convective heat transfer. Because of the complete mathematical analogy between problems of convective heat and mass transfer, a reasonable approach is to set $\bar{f}_h = \bar{f}_v(D_v \to \kappa_a)$; i.e., we may use the ventilation factor for vapor transfer if in (13-57) and (13-58) we replace the Schmidt number $N_{Sc,v} = \nu_a/D_v$ for vapor diffusion by the *Prandtl number* $N_{Pr} = \nu_a/\kappa_a$ for heat diffusion, where κ_a is the thermal diffusivity of air (see Section 13.1.2). Finally then, an approximate expression for the collection kernel for scavenging by thermophoresis is

$$K_{Th} = \frac{4 \pi a B \bar{f}_h k_a(T_\infty - T_a)}{p},$$ (12-97)

for $T_\infty > T_a$ (evaporating drop); for $T_\infty < T_a$, $K_{Th} = 0$. Note that K_{Th} depends on r only through the factor B.

Diffusiophoresis refers to aerosol particle motion induced by concentration gradients in a gaseous mixture. Unlike the case of thermophoresis, both continuum and non-continuum effects contribute to this phenomenon. The continuum contribution is due to *Stephan flow*, which can be regarded as a hydrodynamic flow of the medium which compensates for a diffusive flow of some constituent(s) (such as water vapor). The non-continuum contribution arises from gas slippage along a particle surface due to concentration gradients in some constituent(s); this mechanism is thus analogous to the processes responsible for thermophoresis.

For detailed discussions of diffusiophoresis, see Waldmann and Schmitt (1966) and Hidy and Brock (1971). For our purposes here we merely quote from Waldmann and Schmitt (p. 151) the following approximate result for the diffusiophoretic velocity $\vec{v}_{Df}$ of a particle in stagnant air through which water vapor is diffusing:

$$\vec{v}_{Df} = -\frac{M_w^{1/2}}{(x_v M_w^{1/2} + x_a M_a^{1/2})} \frac{D_v}{x_a} (\nabla x_v)_\infty,$$ (12-98a)

for $N_{Kn} \gtrsim 1$ ($r \lesssim \lambda_a$), and

$$\vec{v}_{Df} = -(1 + \sigma_{va} x_a) \frac{D_v}{x_a} (\nabla x_v)_\infty,$$ (12-98b)

for $N_{Kn} < 1$ ($r > \lambda_a$). The experiments of Schmitt and Waldmann (1960), Schmitt (1961), and Goldsmith and May (1966) showed that the values for $\vec{v}_{Df}$ computed from (12-98a) agree well with observed values for particles of $r \lesssim \lambda_a$. For particles of $r > \lambda_a$, the experiments suggested that $\vec{v}_{Df}$ has to be computed from (12-98b) with $\sigma_{va} \approx -0.26$, i.e. $(1 + \sigma_{va} x_a) \approx 0.74$ for $x_v \lesssim x_a \approx 1$. This value of σ_{va} is supported by the earlier studies of Kramers and Kistenmacher (1943) who proposed on theoretical grounds that for large particles in air through which water vapor diffuses, $\sigma_{va} = (M_w^{1/2} - M_a^{1/2})/(x_v M_w^{1/2} - x_a M_a^{1/2}) \approx -0.21$, from which $1 + \sigma_{va} x_a \approx 0.79$ for $x_a \approx 1$. Thus we may tentatively conclude that one may use (12-98b) with $\sigma_{va} \approx -0.3$ to -0.2.

The diffusiophoretic force $\vec{F}_{Df}$ exerted on an aerosol particle by water vapor diffusing through stagnant air may be obtained by inserting (12-98b) for the diffusiophoretic velocity into the slip flow corrected Stokes drag, $6\pi\eta_a \vec{v}_{Df} r(1 + \alpha N_{Kn})^{-1}$, and setting this drag equal to $\vec{F}_{Df}$. One then finds the diffusiophoretic force on an aerosol particle may be expressed approximately as

$$\vec{F}_{Df} = -\frac{6\pi\eta_a r(1 + \sigma_{va} x_a) D_v(\nabla x_v)_\infty}{(1 + \alpha N_{Kn}) x_a}. \tag{12-99}$$

By introducing the vapor density ρ_v into (12-98b) and assuming $x_v \ll x_a$, we find that also

$$\vec{v}_{Df} \approx (1 + \sigma_{va} x_a) \frac{M_a}{M_w \rho_a} \vec{j}_v, \tag{12-100}$$

where $\vec{j}_v = -D_v \nabla \rho_v$ is the water vapor mass flux vector. Note that the direction of motion induced by diffusiophoresis is the same as that of the vapor flux. Since the diffusiophoretic particle flux vector is $\vec{j}_{Df} = n_0 \vec{v}_{Df}$, where n_0 is the ambient concentration of aerosol particles, the total particle flux $J_{Df} = \int_S \vec{j}_{Df} \cdot d\vec{S}$ becomes after using (13-5) and (12-100)

$$J_{Df} = 4\pi a n_0 D_v(1 + \sigma_{va} x_a) \frac{M_a}{M_w \rho_a}(\rho_{v,\infty} - \rho_{v,a}). \tag{12-101}$$

Finally, an approximate expression for the corresponding collection kernel, including a ventilation factor $\bar{f}_v$ to account for drop motion, is

$$K_{Df} = 4\pi a D_v \bar{f}_v(1 + \sigma_{va} x_a) \frac{M_a}{M_w \rho_a}(\rho_{v,\infty} - \rho_{v,a}), \tag{12-102}$$

for $\rho_{v,\infty} > \rho_{v,a}$ (condensing drop); for $\rho_{v,\infty} < \rho_{v,a}$, $K_{Df} = 0$. Note that this kernel depends on particle size only through the factor $\bar{f}_v$. We see from (12-97) and (12-102) that thermophoresis and diffusiophoresis have opposing effects on the scavenging behavior of a drop.

Young (1974) has carried out a numerical evaluation of expressions similar to (12-90), (12-97), and (12-102) for various atmospheric conditions. His results for the case of a water drop of 10 μm radius evaporating at 98% relative humidity, 600 mb, and $-5\,°C$, and for a drop growing at a supersaturation of 0.3%, 600 mb, and $-5\,°C$, are displayed in Figure 12-6. Note that in both cases the effects of thermophoresis overpower those of diffusiophoresis if $r < 1\,\mu$m. This agrees also with the predictions of Slinn and Hales (1971). Note also that both phoretic scavenging processes depend only slightly on the particle size if $0.01 \leqslant r \leqslant 1\,\mu$m, and that above a certain particle size the phoretic effects dominate Brownian motion. The net effect of all three processes on the scavenging rate of aerosol particles has been computed by Young on the assumption that the individual

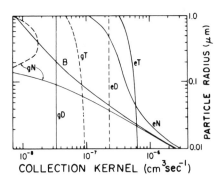

Fig. 12-6.   Collection kernel (cm³ sec⁻¹) for a water drop of $a = 10\,\mu$m in air of 600 mb and $-5\,°C$, collecting aerosol particles by Brownian diffusion (B), thermophoresis (T), or diffusiophoresis (D); (N) is net effect; for 0.3% supersaturation (g), 98% relative humidity (e). (From Young, 1974; by courtesy of Amer. Meteor. Soc., and the author.)

collection kernels may simply be added together. It is questionable whether this assumption is justifiable in the light of the coupling which exists between Brownian diffusion of particles and particle motion due to phoretic forces. On the other hand, the experiments of Goldsmith and May (1966) and the theoretical considerations of Annis and Mason (1975) show that for water vapor diffusing in air under atmospheric conditions, for which $x_v \ll x_a$, the thermo- and diffusiophoretic forces are additive, i.e., thermo- and diffusiophoretic effects are not coupled.

The results of Young, displayed in Figure 12-7, show that by considering convective Brownian diffusion and phoretic effects only, the collection kernel, and hence the scavenging efficiency of a cloud drop of given size, decreases with increasing particle size; also, for an aerosol particle of given size the scavenging efficiency increases with increasing drop size. The values shown in Figure 12-7

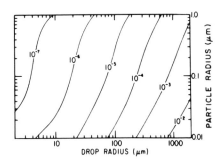

Fig. 12-7.   Collection kernel (cm³ sec⁻¹) for a water drop in air collecting aerosol particles by Brownian diffusion and phoretic forces as a function of aerosol and drop radii for drops evaporating at 600 mb, $-5\,°C$, and 98% relative humidity. (From Young, 1974; by courtesy of Amer. Meteor. Soc., and the author.)

indicate that for particles of $r \geq 1\ \mu$m in concentrations of $10\ cm^{-3}$ and in air of 600 mb, $-5\ °$C, and 98% relative humidity, 10^2 to 10^5 aerosol particles are scavenged by water drops of 10 to $1000\ \mu$m radius within 10 min, in fair agreement with the field observations of Rosinski (see Section 8.3). However, these estimates do not include the effects of inertial impaction, and thus most likely underestimate the number of aerosol particles scavenged.

Due to the complicated shapes of atmospheric ice particles, no theoretical estimates are available for the corresponding rates at which they scavenge aerosol particles as a result of phoretic forces. However, Vittori and Prodi (1967) studied experimentally the scavenging of relatively large aerosol particles by large ice crystals growing by diffusion in a stagnant, water saturated environment. Unfortunately, in these experiments the ice crystals were suspended on small fibers, so that the latent heat released during vapor deposition was very likely conducted away through the structure supporting the crystals. In this manner thermophoresis must have become suppressed to such an extent that diffusiophoresis dominated, since aerosol particles were captured by the ice crystal during its diffusional growth. However, by analogy with the behavior of drops we would expect that under conditions where ice crystals are freely falling, thermophoresis should dominate diffusiophoresis for $r < 1\ \mu$m. Thus, one must expect that in supercooled clouds, where ice crystals grow at the expense of the surrounding supercooled water drops, the dominating thermophoretic effects will tend to drive submicron particles to the droplets.

Finally, we should comment on the experimentally observed dust-free space in the neighborhood of an evaporating water drop (Facy, 1955, 1958, 1960; Goldsmith *et al.*, 1963; Vittori and Prodi, 1967), which implies the dominance of diffusiophoresis over thermophoresis. To properly interpret this observation in the present context one must consider that some of these experiments were carried out with tracer particles of radii larger than $1\ \mu$m, for which diffusiophoresis does overwhelm thermophoresis. In other experiments the energy for vaporization possibly was supplied by modes other than conduction from the surrounding gas, such as by radiation from illumination sources or by transient conduction through the structures used to support the drop. This again would tend to reduce the effect of thermophoresis relative to diffusiophoresis. Both experimental peculiarities favor diffusiophoresis to be the dominant process, explaining the transport of aerosol particles away from the evaporating drop. However, such behavior is not generally expected under natural conditions in the atmosphere.

12.7.3 SCAVENGING BY TURBULENCE

The approximate description of turbulent shear and turbulent inertial coagulation given in Section 12.6.2, and the comparison of these effects with Brownian and gravitational coagulation in Figure 12-5 for spherical particles of $0.1 \leq r \leq 10\ \mu$m, should apply also to the problem of turbulent scavenging of aerosol particles by small drops of radii $\leq 10\ \mu$m. For larger drops with significant fall speeds, one must account also for the possible convective enhancement of the particle flux.

Although no rigorous treatment of this problem exists, several simple and plausible approaches suggest themselves. For example, we may attribute the turbulent shear collection kernel (12-65) to a process of diffusion characterized by the constant diffusion coefficient D_{TS}. By comparing the forms of the collection kernels for Brownian diffusion, (12-46), and turbulent shear, we find $D_{TS} \approx a^2(\varepsilon/\nu)^{1/2}/4\,\pi$, for the case $r \ll a$. Then by assuming the independence of Brownian and turbulent diffusion, we may represent the net effect of both by the diffusion coefficient $D_{net} = D_{TS} + D$. From (12-90) the corresponding collection kernel for convection-enhanced turbulent-shear and Brownian diffusion scavenging is estimated to be $K_{net} \approx 4\,\pi a\,D_{net}\bar{f}(D_{net})$. An approach of this kind has been followed by Williams (1974).

Aside from the modification of the ventilation factor, the preceding example illustrates the superposition assumption which characterizes nearly all formulations of the scavenging problem: one assumes the collection kernel for scavenging by various processes which occur simultaneously is given by the sum of the kernels for the individual processes as they occur in isolation. A variation on this theme has been provided by Saffman and Turner (1956), whose model results suggest the resultant kernel is better represented by taking the square root of the sum of the squares of the individual kernels. We feel it is presently impossible to judge which of these approaches is quantitatively superior. In principle, a more rigorous approach would be to superpose the various forces acting on the particles, and then proceed to work out the resulting collection kernel by determining the particle relative trajectories. Unfortunately, this approach usually involves great computational difficulties, and these may not be worth confronting in view of the fact that even the forces arising from the simultaneous acton of various scavenging mechanisms may not be accurately known.

Let us now make one final observation concerning the influence of turbulence on scavenging. We should note that the fall velocities of drops relative to the particle aerosol will enhance the turbulent scavenging rate not only through ventilation (convective diffusion) effects, but also because of gravitational or inertial capture. As we mentioned in Section 12.6.3, the problem of the enhancement of gravitational coagulation by turbulence has been formulated in a relatively comprehensive manner by de Almeida (1975). Unfortunately, however, de Almeida did not carry out computations for the situation of relevance here, namely for the collection of particles typically with radii $\leqslant 1\,\mu$m by drops of radii $\geqslant 10\,\mu$m. Nevertheless, from the trend of his results for small drop sizes and size ratios (see Section 14.5), it would appear that weak to moderate levels of turbulence, in conjunction with Brownian diffusion, can greatly increase the efficiency of scavenging by gravitational or inertial impaction. This is clearly another problem area in need of further study.

12.7.4 SCAVENGING BY GRAVITATIONAL OR INERTIAL IMPACTION

We are concerned here with the collection process introduced in Section 12.6.3, applied now in the context of small aerosol particles being collected by relatively enormous cloud particles. Assuming a sticking efficiency of unity, which is consistent with the experiments of Weber (1968, 1969), the collection efficiency problem reduces to that of determining the collision efficiency E(r, a). (Whether a captured insoluble particle remains on the surface of a drop or becomes completely immersed depends on the contact angle, defined in Section 5.4. See McDonald (1963b) for a theoretical study of this effect.) Since generally we have $r \ll a$, E(r, a) may be determined by using the superposition scheme of hydro-dynamic interaction (see Sections 12.6.3 and 14.3). Of course, once E(r, a) is known the corresponding collection kernel K_G may be obtained from (12-77).

The first theoretical estimates of E(r, a) were made by Langmuir and Blodgett (1946) and Langmuir (1948). Their computations were later extended and im-proved by Mason (1957) and Fonda and Herne (in Herne, 1960). The method used by these authors is based on a scheme which allows interpolating between potential flow, assumed to characterize the flow past a large collector drop, and Stokes flow, assumed to characterize the flow past a large collector drop. The flow past aerosol particles was considered to have negligible effect. Recently, Beard and Grover (1974) have computed improved values of E(r, a) by using the numerically determined flow fields of Le Clair et al. (1970) (see Section 10.2.6) for spherical drops of $1 \le N_{Re} \le 400$, corresponding to $40 \le a \le 600 \mu m$. Their formulation involves the standard superposition scheme, described in detail in Section 14.3, and they also include the refinement of invoking the slip-flow correction (12-16) for the drag on the aerosol particle. Thus their equation of motion for a particle of mass m and velocity $\vec{v}$ is

$$m\frac{d\vec{v}}{dt} = m\vec{g}^* - \frac{6 \pi \eta_a r}{1 + \alpha N_{Kn}}(\vec{v} - \vec{u}),$$ (12-103)

where $\vec{g}^* = \vec{g}(\rho_P - \rho_a/\rho_P)$ is the buoyancy-corrected acceleration of gravity, and $\vec{u}$ is the flow field past the drop, evaluated at the location of the particle. Beard and Grover worked with a dimensionless form of this equation, given by (14-10) and (14-11). In the latter equation the 'inertia parameter' or *Stokes number* N_S acquires the form $N_S = p^2 \rho_P N_{Re}(1 + \alpha N_{Kn})/9 \rho_a$, where $p = r/a$.

The values of E(r, a) found by Beard and Grover for particle radii $r \ge 1 \mu m$, subject to $r/a \le 0.1$, are intermediate to previously calculated values for the potential and Stokes flow limits, but do not follow the Langmuir interpolation formula mentioned above. For the particle sizes considered E(r, a) was found to increase monotonically with increasing r for $1 \le N_{Re} \le 400$. However, a later extension of these computations down to $r \approx 0.1 \mu m$ by Beard (1974b) revealed a minimum in the curve of E(r, a) versus r, near $r \approx 0.5 \mu m$ for $N_{Re} \ge 20$ and $\rho_P = 1 g cm^{-3}$. Beard's computations for rigid spheres were extended by Grover (1978) to spherical water drops with internal circulation ($1 \le N_{Re} \le 200$), and to particles with $\rho_P = 2 g cm^{-3}$. The results of Grover, shown in Figure 12-8, agree with Beard's computations in revealing a minimum in E(r, a) for $0.3 \le r \le 0.5 \mu m$, and for drops with $N_{Re} = 100$ and 200. Although the decrease in E with decreasing r is readily understandable on considering the effects of particle inertia, the increase in E for decreasing r less than $r = r(E_{min})$ may seem strange at first sight. Beard suggested that for drops with $N_{Re} \ge 20$ the latter finding may be explained on the basis of small particle capture by the standing eddy on the

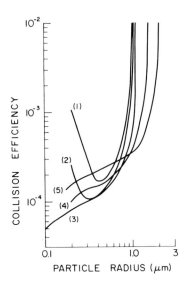

Fig. 12-8. Numerically computed efficiency with which water drops collide by inertial impaction with aerosol particles in air of 10 °C and 900 mb; (1) $N_{Ne} = 200$ ($a = 438\,\mu$m), (2) $N_{Re} = 100$ ($a = 310\,\mu$m), (3) $N_{Re} = 30$ ($a = 173\,\mu$m), (4) $N_{Re} = 10$ ($a = 106\,\mu$m), (5) $N_{Re} = 4$ ($a = 72\,\mu$m). (From Grover, 1978; by courtesy of the author.)

downstream side of the drop, which is aided by gravity pulling the particle towards the drop (*rear capture*). (Note this form of wake capture is quite distinct from another type which involves falling drops of comparable size. In the latter situation the trailing drop becomes drawn into the wake of the leading one and eventually collides with it – see Sections 14.2 and 14.6.) The effectiveness of the rear-capture process should obviously increase with decreasing particle inertia and with increasing eddy size, thus explaining the shape of the collision efficiency curves in Figure 12-8 for $N_{Re} > 30$. Grover's (1978) more refined computations generally support the notion of the existence of a rear-capture process; however, his results indicate that its onset occurs for $N_{Re} \geqslant 30$ rather than for $N_{Re} \geqslant 20$ as suggested by Beard (see Figure 12-8). Note that the curves (4) and (5) cross over curves (1), (2), and (3), indicating that in a narrow particle size range drops with $N_{Re} = 4$ and 10 exhibit a slightly larger collision efficiency than drops of $N_{Re} = 30$, 100, and 200. This result is a reflection of the 'interception effect' discussed in Section 12.6.3, where we showed that for interacting particles of sufficiently small size the collision efficiency is controlled by the p-ratio ($p = r/a$), and increases with increasing p-ratio (i.e., with decreasing drop size for given r).

Theoretical values of E for $N_{Re} \geqslant 400$ have not been computed since numerical solutions for flow fields past drops exclude the phenomena of eddy shedding and drop oscillations which are known to occur for such sizes (recall Section 10.2.6). On the other hand, several experimental studies have been carried out for this Reynolds number range. The results of these experiments provide considerable insight into the behavior of E, although they cannot be considered to uniquely reflect hydrodynamic effects since the experimental setups did not allow a

complete elimination of phoretic and Brownian diffusion effects, and since in
most of these studies the electric charge on the drop and aerosol particles was
not controlled. Thus, the experiments of Oaks (1960) with oil droplets, paraffin
particles, and ammonium chloride particles of 0.5 to 3.7 μm diameter, of Starr
and Mason (1966) and Starr (1967) with pollen of 4 to 12 μm diameter, of Adam
and Semonin (1970) with rod-shaped spores of 0.7 μm diameter and 1.2 μm
length, of Dana (1970) with dye particles of 0.4 to 7.5 μm diameter, and of
Hampl *et al.* (1971) and of Kerker and Hampl (1974) with silver chloride particles
of 0.3 to 1.2 μm diameter, indicate qualitatively that the efficiency with which
these particles are collected by water drops larger than about 600 μm equivalent
radius decreases with increasing drop size. Toulcova and Podzimek (1968) and
Beard (1974b) conjectured this may indicate the existence of a maximum for E
which is determined by the drop size for which eddy shedding begins. Indeed,
one would expect that with increasing turbulence in the wake of a falling drop,
rear capture of aerosol particles would become less likely. The existence of such
a maximum has qualitatively been verified by Starr and Mason (1966) and in a
quantitative manner by Wang and Pruppacher (1977b). The latter experimentally
determined the efficiency with which aerosol particles of almost uniform size
(r $\approx$ 0.25 μm) were captured by drops of 100 $\leq$ a $\leq$ 2500 μm falling through an
aerosol chamber in which the relative humidity was 23% and the temperature
22 °C. The attainment of terminal velocity by the drops was assured by allowing
the largest of them to fall through an enclosed 33 m-long shaft. The results of
Wang and Pruppacher's experiments are plotted in Figure 12-9. Note that the

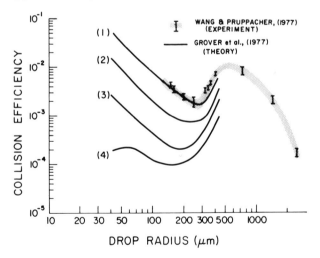

DROP RADIUS (μm)

Fig. 12-9. Variation of the efficiency with which water drops collide with aerosol particles of
r = (0.25 ± 0.03) μm, ρ_P = 1.5 g cm^{-3} in air of 1000 mb, as a function of drop size and relative humidity
ϕ_v of the air. Experimental results are due to inertial forces, phoretic forces, and Brownian diffusion, in
air of ϕ_v = 23%, T = 22°C, and p = 1000 mb. Theory considers inertial and phoretic effects only, (for
such a particle size the collection efficiency due to Brownian diffusion is about one order of
magnitude lower than that due to phoretic effects – see Figure 12-6); (1) ϕ_v = 20%, (2) ϕ_v = 75%,
(3) ϕ_v = 95%, (4) ϕ_v = 100%. (From Wang and Pruppacher, 1977b; by courtesy of Amer. Meteor. Soc.,
and the authors.)

collision efficiency appears to be a maximum for drops of radii between 500 and 600 μm, i.e., of Reynolds number between 300 and 400, in good agreement with the Reynolds number range at which shedding of eddies begins (see Section 10.2.6).

Since no theoretically computed values are available for the efficiency with which aerosol particles collide with ice crystals, it is necessary to rely on experimental observations. A number of field and laboratory studies on the scavenging efficiency of snow crystals have been reported. Although all of these studies involved rather large ice crystals and aerosol particles, so that inertial impaction should have been the dominating scavenging mechanism, Brownian diffusion and phoretic effects must have nevertheless contributed to the collection of particles, in particular since these were collected in an ice-subsaturated atmosphere. Sood and Jackson (1969, 1970) studied 1 to 10 mm diameter ice crystals of various shapes collecting latex particles, NaCl particles, and spores of diameters between 0.1 and 1.01 μm. In analogy with the behavior of water drops, Sood and Jackson found that for a given aerosol particle size the collection efficiency of ice crystals larger than 1 mm diameter decreased with increasing crystal size, while for a given ice crystal size the collection efficiency decreased with decreasing aerosol particle radius to reach a minimum value at $r \approx 0.5 \, \mu$m. Due to rear capture the collection efficiency increased again for aerosol particles of less than 0.5 μm radius. Similar experiments have been carried out more recently by Magono *et al.* (1974). Unfortunately, however, their values for the collection efficiency do not agree well with those of Sood and Jackson. In this regard it should be noted that both of these investigations suffered from a lack of control of the electric charge on the ice crystals and aerosol particles.

A more extensive investigation of the efficiency with which natural snow crystals scavenge aerosol particles has recently been carried out by Knutson *et al.* (1976) for five different types of aerosol particles (B. subtilis, P 1600 pigment, zinc sulfide, polystyrene latex, and anthracene), for $0.25 \leqslant r \leqslant 3.5 \, \mu$m, and for planar type natural ice crystals or snowflakes of radii 1 to 10 mm. From their observations Knutson *et al.* found that, independently of ice crystal shape, the collection efficiency E_c of the ice crystals could be correlated with a correlation coefficient of $\sim$0.8 to the interception parameter d_P/d_c by the relation

$$\log_{10} E_c = 2.477 + 1.366 \log_{10} (d_P/d_c),$$

where d_P is the particle diameter and d_c is the ice crystal diameter. Unfortunately again, neither the electric charge on the ice crystals nor that on the aerosol particles was known.

Qualitatively, one would expect that ice crystals should be more efficient scavengers of aerosol particles than water drops for a given cloud- or precipitation water content, because of their larger surface-to-volume ratio. Field measurements by Graedel and Franey (1976) have verified this expectation.

12.7.5 OVERALL SCAVENGING EFFECTS

An interesting result is obtained when the effects of various scavenging mechanisms are added together. For scavenging processes in the atmosphere, this was first done by Greenfield (1957), who considered Brownian diffusion, turbulent shear diffusion, and inertial impaction. He found that the overall scavenging coefficient exhibited a strong broad minimum for aerosol particles between about 0.1 and 1.0 μm radius. In the literature this minimum is therefore often referred to as the 'Greenfield gap.' It is the result of Brownian diffusion dominating particle capture for $r < 0.1$ μm, and of inertial impaction dominating capture for $r > 1$ μm. Later investigators (e.g., Slinn and Hales (1971), Pilat and Prem (1976), and Wang and Pruppacher (1978), who included contributions from Brownian diffusion, inertial capture, as well as thermo- and diffusiophoresis) have obtained similar results. Figure 12-10 shows the gap as estimated by Slinn and Hales. Results from the more recent gap studies of Grover *et al.* (1977) and Wang *et al.* (1978) are reproduced in Figures 17-8 to 17-10. In these studies the particle trajectory method of Beard and Grover (1974) (see Equation (12-103)) was extended and combined with the particle flux method of Wang *et al.* (1978), discussed in Section 17.5.6. Note from these figures that the phoretic processes tend to fill in the gap quite significantly, but it remains distinct nonetheless. Note also from Figures 17-8 and 17-9 that rear capture of sufficiently small particles by both hydrodynamic and phoretic effects tends to maintain the gap phenomenon.

The theoretical results of Grover *et al.* (1977) are compared with experiments in Figure 12-9. Note from this figure (curves 1 to 3) that, for a given particle size,

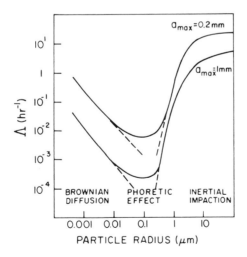

Fig. 12-10. Scavenging coefficient as a function of aerosol particle size for Brownian diffusion, inertial capture, and thermo- and diffusiophoresis; $\Delta T = T_\infty - T = 3\,°C$, precipitation rate $R = 10$ mm hr^{-1}; raindrop size distribution $n(a)\,da = (10^{-4}\,R/6\,\pi a_{max}^7)a^2 \exp(-2\,a/a_{max})\,da$, with R in cm sec^{-1}; drop terminal velocity $V_\infty = 8000\,a$ (sec^{-1}) with a in cm, collision efficiency for inertial impaction based on values of Zimin (1964). (From Slinn and Hales, 1971; by courtesy of the authors.)

E exhibits a minimum not only in a plot of E vs. r for a given a but also in a plot of E vs. a for a given r. The good agreement between the experimental results of Wang and Pruppacher (1977b) and the theoretical predictions of Grover et al. (1977) allows one to conclude that the increase of E with decreasing drop size for $a < a(E_{min})$ is a result of the increasingly pronounced phoretic effects as the flow field past the drop becomes weaker, while the increase of E with increasing drop size for $a > a(E_{min})$ is a result of the increasingly pronounced hydrodynamic effects which allow particles to be captured by the growing, standing eddy in the rear of the drop.

It has been pointed out by McDonald (1964) that the atmosphere does provide a mechanism to 'bridge the Greenfield gap,' in that the aerosol particles of radii between 0.1 and 1 μm are precisely those which most readily serve as cloud condensation and ice nuclei. As such they may be removed from the atmosphere if cloud formation is followed by precipitation. It also seems likely that turbulence and electrical effects will tend to fill in the gap. Thus Grover et al. (1977) and Wang et al. (1978) demonstrated this filling-in effect for the case of scavenging by Brownian diffusion, inertial- and phoretic forces, and because of the presence of electric charges and external electric fields. These results will be discussed in Section 17.5.6. However, at the present time no definitive assessment of the overall problem including turbulence exists.

Let us now briefly consider the problem of obtaining a reasonable upper bound Λ_{max} for the net scavenging coefficient in a precipitating cloud. As we have seen, several processes contribute to the overall effect. In view of the fact that only approximate descriptions exist for most of them, even when they occur in isolation and involve only idealized particles (for example, spheres or infinite cylinders of homogeneous properties), we can expect to achieve only a rough estimate for Λ_{max}.

In spite of the complexities of the problem, an important tentative conclusion may be drawn from the many idealized case studies which have been carried out in recent years (e.g., Slinn and Hales (1971), Dingle and Lee (1973), Crandall et al. (1973), Young (1974), Williams (1974)). These imply that in-cloud scavenging by convective, precipitating systems may be regarded basically as a two-stage process. In the first, all the scavenging mechanisms, and especially nucleation, Brownian and turbulent diffusion, and diffusio- and thermophoresis, serve typically to transfer a major fraction of the aerosol particles (assumed here to be characterized by radii $r \leqslant 1 \mu$m) to the cloud water, predominately the fraction comprised of small droplets and/or ice crystals which together possess most of the total absorbing surface of the cloud water, within a relatively brief time period of the order of 20 min or less. In the second stage, this 'polluted' cloud water is scavenged primarily through inertial capture by relatively large precipitating cloud particles (raindrops, snowflakes, and graupel or soft hail particles). This accretion process can cause a major fraction of the aerosol-containing cloud water to fall out of the cloud in a time period similar to, though generally somewhat larger than, that for stage one. Thus, net fractional depletions of aerosol particle concentrations of the order of unity may occur within the span of one hour, corresponding to an overall scavenging rate of 10^{-3} to 10^{-4} sec^{-1}.

This overall theoretical conclusion is quite consistent with the observation, based on experiments in the field, that in-cloud scavenging operates with approximately the same efficiency as the processes which convert water vapor to precipitation (Dingle, 1975).

This description of the nature of in-cloud scavenging, in which the second stage is generally rate-limiting for the entire process, suggests that a reasonable upper bound for the effective scavenging rate is given by the accretion rate of small cloud particles by precipitation, with the former assumed to have already absorbed most of the aerosol particle mass through stage-one processes. We shall proceed in the sequel to obtain a simple estimate of this kind. Clearly, such a formulation will also represent an upper bound for the case of below-cloud scavenging ('washout') as well, since the efficiency of the accretion process increases rapidly with the size of the collected particles. We shall also ignore possible evaporation losses between cloud and ground. Though evaporation typically reduces the water flux at the ground to about half of that at cloud base, usually only a small fraction of this water loss is caused by drops which evaporate completely, allowing their captured particles to escape again into the air, before being 'recycled' through further drop collection and breakup events. Finally, we shall consider only the case of a 'warm' cloud with no ice phase, since terminal fall velocities and collision cross sections for most types of cloud ice particles are not well known.

Our objective then is to evaluate (12-80) for a collection kernel of the form of (12-77), all in the context of the collection of droplets of radius a' by precipitation drops of radius $a \gg a'$. An evaluation of the collection kernel for the drop-droplet accretion process is discussed in Section 15.2.3.2, and the resulting plot of $K(a', a)$ versus for various a'/a is shown in Figure 15-8. An interesting feature of the curves is a surprisingly weak dependence of K on a'/a. Thus, for $10 \leq a \leq 500 \, \mu m$, K varies over about eight orders of magnitude for fixed a'/a, whereas for fixed a it changes by less than one order of magnitude for $0.2 \leq a'/a \leq 0.9$. This implies that Λ_{max} is only a weak function of a'.

Let us now assume a steady state rain spectrum according to the empirical description of Marshall and Palmer (1948), given by (2-11) but expressed here in different units:

$$n(a) = n_0 \exp(-\alpha a), \tag{12-104}$$

where a is in cm, $n_0 = 1.6 \times 10^5 \, m^{-3} \, cm^{-1}$, $\alpha = 82 \, R^{-0.21} \, cm^{-1}$, and where R is the rain rate in mm hr^{-1}. In view of the results shown in Figure 15-8, we may now combine (12-77), (12-80), and (12-104) to obtain the following simple estimate for the maximum possible scavenging coefficient, which is independent of both a' and t:

$$\Lambda_{max} \approx \pi n_0 \bar{E} \int_0^\infty a^2 U_\infty(a) \, e^{-\alpha a} \, da, \tag{12-105}$$

where $U_\infty(a)$ denotes the drop terminal velocity. To obtain this expression we have made use of the fact that K is well represented by its form for $a'/a \ll 1$; $\bar{E}$ thus represents a characteristic collision efficiency for drop-droplet interactions, and is approximately unity (see Figure 14-7). Also, because of the factors a^2 and $e^{-\alpha a}$ in (12-105), we may represent $U_\infty(a)$ adequately by its variation over the range $100 \leqslant a \leqslant 500 \ \mu m$, namely $U_\infty \approx ca$, where c is a function of p and T (see Figure 10-20).

We may eliminate reference to the parameters n_0 and c by introducing the definition of rain rate. When expressed as a mass flux of rain, the rain rate is $R' = (4 \pi \rho_w/3) \int_0^\infty a^3 n(a) U_\infty(a) \, da$, if we assume the absence of an appreciable updraft. If R' is given in cgs units (i.e., as $g \, cm^{-2} \, sec^{-1}$), then the rain rate in $mm \, hr^{-1}$ is $R = 3.6 \times 10^4 \, R'/\rho_w$ (cf. (2-9)). Therefore, we find

$$\frac{\Lambda_{max}}{R} \approx \frac{\bar{E}}{4.8 \times 10^4} \frac{\displaystyle\int_0^\infty a^3 e^{-\alpha a} \, da}{\displaystyle\int_0^\infty a^4 e^{-\alpha a} \, da} = \frac{\bar{E}}{4.8 \times 10^4}\left(\frac{\alpha}{4}\right),$$

which on substituting $\alpha = 82 \, R^{-0.21}$ becomes

$$\Lambda_{max}(sec^{-1}) \approx 4.2 \times 10^{-4} \, \bar{E} R^{0.79}. \tag{12-106}$$

A similar dependence of Λ on R for the rain scavenging of large particles is evident from the early numerical calculations of Chamberlain (1953). Furthermore, excellent agreement between (12-106) and the results of a numerical computation by Crandall $et \ al.$ (1973) of $\Lambda(R, a')$ vs R for $a' \geqslant 10 \ \mu m$ can be achieved by choosing $\bar{E} = 0.83$.

It is interesting to consider whether there are any conditions under which one might expect a significantly larger scavenging coefficient than (12-106). This might be expected to occur in a situation where, for example, strong turbulence and/or electrical forces cause the rapid 'self-collection' of drops of similar size, in addition to an enhanced accretion rate of small drops by large ones. In this case the particle-to-water mixing ratio would be a weaker function of drop size, so that the scavenging rate could be estimated simply by the rate of total water depletion from the precipitating volume. For example, if we suppose this volume is characterized by a vertical extent H and a horizontal cross section A, and that its liquid water content is w_L, then it contains a water mass of order AHw_L. If the rain rate is R', this amount of mass evidently will cross the bottom surface of the volume in a time τ, where $AR'\tau \approx AHw_L$. Hence, in this case we estimate the scavenging rate to be $\Lambda'_{max} \approx \tau^{-1} \approx R'/Hw_L = \bar{v}/H$, where $\bar{v} \equiv R'/w_L$ is the mass-weighted average velocity of precipitation. For the distribution (12-104) we find $w_L(g \, m^{-3}) = 8.9 \times 10^{-2} R^{0.84}$, so that $\bar{v}(m \, sec^{-1}) = 3.1 R^{0.16}$ and

$$\Lambda'_{max} \approx \frac{3 \, R^{0.16}}{H}, \tag{12-107}$$

for H in meters.

It may seem puzzling to note from (12-106) and (12-107) that Λ_{max} exceeds Λ'_{max} if $R^{0.6} \gtrsim 10^4/H$. This points up the fact that for sufficiently large rain rates, the accretion of particle-laden cloud water by rain occurs in a time period which is smaller than the time needed for fallout of the rain from the cloud volume. For such rain rates, (12-107) provides a better (and smaller) estimate for the scavenging coefficient. As an example, if $H = 1$ km then (12-106) becomes unrealistically large for $R \gtrsim 50$ mm hr^{-1}.

12.8 Explanations for the Observed Size Distribution of the Atmospheric Aerosol

From our studies of the mechanics of aerosol particles we are now in a position to consider some models which have been put forth to explain the observed regularities in the size distributions of tropospheric aerosols (recall Section 8.2.10).

12.8.1 QUASI-STATIONARY DISTRIBUTIONS (QSD)

A particularly simple and physically appealing way of dealing with the problem of the steady state aerosol particle distribution n(r) was introduced by Friedlander (1960a, b), who proposed the theory of *quasi-stationary distributions* (QSD). This theory is based on the assumption that for $r \gtrsim 0.1\,\mu$m the aerosol has attained a state of dynamic equilibrium between Brownian coagulation and gravitational sedimentation. The theory further assumes that the form of n(r) is completely determined by the two parameters characterizing the process rates for coagulation and sedimentation, and by the rate ε_m at which matter enters the upper end of the spectrum by coagulation of smaller particles. The QSD theory is somewhat analogous to the theory of turbulence in the inertial subrange, discussed in Section 12.6.2. There the assumption was made that $v_\lambda = v_\lambda(\lambda, \varepsilon)$, where ε measures the energy flow rate down the eddy size spectrum. Similarly, according to QSD an aerosol in dynamic equilibrium is characterized principally by ε_m, the flow rate of matter passing up the aerosol size spectrum.

Friedlander delineates two subranges for n. At the lower end of the equilibrium range, the 'coagulation subrange' for which $0.1 \leqslant r \leqslant 0.5\,\mu$m, he assumes sedimentation to be negligible. Since the concentration of particles in this subrange is much larger than in the range $r > 1\,\mu$m, practically all the matter being transferred up the spectrum over the coagulation subrange will do so by Brownian coagulation rather than by inertial impaction with the larger sedimentation particles. Therefore, for the coagulation subrange it is reasonable to assume that $n = n(r, C, \varepsilon_m)$, where $C = 2 kT/3\,\eta_a$ is the characteristic coagulation parameter (cf. (12-51)). If L^3 denotes a characteristic air volume and ℓ denotes a characteristic spectral length (particle radius), then on dimensional grounds one finds for the units of n, r, C, and ε_m: $[n] = L^{-3}\ell^{-1}$, $[r] = \ell$, $[C] = L^3 t^{-1}$, $[\varepsilon_m] = \ell^3 L^{-3} t^{-1}$, where t represents time. It therefore follows from the constraint of dimensional consistency that

$$n = B_1 \left(\frac{\varepsilon_m}{C}\right)^{1/2} r^{-5/2}, \quad 0.1 \le r \le 0.5 \,\mu m, \tag{12-108}$$

where B_1 is a dimensionless constant.

The second subrange is considered to be confined to the upper end of the spectrum, $r \ge 5 \,\mu m$, where it may be assumed that Brownian coagulation is negligible. Thus, matter entering this subrange is lost by sedimentation without significant further transfer within the range by coagulation. Therefore, for this subrange n is assumed to be a function of r, ε_m, and the characteristic sedimentation parameter $A = (2/9)(g/\eta_a)(\rho_P - \rho_a)$ (cf. (12-22)). Since $[A] = L\ell^{-2}t^{-1}$, by dimensional analysis we find

$$n = B_2 \left(\frac{\varepsilon_m}{A}\right)^{3/4} r^{-19/4}, \quad r \ge 5 \,\mu m, \tag{12-109}$$

where B_2 is another dimensionless constant.

The physical basis of this second spectral form appears not be as sound as that for the coagulation subrange. Particle loss by sedimentation from an aerosol volume element requires a vertical gradient of particle concentration, and the characteristic length for this gradient will depend in part on some additional transport mechanism such as turbulent diffusion. The model leading to (12-109) misrepresents the physics of this situation by ignoring the possibility of such an additional independent characteristic length.

Although the power laws obtained for the two particle subranges agree qualitatively with aerosol spectra, it should be recalled from Section 8.2.10 that in the lower troposphere observed particle size spectra are usually better represented by $n \sim r^{-4}$, over a range as large as $0.1 < r < 100 \,\mu m$ (see Section 8.2.10).

In another attempt to check the QSD, Friedlander used the sedimentation subrange to obtain an estimate for ε_m. Assuming $B_2 \approx 1$, which is consistent with a principle of dimensional analysis given by Bridgeman (1931), (12-109) gives

$$\varepsilon_m \sim A n^{4/3} r^{19/3}. \tag{12-110}$$

On substituting data given by Junge (1953) for n and r, Friedlander obtained $\varepsilon_m \approx 10^{-13} \, cm^3 \, cm^{-3} \, sec^{-1}$. Since ε_m represents approximately the rate at which matter enters the upper end of the spectrum by Brownian coagulation of Aitken particles, the time scale for 'processing' Aitken particles by coagulation ($\equiv$ mean residence time of matter in the Aitken range) may be estimated by dividing the volumetic concentration of Aitken particles by ε_m. From the data of Junge (1953) the volumetic concentration is of the order of $10^{-11} \, cm^3 \, cm^{-3} \, sec^{-1}$, corresponding to a characteristic residence time of the order of 10^2 sec. This is a very short time (for example, the half life for Aitken particles of $r = 0.01 \,\mu m$ coagulating by Brownian motion is one order of magnitude larger (recall Section 12.5)) and indicates that the estimate for ε_m is somewhat high. This result is not surprising considering the simplicity of the QSD theory. A numerical study by Storebo (1972) on steady state aerosol distributions indicates that the residence time may easily vary from several minutes to 10^2 hr for realizable conditions, depending primarily on the characteristics of the source of particles at the small size end of the spectrum.

Quasi-stationary distributions arising from a dynamic balance between two or more processes (such as Brownian and turbulent coagulation, gravitational sedimentation, gas to particle conversion, and interactions between aerosol particles and clouds) cannot occur in the atmosphere unless the aerosol source and sink processes operate with time scales which are short compared to meteorological time scales. Model calculations of Junge and Abel (1965) show that the time needed for establishing a steady state among the various aerosol source, modification, and removal processes on a large scale may be as much as 50 days. This is slow compared to the pace of changes induced by varying meteorological conditions, which may easily occur with a time scale of the order of 1 day. On this basis Junge (1969b) concluded that the QSD theory cannot explain the observed r^{-4} dependency of aerosol concentration on size in the lower troposphere for $r > 0.1 \mu m$.

However, under special conditions it is presumably possible for QSD to exist. For example, aerosols in very clean air masses above the planetary boundary layer (i.e., the tropospheric background aerosol; see Section 8.2.8) seem to show a tendency for concentrations to vary like r^{-5} (r^{-k} with $4.5 \leqslant k \leqslant 5$). This is close to (12-109), which may thus approximately describe a global background QSD (see also Section 12.8.5). Also, the upper end of the sea salt particle spectrum within the planetary boundary layer over the ocean may be a QSD between salt particle production and vertical transport by sedimentation and turbulent diffusion (Toba, 1965a, b).

12.8.2 SELF-PRESERVING DISTRIBUTIONS (SPD)

Another plausible approach for explaining the observed regularities of size distributions for tropospheric aerosols involves the notion that they represent asymptotic solutions to the coagulation equation, rather than equilibrium solutions as in the QSD theory. It seems reasonable to expect that as time progresses, a coagulating aerosol might lose its 'birth marks' and acquire a size distribution independent of its initial form. It is also reasonable to anticipate that such asymptotic solutions should have relatively simple forms, which might be investigated by the use of similarity transformations, i.e., a transformation of variables which will reduce the coagulation equation to an equation in only one independent variable. The single variable would then suffice to describe the form of the asymptotic distribution. Following the example of the self-preserving hypothesis used in the theory of turbulence (see, for example, Townsend, 1956), Friedlander (1961) introduced a similarity transformation which forms the basis of his theory of *self-preserving distributions* (SPD). Further development and testing of the theory has been reported in a series of subsequent articles (Swift and Friedlander, 1964; Hidy, 1965; Friedlander and Wang, 1966; Wang and Friedlander, 1967, Friedlander and Hidy, 1969).

The similarity transformation for $n(v, t)$ in the SPD theory is as follows:

$$n(v, t) = g(t)\psi_1\left(\frac{v}{v_+(t)}\right),$$ (12-111)

where g and v_+ are functions of time, and it is assumed that ψ_1, the dimensionless 'shape' of the distribution, does not change with time. The functions g and v_+ can be evaluated to within a constant from any two integral functions of n, such as the zeroth and first moments of the distribution (i.e., the total number of particles per unit volume of air, N, and the volumetric concentration or volume fraction of dispersed phase, ϕ). Thus we have $N = \int_0^\infty n(v, t)\, dv = v_+ g c_1$, and $\phi = \int_0^\infty v n(v, t)\, dv = v_+^2 g c_2$, where c_1 and c_2 are constants since ψ_1 is assumed to be independent of time. On substituting the resulting expressions for v_+ and g into (12-111), we obtain

$$n(v, t) = \frac{c_2 N^2}{c_1^2}\psi_1\left(\frac{c_2 N v}{c_1 \phi}\right) = \frac{N^2}{\phi}\psi(\eta),$$ (12-112a)

where $$\eta = \frac{Nv}{\phi},$$ (12-112b)

and ψ is another suitable dimensionless distribution function. This representation also makes sense on dimensional grounds, since $\phi/N \sim \ell^3$ is a characteristic spectral volume, namely the average particle volume.

The conjecture that (12-111) or (12-112) constitutes a solution can only be tested for a specific kernel K by substitution into the coagulation equation. The SPD theory was developed to apply to situations in which there is no addition or removal of aerosol mass in any elementary volume element (as, for example, by sedimentation), so that a basic condition for the validity of (12-112) is that ϕ be constant in time.

Physically, it is obvious that (12-48) must satisfy the condition $\phi = $ constant. It can also be easily demonstrated by integrating (12-48) multiplied by v:

$$\frac{d\phi}{dt} = \frac{d}{dt}\int_0^\infty dv\, v n(v, t) = \frac{1}{2}\int_0^\infty dv\, v \int_0^v K(u, v - u)n(u, t)n(v - u, t)\, du$$

$$-\int_0^\infty dv\, v n(v, t)\int_0^\infty K(u, v)n(u, t)\, du.$$

If the order of integration is interchanged in the first integral on the right (i.e., $\int_0^\infty dv \int_0^v du\, F(u, v) = \int_0^\infty du \int_u^\infty dv\, F(u, v)$), and this is followed by the substitution $v' = v - u$, the two integrals on the right will cancel, so that $\phi = $ constant as expected.

We would like now to show that (12-112) represents a solution to (12-48) for the case that K is a homogeneous function of its arguments, i.e., $K(au, av) = a^b K(u, v)$. This includes the important cases of Brownian coagulation without the slip correction ($b = 0$), and laminar shear coagulation ($b = 1$). In order to facilitate our discussion here and again in Chapter 15 where more elaborate solutions and solution methods are presented, let us first introduce a dimensionless form of the coagulation equation. Following Scott (1968) and Drake (1972), we write

$$u \equiv v_0 y, \quad v \equiv v_0 x, \quad \tau \equiv K_0 N(0)t, \quad K(u, v) \equiv K_0 \alpha(y, x),$$

$$\tag{12-113}$$

$$v_0 n(v, t) \equiv N(0)f(x, \tau), \quad v_0 n(v, 0) = N(0)f(x, 0) \equiv N(0)f_0(x),$$

where $N(0)$ is the initial total number density, v_0 the initial mean particle volume ($v_0 = \phi/N(0)$), K_0 a normalizing factor with dimensions of volume per unit time, τ the dimensionless time, $f(x, \tau)$ the dimensionless concentration, $\alpha(x, y)$ ($= \alpha(y, x)$) the dimensionless collection kernel, and x and y are dimensionless particle volumes. Substituting (12-113) into (12-48) gives for the dimensionless coagulation equation

$$\frac{\partial f(x, \tau)}{\partial \tau} = \frac{1}{2} \int_0^x \alpha(x - y, y)f(x - y, \tau)f(y, \tau) \, dy - f(x, \tau) \int_0^\infty \alpha(x, y)f(y, \tau) \, dy.$$

$$\tag{12-114}$$

If we denote the moments of f by $M_n(\tau)$, i.e.,

$$M_n(\tau) \equiv \int_0^\infty x^n f(x, \tau) \, dx, \tag{12-115}$$

we find from the condition of aerosol mass conservation ($\phi = \text{const.} = v_0 N(0)$) that $M_1(\tau) = 1$. Therefore, a similarity transformation for $f(x, \tau)$ which is completely equivalent to (12-112) is as follows:

$$f(x, t) = M_0^2(\tau)\psi(\eta), \tag{12-116a}$$

where

$$\eta = M_0(\tau)x. \tag{12-116b}$$

Now let us proceed to show that this transformation is indeed a solution to the coagulation equation for homogeneous kernels. From (12-116a) we have

$$\frac{\partial f}{\partial t} = \left(2\,\psi + \eta \frac{d\psi}{d\eta}\right)M_0 \frac{dM_0}{d\tau}. \tag{12-117}$$

Evidently, an expression for $dM_0/d\tau$ is needed. An ordinary integro-differential equation for any of the moments is easily obtained by integration of the coagulation equation in the manner discussed above for the first moment. The result is

$$\frac{dM_n}{d\tau} = \frac{1}{2} \int_0^\infty \int_0^\infty [(x + y)^n - x^n - y^n]\alpha(x, y)f(x, \tau)f(y, \tau) \, dx \, dy. \tag{12-118}$$

Therefore, we have

$$
\frac{dM_0}{d\tau} = -\frac{1}{2}\int_0^\infty\int_0^\infty \alpha(x, y)f(x, \tau)f(y, \tau)\,dx\,dy
$$

$$
= -\frac{M_0^{2-b}}{2}\int_0^\infty\int_0^\infty \alpha(\eta, \xi)\psi(\eta)\psi(\xi)\,d\eta\,d\xi, \tag{12-119}
$$

in view of the homogeneity of α. Similarly, the right side of (12-114) in similarity variables is

$$
M_0^{3-b}\left\{0.5\int_0^\eta \alpha(\eta - \xi, \xi)\psi(\eta - \xi)\psi(\xi)\,d\xi - \psi(\eta)\int_0^\infty \alpha(\eta, \xi)\psi(\xi)\,d\xi\right\}.
$$

On combining these expressions we see that the similarity transformation does work (i.e., it is successful in eliminating reference to τ), and reduces the coagulation equation to the following ordinary integro-differential equation:

$$
\left(\eta\frac{d\psi}{d\eta} + 2\,\psi\right)\int_0^\infty\int_0^\infty \alpha(\eta, \xi)\psi(\eta)\psi(\xi)\,d\eta\,d\xi
$$

$$
= 2\int_0^\infty \alpha(\eta, \xi)\psi(\eta)\psi(\xi)\,d\xi - \int_0^\eta \alpha(\eta - \xi, \xi)\psi(\eta - \xi)\psi(\xi)\,d\xi \tag{12-120}
$$

(Friedlander and Wang, 1966). Two accompanying integral constraints, corresponding to $M_0(0) = M_1(\tau) = 1$, are as follows:

$$
\int_0^\infty \eta\psi(\eta)\,d\eta = \int_0^\infty \psi(\eta)\,d\eta = 1. \tag{12-121}
$$

In order to gain some idea of the spectral shapes predicted by the SPD theory, let us now solve (12-120) for the simplest case of Brownian coagulation with a constant collision kernel. For $\alpha = $ constant (12-120) reduces to

$$
\eta\frac{d\psi}{d\eta} = -\int_0^\eta \psi(\eta - \xi)\psi(\xi)\,d\xi. \tag{12-122}
$$

The convolution form suggests the use of Laplace transforms, by which the following solution is readily obtained:

$$
\psi = e^{-\eta}. \tag{12-123}
$$

This also satisfies (12-121). The corresponding spectrum function is

$$
f(x, \tau) = M_0^2(\tau)\,e^{-M_0(\tau)x}. \tag{12-124}
$$

To proceed further we must choose an explicit representation for τ, or, in other words, K_0 (see (12-113)). From (12-51) the kernel is $K = 16\,\pi Dr = K_0\alpha$. In order to obtain the best correspondence with the notation for Smoluchowski's discrete solution, (12-43), we choose $K_0 = 8\pi Dr$, and $\alpha = 2$. Then from (12-118) we have $dM_0/d\tau = -M_0^2$, or $M_0 = (1 + \tau)^{-1}$, so that the spectrum function is

$$f(x, \tau) = \frac{e^{-x/(1+\tau)}}{(1 + \tau)^2}. \tag{12-125}$$

This result represents a particular asymptotic solution to the coagulation equation. It corresponds to the initial spectrum $f(x, 0) = \exp(-x)$, but it is expected that any initial distribution would approach (12-125) for large τ. See Wang (1966) for a proof of this assertion for the case of constant α.

Actually, the solution (12-125) was obtained in a similar way some time ago by Schumann (1940), who also conjectured it would be approached asymptotically after a long time, no matter what the initial distribution might be. In support of his conjecture he showed how an initially monodisperse distribution would, by a discrete growth process, adopt a form like (12-125) for small steps and large times. We can easily carry through a similar procedure here, and show that Smoluchowski's discrete solution, (12-43), approaches the self-similar form of (12-123) for large times. It suffices for this purpose to find the limit of $f_k(\tau)/(\Sigma f_k)^2$ as $\tau \to \infty$ for fixed $\eta = k(\Sigma f_k)$ (recall (12-116), and note that $k = v/v_0 = x$). Proceeding in this manner, we find for fixed $\eta = k(1 + \tau)^{-1}$ (recall (12-42)) that

$$\lim_{\tau \to \infty} \left[\frac{f_k(\tau)}{(\Sigma f_k)^2} \right] = \lim_{\tau \to \infty} \left\{ (1 + \tau)^2 \left[\frac{\tau^{k-1}}{(\tau + 1)^{k+1}} \right] \right\}$$

$$= \lim_{\tau \to \infty} \left\{ \left(1 + \frac{1}{\tau} \right)^{1-\eta} \left[\left(1 + \frac{1}{\tau} \right)^\tau \right]^{-\eta} \right\} = e^{-\eta}, \tag{12-126}$$

in agreement with (12-123).

An interesting feature of (12-125) is the indication that SPD due to Brownian coagulation decline faster at large sizes than observed tropospheric spectra. The inclusion of size dependence and a slip correction for the Brownian kernel does not alter this conclusion, as shown by Hidy (1965). He carried out numerical calculations of $n(r, t)$ using (12-51) for $0 \le N_{Kn,0} \le 2$, where $N_{Kn,0}$ is the Knudsen number for initially uniform distributions, and for a variety of initial distributions without slip ($N_{Kn,0} = 0$). In all cases the solutions converged toward SPD, with the exact shape depending on $N_{Kn,0}$. In general, however, the self-similar shape for size-dependent continuum Brownian coagulation was found to approach $e^{-\eta}$ for $\eta > 1$.

Evidently then, the shape of Brownian coagulation SPD does not conform well with tropospheric aerosol distributions. This point has been emphasized by Junge (1969b), who plotted Hidy's SPD for $N_{Kn} = 1$, corresponding to particle radii near $0.1\ \mu m$, on a graph with coordinates $\partial N'/\partial \log_{10} r$ versus $\log_{10} r$, for various values of time. The quantity $N'(r, t)$ is the number of particles per unit volume with radii less than r, i.e., $N'(r, t) = \int_0^r n(r, t)\, dr$. Therefore, we have $\partial N'/\partial r = n(r, t) = n(v, t)\, dv/dr = 4\pi r^2 n(v, t)$, or $\partial N'/\partial \log_{10} r = 3 (\ln 10) vn(v, t)$. Then, since $vn(v, t) = N(0)xf(x, \tau) = N(t)\eta\psi(\eta)$, where $N(t) = N'(\infty, t)$ is the total particle concentration at time t (note also that $N'(r, t) = N(t) - N(r, t)$, where

$N(r, t)$ is given by (8-26)), the desired form for plotting is $\partial N' / \partial \log_{10} r \approx$ $6.9 \, N(t) \eta \psi(\eta)$, where $\eta \psi(\eta)$ represents Hidy's numerical solution for the SPD. The result of this computation is shown in Figure 12-11, where it is assumed that at $\tau = 3$ and $r = 0.1 \, \mu m$, $\partial N' / \partial \log_{10} r$ ($= 4.5 \times 10^3 \, cm^{-3}$) is maximal; this concentration is in accordance with observations under moderately polluted conditions. It is seen that the SPD is approached by $t = 13.4$ days, and subsequently changes with time as indicated in the figure. It is interesting to note that the slope of the envelope curve for the SPD follows an r^{-3} law (which we recall from Chapter 8 is similar to the observed spectral shapes in the lower troposphere). This is a consequence of aerosol volume conservation ($\phi = $ const.), which causes radii for different times to vary as $N^{-1/3}$; i.e., for a time shift of δt the ordinate in Figure 12-11 shifts by $(\partial \log_{10} N / \partial t) \delta t$, whereas since $\eta \sim r^3 N / \phi$, for constant ϕ the abscissa shifts by $-(\partial \log_{10} N / \partial t)(\delta t / 3)$. From the figure it is clear that the shape of the Brownian coagulation SPD differs from observed distributions, especially with respect to the steep slope beyond the maximum. Also, just as for QSD, the time for establishing Brownian SPD is long compared to meteorological time scales. Therefore, although there remains a need for research into the possibilities of SPD for other mechanisms, we may conclude that the occurrence of SPD for Brownian coagulation is unlikely in the tropospheric aerosol.

A similar conclusion was implicit in the earlier studies of in-cloud scavenging by Junge and Abel (1965). The results of their computations are displayed in

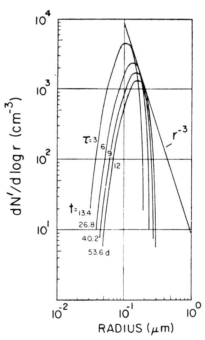

Fig. 12-11. Variation with time of aerosol size distribution; at $\tau = t = 0$ aerosol is monodisperse, $r = 0.04 \, \mu m$, and $N = 3.55 \times 10^4 \, cm^{-3}$; at $\tau = 3$ or $t = 13.4$ d the SPD is approached and subsequently changes with time as indicated. (From Junge, 1969b; by courtesy of Amer. Meteor. Soc., and the author.)

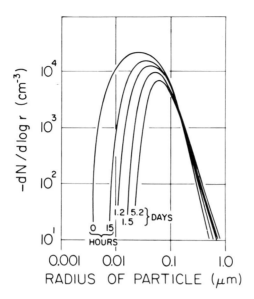

Fig. 12-12.   Variation with time of an average atmospheric aerosol particle size distribution due to sedimentation, capture by cloud and raindrops, and due to their involvement in condensation. (From Junge and Abel, 1965; by courtesy of the authors.)

Figure 12-12 in terms of the concentration $N(r, t)$ of particles with radii $\geqslant r$ (cf. (8-26) and (8-28)). These results are based on the assumptions: (1) that the troposphere has a height of 8 km, is uniformly mixed, and is filled initially with an aerosol of size distribution specified by curve $t = 0$ in Figure 12-12; (2) that a uniform fraction 0.2 of the troposphere is filled with clouds which are composed of uniformly sized drops of radius 10 μm present in a number concentration of 200 cm^{-3}; (3) that the clouds have an average base height of 1.5 km and an average depth of 4 km; (4) that all clouds evaporate after a time period of 1.5 hr, after which they reform; (5) that the troposphere undergoes 10 such evaporation-condensation cycles for each precipitation event; (6) that the mean annual global rainfall rate is 1000 mm, and that rain is composed of raindrops of 0.8 mm in diameter. From their computations, Junge and Abel concluded that neither Brownian coagulation, nor condensation, nor interaction of aerosol particles with cloud and raindrops significantly affects the size distribution of aerosol particles of radii larger than 0.1 μm, at least during time periods up to a week.

12.8.3   QUASI-STATIONARY SELF-PRESERVING DISTRIBUTIONS

Let us suppose the coagulation equation is extended to include possible sources and sinks of particles, so that in general ϕ is no longer constant. Liu and Whitby (1968) investigated the spectral form which could result in this situation if the following restrictions were to apply: (1) dynamic equilibrium exists within some subrange, so that $\partial n/\partial t = 0$; (2) in that subrange a self-similar form ψ for n exists, so that ϕ is constant or nearly so over the time interval of interest.

These strong constraints suffice to specify ψ and n to within a constant. Thus, the governing equation for ψ is simply

$$2\,\psi + \eta \frac{\partial \psi}{\partial \eta} = 0, \tag{12-127}$$

as can be seen from (12-117). The solution is $\psi = C_1 \eta^{-2}$, where C_1 is a dimensionless constant. The corresponding solution for n is, from (12-112), $n(v) = C_1 \phi v^{-2}$, or

$$n(r) = \frac{C_1 \phi}{v^2} \frac{dv}{dr} = \frac{C_2 \phi}{r^4}, \tag{12-128}$$

where C_2 is another dimensionless constant (Liu and Whitby, 1968).

Of course, this simple result is in good agreement with observations, and it is tempting to regard this as a demonstration that the tropospheric aerosol is a quasi-stationary self-preserving distribution. However, it must be remembered that the r^{-4} law is not rigorous experimentally, and that there is no proof of the existence of SPD for general coagulating mechanisms. The simple expression (12-128) is just the envelope curve of the assumed SPD (both time independent and self-preserving), and is a result of an almost overdetermined formulation.

12.8.4 STATISTICAL DISTRIBUTIONS

Over portions of the spectrum where $n(r) \sim r^{-4}$, the volume distribution $(\partial V/\partial \log_{10} r) = (4\,\pi r^3/3)(\partial N'/\partial \log_{10} r)$ is constant. Thus one may say there is a tendency for atmospheric aerosols to form quasi-constant log volume distributions, which in turn may be approximated by broad log normal volume distributions. Noting that mechanical processes such as grinding form log-normal volume distributions, Junge (1969a, b) has suggested it is not unreasonable to expect a large number of independent aerosol sources could produce log-normal distributions also, and that this could provide an explanation for the observations: (1) that the r^{-4} distribution is only an approximation for average

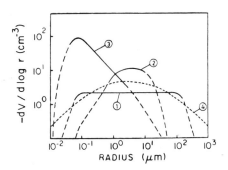

Fig. 12-13. Log-volume distributions for atmospheric aerosol particles under typical conditions. (1) Continental surface air, $\alpha = 4$; (2) maritime surface air, $\alpha \approx 3$, but not constant; (3) upper tropospheric air, $\alpha = 5$; (4) log-normal distribution, $\alpha \approx 4$, but not quite constant. (From Junge, 1969b; by courtesy of Amer. Meteor. Soc., and the author.)

conditions with irregular deviations in individual cases, (2) that the r^{-4} distribution is usually best realized in polluted areas where the statistics are better due to a large number of sources, and (3) that the deviations become more pronounced if one or only a few sources dominate, as in the case, for instance, of aerosols within the lowest 2 to 3 km over the ocean where sea salt particles dominate. These ideas are schematically illustrated in Figure 12-13. Although the statistical explanation for the form $n \sim r^{-4}$ seems quite plausible, it remains to be demonstrated quantitatively under what conditions log-normal distributions will result.

12.8.5 POWER LAW SOLUTIONS FOR A SOURCE-ENHANCED AEROSOL

Distributions somewhat like measured ones of the form $n(r) \sim r^{-k}$ (k = constant) have been obtained by numerical integration of the coagulation equation with sources (e.g., Walter (1973), Storebo (1972), Takahashi and Kasahara (1968), Mockros et al. (1967), Quon and Mockros (1965)), and by analytical solution of simple models based on a related growth equation (Brock, 1971). In addition, there is an analytical solution for the steady state case with a kernel of the form $K(u, v) \sim u^{\beta}v^{\beta}$, where β is a constant between 0 and 1 (Klett, 1975). Because it includes some cases considered by others and provides an illustrative exact solution, we shall now turn to a brief discussion of this last theoretical model.

Let us modify the coagulation equation (12-48) by including a contribution to $\partial n / \partial t$ from a local particle source term, $I(v, t)$, due to unspecified processes. For example, $I(v, t)$ may describe gas-to-particle conversion; alternatively, if we are interested in describing n over some size interval where a particular coagulation mechanism is known to be dominant, $I(v, t)$ can be regarded as describing the entry of particles into the lower end of this range, caused by other coagulation processes which are controlling for still smaller particle sizes. We shall assume a class of particle sources which can be represented by steady gamma distributions, i.e., $I = I_0 J(x)/v_0$, where I_0 is a constant, v_0 is the average source particle volume, $x = v/v_0$, and

$$J(x) = \frac{(p + 1)^{p+1}x^p e^{-(p+1)x}}{\Gamma(p + 1)}, \tag{12-129}$$

where $\Gamma(p)$ is the gamma function (given by A.15-18) and p is a positive real number. This represents a fairly wide range of reasonable unimodal distributions. In particular, we may note the limiting forms $\lim_{p \to 0} J(x) = e^{-x}$ and $\lim_{p \to \infty} J(x) = \delta(x - 1)$, the Dirac delta function (corresponding to a constant feed rate of single particles of volume v_0).

In terms of the normalization of (12-113), the source-enhanced, dimensionless coagulation equation for the steady state and with $\alpha(x, y) = x^{\beta}y^{\beta}$ may be expressed as

$$x^{\beta}f(x) = \frac{v_0^{\beta}M_0}{2 M_{\beta}} \int_0^x (x - y)^{\beta}y^{\beta}f(x - y)f(y) \, dy + \frac{I_0 J(x)v_0^{\beta}}{K_0 M_0 M_{\beta}} \tag{12-130}$$

(cf. (12-114) and (12-115)). The convolution form suggests the use of Laplace transforms. (Details of solution techniques of the type needed for this problem are discussed at greater length in Section 15.2.2 and A-15.2.2.) Denoting the Laplace transforms of $g(x) \equiv x^\beta f(x)$ and $J(x)$ by $\bar{g}(s)$ and $\bar{J}(s)$, respectively, we find

$$\bar{g}(s) = \frac{M_\beta}{M_0 v_0^\beta}\left[1 - \left(1 - \frac{2\,I_0 v_0^{2\beta}}{K_0 M_\beta^2}\,\bar{J}(s)\right)^{1/2}\right] = \frac{1}{M_0}\left(\frac{2\,I_0}{K_0}\right)^{1/2}\{1 - [1 - \bar{J}(s)]^{1/2}\}. \quad (12\text{-}131)$$

The last step is possible since $2\,I_0 v_0^{2\beta} = K_0 M_\beta^2$, as can be seen by setting $n = 0$ and $dM_n/d\tau = 0$ in the source-enhanced moment equation, which is obtained by adding the term $(I_0 v_0^{2\beta}/K_0)\int_0^\infty x^n J(x)\,dx$ to the right side of (12-118). Since $\bar{J}(s) = (p+1)^{p+1}/(s+p+1)^{p+1}$, the radical in (12-131) may be expanded in a binomial series, which leads to the solution

$$f(x) = \left(\frac{2\,I_0}{K_0}\right)^{1/2}\frac{x^{-\beta}}{M_0}\left\{\frac{J(x)}{2} + \sum_{k=2}^{\infty}\frac{(2k-3)!!}{2^k\,k!}\,L^{-1}[\bar{J}^k(s)]\right\}, \quad (12\text{-}132a)$$

where

$$L^{-1}[\bar{J}^k(s)] = \frac{(p+1)^{(p+1)k}\,e^{-(p+1)x}\,x^{p+(k-1)(p+1)}}{\Gamma[k(p+1)]}, \quad (12\text{-}132b)$$

and where $(2k-3)!! = (2k-3)(2k-1)\cdots 3\cdot 1$. A plot of this solution is given in Figure 12-14, where it can be seen that the influence of the source on the spectral shape soon becomes secondary to that of the collection kernel. The asymptotic form of $f(x)$ for finite p,

$$f(x) = \frac{1}{2\,M_0}\left(\frac{2\,I_0}{\pi K_0}\right)^{1/2}x^{-(\frac{3}{2}+\beta)}, \qquad x \gg 1 \quad (12\text{-}133)$$

is thus achieved for $x \geq 6$ even for the rather sharply peaked source distribution shown in the figure.

From (12-133) and (12-113) we see that for $\beta = 0$ and $v \gg v_0$ the solution is $n(v) \approx (v_0 I_0/2\,\pi K_0)^{1/2}v^{-3/2}$; this is equivalent to the numerical solution of Quon and

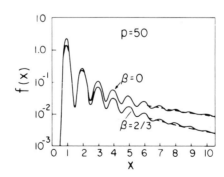

Fig. 12-14. Dimensionless size spectrum $f(x)$ for $\beta = 0$ (constant collision kernel), $\beta = 2/3$ (gravitational collection), and for $p = 50$ (sharply-peaked source distribution). Dashed lines represent the corresponding asymptotic power-law forms of $f(x)$, given by (12-133). (From Klett, 1975; by courtesy of Amer. Meteor. Soc.)

Mockros (1965) for the case of a constant input rate of single particles. Since $n(v) = 4 r^2 n(v)$ and $v_0 I_0 = \varepsilon_m$, the rate of aerosol particle volume input to the system, the solution may also be expressed in the form $n(r) \approx (3/2 \pi)(3/2)^{1/2}$ $(\varepsilon_m/K_0)^{1/2} r^{-5/2}$, which is just Friedlander's QSD in the Brownian coagulation regime (cf. (12-108)). For $\beta = 2/3$ the kernel coincides with (12-75), so that the resulting solution, $n(r) \sim r^{-9/2}$, provides an approximate description of the effects of gravitational coagulation. The solution for this case is plotted in Figure 8-17, and can be seen to conform reasonably well with the background aerosol spectrum in the upper troposphere, where such coagulation would be expected to dominate for $r \geqslant 1 \ \mu m$.

CHAPTER 13

DIFFUSION GROWTH AND EVAPORATION OF WATER DROPS AND ICE CRYSTALS

Immediately following their formation through heterogeneous nucleation, cloud particles proceed to grow by the process of vapor diffusion. Later on they may also experience growth by the mechanisms of collision and subsequent coalescence or sticking. In the present chapter we shall describe the individual and collective growth (and evaporation) of cloud particles by vapor diffusion; collisional growth (and breakup) are considered in Chapters 15 and 16.

13.1 Laws for Diffusion of Water Vapor and Heat

13.1.1 DIFFUSION WATER VAPOR

In Sections 12.5 and 12.7.1 we discussed the problems of aerosol particle diffusion to stationary and falling drops, respectively. An adequate treatment of water vapor diffusion to drops or ice crystals can be based on a strictly analogous formulation. Thus, if we let $\vec{j}_v$ denote the flux density vector of water vapor mass, then for moist air moving with velocity $\vec{u}$ relative to the drop or crystal under study, $\vec{j}_v$ may be represented as the sum of contributions due to diffusion and convection:

$$\vec{j}_v = -D_v \nabla \rho_v + \rho_v \vec{u}, \tag{13-1}$$

where D_v is the diffusion coefficient or diffusivity for water vapor, and ρ_v is the water vapor density. Then from the continuity equation, $\partial \rho_v / \partial t = -\nabla \cdot \vec{j}_v$, we obtain the convective diffusion equation for water vapor:

$$\frac{\partial \rho_v}{\partial t} + \vec{u} \cdot \nabla \rho_v = D_v \nabla^2 \rho_v. \tag{13-2}$$

To obtain (13-2) we have made the usual assumptions that $\nabla \cdot \vec{u} = 0$ (recall the discussion just before (10-2)), and that D_v is constant over the region of interest (in applications, this means D_v must not vary over distances of the order of the size of the growing or evaporating drop or ice crystal). We have also made the implicit assumption that the total air density ρ is constant in the vicinity of the drop or crystal; otherwise we would have had to express the diffusive transport of vapor in terms of the concentration gradient of the mixing ratio $w_v = \rho_v / \rho_a$, or in terms of ρ_v / ρ.

The diffusivity D_v of water vapor in air has been experimentally determined only for temperatures warmer than $0\,°C$, and in this range the experimental results scatter quite strongly. The unreliability of past experimental values for D_v has also been stressed by Ranz and Marshall (1952), Reid and Sherwood (1966), Thorpe and Mason (1966), and Beard and Pruppacher (1971a), all of whom pointed out that the conventionally used values for D_v tabulated in the Smithsonian Meteorological Tables (1968) may be too high by as much as 10%. In a recent survey Marreno and Mason (1972) pointed out that probably the best experimental values for D_v at temperatures above $0\,°C$ are those of O'Connell et al. (1969). Following an extrapolation procedure suggested by E. A. Mason (1975, private comm.), Hall and Pruppacher (1976) arrived at the following best estimate relation for the diffusivity of water vapor in air for temperatures between -40 and $40\,°C$:

$$D_v = 0.211 \left(\frac{T}{T_0}\right)^{1.94} \left(\frac{p_0}{p}\right), \tag{13-3}$$

with $T_0 = 273.15\,°K$, $p_0 = 1013.25$ mb, and D_v in $cm^2\ sec^{-1}$.

For a fixed cloud particle surface (ignoring any tangential motion of the surface, and ignoring the rate of change of particle size due to evaporation or condensation), the usual flow boundary condition at the surface (S) taken to accompany (13-2) is $\tilde{u}|_S = 0$. However, when we recognize that $\tilde{u}$ is the mass-average velocity $(\rho_a \tilde{u}_a + \rho_v \tilde{u}_v)/\rho$ of the velocities of vapor $(\tilde{u}_v)$ and dry air $(\tilde{u}_a)$, we can appreciate that $\tilde{u}_v$ does not vanish at the surface, and that the proper boundary condition is $\tilde{u}_a|_S = 0$. This latter boundary condition leads, from (13-1), to the surface flux relation $\vec{j}_v|_S = \rho_v \tilde{u}_v|_S = -[D_v(1 - \rho_v/\rho)^{-1}\nabla\rho_v]_S$; this differs by a factor of $(1 - \rho_v/\rho)^{-1}$ from the more familiar statement $\vec{j}_v|_S = -[D_v\nabla\rho_v]_S$, which arises from the former boundary condition $\tilde{u}|_S = 0$. On the other hand, since usually $\rho_v/\rho = O(10^{-3})$, we are justified in ignoring this refinement, and the complications it entails. Therefore, we shall henceforth employ the simpler formulation expressed by (13-1), (13-2), and the condition that the total air velocity (or its radial component for a drop with internal circulation) must vanish at the particle surface.

Additional complications may arise if there are gradients in temperature or pressure. Then the phenomena of thermal- and pressure diffusion of vapor will also occur. Discussions of these effects may be found, for example, in Bird et al. (1960) and Hidy and Brock (1970). Fortunately, these effects are of negligible importance for circumstances of interest in cloud microphysics.

We shall now explore the possibility of simplifying (13-2) by dropping the $\partial\rho_v/\partial t$ term. For this purpose let us consider the initial-value problem of radially symmetric diffusion from a motionless drop $(\tilde{u} = 0)$ of radius a, subject to the boundary condition $\rho_v(a, t) = \rho_{v,a} = $ constant, and the initial condition $\rho_v(r, 0) = \rho_{v,\infty} = $ constant for $r > a$. This problem is mathematically equivalent to the one of particle diffusion to a sphere discussed in Section 12-5; in fact, the solution is given by (12-35) with the replacements $r_{12} \to a$, $D_{12} \to D_v$, $n_2 \to \rho_v - \rho_{v,a}$, and $n_{2,\infty} \to \rho_{v,\infty} - \rho_{v,a}$:

$$\rho_v(r, t) = \rho_{v,\infty} + (\rho_{v,a} - \rho_{v,\infty})\frac{a}{r}\left[1 - \mathrm{erf}\left(\frac{r - a}{2\sqrt{D_v t}}\right)\right]. \tag{13-4}$$

From this expression the vapor flux at the drop surface is found to be

$$\vec{j}_v \cdot \hat{e}_r|_{r=a} = -D_v \left(\frac{\partial \rho_v}{\partial r}\right)_{r=a} = \frac{D_v(\rho_{v,a} - \rho_{v,\infty})}{a}\left(1 + \frac{a}{\sqrt{\pi D_v t}}\right). \tag{13-5}$$

Thus it can be seen that the steady state description will be valid for times $t \gg t_c = a^2/\pi D_v$; for $T = 0\,°C$ and $p = 1$ atm., $t_c < 1.7 \times 10^{-6}$ sec for drops smaller than $100\ \mu m$ radius. Since in the atmosphere significant changes in vapor density fields occur over times much larger than this, we may justifiably ignore the non-steady state contribution to the diffusional growth or evaporation of cloud particles under natural conditions. (We have ignored the intrinsic unsteadiness due to the rate of contraction or expansion of an evaporating or growing drop. This effect may also be shown to be generally negligible for $t \gg t_c$ (e.g., see Section 4.2 of Hidy and Brock (1970)).)

From (13-5) we see that the steady state description of the rate of change of drop mass m for a motionless drop, $(dm/dt)_0$, is given by

$$\left(\frac{dm}{dt}\right)_0 = -\int_S \vec{j}_v \cdot \hat{e}_r\, dS = 4\,\pi a\, D_v(\rho_{v,\infty} - \rho_{v,a}), \tag{13-6}$$

where $S\ (= 4\,\pi a^2)$ denotes the drop surface area (Maxwell, 1890). An alternative expression in terms of vapor pressure e and temperature T may be obtained by substituting the equation of state (4-24) into (13-6):

$$\left(\frac{dm}{dt}\right)_0 = \frac{4\,\pi a\, D_v M_w}{\mathscr{R}}\left(\frac{e_\infty}{T_\infty} - \frac{e_a}{T_a}\right). \tag{13-7}$$

The description of diffusional growth or evaporation provided by (13-6) and (13-7) must break down for very small drops. This happens because the assumption that the moist air is a continuous field right up to the drop surface, which is implicit in the formulation presented so far, becomes quite unrealistic for drops with radii comparable to the mean free path λ of air molecules. An intuitive, semi-empirical extension of the continuous description to account for this effect has been carried out by Schäfer (1932) and later in much more detail by Fuchs (1959); both of these treatments follow Langmuir (1918), who pointed out the existence of a rapid change in vapor concentration at the surface of an evaporating drop. (An analogous discontinuity in temperature had been known to exist earlier (Lazarev, 1912).) Since Fuchs's method is often used and produces results of reasonable accuracy (see, e.g., Fuchs, 1959; Fukuta and Walter, 1970), we shall now present a version of it, following Fitzgerald (1972). A discussion of more rigorous treatments of the discrete field problem may be found in Hidy and Brock (1970).

Fuchs assumed the diffusion equation and its solution are valid only for distances greater than the 'vapor jump' length $\Delta_v \approx \lambda$ from the drop surface. Within the layer $a \leq r \leq a + \Delta_v$, vapor transport is assumed to occur according to an elementary gas kinetic mechanism. The condition of continuity of vapor flux across the surface $r = a + \Delta_v$ may be invoked to complete the description. Proceeding in this manner, for the region $r \geq a + \Delta_v$ we must solve the steady state diffusion (Laplace's equation); assuming axial symmetry, this is

$$\nabla \rho_v = \frac{1}{r} \frac{d^2}{dr^2} (r\rho_v) = 0, \tag{13-8}$$

subject to the boundary condition $\rho_v = \rho_{v,\infty}$ for $r \to \infty$. The appropriate solution is simply

$$\rho_v(r) = \rho_{v,\infty} + \frac{A}{r}, \tag{13-9}$$

where A is a constant to be determined.

For the region $a \leqslant r \leqslant a + \Delta_v$, we assume the vapor flux may be described according to the gas kinetic expressions presented in Section 5.9. If we assume the vapor density at $r = a + \Delta_v$ measures the concentration of molecules which strike the surface (which is reasonable since on the average these molecules will have suffered their last collision about one mean free path above the surface), then from (5-49) and (5-53) we estimate the condensation flux of vapor mass per unit area of the surface to be $\alpha_c \dot{m}_w N_v \bar{v}_v / 4 = \alpha_c \bar{v}_v \rho_v (a + \Delta_v) / 4$, where $\dot{m}_w$ is the mass of the water molecule, $\bar{v}_v$ is the average thermal velocity of vapor molecules, N_v is the concentration of vapor molecules at $r = a + \Delta_v$, and α_c is the condensation coefficient. Similarly, we estimate the evaporation flux per unit area of surface to be $\alpha_e \bar{v}_v \rho_v(a)/4$, where α_e is the evaporation coefficient (we have ignored any effects due to temperature gradients or incomplete thermal accommodation). On setting $\alpha_e = \alpha_c$, which is justified by the absence of contrary experimental evidence, we may therefore express the net evaporation flux of water mass from the drop surface as follows:

$$J_v(a) = \pi a^2 \alpha_c \bar{v}_v [\rho_v(a) - \rho_v(a + \Delta_v)]. \tag{13-10}$$

For a steady state description (13-10) must also be the flux $J_v(a + \Delta_v)$ at $r = a + \Delta_v$. On the other hand, from (13-9) we have an alternate expression for $J_v(a + \Delta_v)$, viz.,

$$J_v(a + \Delta_v) = -4\pi(a + \Delta_v)^2 D_v \left(\frac{\partial \rho_v}{\partial r}\right)_{a+\Delta_v} = 4\pi D_v A. \tag{13-11}$$

Therefore, from (13-9)–(13-11) we determine the constant A to be

$$A = \frac{a(\rho_{v,a} - \rho_{v,\infty})}{\left(\dfrac{a}{a + \Delta_v} + \dfrac{4 D_v}{a\alpha_c \bar{v}_v}\right)}. \tag{13-12}$$

Then since $(dm/dt)_0 = -J_v(a)$, we may also express our result in the form of (13-6) or (13-7), where now D_v is replaced by a modified diffusivity given by

$$D_v' = \frac{D_v}{\left[\dfrac{a}{a + \Delta_v} + \dfrac{D_v}{a\alpha_c} \left(\dfrac{2\pi M_w}{\mathcal{R} T_a}\right)^{1/2}\right]} \tag{13-13}$$

(Okuyama and Zung, 1967; Fitzgerald, 1972). To obtain this result we have used (5-50) for $\bar{v}_v$. Equation (13-13) coincides with an expression derived by Fukuta and Walter (1970) on setting $\Delta_v = 0$; also, for the case that $\alpha_c = 1$ and $\Delta_v = C\lambda$, where $C \approx 0.7$ is the 'Cunningham constant,' it reduces to a result obtained by Langmuir (1944).

The magnitude of this correction for the diffusivity (and the corresponding correction for the thermal conductivity, which we shall discuss in the next section) can be seen from Table 13-1. The table indicates the usual continuum theory of Maxwell strongly overestimates the growth rate for submicron particles. The difference in growth rates is illustrated further in Figure 13-1, which presents the results of a computation by Fukuta and Walter (1970) for the case of a single drop growing by diffusion at 10 °C, 1000 mb, a supersaturation of 1%, and a thermal accommodation coefficient of unity. The figure shows that the ratio of droplet masses, $x_m = m(D_v)/m(D_v')$, computed from the Maxwell theory and the modified theory using D_v' of (13-13) (with $\Delta_c = 0$) reduces to $x_m \approx 2$ within the first 20 sec after the start of growth. For longer times x_m decreases until finally the difference between the predictions of the two models becomes negligible

TABLE 13-1

Diffusivity of water vapor and thermal conductivity of air computed from (13-13) and (13-20) for 10°C and 800 mb assuming $\alpha_c = 0.036$, $\alpha_T = 0.7$, $D_v = 0.30$ cm^2 sec^{-1}, $k_a = 5.97 \times 10^{-5}$ cal cm^{-1}sec^{-1}°C^{-1}, $\lambda = 8 \times 10^{-6}$ cm, $\Delta_T = 2.16 \times 10^{-5}$ cm, $\Delta_v = 1.3 \lambda$.

Drop radius	0.01	0.1	1.0	10.0	∞
D_v' (cm^2 sec^{-1})	5.2×10^{-4}	5.2×10^{-3}	4.5×10^{-2}	0.19	0.30
k_a' (cal cm^{-1} sec^{-1}°C^{-1})	1.9×10^{-6}	1.7×10^{-5}	5.2×10^{-5}	5.9×10^{-5}	6.0

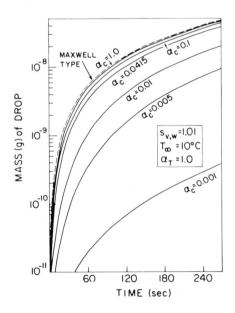

Fig. 13-1.   Influence of kinetic effects on the diffusion growth of small drops, for various α_c. (From Fukuta and Walter, 1970; by courtesy of Amer. Meteor. Soc., and the authors.)

($x_m \approx 1$). Since, as we shall see below, growth times much longer than 20 sec are generally needed to establish the spectral characteristics of a drop population growing by diffusion, the results shown in Figure 13-1 suggest that the correction of the continuum diffusion model for gas kinetic effects is not likely to make a significant quantitative difference in the predictions of growth models for natural conditions. (Also, the practical advantages of the corrected model are further narrowed by the present uncertainties in the known values of Δ_v, D_v, and α_c.)

13.1.2 DIFFUSION OF HEAT

Water vapor transport by diffusion to or from a cloud particle necessarily involves a substantial flow of heat as well, owing to the release or absorption of heat of phase change. The resulting temperature difference between the particle and its local environment causes a flow of sensible heat by the familiar process of thermal diffusion or heat conduction. The flux density vector $\vec{j}_h$ for heat transport by this process in given by Fourier's law:

$$\vec{j}_h = k\nabla T, \tag{13-14}$$

where k is the thermal conductivity of medium through which heat is transported.

From (13-14), and for the physical conditions of relevance in cloud microphysics, we may easily derive an equation for the diffusion of heat which is analogous to (13-2). From the meaning of $\vec{j}_h$, the conductive heat change experienced by a moving volume element δV of air in time dt is just $-\nabla \cdot \vec{j}_h \delta V \, dt$. Since the heated or cooled air in the vicinity of a cloud particle is free to expand or contract, the heat exchange may be assumed to occur at constant pressure. Then from (4-6) the corresponding enthalpy change of the considered volume element is just $d(h\rho\delta V) = -\nabla \cdot \vec{j}_h \delta V \, dt$, where h is the enthalpy of air per unit mass of air. Also, since $dh = c_p \, dT$ from (4-12), and considering that $d(\rho\delta V) = 0$ by conservation of mass, we arrive at the desired result:

$$\frac{dT}{dt} = \frac{\partial T}{\partial t} + \vec{u} \cdot \nabla T = -\frac{\nabla \cdot \vec{j}_h}{\rho c_p} = \kappa \nabla^2 T, \tag{13-15}$$

where $\kappa \equiv k/\rho c_p$ is called the thermal diffusivity, with ρ and c_p being the density and specific heat of air, and where (13-14) has been inserted for $\vec{j}_h$. To obtain (13-15) we have assumed k to be a constant over the region of interest (within a few radii of the given cloud particle). We have also ignored heat changes arising from radiation, and from frictional dissipation of air motion (or drop internal motion). Heating by frictional dissipation is always of negligible importance for individual cloud microphysical processes (see, for example, §55 of Landau and Lifschitz, 1959). Radiative heat exchange is also negligible in the context of interest in this chapter, since temperature differences between particles and the local environment are always quite small.

A survey of the best experimentally and theoretically determined values for the thermal conductivity of dry air, k_a, and water vapor, k_v, has been given by Beard and Pruppacher (1971a). They suggested that the temperature variation for k_a and k_v can be expressed adequately by the relations

$$k_a = (5.69 + 0.017 \text{ T}) \times 10^{-5}, \tag{13-16}$$

$$k_v = (3.78 + 0.020 \text{ T}) \times 10^{-5}, \tag{13-17}$$

with T in °C and k in cal cm^{-1} sec^{-1} °C^{-1}. One may then use the Mason-Sexena formula (Bird *et al.*, 1960) to find the thermal conductivity of moist air, viz.,

$$k = k_a[1 - (\gamma_1 - \gamma_2 k_v)/k_a)x_v], \tag{13-18}$$

where x_v is the mole fraction for water vapor in moist air and $\gamma_1 \approx 1.17$, $\gamma_2 \approx 1.02$. Since for typical atmospheric conditions $x_v \ll 1$ we see that $k \approx k_a$.

From the identical forms of (13-2) and (13-15) we can borrow the argument of the previous section concerning the validity of assuming a steady state: this will be permissible for times $t \gg t_c' = a^2/\pi\kappa$. Since $\kappa \approx D_v$, then $t_c' \approx t_c$ and so we can generally ignore the non-steady state contribution to the conductive heat flow to or from evaporating or growing cloud particles. Therefore, from the same mathematical arguments as used in the previous section we conclude the rate of conductive heat transfer to a motionless drop may be expressed as

$$\left(\frac{dq}{dt}\right)_0 = 4\pi a \, k_a (T_\infty - T_a), \tag{13-19}$$

(cf. (13-6)).

We may also correct k for gas kinetic effects in exactly the same manner as for the previous case of vapor diffusion. The only difference is in the replacement of the vapor mass flux per unit area, $\alpha_c \rho_v \bar{v}_v/4$, by the heat flux per unit area, $\alpha_T \rho \bar{v}_a c_{pa}/4$, where α_T is the thermal accommodation coefficient (see Section 5.9), and c_p is the specific heat of air. The modified form for the thermal conductivity is then given by

$$k_a' = \frac{k_a}{\left[\dfrac{a}{a + \Delta_T} + \dfrac{k_a}{a\alpha_T\rho c_{pa}}\left(\dfrac{2\pi M_a}{\mathscr{R} T_a}\right)^{1/2}\right]} \tag{13-20}$$

(Fitzgerald, 1972) (cf. (13-13)). The 'thermal jump' distance Δ_T is analogous to Δ_v in (13-13). If $\Delta_T = 0$ and c_p is replaced by $c_{pa} + R_a/2$, (13-20) coincides with an expression obtained by Fukuta and Walter (1970). It can also be made to agree with a result derived by Carstens (1972) and Carstens *et al.* (1974) by making the replacements $c_{pa} \to c_{pa} + R_a/2$ and $\alpha_T \to \alpha_T(1 - \alpha_T/2)^{-1}$. These differences arise from slightly different modeling approaches, and are of little consequence. An evaluation of (13-20) is given in Table 13-1. The previous discussion of the significance of the differences between D_v' and D_v applies also to the present case of k' versus k.

13.2 Diffusional Growth of Aqueous Solution Drops

13.2.1 GROWTH OF AN INDIVIDUAL STATIONARY DROP

We shall now formulate a governing equation for the diffusional growth (or evaporation) of a single stationary drop in a motionless atmosphere. From (13-7), the rate at which such a drop changes its radius a may be expressed as

$$a \frac{da}{dt} = \frac{D'_v M_w}{\mathcal{R}} \left(\frac{e_\infty}{T_\infty} - \frac{e_a}{T_a} \right), \tag{13-21}$$

where we have replaced D_v by D'_v to include the correction for gas kinetic effects. An expression for the temperature T_a at the drop surface can be obtained by considering the coupling of the rates of change of heat and mass through latent heat release:

$$\left(\frac{dq}{dt} \right)_0 = -L_e \left(\frac{dm}{dt} \right)_0, \tag{13-22}$$

where L_e is the specific latent heat of evaporation. Then on substituting (13-19) (with $k_a \rightarrow k'_a$) into this equation we find

$$T_a = T_\infty + \frac{L_e \rho''_s}{k'_a} a \frac{da}{dt}, \tag{13-23}$$

where ρ''_s is the density of the aqueous solution drop. As expected, we see that an evaporating drop $(da/dt < 0)$ is predicted to be cooler than its environment, and vice versa.

For the sequel it is convenient to write (13-23) in the form

$$\frac{T_a}{T_\infty} = 1 + \delta, \quad \delta = \frac{L_e \rho''_s}{T_\infty k'_a} a \frac{da}{dt}. \tag{13-24}$$

For usual conditions of drop growth, $\delta \leqslant 10^{-5}$ (Neilburger and Chien, 1960), which means that the heat released by condensation is very efficiently dissipated by conduction. One might therefore expect that the heating of the drop by release of latent heat could be ignored, as was assumed in an early study by Houghton (1933). However, this turns out not to be the case; Neiburger and Chien showed that the neglect of temperature differences between the drop and its environment leads to large errors for all sizes of drops and condensation nuclei.

To integrate (13-21) we also need an equation for the vapor pressure e_a at the surface of the solution drop. An adequate expression for this purpose, including both the drop curvature and solution effects, is given by (6-26b). However, this still does not complete the formulation, since (6-26b) expresses $e_a(T_a)$ as a function of $e_{sat,w}(T_a)$. To close the system of equations, we must also invoke the Clausius-Clapeyron equation, (4-92), in order to relate $e_{sat,w}(T_a)$ to $e_{sat,w}(T_\infty)$, which is a function only of the environmental temperature T_∞. On integrating (4-92), we obtain

$$e_{sat,w}(T_a) = e_{sat,w}(T_a) \exp \left[\frac{L_e M_w}{\mathcal{R}} \left(\frac{T_a - T_\infty}{T_a T_\infty} \right) \right]. \tag{13-25}$$

Finally, on combining (13-21), (13-24), (6-26b), and (13-25), we obtain the desired governing form:

$$a\frac{da}{dt} = \frac{D_v'M_we_{sat,w}(T_\infty)}{\rho_s''\mathcal{R}T_\infty}\left\{S_{v,w} - \frac{1}{(1+\delta)}\exp\left[\frac{L_eM_w}{\mathcal{R}T_\infty}\left(\frac{\delta}{1+\delta}\right)\right.\right.$$
$$\left.\left. + \frac{2M_w\sigma_{s/a}}{\mathcal{R}T_\infty(1+\delta)\rho_wa} - \frac{\nu\Phi_sm_sM_w/M_s}{(4\pi a^3\rho_s''/3) - m_s}\right]\right\}. \tag{13-26}$$

This result applies to the case of condensation on a water-soluble nucleus of mass m_s; for the case of condensation on a mixed particle of soluble mass-fraction ε_m, the last term in (13-26) must be replaced by $\nu\Phi_s\varepsilon_mM_w\rho_{NT}r_N^3/M_s\rho_w(a^3 - r_N^3)$ (cf. (6-35)).

The result (13-26) is rather cumbersome and, worse yet, is an implicit equation due to the dependence of δ on da/dt. Fortunately, several simplifications are possible. Thus, since $\delta \ll 1$ we may set $(1+\delta)^{-1} \approx 1$, and write $\exp[L_eM_w\delta/\mathcal{R}T_\infty(1+\delta)] \approx 1 + L_eM_w\delta/\mathcal{R}T_\infty$; also, on defining

$$y = \frac{2\sigma_{s/a}M_w}{\mathcal{R}T_\infty\rho_wa} - \frac{\nu\Phi_sm_sM_w/M_s}{(4\pi\rho_s''a^3/3) - m_s}, \tag{13-27}$$

we may write $e^y \approx 1 + y$, since both the curvature and solute contributions to the equilibrium vapor pressure over a solution drop are generally small for $a \gtrsim 1\,\mu m$ (see Figure 6-2). With these approximations and on setting $\rho_s'' \approx \rho_w$ (Section 4.7), and $L_e \approx L_{e,0}$ (Section 4.8), (13-16) reduces to

$$a\frac{da}{dt} \approx \frac{S_{v,w} - y}{\dfrac{\rho_w\mathcal{R}T_\infty}{e_{sat,w}(T_\infty)D_v'M_w} + \dfrac{L_{e,0}\rho_w}{k_a'T_\infty}\left(\dfrac{L_{e,0}M_w}{T_\infty\mathcal{R}} - 1\right)}, \tag{13-28}$$

where the supersaturation $s_{v,w} + 1 = e_\infty/e_{sat,w}(T_\infty)$. Equation (13-28) with the replacements $D_v' \to D_v$ and $k_a' \to k_a$ is equivalent to one used by Howell (1949). For realistic conditions of growth it agrees with (13-26) to within a few percent (for the same diffusivities Langmuir (1944) derived a similar equation, except for the omission of the solute effect, which is the second term in y).

The growth histories of individual solution drops calculated from an equation almost identical to (13-28) are shown in Figure 13-2, for $\alpha_c = 0.045$ and an environment characterized by $T = 20\,°C$ and a constant supersaturation of $s_{v,w} = 1\%$. It can be seen there is a tendency for the more dilute solution drops, of smaller initial radii since they contain less salt, to catch up to the size of the more concentrated drops. This happens because both the curvature and solute effects rapidly become negligible with increasing a, so that approximately $y \approx 0$ and hence $da/dt \approx C/a$, or $a^2 \approx a_0^2 + Ct$, where C is a constant. The prediction of a parabolic growth law for relatively large and pure water drops has been verified by the experiments of Houghton (1933).

13.2.2 DIFFUSIONAL GROWTH OF A POPULATION OF SOLUTION DROPS OF NEGLIGIBLE FALL VELOCITY

The growth equations derived in the previous section may be used to describe the evolution in size of a given population of initially highly concentrated aqueous solution drops. For this purpose we must make a distinction between

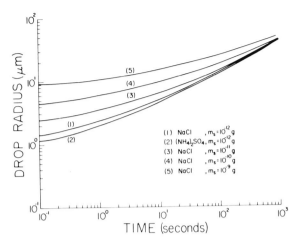

Fig. 13-2.   Diffusional growth rate of individual aqueous solution drops as a function of time at 1% supersaturation and 20 °C; $\alpha_c = 0.045$; the initial drop radius corresponds to that of a salt saturated solution drop in equilibrium with the relative humidity at which the salt deliquesces. (Based on data of Low (1971).)

drop growth in cumuliform clouds where vertical velocities are appreciable and where the rate of cooling can be described by a saturation adiabatic ascent corrected for entrainment (see Chapter 11), and stratiform clouds where the vertical velocities are small and cooling may be considered isobaric. We shall first consider diffusional drop growth in cumuliform clouds since such growth has been treated quite extensively in the literature. Subsequently we shall briefly summarize the few studies that have been carried out on diffusional drop growth in stratiform clouds.

13.2.2.1 Condensation Growth in Cumuliform Clouds

As we shall show below in Section 13.3, the diffusional growth or evaporation of drops with $a \lesssim 50\ \mu$m falling at terminal velocity is essentially unaffected by the air flow around them. Hence the characteristic shapes of drop spectra which evolve through diffusion growth can be established without considering ventilation effects. It is also reasonable to omit consideration of vapor field interactions between pairs of growing drops. This is justified mainly by the fact that nearest neighbor distances of drops are commonly $\geqslant 10^2\ \bar{a}$, where $\bar{a}$ is the average drop radius, even for clouds which have already experienced substantial diffusion growth subsequent to nucleation (see Section 14.1).

Although a complete description of the evolution of a drop spectrum by diffusion growth in a cloud updraft is fairly complicated and requires the simultaneous numerical solution of several differential equations, some important features of the process are not only easy to understand, but are even amenable to a simple approximate analysis. As a parcel of air rises and cools by expansion, its humidity will increase rapidly (nearly in proportion to its upward velocity, as shown below). Eventually the largest and most hygroscopic of the suspended condensation nuclei will deliquesce to solution drops, which will proceed to grow toward their critical radius for activation and subsequent rapid growth (see Section 6.5). The parcel supersaturation will continue to increase for a while longer as more nuclei are activated, until finally the growing drop population is able to absorb excess vapor as fast as it can be released by expansion. Beyond this point the supersaturation will fall, the activated drops will tend to approach a fairly uniform size (since $da/dt \sim 1/a$), and the un-activated solution drops will tend to evaporate.

From this physical picture we would expect to find a strong correlation between the concentration of drops produced by diffusion growth and the maximum supersaturation achieved in the expanding parcel. Approximate theoretical correlations of this nature have been established by Squires (1958a) and Twomey (1959c). We shall now turn to a derivation of Twomey's equation, since this model serves as a good example to illustrate some of the physics of condensation growth. Furthermore, the results are widely used, owing to their simplicity and surprising accuracy.

We first require a description of the rate of change of supersaturation with height. This is readily obtained by substituting (11-19) and (11-22) into (11-21), for the case $\mu = 0$ (entrainment effects are ignored in Twomey's model). The result is

$$\frac{ds_{v,w}}{dt} = \left(\frac{\varepsilon L_e g}{R_a T^2 c_{pa}} - \frac{g}{R_a T} \right) W - \left(\frac{p}{\varepsilon e_{sat,w}} + \frac{\varepsilon L_e^2}{R_a T^2 c_{pa}} \right) \frac{dw_L}{dt}, \qquad (13\text{-}29)$$

where we have made the approximation $1 + s_{v,w} \approx 1$ on the right side of (11-21) (from Figure 2-2 we see that usually $s_{v,w} \lesssim 10^{-2}$ in clouds). Equation (13-29) shows how the supersaturation increases nearly in proportion to W, in the absence of condensation, and how it is decreased by the production of liquid water. Although we could proceed directly with (13-29), a slightly modified form which replaces w_L (g water/g air) (see Chapter 11) by the density ρ_L (g water/m^3 air) of condensed water is more convenient. Since $\rho_L = \rho w_L$, where ρ (g air/m^3 air) is the density of moist air, we have $dw_L/dt = \rho^{-1} (d\rho_L/dt - w_L \, d\rho/dt) \approx \rho^{-1} d\rho_L/dt$; with this result and the ideal gas law we obtain the desired result:

$$\frac{ds_{v,w}}{dt} = A_1 W - A_2 \frac{d\rho_L}{dt}, \qquad (13\text{-}30)$$

where

$$A_1 = \frac{\varepsilon L_e g}{R_a T^2 c_{pa}} - \frac{g}{R_a T}, \qquad A_2 = \frac{R_a T}{\varepsilon e_{sat,w}} + \frac{\varepsilon L_e^2}{p T c_{pa}}. \qquad (13\text{-}31)$$

We next make use of (13-28), ignoring the relatively small curvature and solute terms for activated drops by setting $y = 0$:

$$a \frac{da}{dt} \approx A_3 s, \qquad (13\text{-}32)$$

where A_3 is the denominator in (13-28), and we have omitted the subscripts on $s_{v,w}$ for brevity. Noting that the characteristic time for substantial drop growth is short compared to the time for significant changes in A_3, (13-32) may be formally integrated to yield

$$a^2(t) - a^2(\tau) \approx a^2(t) \approx 2 A_3 \int\limits_{\tau}^{t} s(t') \, dt', \tag{13-33}$$

where τ is the activation time for the considered drop. From (13-32) and (13-33) the rate of mass increase for the drop may be expressed as

$$\frac{dm}{dt} \approx 2 \pi \rho_w (2 A_3)^{3/2} s(t) \left[\int\limits_{\tau}^{t} s(t') \, dt' \right]^{1/2}. \tag{13-34}$$

Now let $n(s) \, ds$ denote the concentration of nuclei activated on the interval $(s, s + ds)$; assuming (9-1), we thus have $\int_0^s n(s') \, ds' = Cs^k$, so that

$$n(s) = kCs^{k-1}. \tag{13-35}$$

With (13-34) and $n(s)$ we can now express $d\rho_L/dt$ as follows:

$$\frac{d\rho_L}{dt} \approx 2 \pi \rho_w (2 A_3)^{3/2} s(t) \int\limits_0^s n(s') \left[\int\limits_{\tau(s')}^{t} s(t') \, dt' \right]^{1/2} ds'. \tag{13-36}$$

If (13-36) is substituted into (13-30), an approximate governing equation for s results which, unfortunately, cannot be solved analytically. However, from our previous description of the condensation process, we can expect to obtain a reasonable estimate of the final drop concentration by determining just the maximum supersaturation s_{max}. And from (13-30) and (13-36) it is clear that an analytical lower bound approximation to $\int_\tau^t s \, dt$ will yield an analytical upper bound for s_{max} (by the setting $ds/dt = 0$ in (13-30) for $s = s_{max}$). Reasoning thus, Twomey obtained the following lower bound estimate:

$$\frac{(s^2 - s'^2)}{2 A_1 W} < \int\limits_{\tau(s')}^{t} s(t') \, dt'. \tag{13-37}$$

The basis for this choice can be seen from Figure 13-3, which illustrates schematically the curve of $s = s(t)$. In the absence of condensation, the supersaturation would follow the straight line $s = A_1 Wt$ shown in the figure. The triangle ABC, with its hypotenuse parallel to $s = A_1 Wt$, has an area $s^2/2 A_1 W$, and this is clearly less than $\int_0^B s \, dt$; hence (13-37) follows.

On substituting (13-35) and (13-37) into (13-36), we see we must calculate the integral I:

$$I = \int\limits_0^s s'^{k-1} (s^2 - s'^2)^{1/2} \, ds'. \tag{13-38}$$

Letting $y = (s'/s)^2$, I becomes

$$I = \frac{s^{k+1}}{2} \int\limits_0^1 y^{(k/2)-1} (1 - y)^{1/2} \, dy = \frac{s^{k+1}}{2} B \left(\frac{k}{2}, \frac{3}{2} \right), \tag{13-39}$$

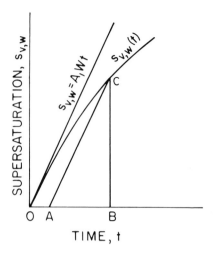

Fig. 13-3. Schematic time variation of the supersaturation in a rising air parcel. (From Twomey, 1959c; by courtesy of *Geophys. pura e appl.* and the author.)

where $B(u, v) \equiv \int_0^1 x^{u-1}(1-x)^{v-1} dx$ is the beta function. Finally, on substituting (13-39) and (13-36) into (13-30) and setting $ds_{v,w}/dt = 0$, we obtain the desired estimate of $(s_{v,w})_{max}$:

$$(s_{v,w})_{max} \lesssim C^{1/(k+2)} \left[\frac{A_1^{3/2} W^{3/2}}{2 \pi \rho_w A_2 A_3^{3/2} k B \left(\frac{k}{2}, \frac{3}{2}\right)} \right]^{1/(k+2)}$$

$$\approx C^{1/(k+2)} \left[\frac{6.9 \times 10^{-2} W^{3/2}}{k B \left(\frac{k}{2}, \frac{3}{2}\right)} \right]^{1/(k+2)}, \tag{13-40}$$

where the numerical evaluation has been carried out for $T = 10\,°C$ and $p = 800$ mb, W is in cm sec^{-1}, C is in cm^{-3}, and $(s_{v,w})_{max}$ is expressed as a percent. The corresponding estimate $N_{max} \approx C(s_{v,w})_{max}^k$ of the cloud drop population is

$$N_{max}(cm^{-3}) \approx C^{2/(k+2)} \left[\frac{6.9 \times 10^{-2} W^{3/2}}{k B \left(\frac{k}{2}, \frac{3}{2}\right)} \right]^{k/(k+2)}. \tag{13-41}$$

The success of this model is illustrated in Figure 13-4, which exhibits a correlation coefficient in excess of 90% between the observed and computed values (from 13-41) for the drop concentration in the bases of small to moderate, non-participating cumulus clouds.

Although the above simple model is apparently capable of predicting quite well the total droplet population soon after the onset of diffusion growth, it tells us nothing about the variation of drop spectral shape with time. Such detailed behavior can be studied only through numerical solution of the full set of governing equations for a given cloud model. Several such studies using different models have been carried out. For example, Howell (1949), Squires (1952), Mordy (1959), Neiburger and Chien (1960), Warner (1969b), Bartlett and Jonas

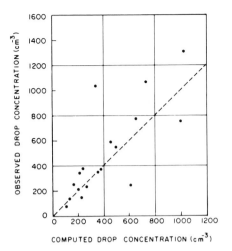

Fig. 13-4. Comparison of observed mean cloud drop concentration with cloud drop concentration computed from CCN spectra for an updraft of $3 \, \mathrm{m \, sec^{-1}}$. Dashed line represents exact agreement between observation and computation. (From Twomey and Warner, 1967; by courtesy of Amer. Meteor. Soc., and the authors.)

(1972), and Fitzgerald (1974) used closed, adiabatic parcel models. Of these, only Bartlett and Jonas included some effects of turbulence. Chen (1971) employed a parcel model which considered entrainment of air devoid of aerosol particles, while Mason and Chien (1962), and Warner (1973) allowed for the entrainment of air containing active CCN. Warner's study also included a comparison of the cases of no entrainment, and entrainment with and without aerosol particles. On the other hand, his study was restricted to drop spectral evolution in maritime clouds only. A more extensive comparative study has been carried out recently by Lee and Pruppacher (1977), who considered the evolution of drop spectra in both maritime and continental clouds. In their study, which we shall discuss further below, they employed the four most often used, simple parcel models: (1) a closed parcel for which the vertical velocity is prescribed; (2) a closed parcel for which the vertical velocity is one of the computed variables; (3) an open parcel which computes vertical velocity and entrains air devoid of CCN; (4) an open parcel which computes vertical velocity and entrains active CCN. The effects of turbulence were not considered, except indirectly through the occurrence of entrainment.

A first step in calculations of this type is to specify the chemical composition and size distribution of the dry aerosol particles on which the drops condense. In this regard, it is important to recall from Chapter 6 that (13-28) does not describe the deliquescence of an aerosol particle, i.e., its transformation from a dry particle into a saturated solution drop of equilibrium size. In order to get around this problem, one generally assumes that each aerosol particle of the considered population of dry particles acquires its equilibrium size at the specified initial ambient relative humidity (assumed to lie between 90 and 99% to insure deliquescence of all water-soluble components in the aerosol particle) in a time period which is short compared to the time over which significant changes occur in the atmospheric environment. This assumption has experimentally been verified by Zebel (1956), Orr et al. (1958), and Winkler (1967), who also demonstrated that the rate determining (slower) step in this transformation is the growth of the saturated solution drop to its new equilibrium size, rather than the deliquescence of the dry aerosol particle. The time required for a saturated solution drop to grow to a new equilibrium size can readily be computed from (13-28). For NaCl dissolved in the drop, Lee and Pruppacher (1977) found that a solution drop in equilibrium with an environment, initially of 70% relative humidity, grows to within 10% of the equilibrium size corresponding to a suddenly-imposed 99% humidity in about 2×10^{-4} sec, 2×10^{-2} sec, and 3 sec for drops containing a salt mass equivalent to a NaCl particle of 0.01 μm, 0.1 μm, and 1.0 μm radius, respectively, in good agreement with the experimental results of the authors mentioned above. Similar but less accurate computations were made earlier by Zebel (1956). Such computations show that particles of sizes typically involved in the condensation process ($r \leqslant 1 \mu$m) are transformed into solution drops of equilibrium size in times which are usually short even compared to the turbulence microscale time τ_k (from the discussion immediately following (12-66) we see that τ_k typically lies in the interval $4 \times 10^{-2} \leqslant \tau_k \leqslant 3 \times 10^{-1}$ sec). Since τ_k provides a lower bound estimate of the quickest fluctuation the CCN can be subjected to by the environment, we see that the assumption of a negligible adjustment time for the conversion from a dry particle to an equilibrium drop is well founded.

Let us now consider a vertically rising, and thus cooling, air parcel in which a population of aerosol particles of given composition and size distribution is contained. From the previous paragraph we may assume that all water-soluble components go into solution when the relative humidity in the parcel reaches the critical value for deliquescence of the particular salts, and that when the relative humidity has reached 99% all solution drops will have reached their corresponding equilibrium size. If we divide the original spectrum of dry particles into small size intervals, we may readily compute the corresponding equilibrium size distribution of solution drops by using (6-26a) if the aerosol particles consist of water-soluble material only, or by using (6-35) or (6-36) if the aerosol is mixed. The evolution in time of this drop size distribution may now be determined from (13-28) if the parcel's temperature, humidity, and vertical velocity are known as functions of time. Relations which describe the variation with time (and thus with height) of these parcel parameters have been derived in Chapter 11. Thus for an entraining parcel we may solve (13-28) for each size interval simultaneously with (11-18)–(11-24), (11-25) or (11-27), (11-26) or its equivalent for a spherical thermal (the calculations presented here are based on the 'plume' geometry, i.e., (11-25) and (11-26), (11-29), and (11-30). For the closed adiabatic parcel for which the vertical velocity is a dependent variable, the only changes from the previous case consist in setting $\mu = 0$ and omitting (11-24). For the case that the vertical velocity is prescribed we must also omit (11-18).

As we have seen from Chapter 8, there are significant differences between maritime and continental aerosols. For their computations Lee and Pruppacher used a Junge power law form (Section 8.2.10) as given in Figure 13-5 to represent

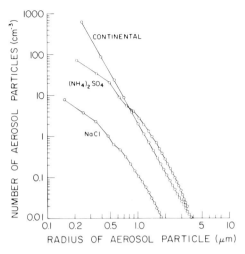

Fig. 13-5.   Initial size distribution of dry aerosol particles used by Lee and Pruppacher (1977); total number of particles: 750 cm^{-3} (continental case); 136 cm^{-3} (NH$_4$)$_2$SO$_4$, 18 cm^{-3} NaCl (maritime case). (From Lee and Pruppacher, 1977; by courtesy of *Pure and Applied Geophys.*, and the authors.)

a typical dry continental aerosol. The particles were assumed to consist of 70% (NH$_4$)$_2$SO$_4$ and 30% water-insoluble material (silicates). The figure also shows the assumed maritime spectrum, which follows the Meszaros and Vissy (1974) form given in Figure 8-18. As indicated in Figure 13-5, the maritime aerosol is a superposition of (NH$_4$)$_2$SO$_4$ and NaCl particles, the relative proportions being in accordance with the observations of Meszaros and Vissy. The corresponding equilibrium drop size distributions at 99% relative humidity, the initial value chosen at cloud base, are shown in Figure 13-6.

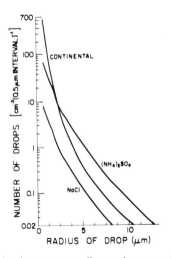

Fig. 13-6.   Initial drop size distribution corresponding to dry aerosol particle distribution in Figure 13-5, each drop being at equilibrium with an environment of 99% relative humidity. (From Lee and Pruppacher, 1977; by courtesy of *Pure and Applied Geophys.*, and the authors.)

TABLE 13-2

Initial sounding for continental and maritime cumulus case considered by Lee and Pruppacher
(1977).

	Continental Case			Maritime Case	
Pressure (mb)	Temperature (°C)	Relative humidity (%)		Temperature (°C)	Relative humidity (%)
900	20.00	90		20.00	95
850	16.10	88		17.07	88
700	3.24	80		7.34	80
500	−17.71	67		−8.77	67
450	−23.94	63		−0.48	0
400	−15.19	0		−	−

For both cloud types Lee and Pruppacher also assumed an initial vertical velocity at cloud base of $1 \, \mathrm{m \, sec^{-1}}$, and an initial updraft radius of 1 km. The entrainment parameter was specified by (11-26) with $C = 0.2$. The assumed environmental conditions of temperature and humidity (considered constant in time) are given in Table 13-2.

Some results of the computations are summarized in Figures 13-7 to 13-13. Figure 13-7 illustrates the behavior we anticipated earlier, namely that the largest hygroscopic nuclei become activated drops which quickly grow to visible size, while the smaller particles produce unactivated drops which undergo partial evaporation after the peak in supersaturation has been achieved. Figure 13-8 points up a fault of the closed, adiabatic model which is easy to understand: the

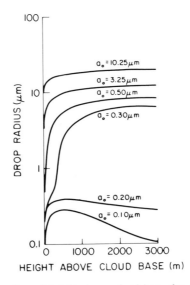

Fig. 13-7. Drop size as a function of height above cloud base, for continental cumulus case with entrainment of air but no nuclei; a_0 is the initial drop size at 99% relative humidity. (From Lee and Pruppacher, 1977; by courtesy of *Pure and Applied Geophys.*, and the authors.)

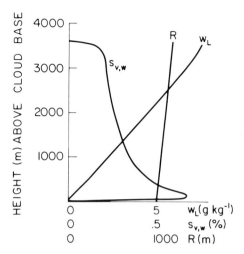

Fig. 13-8.   Continental cumulus case. Variation of supersaturation $s_{v,w}$, liquid water content w_L, and updraft radius R with height above cloud base. Adiabatic model for a constant updraft of 1 m sec^{-1}. (From Lee and Pruppacher, 1977; by courtesy of *Pure and Applied Geophys.*, and the authors.)

lack of mixing with drier air yields an overprediction of liquid water content (cf. Figure 2-15). Also, for the case shown of a specified updraft speed, the peak supersaturation is reached within a few tens of seconds after the air becomes saturated. This implies, in contrast to observations, that the main features of the drop spectrum are determined within a few tens of meters above cloud base. For the adiabatic model in which the updraft is computed (not shown here), the supersaturation more realistically decreases rather slowly with height; however, the computed updraft speed, like w_L, assumes unrealistically large values since no mass exchange with the environment is allowed. Although Figure 13-8 shows results only for the continental cumulus case, very similar behavior holds for the maritime case as well: The features shown are not sensitive functions of the CCN spectrum, but rather depend mainly on the dynamics of the adiabatic parcel model.

The corresponding microphysics of the adiabatic model is also faulty, as Figure 13-9 reveals. The predicted drop size distribution has a single mode at all heights, whereas observations often reveal a double mode distribution, especially at higher levels (recall Figures 2-3 and 2-7b, c). (We should point out that in early studies on condensation growth, the double mode obtained in the *total* particle size distribution, i.e., activated *and* unactivated drops, often was incorrectly interpreted as the observed double mode in the drop size distribution.) Also, Figure 13-9 shows that the originally rather broad aerosol distribution develops into a narrow drop size distribution. This distribution merely shifts as a whole to larger sizes with increasing height in the cloud, thus creating an increasingly pronounced gap between the unactivated particles and the drops. This is also quite in contrast to observations, which reveal the presence of large numbers of small drops ($a \leqslant 5\ \mu$m). Again, the behavior shown in Figure 13-9 holds for the maritime cloud case as well.

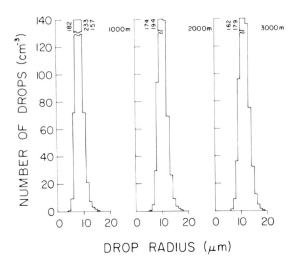

Fig. 13-9. Continental cumulus case. Variation of drop size spectrum with height above cloud base. Adiabatic model for a constant updraft of 1 m sec^{-1}; numbers on top of histograms refer to number of drops in histogram interval. (From Lee and Pruppacher, 1977; by courtesy of *Pure and Applied Geophys.*, and the authors.)

As might be expected, the inclusion of entrainment produces more realistic profiles of updraft speed and radius, liquid water content, and supersaturation (Figures 13-10 and 13-11). However, if the entrained air is devoid of active CCN, the drop spectral shape suffers from the same deficiencies as before: it is still very narrow, and has only a single maximum. The principal change is an overall shift to smaller sizes, reflecting the relative dryness due to mixing with air outside the 'cloud.'

We would expect much better results if active CCN were entrained, since this would provide a continuous source of smaller drops such as are generally observed. This expectation is borne out fairly well by the computed results shown in Figures 13-12 and 13-13. Note that much broad spectra are produced, and that they become bimodal with increasing height.

Nevertheless, condensation models of the type discussed so far are still deficient in various respects, as pointed out by Warner (1973) and Mason and Jonas (1974): (1) In continental clouds the predicted number of drops of radii larger than 20 μm is still too small to account for the initiation of precipitation by the collision and coalescence process (see Chapter 14). (2) In maritime clouds a broad drop size distribution with drops of radii up to 30 μm is predicted largely as a result of the relatively broad initial aerosol size distributions typically chosen. Although such distributions agree well with the recent observations of Meszaros and Vissy (1974) taken at ship level above the ocean surface, considerably fewer particles with r $\approx$ 1 μm are likely to be present at the level of cloud formation. (3) The shape of the predicted drop size distributions is still lopsided as compared to many observed spectra, in that at lower levels of the cloud the drop concentration maximum at the smaller size end of the spectrum is too low relative to that of the second mode for the larger drops.

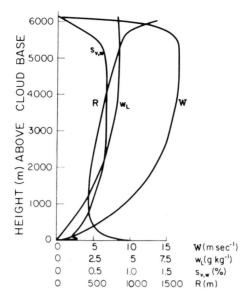

Fig. 13-10. Continental cumulus case. Variation of updraft W, supersaturation $s_{v,w}$, liquid water content w_L, and updraft radius R with height above cloud base. Entrainment model, aerosol particles entrained. (From Lee and Pruppacher, 1977; by courtesy of *Pure and Applied Geophys.*, and the authors.)

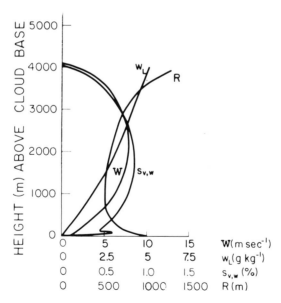

Fig. 13-11. Maritime cumulus case. Variation of updraft W, supersaturation $s_{v,w}$, liquid water content w_L, and updraft radius R with height above cloud base. Entrainment model, aerosol particles entrained. (From Lee and Pruppacher, 1977; by courtesy of *Pure and Applied Geophys.*, and the authors.)

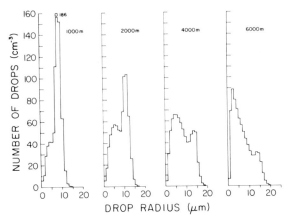

Fig. 13-12. Continental cumulus case. Variation of drop size spectrum with height above cloud base. Entrainment model, aerosol particles entrained; numbers on top of histogram refer to number of drops in histogram interval. (From Lee and Pruppacher, 1977; by courtesy of *Pure and Applied Geophys.*, and the authors.)

In attempting to identify some possible causes for the above mentioned discrepancies, one pervasive and artificial feature of all the condensation models discussed so far comes to mind, namely the fact that they basically involve the lifting of just a single mass of air. It is well known from observations, especially those made with the aid of time lapse photography, that real convective clouds do not grow in this manner. Rather, their development is characterized, at least in its early stages, by a succession of lifting and sinking motions. It is reasonable to expect a strong cumulative effect of such percolating motion, with its cycles of mixing, evaporation, and further condensation, on the evolving cloud drop

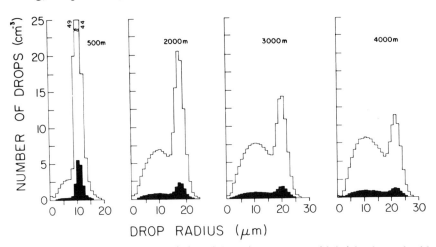

Fig. 13-13. Maritime cumulus case. Variation of drop size spectrum with height above cloud base. Entrainment model; aerosol particles entrained. The shaded area of the histogram refers to NaCl particles, the unshaded area to $(NH_4)_2SO_4$ particles; numbers on top of histogram refer to number of drops in histogram interval. (From Lee and Pruppacher, 1977; by courtesy of *Pure and Applied Geophys.*, and the authors.)

spectra. This expectation is borne out by the condensation model of Mason and Jonas (1974), which, in contrast to all other studies, includes at least a simplified simulation of the observed percolating motion. Thus in their model a single spherical 'thermal' is allowed to rise and then sink back again under the influence of evaporative cooling resulting from the entrainment of drier environmental air. Following this a second thermal is released to raise through the residue of the first. In consequence, though most of the droplets in the subsiding first thermal evaporate, a few of the largest survive to be caught up and experience further growth in the second thermal. In this way it is possible to produce a few large drops without invoking the presence of large hygroscopic particles. Also, the subsidence of the first thermal boosts the relative concentration of small droplets to more realistic values.

Aside from the novel feature of providing for a succession of thermals, the computational details of the model are essentially the same as those of Mason and Chien (1962), Warner (1973), and Wang and Pruppacher (1977), described above. The variations with height and time of W, $s_{v,w}$, w_L, and $w_L/w_{L,ad}$ (the ratio of the actual liquid water content to that which would be produced by the adiabatic lifting of a closed parcel) are shown in Figure 13-14 for the case of a maritime cumulus cloud. In these computations the temperature and pressure at condensation level were taken respectively as 10 °C and 900 mb, and the initial thermal radius, vertical velocity, and excess virtual temperature as 350 m, 1.0 m sec^{-1}, and 0.1 °C, respectively. The lapse rate of the environment was 7 °C km^{-1} and its humidity was held constant at 85%. The computed drop spectra resulting after the ascent of the second thermal are given in Figure 13-15 for heights of 150 and 1400 m above cloud base. Also included in this figure are

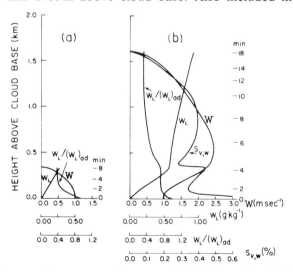

Fig. 13-14.   Maritime cumulus case; entrainment model. Variation of updraft W, supersaturation $s_{v,w}$, liquid water content w_L, and $w_L/(w_L)_{ad}$ with height above cloud base. (a) During ascent of first thermal, (b) during ascent of second thermal through residue of first thermal. Numbers on right of diagrams show the time elapsed since the start of each stage. (From Mason and Jonas, 1974; by courtesy of *Quart. J. Roy. Meteor. Soc.*)

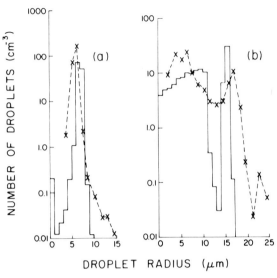

NUMBER OF DROPLETS (cm⁻³)

DROPLET RADIUS (μm)

Fig. 13-15.   Maritime cumulus case; entrainment model. (a) Comparison of drop spectrum predicted by model after ascent of second thermal (150 m above cloud base) with drop spectrum observed by Warner at a similar height in a similar sized cloud of similar total drop concentration. (b) Comparison of computed drop size spectrum at 1400 m above cloud base with mean of two drop size spectra observed by Warner near top of a cloud 1.4 km deep, and of a similar total drop concentration. (From Mason and Jonas, 1974; by courtesy of *Quart. J. Roy. Meteor. Soc.*)

representative observed maritime spectra (Warner, 1969a, b) for clouds with total drop concentrations similar to those in the model cloud. The size distribution of NaCl particles assumed in the model calculations is given in Table 13-3, the total number of particles being 66 (mg air)$^{-1}$ ≈ 66 cm^{-3}.

A comparison between Figures 13-14 and 2-15 shows that $w_L/w_{L,ad}$ according to the model is significantly larger than observed values. This discrepancy has been emphasized by Warner (1975b), who pointed out also that the model results would only deteriorate in this respect if the number of successive thermals were allowed to increase, since each of these would deposit additional moisture in the air to be traversed by the next in the series (for example, note from Figure 13-14b that $w_L/w_{L,ad} \approx 0.8$ in the lowest 400 m of the cloud, which can be seen from Figure 13-14a to be the height achieved by the first thermal). Against this criticism Mason (1975) has argued that the observed values of $w_L/w_{L,ad}$ represent averages over cloud regions which are probably composed largely of the residues of earlier thermals, and that therefore one would expect $w_L/w_{L,ad}$ to be significantly larger in active growing thermals such as are simulated by the Mason-Jonas model.

In spite of the likelihood that the Mason-Jonas cloud is unrealistically wet, its microphysical properties seem to agree better with observations than do those of the other condensation models. Thus we see from Figure 13-15 that the predicted droplet spectra agree quite well with observations. The model drop spectrum for the maritime cloud was found to broaden rapidly, and produced drops with

TABLE 13-3

Initial aerosol particle size distribution for maritime case considered by Mason and Jonas (1974).

m_s (g)	5×10^{-16}	10^{-15}	3×10^{-15}	10^{-14}	3×10^{-14}
r_N (μm)	0.038	0.048	0.069	0.10	0.15
number (cm^{-3})	38	19	6	1.9	0.6
m_s (g)	10^{-13}	3×10^{-13}	10^{-12}	3×10^{-12}	10^{-11}
r_N (μm)	0.22	0.32	0.48	0.69	1.03
number (cm^{-3})	0.19	60×10^{-3}	19×10^{-3}	6×10^{-3}	1.9×10^{-3}

$a \geqslant 25 \mu$m in concentrations of about 300 m^{-3}, and drops with $a \geqslant 20 \mu$m in concentrations of about 10 cm^{-3}, within 30 min; as we shall see in Chapter 15, this is sufficient to continue growth by collision and coalescence. It is particularly noteworthy that the production of these relatively large drops was accomplished without assuming as large a population of large and giant nuclei as has been found necessary in the other condensation models.

Similar model computations by Mason and Jonas for the case of a continental type cloud produced total drop concentrations about an order of magnitude larger than for the maritime cloud, but with very few drops of even 20 μm radius. In this respect the results are quite similar to those obtained from the other condensation models. The qualitative differences between the predicted drop spectra for the maritime- and continental type clouds are generally consistent with observations (see Section 2.1.1), and reflect the greater colloidal stability of continental clouds. However, as with the other condensation growth simulations, the Mason-Jonas model somewhat underestimates the observed capacity of continental type clouds to produce precipitation. As we shall see in Section 15.4, more realistic results evincing a reduced colloidal stability have been obtained by an extension of their model to include simultaneous condensation and coalescence (Jonas and Mason, 1974).

We have noted that condensation growth simulations, and especially those which involve the lifting of a single air mass, produce drop spectra which are unrealistically narrow. It has been argued that this shortcoming may be due to their neglect of the existence of turbulence-induced fluctuations in the local supersaturation in updrafts. The notion that turbulence may significantly increase the dispersion in drop sizes has been pursued especially by Russian and Chinese workers (e.g., Belyayev (1961), Mazin (1965), Sedunov (1965), Levin and Sedunov (1966), Jaw Jeou-Jang (1966), Wen Ching-Sung (1966), Stepanov (1975, 1976)). As one representative result of these efforts, we note that Mazin (1965) predicted turbulence can produce in 5 min a standard deviation of 4 μm in the radius of drops initially centered about 10 μm radius.

On the other hand, quite different results have been obtained by Bartlett (1968), Warner (1969b), and Bartlett and Jonas (1972). They have found that the turbulence-induced spread in radii of drops at any given level is quite small, in fact generally less than 0.2 μm. According to these authors, the key argument for understanding this result is based on the fact that the supersaturations and updrafts which a growing drop encounters are closely correlated to each other. Thus, a droplet which experiences a high supersaturation – and therefore grows rapidly – is at the same time likely to be in a strong updraft which restricts the drop's growth between any two given levels to a relatively short time. Conversely, low supersaturations are associated with small updrafts and longer growth times. It appears that the resulting significant distinction to be drawn between calculations of drop-size dispersions after a given time and those at a given level have generally been overlooked in the studies which have predicted large dispersions. For example, Sedunov (1965) obtained an estimate of the drop-size distribution at a given time, but refers to it as the distribution at a given level. Another fault of much of the Russian and Chinese work is that it generally involves linearized versions of the full governing equations, this being done to permit the extraction of analytical solutions. Unfortunately, some of the physics is unavoidably lost by such simplifications.

We shall now describe briefly the turbulent updraft model of Bartlett and Jonas (1972). They assumed that: (1) the turbulence is homogeneous and stationary; (2) the velocities of all particles in the parcel under study are, at any instant, normally distributed about a mean velocity $\bar{W}$ (which remains constant, or only varies slowly), with a standard deviation of σ_W (thus their simulations do not account for turbulence exchange between adjacent parcels); (3) after a given time interval the distribution of velocity components of a set of particles which originally had the same velocity depends only on the characteristics of the turbulence field, the initial velocity, and the time interval.

In accordance with these assumptions the vertical velocity $W_{t+\tau}$ of a typical parcel at time $t + \tau$, in a cloud where the turbulence energy dissipation rate is ε, can be expressed in terms of its velocity W_t at time t by means of the following relations (Jonas and Bartlett, 1972):

$$W_{t+\tau} = \bar{W} + f_\tau(W_t - \bar{W}) + n\sigma_\tau, \tag{13-42}$$

where

$$\tau_\tau^2 = \sigma_W^2(1 - f_\tau^2), \tag{13-43}$$

and

$$f_\tau = \exp(-\varepsilon\tau/2\,\sigma_W^2). \tag{13-44}$$

In (13-42), n is a number drawn at random from a set which is normally distributed with a mean of zero and unit standard deviation. Each sequence of random numbers produces a realization of the vertical velocity of a typical parcel of air at intervals of τ.

With this extension of the closed adiabatic parcel condensation model, described above, and assuming $\bar{W} = 1$ m sec^{-1}, $\varepsilon = 100$ cm^2 sec^{-3}, an initial temperature and pressure of 10 °C and 900 mb, an initial drop concentration of 10^2 cm^{-3}, and initially monodisperse drops of 5 μm radius, Bartlett and Jonas found a variation of drop radius with height in the cloud as given in Table 13-4. The predicted spread in sizes is seen to be quite small at all levels. Furthermore, similar computations using $\varepsilon = 10^3$ cm^2 sec^{-3} only doubled the corresponding

TABLE 13-4

Variation of cloud drop radius with height in cloud with a turbulent updraft for $\varepsilon = 100 \, \mathrm{cm^2 \, sec^{-2}}$. (From Bartlett and Jonas, 1972; by courtesy of *Quart. J. Roy. Meteor. Soc.*)

Height above starting level (m)	Mean droplet radius (μm)	Standard deviation (μm)
0	5.00	0.00
300	11.39	0.26
600	14.05	0.13
1 200	17.40	0.09
1 800	19.67	0.06
2 400	21.39	0.01
3 000	22.75	0.01

standard deviations at the various heights. Hence it appears unlikely that the neglect of turbulence in the usual condensation models is of any consequence.

13.2.2.2 *Condensation Growth in Stratiform Clouds*

A somewhat different overall theoretical approach from the previous section is appropriate for the problem of describing drop spectral evolution in stratiform clouds, since they generally have longer lifetimes and much weaker updrafts (and hence have smaller vertical extent) than cumuliform clouds. These characteristic features should especially enhance the role played by turbulent exchange of air between the cloud and its environment. Hence it is perhaps not surprising that the model computations of Neiburger and Chien (1960), for which an equation similar to (13-26), combined with an isobaric cooling law and the assumption of a Junge-type size distribution of NaCl particles within a closed air parcel, produced rather unrealistic results. For example, the peak saturation in their simulation was reached only a few tens of seconds after the air parcel became saturated, so that the main characteristics of the drop-size spectrum were determined at an unrealistically early stage. Also, the computed size distribution was considerably narrower than those observed in stratiform clouds.

The first attempt to describe the drop spectra in stratus clouds in terms of a balance between condensation growth and turbulent transport of drops to the cloud boundaries, where they evaporate, was made by Best (1951b, 1952). In his first paper Best obtained an order of magnitude estimate of the mean lifetime of cloud drops subjected to turbulent diffusion. To accomplish this he applied a formula of Sutton (1932) for the standard deviation σ_P of airborne smoke particles from their mean path, namely $2 \sigma_P^2 = c^2(ut)^m$, where u is the wind velocity, t denotes time, c has a value of about 0.08, and m is 1.75 (cgs units). By assuming σ_P can be taken to represent half the cloud thickness, and adopting the value $u = 10 \, \mathrm{m \, sec^{-1}}$, Best found that the resulting values of t, interpreted as the mean lifetime of the drops, were consistent with their mean size as computed from an equation similar to (13-28) and assuming a constant supersaturation of

0.05%. In his second paper Best claims to have accounted as well for the observed drop spectral shapes, but the mathematical development presented in the paper appears to be flawed by several errors.

A more straightforward and complete treatment of this diffusional transport model has been provided by Mason (1952b, 1960b). Mason assumed an idealized cloud with boundaries at $\pm z_0$, and with turbulent eddying motions on a scale $\ell \ll z_0$, so that classical diffusion theory could be applied. Then in terms of the diffusion problem described in Section 12.2, the probability density $w(z, t) = n(z, t)/N$ for finding drops at distance z from the origin $z = 0$ at time t, assuming they were introduced at $z = 0$ and $t = 0$, is governed by the equation

$$\frac{\partial w}{\partial t} = D_e \frac{\partial^2 w}{\partial z^2}, \tag{13-45}$$

where D_e is the effective eddy diffusion coefficient. Drops which arrive at the boundaries are assumed lost, so that the boundary conditions for (13-45) are $w(\pm z_0, t) = 0$. Also, according to the above description the initial condition may be expressed as $w(z, 0) = \delta(z)$, where δ is the delta function.

The solution to this problem may be obtained most easily by means of the procedure described in Appendix A-12.4, i.e., by simulating the boundary conditions with an extended initial distribution, so that the formal solution form of (A.12-7) may be applied. The appropriate simulation is easily obtained by inspection. Thus, the boundary conditions $w(\pm z_0, t) = 0$ may be maintained by introducing the respective source terms $-\delta(z + 2 z_0)$ and $-\delta(z - 2 z_0)$ at the boundaries of the extended region $-2 z_0 \leq z \leq 2 z_0$. A further extension to $-4 z_0 \leq z \leq 4 z_0$ may be obtained by introducing the sources $\delta(z + 4 z_0)$ and $\delta(z - 4 z_0)$, and so forth. Thus the initial distribution for the equivalent problem of diffusion in the infinite space $-\infty < z < \infty$ is

$$w(z, 0) = \delta(z) + \sum_{n=1}^{\infty} (-1)^n [\delta(z + 2 n z_0) + \delta(z - 2 n z_0)]. \tag{13-46}$$

On substituting (13-46) into (A.12-7) (with $w(z, t)$ replacing $n'(z, t)$), the following solution is immediately obtained for the region $-z_0 \leq z \leq z_0$:

$$w(z, t) = \frac{1}{(4 \pi D_e t)^{1/2}} \left\{ e^{-z^2/4 D_e t} + \sum_{n=0}^{\infty} (-1)^n [e^{-(z+2 n z_0)^2/4 D_e t} + e^{-(z-2 n z_0)^2/4 D_e t}] \right\}. \tag{13-47}$$

From (13-47) the probability that the drops may be found somewhere in the cloud at time t is

$$P(t) = \int_{-z_0}^{z_0} w(z, t) \, dz = \text{erf} (x) + \sum_{n=1}^{\infty} (-1)^n \{\text{erf} [(2n + 1)x] - \text{erf} [(2n - 1)x]\}, \tag{13-48}$$

where $x = z_0/(4 D_e t)^{1/2}$. Thus if we denote the half-life of the drops by λ, then $P(\lambda) = 0.5$, and from (13-48) the corresponding x is $x_\lambda = 0.81$. From this the fraction of drops $P(n\lambda)$ which remain in the cloud for any other time $n\lambda$ can be determined from (13-48) by setting $x = x_n = 0.81 \, n^{-1/2}$.

Having obtained the values of $P(n\lambda)$, Mason proceeded next with a step by step computation of the fraction of drops (the same fraction for all sizes) removed from the cloud in a given time interval, assuming that lost drops are replaced by an equal number of unactivated nuclei in order to maintain a constant total concentration. Thus the drops present at the beginning of a time step will include those surviving previous time intervals plus those activated during the last time step. Corresponding to this time history of the drop population, a cumulative size distribution was computed by integrating an equation of the type given by (13-28) over the lifetimes belonging to each fraction of the total population. This was done under the assumptions of a constant supersaturation $s_{v,w} = 0.05\%$, and an initial NaCl particle size distribution specified as follows: 270 cm^{-3} of particles of mass $m_s = 10^{-14}$ g, 27 cm^{-3} of $m_s = 10^{-13}$ g, and 3 cm^{-3} of $m_s = 10^{-12}$ g. Also, a half-life of $\lambda = 10^3$ sec was specified, based on the above mentioned smoke dispersion formula of Sutton (1932) and the assumptions of a cloud 150 m thick in a region where the horizontal wind speed is 5 m sec^{-1}.

The cumulative drop size distribution obtained in this manner is shown in Figure 13-16 for a growth time of 6000 sec, i.e., 1 hr 40 min. At this time the median drop radius is 4.8 μm and the liquid water content is 0.4 g m^{-3}. The characteristic spectral shape, which is established after about 2000 sec, is seen to agree fairly well with the drop spectrum observed by Neiburger (1949) near the middle of a California stratus layer with a similar total drop concentration. The extent of agreement suggests the theoretical model is basically sound, in spite of the fact that it is obviously quite idealized and requires the somewhat arbitrary specification of a number of parameters. On the other hand, Mason's model is insufficiently general to account for the double mode observed in some stratus layers (see Figure 2-7b).

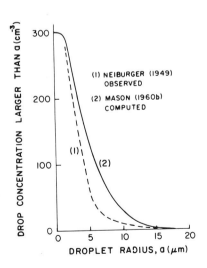

Fig. 13-16. Comparison of computed and observed cumulative size distributions of droplets in stratus cloud. (From Mason, 1960b; by courtesy of Amer. Meteor. Soc., and the author.)

13.2.3 STEADY STATE EVAPORATION OF WATER DROPS FALLING IN SUBSATURATED AIR

In this section we shall consider the ventilating effect of a drop's motion on its rate of diffusional growth or evaporation. The results presented here will justify our previous neglect of the ventilation effect for the small sizes which are involved in the early stages of evolution of cloud drop spectra.

From the discussions of Sections 12.7.1 and 13.1.1, it is clear that our problem here is to solve the steady state convective diffusion equation,

$$\bar{u} \cdot \nabla \rho_v = D_v \nabla^2 \rho_v, \tag{13-49}$$

for a spherical drop of radius a past which moist air containing water vapor of density ρ_v flows with velocity $\bar{u}$. There are several conventional ways of describing the resulting convective enhancement of the diffusional growth or evaporation rates. Thus in the cloud physics literature one generally uses the *mean ventilation coefficient*, $\bar{f}_v$, defined as the ratio of the water mass fluxes to or from the drop for the cases of a moving and a motionless drop, viz.,

$$\bar{f}_v \equiv \frac{dm/dt}{(dm/dt)_0}, \tag{13-50}$$

where $(dm/dt)_0$ denotes the mass flux for the motionless case of pure diffusion, and is given by (13-6) for a growing drop (the same expression but with opposite sign applies to the case of an evaporating drop). A closely related quantity used in the chemical engineering literature is the *mean mass transfer coefficient*, $\bar{k}_v$, defined by

$$\bar{k}_v \equiv \frac{\int_S \left[-D_v \left(\frac{\partial \rho_v}{\partial r} \right) \right] dS}{4 \pi a^2 (\rho_{v,a} - \rho_{v,\infty})}, \tag{13-51}$$

where S denotes the drop surface. On comparing (13-6), (13-50), and (13-51), we see that also

$$\bar{k}_v = \frac{D_v \bar{f}_v}{a}. \tag{13-52}$$

Finally, another useful dimensionless measure found in the chemical engineering literature is the *mean Sherwood number*, $\bar{N}_{Sh}$, defined in terms of $\bar{k}_v$ as

$$\bar{N}_{Sh} \equiv \frac{2 \bar{k}_v a}{D_v} = 2 \bar{f}_v. \tag{13-53}$$

From this last result and (13-50), we see that $\bar{N}_{Sh} = 2$ for a motionless drop.

Of course to solve (13-49) for ρ_v requires specifying $\bar{u}$, and generally this can be accomplished only through a numerical solution of the Navier-Stokes equation for $\bar{u}$; hence for nearly all cases of interest (13-49) must be solved by numerical methods. To do this one generally first renders (13-49) dimensionless by introducing the variables $r' \equiv r/a$, $\rho_v' \equiv (\rho_v - \rho_\infty)/(\rho_{v,a} - \rho_{v,\infty})$, and $\bar{u}' \equiv$

$\tilde{u}/U_\infty$; in terms of these (13-49) becomes

$$\tilde{u}' \cdot \nabla' \rho_v' = \frac{2}{N_{Pe,v}} \nabla'^2 \rho_v, \tag{13-54}$$

where $\nabla' = a\nabla$ and $N_{Pe,v} = 2 U_\infty a/D_v$ is the Peclet number for vapor transport (cf. (12-84)). An axially symmetric numerical solution of (13-54) may be carried out in spherical coordinates (r', θ, ϕ) in conjunction with (10-71) by writing $\tilde{u}' = \hat{e}_\phi \times (\nabla \psi'/r' \sin \theta')$ (cf. (A.10-18)), and by imposing the following boundary conditions: (1) On the sphere surface $(r' = 1)$, $\psi' = 0$, $\zeta' = E'^2 \psi'/\sin \theta$, and $\rho_v' = 1$; (2) along the symmetry axis $(\theta = 0, \pi)$, $\psi' = 0$, $\zeta' = 0$, $\partial \rho_v'/\partial \theta = 0$; (3) far from the sphere surface $(r' = r_\infty')$, $\psi' = \frac{1}{2}r_\infty'^2 \sin^2 \theta$, $\zeta' = 0$, and $\rho_v' = 0$. From the solution for ρ_v' one may then also express the ventilation effect in terms of a local Sherwood number, $N_{Sh}(\theta)$, which is related to the mean Sherwood number according to

$$\bar{N}_{Sh} = \frac{1}{2} \int_0^\pi N_{Sh}(\theta) \sin \theta \, d\theta. \tag{13-55}$$

On comparing (13-51), (13-53), and (13-55), we thus find that

$$N_{Sh}(\theta) = -2 \left(\frac{\partial \rho_v'(r', \theta)}{\partial r'} \right)_{r'=1}. \tag{13-56}$$

Numerical solutions as outlined above have been obtained by Woo (1971) and Woo and Hamielec (1971) for an evaporating drop in air with a Schmidt number $N_{Sc,v} = 0.71$ (cf. (12-85b)). Some resulting plots of $N_{Sh}(\theta)$ versus θ are shown in Figure 13-17 for various $N_{Re} \gg 1$. It is seen that the effect of ventilation varies strongly with the angle θ from the forward stagnation point, and is smallest near the location of flow separation from the drop. Also, the ventilation effect, and hence the rate of evaporation, is greatest on the upstream side of the drop, as would be expected.

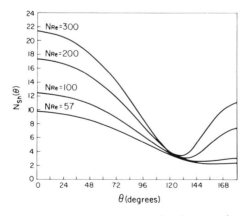

Fig. 13-17. Variation of local Sherwood number for ventilated water sphere in air ($N_{Sc,v} = 0.71$) as a function of angle θ from foreward stagnation point on sphere. (From Woo and Hamielec, 1971; by courtesy of Amer. Meteor. Soc., and the authors.)

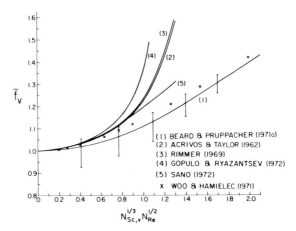

Fig. 13-18. Comparison of experimentally determined ventilation coefficient for water drops in air at low Reynolds numbers with numerical results of Woo and Hamielec (1971), and with various analytical results. (From Beard and Pruppacher, 1971; by courtesy of Amer. Meteor. Soc., and the authors.)

We have shown in Section 12.7.1 how boundary layer theory leads to the prediction that the convective enhancement of the mass or particle flux to a sphere is proportional to $N_{Sc}^{1/3} N_{Re}^{1/2}$ for $N_{Re} \gg 1$ (cf. (12-89)). The rate at which this asymptotic behavior is approached is indicated by Figures 13-18 and 13-19, where the numerical results of Woo and Hamielec for $\bar{f}_v = N_{Sh}/2$ are plotted for various conditions. Recalling that $N_{Sc,v} = 0.71$ in their computations, we see that

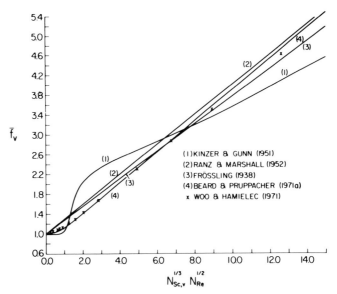

Fig. 13-19. Comparison of various experimentally determined ventilation coefficients for water drops in air at moderate Reynolds numbers with numerical results of Woo and Hamielec (1971). (From Beard and Pruppacher, 1971a; by courtesy of Amer. Meteor. Soc., and the authors.)

the expected linear dependence of the ventilation rate on $N_{Sc}^{1/3}N_{Re}^{1/2}$ begins near $N_{Re} \approx 3$. For larger values the ventilation coefficient thus increases nearly as $N_{Re}^{1/2}$; for example, at $N_{Re} = 250$ ($a = 490\,\mu m$), $\bar{f}_v \approx 5$; i.e., the evaporation rate is 5 times what it would be if the drop were stationary.

Figures 13-18 and 13-19 also include comparisons with experiment and various analytical results. It can be seen that the wind tunnel measurements of drop evaporation by Beard and Pruppacher (1971a) are in excellent agreement with the numerical results of Woo, and Woo and Hamielec. The earlier experimental investigations can be seen to have produced some conflicting results. These can be traced to various experimental deficiencies. For example, Frössling (1938) and Ranz and Marshall (1952) studied drops which were not freely suspended in an air stream, but rather were attached to supports, while Kinzer and Gunn's (1951) experiment suffered from inaccuracies in the assumed or determined values of vapor diffusivity, drop surface temperatures, and terminal velocities.

The values of $\bar{f}_v$ obtained by Beard and Pruppacher may be closely approximated by the following empirical expressions:

$$\bar{f}_v = 1.00 + 0.108(N_{Sc,v}^{1/3}N_{Re}^{1/2})^2, \tag{13-57}$$

for $N_{Sc,v}^{1/3}N_{Re}^{1/2} < 1.4$, i.e., for $N_{Re} \lesssim 2.5$ (with $N_{Sc,v} = 0.71$), and

$$\bar{f}_v = 0.78 + 0.308\,N_{Sc,v}^{1/3}N_{Re}^{1/2}, \tag{13-58}$$

for $N_{Sc,v}^{1/3}N_{Re}^{1/2} > 1.4$ ($N_{Re} > 2.5$). An extrapolation of (13-58) to drops of sizes $a_0 \lesssim 2\,mm$ is given in Figure 13-20. Considering the fact that vortex shedding from the rear of such drops most likely slightly enhances the rate of vapor transport from the drop to the environment, and considering that the deformation of drops of $a_0 \gtrsim 500\,\mu m$ reduces the fraction of the drop area which is effectively ventilated, one may assume – as was done in the extrapolation shown in the figure – that these two opposing effects approximately cancel each other.

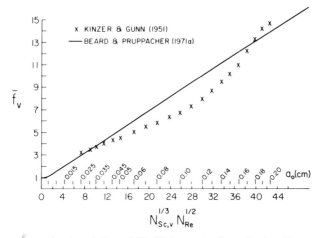

Fig. 13-20. Comparison of extrapolation of (13-58) for water drops in air with experimental results of Kinzer and Gunn (1951).

This assumption is supported by the fair agreement shown in the figure between the extrapolation of (13-58) for spherical drops and the experimental results of Kinzer and Gunn for deformed drops falling in air.

Note also from Figure 13-20 that $\bar{f}_v \approx 1$ for drops of $a \lesssim 50 \, \mu\text{m}$. This result justifies the neglect of ventilation effects in studies of drop spectral evolution by condensation growth.

We see from (13-57) that $\bar{f}_v$ no longer varies approximately linearly with $N_{Sc,v}^{1/3} N_{Re}^{1/2}$ for $N_{Re} \lesssim 2.5$. This is of course what we would expect, since boundary layer theory becomes inapplicable for $N_{Re} \lesssim O(1)$. However, for such decreasing N_{Re} the flow becomes increasingly linear and simple, so that various analytical approximations to $\bar{f}_v$ become possible. For example, (13-54) has been solved by the method of matched asymptotic expansions (see case 4 of Section 10.2.4) for small N_{Pe} and N_{Re} by Acrivos and Taylor (1962), Rimmer (1969), Gupalo and Ryazantsev (1972), and Sano (1972). The differences in these studies are due to the assumed form for $\vec{u}'$. Thus, Acrivos and Taylor assumed Stokes flow; Rimmer, and Gupalo and Ryazantsev assumed 'Proudman-Pearson flow,' expressed by (10-48); and Sano assumed potential flow. As one representative example, we shall quote the series expression obtained by Acrivos and Taylor:

$$\bar{N}_{Sh} = 2 + \frac{N_{Pe}}{2} + \frac{N_{Pe}^2 \ln N_{Pe}}{4} + 0.03404 \, N_{Pe}^2 + \frac{N_{Pe}^3 \ln N_{Pe}}{16} + \cdots. \qquad (13-59)$$

This result and some others are plotted in Figure 13-18. Note that the experimental, numerical, and analytical results merge as $\bar{f}_v \to 1$ with $N_{Re} \to 0$. Note also that the analytical results begin to diverge noticeably from the experimental and numerical results as $N_{Re} \gtrsim 0.1$.

To determine the rate of evaporation of a falling drop we must also take into account the convective-diffusional transfer of heat to the drop from the environment. Because of the complete mathematical analogy between problems of convective heat and mass transfer (cf. (13-2) and (13-15)), we may express the ventilation effect on heat transfer in terms of a mean ventilation coefficient for heat, $\bar{f}_h$, which is obtained from $\bar{f}_v$ merely by replacing D_v with κ_a, i.e., $\bar{f}_h = \bar{f}_v(D_v \to \kappa_a)$. Similarly, corresponding to the mean Sherwood number for mass transfer there is the *mean Nusselt number*, $\bar{N}_{Nu}$, for heat transfer, given by $\bar{N}_{Nu} = \bar{N}_{Sh} (D_v \to \kappa_a)$.

Because the ventilation effect shows a strong local variation over the drop surface (Figure 13-17), there must be a non-zero temperature gradient at the surface. This gradient will be degraded to some extent by the drop internal circulation, but unfortunately there are no computations available with which to estimate the resultant steady state gradient. For lack of any information on this point it is customary to assume the local cooling effect is negligible, so that the drop has an isothermal surface at temperature T_a. Simple order of magnitude estimates indicate this is probably a very good assumption, because of the large thermal conductivity of water and the expected small differences in the heat flux around the drop surface.

In order to determine T_a for an evaporating drop we first write, in analogy to

(13-22),

$$\frac{dq}{dt} = \bar{f}_h \left(\frac{dq}{dt}\right)_0 = -L_e \frac{dm}{dt} = L_e \bar{f}_v \left(\frac{dm}{dt}\right)_0,$$ (13-60)

for the coupling of the rates of heat and mass exchange for the ventilated drop. Then on substituting (13-7) and (13-19) into this equation, we find that

$$T_a = T_\infty - \frac{L_e D_v M_w}{k_a \mathcal{R}} \left(\frac{e_a}{T_a} - \frac{e_\infty}{T_\infty}\right)\left(\frac{\bar{f}_v}{\bar{f}_h}\right),$$ (13-61)

where $e_a = e_a(e_{sat,w}(T_\infty))$ through (6-26b) and (13-25).

With T_a available the drop evaporation rate may be determined from (13-60) and (13-7), viz.,

$$a \frac{da}{dt} = -\frac{M_w D_v}{\rho_w \mathcal{R}} \left(\frac{e_a}{T_a} - \frac{e_\infty}{T_\infty}\right) \bar{f}_v.$$ (13-62)

Computations of drop evaporation based on (13-61) and (13-62) are shown in Figures 13-21 and 13-22 for different environmental conditions. (Of course (13-62) may also be used to experimentally determine $\bar{f}_v$ from observed drop evaporation rates $a\, da/dt$.) From Figure 13-21 we see that, as expected, the fate of drops falling from cloud base depends strongly on the prevailing temperature and humidity structure of the atmosphere. Similarly, Figure 13-22 shows that the probability of a drop surviving a fall through a layer of air 300 m thick, of 90% relative humidity and at $T = 0\,°C$ and $p = 765$ mb, is a sensitive function of its initial size. Thus, drops of $a > 150\,\mu m$ experience a negligible change in size, while drops of $a < 100\,\mu m$ do not survive the fall.

Another related quantity of interest is the time required for a falling drop to achieve its quasi-steady state temperature difference between itself and the environment. Let us consider this adaptation time for the case of a drop of initial temperature $T_{a,0} = T_a$ (t = 0) placed abruptly in an environment at $T_\infty > T_{a,0}$ and

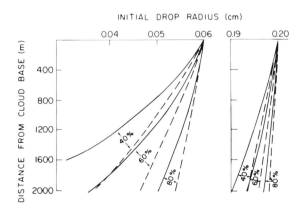

Fig. 13-21. Effect of humidity and thermal structure of ambient air on rate of evaporation of rain and drizzle drops. Solid line: lapse rate 1 °C/100 m; dashed line: isothermal atmosphere; cloud base temperature 5 °C. (From Kühme, 1968; by courtesy of the author.)

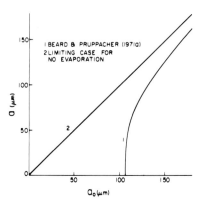

Fig. 13-22.   Final size a of various drops of initial size a_0 after an isobaric and isothermal excursion of 300 m into air of 90% relative humidity, at 0 °C and 765 mb. (From Beard and Pruppacher, 1971a; by courtesy of Amer. Meteor. Soc., and the authors.)

with a relative humidity $\phi_v < 1$. Then because of evaporative cooling the instantaneous drop temperature $T_a(t)$ satisfies $T_a < T_\infty$, so that there is a conductive heat flux from the air to the drop given by $4\pi\bar{f}_h a k_a(T_\infty - T_a)$ (cf. (3-19)). At the same time the rate at which heat is lost from the drop due to its evaporation is $-L_e\, dm/dt$ (>0 since $dm/dt < 0$). Thus we find that the total rate at which the drop gains heat is given by

$$\left(\frac{dq}{dt}\right)_{\text{total}} = 4\pi\bar{f}_h a k_a(T_\infty - T_a) + L_e\frac{dm}{dt}. \tag{13-63}$$

This total heat flux to the drop must cause the temperature difference between the drop and its environment to change at a rate determined by the obvious heat balance relation $(dq/dt)_{\text{total}} = mc_w\, d(T_a - T_\infty)/dt$, where c_w is the specific heat capacity of water and m is the mass of the drop. From this relation, (13-63), and (13-50) and (13-16) for dm/dt, we obtain the governing equation for temperature adaptation:

$$\frac{a^2\rho_w c_w}{3}\frac{d}{dt}(T_\infty - T_a) = -k_a(T_\infty - T_a)\bar{f}_h - L_e D_v(\rho_{v,\infty} - \rho_{v,a})\bar{f}_v. \tag{13-64}$$

Let us now assume that the drop is sufficiently large that curvature effects are negligible; then we may set $\rho_{v,a} = \rho_{v,\text{sat}}(T_a)$. If we make the further approximation that $[\rho_{v,\text{sat}}(T_\infty) - \rho_{v,\text{sat}}(T_a)]/(T_\infty - T_a) \approx (d\rho_v/dT)_{\text{sat}}$, which represents the mean slope of the saturation vapor density curve over the interval $T_a \leqslant T \leqslant T_\infty$, then (13-64) may be expressed in the form

$$\frac{d(T_\infty - T_a)}{dt} = -A(T_\infty - T_a) + B, \tag{13-65}$$

where

$$A = \frac{3}{a^2\rho_w c_w}\left[k_a\bar{f}_h + L_e D_v\left(\overline{\frac{d\rho_v}{dT}}\right)_{\text{sat}}\bar{f}_v\right], \tag{13-66a}$$

and

$$B = \frac{3 \, D_v L_e \bar{f}_v}{a^2 \rho_w c_{w'}} [(1 - \phi_v) \rho_{v,sat}(T_\infty)], \tag{13-66b}$$

where $\phi_v = \rho_{v,\infty}/\rho_{v,sat}(T_\infty)$. An approximate solution of (13-65)–(13-66) is

$$T_\infty - T_a(t) - \delta = (T_\infty - T_{a,0} - \delta)e^{-t/\tau}, \tag{13-67}$$

where $\delta = \bar{B}/\bar{A}$ is the steady state temperature difference between the drop and its environment (i.e., $T_\infty - T_a$ ($t = \infty$) $= \delta$), and where the relaxation time τ is

$$\tau = \frac{1}{\bar{A}} = \frac{a^2 \bar{\rho}_w \bar{c}_w}{3 \left[k_a \bar{f}_h + \bar{L}_e \bar{D}_v \left(\dfrac{\overline{d\rho_v}}{dT}\right)_{sat} \bar{f}_v \right]} \tag{13-68}$$

(Kinzer and Gunn, 1951). The bars over $\bar{A}$, $\bar{B}$, $\bar{\rho}_w$, $\bar{c}_w$, $\bar{L}_e$, and $\bar{D}_v$ denote averages over the integration interval.

A plot of $\tau = \tau(a)$ is given in Figure 13-23 for three different drop tempera-tures. To obtain these curves it has been assumed that $\bar{f}_h = \bar{f}_v$; also, $(d\rho_v/dT)_{sat}$ has been evaluated for $T = T_a$. It can be seen that, as expected, larger drops have larger adaptation times. The only experimental results which are available for comparison with the theoretical values are those of Kinzer and Gunn (1951). They studied drops of $a_0 = 1.35$ mm radius cooling from 22.4 °C to 14.9 °C while falling in subsaturated air. The observed adaptation time was 4.4 sec, at which time the drop temperature was about 17.6 °C. For these data (13-68) predicts $\tau \approx 4.6$ sec, in surprisingly good agreement with the experimental value, considering the approximations involved in (13-68) and the uncertainties in Kinzer and Gunn's experiments. This good agreement suggests the neglect of local unsteadiness (terms involving $\partial/\partial t$) in the derivation of (13-68) are of little consequence. This indication has been verified quantitatively by Watts and Farhi (1975), who solved for τ including the effect of local time dependence.

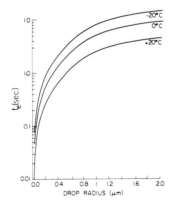

Fig. 13-23.   Variation of the e-folding adaptation time for evaporating, ventilated water drops falling at terminal velocity in subsaturated air.

13.3 Diffusional Growth of Ice Crystals

13.3.1 GROWTH OF A STATIONARY ICE CRYSTAL

The diffusional growth of simple ice crystals can be treated in the same manner as for drops, by making use of an analogy between the governing equations and boundary conditions for electrostatic and diffusion problems (Stefan, 1873; Jeffreys, 1918). To understand this approach, recall that the electrostatic potential function Φ outside a charged conducting body (assumed to be the only source for Φ) satisfies Laplace's equation, i.e., $\nabla^2\Phi = 0$, and that the boundary conditions on Φ are $\Phi = \Phi_S$ = constant on the conductor and $\Phi = \Phi_\infty$ = constant at infinity. Similarly, if we now consider a stationary growing or sublimating ice particle of the same geometry as the conducting body, then in the steady state the vapor density field ρ_v also satisfies Laplace's equation and, if the particle surface is at a uniform temperature T_S, the boundary conditions are similarly given by $\rho_v = \rho_{v,S} = \rho_{v,sat}(T_S)$ = constant on the particle (ignoring the dependence of $\rho_{v,sat}$ on local curvature) and $\rho_v = \rho_{v,\infty}$ = constant at infinity. Therefore, with this complete analogy we can borrow known solutions for electrostatic problems and use them to describe the diffusional growth of ice particles of corresponding geometry.

In particular, the growth rate of the particle can be written down immediately if the capacitance C of the corresponding conductor is known. Thus by integrating Gauss's law, (17-2), over the surface S of the conductor we have, with $\vec{E} = -\nabla\Phi$,

$$\int_S \nabla\Phi \cdot \hat{n}\, dS = -4\,\pi Q = -4\,\pi C(\Phi_S - \Phi_\infty), \qquad (13\text{-}69)$$

where Q is the total charge on the conductor, and where the last form on the right side of (13-69) follows from the definition of capacitance. Then from the analogy between the Φ and ρ_v fields and the physical meaning of $-D_v\nabla\rho_v|_S$ as the vapor mass flux vector at the ice particle surface, we immediately conclude that the growth rate of the particle is given by

$$\left(\frac{dm}{dt}\right)_0 = \int_S D_v\nabla\rho_v \cdot \hat{n}\, dS = -4\,\pi D_v C(\rho_{v,s} - \rho_{v,\infty}). \qquad (13\text{-}70)$$

A comparison of (13-70) and (13-6) shows that they are formally equivalent, with the capacitance of the particle or crystal replacing the radius of the sphere. Therefore we may immediately write down a growth rate equation for the crystal in a form analogous to (13-28), but expressed in terms of dm/dt and with C replacing a; for negligible curvature effects ($y \approx 0$), the result is

$$\left(\frac{dm}{dt}\right)_0 \approx \frac{4\,\pi C s_{v,i}}{\dfrac{\mathscr{R}T_\infty}{e_{sat,i}(T_\infty)D_v'M_w} + \dfrac{L_s}{k_a'T_\infty}\left(\dfrac{L_s M_w}{\mathscr{R}T_\infty} - 1\right)}, \qquad (13\text{-}71)$$

where now the subscript i refers to the ice phase, and L_s is the specific latent heat of sublimation.

In order to apply (13-71) to a particular crystal form we must specify C, which is a function only of the crystal geometry. For example, in the simplest case of a spherical particle of radius a, $C = a$ (in electrostatic units), and we recover the previous description. As we recall from Section 10.4, the hydrodynamic behavior of many simple crystals can be described adequately in terms of the behavior of very simple geometric forms, such as finite circular cylinders, thin circular disks, and prolate or oblate spheroids. Similar idealizations work well in the present context also, and so we shall now quote the capacitances for some simple geometries (in e.s.u.). To model the shape of a simple thin hexagonal ice plate we use the idealization of a circular disk of radius a, for which

$$C = \frac{2a}{\pi}.$$ (13-72)

The shape of simple ice plates of various thicknesses may be modeled by an oblate spheroid of semi-major and minor axes lengths a and b, for which

$$C = \frac{ae}{\sin^{-1}e}, \quad e = \left(1 - \frac{b^2}{a^2}\right)^{1/2}.$$ (13-73)

The shape of a columnar crystal may be modeled by a prolate spheroid of semi-major and minor axes a and b, for which

$$C = \frac{A}{\ln\left[(a+A)/b\right]}, \quad A = (a^2 - b^2)^{1/2}.$$ (13-74)

Finally, for $b \ll a$, (13-74) transforms into

$$C = \frac{a}{\ln(2a/b)},$$ (13-75)

which is applicable to a long, thin needle.

Measurements by McDonald (1963c) and Podzimek (1966) of the capacitances of metal models of snow crystals agree surprisingly well with the above theoretical values for the idealized forms. Thus, the capacitance of a simple hexagonal plate (ice crystal shape P1a in the nomenclature of Figure 2-27) is within 4% of the value for a circular disk of the same area. Similarly, Podzimek compared the capacitance C_0 of a prolate spheroid and that (C) of a hexagonal prism of the same maximum dimensions, and found that $C/C_0 = 1.116$ for a solid short prism, and $C/C_0 = 1.099$ for a short prism with hourglass-like hollow ends. The capacitances C of various dendritic models were found to be only a little less than that (C_0) of a hexagonal plate having the same maximum dimensions. For example, referring again to the crystal forms shown in Figure 2-27, McDonald found that: (1) for crystal shape P1c, $C/C_0 = 0.98$; (2) for shape P2a, $C/C_0 = 0.97$, decreasing to 0.91 with decreasing end plate size; (3) for shape P2g, $C/C_0 = 0.97$, decreasing to 0.80 with increasing openness of the dendrite; (4) for shape P1d, $C/C_0 = 0.77$. Similar values were also obtained by Podzimek. The observed insensitivity of C to such large variations in crystal surface area for crystals of the same maximum dimensions is apparently due to the increasing edge length with decreasing area; such edges of high curvature can store relatively large amounts of charge and hence compensate for the loss of surface.

It is of interest to note that (13-71) can be used as a basic model for the Wegener-Bergeron-Findeisen mechanism of precipitation formation, described in Chapter 1. In order to apply (13-71) to a study of this mechanism, according to which ice crystals grow by vapor diffusion at the expense of supercooled drops, we shall assume a water saturated environment corresponding to a preponderance of supercooled drops. This condition specifies the supersaturation as a function of temperature according to the relation $s_{v,i} = (e_{sat,w}/e_{sat,i}) - 1 > 0$. The resulting normalized mass growth rate, $(4 \pi C)^{-1} dm/dt$, calculated from (13-71) is displayed as a function of temperature in Figure 13-24 for two pressure levels. Note that the growth curves each have a single maximum, near $-14\,°C$ for the 1000 mb curve and $-17\,°C$ for the 500 mb curve. From our discussions in Section 4.9 we might have expected the maxima to occur near $-12\,°C$, since at that temperature the difference $e_{sat,w} - e_{sat,i}$ is a maximum (see Figure 4-5). The reason the actual maxima are shifted to lower temperatures is that local heating from latent heat release causes the vapor pressure difference between the crystal surface and its environment to be reduced slightly. The system must therefore cool below $-12\,°C$ before the vapor flux and hence the growth rate can achieve its maximum value.

Finally, it should be recalled from Section 2.2 that the growth habit of ice crystals is strongly dependent on both temperature and vapor pressure. This implies that C is also indirectly a function of these variables, through its dependence on crystal geometry. Hence (13-71) can provide a complete description of crystal growth only if $C = C(T, e)$ is known (see also Section 13.3.3).

13.3.2 GROWTH OF A VENTILATED ICE CRYSTAL

If an ice crystal has grown by vapor diffusion to a size at which it has an appreciable fall velocity, it is necessary to take into account the effect of

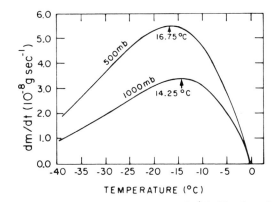

Fig. 13-24. Normalized diffusion mass growth rate $(4 \pi C)^{-1} dm/dt$ of an ice crystal in a water saturated environment as a function of temperature. (From *Elements of Cloud Physics* by H. R. Byers, copyrighted Univ. Chicago Press, 1965.)

ventilation on the diffusion of water vapor and heat. This may be done through methods which are entirely analogous to those we have discussed in Section 13.2.3 for the case of spherical water drops, the only additional complication in the present application being the more complex geometry of ice particles. Fortunately, however, experiments on heat and mass transfer from bodies of various shape (e.g., Pasternak and Gauvin (1960), Brenner (1963), Skelland and Cornish (1963)) have shown that most of the complications arising from ice crystal geometry may be effectively circumvented if the ventilation effect is described in terms of a Sherwood number or ventilation coefficient which depends on a particular characteristic length, L*.

The appropriate definition for L* follows from a recognition of the physical fact that the degree of ventilation is controlled by that portion of the particle's area which is most directly exposed to the oncoming flow. Thus we define L* as the ratio of the total surface area Ω of the body to the perimeter P of its area projected in the flow direction:

$$L^* \equiv \frac{\Omega}{P}. \tag{13-76}$$

As a trivial example, we see that for a sphere $L^* = 2\,a$. The corresponding mean Sherwood number is (cf. (13-53))

$$N_{Sh,L^*} \equiv \frac{\bar{k}_{v,\Omega} L^*}{D_v}, \tag{13-77}$$

where, in analogy with (13-51), the mean mass transfer coefficient is defined by

$$\bar{k}_{v,\Omega} \equiv \frac{-dm/dt}{\Omega(\rho_{v,s} - \rho_{v,\infty})}. \tag{13-78}$$

(Note that $\bar{k}_{v,\Omega} \geq 0$ for particle growth or evaporation.)

From (13-76)–(13-78) the mass rate of change may be expressed in terms of the mean Sherwood number as

$$\frac{dm}{dt} = -PD_v(\rho_{v,s} - \rho_{v,\infty})\bar{N}_{Sh,L^*}. \tag{13-79}$$

On comparing (13-79) with the corresponding expression for a stationary crystal, (13-70), we see that the ventilation coefficient, $\bar{f}_{v,L^*}$, defined as in (13-50), is

$$\bar{f}_{v,L^*} \equiv \frac{dm/dt}{(dm/dt)_0} = \frac{P}{4\pi C}\bar{N}_{Sh,L^*}. \tag{13-80}$$

From this equation we see that for a stationary particle,

$$\bar{f}_{v,L^*} = 1, \quad (\bar{N}_{Sh,L^*})_0 = \frac{4\pi C}{P}. \tag{13-81}$$

In order to make use of (13-79), given $\bar{N}_{Sh,L^*}$, we need expressions for the perimeter P. For a sphere, $P = 2\pi a$; this is also the appropriate expression for a disk or oblate spheroid falling with its maximum dimension normal to the flow direction. For a prolate spheroid in the falling mode of a columnar ice crystal (with its longest extension, $L = 2a$, perpendicular to the direction of fall), $P = \pi L y$ with $y = -0.25\,e^2 - 0.0469\,e^4 - 0.0195\,e^6 - 0.0107\,e^8 - 0.0067\,e^{10} - \cdots$, where $e = [(1 - (b/a)^2]^{1/2}$ is the eccentricity of the spheroid.

Since from (13-81) we see that the Sherwood number for a stationary particle depends on its shape, while the ventilation factor does not, it is more convenient to express the mass rate of change in terms of $\bar{f}_{v,L^*}$ rather than $\bar{N}_{Sh,L^*}$ as in (13-79). From (13-76), (13-77), and (13-79) we find the desired alternative form:

$$\frac{dm}{dt} = -4\,\pi C D_v (\rho_{v,s} - \rho_{v,\infty})\bar{f}_{v,L^*}. \qquad (13\text{-}82)$$

Note that if $\bar{f}_{v,L^*}$ is given (most likely from experiment), then we no longer need to know P, but rather C.

Unfortunately, very few studies have been carried out to determine $\bar{f}_{v,L^*}$ for ice crystal-shaped bodies. However, Thorpe and Mason (1966) measured dm/dt for a hexagonal ice plate and found (from (13-82)) that $\bar{f}_{v,L^*}$ could be expressed as

$$\bar{f}_{v,L^*} = 0.65 + 0.44\,N_{Sc,v}^{1/3} N_{Re,L^*}^{1/2}. \qquad (13\text{-}83)$$

Also, theoretical estimates of $\bar{f}_{v,L^*}$ have been made by Pitter et al. (1974) and Masliyah and Epstein (1971) for oblate spheroidal shaped ice particles by means of numerical solutions to the convective diffusion equation (13-49). The formulation was strictly analogous to the description given just after (13-54), the only difference being the use of oblate spheroidal rather than spherical coordinates. From their numerical solutions Pitter et al. obtained the estimates

$$\bar{f}_{v,L^*} = 1 + 0.142\,X^2 + 0.054\,X^4 \ln(0.893\,X^2), \qquad (13\text{-}84)$$

for $X \equiv N_{Sc,v}^{1/3} N_{Re,L^*}^{1/2} \leq 0.71$ or $N_{Re,L^*} \leq 0.63$ with $N_{Sc,v} = 0.71$, and

$$\bar{f}_{v,L^*} = 0.937 + 0.178\,X, \qquad (13\text{-}85)$$

for $X > 0.71$ or $N_{Re,L^*} > 0.63$. Also, for $X \leq 1$ there is a corresponding analytical estimate, obtained by the method of matched asymptotic expansions (Brenner, 1963):

$$\bar{f}_{v,L^*} = 1 + \frac{1}{2\,\pi}\,N_{Pe,L^*} + \frac{2}{3\,\pi}\,N_{Pe,L^*}^2 \ln N_{Pe,L^*}, \qquad (13\text{-}86)$$

where $N_{Pe,L^*} = U_\infty L^*/D_v$ and $L^* = 2a$.

A comparison of these various results (13-83)–(13-86) is given in Figure 13-25. Note that $\bar{f}_{v,L^*}$ varies linearly with X for $X \gtrsim 1$, as would be expected from boundary layer theory. Note also that there is only fair agreement between the numerical results for $\bar{f}_{v,L^*}$ and the experimental values of Thorpe and Mason. This is due in part to the intrinsic differences in geometry (hexagonal plate versus oblate spheroid), but probably the most important cause of the discrepancies shown is a different thickness to diameter ratio for the hexagonal plate than for the oblate spheroids (unfortunately, Thorpe and Mason did not report this ratio). Finally, a comparison of Figures 13-25 and 13-18 shows that for a given X the overall effect of ventilation on vapor diffusion increases with

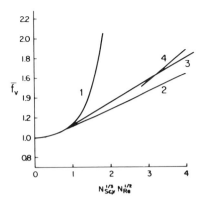

Fig. 13-25.   Ventilation coefficient for an oblate spheroid of ice in air ($N_{Sc,v} = 0.71$). (1) Analytical result of Brenner (1963) for infinitely thin circular disk, (2) Numerical results of Pitter *et al.* (1974) for spheroid of axis ratio 0.05, (3) Numerical result of Masliyah and Epstein (1971) for spheroid of axis ratio 0.2, (4) Experimental result of Thorpe and Mason (1966) for hexagonal ice crystals. (From Pitter *et al.*, 1974; by courtesy of Amer. Meteor. Soc., and the authors.)

increasing axis ratio of the oblate spheroid, and is largest for a sphere. On the other hand, the solution of Pitter *et al.* shows, as would be expected, that the local ventilation effect is greater near the waist of a thin oblate spheroid than anywhere on a sphere surface.

There are unfortunately no theoretical or experimental studies available from which one may directly obtain $\bar{f}_{v,L*}$ for columnar ice crystals of finite length. However, the experiments of Skelland and Cornish (1963) and Pasternak and Gauvin (1960) indicate that the mean Sherwood numbers for finite circular cylinders are essentially the same as those for disks and spheres, so long as the Reynolds numbers based on the length L* are the same for each geometry. (This, of course, is the fortunate circumstance we alluded to at the beginning of this section.) Following this lead, Hall and Pruppacher (1977) have correlated the numerical results for spheres and oblate spheroids of axis ratio 0.05 and 0.2 to obtain the estimates

$$\bar{f}_{v,L*} = 1 + 0.14\,X^2, \quad X = N_{Sc,v}^{1/3}N_{Re,L*}^{1/2} < 1.0, \tag{13-87a}$$

$$\bar{f}_{v,L*} = 0.86 + 0.28\,X, \quad X \geqslant 1.0. \tag{13-87b}$$

For $N_{Re,L*} \leqslant 100$, (13-87) agrees with the individual data to within 8%, again showing that the use of L* does effectively minimize the dependence on geometry.

Hall and Pruppacher have also used (13-82) and (13-87) to determine the growth rate of various ice crystals falling at terminal velocity, on the assumption that the crystals obey the dimensional relationships specified by Auer and Veal (1970) (see Section 2.2.1) for an environment at water saturation. The results, shown in Figure 13-26, show that the differences in growth rates for the various crystal types are small for early growth times, but become increasingly significant with increasing growth time. This is in marked contrast to the case of water drop diffusional growth (recall Figures 13-2 and 13-7), and reflects the fact

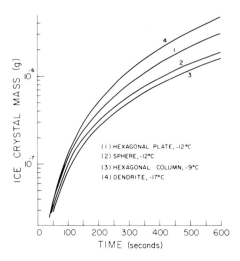

Fig. 13-26. Diffusional growth rate of ice crystals of various shapes in a water saturated environment, as a function of time; initial ice crystal radius 3 μm; for $\alpha_c = 0.1$ and for crystal densities as given in Table 2-3.

that the crystal geometry can increasingly 'assert itself' as the crystal size increases.

Several other studies on crystal growth have been carried out which, in addition to using equations similar to (13-82) and (13-87), have also taken into account the different crystal growth habits for different temperatures; i.e., they have included the specification of $C = C(T)$ (Koenig (1971), Jayaweera (1971), Middleton (1971), and Houghton (1972)). Unfortunately, a comparison of these studies is complicated by the different expressions used for the ventilation coefficients, and the different assumed dimensional relationships for the crystals. For purposes of illustrating the characteristic features of these studies we shall therefore restrict ourselves to a discussion of Middleton's results, shown in Figures 13-27 and 13-28. Middleton used Frössling's (1938) ventilation coefficients for a sphere (see Figure 13-19) and Auer and Veal's (1970) formulae relating crystal dimensions (see Section 2.2.1). The figures show that the crystal growth rate is greatest at temperatures where needles form and where growth is dendritic, while the slowest growth occurs at temperatures favoring the development of solid columns and plates (refer to Figure 2-27 for a translation of the symbols designating crystal type in Figures 13-27 and 13-28). These differences increase with time, since dm/dt increases with crystal size. Thus, after 10 min an ice crystal growing dendritically at $-15\,^{\circ}$C has acquired a mass which is almost six times that of a crystal growing as a solid column near $-8\,^{\circ}$C. Also, if we recall from Section 2.2.2 that planar and columnar ice crystals must respectively grow by diffusion to diameters of 300 and 50 μm before they can commence riming, we see from Figure 13-28 that this implies a hexagonal plate of type C1g or P1a must grow by diffusion at water saturation and $-12\,^{\circ}$C for about 6 min before riming is possible. Similarly, a columnar crystal must grow for about 6 min at water saturation and $-9\,^{\circ}$C before it may commence riming.

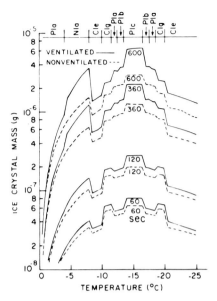

Fig. 13-27.   Mass of ice crystals grown by vapor diffusion in a water saturated environment as a function of growth time (sec) and temperature, including (solid line) and excluding (dashed line) ventilation, at 700 mb. (From Middleton, 1971; by courtesy of the author.)

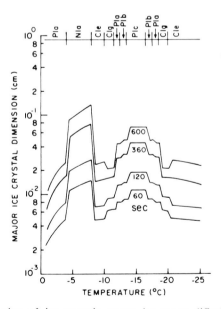

Fig. 13-28.   Major dimension of ice crystals grown by vapor diffusion in a water saturated environment as a function of growth time and temperature, at 700 mb. (From Middleton, 1971; by courtesy of the author.)

Of course, the same set of equations can serve to describe ice crystal evaporation in an ice-subsaturated environment. Some early calculations of this nature by Braham (1967) and Braham and Spyers-Duran (1967) indicated that ice crystals falling from cirrus clouds may survive for considerable distances (greater than a few kilometers) in the ice-subsaturated air below the clouds. This problem has recently been studied more comprehensively by Hall and Pruppacher (1977), who included the effects of radiational heat transfer between the ice particle and its environment. By using maximum and minimum limits for the upward and downward radiation fluxes at cirrus cloud levels, and by considering the emission and absorption properties of ice, they concluded that radiative heat transfer affects the survival distance of columnar ice crystals falling from cirrus clouds by less than 10% if the relative humidity of the environmental air is less than 70%. From a consideration of these radiative effects and a wide range of values for the initial size and bulk density of the ice particle, they showed that ice crystals falling from cirrus clouds can survive fall distances of up to 2 km when the relative humidity is below 70% in a typical midlatitude atmosphere. Larger survival distances are possible only if the ambient air has a relative humidity larger than 70%. A comparison of their theoretical results with a field study by Heymsfield (1973) of the survival distance of cirrus ice particles is shown in Figure 13-29. Note that the theoretical predictions for $\rho_c = 0.75 \text{ g cm}^{-3}$

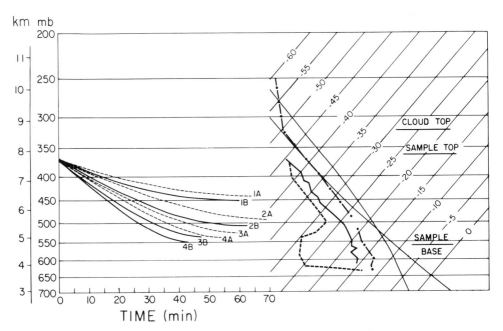

Fig. 13-29. Comparison between survival distance of cirrus ice particles theoretically computed by Hall and Pruppacher (1977), and observed by Heymsfield (1973) on March 4, 1972 above Salem (Ill.). On right: $-\cdot-\cdot-\cdot$ radiosonde data of Salem, ———— temperature measured by aircraft in situ. On left: computed fall distances, (1) $\rho_c = 0.3 \text{ g cm}^{-3}$, (2) $\rho_c = 0.5 \text{ g cm}^{-3}$, (3) $\rho_c = 0.75 \text{ g cm}^{-3}$, (4) $\rho_c = 0.9 \text{ g cm}^{-3}$; (A) column, initial size $800 \times 164 \, \mu\text{m}$, (B) sphere, initial radius $160 \, \mu\text{m}$; $\alpha_c = 0.014$. (From Hall and Pruppacher, 1977; by courtesy of Amer. Meteor. Soc., and the authors.)

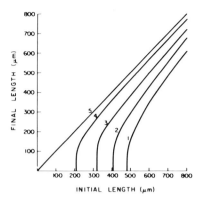

Fig. 13-30. Final length of columnar ice crystals of given initial length after an isobaric and isothermal excursion of 200 m into air of (1) 30%, (2) 50%, (3) 70%, (4) 90% relative humidity, at $-32.7\,°C$ and 400 mb, for $\alpha_c = 0.1$, (5) limiting case of no evaporation; crystal width w and length L are related by $w = 0.0627\, L^{0.53}$. (From Hall and Pruppacher, 1977; by courtesy of Amer. Meteor. Soc., and the authors.)

and $\rho_c = 0.9\, \text{g cm}^{-3}$ agree satisfactorily with the lowest level (sample base) at which cirrus ice particles were detected.

Hall and Pruppacher also determined the final length of columnar crystals, of given initial length and width, which traverse a 200 m thick layer of air at $-32.7\,°C$ and 400 mb, for various relative humidities. As would be expected, the results, given in Figure 13-30, are qualitatively similar to those shown in Figure 13-22 for the case of isobaric and isothermal excursions of water drops. Thus for the stated conditions the initial crystal lengths must be greater than 210, 320, 410, and 480 μm in order to survive the 200 m distance if the relative humidities are 90, 70, 50, and 30%, respectively.

Most experimental studies on the diffusional growth rate of ice crystals have been carried out with experimental setups which did not allow the ice crystals to fall freely in air. They also generally provided growth rates for only a very limited temperature range, and for widely different growth times (for a summary of earlier work see Jayaweera, 1971). Nevertheless, the earlier experimental studies were consistent in showing that the mass growth rate is a function of temperature, being highest near $-6\,°C$ and $-15\,°C$ and lowest near to -8 to $-9\,°C$ and at temperatures below $-20\,°C$. These qualitative results are supportive of the theoretical results shown in Figures 13-27 and 13-28. Further, more quantitative confirmation of such theoretical studies comes from the experiments of Fukuta (1969), Ryan et al. (1974, 1976), and Gallily and Michaeli (1976). The measurements of Ryan et al. (1976) are shown in Figure 13-31. A comparison of this figure and Figure 13-27 reveals a general good agreement between the theoretical and experimental values of m(T, t) after a growth time of t = 150 sec. Unfortunately, the growth times realized in laboratory studies have been relatively short ($\leqslant 150$ sec). However, ice crystal growth studies for somewhat longer growth times ($\leqslant 300$ sec) have been carried out by Hoffer and Warburton (1970) and Davis and Auer (1974) in clouds which were seeded with AgI. Subsequent to the seeding, ice crystals were sampled at various time

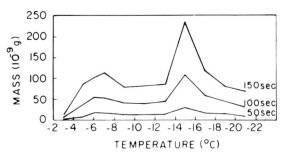

Fig. 13-31. Experimentally determined variation of the mass of ice crystals growing by diffusion of vapor in a water saturated environment, as a function of growth time and temperature. (From Ryan *et al.*, 1976; by courtesy of Amer. Meteor. Soc., and the authors.)

intervals, and the growth rates thus determined. Although the temperature and humidity conditions are considerably less accurately defined during such field studies than in laboratory experiments, the observed growth rates were found to be within 15% of the rates predicted by the theory.

13.3.3   GROWTH RATE OF ICE CRYSTAL FACES – ICE CRYSTAL HABIT CHANGE

The fact that the theoretical framework for crystal growth given by (13-82) and, for example, (13-87), is incomplete without a specification of $C = C(T, e)$ indicates clearly that a consideration of vapor and heat exchange between the crystal and its environment cannot provide by itself a complete description of crystal growth mechanisms. The reason for this is that, in addition to the driving forces of vapor and heat gradients, there are also crystal surface forces which control the incorporation of water molecules into the ice lattice. From the observed changes in crystal growth habit with temperature, these latter forces are evidently sensitive functions of temperature, and are different for differential crystal faces.

   The extent to which the surface forces alone control the growth habit of ice crystals has been studied experimentally by Lamb and Hobbs (1971). They measured the 'linear' growth rates of the basal and the prism faces of ice crystals under conditions which insured that the growth of the crystals was controlled by molecular events taking place at the crystal surface, and not by the rate of supply of water molecules from the vapor phase or by the rate at which latent heat of deposition was removed. For this purpose the measurements were carried out in pure water vapor at a constant excess vapor pressure of 1.3×10^{-2} mb; such low pressures reduce the vapor and heat gradients to negligible levels. The results of this study are shown in Figure 13-32, where it can be seen that the growth rates of the prism- and basal faces vary strongly with temperature, decreasing rapidly at first with decreasing temperature, and subsequently increasing sharply to a maximum followed by a monotonic decrease with further decrease in temperature. Note also that the growth rate curves are shifted with respect to each other, crossing at -5.3 and $-9.5\,°C$. At temperatures from 0 to $-5.3\,°C$ and at temperatures from -9.5 to near $-22\,°C$, the linear growth rate of

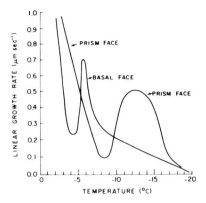

Fig. 13-32. Best fit curves to experimental measurements of the linear growth rates on the basal and prism faces of ice, at constant vapor pressure excess of 1.3×10^{-2} mb. (From Lamb and Hobbs, 1971, with changes.)

the prism face is larger than that of the basal face, implying a-axis growth and thus the development of plate-like ice crystals. On the other hand, at temperatures between -5.3 and $-9.5\,°C$ and at temperatures below about $-22\,°C$ the basal face has a linear growth rate which is larger than that of the prism face, implying c-axis growth and thus the development of columnar ice crystals. These results are in reasonable agreement with the observed growth habit behavior of ice crystals in air under natural conditions, which demonstrates the dominance of surface kinetic effects in controlling the basic growth habit.

Given the linear growth rates G_B and G_P of the basal and prism faces, the mass growth rate of the crystal can evidently be determined from the relation

$$\frac{dm}{dt} = \rho_c[G_B(T)\Omega^{(B)} + G_P(T)\Omega^{(P)}], \qquad (13\text{-}88)$$

where ρ_c is the density of the crystal and $\Omega^{(B)}$ and $\Omega^{(P)}$ are the total respective areas of the basal and prism faces. From Figure 5-9 we easily find $\Omega^{(B)} = 4\sqrt{3}\,a'^2$ and $\Omega^{(P)} = 4\sqrt{3}\,a'H$, where H is the height of the hexagonal prismatic ice crystal, and a' is the radius of the circle inscribed in the hexagonal base of the ice prism. Since in an environment of pure water vapor the linear growth rates are constant with time at any one temperature, $a' = G_P(T)t$ and $H = 2\,G_B(T)t$. On combining these relations with (13-88) we obtain

$$\frac{dm}{dt} = 12\sqrt{3}\,\rho_c G_P^2 G_B t^2. \qquad (13\text{-}89)$$

The growth rate described by (13-89) was referred to by Lamb and Hobbs (1971) as 'inherent,' since it is governed by kinetic processes on the surface of ice and not by diffusion of water vapor or heat through air. With their experimental values of G_P and G_B Lamb and Hobbs evaluated (13-89) and obtained the results shown in Figure 13-33. We see that the 'inherent' growth rate attains peak values near -6 and $-12\,°C$, which are separated by a minimum near $-8\,°C$. Since these results are quite similar to those for a crystal growing at water saturation in air,

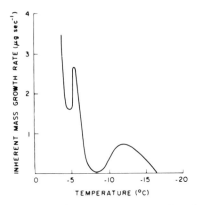

Fig. 13-33. 'Inherent' mass growth rate of ice after 1000 sec as a function of temperature at constant excess vapor pressure of 1.3×10^{-2} mb, based on the linear growth rates given in Figure 13-32. (From Lamb and Hobbs, 1971; by courtesy of Amer. Meteor. Soc., and the authors.)

we again conclude that surface kinetic effects must largely control the growth habit of ice crystals.

As for the modifications expected from the effects of finite vapor and heat gradients, we may first note that the second maximum near $-12\,°C$ in Figure 13-33 should be enhanced relative to the first by virtue of the fact that $e_{sat,w} - e_{sat,i}$ attains a maximum value for $T \approx -12\,°C$ (see Figure 4-5). Also, once the crystal habit has been decided by surface kinetic effects at an early stage of crystal development, the vapor flux to the crystal will not significantly alter its geometry, at least for low to moderate vapor density excess, since the excess water mass arriving at the corners and edges of the crystal can be effectively redistributed by surface diffusion. However, at sufficiently high vapor density excess, surface diffusion will no longer suffice to compensate for the large nonuniform vapor deposition. Growth will then occur preferentially at the corners (where the liberated latent heat is also most effectively dissipated), and this will result in the formation of sector plates, dendrites, stellars, etc.

In attributing the characteristics of the mass growth of an ice crystal to the characteristics of the linear growth rate of its crystallographic faces, we of course have really done nothing more than exchange one problem for another. Thus, for a further understanding of the mass growth rate behavior of ice crystals we must attempt to gain further insight into the linear growth rate mechanism. For this purpose we may note the experimental results of Hallett (1961) and Kobayashi (1967), who determined the rate of propagation of steps at the surface of ice crystals. They found that steps of heights between 200 to 1000 Å have propagation velocities which are inversely proportional to the step height. Furthermore, they increase with increasing excess vapor density, and exhibit at constant excess vapor density a characteristic variation with temperature. If the results of Kobayashi are corrected for the effect of heating of the crystal surface by latent heat release (according to Ryan and Macklin (1969) this amounts to a surface temperature correction of $+4\,°C$), the measurements of Hallett and of Kobayashi agree with each other in suggesting that at constant excess vapor pressure the step propagation velocity on the basal face of an ice crystal decreases with increasing temperature in the interval 0 to $-4\,°C$ to reach a minimum near $-4\,°C$, then sharply increases in the interval -4 to $-6\,°C$ to reach a maximum near $-6\,°C$, and subsequently decreases monotonically towards $-20\,°C$. This behavior is indicated in Figure 13-34. Note from a comparison of this figure with Figure 13-32 that the temperature variation of the step propagation velocity on the basal face of ice is practically identical to that of the linear growth rate of the basal face. Unfortunately, no similar experimental study is available for the step propagation velocity on the prism face of ice.

It is of course reasonable to find that the linear growth rate G of a crystallographic face of ice is proportional to the propagation velocity v_0 on that face, since the rate at which the face grows depends directly on how fast the successive molecular layers can be formed. On the other hand, this argument implicitly assumes there is no time lag in the initiation of each new successive layer. Since rapid step regeneration by nucleation is improbable at the low supersaturations prevailing under natural conditions, another mechanism must be responsible. Perhaps the most likely candidate is that of Frank (1949), who proposed the emergence of screw dislocations at the crystal surface (recall Section 5.10 and Figure 5-10). If we assume now that a step is part of a screw dislocation, and that the latter can be represented by an Archimedes spiral, i.e., $r(\theta) = K\theta$, where K is a constant and $\vec{r} = \vec{r}(r, \theta)$ is the radius vector from the dislocation center to any point on the spiral, then for each full rotation of the spiral another step will be released without delay from the spiral center and allowed to propagate across the crystal face. Furthermore, in this case the frequency of step regeneration will be $f = \omega/2\,\pi$, where $\omega = d\theta/dt = K^{-1}\,dr/dt = v_0/K$; hence, the linear growth rate of the crystals face will be

$$G = fh = \frac{v_0 h}{2\,\pi K},\tag{13-90}$$

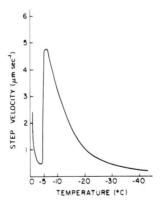

Fig. 13-34.   Variation with temperature of the propagation velocity of steps 250 Å high on the basal face of ice; excess vapor density $0.25 \times 10^{-6}\,g\,cm^{-3}$. (From Hallett, 1961; by courtesy of *Phil. Mag.*, and the author.)

where h is the step height. Thus, we find $G \propto v_0$ for a screw dislocation growth process, and also that the spacing between the steps is given by $\Delta x = 2 \pi K$.

Hallett (1961) interpreted his finding that $v_0 \propto h^{-1}$ as an indication that steps grow largely by surface diffusion, using arguments similar to those presented in Section 5.7.3 for estimating the bunching time of monolayers on an ice surface. Thus if d_S is the average migration distance which adsorbed water molecules travel by surface diffusion before re-evaporating, then it is reasonable to assume that a step of height h and unit length will receive material by surface diffusion at the rate $2 w_{net,i} d_S$ and by direct deposition from the vapor at the rate $w_{net,i} h$, where $w_{net,i}$, given by (5-56), is the net flux to the surface (assumed to be uniform over the surface). Hence, if ρ_i is the density of ice the velocity of the step is

$$v_0 = \frac{w_{net,i}}{h \rho_i} (2 d_S + h). \tag{13-91}$$

Hence, if $d_S \gg h$ the step will grow by surface diffusion such that $v_0 \propto h^{-1}$, which is consistent with Hallett's measurements. Also, given this result we find from (13-90) that the growth rate has the form $G \approx w_{net,i} d_S / \pi K$, i.e., it is independent of the step height and velocity, and the migration distance d_S emerges as the key surface parameter.

Mason et al. (1963) estimated d_S as one half the critical separation for which the velocity of approach of two neighboring growth layers slows down (see Section 5.7.3). If their measured values are corrected for crystal surface heating from latent heat release (Ryan and Macklin, 1969), excellent agreement among the temperature variations of $d_S^{(B)}$, $v_0^{(B)}$, and G_B for the basal plane is found. Although no experiments have yet been carried out to determine d_S on the prism plane of ice, a similar agreement between $d_S^{(P)}$, $v_0^{(P)}$, and G_P is expected. The available evidence therefore suggests that on the temperature intervals 0 to $-5.5\,°C$ and -9.5 to $-22\,°C$, in which plate-like ice crystals form preferentially, the inequalities $d_S^{(P)} > d_S^{(B)}$, $v_0^{(P)} > v_0^{(B)}$, and $G_P > G_B$ hold; similarly, on the interval -5.5 to $-9.5\,°C$ and below about $-22\,°C$, where columnar crystals are favored, one should find $d_S^{(B)} > d_S^{(P)}$, $v_0^{(B)} > v_0^{(P)}$, and $G_B > G_P$.

We have described how, because of the dominance of surface diffusive transport of water molecules over their direct arrival from the vapor, d_S appears to control the linear growth rates and step propagation velocities on the surface of an ice crystal. Therefore, for a deeper understanding of the temperature variation of crystal growth habits it is necessary to turn to a study of the dependence of d_S on the physics of molecular processes on crystal surfaces. Unfortunately, a detailed consideration of these processes is beyond the scope of this book. However, we can provide a brief indication of the relevant concepts, following to some extent the discussion given in Section 9.1.3.1. Thus, if ΔF_{des} is the energy of desorption per molecule at a site on the surface, the probability per unit time that an adsorbed molecule will leave the surface (evaporate) is $\nu \exp(-\Delta F_{des}/kT)$, where ν is the natural vibration frequency of the molecule normal to the surface at the site. Similarly, the probability per unit time that the molecule will jump to a neighboring adsorption site is $\nu \exp(-\Delta F_{sd}/kT)$, where ΔF_{sd} is the activation energy for surface diffusion of the molecule. Therefore, the number of molecular jumps before evaporation is $N = \exp[(\Delta F_{des} - \Delta F_{sd})/kT]$. Finally, the migration of the molecule on the surface will be a random walk, so that $d_S = d_0\sqrt{N}$, where d_0 is the molecular jump distance; hence we find that

$$d_S = d_0 \exp[(\Delta F_{des} - \Delta F_{sd})/2\,kT] \tag{13-92}$$

(Burton et al., 1951).

Unfortunately, (13-92) appears to conflict sharply with the observed dependence of d_S on temperature: Since it is generally expected that $\Delta F_{des} > \Delta F_{sd}$, we find that d_S, and hence the linear growth rates, are predicted to decrease with increasing temperature. Several more or less plausible ways around this clear contradiction of the experimental results have been proposed. For example, Mason et al. (1963) suggested that ΔF_{des} and ΔF_{sd} must be strong functions of temperature which, moreover, must contain discontinuities at the appropriate temperatures which would generate the observed maxima and minima in the growth curves (as seen, for example, in Figure 13-32). On the other hand, Hobbs and Scott (1965) explored the possibility that adsorbed molecules might hinder the subsequent adsorption of further molecules in such a way as to define a new effective 'molecular collection distance' $d_{S'}$, which might have a better temperature dependence than d_S. For this purpose Hobbs and Scott applied the Langmuir (1918) monolayer adsorption model (see Section 5.6). Their results gave some improvement in the predicted average trend of growth variables with temperature, but the overall effect of the assumed monolayer was rather weak. Lamb and Scott (1974) have obtained somewhat better results for the average growth trends by extending the model of Hobbs and Scott to account for multilayer adsorption. For this purpose they used the BET adsorption model (Section 5.6).

We have said the adsorption models attempt to describe the average growth trends with temperature. By this we mean that they involve a certain conceptual interpretation of the observed temperature dependence of the growth rates, namely that they may be regarded as a superposition of two distinct types of behavior: (1) a strong average trend with temperature, and (2) a nearly discontinuous change in growth rates, which generate the observed maxima and minima. The adsorption models are concerned with the former behavior. Several suggestions have also been made concerning the latter behavior. For example, Ryan and Macklin (1969) and Michaeli (1971) suggested that such discontinuities may originate with changes in the mode of surface diffusion. Thus, they obtained evidence that such diffusion may occur via interstitial migration mechanisms (as assumed in the derivation of (13-92)), via lattice vacancy mechanisms, and, close to 0 °C, via a mechanism involving the 'quasi-liquid' layer on ice (see Section 5.7.3). How these mechanisms might depend on temperature in a manner specific to a given crystallographic face still remains to be resolved.

CHAPTER 14

CLOUD PARTICLE INTERACTIONS – COLLISION, COALESCENCE, AND BREAKUP

In Chapter 10 we discussed the behavior of isolated cloud particles in some detail. Now we shall consider their hydrodynamic interactions, with a view to providing a quantitative assessment of the processes of particle growth by collision and coalescence, and of collisional breakup. We shall first treat the collision problem for drops of radii less than about 500 μm which, in accordance with our previous description of drop distortion in Section 10.3.2, may be regarded as rigid spheres (at least when falling in isolation). This will be followed by a somewhat briefer discussion of the phenomena of drop coalescence and breakup. Finally, we shall consider water drop-ice crystal and ice crystal-ice crystal interactions, which lead to the formation of graupel, hail, and snow particles.

14.1 The Basic Model for Drop Collisions

In any cloud the effectiveness of the drop interaction process will be determined in part by such factors as uplift speed and entrainment, droplet charge and external electric field, and turbulence. For simplicity we shall consider first the basic problem of determining the interaction of droplets falling through otherwise calm air under the influence of gravity and the hydrodynamic forces associated with their motion. Turbulence-enhanced gravitational collision is discussed in Section 14.5 and Appendix A-14.5, while the effects of electric fields and charges are discussed in Chapter 17.

Although in principle the collision problem is a many-body problem, the droplet concentrations in natural clouds are low enough so that only the interactions of pairs of droplets need be considered. This can be appreciated by the following very simple argument (a more detailed analysis is given in Appendix A-14.1): The liquid water content of clouds, w_L, typically equal to 1 cm^3 of water per cubic meter of air, may be approximated by the ratio of the volume of the average cloud droplet to the volume of a sphere whose radius is S, the average distance between droplets; thus

$$a/S \sim w_L^{1/3} \sim 10^{-2}, \tag{14-1}$$

showing that clouds are rather disperse aerosols and that therefore two body interactions should dominate.

It is clear that such interactions must occur extensively if rain is to form.

From Chapter 13 we know that the initial stage of droplet growth by vapor diffusion produces a fairly uniform distribution, with droplets generally of 1 to 10 μm radius, thus possessing only 10^{-6} or so of the mass of a typical 1 mm diameter raindrop. Further broadening of the spectrum toward precipitation-sized droplets (conveniently defined as those with radii $\geqslant 100\,\mu$m) is possible only when conditions in the cloud are such that droplets can collide and coalesce.

In order to avoid excessive mathematical complications, we shall also assume that interacting drops which are spherical in shape when falling in isolation remain so for all separations of their surfaces. As we shall see, the reasonableness of this assumption is borne out by the general good agreement between the resulting predicted and observed effective collision cross sections. The shape distortions which may occur when the drops are in close proximity may thus be considered of secondary importance in most cases of interest.

In consequence of these simplifying assumptions, the basic physical model for the collision problem reduces to one of two rigid spheres, initially widely separated in the vertical, falling under gravity in an otherwise undisturbed and unbounded fluid.

14.2   Definition of Collision Efficiency

The essential piece of information to be extracted from a study of the collision model described above is the effective cross section for collision of the two spheres. Thus, the goal is to find y_c, the initial horizontal offset of the center of the lower (smaller) sphere of radius a_2, from the vertical line through the center of the upper (larger) sphere of radius a_1 such that a grazing trajectory will result (see Figure 14-1). Then y_c/a_1, called the *linear collision efficiency*, is a simple dimensionless measure of the tendency for collision. Another measure is the *collision efficiency*, E, defined here to be the ratio of the actual collision cross section πy_c^2 to the geometric cross section $\pi(a_1 + a_2)^2$:

$$E \equiv \frac{y_c^2}{(a_1 + a_2)^2}. \tag{14-2}$$

Thus on introducing the size ratio $p \equiv a_2/a_1$, we have the simple relationship

$$y_c/a_1 = (1 + p)E^{1/2}. \tag{14-3}$$

(An alternative definition for E, which also appears often in the literature, is $E \equiv y_c^2/a_1^2$. We shall use the definition (14-2) exclusively.)

Let us now consider the dependence of E on the physical parameters characterizing the problem. The initial vertical separation z_0 for unequal droplets is taken to be large enough so that they fall independently in the beginning; thus E will not depend on z_0. For equal droplets this condition must be altered slightly, and we start with a vertical separation such that interaction is at a threshold. We can repeat trials with varying separation, and look for the limiting value of E as the separation increases. In practice it has been found that E is generally insensitive to z_0 if $z_0 \geqslant 10^2\,a_1$, for any size ratio a_2/a_1.

Therefore, assuming we start with a large enough z_0, the only independent parameters which could possibly play a role in the interaction process are a_1, a_2, the kinematic viscosity ν and density ρ of the fluid (air), the density ρ' of the spheres, the acceleration g of gravity, and y, the initial horizontal offset. A collision ($y = y_c$) may therefore be described by some relation of the form $f_1(a_1, a_2, \nu, g, \rho, \rho', y_c) = 0$. By applying the π-theorem of dimensional analysis (which states that in an equivalent dimensionless formulation of a physical problem the number of independent dimensionless parameters equals the number of independent dimensional parameters occurring in the original problem, less the number of dimensions involved – see, for example, Kline (1965)), this relationship may be expressed equivalently in terms of four dimensionless combinations of the seven parameters. For example, we may write $f_2(E, a_2/a_1, \rho/\rho', ga_1^3/\nu^2) = 0$; or, explicitly in terms of E

$$E = E(a_2/a_1, \rho/\rho', ga_1^3/\nu^2). \tag{14-4}$$

Since ρ/ρ', g, and ν are fixed in the case of water droplets falling in air, we have for water drops in air

$$E = E(a_2/a_1, a_1). \tag{14-5}$$

In words, the collision process for cloud droplets in air can be described by a one parameter set of curves. The usual display is of E versus a_2/a_1 with a_1 as the curve label. (Of course, ν and ρ' will vary with height and this will also affect the value of E. For example, a recent calculation by de Almeida (1977) gave a 22% increase in E for a change in ambient conditions from 995 mb, 14 °C to 800 mb, 25 °C.)

One may anticipate the qualitative shape of the curves described by (14-5). The tendency for the small sphere to move with the flow around the large sphere is depicted schematically in Figure 14-1. As a consequence of this tendency for deflection by viscous forces from a collision trajectory, one can anticipate that generally $E < 1$ (except possibly for $a_2/a_1 \rightarrow 1$; see below). Since the forces of

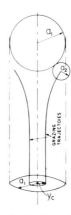

Fig. 14-1.   Schematic representation of the hydrodynamic interaction of a pair of spheres; y_c is the critical, horizontal offset for a grazing trajectory of the small sphere.

deflection must become less effective as the inertia of the spheres increases, E should be a montonic increasing function of a_1. For the same reason, E also should be a monotonic increasing function of a_2/a_1, at least for $a_2/a_1 \ll 1$. However, the situation for a_2/a_1 approaching unity is not obvious. For example, as $a_2/a_1 \to 1$ the relative velocity of approach decreases so that small forces of deflection have a relatively long time to operate and hence prevent a collision. This effect tends to reduce E. On the other hand, there is the possibility of 'wake capture' as the trailing sphere falls into the wake of the leading sphere, and thus encounters less resistance to motion than the latter. This effect tends to increase E. (Another form of 'wake capture' for the case of $a_2/a_1 \ll 1$ is discussed in Chapter 12.) The overall result can be determined only by elaborate calculations, and, as we shall see below, there is not yet complete agreement even as to the qualitative outcome. However, since the strength and size of the wake increases with the Reynolds number, one at least may expect the wake capture effect to increase strongly with increasing a_1. In summary then, we may expect to find $E \ll 1$ for sufficiently small a_1, and for $a_2/a_1 \ll 1$ with arbitrary a_1. Also, E should increase with a_1 for fixed a_2/a_1, and there is a possibility of a greater than geometric cross section ($E > 1$) by wake capture as $a_2/a_1 \to 1$ for sufficiently large a_1.

Note that if one hopes to simulate the collision process in the laboratory (which is desirable since it is difficult to observe the collision of small droplets in air), it is sufficient that a_2/a_1, ρ/ρ', and a_1^3/ν^2 have the same values in the laboratory as in the natural cloud. Unfortunately, it has not been possible to model this way. To be a useful operation, the simulation should employ easily observable (large) spheres, and high viscosities in order to achieve low velocities. But then the density ratio ρ/ρ' cannot be preserved; whereas in clouds this ratio is about 10^{-3}, in model experiments it is generally larger than about 5×10^{-2}. However, we shall see later there is some evidence that practical similitude may be achieved so long as $\rho/\rho' \lesssim 10^{-1}$ in the laboratory, and this less severe modeling constraint can be met.

From (10-108) and (10-110) it can be seen the parameter ga_1^3/ν^2 is a function only of ρ/ρ' and the terminal velocity Reynolds number $N_{Re,1}$ of sphere a_1 in isolation. Therefore, an expression equivalent to (14-5) for E is

$$E = E(a_2/a_1, \rho/\rho', N_{Re,1}). \tag{14-6}$$

This is the more usual way of describing the dependence of E.

14.3   The Superposition Method

An account of the two sphere problem which directly incorporates results from the theory of flow past a single sphere obviously would be desirable. Such a scheme is known as the method of superposition, according to which each sphere is assumed to move in a flow field generated by the other sphere falling in isolation. More explicitly, if spheres 1 and 2 move with the instantaneous velocities $\vec{v}_1$ and $\vec{v}_2$, respectively, the force $\vec{F}_{1,sup}$ on sphere 1 is assumed to have the form (cf. (10-39) and (10-109))

$$\vec{F}_{1,\text{sup}} = -6\,\pi a_1 \eta \left(\frac{C_D N_{Re}}{24}\right)_1 (\vec{v}_1 - \vec{u}_2), \tag{14-7}$$

where $\vec{u}_2$ is the velocity field, due to the motion of sphere 2, which would exist at the location of the center of sphere 1 if it were absent. The corresponding expression for the force on sphere 2 is obtained by interchanging 1 and 2 in (14-7). The quantity $(C_D N_{Re}/24)_1$, a function only of the Reynolds number for sphere 1, evidently should be based on the relative velocity $(\vec{v}_1 - \vec{u}_2)$, although sometimes it is based instead on $\vec{V}_{\infty,1}$, the terminal velocity of sphere 1 in isolation.

As would be expected, and as is demonstrated in the example presented in Appendix A-14.3 on the use of the superposition method in the Stokes regime, superposition does not give a very accurate account of the interaction for close separations. However, if the relative velocity of the spheres is sufficiently large, they may pass through the region of close proximity so rapidly that the inaccurate description of the forces acting there is of little consequence. In general the method of superposition becomes increasingly accurate with decreasing a_2/a_1.

Langmuir (1948) was one of the first to adopt the superposition scheme to the collision problem. He restricted his investigation to small values of a_2/a_1, so that sphere 2 could be regarded as a mass point moving in the flow around sphere 1. Two extreme cases for the flow past sphere 1 were considered, namely potential flow for large $N_{Re,1} = 2\,a_1 V_{\infty,1}/\nu$, and Stokes flow for small $N_{Re,1}$. A recent calculation of a similar nature is that of Beard and Grover (1974), who used the method to determine E for small raindrops colliding with micron sized particles. They employed the numerical solution of Le Clair et al. (1970) for the flow field past an isolated sphere.

The method also has been used to investigate the interaction of spheres of comparable size by Pearcey and Hill (1956), Shafrir and Neiburger (1963), Plumlee and Semonin (1965), Neiburger (1967), Shafrir and Gal-Chen (1971), Lin and Lee (1975), and Schlamp et al. (1976). For the single sphere flow field, Pearcey and Hill chose Goldstein's (1929) complete solution for Oseen flow. Shafrir and Neiburger, like Langmuir, employed two different Reynolds number regimes: (a) Stokes flow for $N_{Re,1} \leqslant 0.4$ ($a_1 \leqslant 30\,\mu$m), and (b) a modified form of Jenson's (1959) numerical solution for $N_{Re,1} > 0.4$. Plumlee and Semonin used the second approximation to the inner expansion of the flow field obtained by Proudman and Pearson (1957). Shafrir and Gal-Chen used a numerical solution by Rimon and Cheng (1969), Lin and Lee used their own numerical solution (Lin and Lee (1973)), and Schlamp et al. (1976) used Le Clair's (1970) solution.

Let us consider briefly the form of the equations of motion for the droplets. Assuming the method of superposition correctly describes the hydrodynamic forces, Newton's second law for sphere 2 of mass m_2 becomes

$$m_2 \frac{d\vec{v}_2}{dt} = m_2 \vec{g}^* + \vec{F}_2, \tag{14-8}$$

where $\hat{g}^*[=(\rho'-\rho)\hat{g}/\rho']$ is the gravitational acceleration corrected for buoyancy, and $\vec{F}_2$ is given by (14-7) (with subscripts 1 and 2 interchanged). It is convenient to render these equations dimensionless by using a_1 for the unit of length, and the terminal velocity of sphere 1, $V_{\infty,1}$, as the unit of velocity; i.e., we introduce the following dimensionless quantities:

$$\vec{v}_2' \equiv \vec{v}_2/V_{\infty,1}, \quad \vec{u}_1' \equiv \vec{u}_1/V_{\infty,1}, \quad t' \equiv tV_{\infty,1}/a_1. \tag{14-9}$$

Then in place of (14-8) we have

$$\frac{d\vec{v}_2'}{dt'} = \frac{\hat{g}^*}{N_{Fr}} - \frac{\vec{v}_2' - \vec{u}_1'}{N_S}, \tag{14-10}$$

where $N_{Fr} \equiv V_{\infty,1}^2/g^*a_1$ is the *Froude number* for sphere 1, $\hat{g}^*$ is a unit vector in the direction of gravity, and

$$N_S = \frac{2\,a_2^2\rho'V_{\infty,1}}{9\,\eta a_1}\left(\frac{24}{C_D N_{Re}}\right)_2. \tag{14-11}$$

The quantity N_S often is called the 'inertia parameter' or *Stokes number*. It provides a measure of the ability of sphere 2 to persist in its state of motion in a viscous fluid. For example, in the limit of zero Reynolds number it is directly proportional to the 'range' or penetration distance of sphere 2 injected with its Stokes terminal velocity, $V_{S,2}$, into a fluid at rest (the law governing the deceleration of sphere 2 is assumed to be the obvious one involving the steady state Stokes drag):

$$-6\,\pi a_2 \eta v_2 = m_2\frac{dv_2}{dt} = m_2 v_2 \frac{dv_2}{dx}, \tag{14-12}$$

so that the 'range' x_s is

$$x_s = \frac{m_2 V_{S,2}}{6\,\pi a_2}. \tag{14-13}$$

Then since $V_{S,2} = p^2 V_{S,1}$, where $p \equiv a_1/a_2$, we have the relationship

$$\frac{x_s}{a_1} = \frac{2\,a_2^2 p^2 \rho' V_{S,1}}{9 a_1 \eta} = \lim_{N_{Re,1}\to 0} p^2 N_S. \tag{14-14}$$

Other related parameters which often are employed in the collision problem are $q \equiv N_S^{-1}$ (Shafrir and Neiburger (1963)), $N_S^* \equiv x_s/a_1$ (Klett and Davis (1973)), and I (Hocking (1959), Hocking and Jonas (1970)), where

$$I = \lim_{N_{Re,1}\to 0} \frac{N_S}{p^2} = \frac{4\rho'^2 a_1^3 g^2}{81\,\eta^2}. \tag{14-15}$$

From the definition of N_S we may expect increasing viscous deflection of sphere 2 from a collision trajectory with sphere 1 (and thus decreasing values for E) for decreasing N_S, I, N_S^*, or increasing q.

Also, by rewriting N_S as

$$N_S = \frac{2}{9}p^2\left(\frac{\rho'}{\rho}\right)N_{Re,1}\left(\frac{24}{C_D N_{Re}}\right)_2, \tag{14-16}$$

and assuming that $(24/C_D N_{Re})_2$ is calculated according to the terminal velocity of sphere 2, we see that

$$N_S = N_S(p, \rho'/\rho, N_{Re,1}), \tag{14-17}$$

since there is a unique relation between sphere size and Reynolds number for a given viscosity and density ratio. Similarly, for the Froude number we have

$$N_{Fr} = \frac{V_{\infty,1}^2}{g^* a_1} = I\left(\frac{C_D N_{Re}}{24}\right)_1^2 = N_{Fr}(p, \rho'/\rho, N_{Re,1}). \tag{14-18}$$

Consequently, from (14-10) we infer the dimensionless sphere trajectories depend on the values of p, ρ'/ρ, and $N_{Re,1}$. This is consistent with (14-6) for the dependence of E.

14.4 The Boundary Value Problem Approach

14.4.1 THE QUASI-STATIONARY ASSUMPTION

Since two interacting spherical cloud drops generally fall with unequal velocities, the flow must be intrinsically unsteady. The method of superposition simply ignores this difficulty by employing steady state flow fields for a single sphere. Even more rigorous formulations of the two sphere problem, which attempt to take into consideration the no slip boundary conditions on both spheres, must likewise ignore the fluid unsteadiness in order to avoid overwhelming mathematical difficulties. The general boundary value formulation deals with the unsteadiness of the flow as follows: The conditions of fall are assumed to be such that the motion is 'quasi-stationary,' meaning the instantaneous velocities of the spheres at any moment define boundary values for the velocity field generated by the steady state equation of motion.

This quasi-stationary assumption would be unrealistic if the spheres experienced very large accelerations, for then the past history of the flow field would become important. As an extreme example, if the spheres were suddenly stopped in their motion the assumption would predict zero velocity immediately everywhere in the fluid, which obviously would be incorrect. Since one would expect the acceleration of a falling sphere to vary inversely with its inertia relative to a similar volume of fluid, the quasi-stationary assumption should be applicable when $\rho/\rho' \ll 1$.

Hocking (1959) has given a somewhat different argument leading to the same conclusion by considering the order of magnitude of terms in the Stokes equation for the case of an accelerating spherical particle in a fluid otherwise at rest. Suppose the particle has radius a, velocity U, and density ρ'. Then, as we have seen, the viscous force on the particle will be of order $\rho \nu U a$ for N_{Re} of the order of unity or smaller. Therefore, by Newton's second law the acceleration of the particle will be of order $\rho \nu U / \rho' a^2$ or smaller. The local acceleration of the fluid, $\partial \tilde{u}/\partial t$, will therefore be of this order or smaller. But the fluid viscous acceleration, $\nu \nabla^2 \tilde{u}$, is of order $\nu U/a^2$. Consequently, we find $|\partial \tilde{u}/\partial t|/|\nu \nabla^2 \tilde{u}| \lesssim \rho/\rho'$, so that the time derivative term in the equation of motion may be neglected if $\rho/\rho' \ll 1$. A similar argument may be made for the case $N_{Re} \gg 1$. Then the force on the particle will be of order $\rho C_D a^2 U^2$, and the local acceleration of the fluid of order $\rho C_D U^2/\rho'a$ or less. Hence we find in this case $|\partial \tilde{u}/\partial t|/|\tilde{u} \cdot \nabla \tilde{u}| \lesssim C_D \rho/\rho'$, and again the term $\partial \tilde{u}/\partial t$ becomes relatively unimportant if $\rho/\rho' \ll 1$.

We conclude the quasi-stationary assumption may be used to calculate theoretical collision efficiencies for water droplets in air, since $\rho/\rho' \approx 10^{-3}$ in that case. Also, on the basis of the fairly good agreement obtained between theoretical and experimental trajectories of pairs of equal spheres falling along their line of center for $\rho/\rho' \approx 10^{-1}$ (see Section 14.4.4), it appears the quasi-stationary assumption should remain valid for the relatively large density ratios one encounters in model simulations of cloud droplet interactions.

14.4.2 TWO SPHERES IN STEADY STOKES FLOW

From the preceding discussion it appears reasonable to apply time-independent Stokes flow theory to the collision problem for Reynolds numbers which are small compared to unity. Traditionally, such treatment has been assumed valid for droplet radii less than 30 microns, corresponding to $N_{Re} \leq 0.4$. Strictly speaking, of course, the Stokes theory applies rigorously only for $N_{Re} = 0$, and becomes progressively more inaccurate with increasing N_{Re}.

Assuming steady Stokes flow, the equations to be solved are

$$\nabla p = \eta \nabla^2 \tilde{u}, \tag{14-19}$$

$$\nabla \cdot \tilde{u} = 0. \tag{14-20}$$

The velocity must vanish at infinity and match that of the spheres on their surfaces. In general, the spheres will have velocity components along and perpendicular to their line of centers. The latter components will create shears causing the spheres to rotate with angular velocity vectors perpendicular to the plane containing the direction of gravity and the line of centers. The flow will maintain symmetry with respect to that plane.

As the governing equations are linear, the complete solution may be built up out of special cases which are easier to treat. For example, the problem of general motion for two spheres may be considered solved when solutions to the following three special problems are available: problem-1 when one sphere is stationary, while the other moves toward or away from it along the line of centers; problem-2 when neither sphere translates, but one rotates about a diameter perpendicular to the line of centers; problem-3 when one sphere is stationary while the other moves perpendicular to the line of centers. Of course, this decomposition of the general problem is not unique. Historically, the first 2-sphere problem to be solved was that for equal parallel velocities along the line of centers by Stimson and Jeffery (1926). Later, the case of antiparallel motion along the line of centers was treated by Maude (1961) and Brenner (1961). These two axisymmetric problems are equivalent to case-1 above.

The solutions for problems 1, 2, and 3 are outlined in Appendix A-14.4.2. The solution for problem-1 leads to the results plotted in Figures 14-2a and 14-2b,

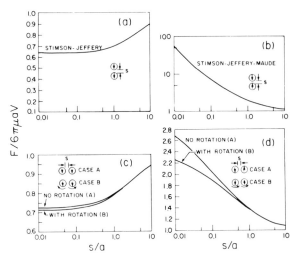

Fig. 14-2. Theoretical resistance coefficients for equal sized spheres in Stokes flow. (a) Parallel motion along the line of sphere centers, (b) antiparallel motion along the line of centers, (c) parallel motion perpendicular to the line of centers, (d) antiparallel motion perpendicular to the line of centers. (From M. H. Davis, 1966, with changes.)

which show the drag forces on identical spheres moving with equal parallel and antiparallel velocities along their line of centers. The forces are normalized with respect to the single sphere Stokes drag, so that they approach unity for large separations. For equal parallel motion the forces on the spheres are identical, which is indicative of the absence of any wake in Stokes flow. For equal antiparallel motion the forces are equal and opposite, and become singular like s^{-1} as the separation s between their surfaces approaches zero. This is generally true for any size ratio, and means that rigid spheres cannot be brought into contact according to the continuum theory of viscous flow. Physically, of course, the continuum theory will break down as the separation approaches the mean free path of air molecules. The effect of this breakdown on the collision problem is discussed in the next section.

The solutions for problems 2 and 3 lead to the results plotted in Figures 14-2c and 14-2d. Figure 14-2c shows the drag forces experienced by equal spheres falling with equal parallel velocities perpendicular to their line of centers with and without rotation. The angular velocities $\vec{\Omega}$ of the spheres for the case of rotation are determined by the condition that the spheres experience no torque $\vec{\Gamma}$; i.e., Ω_1 and Ω_2 are found by setting $\Gamma_1 = \Gamma_2 = 0$ in (A.14-41). Similarly, Figure 14-2d shows the drag forces for equal antiparallel motion perpendicular to the line of centers, with and without rotation. Note the forces for antiparallel motion appear to become singular with vanishing separation, as in the case of axisymmetric motion. In fact, O'Neil (1964) has shown they do become singular, like log s, for s → 0.

14.4.3 THE SLIP-FLOW CORRECTION IN STOKES FLOW

As we mentioned above, droplet collisions are theoretically impossible in Stokes flow since the resistance to the approach of the spheres varies inversely with the gap between them, for separations small compared to their radii. Therefore, in the collision efficiency calculations which are based on rigorous Stokes flow (Davis and Sartor, 1967; Hocking and Jonas, 1970), it is assumed that a collision has occurred if the computed gaps become less than a given small fraction, $s' \equiv s/a_1$, of the radius of the larger droplet. Of course, such an approach can be useful only if the computed collision efficiencies are relatively insensitive to the choice of s'. This turns out to be the case for $a_1 \geqslant 30 \, \mu$m. However, for smaller sizes the choice of s' noticeably affects the results.

This uncertainty brought about by the arbitrary choice of a collision criterion has been removed by Davis (1972), Jonas (1972), and Hocking (1973), through application of the theory of slip-flow. The assumption that the air gap between the droplets is a continuum will break down progressively as the separation becomes of the order of ten times the mean free path of air molecules or less. The relevant manifestation of the onset of non-continuum flow will be that the effective viscosity will tend to decrease as the Knudsen number based on the separation of the droplets becomes larger than about 0.1. (This effect applies only to the component of antiparallel motion, which causes the air to be squeezed out from between the droplets.) The theory of slip-flow enables one to determine approximately the form of the effective viscosity.

According to slip-flow theory, the tangential component of the fluid velocity at a body surface is less than the surface velocity, with the velocity difference being proportional to the local tangential stress in the fluid. The factor of proportionality between the velocity slip and the local stress, the 'coefficient of external friction,' β, then can be related to the mean free path. This leads directly to a new boundary condition, which must replace the continuum boundary condition of no slip (some further details of slip-flow modification of the Stokes flow problem are given in Appendix A-14.4.3).

The analysis of Davis (1972) yields forces which are functions of the dimensionless parameter $C \equiv \eta/\beta c$, where c is the length scale factor appearing in the transformation from cylindrical to bispherical coordinates (see A.14-22). It is necessary to provide representative values of C to complete the force description. It is also of interest to determine the dependence of C on the mean free path λ. For the case of low pressure gas flow through glass capillaries, Davis found that good agreement between experimental data and the slip-flow analysis could be obtained by setting $\eta/\beta = A\lambda$, where A is a number in the range 1.3 to 1.5. Therefore, by analogy he conjectured the dependence $C = A(\lambda/c)$. Then for purposes of presenting results a new quantity k was defined as $k \equiv C(c/a_2) = A(\lambda/a_2)$. The advantage of k over C is that the former is fixed for a given pair of spheres, while the latter varies with separation. Since the cross-over in dominance from Brownian to gravitational coagulation occurs for radii of a few microns, we see it is sufficient to evaluate the forces for $k < 0.1$.

Over a wide range of separations and size ratios, the calculations produce the

TABLE 14-1

Effect of slip on forces of interaction between equal sized spheres in Stokes flow; equal parallel motion along line of centers for $s'' = 0.01$. (From M. H. Davis, 1972; by courtesy of Amer. Meteor. Soc., and the author.)

$a_1/a_2 = 1.0$		$a_1/a_2 = 5.0$	
k	$F_1/F_{s,1}$	$F_1/F_{s,1}$	$F_2/F_{s,2}$
0.01	0.992	0.998	0.992
0.02	0.985	0.996	0.984
0.05	0.961	0.991	0.966

following results: 1) the effect of slip on forces for equal parallel motion along the line of centers is, as expected, insignificant for $k < 0.1$ (see Table 14-1; F_S is the Stokesian value); 2) forces on two spheres in equal antiparallel motion along the line of centers are reduced significantly at close separations $s'' \equiv s/a_2$ for finite k. Further, if $F_{S,i}$ denotes the Stokesian value for sphere i, then an approximate representation of the forces is as follows:

$$F_i \approx F_{S,i} f, \tag{14-21}$$

where

$$f = \left(1 + \frac{k}{s''}\right)^{-1} = \left(1 + A \frac{\lambda'}{s''}\right)^{-1} \tag{14-22}$$

and $\lambda' = \lambda/a_2$. Table 14-2 shows this representation (which of course has the familiar form of the Cunningham slip-correction factor; cf. Section 10.3.5) is reasonably good over a wide range of parameter values.

It can be seen that the form of f is such that the s^{-1} behavior of the Stokesian force coefficients for approaching spheres at close separations is eliminated. Therefore, if the force components that refer to antiparallel motion along the line of centers are each multiplied by the factor f to allow for gas kinetic effects

TABLE 14-2

Effect of slip on forces of interaction between equal sized spheres in Stokes flow; equal antiparallel motion along line of centers. (From M. H. Davis, 1972; by courtesy of Amer. Meteor. Soc., and the author.)

k	$s'' = 0.10$		$s'' = 0.01$		$s'' = 0.001$	
	$(F_1/F_{s,1})$	f	$(F_1/F_{s,1})$	f	$(F_1/F_{s,1})$	f
0.002	0.985	0.980	0.754	0.833	0.260	0.333
0.005	0.944	0.952	0.586	0.667	0.152	0.167
0.010	0.888	0.909	0.451	0.500	0.097	0.091
0.020	0.804	0.833	0.329	0.333	0.061	0.048

in the region between the droplets, the need for imposing an arbitrary minimum gap in collision efficiency calculations is eliminated.

A related problem which has not yet been solved is the determination of the slip-flow modification of forces experienced by two spheres moving with antiparallel velocities perpendicular to their line of centers at close separations. However, Davis (1972) conjectured that the resulting corrections would have little effect on droplet trajectories. We agree because, as we discussed in the previous section, the forces in question become singular at a slower rate with vanishing s (like log s) than those for antiparallel line of centers motion.

14.4.4 TWO SPHERES IN MODIFIED OSEEN FLOW

We have seen that great mathematical complications attend the basic collision model in the theory of accretion growth of cloud droplets. This is true primarily because of the nonlinearity of the convective acceleration terms in the governing Navier-Stokes equations, and because of the problem of satisfying the flow boundary conditions on two separate surfaces. The superposition method provides a means of estimating the effects of nonlinear inertial accelerations in the fluid even for quite large Reynolds numbers, but only at the cost of completely avoiding the boundary value problem. On the other hand, as we have seen, the Stokes equations are just simple enough to permit a complete boundary analysis of the problem, but their use implies the complete omission of all fluid inertial effects.

Unfortunately, there is evidence that the applicable range of the Stokes-flow model may be less than is often assumed. We have said that traditionally the model has been assumed valid for $a_1 \lesssim 30 \, \mu$m, corresponding to $N_{Re} \lesssim 0.4$. But while it is true that for single spheres of such radii the Stokes drag closely approximates the actual drag, for two spheres the effect of inertial accelerations in the fluid can cause a significant differential rate of fall even for considerably smaller sizes. This is especially true for spheres of comparable size. For example, Steinberger *et al.* (1968) observed the motion of pairs of equal spheres falling along their line of centers (for which case any relative motion is due entirely to non-Stokes behavior of the fluid), and found a significant acceleration effect even for Reynolds numbers as small as 0.05, corresponding to droplet radii of 15 μm. Similar behavior has been observed in the model experiments of Pshenai-Severin (1957, 1958), Schotland (1957), Telford and Cottis (1964), and Horguani (1965).

From the foregoing it would appear useful to obtain an analysis of the collision problem which does not assume that either the superposition scheme or the Stokes approximation are adequate to describe the flow. We now outline such a treatment, due to Klett and Davis (1973). It is an approximate boundary value analysis, based on the Oseen equations of motion as modified by Carrier (see Section 10.2.4).

For the two-sphere problem, Carrier's method must be generalized somewhat, since four variable Reynolds numbers are involved: with sphere ℓ ($\ell = 1, 2$) there are associated two characteristic Reynolds numbers, one (R_ℓ) based on its velocity U_ℓ and radius a_ℓ, and the other (R'_ℓ) on U_ℓ and the center-to-center distance d to the other sphere. The latter Reynolds number is the one involved in the inertial correction to the force on a given sphere due to the presence of the other sphere. Typically, it will vary over an order of magnitude for those separations which give rise to a significant hydrodynamic interaction.

The analysis proceeds by first selecting for each sphere a primitive solution to an Oseen governing equation. For each such equation, the constant velocity appearing in the acceleration term is taken to be the velocity of the respective sphere through the fluid. Next, the sum of the two fields is made to satisfy the no-slip condition on the average over the surface of the spheres. The expressions for the forces are then obtained from this approximate solution, and these are modified further by making the replacements $R_\ell \to C_\ell R_\ell$ and $R'_\ell \to C'_\ell R'_\ell$, where

$$C_\ell = 1 - 0.08 \log (1 + 50\, R_\ell), \quad R_\ell \leq 2,$$

$$C'_\ell = 0.43.$$

(14-23)

The 'Carrier constants' C_ℓ ensure agreement to within 1% between the numerically determined drag values of Le Clair et al. (1970) and the Carrier-Oseen drag on sphere ℓ falling in isolation for $R \leq 2$ ($N_{Re} \leq 4$). The constants C'_ℓ have the value suggested originally by Carrier.

Details of the analysis are discussed in Appendix A-14.4.4. Some results of the computations are plotted in Figure 14-3, where the velocity difference between two equal spheres falling along their line of centers is given as a function of the distance d between the sphere centers. Also, a comparison is made with the theoretical predictions of Oseen (1927) and Stimson and Jeffery (1926), and with the experimental results of Steinberger et al. (1968), which covered the Reynolds number interval $0.06 \leq N_{Re,2} \leq 0.216$ ($N_{Re,2} = 2\,R_2$), corresponding roughly to cloud droplets from 15-22 μm in radius. In all cases observed by Steinberger et al., both spheres continually accelerated as they fell, and the upper sphere fell

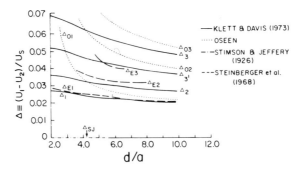

Fig. 14-3. Velocity difference between two spheres of equal size falling along their line of centers, according to Klett and Davis (Δ_1, Δ_2, Δ_3, Δ'_3), Oseen (Δ_{01}, Δ_{02}, Δ_{03}), Stimson and Jeffery (Δ_{SJ}), and the experiments of Steinberger et al. (Δ_{E1}, Δ_{E2}, Δ_{E3}). Δ_1: $0.060 \leq N_{Re} \leq 0.086$, Δ_2: $0.082 \leq N_{Re} \leq 0.0118$, Δ'_3: $0.120 \leq N_{Re} \leq 0.174$, Δ_3: $0.168 \leq N_{Re} \leq 0.240$, Δ_{01}: $0.060 \leq N_{Re} \leq 0.086$, Δ_{02}: $0.082 \leq N_{Re} \leq 0.124$, Δ_{03}: $0.170 \leq N_{Re} \leq 0.250$, Δ_{SJ}: Stokes flow, Δ_{E1}: $0.060 \leq N_{Re} \leq 0.086$, Δ_{E2}: $0.080 \leq N_{Re} \leq 0.150$, Δ_{E3}: $0.170 \leq N_{Re} \leq 0.216$. (From Klett and Davis, 1973; by courtesy of Amer. Meteor. Soc., and the authors.)

faster and accelerated more than the lower one. Because the velocities changed as each experimental run proceeded, a range of values of $N_{Re,2}$ was generated. They grouped their results according to the ranges $0.060 \leqslant N_{Re,2} \leqslant 0.086$, $0.086 \leqslant N_{Re,2} \leqslant 0.150$, and $0.170 \leqslant N_{Re,2} \leqslant 0.216$.

We note from Figure 14-3 that the model of Stimson and Jeffery is obviously the most deficient, predicting zero velocity difference independently of separation and Reynolds number. Of course, this unrealistic result is due to the symmetry of the Stokes flow field (there is no wake), which yields equal forces on the spheres. The Klett-Davis formulation is seen to be the most successful of the three in predicting the magnitude of the velocity difference and its trend toward higher values with both increasing Reynolds number and decreasing separation. Oseen's model (essentially equivalent to the superposition method applied with single sphere Oseen flow fields) seriously overestimates these trends.

Another noteworthy feature of the motion observed by Steinberger et al., but not revealed in Figure 14-3, is that the drag force for each sphere has a Reynolds number dependence. This is in agreement with the Klett-Davis formulation, but is at odds with Oseen's model, which predicts no such dependence for the upper sphere. Of course in the Stimson-Jeffery model all Reynolds numbers are effectively equal to zero.

The experiments of Steinberger et al. were conducted with both steel and tungsten carbide spheres in oil, for which the density ratios were $\rho/\rho' = 0.11$ and 0.06, respectively. They found that for a given Reynolds number range the results were independent of the kind of spheres used. Therefore, one can conclude that the density ratio was small enough not to play a noticeable role in the evolution of trajectories. This means the results shown are relevant to the case of water drops in air.

Additional comparisons between theory and experiment are made in Section 14.6, where we discuss collision efficiencies.

14.5 Enhancement of Gravitational Collection by Turbulence

Since some degree of turbulence is always present in clouds, it is important to consider its effect on the collision process. A major problem of this class is to determine how the low intensity turbulence in young clouds affects the early stages of the evolution of droplet spectra through the enhancement (or suppression) of gravitational collection. This very difficult problem has been dealt with in a relatively comprehensive manner only recently by de Almeida (1975, 1976). Details of de Almeida's analysis are given in Appendix A-14.5. With his model it is possible to determine approximately the trajectories of two interacting drops in a turbulent medium.

The presence of turbulence evidently requires a generalization of the notion of collision efficiency, since turbulent fluctuations impart a degree of randomness to the trajectories of droplets as they interact hydrodynamically. Thus, whereas in the non-turbulent case a collision is assured if the initial horizontal offset distance y satisfies $y \leqslant y_c$ and is impossible for $y > y_c$ (recall Figure 14-1), in the present case there is a finite probability $P(y)$ for a collision at any y. This distinction between the previous deterministic and present probabilistic problem is illustrated in Figure 14-4.

Evidently, $P(y)$ can be determined only by the laborious procedure of making repeated trials for a given set of initial conditions. Thus, de Almeida obtained $P(y)$ as follows:

$$P(y) = \frac{n(y)}{T}. \qquad (14\text{-}24)$$

Here $n(y)$ denotes the number of successes (collisions) out of T trials for a pair of spheres starting their motion with a given off-center horizontal separation y. After experimenting with various values of T, de Almeida decided on the choice $T = 200$ for his entire set of computations.

A somewhat analogous procedure is involved even in the non-turbulent case, since a sequence of trials is required to produce a converging sequence of estimates of y_c. In this case the error in the collision efficiency E is, from (14-2),

$$\Delta E = \frac{2\, y_c \Delta y}{(a_1 + a_2)^2} \leqslant \frac{2\, \Delta y_c}{a_1 + a_2}, \qquad (14\text{-}25)$$

where Δy_c is the error in the trial and error solution for the critical horizontal separation y_c. (As we have discussed in Section 14.2, and as we shall see further below, the inequality in (14-25) may break down for $a_2/a_1 \to 1$, due to the wake effect.)

Given $P(y)$, the total effective cross section for collisions is just $\int_0^\infty 2\pi y P(y)\,dy$. Therefore, the appropriate generalization of E to include the case of turbulence is

$$E = \frac{2}{(a_1 + a_2)^2} \int\limits_0^\infty y P(y)\,dy. \qquad (14\text{-}26)$$

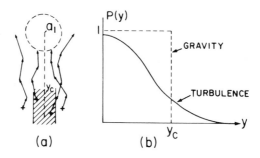

Fig. 14-4. (a) Typical drop-droplet relative trajectories in turbulent air, (b) collision probability curve (solid line) compared with a typical deterministic case (dashed line). (From de Almeida, 1975; by courtesy of the author.)

14.6 Theoretical Collision Efficiencies of Water Drops in Air

14.6.1 THE CASE OF CALM AIR

Let us consider first the non-turbulent collision efficiencies E of small droplets with radii of 30 μm or less. As we discussed in the previous sections, it is this realm which has been treated most rigorously, through application of the model of slip-corrected Stokes flow (Davis (1972), Jonas (1972), Hocking (1973)). Hence the resultant values of E should be the best available, subject to the proviso that the radius ratio $p \equiv a_2/a_1 \lesssim 0.6$ so that fluid inertial effects, and especially the wake capture phenomenon, will be of negligible importance. Slip-flow corrected values of E computed by Jonas (1972) are shown in Figure 14-5, where they are compared with a representative example of computations (Hocking and Jonas, 1970) based on continuum Stokes flow and the assumption that a collision occurs whenever the separation of the sphere surfaces becomes less than $10^{-4} a_1$. The effect of slip is seen to be considerable for $a_1 \lesssim 20 \mu$m, but to have little effect on drops as large as 30 μm.

Figure 14-5 also includes the corresponding values of E obtained by setting all Reynolds numbers equal to zero in the modified Oseen flow model of Klett and Davis (1973). This provides a measure of the accuracy of the purely viscous forces according to their formulation. It can be seen that the Klett-Davis values agree better with the slip-corrected E's than do those that follow from the continuum Stokes flow model.

Computations for a given zero-Reynolds number model show that for $p = 1$, E is independent of the size of the spheres. This behavior can be seen in Figure 14-5 by the tendency of each set of curves to converge to a single point near

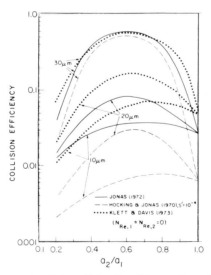

Fig. 14-5. Theoretically computed collision efficiencies of interacting spheres in Stokes flow, and in modified Oseen flow (zero Reynolds number limit) for spheres of various radii (given by the label of each curve). (From Jonas, 1972, with changes.)

p = 1. This (unphysical) behavior may be explained by noting that in the limit
p → 1, the sphere accelerations become vanishingly small since no fluid inertial
effects are permitted. Hence dynamical effects vanish and so geometric similarity implies dynamic similarity as well, no matter what the absolute size of the
equal spheres may be.

An indication of the importance of wake capture and other fluid inertial effects
is provided by Figure 14-6, where the collision efficiencies derived from the
Stokes flow model of Jonas (1972) are compared with those computed from the
Klett-Davis model with non-zero Reynolds numbers, and with those following
from the superposition models of Lin and Lee (1975) and Schlamp *et al.* (1976).
It can be seen that the effect of fluid inertia on drop collisions is most
pronounced for p-ratios near unity, where there is a marked tendency for an
increase in E due to wake capture.

The trends of the collision efficiency curves in Figure 14-6 for all three of the
most recent models which include inertial effects are in accord with what was
anticipated qualitatively in Section 14.2. Quantitatively, however, the agreement
among the values of E computed by these models is only fair. In particular, the

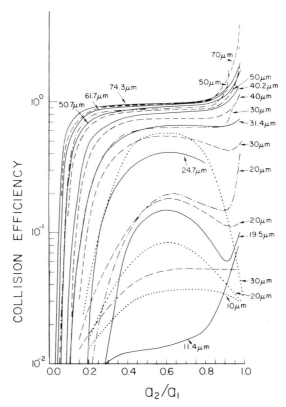

Fig. 14-6. Theoretical collision efficiencies of spherical water drops in calm air as a function of
p-ratio and of collector drop radius (given by label of each curve); ——— Schlamp *et al.* (1976), - - - -
Klett and Davis (1973), -·-·- Lin and Lee (1975), ····· Jonas (1972).

E values deriving from the superposition approach are somewhat larger than those computed from the approximate boundary value analysis of Klett and Davis, for $a_1 \gtrsim 30 \, \mu$m and all p-ratios. The differences, which are especially large for p near unity, are most likely due to the following inherent deficiencies of the superposition model (cf., Klett, 1976): Since under superposition the individual flow fields do not interact, the strength of wake formation behind the leading sphere of two spheres falling in close proximity will be overestimated. This effect will be enhanced also by the underestimate of the strength of viscous interaction between the spheres. This leads to spuriously low drag, hence higher velocities, and hence stronger wakes. In short, the deficiencies of superposition lead one to expect overestimated wake capture, and underestimated deflection by viscous forces from collision trajectories.

Finally, Figure 14-7 depicts some representative results in another fashion, with E plotted as a function of the collected drop radius a_2, for various collector drop radii a_1. This manner of presentation shows clearly that E is near unity for $a_1 \gtrsim 40 \, \mu$m and $a_2 \gtrsim 15 \, \mu$m, unless $a_1 \approx a_2$, in which case $E > 1$.

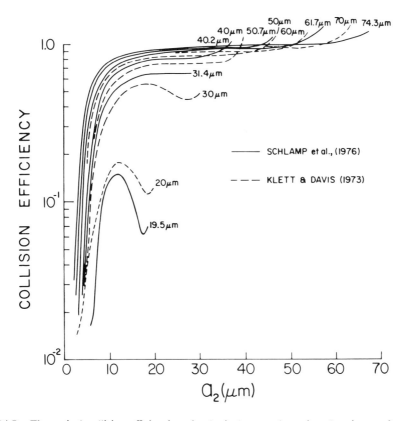

Fig. 14-7.   Theoretical collision efficiencies of spherical water drops in calm air as a function of small drop radius and of large drop radius (given by label of each curve).

14.6.2 THE CASE OF TURBULENT AIR

The study of de Almeida (1975, 1976), referred to earlier, is the best presently available for predicting the influence of turbulence on the collision growth process. de Almeida carried out numerical evaluations of his formulation (outlined in Appendix A-14.5) for two cases of weak turbulence, characterized by the energy dissipation rates $\varepsilon = 1 \text{ cm}^2 \text{ sec}^{-3}$ and $\varepsilon = 10 \text{ cm}^2 \text{ sec}^{-3}$. Such values are probably appropriate for the early stages of cloud development. The results of his collision efficiency computations are summarized in Figure 14-8. It is seen that E for small collector drops increases sharply as ε increases from 0 to $1 \text{ cm}^2 \text{ sec}^{-3}$, the effect being most noticeable for $a_1 \lesssim 30 \mu\text{m}$. Larger drops are not greatly affected by the turbulence levels studied; presumably this is a consequence of their larger inertia.

A somewhat surprising result is that for many droplet pairs E is smaller for $\varepsilon = 10$ than for $\varepsilon = 1 \text{ cm}^2 \text{ sec}^{-3}$. This trend is especially noticeable for p near unity. Inspection of Figure 14-9, in which P(y) (Equation (14-24)) is plotted vs. y, suggests a possible explanation of this behavior: For drops of comparable size, P(y) remains small and nearly constant. This means that turbulence effectively interferes with the interaction of such drop pairs. Their relative velocity is small so that turbulent fluctuations can seriously disrupt their weak interaction. The capacity for disrupting the drop interaction naturally grows progressively with increasing ε.

A more troublesome result also appears on inspection of Figure 14-8. As discussed in Appendix A-14.5, de Almeida has incorporated the Klett-Davis force description for drop interactions in his turbulence model, so that for $\varepsilon = 0$ his E values should agree with those of Klett-Davis shown in Figure 14-6. Indeed, for $a_1 \gtrsim 30 \mu\text{m}$ and $p \lesssim 0.8$, the results are in good agreement. However, for $a_1 \lesssim 25 \mu\text{m}$ de Almeida's E values are significantly smaller. In addition, the collision efficiencies of de Almeida do not exhibit the marked increase for p near unity which is so clearly evident in Figure 14-6. de Almeida attributes these differences to the occurrence of numerical instabilities in the numerical schemes

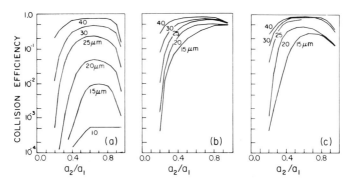

Fig. 14-8. Theoretical collision efficiencies of spherical water drops in turbulent air as a function of p-ratio and of collector drop radius (given by label of each curve). (From de Almeida, 1975; by courtesy of the author.)

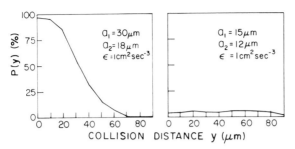

Fig. 14-9. Turbulent probability curves for interacting spherical water drops in air. (From de Almeida, 1975; by courtesy of the author.)

used by Klett-Davis and others to integrate the drop trajectory equations. However, in view of the fact that there is now considerable experimental evidence for the wake capture effect and the occurrence of $E > 1$, it is not clear that de Almeida's explanation is correct. Evidently, this is another point in need of further clarification.

Finally, we should introduce a general note of caution regarding the results shown in Figure 14-8. There are, unfortunately, two possibly seriously erroneous features of de Almeida's model (for details, see Appendix A-14.5). One is that the turbulent velocity correlation used in the model cannot be valid for the small drop separations which are most important in the collision problem. The second problem, which is likely more serious, is that the basic governing form for the equation of motion of the drops is not valid in principle for situations like the one of interest, in which the particles in the turbulent medium have a density large compared to that of the medium. The violation of this condition probably implies that the effect of turbulence on the drop collision problem is overestimated in de Almeida's model, since the drops' resistance to turbulent entrainment is apparently underestimated. In view of these problems, the results shown in Figure 14-8 should tentatively be regarded as exploratory rather than definitive. Nevertheless, we reiterate they are the best presently available.

Before closing this section on theoretical determinations of E, we should comment on the problem of extending the calculations to larger drop sizes. In general, for large falling drops with $N_{Re} \gg 1$ the strong nonlinearity of their hydrodynamic interaction makes the superposition method the only mathematically feasible approach known for estimating E. However, even this method has its limitations in implementation. We recall from Chapter 10 that for sufficiently large drops, such complications as wake oscillations, eddy shedding, and shape deformations occur. Since there are no numerical solutions in existence for flows past single drops which reproduce these features, one must resort instead to the use of steady state numerical flows past rigid spheres or oblate spheroids. Since such flows become increasingly artificial for $N_{Re} \gtrsim 10^2$ ($a_0 \gtrsim 300 \ \mu m$), one must expect the same for the corresponding estimates of E.

14.7 Experimental Verification

The many uncertainties inherent in the theoretical models for determining collision efficiencies make it especially important to check the computations against measurements. Unfortunately, however, the experimental approach is also beset with great difficulties. One major obstacle lies in the fact that in reality there is no clean conceptual division between the processes of 'collision' and 'coalescence.' When a pair of drops is allowed to interact, generally what is observed is either a coalescence or a non-coalescence event; in the latter case, it is usually not possible to say whether the drops actually collided but did not coalesce, or whether they simply experienced a 'near miss.' Thus, the experimentally accessible quantity is the *collection efficiency*, E_c, which is the ratio of the actual cross section for drop coalescence to the geometric cross section. This may be regarded as equivalent to the collision efficiency, as we have defined it, only if coalescence necessarily follows whenever the center-to-center separation of the two interacting drops becomes less than the sum of their (undistorted) radii.

To determine E_c one may either observe (usually photographically) the trajectories of interacting drops, or measure the rate at which a drop grows as it falls through a cloud of smaller drops. Suppose, for example, that a drop of radius a_1 and fall speed $V_{\infty,1}$ grows by collecting drops each of radius a_2 and fall speed $V_{\infty,2}$; then we have $E_c = (dm/dt)/\pi(a_1 + a_2)^2(V_{\infty,1} - V_{\infty,2})w_L$, where dm/dt is the mass growth rate of the drop, and w_L is the liquid water content of the cloud of a_2-drops. Unfortunately, it is experimentally quite difficult to determine accurately drop sizes, trajectories, growth rates, and fall speeds, and to produce a homogeneous cloud of known liquid water content. Hence, experimental values of E_c are commonly subject to errors of 10 to 20%.

Telford *et al.* (1955), Telford and Thorndyke (1961), Woods and Mason (1965, 1966), Beard and Pruppacher (1968), and Abbott (1974) have experimentally determined the collection efficiency of pairs of water drops in air with p-ratios close to unity, while Kinzer and Cobb (1958), Picknet (1960), Woods and Mason (1964), and Beard and Pruppacher (1971b) determined E_c for drop pairs of low p-ratios. A few experimental results are also available for pairs of small drops with intermediate p-ratios (Jonas and Goldsmith, 1972). In Figure 14-10, a comparison is made between theoretical collision efficiencies and experimental collection efficiencies for $p \approx 1$. The agreement between theory and the measurements of Abbott (1974) and Telford *et al.* (1955) is quite good. Unfortunately, however, the results of Woods and Mason (1965) and Beard and Pruppacher (1968) do not support this agreement; in fact, they observed no strong wake capture effect, and found $E_c < 1$ in all cases. The reasons for these different outcomes have not been clarified.

A comparison between theory and experiment for the case of small p-ratios $(0.02 \lesssim p \lesssim 0.2)$ is given in Figure 14-11. The computed values of Klett and Davis (1973) appear to be in good agreement with experiment, while the results of Schlamp *et al.* (1976) appear somewhat high.

While no direct measurements of collection efficiencies in turbulent clouds have been made, some related experiments which bear on the problem have

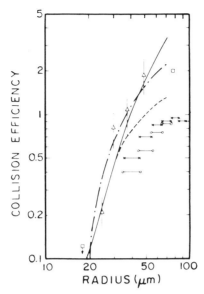

Fig. 14-10. Comparison of theoretical collision efficiencies with experimentally measured collection efficiencies for water drops of nearly equal size in air ····· Schlamp *et al.* (1976), p = 0.97, theory; ———— Klett and Davis (1973), p = 0.98, --- Klett and Davis (1973), p = 0.90, theory; ×—× Woods and Mason (1965) experiment; ○—○ Beard and Pruppacher (1968), experiment; □ Telford *et al.* (1955), experiment; △ Abbott (1974), experiment. (From Abbott, 1974; with changes.)

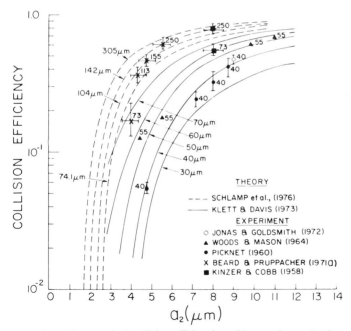

Fig. 14-11. Comparison of theoretical collision efficiencies with experimentally determined collection efficiencies, for water drop pairs of small p-ratio. Label of each curve and data point gives large drop radius in μm.

been carried out (Woods *et al.* (1972), Jonas and Goldsmith (1972)). Both groups measured E_c for drops in an approximately steady, laminar shear flow. As we discussed in Section 12.6.2, small aerosol particles interact on such small length and time scales that the effects of any turbulence which may be present can be modeled roughly through replacement of the turbulent velocity field by a much simpler linear shear field. This same basic modeling approach, including the selection of a representative shear strength near the Kolmogorov microscale value, is thought to be reasonable also for the problem of small droplet collisions in a turbulent cloud (see, for example, Tennekes and Woods, 1973).

Woods *et al.* and Jonas and Goldsmith measured the effect of shear on E_c for drops of $10 \lesssim a_1 \lesssim 40 \, \mu m$ colliding with drops of $9 \lesssim a_2 \lesssim 9.5 \, \mu m$, and for linear shears in the horizontal wind varying up to $27 \, sec^{-1}$. Since the results for the case of no shear appear somewhat low, we shall only discuss the qualitative trends which were observed. It was found that E_c values for drops of radius $a_1 \lesssim 25 \, \mu m$ are much larger in shear flows than in still air, while for larger drops the effects of shear were much weaker. The effects were most pronounced for drops of comparable size. To this extent the experiments support the theoretical predictions of de Almeida (1975). However, the experiments also indicated that the increase in E_c due to shear varies approximately linearly once a threshold value of shear has been exceeded. This is in conflict with de Almeida's prediction that an increase in the turbulent energy dissipation rate from 1 to $10 \, cm^2 \, sec^{-3}$ leads in most cases to a decrease in E_c (cf. Figures 14-8b and c).

A theoretical treatment by Jonas and Goldsmith of the effect of shear flow on small droplet collisions failed to reveal any significant enhancement in E_c for shears of the magnitude of those employed in the experiments. This failure of the theory may be due to the fact that Stokes flow was assumed. An effect of shear is to induce a flux of droplets past one another. As suggested by Manton (1974), this indicates the possibility that some droplets might therefore be expected to intersect the wakes of neighboring ones, and hence experience wake capture. Of course, such a possible mechanism for the enhancement of E_c is precluded by the assumption of Stokes flow, wherein no wakes can occur.

The plausibility of shear-induced wake capture is supported by measurements of the distance z_c of the wake interaction region behind falling drops of comparable size. For example, in the model experiments of Steinberger *et al.* (1968), discussed in Section 14.4.4, it was found that $z_c \gtrsim 12$ radii, even for a Reynolds number as low as 0.06 (equivalent to a drop of $a \approx 15 \, \mu m$). Not surprisingly, z_c is observed to increase strongly with N_{Re}. Thus Eaton (1970) found $z_c \gtrsim 20 \, a$ for roughly equal-sized drops of $a = 150$ to $250 \, \mu m$ falling in air. Cataneo *et al.* (1971) observed much larger values of $z_c \gtrsim 2 \times 10^2 \, a$ and $3 \times 10^2 \, a$ for equal water drops of $a = 57.5$ and $350 \, \mu m$, respectively. Finally, List and Hand (1971) found a maximum value of $z_c \approx 2.7 \times 10^3 \, a$ for drops of $a = 1.45 \, mm$.

Manton (1974) has provided a theoretical description of shear induced wake capture which agrees with the qualitative features found in the experiments. However, his model is rather artificial, and greatly oversimplifies the hydro-dynamics of the problem. For example, the wake interaction region is assumed to be undisturbed by the shear, and the interaction of the drops in close proximity is assumed independent of the shear.

In conclusion, it is clear that our present knowledge of drop collection efficiencies in turbulent clouds is still inadequate, and that some substantial differences between the current theories and experiments still need to be resolved. On the other hand, there is general good agreement between theoretical E and experimental E_c for drops of $a \leqslant 100 \ \mu$m falling in still air. Hence, for this size range the evidence suggests that shape distortions of interacting drops are negligible, and that the probability of coalescence following a collision is near unity. In the next section we shall see that this conclusion appears to be justified by further independent experiments.

14.8 Coalescence of Water Drops in Air

It is well known that not all drop collisions result in a permanent union by coalescence. Rather, there are two additional possibilities with which we must be concerned: (1) The drops may bounce apart before surface contact is made, owing to the presence of an air film trapped between their surfaces. (2) The drops may disrupt following temporary coalescence; as we shall see, this latter behavior may be explained satisfactorily in terms of the relative magnitudes of the surface energy and the rotational kinetic energy of the coalesced drop pair.

Unfortunately, the drop coalescence problem is in general quite complex, and not nearly so well in hand as the collision problem. For example, there are no theoretical treatments which incorporate accurately the large amplitude surface distortions which may occur, and thereupon inhibit the rate of air film drainage between the drops. Similarly, it has proven very difficult to conduct experiments which can faithfully reflect natural conditions, and at the same time provide sufficient control and resolution for the parameters of interest.

A wide variety of experimental arrangements have been employed in the coalescence problem. For example, Lindblad (1964) and Semonin (1966) studied the coalescence between two mm-size water drops, artificially held quasi-fixed at the end of capillaries which could be moved at a variable relative velocity; Magono and Nakamura (1959), Schotland (1960), and Jayaratne and Mason (1964) studied the conditions for coalescence between drops of $29 \ \mu$m $\leqslant a_0 \leqslant 400 \ \mu$m and a very large stationary, plane or hemispherical water surface; Whelpdale and List (1969), Whelpdale (1970), and List and Whelpdale (1971) observed coalescences between a moving $35 \ \mu$m radius drop and stationary larger drops of 500 m to 1750 m radius; Prokhorov (1951, 1954), Grover et al. (1960), Nakamura (1964), Adam et al. (1968), and Park (1971) studied the coalescence of freely moving drops of 3.1 mm to $25 \ \mu$m radius colliding with each other at various imparted relative velocities; Montgomery (1971), Nelson and Gokhale (1973), and Spengler and Gokhale (1973a, b) studied the coalescence of mm-size drops in a wind tunnel where the drops could be freely suspended in the vertical air stream of the tunnel, while Neiburger et al. (1972) and Levin et al. (1973) used a wind tunnel to infer the coalescence efficiency of water drops with radii between 45 and $120 \ \mu$m from their growth rates.

Unfortunately, most of these experiments were not carried out under natural conditions. For a proper simulation both interacting drops should fall freely at the relative velocities following from their size difference, or from wake capture effects. Only in this way can the flows controlling drop deformations and relative trajectories for close separations be represented accurately. For this reason, experiments with fixed large drops should best be regarded as primarily exploratory and qualitative in nature. The use of stationary flat water targets is even less realistic, except possibly for the case of very small radius ratios, as film drainage rates and the forces resisting deformation can be expected to depend strongly on drop curvature. Serious shortcomings are also inherent in those studies which involve the measurement of drop growth rates, since it is difficult to ensure that no mass gain or loss occurs by diffusional growth or evaporation.

These experimental difficulties are exacerbated by the sensitivity of the coalescence phenomenon to such influences as turbulence, surface contaminants, and electric fields and charges. Because of these problems, our discussions in this section are limited primarily to just the qualitative trends which can be gleaned from the small number of studies which appear to be the most comprehensive, and which yield the most mutually consistent results. The effects of turbulence and contaminants are not considered. Some discussion of the effects of electric fields and charges is presented in Chapter 17.

14.8.1 THE REBOUND PROBLEM

It is now widely recognized that coalescence, if only temporary, will proceed once the drop surfaces make contact, since surface energy is lowered by the destruction of surface area (some apparent exceptions to this consensus include Gunn (1965) and Cotton and Gokhale (1967)). Therefore, the drop rebound problem is primarily the problem of air film drainage. Many experiments have confirmed that this drainage is hindered in particular when the two approaching drops are large enough to deform easily. Then local flattening of their surfaces can strongly impede the expulsion of the intervening air. In addition to a dependence on drop size, the rate of air film drainage is controlled also by the relative velocity $\vec{v}_r$ of the drop pair, and by the impact angle θ between $\vec{v}_r$ and their line of centers at impact. Finally, many experiments have shown that the probability of coalescence rises sharply once the air film has thinned locally to a thickness $s \lesssim 0.1\,\mu$m. At such distances, attractive van der Waal's forces vary like s^{-3} (e.g., Adamson, 1967), and will likely be of sufficient strength to induce coalescence, especially with the help of small, random surface perturbations.

The most comprehensive set of observations on drop coalescence versus rebound and disruption for freely moving water drops with various imposed relative velocities has been carried out by Park (1970), and is summarized in Figure 14-12. From Figures 14-12a, b we notice that equal sized drops rebound at high impact angles only above a certain critical relative velocity. Considering now that numerical calculations of drop relative trajectories due to wake capture show that for roughly equal sized drops $v_r \approx V_\infty/10$ near impact, where V_∞ is the terminal velocity of either drop falling in isolation, we estimate $v_r \lesssim 30$ cm sec^{-1}

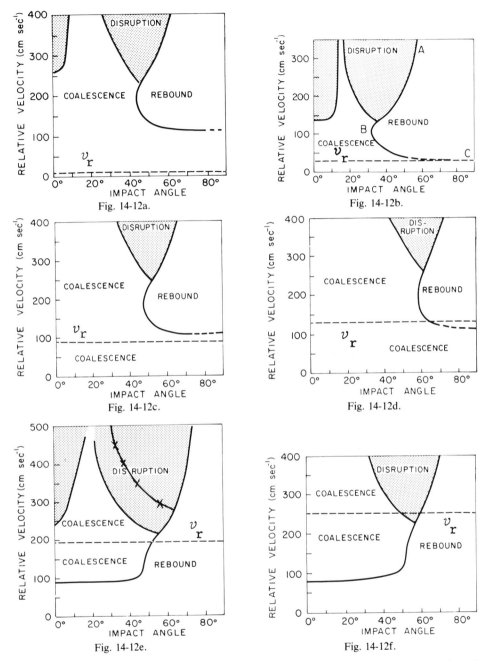

Fig. 14-12a.

Fig. 14-12b.

Fig. 14-12c.

Fig. 14-12d.

Fig. 14-12e.

Fig. 14-12f.

Fig. 14-12. Experimentally determined conditions for permanent coalescence, rebound, and temporary coalescence (disruption) of water drops in air for various relative velocities imposed on the drops. (a) $a_1 \approx a_2 = 100\ \mu m$; (b) $a_1 \approx a_2 = 350\ \mu m$; (c) $a_1 = 200\ \mu m$, $a_2 = 100\ \mu m$; (d) $a_1 = 225\ \mu m$, $a_2 = 75\ \mu m$; (e) $a_1 = 450\ \mu m$, $a_2 = 225\ \mu m$; (f) $a_1 = 450\ \mu m$, $a_2 = 150\ \mu m$. In Figures 14-12a, b v_r is the estimated relative velocity near impact of the two interacting drops. In Figures 14-12c-f v_r is the difference in terminal velocity of the two interacting drops; ✕✕✕✕ computed from (14-31) and (14-32). (From Park, 1970, with changes.)

for drops of radius $a \lesssim 350 \, \mu\text{m}$. Hence from Park's results we conclude that equal drops with $a \lesssim 350 \, \mu\text{m}$ should coalesce on impact, whatever their contact angle.

Figures 14-12c, d suggest the coalescence behavior of unequal sized drops has a similar character to that just discussed, at least for larger drop radii in the interval $150 \lesssim a_1 \lesssim 300 \, \mu\text{m}$. We see that at sufficiently low relative velocities $v_r \approx V_{\infty,1} - V_{\infty,2}$ no rebound occurs for any angle of impact. In fact, the figures suggest the conclusion that colliding drops of $a \lesssim 150 \, \mu\text{m}$ ($V_\infty \lesssim 115 \, \text{cm sec}^{-1}$) will suffer no coalescence problem for any p-ratio or impact angle. Furthermore, Figures 14-12c, d imply that drops of $150 \lesssim a \lesssim 300 \, \mu\text{m}$ ($115 \lesssim V_\infty \lesssim 245 \, \text{cm sec}^{-1}$) will coalesce for all p-ratios if the impact angle is less than about $60°$. At higher angles the drops will rebound.

The situation changes abruptly for drops $a \gtrsim 400 \, \mu\text{m}$ (Figures 14-12e, f). Notice that drops in this size range require, at all angles of incidence, a minimum critical relative velocity for coalescence to occur. Below this critical value drops rebound, the rebounding being particularly pronounced for angles of impact larger than $50°$. However, since drops of $a \gtrsim 400 \, \mu\text{m}$ have $V_\infty \gtrsim 330 \, \text{cm sec}^{-1}$, coalescence seems to be guaranteed for all drop pairs with $p \lesssim 0.75$ ($v_r \gtrsim 90 \, \text{cm sec}^{-1}$) colliding at angles of impact less than $50°$. Such behavior would also be consistent with the observations of List and Whelpdale (1971).

Park did not attempt to extract a coalescence efficiency or probability function from his measurements. However, List and Whelpdale have provided a simple estimate of the coalescence efficiency pertaining to their experimental setup. As noted earlier, they studied the coalescence between a moving $35 \, \mu\text{m}$ radius drop and stationary larger drops of $500 \lesssim a_0 \lesssim 1750 \, \mu\text{m}$. On projecting the small drops toward the larger ones, they observed nearly geometric collision cross sections; i.e., the small drops moved rectilinearly until impact. (Such behavior would be expected, since there was no flow around the large drops. If both drops had been falling freely, some flow deflection of the smaller one would have occurred; however, this effect should not be very significant for the drop sizes considered in the experiment.) List and Whelpdale observed rebounding only when the impact angle was near $90°$, i.e., for grazing collision trajectories. Consequently, the observed cross section for coalescence was roughly just πa_1^2, which on comparison with the geometric collision cross section of $\pi(a_1 + a_2)^2$ led to the following estimate of the coalescence efficiency, E_{coa}:

$$E_{coa} \approx \frac{a_1^2}{(a_1 + a_2)^2} = \frac{1}{(1 + p)^2}. \tag{14-27}$$

It should be stressed that this result was obtained only for $a_1 \gtrsim 400 \, \mu\text{m}$. In fact, from Park's results we expect $E_{coa} \approx 1$ for smaller a_1, since he observed essentially no rebound under natural conditions for such sizes. It appears that within its range of applicability (14-27) should somewhat overestimate the actual coalescence efficiency, because List and Whelpdale, and especially Park, also observed rebound at impact angles less than $90°$.

Levin *et al.* (1973) have reported values of E_{coa} significantly less than unity for $a_1 < 114 \, \mu\text{m}$. These results appear to extend the trend found by List and

Whelpdale to smaller drop sizes, and are thus in conflict with the findings of Park. However, since Park's results are strongly supported by the good agreement between several independent sets of calculations of collision efficiencies and measurements of collection efficiencies for $a_1 \lesssim 100 \ \mu$m, it appears likely that the drop growth rates observed by Levin et al., from which they inferred values of E_{coa}, may have been affected by evaporation and/or electric effects. Some support for this conclusion comes from the recent experiments of Levin and Machnes (1976); they found that with decreasing a_1, E_{coa} of drops of a given p-ratio tended to increase towards unity for $1700 \ \mu$m $\leqslant a_1 \leqslant 200 \ \mu$m. However, the values observed for E_{coa} were considerably smaller than those expected from Park's results.

Of course, since Park's measurements were confined to $a_1 \geqslant 100 \ \mu$m, we have had to extrapolate the trends observed by him in order to arrive at the conclusion that there is evidently no coalescence problem for $a_1 < 100 \ \mu$m as well as for $100 \leqslant a_1 \leqslant 150 \ \mu$m. However, there is a simple physical justification for this extrapolation which we should re-emphasize: Surface tension forces increase the resistance to drop deformation with decreasing size. Therefore, since such deformations inhibit air film drainage between interacting drops and thus increase the likelihood of rebound, drops of $a_1 < 100 \ \mu$m should experience no coalescence problem if those of sizes $100 \leqslant a_1 \leqslant 150 \ \mu$m do not. That is to say, for such drops one expects the coalescence problem to reduce to the collision problem. And, as we have seen, there is ample direct and indirect evidence that this is in fact the case.

The most comprehensive theoretical treatment of the effect of drop deformation on the rebound problem is that of Foote (1971). He carried out a numerical simulation of the collision and rebound of equal drops moving with equal antiparallel velocities along their line of centers. Unfortunately, however, in his simulation Foote ignored the air flow around the drops, so that in principle they should have remained spherical up to the moment of contact; i.e., it was as though they were colliding in a vacuum. Thus in principle a rigorous treatment of the air drainage problem was excluded by his formulation. As it turned out, however, for very small separations some local deformation occurred as a result of the 'upstream' propagation of the influence of the boundary condition of zero normal drop flow at the collision interface. The deformation before impact was thus 'numerical' rather than physical; nevertheless, it probably represented a reasonable qualitative description of natural deformations, especially for drops large enough to flatten significantly.

Given this pre-collision deformation, Foote proceeded to study the drainage problem which would result if an air film did in fact lie between the surfaces. The model used for this purpose was one of two approaching disks in a viscous fluid. The disk radii and the velocity of approach were described in accordance with the results of the prior simulation of the drop collision and rebound (coalescence not being allowed to occur in the prior numerical simulation). The water flow velocity at the disk surfaces was prescribed in several ways to explore various effects. Some interesting results of this work include the following: (1) Drop internal circulations of naturally-occurring magnitudes gave negligible assistance to the drainage of air. (2) The local water flow set up as the drops locally flattened on impact substantially promoted the drainage of the air film. (3) Any realistic reduction in the effective air viscosity due to gas kinetic effects at small separations (recall Section 14.4.3) did not make a significant difference in the rate of film drainage.

In general, it turned out to be difficult to reduce the air film to a thickness of a few tenths of a micron, where coalescence might be expected to commence, in a time period less than that required for rebound (which incidentally was often close to the drop oscillation time corresponding to (10-91)). However, this tendency could be overcome by assuming sufficiently large initial relative velocities. For example, Foote found that drops of $a = 595$ μm will coalesce only if their relative velocity is larger than about $40 \, \text{cm sec}^{-1}$. This result appears consistent with the observations of Park (Figures 14-12e, f). (We may note also that for drops of this size the relative velocity near impact induced by wake capture is $v_r \approx 50 \, \text{cm sec}^{-1}$, so that coalescence would in fact be expected.)

Also, as would be expected on the basis of dimensional arguments, it was found that the results of the collision and rebound dynamics could be expressed in terms of the collisional Weber number, $N'_{We} \equiv \rho_w d v_r^2 / \sigma$, where d is the drop diameter. Obviously, the smaller N'_{We}, the less the deformation and hence the less likely the occurrence of rebound. Foote did not find a minimum value for N'_{We} below which coalescence should always occur, because of the above mentioned difficulties of properly representing the onset of drop deformations.

An indication of how important it is to represent the initial deformations accurately may be provided by noting that the force required to squeeze out viscous fluid between two disks with surface separation s varies as s^{-3} (e.g., see § 20 of Landau and Lifschitz (1959)), whereas it varies as s^{-1} for the case of two spheres. This indicates that the computed insensitivity of air film drainage rates to the breakdown in viscosity at close separations (result (3) above) was due to the assumption of the particular geometry, and that drainage rates for small drops could be expected to be much higher than those computed.

It is obvious that the drop coalescence-rebound problem is yet another area in need of further experimental and theoretical study.

14.8.2   DISRUPTION FOLLOWING COALESCENCE

Drop break-up following coalescence of two drops has been studied experimentally by Blanchard (1948, 1949, 1950, 1962), Cotton and Gokhale (1967), List et al. (1970), Montgomery (1971), Park (1970), Whelpdale and List (1971), Brazier-Smith et al. (1972, 1973), Spengler and Gokhale (1973), and McTaggart-Cowan and List (1975). These experiments have demonstrated that during the early stages of coalescence the coalescing drop-drop system is often highly unstable and may break up. Bag-type, dumbbell-type, neck-type, sheet-type, and disk-type break-up (McTaggart-Cowan and List (1975)) have been observed. The wind tunnel studies of List et al. (1970) with freely falling water drops of $1.0 \leq a_0 \leq 2.3$ mm showed that neck-type break-up produced up to 10 fragments, with about four fragments on the average, and that the two temporarily coalescing drops 'reappeared' as the largest two fragments with relatively little mass change while the smaller fragments were due to the disruption of the neck temporarily joining the two coalescing drops. The size distribution of the fragments formed after break-up of a temporarily coalesced 3.5 and 2.1 mm diameter drop pair is illustrated by Figure 14-13. Note that the two modes in the size distribution correspond to the size of the two drops before coalescence. Note also that when more than two fragments are formed a third fragment size mode appears at a size smaller than either of the two original drops. Brazier-Smith *et al.* studied somewhat smaller freely falling drops of $150 \leq a_0 \leq 750\,\mu m$ and $0.4 \leq a_2/a_1 \leq 1.0$. The number of satellite drops (drops in addition to the two main fragments) varied from one to eight, and was three on the average. The fraction of the total drop mass used to form these satellites was found to be approximately $0.12\,\gamma^3/(1 + \gamma^3)^2$, where $\gamma = a_1/a_2 = p^{-1}$.

Brazier-Smith *et al.* also presented an appealingly simple theoretical explanation for their coalescence observations. (A similar explanation has been given by Park (1971), but with less elaboration.) They suggested that breakup following coalescence will occur if the rotational kinetic energy of the coalesced drop pair exceeds the surface energy required to reform the original two drops from the coalesced system.

The drops studied by Brazier-Smith *et al.* had relative velocities ranging from about 0.3 to 3.0 m sec^{-1}. For the drop sizes and relative velocities considered, the

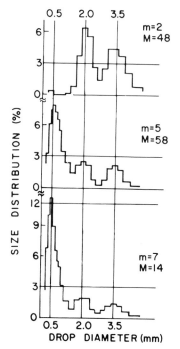

Fig. 14-13. Experimentally determined drop size distribution found after M collisions between drops of 3.5 mm and 2.1 mm diameter for various fragment numbers m. (From List *et al.*, 1970, with changes.)

reasonable assumption was made that the air should exert a negligible influence on both the trajectories and the initial coalescence process. Hence the angular momentum of the coalesced system could be found from elementary mechanical arguments, given the initial velocity v_r on the small drop relative to the larger one, and the initial perpendicular distance X between the center of the large drop and the undeflected trajectory of the center of the smaller one. In fact, the angular momentum L of the coalesced pair about its center of mass may be written down immediately from the theory of the two-body problem (e.g., see Ch. 3 of Goldstein (1965)): $L = \mu v_r X$, where $\mu = m_1 m_2 / (m_1 + m_2)$ is the *reduced mass* of the system comprised of masses m_1 and m_2 of the two drops. Hence, if the coalesced pair is a sphere of radius $a = (a_1^3 + a_2^3)^{1/3}$, then its angular momentum must be

$$L = \frac{4 \pi \rho_w v_r X a_1^3 a_2^3}{3(a_1^3 + a_2^3)}. \tag{14-28}$$

The moment of inertia about the center of the coalesced pair is $I = 8 \pi a^5 \rho_w / 15$, so that its rotational kinetic energy is

$$E_{rot} = \frac{L^2}{2 I} = \frac{5 \pi \rho_w v_r^2 X^2 a_1^6 a_2^6}{3(a_1^3 + a_2^3)^{11/3}}. \tag{14-29}$$

On the other hand, the surface energy required to form two drops of radius a_1 and a_2 from the one of radius a is

$$E_{surf} = 4 \pi \sigma [a_1^2 + a_2^2 - (a_1^3 + a_2^3)^{2/3}]. \tag{14-30}$$

Hence from (14-29) and (14-30) disruption following coalescence should occur for $X > X_c$, where X_c is the impact parameter for which $E_{rot} = E_{surf}$; i.e.,

$$\frac{X_c}{a_1 + a_2} = \left[\frac{2.4 \, \sigma}{\rho_w v_r^2 a_2} f(\gamma) \right]^{1/2}, \tag{14-31}$$

where

$$f(\gamma) = \frac{[1 + \gamma^2 - (1 + \gamma^3)^{2/3}](1 + \gamma^3)^{11/3}}{\gamma^6 (1 + \gamma)^2}, \tag{14-32}$$

and where $\gamma = a_1 / a_2 = p^{-1}$. Note that the left side of (14-31), which is the effective coalesce efficiency for the interaction mode under consideration, is a function only of the collisional Weber number and the radius ratio of the interacting drops.

Brazier-Smith *et al.* found that (14-31) and (14-32) accurately predict their experimental results. The theory also produces fair agreement with Park's observations. For example, for $a_1 = 450 \, \mu m$ and $a_2 = 225 \, \mu m$ we find $f(\gamma) = 3.68$; on substituting this value into (14-31) and noting that $X_c = a_1 \sin \theta$, where θ is the impact angle, we obtain the curve of $v_r = v_r(\theta)$ shown in Figure 14-12e. It can be seen that the theory predicts somewhat larger values of v_r for disruption than were observed by Park; however, the slopes of the theoretical and experimental curves are in good agreement. Note that if the drops fall at their terminal velocities, then $v_r \approx 190 \, cm \, sec^{-1}$, so that from Figure 14-12e no disruption is predicted.

14.9   Collisions of Ice Crystals with Water Drops

In clouds with temperatures lower than 0 °C both supercooled water drops and ice crystals may be present. It is clear that in this case essentially two different types of hydrodynamic interactions may occur, depending on whether the flow past the ice crystal or the drop dominates. Thus far only the former case has been studied.

Due to the complicated shapes of snow crystals, the simple approach of the superposition method has proven to be the only feasible means for describing the interactions between drops and ice crystals. For example, Pitter and Pruppacher (1974) have used the method to study the collisions of small supercooled drops with simple hexagonal ice plates. A numerical solution for the flow past a thin oblate spheroid was used as an approximation to the actual flow past the crystal. Similarly, Schlamp et al. (1975) used numerical flows past infinite cylinders in conjunction with superposition to estimate collision efficiencies of drops colliding with hexagonal columnar ice crystals. In both studies the flow past the drop was assumed to be given by the numerical solution of Le Clair et al. (1970) for the flow past a liquid sphere, including the effects of internal circulation. For the drop-planar crystal problem, the form of (14-7) remains valid, it being understood that the radius a_1 now refers to the semi-major axis of the oblate spheroid. For the drop-columnar crystal problem, we see from (10-27) that the required modification of (14-7) is $\vec{F}_1 = -\rho_a v_1 C_{D,1} L a_1 (\vec{v}_1 - \vec{u}_2) = 0.5 L \eta_a C_{D,1} N_{Re,1} (\vec{v}_1 - \vec{u}_2)$, where L and a_1 are the crystal length and radius, respectively.

Earlier, less accurate computations by Wilkins and Auer (1970) and Ono (1969) were based on the computations of Ranz and Wong (1952) for inviscid flow past disks, and on the results of Davies and Peetz (1956), who studied the interaction between small droplets and an infinite cylinder in Oseen flow for $N_{Re} \leq 0.2$, and in flow numerically determined by Thom (1933) for $N_{Re} = 10$. Unfortunately, the use of inviscid flow is not well justified since, as we know from Chapter 10, ice crystals typically have $N_{Re} \leq 100$. Also, the early numerical computations of Thom cannot be considered reliable.

Pitter and Pruppacher (1974) and Pitter (1977) computed collision efficiencies for drops of $1 \leq a_2 \leq 55 \mu m$ colliding with oblate spheroidal ice plates of axis ratio $b_1/a_1 = 0.05$ and of $147 \leq a_1 \leq 404 \mu m$; the results are shown in Figure 14-14. The cutoffs of zero E for both large and small drop sizes are probably the most notable features shown, being in contrast to the behavior found for drop-drop collisions. The cutoff for $a_2 \leq 10 \mu m$ is in excellent agreement with the field observations of Harimaya (1975) for simple plate-like and dendritic crystals, summarized in Figure 14-15. Harimaya found that the number of drops frozen on the crystals decreased to zero for $a_2 \approx 5 \mu m$. Note also from Figure 14-15 that there is also an upper size limit for the collected drops, in qualitative agreement with Pitter and Pruppacher.

Another interesting feature of Figure 14-14 is the evidence that there is a minimum crystal size below which drops cannot be collected. The computations predict this size is $a_1 \approx 150 \mu m$, which is in excellent agreement with the observations of Harimaya (1975), Wilkins and Auer (1970), and Ono (1969). They

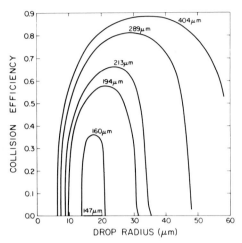

Fig. 14-14. Theoretical efficiency with which thin oblate spheroids of ice of semi-major axis $a_1 = 147\ \mu m$ ($N_{Re} = 2.0$), $a_1 = 160\ \mu m$ ($N_{Re} = 2.5$), $a_1 = 194\ \mu m$ ($N_{Re} = 4$), $a_1 = 213\ \mu m$ ($N_{Re} = 5$), $a_1 = 289\ \mu m$ ($N_{Re} = 10$), and $a_1 = 404\ \mu m$ ($N_{Re} = 20$), in air of $-10\ °C$ and 700 mb, collide with spherical, supercooled water drops of various radii. (From Pitter, 1977; by courtesy of Amer. Meteor. Soc., and the author.)

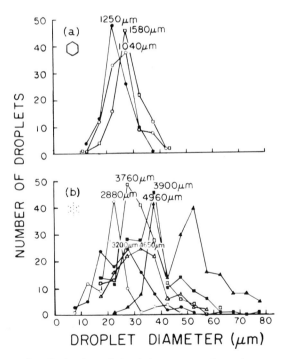

Fig. 14-15. Observed size distribution of cloud drops accreted on planar snow crystals. Number labelling curve is diameter of collector crystal. (From Harimaya, 1975; by courtesy of *J. Meteor. Soc., Japan.*)

found that simple ice hexagonal ice plates must grow by vapor diffusion to a size larger than about 150 μm in radius before they may commence riming. Figure 14-16 illustrates this fact and also shows that the onset for riming shifts to progressively larger sizes with increasing dendritic shape of the planar snow crystal. The latter result is expected since the fall velocity of a crystal of given size decreases with increasing dendritic features (see Section 10.5).

Lastly, the studies of Pitter and Pruppacher show that there is often a circular region concentric with the crystal center where no drop collisions can occur. The portion of the crystal on which collisions may occur was thus predicted to have an annular shape. This result is in excellent agreement with the field observation of Wilkins and Auer (1974), Zikmunda and Vali (1972), Hobbs et al. (1972), and Knight and Knight (1973), who observed that lightly rimed, natural planar snow crystals preferentially rime near the crystal edges (see Plate 3, pp. 48–49). This behavior was explained by Pitter and Pruppacher in terms of an 'air pillow' beneath the central portion of the falling ice crystal plate, inside of which drops may accelerate to the terminal fall velocity of the ice plate before they collide with it, giving the viscous forces an opportunity to move the drop around the falling crystal.

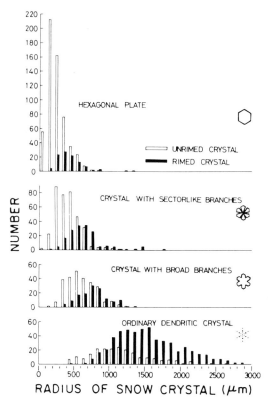

Fig. 14-16. Observed relationship between the onset of riming and the radius of planar snow crystals; open columns represent unrimed crystals; solid columns represent rimed crystals. (From Harimaya, 1975; by courtesy of *J. Meteor. Soc., Japan.*)

Only a few laboratory studies are available on the efficiency with which planar ice crystals collect supercooled water drops. Sasyo (1971) and Sasyo and Tokune (1973) made model experiments to determine the trajectory of water drops carried by an air stream past a rigidly fixed hexagonal bluff body, and to determine the efficiency with which such a body collects water drops. Unfortunately, the target sizes did not correspond to the air stream velocities chosen to represent the air flow past an ice crystal of given size at terminal velocity. Also, the Reynolds numbers at which the study was carried out were larger than 100, implying that the model had an unsteady wake due to shedding of the eddy at the downstream side of the target. The same shortcomings characterize the model experiments of Kajikawa (1974), although he more realistically allowed his models to fall freely. Fortunately, Kajikawa also carried out experiments with natural ice crystals (hexagonal plates) of $500 \leqslant a_1 \leqslant 200 \ \mu$m, which settled at their terminal velocity ($N_{Re} < 100$) through a cloud of supercooled water drops (of radius 2.5 to 17.5 μm). The results of this study are given in Figure 14-17. Note that within the accuracy of the measurements the magnitude and trend of E_c with collector size agrees with the predictions of Pitter and Pruppacher for $a_2 \geqslant 10 \ \mu$m. However, the reversal of the trend of E_c with ice collector size for smaller drops is in conflict with both the theoretical results and Kajikawa's own model experiments. Unfortunately, no check was made to see whether the cloud drops or snow crystals studied by Kajikawa were electrically charged.

Let us now turn to the case of supercooled drops colliding with columnar ice crystals. In the theoretical study referred to earlier, Schlamp et al. (1975) computed collision efficiencies for drops of $2 \leqslant a_2 \leqslant 134 \ \mu$m colliding with circu-

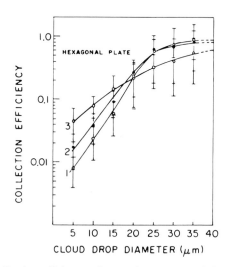

Fig. 14-17. Observed collection efficiency of natural snow crystal (hexagonal plates) for supercooled water drops. (1) $V_r = 45 \ \text{cm sec}^{-1}$, $400 \ \mu\text{m} \leqslant a_1 \leqslant 500 \ \mu\text{m}$; (2) $V_r = 37 \ \text{cm sec}^{-1}$, $300 \ \mu\text{m} \leqslant a_1 \leqslant 400 \ \mu\text{m}$; (3) $V_r = 25 \ \text{cm sec}^{-1}$, $200 \leqslant a_1 \leqslant 300 \ \text{cm sec}^{-1}$. (From Kajikawa, 1974; by courtesy of *J. Meteor. Soc., Japan.*)

lar cylindrical ice crystals of length L and radius a_1, where $67.1 \leqslant L \leqslant 2440\,\mu$m and $23.5 \leqslant a_1 \leqslant 146.4\,\mu$m $(0.2 \leqslant N_{Re} \leqslant 20)$. As we have said, the flow past the crystal was taken to be that past an infinite cylinder. This assumption is justified (at least for the lower half of the falling cylinder) by the observations of Ono (1969), Zikmunda and Vali (1972), and by Iwai (1973) who showed that, in contrast to plate-like ice crystals, columnar crystals rime quite uniformly over their surface (see Plate 4, p. 49). However, in an effort to minimize the errors from neglecting 'end effects,' Schlamp et al. employed drag coefficients determined experimentally for finite cylinders (Jayaweera and Cottis (1969), Kajikawa (1971)), rather than theoretical ones based on infinite cylinder calculations.

The computations are summarized in Figure 14-18. The results are seen to be qualitatively similar to those shown in Figure 14-14, except for the wake capture effect which causes E to increase sharply for $a_1 \gtrsim 77\,\mu$m. The cutoff for $a_2 \lesssim 10\,\mu$m is in good agreement with the field observations of Harimaya (1975) on lightly rimed columnar snow crystals, analogously to the case of planar ice crystals. The minimum ice crystal radius below which $E = 0$ is estimated to be $a_1 \approx 25\,\mu$m. This prediction is in good agreement with the field observation of Iwai (1973), Hobbs et al. (1971), and Ono (1969), who found that columnar ice crystals rarely are rimed if their diameters are smaller than about $50\,\mu$m. Needle shaped ice crystals begin riming at somewhat smaller diameters. Thus, the

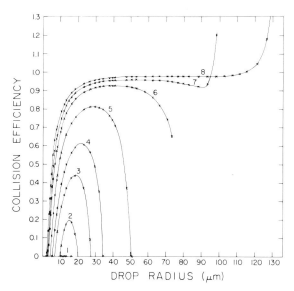

Fig. 14-18. Theoretical efficiency with which a columnar ice crystal of circular cylindrical shape (length L and radius a_1) in air ($-8\,^\circ$C and 800 mb) collides with spherical supercooled water drops of radius a_2. (1) $a_1 = 23.5\,\mu$m, $L = 67.1\,\mu$m ($N_{Re,1} = 0.2$), (2) $a_1 = 32.7\,\mu$m, $L = 93.3\,\mu$m ($N_{Re,1} = 0.5$), (3) $a_1 = 36.6\,\mu$m, $L = 112.6\,\mu$m ($N_{Re,1} = 0.7$), (4) $a_1 = 41.5\,\mu$m, $L = 138.3\,\mu$m ($N_{Re,1} = 1.0$), (5) $a_1 = 53.4\,\mu$m, $L = 237.4\,\mu$m ($N_{Re,1} = 2.0$), (6) $a_1 = 77.2\,\mu$m, $L = 514.9\,\mu$m ($N_{Re,1} = 5.0$), (7) $a_1 = 106.7\,\mu$m, $L = 1067\,\mu$m ($N_{Re,1} = 10$), (8) $a_1 = 146.4\,\mu$m, $L = 2440\,\mu$m ($N_{Re,1} = 20$). (From Schlamp et al., 1975; by courtesy of Amer. Meteor. Soc., and the authors.)

theoretical computations of Schlamp and Pruppacher (1977) predict needle diameters of 30 to 40 μm for riming onset, in a good agreement with the field observations of Reinking (1976). Unfortunately, there are no experimental values of E available with which to compare the theoretical results.

It should be stressed at this point that the collision efficiencies discussed above only apply to ice crystals in their initial stages of riming, i.e., as long as their shape is still that assumed for the computations. During later stages, the shape changes from that of a planar or columnar hexagonal crystal into that of a spherical, conical, or irregular graupel particle and eventually, if riming proceeds further, into that of a hailstone. As might be expected, there are no theoretical values of E available for such irregular shapes, not only because of the complicated surface geometry, but also because such particles often fall with spinning and tumbling motions. Qualitative experimental studies of the efficiency with which cloud drops are collected by artificial hailstones held fixed in a wind tunnel have been carried out by Macklin and Bailey (1966, 1968). More representative experiments have recently been carried out by Pflaum and Pruppacher (1976) and Pflaum et al. (1978), who studied the growth of graupel particles freely suspended in the vertical air stream of a wind tunnel. Their preliminary results indicated that under 'dry-growth' conditions (see Section 16.1) rimed spherical ice particles, which generally have a rough surface texture (protuberances, etc.) and exhibit various oscillatory-, spinning-, and helical motions (Section 10.5), have a collision efficiency which is considerably lower than that of a smooth sphere falling straight.

In closing this section we must briefly touch upon the retention efficiency of drops which collide with ice crystals. Since cloud drops which come into contact with an ice particle are supercooled, they immediately begin their transformation to ice. In the dry-growth regime (see Section 16.1), in which an ice particle acquires drops sufficiently slowly for all the acquired water to freeze, the collected drops tend to be retained by the ice particle. Earlier wind tunnel studies of List (1959a, b, 1960a, b) indicated that also in the spongy- or wet-growth regime (see Section 16.1), in which an ice particle acquires drops too fast for all of the water to freeze immediately, up to 70% of the unfrozen water is accommodated and retained in the dendritic ice mesh of the spongy ice deposit, implying that little, if any, of the water collected by the ice particle will be lost through shedding. However, more recent experiments of Bailey and Macklin (1968a), of Carras and Macklin (1973), Joe (1975), and of List et al. (1976) appear to give some evidence that at sufficiently high impact velocities of drops on riming cylinders and artificial hailstones, water retention is limited, and water may be shed by drop bouncing, detachment of water sheets, or by splashing of impacting drops. Unfortunately, however, little attention was given to the possibility that the observed reduced collection efficiency may also have been the result of a lowering of the collision efficiency due to the presence of surface roughness elements.

14.10   Collisions of Ice Crystals with Ice Crystals

The crystal aggregation mechanism which forms snowflakes is known to be a strong function of air temperature. For example, Dobrowolski (1903) observed 283 aggregation snowfall episodes, of which 83% occurred between +1 and −5 °C, 9% between −5 and −10 °C, and only 8% at temperatures less than −10 °C. Similarly, Magono (1953, 1960) found that snowflakes had their largest dimensions at temperatures near −1 °C, and that aggregation was mostly confined to temperatures warmer than −8 to −10 °C. These observations were confirmed by Hobbs *et al.* (1974b) and Rogers (1974a, b), who also found a second snowflake diameter maximum at temperatures between −12 and −17 °C, in addition to the main maximum near 0 °C (see Figure 14-19). The field observations of Hobbs *et al.* and of Jiusto and Weickmann (1973) demonstrate that most snowflakes are aggregates of planar snow crystals with dendritic habit features. However, aggregates (bundles) of needles are also observed. Aggregates of simple, thick ice-plates and short columnar ice crystals are rare.

Radar studies of the 'bright band' have provided indirect evidence for the preferential aggregation of snow crystals near 0 °C. The 'bright band' is a horizontal layer of intense radar reflectivity located at about the melting level. This large reflectivity is due in part to aggregated snow crystals, which provide relatively large scattering centers for radar waves (the radar reflectivity is proportional to the sixth power of the scattering particle diameter). However, there are also other factors which contribute to the bright band, including the presence of partly melted ice particles which constitute particularly efficient scattering centers (the radar reflectivity is proportional to the refractive index of the scattering particle), and the fact that the marked increase in the terminal velocity of particles which have just melted causes a pronounced decrease in particle concentration near the melting level (the reflectivity is proportional to the concentration of scattering centers).

Observations show that on contact ice crystals 'stick' to each other by forming an ice bond across the surface of contact if the air temperature is relatively close

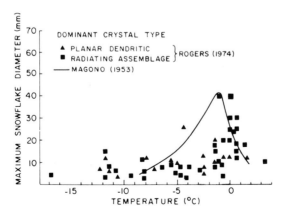

Fig. 14-19.   Maximum observed snowflake diameters as a function of air temperature for two types of snowflake composition. (From Rogers, 1974b; by courtesy of the author.)

to 0 °C, or interlock with each other if the crystals have dendritic features. The 'interlocking mechanism' is expected to occur preferentially at temperatures between −12 and −17 °C and at relatively high ice supersaturations (Ohtake, 1970), since under these conditions dendritic features are most favored (recall Section 2.2). The 'sticking mechanism' is most efficient at temperatures near 0 °C where a pseudo-liquid film is present to promote the formation of an ice neck between the particles (recall Section 5.7.3). Hobbs (1965) concluded from a study of the sintering of ice spheres that the rate of ice-neck formation is sufficiently fast for ice crystal aggregation to occur in clouds at temperatures as low as −20 °C.

There have been relatively few experimental studies of the collection efficiency of ice crystal-ice crystal interactions (and no theoretical ones, owing to the complexity of the problem). Unfortunately, most of these (Hallgren and Hosler (1960), Hosler and Hallgren (1960), Latham and Saunders (1970)) considered only the interactions between small crystals and fixed, large spherical ice targets; thus the consequences of complex crystal geometry and irregular falling motions were not explored. To date only Rogers (1974) has considered the collection efficiency of freely falling snow crystals; the values were inferred from field studies on snowflakes.

The results of the above mentioned studies are summarized in Figure 14-20. The wide scatter shown can, to a certain extent, be attributed to differences in the crystal shapes studied. Thus, both Rogers and Latham and Saunders studied exclusively the collection of plate-like ice crystals, while Hallgren and Hosler, and Hosler and Hallgren dealt with crystal shapes characteristic of the temperature range at which the experiment was carried out. Even so, one might have expected that all results should exhibit the same trend of E_c with temperature. However, although the observations of Rogers and of Hallgren and Hosler

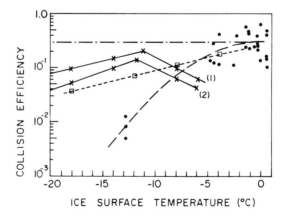

Fig. 14-20. Experimentally determined efficiency with which snow crystals and ice spheres collect micron sized ice crystals, as a function of temperature of collector ice particle. ----- Latham and Saunders (1971), $a_1 = 1000 \mu m$, spheres; ———— Rogers (1974b), $a_1 = 500 \mu m$, snowflake; ---- Hallgren and Hosler (1960), $a_1 = 85 \mu m$, ice sphere; ✕ Hosler and Hallgren (1960), (1) $a_1 = 180 \mu m$, (2) $a_1 = 63.5 \mu m$, ice spheres.

showed E_c is largest near 0 °C and decreases with decreasing temperature, as expected from the behavior of the pseudo-liquid film on ice with temperature, the values for E_c found by Latham and Saunders exhibit no temperature dependence. The maximum in E_c found by Hosler and Hallgren near −11 °C may be the result of a combination of effects, due to crystal shape and the pseudo-liquid film. Despite these discrepancies, Figure 14-20 demonstrates that the efficiency with which ice crystals are collected by ice crystals is rather low except near 0 °C. Primarily, however, the figure points up the need for further more accurate collection efficiency values.

CHAPTER 15

GROWTH OF CLOUD DROPS BY COLLISION
AND COALESCENCE

As we have seen in Chapter 1, it has long been established that the presence of ice is not necessary for precipitation formation in tropical cumuli. Also, radar observations of clouds outside the tropics have shown that the formation of echoes, indicating the presence of precipitation, can occur at temperatures warmer than 0 °C. In such cases the Wegener-Bergeron-Findeisen mechanism of precipitation formation (Section 13.3.1) is absent, and the flow of water up the spectrum from small droplets to rain must occur by the process of collision and coalescence. This is also often referred to as the collection process, or as the 'warm rain' process. The latter designation is somewhat inappropriate, since collection growth occurs also in clouds colder than 0 °C. For example, Braham (1964) found evidence of collection growth of supercooled drops in summer cumuli over the central U.S.

In this chapter we shall discuss quantitatively the evolution of drop spectra by collection growth. We shall first consider the 'continuous growth' model, according to which all large drops of the same size grow at the same continuous rate. This relatively simple model was the first to be applied to the problem of precipitation development, and is capable of reasonable accuracy in describing some aspects of collection growth. However, it generally overestimates the time required to form precipitation, since it does not account for the fact that some small fraction of larger drops will experience by chance a greater than average frequency of collection events, and will thus grow faster than the continuous model predicts. We shall therefore devote most of our attention to the 'stochastic growth' model, which takes this probabilistic aspect of collection growth into account.

15.1 Continuous Model for Collection Growth

We have already briefly introduced the continuous growth model in Section 14.7, where we discussed its application to the experimental problem of determining the collection efficiency E_c by measurement of drop growth rates. Thus, if a drop of radius a_1, fall speed $V_{\infty,1}$, and mass m_1 falls through a cloud of liquid water content w_L, containing uniform drops of radius $a_2 < a_1$ and fall speed $V_{\infty,2}$, then according to the continuous growth model the growth rate of the large drop is

$$\frac{dm_1}{dt} = E_c \pi (a_1 + a_2)^2 (V_{\infty,1} - V_{\infty,2}) w_L = K(a_1, a_2) w_L, \qquad (15\text{-}1)$$

where

$$K(a_1, a_2) = E_c \pi (a_1 + a_2)^2 (V_{\infty,1} - V_{\infty,2}), \tag{15-2}$$

is the collection kernel for hydrodynamic capture (cf. (12-77)). Equation (15-1) follows from the assumption that the water associated with the small drops is distributed continuously and uniformly. Furthermore, if more than one a_1-drop falls through the homogeneous cloud, each one is assumed to grow at the rate specified by (15-1).

An evaluation of $K(a_1, a_2)$ for the case of no turbulence is given in Section 15.2.3.2. There it is shown that K varies approximately as a_1^6 for $10 \leqslant a_1 \leqslant 50 \ \mu m$, and as a_1^3 for $a_1 > 50 \ \mu m$. On comparison with the weaker size dependence, $dm/dt \sim a$, found for the diffusion growth of a small drop (Section 13.2), we conclude that the relative importance of collection growth increases sharply with drop size. In fact, a crossover in dominance from diffusion to collection growth generally occurs for $10 \leqslant a_1 \leqslant 20 \ \mu m$.

We may easily generalize (15-1) to apply to a situation in which an a_1-drop falls through a polydisperse cloud of smaller drops, distributed in size according to the spectrum n(a). Then in place of (15-1) we have

$$\frac{dm_1}{dt} = \frac{4 \pi \rho_w}{3} \int K(a_1, a_2) n(a_2) a_2^3 \, da_2. \tag{15-3}$$

Since $dm_1/dt = 4 \pi \rho_w a_1^2 \, da_1/dt$, and assuming $E_c = E = y_c^2/(a_1 + a_2)^2$ (Equation (14-2)), then from (15-2) and (15-3) an alternative description in terms of the radius growth rate is

$$a_1^2 \frac{da_1}{dt} = \frac{\pi}{3} \int y_c^2(a_1, a_2)[V_{\infty,1} - V_{\infty,2}] a_2^3 n(a_2) \, da_2. \tag{15-4}$$

Several studies of (15-4) are in the literature (e.g., Telford (1955), Twomey (1964), Braham (1968), and Chin and Neiburger (1972)). Chin and Neiburger computed the drop growth rates in a cloud with a Khrgian-Mazin spectrum (Equation (2-2)), and in a monodisperse cloud with the same w_L and mean volume radius. The results, shown in Figure 15-1, demonstrate that the growth

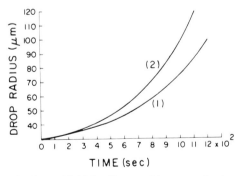

Fig. 15-1. Size variation of a drop of initial radius $a_0 = 30 \ \mu m$ growing by collision and coalescence in: (1) a monodisperse cloud with $\bar{a}_{2,vol} = 10 \ \mu m$, $w_L = 1 \ g \ m^{-3}$; (2) a cloud with a Khrgian-Mazin drop size distribution, $\bar{a}_{2,vol} = 10 \ \mu m$, $\bar{a}_2 = 7.5 \ \mu m$, $w_L = 1 \ g \ m^{-3}$. (From Chin and Neiburger, 1972; by courtesy of Amer. Meteor. Soc., and the authors.)

rate is significantly smaller in the monodisperse cloud, even though it is comprised of drops which are larger than three-fourths of those in the poly-disperse cloud. This example shows that the growth rate is a sensitive function of the drop size distribution as well as the liquid water content.

Braham similarly compared the growth rate in typical maritime and continen-tal cumuli, using the spectra depicted in Figure 15-2. Note that for the continen-tal cloud there are about ten per liter of drops centered in a 5 μm interval about 30 μm radius, and a similar number centered about 40 μm for the maritime cloud. Since raindrop concentrations are typically of order one per liter or less, it is reasonable to use these 30 and 40 μm radius drops as representative of those which ultimately evolve to precipitation size for the two cloud types. The results of the computations are shown in Figure 15-3. We see that drops present in concentrations of ten per liter are predicted to grow much faster in maritime-type clouds than in continental-type clouds, in agreement with the observation that maritime clouds precipitate more often than continental clouds.

The continuous growth model also permits a simple assessment of the effect of an updraft on the development of precipitation. For example, assuming $a_2 \ll a_1$, (15-1) may be written in the form

$$\frac{da_1}{dt} = \frac{E_c V_{\infty,1} w_L}{4 \rho_w}. \tag{15-5}$$

Now suppose there is an updraft of strength $W > V_{\infty,1}$, so that the upward velocity of the a_1-drop relative to the ground is $W - V_{\infty,1} = dz/dt$, where $z(t)$ measures the height of the drop. Then (15-5) may also be expressed as follows:

$$\frac{da_1}{dz} = \frac{E_c V_{\infty,1} w_L}{4 \rho_w (W - V_{\infty,1})}. \tag{15-6}$$

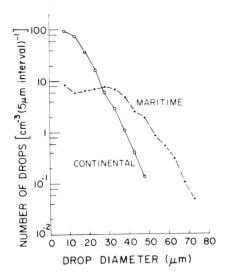

Fig. 15-2.   Cloud droplet spectra for maritime-type and continental-type cumulus clouds used for the drop growth rate computations shown in Figure 15-3. (From Braham, 1968; by courtesy of Amer. Meteor. Soc., and the author.)

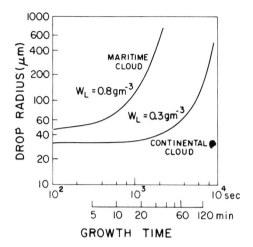

Fig. 15-3. Size variation of a drop growing by collision and coalescence in a maritime- and a continental-type cloud. (From Braham, 1968; by courtesy of Amer. Meteor. Soc., and the author.)

For a constant updraft the formal solution is

$$z(a_1) - z_0 = \frac{4\,\rho_w}{w_L} \left[W \int_{a_{1,0}}^{a_1} \frac{da_1}{V_{\infty,1}(a_1)E_c(a_1)} - \int_{a_{1,0}}^{a_1} \frac{da_1}{E_c(a_1)} \right], \qquad (15\text{-}7)$$

where $a_{1,0} = a_1(z_0)$. This equation has been the basis of several studies (e.g., Langmuir (1948), Bowen (1950), Ludlam (1951), Mason (1952b, 1959), and East (1957)). Figure 15-4 shows as an example some results obtained by Bowen. In this calculation a drop, originating by the chance coalescence of two $10\,\mu$m

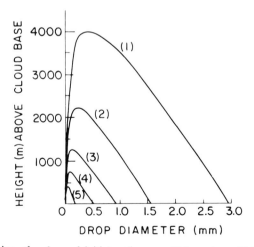

Fig. 15-4. Size variation of a drop of initial radius $a_0 = 12.6\,\mu$m by collision and coalescence in clouds of various constant updraft speeds W: (1) 200, (2) 100, (3) 50, (4) 25, (5) 10 cm sec^{-1}; for $a_2 = 10\,\mu$m, $w_L = 1$ g m^{-3}. Time to reach indicated size after return to cloud base is 60, 60, 70, 85, 115 min for (1), (2), (3), (4), and (5), respectively. (From Bowen, 1950; by courtesy of *Australian J. Sci. Res.*, and the author.)

radius cloud droplets near the cloud base, is assumed to grow further by collection of 10 μm radius drops in accordance with the collision efficiency as given by Langmuir (1948). The figure demonstrates that strong updrafts reduce the growth time of drops considerably, and produce larger drops than weak updrafts.

One feature common to all the examples discussed so far is the prediction of growth times for precipitation-sized drops which are much longer (by a factor of two or more) than the times which are often observed to be necessary. As we indicated earlier, this fault arises from the neglect of the stochastic aspect of collection growth. Since raindrop concentrations are typically 10^5–10^6 times smaller than cloud drop concentrations, one would expect that the fate of the 'favored' small fraction of drops which happen by chance to grow much faster than the average rate should be quite important in the overall process of precipitation development. Many calculations have borne out this expectation. As an example, Figure 15-5 presents a comparison by Twomey (1964) of the development of a drop spectrum according to the continuous and stochastic growth models. For drop concentrations at the level of 100 m^{-3} μm^{-1}, the growth rate in the stochastic mode is seen to be almost ten times faster than in the continuous mode. It is for this reason that in the remainder of this chapter we shall concern ourselves exclusively with the stochastic description of the collision-coalescence process.

15.2   Stochastic Model for Collection Growth

As the reader has probably anticipated, the kinetic coagulation equation, (12-45) or (12-48), is generally taken as the basic governing form for stochastic collec-

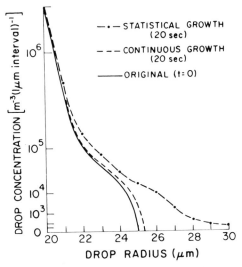

Fig. 15-5. Variation of drop size distribution with time due to continuous- and stochastic growth of cloud drops by collision and coalescence. (From Twomey, 1964; by courtesy of Amer. Meteor. Soc., and the author.)

tion growth. (In the more recent literature on the subject, the coagulation equation is also variously referred to as the scalar transport equation, the kinetic equation, the collection equation, the stochastic coagulation equation, and the stochastic collection equation (SCE); we shall employ this last abbreviated designation henceforth.) Telford (1955) was the first to introduce this approach to the drop collection problem, using a version of the SCE which applies to an idealized cloud consisting initially of just two drop sizes, and for which the collision kernel is a constant. The systematic application of the SCE to the collection problem has occurred only relatively recently, since the appearance of Telford's paper. In part this reflects the rise of computer technology, which was requisite for coping with realistic collection growth problems.

In the sequel we shall consider first the suitability of the SCE for the problem at hand. Having established the conditions under which its use is appropriate, we shall then develop and discuss some of the few known exact solutions. For the most part these solutions do not correspond to very physical collection processes. Even so, they provide useful clues to the behavior in more realistic situations, and also serve as standards against which to judge the accuracy of various numerical integration schemes. We next discuss a few useful or potentially useful approximation techniques which have been applied to the SCE. Finally, we shall turn to some numerical methods of solution, including the Monte Carlo approach, and discuss representative results obtained thereby.

15.2.1 COMPLETENESS OF THE SCE

Let us imagine an idealized cloud which is spatially homogeneous or 'well-mixed' at all times, and which contains drops whose masses are multiples of some unit mass. Let n_k be the number of drops per unit volume containing k units of mass, and let K_{ik} denote the collection kernel which describes the collection rate of i and k drops. Then the discrete SCE for this situation has exactly the form of (12-45). If the cloud volume is V, then $N_k \equiv n_k V$ is the total number of drops of size k, and $A_{ik} \equiv K_{ik}/V$ is the probability per unit time of coalescences between any pair of i and k drops (by the 'well-mixed' assumption). By introducing V into (12-45) we thus arrive at the SCE for $N_k(t)$:

$$\frac{dN_k}{dt} = \frac{1}{2}\sum_{i=1}^{k-1} A_{i,k-i}N_iN_{k-i} - N_k\sum_{i=1}^{\infty} A_{ik}N_i. \tag{15-8}$$

We would like to determine whether (15-8) is suitable for studying the spectral evolution in our ideal cloud. As it happens, only a little reflection suffices to raise some doubts. For example, if we imagine several successive 'runs' or realization of the collection process for a given set of probabilities for binary interactions and a given initial distribution, we would expect to see at least slightly different outcomes for the spectrum from one run to the next. However, (15-8) can only produce a unique spectrum once A_{ik} and $N_k(0)$ are specified. Thus it is natural to suspect that (15-8) is not stochastically complete (i.e., does not describe the probabilities of all possible histories of drop growth), and that the solution to (15-8) must represent only some sort of average spectrum. Such stochastic incompleteness might well be expected to cause trouble, since for most applications in cloud physics we are primarily interested in the long tail of the spectrum where fluctuations are relatively strong, there being relatively few large particles in the system.

This problem of stochastic completeness has been debated extensively (e.g., Scott (1967, 1968), Berry (1967, 1968), Warshaw (1967, 1968), Slinn and Gibbs (1971), Long (1971, 1972), Chin and Neiburger (1972), Gillespie (1972, 1975a), and Bayewitz *et al.* (1974)). We recommend Gillespie's (1972) paper as the most thorough and clear discussion of the general problem. Bayewitz *et al.* (1974) present what is apparently the only known rigorous solution to the full stochastic model which allows for spectrum fluctuations, accomplished under the assumption of a constant collection kernel. Their work provides an interesting concrete example of the extent of the incompleteness of the SCE, and is presented below in Section 15.2.1.2. But first we shall outline Gillespie's (1975a) relatively simple analysis of the problem, which appears to us to delineate very well the salient features with a minimum of effort.

15.2.1.1 *Three Models for Collection Growth*

Let us further simplify our well-mixed cloud so that at time $t = 0$ it consists entirely of N 'drops' each having mass m_0 and N' 'droplets' each of mass μ. We further assume the conditions $m_0 > \mu$ and $N \ll N'$, and that coalescences are possible only between drops and droplets. Thus the masses of the drops are increasing, but the number N of drops remains constant. Because of the condition $N \ll N'$, we assume the number N' of droplets also remains constant. (This is the cloud model used by Telford (1955).)

Our goal is to describe the growth of the N drops. For this purpose we need the collection kernel, which in the present case can be a function only of the drop mass m: $A_{ik} \rightarrow A(m)$. Gillespie (1975a) shows that three growth models are possible, depending upon the physical interpretation given the quantity $A(m)N'$ dt, where dt is an infinitesimal time interval. These models are:

(1) the *continuous model*, in which

$$A(m)N' \, dt = \text{number of droplets which } any \\ \text{drop of mass m will collect in dt;} \qquad (15\text{-}9)$$

(2) the *quasi-stochastic model*, in which

$$A(m)N' \, dt = \text{fraction of the drops of mass m} \\ \text{which will collect a droplet in dt;} \qquad (15\text{-}10)$$

(3) the *pure stochastic model*, in which

$$A(m)N' \, dt = \text{probability that any drop of mass m} \\ \text{will collect a droplet in dt.} \qquad (15\text{-}11)$$

We shall now proceed to explore the consequences of each of these models under the assumption $A(m) = A$, a constant.

Since in the continuous model all drops start with the same mass m_0 and grow at the same rate, the state of the drops may be specified very simply by the function $M(t) \equiv$ mass of any drop at time t. From (15-9) we have $dM = \mu AN' \, dt$, so that

$$M(t) = m_0 + \mu AN't. \tag{15-12}$$

According to the quasi-stochastic model only a certain fraction of the m-drops will collect a droplet in dt. Thus, the m-drops do not grow in unison, and the definition of M(t) given above no longer applies. Instead, for the quasi-stochastic model we define the function $N_m(t) \equiv$ number of m-drops at time t, where $m = m_0, m_0 + \mu, \ldots$. Simple bookkeeping enables us to write down the governing equation for $N_m(t)$: From (15-10), in time $(t, t + dt)$ exactly $N_{m-\mu}(t)AN' \, dt$ drops of mass $m - \mu$ will each collect a droplet and so become drops of mass m, while exactly $N_m(t)AN' \, dt$ m-drops will each collect a droplet and so become drops of mass $m + \mu$. Therefore, $dN_m = (N_{m-\mu} - N_m)AN' \, dt$, or

$$\frac{dN_m}{dt} = AN'(N_{m-\mu} - N_m). \tag{15-13}$$

This is just the SCE as it appears for our simple drop-droplet cloud; it is also the equation considered by Telford (1955).

The solution to (15-13) may be obtained by noting it comprises a set of coupled, linear, first order differential equations. These may be solved sequentially, subject to the initial condition $N_m(0) = N'$ if $m = m_0$, and 0 if $m \neq m_0$. The result, which may be verified by direct substitution, is

$$N_{m_0+k\mu}(t) = \frac{N(AN't)^k \exp(-AN't)}{k!}, \quad k = 0, 1, 2, \ldots. \tag{15-14}$$

Thus the quasi-stochastic model yields a discrete mass spectrum. The average drop mass $M_1(t)$ at time t is, letting $x \equiv AN't$,

$$M_1(t) \equiv \frac{1}{N} \sum_{m=m_0}^{\infty} mN_m(t) = \sum_{k=0}^{\infty} (m_0 + k\mu)x^k e^{-x}/k! = M(t). \tag{15-15}$$

This is what we would expect: The average drop mass coincides with the mass of every drop in the continuous model. Another quantity of interest is the width $\Delta(t)$ of the distribution, as measured by the root mean square (rms) deviation: $\Delta(t) \equiv [M_2(t) - M_1^2(t)]^{1/2}$, where M_2 denotes the second moment of the distribution. A simple calculation yields $\Delta(t) = \mu(AN't)^{1/2}$, which contrasts with the zero width in the continuous model.

The continuous model requires each m-drop to collect a definite number of droplets in a given time interval. As we have seen, the quasi-stochastic model requires only that all m-drops *together* collect droplets at a definite rate. But this is still too restrictive: by the probabilistic nature of the collection process, there should be fluctuations in the number of droplets collected by any group of drops as well as by any individual drop. Such fluctuations are permitted in the pure stochastic model. In consequence, we cannot predict exactly how many drops of a particular size there will be at any time t, so that the definition of the spectrum given above for $N_m(t)$ no longer applies. However, from (15-11) we have a means of predicting the probability of finding a given number of m-drops at time t. Therefore, an appropriate state function in the pure stochastic model is

$$P(n, m; t) \equiv probability \text{ that exactly n drops} \qquad (15\text{-}16)$$
$$\text{have mass m at time t,}$$

where $n = 0, 1, \ldots, N$ and $m = m_0, m_0 + \mu, \ldots$.

In order to calculate $P(n, m; t)$, consider first the probability $\mathscr{P}(k, t)$ that any given drop will collect exactly k droplets in time t; this is just the familiar Poisson distribution (see Gillespie (1975a) or Feller (1967) for details):

$$\mathscr{P}(k, t) = \frac{(AN't)^k \exp(-AN't)}{k!}. \qquad (15\text{-}17)$$

Then, since each drop collects droplets independently of the other drops, the probability that a particular selection of n drops will each collect exactly k droplets in time t while the remaining $N - n$ drops will not is $\mathscr{P}^n(k, t)[1 - \mathscr{P}(k, t)]^{N-n}$. And, since the number of distinct ways of selecting two groups of n drops and $N - n$ drops from a set of N drops is $\binom{N}{n} = N![n!(N - n)!]^{-1}$, the probability that exactly n of the N drops will collect exactly k droplets in time t is just $\binom{N}{n}\mathscr{P}^n(k, t)[1 - \mathscr{P}(k, t)]^{N-n}$. Therefore, we obtain

$$P(n, m_0 + k\mu; t) = \frac{N!}{n!(N - n)!} \mathscr{P}^n(k, t)[1 - \mathscr{P}(k, t)]^{N-n}, \qquad (15\text{-}18)$$

where $n = 0, 1, \ldots, N$ and $k = 0, 1, \ldots$. Equations (15-17) and (15-18) constitute the solution for $P(n, m; t)$ for the simple drop-droplet cloud.

The first moment of $P(n, m; t)$ with respect to n has the physical meaning of being the average number $N_1(m; t)$ of m-drops in the cloud at time t:

$$N_1(m; t) \equiv \sum_{n=0}^{N} nP(n, m; t) = \sum_{n=1}^{N} \frac{nN!}{n!(N - n)!} \mathscr{P}^n(k, t)[1 - \mathscr{P}(k, t)]^{N-n}$$
$$= N\mathscr{P}(k, t) = N_m(t). \qquad (15\text{-}19)$$

So for our simple cloud the average spectrum as defined above coincides with the solution to the SCE. Also of interest is the expected uncertainty associated with $N_1(m; t)$; i.e., we would like to know to what extent the actual number of m-drops in a particular realization can be expected to deviate from $N_m(t)$. In analogy with our previous choice of $\Delta(t)$ as the width of the SCE spectrum, we now choose the rms deviation of $P(n, m; t)$ with respect to n: $\Delta(m; t) \equiv [N_2(m; t) - N_1^2(m; t)]^{1/2}$, where $N_2(m; t)$ is the second moment of $P(n, m; t)$ with respect to n. A straightforward calculation gives the result $\Delta(m; t) = [N_1(m; t)]^{1/2}[1 - N_1(m; t)/N]^{1/2}$. As the second factor on the right side approaches unity for $t \to \infty$, we conclude we may reasonably expect to find roughly between

$$N_m(t) - [N_m(t)]^{1/2} \quad \text{and} \quad N_m(t) + [N_m(t)]^{1/2} \qquad (15\text{-}20)$$

drops of mass in the cloud at time t. Thus, in the present simple case the solution to the SCE provides not only the mean spectrum, but also a measure of the fluctuations about the mean.

The result (15-20) shows that for the simple 'drop-droplet' cloud the SCE is more 'stochastically complete' than its usual derivation via the quasi-stochastic interpretation would suggest. The important question arises as to whether or not the same holds true in the more realistic case in which the collection kernel is size dependent and drops of all sizes are present. A partial answer is provided in Gillespie's (1972) study of the general problem, in which it is shown that if (1) certain correlations can be neglected, and (2) coalescences between drops of the same size are prohibited, then the SCE does indeed determine the mean spectrum $N_1(m;t)$. Also, the function $P(n, m; t)$ then tends to the Poisson form

$$P(n, m; t) \xrightarrow[t \to \infty]{} \frac{N_1^n(m; t) e^{-N_1(m;t)}}{n!}, \tag{15-21}$$

and in particular $\Delta(m; t) \to \sqrt{N_1(m; t)}$, so that the result (15-20) still holds. Gillespie's analysis also provides a simple estimate of the time interval $\tau(m)$ after which the result (15-20) may be assumed applicable. This is given by the implicit relation

$$\int_0^{\tau(m)} \sum_{m'=1}^{\infty} N_{m'}(t) A_{mm'} \, dt = \ln 2. \tag{15-22}$$

The same general conclusion that (15-20) is a valid estimate of the spectrum fluctuations was arrived at earlier by Scott (1967); however, his analysis erroneously implies the equivalent of assuming $\tau(m) = 0$.

However, without assumptions (1) and (2) the situation is unclear. This is so primarily because in general the state of a cloud in the pure stochastic model cannot be determined completely by just the function $P(n, m; t)$. The existence of particle correlations means that various conditional probabilities must be specified also, such as the probability that n drops of mass m are present, given that there are also n' drops of mass m', etc. Such correlations are bound to occur in real clouds, partly because they are not well-mixed, as we have assumed for the 'drop-droplet' cloud: as droplets in a given region coalesce, there will be a corresponding decrease in the number available in that region for further coalescence. In addition, measurements show strong spatial inhomogeneities in cloud liquid water content.

Some indication of the effect of particle correlations is included in the study of Bayewitz et al. (1974), to which we now turn.

15.2.1.2 Correlations in a Stochastic Coalescence Process

Bayewitz et al. consider a well-mixed cloud containing N_k drops of size k. The probability per unit time of coalescences between any pair of drops is taken as a constant, $A_{ik} = A$, as in the previous section. The authors proceed to set up and solve the governing equation which describes all possible histories of drop growth over the full range of drop sizes (an outline of the analysis is given in Appendix A-15.2.1.2). The results show the SCE produces total particle counts in excellent agreement with the true stochastic averages, even for very small initial populations, at least for collection kernels which are not strongly size dependent.

As we have noted, real clouds are not well-mixed. Bayewitz *et al.* go on to consider the consequences of poor mixing by adopting the following approach: A hypothetical large cloud is imagined to be partitioned into many small compartments of volume V_0, with the understanding that drops can coalesce only if they occupy the same compartment. By making V_0 sufficiently small, correlations are introduced which are perhaps similar to those occurring in a real cloud. The effects of poor mixing as simulated by this partitioning model are as follows: (1) If we are interested only in the total number of drops, and not their size distribution, then we find that for initial populations of as few as ten drops the results of the SCE match the true stochastic averages. (2) If we focus instead on the size distribution of the coalescing drops, we find that either for small populations or for systems partitioned into small isolated compartments the results of the SCE may differ significantly from the true stochastic averages, especially in the large-particle tail of the distribution. Additionally, Bayewitz *et al.* make the plausible conjecture that the correlation effects which produce these differences would be enhanced by a more realistic size-dependent collection kernel.

Since precipitation formation is an example of a collection process in an aerosol which is not well-mixed, and in which particles in the tail of the distribution generally appear to play a major, if not dominant, role, it is tempting to infer that the standard SCE probably does not simulate the production of rain with a high degree of accuracy, and that it may be better in some instances to turn to other simulation techniques, particularly the Monte Carlo method (discussed below in Section 15.2.4.3). On the other hand, it also seems plausible that in some cases the flow rate of water up the spectrum toward rain is controlled predominantly by the self-collection of cloud droplets in the mid-range of the spectrum, in which case the SCE may be used with confidence. (Some light is shed on this question below in Section 15.2.5; there we see that according to a case study by Berry and Reinhardt (1974a) based on the SCE, the self-collection of cloud droplets ('auto-conversion') is less important than the self-collection of large drops ('large hydrometeor self-collection') and the collection of small droplets by large drops ('accretion'). This tends to support the notion that the SCE is not suitable for very accurate simulations of rain formation.) Obviously, this is an area where further research is needed.

In addition to the need for further studies of stochastic completeness, such matters as the actual local fluctuations in cloud liquid water content and the efficiency of turbulent mixing should be explored, in order to ascertain the degree to which a cloud approximates a 'well-mixed' aerosol. In this connection we should note a recent paper by Twomey (1976), in which otherwise unpublished measurements by J. Warner are presented showing that cumulus cloud liquid water contents of two to three times the average occur with a frequency of 0.5%. Twomey uses the SCE and the assumption that the collection kernel $K(u, v)$ for drops of volume u and v is homogeneous of degree α $(1 \leqslant \alpha \leqslant 2)$ in w_L $(K(w_L u, w_L v) = w_L^\alpha K(u, v))$ to show that such fluctuations in w_L greatly increase the production rate of large drops. As an example, Twomey estimates that if only one percent of the cloud volume can maintain $w_L = 3\,g\,m^{-3}$, a reasonable value for a saturated adiabatic process, for just a few minutes, with the remainder of the cloud being drier by about a factor of ten (due to turbulent mixing with the environment), then those few very wet pockets of air can produce large drops ($a > 41\,\mu$m) so quickly that when averaged over the entire cloud volume, their concentration will be nine orders of magnitude larger ($10^{-6}\,cm^{-3}$ vs. $10^{-15}\,cm^{-3}$) than if the cloud water were uniformly distributed.

It is worth noting that Twomey's study makes use of solutions to the SCE; thus the SCE can be used to estimate at least some of the effects of poor mixing. This fact, along with the other virtues of the SCE, namely that it is relatively simple and evinces a fairly close approach to stochastic completeness in case of minimal particle correlations (Equation (15-20)), appears to justify its use in studies of collection growth at the present time.

15.2.2   EXACT SOLUTIONS TO THE SCE

In this section we shall be concerned with the continuous and dimensionless version of the SCE given by (12-114):

$$\frac{\partial f(x, \tau)}{\partial \tau} = \frac{1}{2} \int_0^x \alpha(x - y, y) f(x - y, \tau) f(y, \tau)\, dy - f(x, \tau)$$

$$\times \int_0^\infty \alpha(x, y) f(y, \tau)\, dy. \qquad (15\text{-}23)$$

The only known exact solution to the discrete SCE which has any practical value, in the sense of simulating real coagulation processes, is just that of Smoluchowski (1916), (12-43), for the case of a constant collection kernel. (Another solution (McLeod, 1962) has been found for the case $K_{ij} \sim ij$, but this choice of kernels appears quite unrealistic (see the discussion below on the xy kernel).) Also, in the steady state there is an exact solution (Klett, 1975) to the continuous SCE enhanced by a particle source term, for $\alpha = x^\beta y^\beta$ ($\beta < 1$). This solution has already been discussed in Section 12.8.5.

All the known solutions to (15-23) correspond to special cases of the following kernel:

$$\alpha(x, y) = A + B(x + y) + Cxy. \qquad (15\text{-}24)$$

In our development and/or discussion of these solutions, we shall follow the work of Drake (1972a) and Drake and Wright (1972). In the latter paper, families of exact solutions are constructed for the following subclasses of (15-24): (1) $C = 0$ and $B = aA$ for arbitrary a on the interval $0 \le a \le \infty$, (2) $A = 0$ and $C = 2\,aB$ for $0 \le a \le \infty$, (3) $A = 1$, $B = a$, and $C = a^2$ for $0 \le a \le \infty$. Here we shall consider only the first subclass, because of the unphysical behavior associated with the presence of the xy term.

Before going ahead with the construction of solutions, let us consider briefly the essential difficulty with the choice $C \neq 0$: it is that solutions based on the xy kernel can exist only for a finite time interval (McLeod (1964)). This behavior is easily demonstrated by use of the moment equation (12-118). For example, for $\alpha = Cxy$ the equation for the zeroth moment is $dM_0/d\tau = -C/2$, so that $M_0(\tau) = 1 - C\tau/2$. Thus, coalescence apparently stops at least by $\tau = \tau_{max} = 2\,C^{-1}$. Similarly, the equation for the second moment is $dM_2/d\tau = CM_2^2$, so that $M_2(\tau) = M_2(0)[1 - CM_2(0)\tau]^{-1}$. Therefore, the second moment becomes infinite when $\tau = \tau_\infty = [CM_2(0)]^{-1}$, a time which depends on the initial spectrum, $f(x, 0)$. Let us assume, for example, that $f(x, 0) = \delta(x - 1)$; then $M_2(0) = 1$ and $\tau_\infty = C^{-1} = \tau_{max}/2$. The singularity occurs sooner if the initial spectrum is more spread out; thus, if $f(x, 0) = e^{-x}$, then $M_2(0) = 2$ and $\tau_\infty = (2\,C)^{-1} = \tau_{max}/4$. These examples should suffice to illustrate how the xy kernel leads to unreasonable behavior in a finite time, such as the occurrence of an infinite radar reflectivity ($\Sigma\, na^6 \sim M_2$), followed by the apparent collapse of the spectrum.

The solution of (15-23) and (15-24) with $C = 0$ is facilitated by introducing another time variable (Martynov and Bakonov, 1961):

$$T \equiv 1 - M_0(\tau). \tag{15-25}$$

As coagulation proceeds, M_0 slowly decreases from unity to zero. Thus, corresponding to $0 \leq \tau < \infty$ we have $0 \leq T \leq 1$, which makes T a natural measure of the progress of coagulation. Let us define also

$$g(x, T) \equiv f(x, \tau), \tag{15-26}$$

so that

$$\frac{\partial f}{\partial \tau} = \frac{\partial g}{\partial T} \frac{dT}{d\tau} = -\frac{\partial g}{\partial T} \frac{dM_0}{d\tau}. \tag{15-27}$$

Then from (15-23) and the moment equation for $M_0(\tau)$, the governing equation for $g(x, T)$ is found to be

$$D(T) \frac{\partial g(x, T)}{\partial T} = -2\, g(x, T) \int_0^\infty \alpha(x, y)g(y, T)\, dy$$

$$+ \int_0^x \alpha(x - y, y)g(x - y, T)g(y, T)\, dy, \tag{15-28}$$

where

$$D(T) = \int_0^\infty \int_0^\infty \alpha(x, y)g(y, T)g(x, T)\, dx\, dy. \tag{15-29}$$

On substitution of (15-24) with $C = 0$ into (15-28) and (15-29), we obtain

$$D(T) \frac{\partial g}{\partial T} = -2\, g(A + Bx)(1 - T) - 2\, Bg + (A + Bx)g * g, \tag{15-30}$$

where

$$D(T) = A(1 - T)^2 + 2\, B(1 - T), \tag{15-31}$$

and $g * g \equiv \int_0^x g(x - y)g(y)\,dy$. The appearance of the convolution form $g * g$ in (15-30) indicates the feasibility of a solution via Laplace transforms. Therefore, let us introduce the Laplace transform of $g(x, T)$:

$$L[g(x, T)] \equiv \int_0^\infty g(x, T)e^{-sx}\,dx \equiv \Phi(s, T). \tag{15-32}$$

The inverse transform is

$$L^{-1}[\Phi(s, T)] \equiv \frac{1}{2\pi i} \int_{\gamma - i\infty}^{\gamma + i\infty} \Phi(s, T)e^{sx}\,ds = g(x, T), \tag{15-33}$$

where γ is any number such that $g(x, T)e^{-\gamma x} \to 0$ as $x \to \infty$. Then on taking the Laplace transform of (15-30) and using the well known properties $L(f * g) = L(f)L(g)$ and $L(xf) = -d[L(f)]/ds$, we obtain the transformed equation for Φ:

$$D\frac{\partial \Phi}{\partial T} = -2\,A(1 - T)\Phi + 2\,B(1 - T)\frac{\partial \Phi}{\partial s} - 2\,B\Phi + A\Phi^2 - B\frac{\partial}{\partial s}\,\Phi^2. \tag{15-34}$$

Finally, in terms of a new dependent variable, $\eta(s, T)$, defined as follows:

$$\eta(s, T) \equiv \frac{2\,\Phi(s, T)}{D(T)}, \tag{15-35}$$

(15-34) simplifies to

$$\frac{\partial \eta}{\partial T} = \frac{A}{2}\eta^2 - B\eta\frac{\partial \eta}{\partial s} + \frac{2(1 - T)}{D}B\frac{\partial \eta}{\partial s}. \tag{15-36}$$

We shall now construct the solutions for $g(x, T)$ in terms of the Laplace transform $G(s)$ of the initial spectrum, for the two special cases $A = 1$, $B = 0$, and $A = 0$, $B = 1$.

(1) $A = 1$, $B = 0$ ($\alpha(x, y) = 1$).
  The equation for η is

$$\frac{\partial \eta}{\partial T} = \frac{\eta^2}{2}. \tag{15-37}$$

Therefore,

$$\eta(s, T) = \frac{\eta(s, 0)}{1 - \eta(s, 0)T/2} = \frac{2\,G(s)}{1 - G(s)T}. \tag{15-38}$$

Since $T \le 1$ and $G(s) = \int_0^\infty e^{-sx}g(x)\,dx \le \int_0^\infty g(x)\,dx = 1$, (15-38) may be expanded in a geometric series to yield

$$\eta(s, T) = 2\sum_{k=0}^\infty T^k G^{k+1}(s). \tag{15-39}$$

Finally, since $L^{-1}[\eta(s, T)] = 2\,g(x, T)/(1 - T)^2$, we have (Scott, 1968):

$$g(x, T) = (1 - T)^2 \sum_{k=0}^\infty T^k L^{-1}[G^{k+1}(s)]. \tag{15-40}$$

(2) A = 0, B = 1 ($\alpha(x, y) = x + y$).
   The equation for η is

$$\frac{\partial \eta}{\partial T} + (\eta - 1)\frac{\partial \eta}{\partial s} = 0. \tag{15-41}$$

This equation is a 'quasi-linear' first order partial differential equation, and can be solved in the following manner (e.g., Ames (1965), p. 50): If the form of the differential equation is

$$P\frac{\partial \eta}{\partial T} + Q\frac{\partial \eta}{\partial s} = R, \tag{15-42}$$

where P, Q, and R are functions of s, T, and η, the general solution is F(u, v) = 0, where F is an arbitrary, sufficiently differentiable function and u(s, T, η) = a and v(s, T, η) = b form independent solutions of the Lagrange system

$$\frac{dT}{P} = \frac{ds}{Q} = \frac{d\eta}{R}. \tag{15-43}$$

In the present application this leads to the implicit solution $\eta = F[s - (\eta - 1)T]$. From the initial condition $\eta(s, 0) = F(s) = G(s)$, we find F = G:

$$\eta(s, T) = G[s - (\eta - 1)T]. \tag{15-44}$$

   Now on substitution of (15-33) and (15-35) into this last result we have

$$g(x, T) = \frac{(1 - T)}{2\pi i} \int_{\gamma - i\infty}^{\gamma + i\infty} e^{sx}G[s - (\eta - 1)T]\,ds, \tag{15-45}$$

or, on letting $\sigma = s - (\eta - 1)T = s - [G(\sigma) - 1]T$,

$$g(x, T) = \frac{(1 - T)}{2\pi i} \int_{\gamma - i\infty}^{\gamma + i\infty} e^{\sigma x + GTx - Tx}G\left(1 + \frac{dG}{d\sigma}T\right)d\sigma$$

$$= (1 - T)e^{-Tx}\left\{\sum_{k=0}^{\infty} \frac{(Tx)^k}{k!} L^{-1}[G^{k+1}] + \frac{1}{2\pi i}\right.$$

$$\left. \times \int_{\gamma - i\infty}^{\gamma + i\infty} e^{\sigma x} \sum_{k=0}^{\infty} \frac{(Tx)^k}{k!} G^{k+1}T\frac{dG}{d\sigma}\,d\sigma\right\}. \tag{15-46}$$

The last term may be integrated by parts, so that finally we have (Scott, 1968):

$$g(x, T) = (1 - T)e^{-Tx}\sum_{k=0}^{\infty} \frac{(Tx)^k}{(k + 1)!} L^{-1}[G^{k+1}(s)]. \tag{15-47}$$

   In Appendix A-15.2.2 the general solutions (15-40) and (15-47) are evaluated for two particular choices of the initial spectrum. For the choice g(x, 0) = $\delta(x - 1)$ (a monodisperse cloud) the results are as follows:

$$g(x, T) \underset{(\alpha = 1)}{=} (1 - T)^2 \sum_{k=0}^{\infty} T^k\delta(x - k - 1), \tag{15-48}$$

(Melzak, 1953), and

$$g(x, T) = (1 - T)e^{-Tx} \sum_{k=0}^{\infty} \frac{(Tx)^k}{(k + 1)!} \delta(x - k - 1), \qquad (15\text{-}49)$$
$$(\alpha = x + y)$$

(Golovin, 1963a).

The solution (15-48) must be completely equivalent to Smoluchowski's (12-43), since it is based on a constant collection kernel and the initial condition of a homogeneous aerosol. It is not difficult to demonstrate this equivalence. Thus, the discrete spectrum $g_n(T)$ which counts the number of n-tuple particles composed of n unit-sized (v_0) particles per unit aerosol volume, normalized by $N(0)$, the initial particle density (recall the normalization in (12-113)), is $g_n(T) = \int_{n-\varepsilon}^{n+\varepsilon} g(x, T)\,dx$ for integral n and $\varepsilon < 1$, so that

$$g_n(T) = (1 - T)^2 T^{n-1}. \qquad (15\text{-}50)$$

The dimensionless time τ may be recovered from T through the relationship $dT/d\tau = -dM_0/d\tau = D(T)/2$, or

$$\tau = 2 \int_0^T \frac{dT}{D(T)}. \qquad (15\text{-}51)$$

In the present instance with $D = (1 - T)^2$, this gives $T = \tau/2(1 + \tau/2)^{-1}$. Therefore, we find

$$g_n(T) = f_n(\tau) = \frac{(\tau/2)^{n-1}}{(1 + \tau/2)^{n+1}}, \qquad (15\text{-}52)$$

which is the Smoluchowski solution (12-43), allowance being made for the fact that τ as defined in (12-113) is just twice the value which appears in (12-43).

For our second choice of the initial spectrum we take the family of gamma distributions: $g(x, 0) = f(x, 0) = J(x)$, where $J(x)$ is given by (12-129). In passing we note that the observations of Levin (1954) led him to conclude that (12-129) with $p = 8/3$ gives a satisfactory description of many fog droplet spectra. Also, as Scott (1968) has pointed out, if the initial droplet distribution is nearly Gaussian in radius, then as a distribution in volume it may be approximated very well by some function of the form of (12-129). Given the distribution (12-129), we find that the mean value of x (or the first moment) is $M_1 = 1$, the most probable value is $p/(p + 1)$, and the relative rms dispersion (the square root of the relative variance, var x) is $(M_2 - M_1^2)^{1/2}/M_1 = (1 + p)^{-1/2}$. Finally we note again that (12-129) contains the limiting forms of a monodisperse distribution ($p \to \infty$) and an exponential distribution ($p \to 0$).

The solutions for this choice of initial distribution are as follows:

$$g(x, T) = x^p(1 - T)^2(p + 1)^{(p+1)}e^{-(p+1)x} \sum_{k=0}^{\infty} \frac{T^k(p + 1)^{k(p+1)}x^{k(p+1)}}{\Gamma[(p + 1)(k + 1)]} \qquad (15\text{-}53)$$
$$(\alpha = 1)$$

(Martynov and Bakanov, 1961), and

$$g(x, T) = x^p(1 - T)(p + 1)^{(p+1)}e^{-(T+p+1)x} \sum_{k=0}^{\infty} \frac{(Tx)^k(p + 1)^{k(p+1)}x^{k(p+1)}}{(k + 1)!\,\Gamma[(p + 1)(k + 1)]} \qquad (15\text{-}54)$$
$$(\alpha = x + y)$$

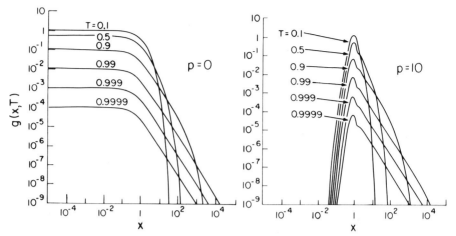

Fig. 15-6. Plots of the evolving drop size spectrum $g(x, T)$ for the kernel $\alpha(x, y) = x + y$. The initial spectrum $f(x, 0)$ is a gamma distribution characterized by the parameter p (Equation (12-129)). (From Drake and Wright, 1972; by courtesy of Amer. Meteor. Soc., and the authors.)

(Scott, 1968). This last solution is probably the most realistic for the drop collection problem, since the collection kernel for drop radii $\geq 50 \ \mu$m is similar to the sum-of-volumes form (see Section 15.2.3.2). A plot of (15-54) for various T and p = 0, 10 is given in Figure 15-6. The 'wiggles' in the curves for p = 10 and x near unity reflect the narrowness of the initial spectrum. Similar behavior occurs in the steady state case of a reinforced aerosol (see Figure 12-14). The curves indicate that for $xT \leq 1$ the form of the evolving spectrum is strongly influenced by the initial spectra, and that the influence is lost for $xT \gg 1$. Also, it turns out that the influence of the initial spectra is retained for larger xT for the sum-of-volumes kernel than for the constant kernel.

Some corresponding curves for the dimensionless liquid water content $xg(x, T)$ are shown in Figure 15-7. The curves, taken from Scott (1968), are labeled with

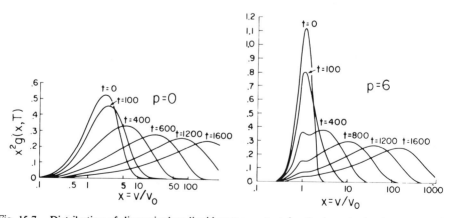

Fig. 15-7. Distribution of dimensionless liquid water content for the kernel $\alpha(x, y) = x + y$, and an initial distribution given by (12-129). (From Scott, 1968; by courtesy of Amer. Meteor. Soc., and the author.)

values of dimensional time t (in seconds). From (12-113) we see this implies a particular choice of K_0 and N(0). Scott assumed the collection kernel has the value $1.80 \times 10^{-4} \, cm^3 \, sec^{-1}$ when the drop volumes are $u_1 = 4.189 \times 10^{-9} \, cm^3$ (radius of 10 μm) and $v_1 = 1.131 \times 10^{-7} \, cm^{-3}$ (radius of 30 μm). This was done to ensure agreement with the collection kernel value found by using the Shafrir-Neiburger (1963) collision efficiency for a 30 μm and 10 μm radius droplet pair. The liquid water content was taken as $1 \, g \, m^{-3}$, and the initial mean particle volume was assumed to be $v_0 = 4.189 \times 10^{-9} \, cm^3$; therefore, $N(0) = 10^{-6} \, v_0^{-1} = 239$ droplets per cm^3. For the constant collection kernel case, $K_0 = 1.80 \times 10^{-4} \, cm^3 \, sec^{-1}$ and so $\tau = N(0) K_0 t = 0.0429 \, t$. For the sum-of-volumes kernel, $K_0 = 1.80 \times 10^{-4} \, v_0 \, (u_1 + v_1)^{-1} \, cm^3 \, sec^{-1}$, so that $\tau = 0.00153 \, t$ in Figure 15-7. The connection between τ and T is given by (15-51). Table 15-1 lists some representative values of t, τ, and T.

As we have mentioned before, the SCE applies in the conservative situation in which there are no mechanisms available to transport particles in or out of volumes of aerosol. Consequently, we have the constraint $M_1(\tau) = \int_0^x xg(x, T) \, dx = 1$, so that the area under the curves of xg versus x is constant in time. This desirable property of equal areas, which shows clearly how the water content is distributed along the range of droplet sizes, may be preserved in a semi-logarithmic plot by noting that $xg \, dx = x^2g \, d \ln x$; thus, in Figure 15-7 the ordinate is linear in the quantity $x^2g(x, T)$.

15.2.3 APPROXIMATION TECHNIQUES FOR THE SCE

We shall first briefly describe five methods of approximation which have been applied to the SCE, but which for practical purposes do not appear to be as

TABLE 15-1

Relationship between T, τ and t for $\alpha(x, y) = 1$ and for $\alpha(x, y) = x + y$. (From data given by Scott, 1968.)

T	$\alpha(x, y) = 1$		$\alpha(x, y) = x + y$	
	τ	t (sec)	τ	t (sec)
0	0	0	0	0
0.05	0.105	2.5	0.053	34
0.10	0.222	5.2	0.105	59
0.20	0.500	12	0.223	146
0.30	0.857	20	0.356	234
0.40	1.333	31	0.511	334
0.50	2	47	0.693	453
0.60	3	70	0.916	599
0.70	4.67	109	1.203	789
0.80	8	186	1.609	1050
0.90	18	420	2.303	1509
0.95	38	885	2.996	1960
0.99	198	4 610	4.605	3020
0.999	1998	46 600	6.908	4520

effective as two other methods we discuss in more detail below. Generally, their shortcomings arise from convergence problems and/or computational difficulties. For a comprehensive discussion of these various methods, see Drake (1972b).

(1) Power series representation. The spectrum is written as a power series; i.e., $f(x, \tau) = \sum_{i=0}^{\infty} a_i(x)\tau^i$, which on substitution into (15-23) leads to recursion relations for the a_i's.

(2) Picard's method of successive approximations. In this approach (15-23) is first integrated with respect to τ, so that $f(x, \tau)$ on the left side of the equation may be set equal to integrals over $f(x, \tau)$ on the right side. Then the nth approximation to $f(x, \tau)$ is obtained by substituting the $n - 1^{th}$ approximation on the right side.

(3) Similarity methods. For special forms of the kernel it has been possible to construct similarity solutions (sometimes referred to as 'self-preserving spectra'). A description and an example of this technique have already been presented in Section 12.8.2.

(4) Perturbation methods. Approximate solutions have been obtained by perturbing the kernel; e.g., $\alpha(x, y) = \alpha_0(x, y) + \alpha_1(x, y)$, where $|\alpha_1| \ll |\alpha_0|$. The corresponding spectrum is $f(x, \tau) = f_0(x, \tau) + f_1(x, \tau)$, where $f_0(x, \tau)$ is the solution for $\alpha = \alpha_0$. Assuming $f_1 \ll f_0$, a linear equation for f_1 results.

(5) Asymptotic approximations via the 'saddlepoint method.' For the kernels $\alpha = 1$, $x + y$, and xy, Scott (1965) has applied the saddlepoint method (e.g., Morse and Feshbach, Vol. I, 1953) for the asymptotic calculation of the integral form of the inverse Laplace transform of the spectrum. Unfortunately, it is difficult to apply the same technique for arbitrary kernels.

15.2.3.1 *Method of Moments*

We now turn to an approximation technique which has proven useful in a variety of applications. This is the *method of moments*, which replaces the problem of solving the SCE by one of solving a set of coupled differential equations for the moments of the distribution. Enukashvili (1964a) used the method of obtain first and second approximations to the spectrum for spatially uniform clouds for the cases of Brownian and gravitational coagulation. Later (1964b), he performed a similar analysis for gravitational coagulation in a spatially heterogeneous cloud in which a specified updraft structure led to an assumed form for drop sedimentation velocities as a function of time, volume, and height. This formulation required an extension of the SCE to include a sedimentation or drift term: If the drop velocities are denoted by $\tilde{c}(v, z, t)$, then the drift current density is $n\tilde{c}$, and so by continuity the additional term appearing on the left side of the SCE is $\nabla \cdot (n\tilde{c})$. Similar calculations were performed by Golovin (1963b, 1965), except that the collection kernel was assumed to have the sum-of-volumes form. Martynov and Bakanov (1961) made use of the moment equations in their search for a universal asymptotic spectrum for the SCE. Wang (1966) and Pich *et al.* (1970) also employed moments in determining various asymptotic spectra for several different collection kernels. Cohen and Vaughan (1971) used the moment equations to solve for the concentration, geometric mean particle volume, and logarithmic variance of an assumed log-normal distribution. More recently, Drake (1972a) has used the method of moments to obtain upper and lower bounds for the asymptotic form of $f(x, \tau)$.

For our first example of the use of moments, we shall follow the formulation of Enukashvili (1964a). Because he employs a special normalization, it is most convenient to start with the dimensional form of the SCE given by (12-48):

$$\frac{\partial n(v, t)}{\partial t} = \frac{1}{2} \int_0^v K(u, v - u)n(v - u, t)n(u, t) \, du$$

$$- n(v, t) \int_0^\infty K(u, v)n(u, t) \, du. \tag{15-55}$$

The moments of $n(v, t)$ are denoted by

$$\bar{M}_m(t) \equiv \int_0^\infty v^m n(v, t) \, dv, \quad m \geq 0. \tag{15-56}$$

(Thus we have $\bar{M}_0 = N$, the total concentration; and $\bar{M}_1 = \phi = w_L$, the volume fraction of water, or the liquid water content.) Then on multiplying (15-55) by $v^n \, dv$ and integrating from 0 to ∞, we get the moment equation (cf. (12-118)):

$$\frac{d\bar{M}_m}{dt} = \frac{1}{2} \int_0^\infty \int_0^\infty (v + u)^m K(u, v)n(u, t)n(v, t) \, du \, dv$$

$$- \int_0^\infty \int_0^\infty v^m K(u, v)n(u, t)n(v, t) \, du \, dv. \tag{15-57}$$

It is assumed that $n(v, t)$ can be developed in the following series:

$$n(v, t) = \sum_{i=0}^\infty a_i(t)\psi_i(v)\rho(v), \tag{15-58}$$

where the $\psi_i(v)$ constitute a complete set of orthogonal polynomials with respect to the weighting function $\rho(v)$, and the $a_i(t)$ are the series coefficients. Because of the orthogonality property, the a_i are obtained by multiplying (15-58) by ψ_j and integrating over v:

$$a_i(t) = \frac{\int_0^\infty \psi_i(v)n(v, t) \, dv}{\int_0^\infty \psi_i^2(v)\rho(v) \, dv}. \tag{15-59}$$

Since $\psi_i(v)$ is a polynomial in v, this integral expression for $a_i(t)$ is a linear combination of the moments.

Now on substituting (15-58) into (15-57), we obtain the desired result:

$$\frac{d\bar{M}_m}{dt} = \sum_{i=0}^\infty \sum_{j=0}^\infty a_i(t)a_j(t)I_{mij}, \tag{15-60}$$

where

$$I_{mij} = -\int_0^\infty \int_0^\infty [v^m - (v+u)^m/2]K(u,v)\psi_i(v)\psi_j(u)\rho(v)\rho(u)\,du\,dv. \qquad (15\text{-}61)$$

Given the kernel $K(u,v)$, the I_{mij} may be evaluated from (15-61). And from (15-59) the a_i may be expressed as a linear combination of the $\bar{M}_m$. Therefore, (15-60) is an infinite set of quasi-linear first-order differential equations fixing the moments of the distribution. Further, the infinite set (15-60) may be made finite by approximating $n(v,t)$ with only the first l terms of (15-58). Then by orthogonality, all the a_i for $i > l$ are zero.

The success of this method depends on having a fast rate of convergence, which depends strongly on the weighting function $\rho(v)$. Since any factor of $n(v,t)$ may be regarded as part of the weighting function, we are free to choose $\rho(v)$ in any way we like. For the most rapid convergence, it is evidently best to choose ρ as close to the unknown function n as possible. Following this line of reasoning. Enukashvili chose for ρ the simple exponential factor of the asymptotic form of n for the case of a constant collection kernel (cf. (12-124)):

$$\rho(v) = \rho'(v,t) = \exp(-v/v_f), \qquad (15\text{-}62)$$

where v_f is the mean droplet volume at time t:

$$v_f(t) = \bar{M}_1(t)/\bar{M}_0(t) = w_L/N(t). \qquad (15\text{-}63)$$

The orthogonal polynomials corresponding to the weighting function e^{-x} are the associated Laguerre polynomials (e.g., Morse and Feshbach, Vol. I, 1953):

$$L_m(x) = e^{-x}\frac{d^m}{dx^m}(x^m e^{-x})$$

$$(L_0 = 1, \quad L_1 = 1 - x, \quad L_2 = x^2 - 4x + 2). \qquad (15\text{-}64)$$

Accordingly, Enukashvili chose the following representation for $n(v,t)$:

$$n(v,t) = e^{-v/v_f}\sum_{i=0}^\infty a_i(t)L_i\left(\frac{v}{v_f}\right). \qquad (15\text{-}65)$$

Since the norm of L_m is $[\int_0^\infty L_m^2(x)e^{-x}\,dx]^{1/2} = m!$, the coefficients a_m are given by

$$a_m(t) = \frac{1}{(m!)^2}\int_0^\infty L_m\left(\frac{v}{v_f}\right)n(v,t)\,d\left(\frac{v}{v_f}\right), \qquad (15\text{-}66)$$

and these are a linear combination of the moments $\tilde{M}_m(t)$ defined by

$$\tilde{M}_m(t) \equiv \frac{\bar{M}_m(t)}{v_f^{n+1}(t)} = \int_0^\infty \left(\frac{v}{v_f}\right)^m n(v,t)\,d\left(\frac{v}{v_f}\right). \qquad (15\text{-}67)$$

The first approximation $n^{(1)}(v,t)$ to $n(v,t)$ is obtained by including the first two terms in (15-65):

$$n^{(1)}(v,t) = e^{-v/v_f}(a_0L_0 + a_1L_1) = \tilde{M}_0(t)e^{-v/v_f}, \qquad (15\text{-}68)$$

since from (15-64), (15-66), and (15-67) we have $a_0(t) = \tilde{M}_0(t)$, $a_1(t) = \tilde{M}_0(t) - \tilde{M}_1(t)$, while from (15-63) and (15-67) we find

$$\tilde{M}_0(t) = \tilde{M}_1(t) = \frac{N(t)}{v_f(t)}. \qquad (15\text{-}69)$$

Therefore, the moment equations (15-60)–(15-61) reduce to

$$\frac{d}{dt}(\tilde{M}_m v_f^{(m+1)}) = a_0^2 I_{m00}, \qquad (15\text{-}70)$$

where

$$I_{m00} = - \int_0^\infty \int_0^\infty [v^m - (v + u)^m/2] K(u, v) e^{-(u+v)/v_f} \, du \, dv. \qquad (15\text{-}71)$$

For every kernel we see that $I_{100} = 0$, so that water is conserved ($\bar{M}_1 = $ constant) in this representation.

For the case of a constant collection kernel, $K = K_0$, (15-70)–(15-71) reproduce the known exact results for the zeroth and second moments:

$$\bar{M}_0(t) = N(t) = \frac{N(0)}{1 + \dfrac{K_0 N(0)t}{2}}, \qquad (15\text{-}72a)$$

$$\bar{M}_2(t) = \bar{M}_2(0) + K_0 w_L^2 t. \qquad (15\text{-}72b)$$

Therefore, the solution $n^{(1)}$ in this case is the asymptotic form of n for arbitrary (but continuous) initial spectra (cf. (12-124)):

$$n^{(1)}_{(K=K_0)}(v, t) = \frac{N^2(t)}{w_L} e^{-N(t)v/w_L}, \qquad (15\text{-}73)$$

For other kernels the form of $n^{(1)}$ will also be the same as in (15-73), but the expression for $N(t)$ will be different. Enukashvili considers two special forms for K:

(1) Gravitational coagulation. The assumed form for K, taken to be valid for droplet radii less than 50 μm, is

$$K_G = k_0(u^{1/3} + v^{1/3})^2 |v^{2/3} - u^{2/3}|, \qquad (15\text{-}74)$$

where $k_0 = 5.9 \times 10^5 \, \text{cm}^{-1} \, \text{sec}^{-1}$. In this expression the factor k_0 includes an assumed constant collection efficiency, and the last factor describes the form of the droplet terminal velocity difference in Stokes flow. For $K = K_G$, (15-71) gives $I_{000} = 1.30 \, k_0 v_f^{10/3}$, so that the equation for $N(t)$ is

$$\frac{dN}{dt} = -1.30 \, k_0 v_f^{4/3} N^2 = -1.30 \, k_0 w_L^{4/3} N^{2/3}; \qquad (15\text{-}75)$$

hence

$$N(t)_{(K=K_G)} = N(0) \left[1 - \frac{2.55 \times 10^5}{N^{1/3}(0)} w_L^{4/3} t \right]^3. \qquad (15\text{-}76)$$

(2) Brownian coagulation. The kernel, omitting the correction for slip, is given

by (12-51):

$$K_B = \frac{2\,kT}{3\,\eta_a}\left[\left(\frac{u}{v}\right)^{1/3} + \left(\frac{v}{u}\right)^{1/3} + 2\right].$$ (15-77)

Accordingly, we obtain $I_{000} = 1.47\,kTv^2/\eta$, and

$$\frac{dN}{dt} = -1.47\,\frac{kT}{\eta_a}\,N^2.$$ (15-78)

The solution is

$$N(t) = N(0)\left[1 + \frac{1.47\,kT}{\eta_a}\,N(0)t\right]^{-1}.$$ (15-79)
$$\scriptstyle K = K_B$$

Comparison of (15-75) and (15-78) indicates that gravitational coagulation as modeled by (15-74) will predominate over Brownian coagulation when the mean droplet volume v_f satisfies the condition

$$v_f^{4/3} > 5.8 \times 10^{-6}\,\frac{kT}{\eta_a},$$ (15-80)

or when the mean radius of the droplets is greater than about 8 μm.

It can be seen from (15-73) that the first approximation to $n(v, t)$ cannot describe the 'spectral broadening' of the large particle tail of the distribution which is brought about when the kernel increases with particle size. However, this does occur to some extent in the second approximation, which includes the first three terms in the expansion (15-65); this is obvious by inspection:

$$n^{(2)}(v, t) = \frac{N(t)}{v_f}\,e^{-v/v_f}\left[1 + \frac{(y-2)}{4}\left(2 - \frac{4\,v}{v_f} + \frac{v^2}{v_f^2}\right)\right],$$ (15-81)

where $y = \tilde{M}_2/\tilde{M}_0$.

We shall now outline Enukashvili's similar analysis (1964b) for sedimenting droplets coagulating according to (15-74) in a cloud updraft region. From the paragraph preceding (15-55) we can see that the relevant generalization of the SCE is

$$\frac{\partial n(v, z, t)}{\partial t} + \frac{\partial}{\partial z}\,[n(v, z, t)c(v, z)] = -n(v, z, t)\int_0^\infty K(u, v)n(u, z, t)\,du$$

$$+ \frac{1}{2}\int_0^v K(u, v - u)n(v - u, z, t)n(u, z, t)\,du,$$ (15-82)

where $n(v, z, t)$ is the concentration of droplets of volume v at height z at time t, and

$$c(v, z) = W(z, t) - \alpha v^\beta$$ (15-83)

is the assumed vertical velocity of droplets in an updraft specified by $W(z, t)$. The constants α and β specify the drag law for the droplets (for complete consistency with (15-74) we note that α and β should be chosen in accordance with the Stokes approximation). Proceeding as before with an expansion of n in orthogonal polynomials we again arrive at an equation like (15-60) for the moments except that: (1) $\bar{M}_m$ is now a function of z as well as t; (2) a sedimentation term

$$\frac{\partial}{\partial z} \sum_{i=0}^{\infty} a_i(z, t) I_{mi}, \tag{15-84}$$

where

$$I_{mi} = \int_0^{\infty} v^m \rho(v) \psi_i(v) c(v, z) \, dv, \tag{15-85}$$

appears on the left side of the equation.

Proceeding in analogy to the previous case, Enukashvili's first approximation to $n(v, z, t)$ is (cf. (15-68)):

$$n^{(1)}(v, z, t) = \tilde{M}_0(z, t) e^{-v/v_f(z,t)}. \tag{15-86}$$

Substitution of this expression into the moment equation yields

$$\frac{\partial N(z, t)}{\partial t} + \frac{\partial}{\partial z} \{ [W - \alpha \Gamma(\beta + 1) v_f^{\beta}] N(z, t) \} = -1.30 \, k_0 v_f^{4/3} N^2(z, t), \tag{15-87}$$

in place of the previous (15-75) (Γ denotes the gamma function), and

$$\frac{\partial w_L(z, t)}{\partial t} + \frac{\partial}{\partial z} \{ [W - \alpha \Gamma(\beta + 2) v_f^{\beta}] w_L(z, t) \} = 0, \tag{15-88}$$

in place of the previous result w_L = constant. These expressions show how the effective transport velocities of total concentration and liquid water content depend on the updraft velocity and mean droplet volume. It can be seen, for example, that sedimentation of liquid water content will occur even when there is no vertical flux of total droplet concentration.

One relatively simple application of (15-87) and (15-88) is to solve for the steady state variations of N and w_L with height above cloud base ($z = 0$). From these one may recover the spectrum also (from (15-86)), and hence the radar reflectivity, which is proportional to $\bar{M}_2 = \int_0^{\infty} v^2 n(v, z) \, dv = 2 \, w_L^2(z)/N(z)$. Proceeding in this way, and assuming $W = W_0$, a constant, (15-87) and (15-88) reduce to

$$W_0 \frac{dN}{dz} = \alpha \Gamma(\beta + 1) \frac{d}{dz} (v_f^{\beta} N) - 1.30 \, k_0 v_f^{4/3} N^2, \tag{15-89}$$

and

$$W_0 \frac{dw_L}{dz} = \alpha \Gamma(\beta + 2) \frac{d}{dz} (v_f^{\beta} w_L). \tag{15-90}$$

The second equation may be integrated directly to give $W_0 w_L - \alpha \Gamma(\beta + 2) v_f^{\beta} w_L = b =$ constant, so that

$$N = \left[\frac{\alpha \Gamma(\beta + 2)^{\beta+1}}{W_0 w_L - b} \right]^{1/\beta}. \tag{15-91}$$

Putting this result into (15-89) and integrating, we obtain

$$\frac{2\,w_L^2}{\beta b\theta} + \frac{2\,w_L}{b}\left[1 + \frac{2}{\beta(\beta+1)}\right]\theta - \frac{2\,\theta^3}{3\,b\beta(\beta+1)} = 1.30\,k_0[\alpha\Gamma(\beta+2)]^{-1/3\beta}z + b_1,$$

(15-92)

where b_1 is the constant of integration and

$$\theta = \left(w_L - \frac{b}{W_0}\right)^{1/2}.$$

Enukashvili has evaluated (15-91)–(15-93) under the assumptions that: (1) α and β are determined according to the Stokes approximation, namely $\alpha = 4.8 \times 10^5\,cm^{-1}\,sec^{-1}$ and $\beta = 2/3$; (2) at cloud base $w_L = 2\,g\,m^{-3}$ and $N = 450\,cm^{-3}$; (3) $W_0 = 1\,m\,sec^{-1}$. The results, shown in Table 15-2, are in fair accord with the experimental data of Zaitsev (1950) and Weickmann and Aufm Kampe (1953) for convective clouds. Also, Enukashvili claims the corresponding prediction that the radar reflectivity increases with height is in qualitative agreement with observations of the lower portions of warm cumulus clouds. (However, for situations in which precipitation has not yet penetrated cloud base, a condition of Enukashvili's model, such a result is rather trivial.)

These examples illustrate how considerable information of direct physical relevance can be provided with relative ease through use of the method of moments. It appears there is much room for further exploitation of this approach as applied, for example, to more realistic descriptions of the collection kernels and/or extensions of the SCE to include other microphysical processes of interest.

15.2.3.2 Polynomial Approximations to the Gravitational Collection Kernel

The most realistic versions of the kernel for gravitational collection are fairly complicated, and thus necessitate a numerical solution of the SCE. On the view that the numerical approach should be used only as a last resort, Long (1974) has searched for simple polynomial approximations to the collection kernel, hoping that analytical solutions based on them could be found.

Though often inevitable at some point when attempting to deal with realistic

TABLE 15-2

Variation of liquid water content (w_L) and total drop concentration (N) with height above cloud base ($z = 0$) for a constant updraft of $W = 1\,m\,sec^{-1}$. (From data given by Enukashvili, 1964b.)

z(m)	0	121	358	504
$w_L\,(gm^{-3})$	2	2.01	2.05	2.1
$10^{-8}\,N\,(m^{-3})$	4.5	3.07	1.25	0.736
z(m)	671	723	767	783
$w_L\,(gm^{-3})$	2.3	2.5	3	4
$10^{-8}\,N\,(m^{-3})$	0.246	0.152	0.092	0.068

problems, numerical methods of solution do have some obvious drawbacks. For example, they tend to be inelegant, in the sense that insight into the physical processes under study is often obscured by a morass of numbers. The dependence of the solutions on various physical parameters is generally not obvious and this tends to limit their usefulness. Also, such calculations are generally time consuming, and, if they are to be incorporated in a larger context, the time requirements may be prohibitive.

For example, consider the three-dimensional, moist, deep-convection model under development at the National Center for Atmospheric Research (NCAR). As described by Drake (1972a), the model considers the following dependent variables: three velocity components, pressure, density, absolute temperature, and the mixing ratios of water vapor, cloud water, cloud ice, rain, and hail. The domain of integration is 50 km by 50 km by 20 km with a mesh size of 500 m in each direction. Even with a large computer system, such as the CDC 7600 presently in use at NCAR, it is not feasible to solve the SCE numerically for, say, 20–40 particle sizes for each of the water categories (cloud water, cloud ice, rain, and hail), at each of the 4×10^5 grid points.

One way around such an impass is to summarize the results of numerical computations through simple 'parameterizations,' wherein the observed dependence of solutions on important physical parameters is approximated by simple algebraic expressions (e.g., Berry and Reinhardt, 1974d). This approach is treated below in Section 15.2.5. Here we shall discuss the limited potential alternative offered by Long. We say potential alternative, because at this writing no analytical solutions have been found for the polynomial kernels he has provided.

The first step in obtaining the polynomial approximations is to evaluate the actual gravitational collection kernel K as accurately as possible. In the expression (15-2) for K we recall the collection efficiency E_c is the product of the collision efficiency E and the coalescence efficiency E_{co}. Long (1974) assumes $E_{co} = 1$ because evidence to the contrary (Woods and Mason, 1964; Whelpdale and List, 1971; Brazier-Smith et al., 1972) covers only a limited range of drop sizes (see Section 14.8.1). For $10 \leqslant a \leqslant 300 \ \mu\text{m}$ Long uses the E values of Shafrir and Gal-Chen (1971) and Klett and Davis (1973). The terminal velocities are evaluated from the approximate formula developed by Long and Manton (1974), which is based on the data of Gunn and Kinzer (1949) and Beard and Pruppacher (1969) obtained at 1013 mb and 20 °C. An altitude correction to the formula is made because the collision efficiencies were calculated for 900 mb and 0 °C. For droplets of radii $<15 \ \mu\text{m}$ this correction is derivable from the Stokes terminal velocity formula, and is due to the change in viscosity of air with temperature. For radii $>1690 \ \mu\text{m}$ Long uses the correction of Foote and du Toit (1969). For intermediate sizes a linear interpolation with respect to the logarithm of the droplet radius is used.

In this manner the curves of K vs. a_1 for various ratios a_2/a_1 are obtained as shown in Figure 15-8. As we noted earlier in Section 12.7.5, the curves indicate a relatively weak dependence of K_G on a_2/a_1. The figure also shows that K_G varies roughly as v^2 for $a \leqslant 50 \ \mu\text{m}$ and as v for $a > 50 \ \mu\text{m}$. This is explained by the

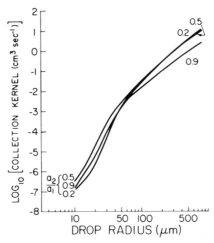

Fig. 15-8.   Collection kernel $K(a_1, a_2)$ based on collision efficiencies of Shafrir and Gal-Chen (1971), and Klett and Davis (1973). (From Long, 1974; by courtesy of Amer. Meteor. Soc., and the author.)

dependence of E and V_∞ on size: For small and intermediate sized-drops, the terminal velocity varies as a^2 and a, respectively. Similarly, for small drops and size ratios, $E \approx 4.5 \times 10^4 \, a^2$ (a in cm), while $E \approx$ constant for intermediate drop sizes.

These results provide the rationale for simple polynomial estimates for K_G of the form v or v^2 (or $(v + u)$ and $(v + u)^2$). The numerical coefficients selected by Long are those which give a minimum rms deviation between the logarithm of the approximating polynomial $P(v, u)$ and the logarithm of $K_G(v, u)$. The deviation between the logarithms of $P(v, u)$ and $K_G(v, u)$ rather than between the functions themselves was chosen because of the large variation in K_G, and because of the presumed importance of representing K_G well in all size intervals. In this fashion Long obtained the following polynomial estimates for K_G:

$$P(v, u) \atop (v>u) = \begin{cases} 9.44 \times 10^9 \, (v^2 + u^2), & 10 \leqslant a \leqslant 50 \; \mu m \\ 5.78 \times 10^3 \, (v + u), & a > 50 \; \mu m \end{cases} \tag{15-93a}$$

and

$$P(v, u) \atop (v>u) = \begin{cases} 1.10 \times 10^{10} \, v^2, & 10 \leqslant a \leqslant 50 \; \mu m \\ 6.33 \times 10^3 \, v, & a > 50 \; \mu m. \end{cases} \tag{15-93b}$$

The merit of the approximations (15-93) has been tested by comparing numerical solutions based on them and the actual K_G; generally, the agreement is quite good. With this evidence of their accuracy, Long has suggested that an attempt should be made to find analytical solutions to the SCE based on either of them. Such solutions could then facilitate the simulation of droplet collection in, for example, multidimensional cloud models. Even if this proves impossible, (15-93) provides simple and concise, and therefore convenient, descriptions of the collection process (for the case of no turbulence) which may be of use in other applications. In passing, we note the restriction $a > 10 \; \mu m$ may be dropped without incurring significant errors (this was in fact done by Long in his computations).

15.2.4 NUMERICAL METHODS FOR THE COLLECTION PROCESS

In spite of the unattractive aspects of numerical methods we have referred to above, they remain the only practical way to cope with the most realistic formulations of the collection problem. A brief description of some of these methods therefore is in order. We shall restrict ourselves here to three methods which appear to have particular merit: the method of Berry (1967) and Reinhardt (1972); the method of Kovetz and Olund (1969); and the Monte Carlo method.

15.2.4.1 *Method of Berry (1967) and Reinhardt (1972)*

Numerical integration of the SCE generally requires replacing the integrals in the continuous version by finite sums, or truncating the infinite summation in the discrete version. In either case the time derivatives are replaced by some suitable finite difference form so that the resulting finite system of nonlinear algebraic equations may be numerically 'stepped along' in time.

For the continuous version of the SCE, the basic procedure is as follows: one discretizes the continuous volume (or mass) axis into 'bins' of width Δv_i, $i = 1, 2, \ldots$, and each bin is characterized by some representative volume v_i ($v_1 < v_2 < \cdots$). One then calculates $\partial n(v_i, t)/\partial t$ at $t = 0$ by numerical evaluation of the integrals on the right side of (15-55), wherein the initial spectrum $n(v_i, 0)$ is inserted. The corresponding spectrum at time Δt is then written as

$$n(v_i, \Delta t) = n(v_i, 0) + \frac{\partial n(v_i, t)}{\partial t}\bigg|_{t=0} \Delta t \ . \tag{15-94}$$

The process can be repeated to obtain the spectrum for every $t > 0$.

A key practical problem which arises is how to subdivide the volume axis into a manageable (small) number of bins, while maintaining adequate resolution over the broad size range of interest (from a few microns to a few millimeters). An effective solution is that of Berry (1965, 1967), who employed an exponential subdivision in which the drop radius a is written as a function of the integer J as follows:

$$a(J) = a_0 e^{(J-1)/J_0}, \tag{15-95}$$

$J = 1, 2, \ldots, J_{max}$; $J_0 =$ unspecified number. The corresponding volume is given by

$$v(J) = \frac{4 \pi a_0^3}{3} e^{3(J-1)/J_0}. \tag{15-96}$$

Here $4 \pi a_0^3/3 \equiv v_0$ is the smallest volume considered and replaces the lower limit zero in the SCE.

To facilitate the discussion slightly, let us now rewrite the SCE in the following form:

$$\frac{\partial n(v)}{\partial t} = \int_0^{v/2} K(v_c, v')n(v_c)n(v') \, dv' - n(v) \int_0^{\infty} K(v, v')n(v') \, dv', \tag{15-97}$$

where $v_c \equiv v - v'$. The symmetry of K permits the change in form of the first (gain) integral. Also, for brevity we have suppressed the dependence of n on time. In terms of the transformation (15-96), the upper limit $v/2$ in the gain integral is specified by the integer J_{up} where $v(J_{up}) = v(J)/2$, or $J_{up} = J - (J_0 \ln 2)/3$. It is convenient to set $(J_0 \ln 2)/3 = 2$, which reduces (15-96) to a geometric progression of sizes, with each drop volume being $\sqrt{2}$ times the preceding one:

$$v(J) = v_0 2^{(J-1)/2}. \tag{15-98}$$

This transformation results in a new distribution function $n(J)$ which is related to $n(v)$ by $n(J) \, dJ = n(v) \, dv$, or

$$n(J) = \frac{\ln 2}{2} \, vn(v). \tag{15-99}$$

Also, from $v_c \equiv v - v'$ and (15-98) we find the number J_c (generally not an integer) which corresponds to v_c is

$$J_c = J + \frac{2}{\ln 2} \ln [1 - 2^{(J'-J)/2}]. \tag{15-100}$$

With these changes the form of (15-97) for the function $n(J)$ is

$$\frac{\partial n(J)}{\partial t} = v(J) \int_1^{J-2} \frac{K(J_c, J')}{v(J_c)} n(J_c) n(J') \, dJ' - n(J) \int_1^{J_{max}} K(J, J') n(J') \, dJ', \tag{15-101}$$

where now J_{max} replaces the upper limit of infinity for the loss integral.

For physical and graphical reasons we have discussed already, there is some advantage in working directly with the distribution of liquid water content per unit $\ln a$ interval, $g(\ln a)$, which is related to $n(v)$ by $g(\ln a) \, d(\ln a) = \rho_w vn(v) \, dv$, or $g(\ln a) = 3 \rho_w v^2 n(v)$. Then on defining $G(J) \equiv g(\ln a)$, we have

$$G(J) = \rho_w J_0 vn(J). \tag{15-102}$$

From (15-101) and (15-102) the governing equation for $G(J)$ follows immediately (Reinhardt, 1972; Berry and Reinhardt, 1974a):

$$\frac{\partial G(J)}{\partial t} = \frac{v(J)}{J_0} \left\{ v(J) \int_1^{J-2} \frac{K(J_c, J')G(J_c)G(J')}{v^2(J_c)v(J')} \, dJ' - \frac{G(J)}{v(J)} \right.$$

$$\left. \times \int_1^{J_{max}} \frac{K(J, J')G(J')}{v(J')} \, dJ' \right\}. \tag{15-103}$$

In Reinhardt's computations a value of G below $10^{-70} \, g \, cm^{-3}$ per unit $\ln a$ is defined as zero, with J_{max} being the integer argument of the smallest G greater than 10^{-70}.

Two types of numerical operations must be carried out in order to evaluate $\partial G/\partial t$ from (15-103). First, since J_c is generally not an integer, the value of $G(J_c)$ must be interpolated. Second, the indicated integrations must be carried out by some suitable process of numerical quadrature. The numerical schemes devised by Reinhardt (1972) apparently are able to accomplish these tasks with considerable accuracy and speed. Thus, Berry and Reinhardt (1974a) claim that calculations performed on the CDC 6600/7600 computer at NCAR show only a 0.15% change in total liquid water content over 1800 iterations for realistic conditions. The time step used was 1.0 sec, and 30 min of real time took 10–15 min of computer time on the 6600 and 2–3 min on the 7600. (However, a note of caution: such accuracy requires a computing system comparable to the one used by Berry and Reinhardt. In applying the method, de Almeida (1975) found that single precision arithmetic on the Univac 1110 computer was not sufficient to render good solutions. In fact, in some cases the errors were so large that the computed results were meaningless.) Reinhardt interpolates on the logarithm of G(J) with a six point Lagrange interpolation formula (see, for example, Collatz, 1960). Over most intervals the numerical quadrature is carried out with three point Lagrange integration coefficients. However, near the zeros of the integrands, which occur when collection kernels containing the drops' relative velocity of approach as a factor are used, four or five coefficients are necessary for high accuracy. Further details are available in the references.

15.2.4.2 *Method of Kovetz and Olund (1969)*

The necessity for interpolation as described in the previous section comes about because drops formed by coalescence generally do not coincide exactly with one of the specified size categories. In the Kovetz and Olund (K-O) formulation, this problem is circumvented by the relatively simple device of redistributing the new droplets into the nearest size classes in such a manner as to preserve both number and volume. This is accomplished through replacement of the discrete SCE by the following governing form:

$$\frac{dn_k}{dt} = \frac{1}{2} \sum_{i=1}^{k} \sum_{j=1}^{k} B_{ijk} K_{ij} n_i n_j - n_k \sum_{i=1}^{\infty} K_{ik} n_k, \tag{15-104}$$

where B_{ijk}, called the redistribution kernel, is given by

$$B_{ijk} = \begin{cases} (r_i^3 + r_j^3 - r_{k-1}^3)/(r_k^3 - r_{k-1}^3), & r_{k-1}^3 < (r_i^3 + r_j^3) \leq r_k^3 \\ (r_{k+1}^3 - r_i^3 - r_j^3)/(r_{k+1}^3 - r_k^3), & r_k^3 < (r_i^3 + r_j^3) < r_{k+1}^3. \\ 0, & \text{otherwise} \end{cases} \tag{15-105}$$

This model for collection has been criticized because it is not strictly equivalent to the SCE and because redistribution tends to force some water into spuriously high size categories, an effect which is referred to an 'anomalous spreading' (Reinhardt, 1972). On the other hand, in a recent paper in defense of the method Scott and Levin (1975a) reiterate the fact that the SCE is itself only an approximate representation of the collection process. They also emphasize the simplicity of the K-O method, and the relative ease with which it can be generalized to include other processes.

Scott and Levin provide a quantitative comparison of the K-O and SCE formulations. For the sum-of-volumes kernel, the numerical techniques of Berry and Reinhardt for the SCE yield results which are essentially indistinguishable from the analytical solution of Golovin (1963a) ((15-54) with p = 0). On the other hand, the K-O solution for the liquid water distribution, g(ln a, t), shows a peak value about 6% lower after 20 min, and a slight extension of the tail. For the case of the more realistic hydrodynamic collection kernel used by Berry (1967), the K-O results are a bit worse. Here there is little anomalous spreading, but the second maximum in the rain-water portion of the spectrum is about 25% lower than according to the SCE.

15.2.4.3 *Monte Carlo Method*

The numerical method of solving the SCE can be regarded as a set of procedural rules, or an algorithm, whereby one uses the SCE to construct the average spectrum at time $t + \Delta t$ from the one at time t. Monte Carlo simulations can provide other such algorithms, which also can accommodate quite general forms for the collection kernel and the initial spectrum. The Monte Carlo algorithm presented by Gillespie (1975b) provides, unlike the SCE algorithm, a rigorous simulation of the stochastic collection process. Therefore, it can produce numerical estimates of both the average of, and the fluctuations in, the developing drop volume (or mass) spectrum. A description of Gillespie's algorithm is given in Appendix A-15.2.4.3.

At this writing, Gillespie's rigorous algorithm has not been applied to specific problems. When it is, it will be most interesting to see whether the SCE, with its no-correlation approximation, is vindicated or not. If not, then Gillespie's numerical algorithm may well stand as the only reliable computational procedure for studying stochastic collection. If the SCE does prove to yield sufficiently accurate information, we would then like to know whether or not the Monte Carlo algorithm offers any computational advantages over algorithms for solving the SCE. One obvious limitation of the Monte Carlo method is that it can account for stochastic collection in only a small portion of a cloud. Gillespie estimates that a run of from 10^3 to 10^5 droplets could be handled with the current state of computer technology, and this corresponds typically to a cloud volume range of from 1 to $10^3 \, cm^3$. This may not be a serious fault, since stochastic collection is a very local process in clouds.

Let us now comment briefly on earlier applications of the Monte Carlo method, following the critique of Gillespie (1975b). The algorithm of Chin (1970) and Chin and Neiburger (1972) applies only to the case of a large drop which falls through a cloud of smaller droplets with a fixed spectrum. This is less general than Gillespie's algorithm, which allows interactions between all drop pairs in a developing spectrum. In addition, their method of choosing the time interval between coalescence events is not completely error-free, as Gillespie's is. The algorithm of Lapidus and Shafrir (1972) abandons the restriction to just drop-droplet coalescences, but does not appear to conform completely to a probabilistic description of collection events. For example, their procedure allows multiple coalescences to occur in a single time step; this will lead to a misrepresentation of correlation effects. The simulation of Robertson (1974) is quite restrictive in scope, since only drop-droplet interactions in a monodisperse droplet cloud are permitted (this constitutes a Monte Carlo simulation of Telford's (1955) early cloud model). His method for picking the time between coalescence events is also not stochastically complete.

15.2.5 PARAMETERIZATION OF THE COLLECTION PROCESS

Kessler (1969) was the first to attempt a general parameterization of the microphysics of the warm rain process. His scheme is based on the assumptions that: (1) the spectrum may be divided into the two categories of 'cloud' droplets having negligible terminal velocities, and 'hydrometeor' drops with significant terminal velocities; and (2) there are two growth processes to consider, namely the self-collection of cloud droplets (auto-conversion), and the collection of small droplets by large drops (accretion). The resulting parameterization is not based on numerical integration of the SCE, and so is considerably simpler than others which are (e.g., Berry, 1968; Simpson and Wiggert, 1969; Cotton, 1972; Berry and Reinhardt, 1974b, d). Here we shall restrict ourselves to a derivation and description of the recent scheme based strictly on the SCE, and elaborated in great detail by Berry and Reinhardt (1974a, b, c, d).

In the first paper of the series, Berry and Reinhardt (1974a) emphasize the need for including the self-collection of large drops, in addition to accretion and auto-conversion, in the collection growth parameterization. They do this through their analysis of the development of the liquid water spectrum which, as we shall see below, tends to evolve to a bimodal form. This leads them to break the spectrum into two parts, one consisting of cloud droplets (S1) and the other primarily of hydrometeors (S2). To do this, the assumption is made that the first three moments of S1 decrease proportionately in time, so that its relative variance is constant (Berry and Reinhardt, 1973). Given this slightly artificial resolution, numerical integration of the SCE elucidates the roles played by the three growth modes as follows: First, the S1 − S1 (auto-conversion) collections serve to add initial water to S2 so that the other modes can operate. The auto-conversion mode soon becomes weak compared to accretion (S1 − S2 collections), which is the primary mechanism for transferring water from S1 to S2. The associated growth rate of the peak of S2 is not large, however, as all drops tend to grow at similar rates. It is the third mode of hydrometeor self-collection (S2 − S2 collections) which turns out to be largely responsible for the emerging position and shape of S2.

In (1974b), Berry and Reinhardt display graphically several numerical solutions of the SCE for a range of initially unimodal (gamma) distributions. An example of the evolution of the liquid water spectrum $g(\ln a, t)$ into a bimodal $(S_1 + S_2)$ form is shown in Figure 15-9. In these calculations, a_f and a_g are the radii corresponding to v_f and v_g which, in the notation of Section 15.2.3.1, are defined as follows:

$$v_f = \bar{M}_1/\bar{M}_0 = w_L/N, \quad v_g = \bar{M}_2/\bar{M}_1 = \bar{M}_2/w_L; \tag{15-106}$$

i.e., v_f is the usual mean volume of the number distribution $n(v)$, while v_g is the mean volume of the water content distribution $vn(v)$. It can be seen from Figure 15-9 that when S_2 is well developed, a_g is close to its peak radius, and so

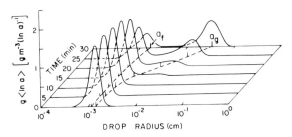

Fig. 15-9.   Time evolution of the liquid water spectrum; $a_f(0) = 12\,\mu$m, var x = 1. (From Berry and Reinhardt, 1974c; by courtesy of Amer. Meteor. Soc., and the authors.)

provides a measure of the size which contributes most to the volume of water present; for this reason a_g is loosely referred to as the 'predominant' radius.

For collection kernels based on the collision efficiencies of Hocking and Jonas (1970) for $a \leqslant 40\,\mu$m and Shafrir and Neiburger (1963) for $a > 40\,\mu$m, for the assumption of a coalescence efficiency of unity, and for initial gamma spectra described by $10 \leqslant a_f(0) \leqslant 18\,\mu$m and $0.5 \leqslant var^{1/2}\,v \leqslant 1.0$ (note var $v = v_g/v_f - 1$), Berry and Reinhardt observed regularities in the spectral evolution which may be summarized in such a way as to yield an effective auto-conversion rate of transfer of cloud droplet water to hydrometeor water. This parameterized auto-conversion rate is based on the time T_2 required for the predominant radius of S2, a_{g2}, to reach 50 μm. In terms of the initial liquid water content w_{L1}, droplet concentration N_1, and relative variance (var v)$_1$ of S1, time T_2 (in seconds) is given approximately by

$$T_2 = 6 \times 10^{-5}[10^3(w_{L1}/N_1)^{1/3}(var\ v)_1^{1/6} - 1.2]^{-1}w_{L1}^{-1}. \qquad (15\text{-}107)$$

At this time, the liquid water content $w_{L,a}$ which is converted to a spectrum centered about a_{g2} (the subscript a referring to auto-conversion), is approximately

$$w_{L,a} = 4 \times 10^{-3}[10^{12}(w_{L1}/N_1)^{4/3}(var\ v)_1^{1/2} - 2.7]w_{L1}. \qquad (15\text{-}108)$$

From these expressions the formula for the average auto-conversion rate is obtained as

$$\left.\frac{dw_{L2}}{dt}\right|_a = -\left.\frac{dw_{L1}}{dt}\right|_a = \frac{w_{La}}{T_2}. \qquad (15\text{-}109)$$

This parameterization is possible because S2 tends to acquire the same shape by time T_2 for all cases considered. The reason for this is that the collection kernel used by Berry and Reinhardt is approximated by the Golovin (1963a) sum-of-volumes form for $a \geqslant 50\,\mu$m (recall (15-93)), so that the asymptotic Golovin spectral form is approached by time T_2, regardless of the initial shape of S1. Of course, this behavior would not hold if the inclusion of other phenomena which can enhance coagulation, such as turbulence, shearing motions, and electric fields and charges, caused the form of the collection kernel to change significantly.

In (1974c), Berry and Reinhardt formulate approximate descriptions of accretion and hydrometeor self-collection. Both processes are parameterized in terms of rate equations. An outline of the analysis is given in Appendix A-15.2.5. In (1974d), Berry and Reinhardt complete the parameterization for accretion and self-collection merely by summing the various rate equations for the fractional changes in v_{f2} and v_{g2}, the mean volumes of the concentration and water content distributions for the hydrometeor spectrum S_2:

$$\frac{1}{v_{f2}}\frac{dv_{f2}}{dt} = b_{cf}w_{L1} + b_{sf}w_{L2}, \tag{15-110a}$$

$$\frac{1}{v_{g2}}\frac{dv_{g2}}{dt} = b_{cg}w_{L1} + b_{sg}w_{L2}. \tag{15-110b}$$

The various expressions for the rate coefficients are given in Appendix A-15.2.5. Unfortunately, they are based upon several rather crude mathematical estimates, so that the net improvement over earlier, much simpler parameterizations may be deceptively small. Also, of course, the parameterization applies only to the case of pure collection, with a kernel which approximates the sum-of-volumes form for $a \geqslant 50 \ \mu$m. Similar remarks apply to the auto-conversion parameterization given in (15-107)–(15-109). Of course, the best way to judge such schemes is through their performance in applications. To our knowledge that remains to be done for the Berry-Reinhardt parameterization.

15.3   Representative Numerical Results for the Collection Process

Telford's (1955) early calculation of stochastic collection for a simple drop-droplet cloud and an assumed constant collection kernel showed that a small fraction of the large drops grows about six times faster than according to the continuous mode. As we have indicated earlier, this work was followed by many detailed numerical integrations of the SCE involving more realistic initial spectra and collection kernels. In this section we shall summarize some of the conclusions which can be drawn from the large body of literature on the subject. More comprehensive summaries of the work done prior to 1971 are available in Mason (1971) and Drake (1972b).

Twomey (1966) considered the following initial spectrum as representative for a typical maritime cumulus cloud: a normal distribution in radius, with a relative dispersion $\sigma/\bar{a} = 0.15$, a liquid water content of $1 \ \text{g m}^{-3}$, and a concentration $N(0) = 50 \ \text{cm}^{-3}$ (hence $\bar{a} \approx 16.5 \ \mu$m and $\sigma \approx 2.5 \ \mu$m). The spectrum was truncated where $n(a) = 0.5 \ \text{cm}^{-3} \ \mu\text{m}^{-1}$ (near $a = 25 \ \mu$m). The collection kernel was based on the collision efficiencies of Hocking (1959) for $a \leqslant 30 \ \mu$m and Shafrir and Neiburger (1963) for $a > 30 \ \mu$m, and a coalescence efficiency of unity was assumed. It was found that 14 min are required to obtain drizzle drops of $100 \ \mu$m radius in concentrations of $100 \ \text{m}^{-3}$; further integration showed such a cloud may be expected to produce rain within 20 to 30 min, a realistically short period of time.

The predicted time to produce rain is considerably longer for a typical continental cumulus cloud. For this case, Twomey selected the initial spectrum as follows: a normal distribution for $a \leqslant 20 \ \mu$m with $\sigma/\bar{a} = 0.15$, $N(0) = 200 \ \mathrm{cm}^{-3}$, and a liquid water content of $1 \ \mathrm{g \ m}^{-3}$ (hence $\bar{a} \approx 10 \ \mu$m and $\sigma \approx 1.5 \ \mu$m). The spectrum was extended slightly to include a few larger drops: at $a = 20 \ \mu$m, $n(a) = 0.04 \ \mathrm{cm}^{-3} \ \mu\mathrm{m}^{-1}$; on the interval $20 \ \mu\mathrm{m} < a \leqslant 21 \ \mu$m, $n(a) = 0.04(21 - a) \ \mathrm{cm}^{-3} \ \mu\mathrm{m}^{-1}$ (a in microns). The time to produce the same concentration of drizzle drops as in the previous case increased to 39 min; this corresponds to more than 1 hour for the production of rain.

These findings are consistent with those of Bartlett (1966, 1970), whose later paper includes results based on the improved collision efficiencies of Hocking and Jonas (1970) for $a \leqslant 30 \ \mu$m. He also investigated the dependence of the time required for producing $40 \ \mu$m radius droplets in concentrations of one per liter on the number of droplets initially present in the $20-25 \ \mu$m radius range. For a continental type cumulus cloud with an initial spectrum like the one selected by Twomey, this time is 16 min, while if the spectrum is extended to include more than one drop/liter for each micron radius interval up to $a = 26 \ \mu$m, it is reduced to 7 min. (For Twomey's initial maritime spectrum, only 4 min are required.) This reinforces the notion that a crucial step for precipitation formation is the production of an adequate number of droplets on the $20-25 \ \mu$m radius range: only if there are a few per liter of such droplets present may precipitation be expected to form in a reasonably short period of time. But since neither condensation growth nor basic hydrodynamic collection in the Stokesian regime appear to be very effective separately in that size range (barring the infrequent occurrence of giant hygroscopic nuclei of mass greater than 10^{-9} g which can grow by condensation to $a = 25 \ \mu$m in less than 5 min), it appears some additional process(es) must be invoked in order to explain rain formation in realistic times for typical conditions.

These general trends are also consistent with the computations of Warshaw (1967, 1968), though for reasons which are not clear some of his calculated growth rates are nearly twice as fast as those found by Twomey for the same cloud model. A Monte Carlo simulation by Kornfeld et al. (1968) agrees very closely with Warshaw's predicted spectrum for a growth period of 2 min.

From Berry's (1967, 1968) and Berry and Reinhardt's (1974b) computations, we expect that for the continental cumulus spectrum truncated at $25 \ \mu$m, the 'predominant radius' a_g (defined in the preceding section) would reach $40 \ \mu$m in a little less than 40 min; for the maritime cloud only 11 min are required. Furthermore, the time taken for a_g to grow from $40 \ \mu$m to drizzle drops of $100 \ \mu$m radius is nearly independent of the shape of the initial spectrum, depending instead primarily on the liquid water content. If this is $1 \ \mathrm{g \ m}^{-3}$, the required time is 4 min; further growth to $400 \ \mu$m takes an additional 6 min.

As we have seen (Section 14.6.1), fluid inertial effects not described by Stokes flow theory enhance the collision cross section of cloud droplets, especially when they are of comparable size. In order to check whether this wake capture phenomenon might serve to close the 'growth gap' referred to above, Ryan (1974) has integrated the SCE using a collection kernel based on the Klett and Davis (1973) collision efficiencies. For $a > 100 \ \mu$m, it was assumed that $E = 1$ (geometric cross section). Also, the usual assumption was made of inevitable coalescence following collision. For the initial spectra, Ryan chose gamma distributions in volume quite similar to Twomey's maritime and continental cumulus spectra: a liquid water content of 1 g m^{-3}, a relative dispersion in radius of 0.2, and a total droplet concentration of either 50 or 200 cm^{-3}.

Ryan has also attempted to explore the consequences of shearing flow arising from background turbulence. His model for this influence is based on the work of Manton (1974) (see Section 14.7). Ryan's modified collision kernel for this effect is expressed as

$$K'(v, u) = K(v, u) + r^* X^2 C, \qquad (15\text{-}111)$$
$$\scriptsize v>u$$

where $K(v, u)$ is the usual collection kernel without shear, X and r^* are respectively the length and width of the effective interaction region in the wake of a falling drop (Ryan sets $r^* = a_v$, the radius of the larger drop of the interacting pair), and C is the effective shear strain rate. On the basis of the experimental results of Jonas and Goldsmith (1972), Ryan makes the estimate $X^2 = 50\, a^2$ for drops of equal size, and assumes it decreases exponentially with decreasing size ratio. The strain rate C is assumed to be 5 sec^{-1}. This leads to

$$K'(v, u) = K(v, u) + 250\, a_v^3 e^{-(a_u/a_v - 1)}. \qquad (15\text{-}112)$$
$$\scriptsize v>u$$

The various results for the 100th largest drop m^{-3} m^{-1} are shown in Figure 15-10. It appears that wake capture and shearing motions can easily more than double relevant droplet growth rates (the slowest growth curve in the figure corresponds closest to the continental cloud growth rates discussed above). It is

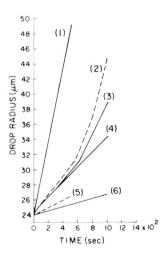

Fig. 15-10. Growth rate of the 100th largest drop m^{-3} (μm)$^{-1}$ as a function of time for various collection kernels with an initial drop concentration of 200 cm^{-3} and $w_L = 1$ g m^{-3}; (1) geometric; (2) Shear & Klett and Davis (1973); (3) Klett and Davis; (4) Davis (1972), Shafrir and Neiburger (1963); (5) Shear & Hocking and Jonas (1970), Shafrir and Neiburger, (6) Hocking and Jonas, Shafrir and Neiburger. (From Ryan, 1974; by courtesy of Amer. Meteor. Soc., and the author.)

also interesting to note that just correcting the Stokes flow collision efficiencies for slip (see Section 14.4.3) leads to a marked increase in growth rates.

These mechanisms of increased growth rates are given firmer footing in the work of de Almeida (1975) (Section 14.5). His treatment of wake capture and turbulent shear is better, at least in principle, in that the effects of velocity correlations, turbulent accelerations, and fluid inertia are superposed at the rigorous, primitive level of forces in the drops' equations of motion.

Of the five initial spectra considered by de Almeida, we shall select for our discussion the two most like those considered already. Both spectra are gamma distributions in volume, each with a liquid water content of $1 \, \text{g m}^{-3}$ as before, but somewhat broader with a relative dispersion in radius of 0.36. From Berry and Reinhardt (1974b), it appears a growth lag of less than 10 min to achieve $a_g \gtrsim 40 \, \mu\text{m}$ should accompany a change from this relative dispersion to one equal to 0.2. The continental type spectrum has $\bar{a} = 10 \, \mu\text{m}$ and $N(0) = 237 \, \text{cm}^{-3}$; the maritime spectrum has $\bar{a} = 14 \, \mu\text{m}$ and $N(0) = 87 \, \text{cm}^{-3}$.

Before presenting results, we should mention another novel refinement in de Almeida's work. In his calculation of the collision kernel from the collision efficiency he replaced the usual relative terminal velocity difference factor by the more realistic average velocity difference determined over the collision trajectory. On the average, the effect is a reduction in the relative velocity term by a factor of about 0.3; the collision kernel is thus reduced by the same amount. In the computations of spectral evolution this shows up approximately as a time lag for a given spectral shape ranging from 1 to 7 min. The effect depends noticeably on the choice of initial distribution.

The evolution of the continental spectrum is shown in Figure 15-11 for two levels of turbulence intensity: $\varepsilon = 0$ and $1 \, \text{cm}^2 \, \text{sec}^{-3}$. The $\varepsilon = 0$ (pure gravitational) case shows very little development in 30 min, in agreement with other similar model calculations. The other case clearly indicates a greatly accelerated growth rate due to turbulence. For $\varepsilon = 1 \, \text{cm}^2 \, \text{sec}^{-3}$, the predominant radius a_g reaches about $600 \, \mu\text{m}$ within 30 min; for $\varepsilon = 10 \, \text{cm}^2 \, \text{sec}^{-3}$ (not shown) the corresponding radius is $a_g = 400 \, \mu\text{m}$. This implies rainfall production in that time period, a very reasonable prediction for the given initial conditions.

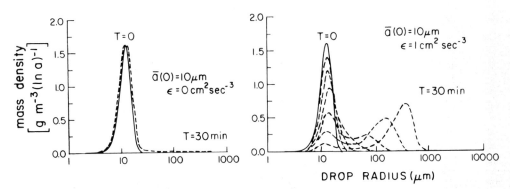

Fig. 15-11.   Evolution of a continental cloud liquid water spectrum for different levels of turbulence. (From de Almeida, 1975; by courtesy of the author.)

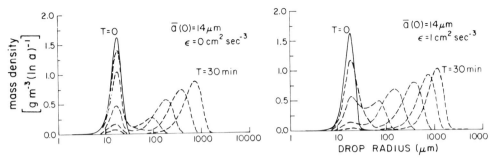

Fig. 15-12.   Evolution of a maritime cloud liquid water spectrum for different levels of turbulence.
(From de Almeida, 1975; by courtesy of the author.)

For the maritime cloud the effect of turbulence is not as great, since initially there are enough large drops for pure gravitational collection to 'take hold.' Even so, on comparing Figures 15-12a and 15-12b one can see that the degree of evolution which takes 30 min to achieve without turbulence takes about 7 min less time when $\varepsilon = 1 \text{ cm}^2 \text{ sec}^{-3}$ (5 min less for $\varepsilon = 10 \text{ cm}^2 \text{ sec}^{-3}$).

It was pointed out in Section 14.6.2 how the effect of weak turbulence appears marginal for $a \geqslant 40 \ \mu$m, but is of importance for smaller sizes, and especially for the range $15 \leqslant a \leqslant 25 \ \mu$m. In some cases the increase in collision efficiencies amounts to two orders of magnitude. As would be expected, and as is indicated by the results shown here, this enhancement of gravitational coagulation is the sort of boost needed to close the 'growth gap' in the theory of warm cloud droplet evolution.

Some discussion bearing on the somewhat surprising decrease in growth rates with increasing turbulence intensity beyond $\varepsilon = 1 \text{ cm}^2 \text{ sec}^{-3}$ is included in Section 14.6.2. For our purposes here, we merely point out that the degree of suppression is small, amounting to a time lag of about 3 min for both the continental and maritime models.

Finally, we note from Figure 15-12 that turbulence suppresses the tendency for the formation of bimodal distributions. Presumably, this is because turbulence strongly enhances the coagulation rate of small droplets among themselves and with larger drops, but has relatively little effect on larger drops. Referring to the previous section, this means the auto-conversion and accretion modes are preferentially strengthened relative to the hydrometeor self-collection mode, and that the transfer of water from S1 to S2 is faster.

15.4   Collection Growth with Condensation and Breakup

Collection growth in a real cloud is naturally subject to the influences of a host of other processes and environmental factors. This has led in the past decade to the development of a series of increasingly complex numerical models for simulating precipitation growth. It is hoped, but not yet certain, that this sequence of modeling efforts will eventually converge to something resembling a complete cloud model, including the relevant coupling of cloud dynamics and microphysics. As the numerical modeling approach of compounding several contributing processes is generally beyond the scope of this book, in this section we would like merely to discuss briefly and qualitatively some recent results of particular relevance. We refer here to models which have explored the influence of condensation and drop breakup on the collection growth process.

Kovetz and Olund (1969), Leighton and Rogers (1974), and Jonas and Mason (1974) have studied drop development by simultaneous condensation and collection. Kovetz and Olund found that even a gentle updraft of 10 cm sec^{-1} causes condensation growth sufficient to significantly promote the collection process. Leighton and Rogers considered the effect of strong updrafts (7.5 to 15 m sec^{-1}). They found that an initial drop spectrum represented by a gamma distribution with p = 3 (see (12-129)) and $\bar{v} = 2.15 \times 10^{-9}$ cm^3 ($a_f = 8 \mu$m and a relative dispersion in radius of 0.2) will evolve to produce appreciable numbers of precipitation-sized drops in just 10–15 min, assuming $w_L \approx 1$ g m^{-3}. This represents a formation time for precipitation which is roughly one-fourth that required by pure collection growth. Figure 15-13 gives an example for the development of an initial cloud drop spectrum by collision and coalescence when condensation is neglected and when it is included. In order to obtain these results Jonas and Mason (1974) incorporated the SCE into a simplified version of their condensation model (see Section 13.2.2.1). For their computations they used the collision efficiencies of Shafrir and Neiburger (1963), Hocking and Jonas (1970), and Jonas (1972). Note from Figure 15-13 that the initial drop spectrum contains about 1 ℓ^{-1} of droplets of $a > 25 \mu$m, and so is just beginning to be susceptible to modification by coalescence. Coalescence acting alone produces drops of 29 μm radius in concentration of ~ 100 m^{-3} after 4 min (and 75 μm after 10 min). The corresponding drop radii are 54 μm and 105 μm if condensation growth is also included. This enhanced growth was found to be a consequence of the fact that continued condensation causes the concentration of small drops in the 8 to 10 μm radius range to be increased some 30-fold, thus accelerating the growth of

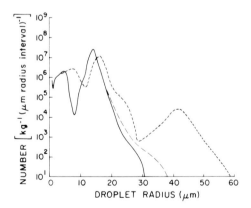

Fig. 15-13. Development of an initial drop spectrum (solid line) by collision and coalescence when condensation is neglected (long dash) and when it is included (short dash). Development in both cases is over a 4-minute period; entrainment parameter $\mu = 1.5 \times 10^{-3}$ sec^{-1} (see Section 11.5), $s_{v,w} = 0.25\%$. (From Jonas and Mason, 1974; by courtesy of *Quart. J. Roy. Meteor. Soc.*)

the largest drops through the capture of these more abundant smaller droplets. Jonas and Mason also found, in agreement with the previously discussed work of Ryan (1974) and de Almeida (1975), that the degree of spectral broadening is sensitive to the values adopted for the collision efficiencies of droplet pairs, especially to those for collector drops of $a < 40 \, \mu$m.

An unrealistic feature of solutions to the SCE is that they predict a continual flow of water mass to larger and larger drop sizes. This is not a serious flaw when we wish only to estimate the time required for the first few precipitation-sized drops to form. However, we cannot expect to obtain realistic theoretical descriptions of quasi-steady state rain spectra, such as exponential distributions of the Marshall-Palmer type (Equation (2-11)), unless we take into account the spectral shaping due to drop breakup as well as coalescence.

The fragmentation of large drops may be induced by collisions ('collisional breakup' – see Section 14.8.2), or by hydrodynamic instabilities ('spontaneous breakup' – see Section 10.3.4). One might expect the latter mode to be relatively unimportant on the basis of the observation that drops large enough to enter the realm of spontaneous breakup (diameter $\geqslant 5$ mm) are very rare. Indeed, model calculations by Young (1975), in which both breakup modes were included, showed that spontaneous breakup is negligible. Furthermore, the spectral shape produced by a balance between spontaneous breakup and coalescence is unrealistically flat, i.e., there is too great a bias toward the larger drop sizes (Srivastava, 1971).

It is a simple matter to demonstrate that there is ample opportunity for the occurrence of collisional breakup in rain. We can show this by estimating the mean free path between collisions for raindrops. Consider, for example, raindrops of radius a_1 in concentration N_1 falling through a region containing drops of radius a_2 ($a_2 < a_1$) in concentration N_2. By elementary arguments, if the N_1-drops move a distance dx relative to the N_2-drops, the change in their concentration will be given approximately by $dN_1 = -N_1[E\pi(a_1 + a_2)^2 N_2]$ dx. But for a small time interval dt we have $dx = (V_{\infty,1} - V_{\infty,2})$ dt, assuming the relative velocity of approach is given by the difference in terminal velocities. Hence the mean free time between collisions is $\tau = [E\pi(a_1 + a_2)^2 N_2(V_{\infty,1} - V_{\infty,2})]^{-1}$, and the mean distance traveled by the N_1-drops between collisions is $L = V_{\infty,1}\tau$ or

$$L = \frac{V_{\infty,1}}{E\pi(a_1 + a_2)^2 N_2(V_{\infty,1} - V_{\infty,2})}. \tag{15-113}$$

List *et al.* (1971) have used (15-113) to estimate that for rain characterized by a Marshall-Palmer distribution, a drop with $a_1 = 1.8$ mm will have $L = 750$ m for collisions with drops equal to or greater than 0.5 mm in radius, and for medium rain with a rain rate of $R = 5$ mm hr^{-1}. The value of L becomes 180 m for heavy rain ($R = 50$ mm hr^{-1}) and 65 m for intense rain ($R = 500$ mm hr^{-1}). These estimates imply a rapid local equilibration between coalescence and collisional breakup may often occur in rain.

Recent computations by List and Gillespie (1976) of the effects of stochastic collection and collisional breakup support the conclusion reached above. They modeled collisional breakup in accordance with the data of McTaggard-Cowan and List (1975) (see also Section 14.8.2), and found that drops with diameters greater than 2 to 3 mm, falling through a typical rain spectrum of smaller drops, suffer breakup in fairly short times (1–5 min in rainfalls of 100 mm hr^{-1}) and therefore vanish from the drop population after a relatively short fall distance. They further found that if the initial spectrum is of the Marshall-Palmer type, the drop spectrum produced by collision, coalescence and breakup will also have an exponential form. However, the final total drop concentration is larger than the initial Marshall-Palmer value, owing to the breakup of large drops. From their computations, List and Gillespie concluded that in precipitation from 'warm' clouds, where no ice particles are present, the raindrop size distribution is likely to be limited to drops of diameters less than 2 to 3 mm, as a result of collisional breakup. On the other hand, in precipitation from 'cold' clouds, which do contain ice particles, raindrops larger than 3 mm in diameter may be present if the melting level in the atmosphere is relatively close to the ground such that the ice particles have sufficient time to melt, but insufficient time to change their size spectrum by collisional breakup.

Numerical models incorporating the features of simultaneous condensation, collection, and breakup have also been developed (e.g., Brazier-Smith *et al.* (1973), Young (1975)). Brazier-Smith *et al.* studied rainfall development in a cloud volume where water is released by condensation at the constant rate J. The distribution of satellite drops from collisional breakup was specified in accordance with the data of Brazier-Smith *et al.* (1972) (see Section 14.8.2). The calculations indicated that the cloud and rainwater contents, and the radar reflectivity, are dependent mainly on J, the distribution of satellite drops and the coalescence efficiency E_{co} being relatively unimportant. On the other hand, the drop size distribution depends strongly upon E_{co} and the production of satellites. This model tends to produce a bimodal spectrum, in conflict with the results of List and Gillespie (1976). The condensation, coalescence, and collisional breakup model of Young (1975) produces a steady state spectrum which, unlike the results of Brazier-Smith *et al.* (1973), is quite insensitive to the distribution of satellite drops. Furthermore, the spectral shape is roughly exponential, not bimodal, and is in fair agreement with the Marshall-Palmer distribution.

CHAPTER 16

MICROPHYSICS OF ICE PARTICLE-DROP INTERACTIONS

In Chapter 13 we discussed the growth of ice particles by vapor diffusion from drops to ice crystals, and in Chapter 14 we described ice particle growth by the accretion of supercooled drops, and the formation of snowflakes by the process of ice crystal aggregation. In this chapter we shall look closer at the microphysics of ice particle-drop interactions, and also at the physics of melting and freezing of individual cloud particles.

16.1 Growth Mode and Structure of Rimed Ice Particles, Graupel, and Hailstones

Ice particles which grow by collisions with supercooled drops have a surface temperature which is higher than that of the air surrounding them, owing to the release of latent heat during the freezing process. This heating-up of the ice particle is counteracted and, at steady state, just balanced by the transfer of heat to the environmental air by conduction, and by evaporation if the air is water vapor subsaturated. As long as the latent heat of freezing is dissipated from the growing ice particle in such a way that its temperature remains below 0 °C, all accreted cloud water must freeze on the ice particle. The particle is then considered to grow in the so-called *dry growth regime* (Ludlam, 1958). With increasing liquid water content of the cloud, increasing drop size, and increasing frequency of collision between drops and the ice particle, the temperature of the growing particle gradually rises. This rise in temperature generally comes to an end when the surface temperature of the ice particle approaches 0 °C. Under such growth conditions not all accreted water is converted to ice, the amount of ice formed being determined by the rate at which heat is dissipated from the particle. The ice particle is now considered to grow in the *wet growth regime* (Ludlam, 1958). The critical conditions for which all the accreted water freezes on the ice particle and acquires a temperature of 0 °C in the solid phase is known as the *Schumann-Ludlam Limit* (SLL); the SLL thus marks the boundary between the two growth regimes.

Schumann (1938) and Ludlam (1958) studied the thermodynamics of the wet growth regime under the assumptions that the growing ice particle would remain solid and shed all excess water, acquired as a liquid film over its surface. However, the wind tunnel studies of List (1959, 1960) and Macklin (1961) demonstrated that little or no shedding occurs in the wet growth regime. Furthermore, the assumption of impervious ice particles was found to be incorrect; rather, they were found to consist of a dense ice framework whose capillaries are filled with water and air bubbles, termed *spongy ice* by List (1965). Such spongy ice may either form directly if the heat exchange between the growing ice deposit and the surroundings is insufficient to freeze all accreted water (the latter being retained in a mesh of ice dendrites), or by intake of unfrozen water into porous ice formed beforehand during growth of the ice particle in the dry growth regime (List, 1965).

Microphotographs of rimed ice cylinders (Macklin and Payne, 1968), and of rimed ice crystals and graupel freely floating in a wind tunnel (Pflaum *et al.*, 1978) show that at low temperatures of the ice deposit and for relatively small drop impact velocities and sizes, supercooled drops tend to freeze rapidly as individual ice spheres, forming loosely woven chains with densities as low as 0.1 to $0.3 \, \text{g cm}^{-3}$. At higher deposit temperatures and for larger drop impact velocities and sizes, the drops become increasingly distorted on impact and tend to pack more closely.

The spreading of supercooled water drops impacting on ice surfaces has been studied by Brownscombe and Hallett (1967) and Macklin and Payne (1969). Macklin and Payne considered air temperatures between -11 and $-22\,°C$, surface temperatures of the ice deposit between -3 and $-20\,°C$, and cloud drops of radii between 17 and 511 μm and impact speeds between 5 and 30 m sec^{-1}. They found that the spreading factor, defined as the final maximum drop radius to the initial radius, varied between 1.3 and 6, depending primarily on impact speed and deposit temperature. Only at the lowest impact speeds and lowest deposit temperatures studied did the drops freeze as hemispheres or truncated spheres. With an increase in either of these parameters, the drops became increasingly flattened. They concluded that the final drop shape is a function of several compensating factors. Three of these are: (1) the kinetic energy which the drop possesses at impact and which acts to distort it, (2) the surface tension of the drop which acts to retain its spherical shape, and (3) the rate at which the drop freezes and thus terminates deformation.

The density ρ_R of rime deposits has been studied by Macklin (1962) and Macklin and Payne (1968). Their observations showed that ρ_R is related to the surface temperature T_s of the ice substrate, and to the radius a and impact velocity v_r of the cloud drops. For deposit temperatures ranging between -5 and $-20\,°C$, impact speeds between 2 and 12 m sec^{-1}, drop radii between 11 and 32 μm, and for cloud liquid water contents between 1 and 7 g m^{-3}, Macklin (1962) found the relation

$$\rho_R = 0.11 \left(-\frac{a v_r}{T_s} \right)^{0.76}, \tag{16-1}$$

with the units T_s (°C), v_r (m sec^{-1}), a (μm), and ρ_R (g cm^{-3}). Equation (16-1) predicts, for example, that $\rho_R = 0.19 \, \text{g cm}^{-3}$ for $a = 20 \, \mu\text{m}$, $v_r = 2$ m sec^{-1}, and $T_s = -20\,°C$; similarly $\rho_R = 0.86 \, \text{g cm}^{-3}$ for $a = 30 \, \mu\text{m}$, $v_r = 5$ m sec^{-1}, and $T_s = -10\,°C$. (For comparison we may note that the density of a structure composed of regularly packed spheres of individual density $0.9 \, \text{g cm}^{-3}$ is $0.67 \, \text{g cm}^{-3}$, while the corresponding figure for randomly packed spheres is 0.57 g cm^{-3}.) If such low density rime becomes soaked with water, its density obviously increases and assumes values between the bulk density of ice and that of water (see Equations (4-71) to (4-73)).

In addition to causing a low bulk density, the beaded, chain-like growth features of rime deposited in the dry growth regime also have been found to significantly reduce the efficiency with which a rimed ice particle collects cloud drops relative to the efficiency of a smooth sphere of the same Reynolds number. This has been shown by recent wind tunnel studies of Pflaum and Pruppacher (1976) with ice particles freely suspended in an air stream. Their experiments also indicated that the rough surface of a riming frozen drop promotes capture of small drops near the rear stagnation point, leading to the buildup of a cone-like cap. This cap, in turn, tends to induce in the riming ice particle an oscillatory motion that further promotes the cone shape through preferential drop capture now on the lower side of the ice particle. Ice plates were found to capture drops preferentially on their lower side and, in particular, along their rim. However, oscillations eventually caused the rimed plate to flip over and to continue growing on what had previously been the upper side. Continued riming caused the ice particle to 'round off,' and to eventually develop into a conical particle in the same manner as outlined above for a frozen drop.

In Section 2.2.2 we described in some detail the characteristic layer structure of rimed ice crystals, graupel, and hailstones, due to a varying concentration and size of trapped air bubbles. From their studies of hailstones produced in a wind tunnel, List and Agnew (1973) concluded that it is the cloud liquid content which controls both the air bubble concentration and size. On the other hand, Carras and Macklin (1975) inferred from their experiments that a more fundamental determinant is the rate at which the collected water freezes. They found that in both the dry and wet growth regimes the bubble concentration increases with increasing freezing rate, which is governed in the dry growth regime by the temperature of the ice deposit and by the size and impact speed of the accreted droplets and, in wet growth regime, by the rate of heat transfer by forced convection away from the riming ice particle. This finding is understandable if we consider that for smaller freezing rates more dissolved air can escape by diffusion and more bubbles can migrate to the surface of the accreting ice particle. Therefore, small freezing rates lead to relatively clear ice, and large freezing rates to relatively opaque ice. Thus, Carras and Macklin typically found bubble concentrations of 10^5 to 10^6 cm^{-3} in the wet growth regime with small freezing rates, and 10^6 to 10^8 cm^{-3} in the dry growth regime with relatively large freezing rates.

In a similar way we can understand how the freezing rate can control the bubble size, by limiting its growth time; thus, the smaller the freezing rate, the larger the bubble size. In the dry growth regime this trend is counteracted somewhat in that, at increasingly warm deposit temperatures, given the relatively slow growth rate and the considerable spreading of the drops on the accreting ice surface, an increasingly larger amount of air is allowed to diffuse away and become unavailable for bubble growth. Typically, Carras and Macklin found bubble sizes of 10 to 50 μm radius in the wet growth regime and 1 to 4 μm radius in the dry growth regime.

Thin sections of rimed ice crystals, graupel, and hailstones viewed in polarized light reveal alternating layers of large concentrations of small ice crystallites and of smaller concentrations of larger ice crystallites. Wind tunnel studies of Aufdermauer et al. (1963), Levi and Aufdermauer (1970), Levi et al. (1970), Macklin and Rye (1974), and Rye and Macklin (1975) showed that in the dry growth regime the mean length of ice crystallites generally decreases from ~8000 to ~250 μm, and the mean width from ~1000 to ~200 μm, as the air temperature (i.e., the temperature of the accreted drops) decreases from −5 to −30 °C. At air temperatures colder than −15 °C the crystallite size also depends on the temperature of the ice deposit such that for each air temperature a critical deposit temperature exists at which the crystallite size rather abruptly decreases to below 50 μm. Drop size and impact velocity of the accreted drops were not found to affect the ice crystallite size. Unfortunately, no similar studies have been carried out for the wet growth regime.

The above mentioned wind tunnel studies also demonstrated that ice crystallites exhibit preferred orientations with respect to their c-axis and the radial growth direction of the accreting ice particle. Thus, Levi et al. (1970) and Levi and Aufdermauer (1970) showed that in the dry growth regime the crystallographic c-axis of individual crystallites tends to be oriented normal to the growing ice surface, i.e., parallel to the radial growth direction of the hailstone, while in the wet growth regime the c-axis tends to be parallel to the growing ice surface, i.e., normal to the radial growth direction of the accreting ice particle. In terms of the frequency distribution of orientation angles α (see Figure 16-1), a distinct maximum in the dry growth regime was thus found near $\alpha = 0°$ for air temperatures warmer than −10 °C, shifting to ~45° as the air temperature decreased below −20 °C. In the wet growth regime a distinct maximum in the frequency distribution was observed near $\alpha = 90°$ for air temperatures warmer than −10 °C, shifting to ~45° as the air temperature decreased to below −20 °C. These results were essentially confirmed by Macklin and Rye (1974), although their frequency distributions were somewhat more complicated than those found by Levi and co-workers. In particular, two (instead of one) frequency maxima were observed for the dry growth regime, one near $\alpha = 0°$, relatively in-

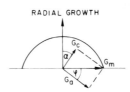

RADIAL GROWTH

Fig. 16-1. Schematic relation between non-rational dendrite growth velocity G_m, its component growth velocity G_a along the basal plane, and its component growth velocity G_c perpendicular to the basal plane of ice. Also shown are the orientation angle α between the radial growth direction of ice and G_c, and the 'splitting angle' φ between G_m and G_a.

dependent of air temperature, and one near $\alpha = 22°$, shifting to higher angles with decreasing air temperature.

It is obvious that information concerning the structural features of ice particles, such as discussed above, is very useful for purposes of interpreting the growth history of natural graupel and hailstones. Attempts to determine the growth history of fallen hailstones in this manner have been made by Knight and Knight (1968a, b), Levi et al. (1970), Macklin et al. (1970), and by Macklin et al. (1976).

In the following three sections we shall learn about the physical bases for the development of the structural features of rimed ice crystals, graupel, and hailstones, discussed in this section. In Section 16.2 we shall show that the freezing of a quantity of supercooled water is essentially a two-stage process. The first stage, although considerably shorter than the second, plays the decisive role in determining the structural features of ice particles which grow by riming. In Sections 16.3 and 16.4 we shall discuss the structure and rate of propagation of ice grown in supercooled water during this first stage of freezing.

16.2 Freezing Time of Water Drops

Observations show that following the collision of an ice particle with a supercooled drop, solidification of the supercooled water proceeds in two major stages. The first stage, controlled by the intrinsic rate of ice propagation in water, is completed within a relatively short time interval. During this stage only a very small portion of the liberated latent heat of freezing is transferred to the drop environment, most of the heat being absorbed instead by the water of the liquid portion of the drop, and warming the latter quickly towards 0 °C. From this one may readily show that for a drop supercooling of ΔT, only a fraction of about $\Delta T/80$ of the drop becomes converted to ice in stage one. Thus, if m_i and m_w respectively denote the masses of ice and water in the drop at the end of stage one, the heat balance for the drop is, assuming no heat loss to the environment,

$$m_i L_m = (m_w c_w + m_i c_i)\Delta T, \tag{16-2}$$

where c_w and c_i are the specific heats of water and ice, and L_m is the latent heat of melting. First we may note that for no realizable supercooling will be the drop freeze entirely during this adiabatic stage; in fact, we see from (16-2) with $m_w = 0$ that the supercooling $\Delta T'$ which would be required for this to occur is $\Delta T' = L_m/c_i \approx 160°C$, with $L_m \approx 80 \, \text{cal g}^{-1}$ and $c_i \approx 0.5 \, \text{cal g}^{-1}°C^{-1}$. Hence we have $m_i c_i \ll m_w c_w$, and so from (16-2),

$$\frac{m_i}{m_w} \approx \frac{c_w \Delta T}{L_m} \approx \frac{\Delta T}{80}. \tag{16-3}$$

The initial stage of freezing is followed by a second stage during which the remainder of the water is frozen. The freezing rate during this stage is one to several orders of magnitude smaller than during the first stage and is controlled partly by the rate at which heat is conducted into the underlying ice surface, and partly by the rate at which heat is dissipated by forced convection into the environment. Although it is clear that the two freezing stages overlap, the total time it takes to freeze a drop may be regarded approximately as the sum of the initial freezing time t_1 and the subsequent freezing time t_2.

Macklin and Payne (1967, 1968) estimated t_1 and t_2 by assuming that the ice substrate surrounded by air is a sphere of radius a and has an initial temperature T_d. In their model every impinging drop is assumed to spread uniformly over the ice substrate into a thin layer of water of thickness δa. The latent heat of freezing is considered to flow in a radial direction only. A crude estimate of the initial freezing time t_1 can then be made from

$$t_1 \approx \delta a/G, \tag{16-4}$$

where G is the growth rate of ice dendrites in supercooled water. (Note we have not used just a fraction (m_i/m_w) of the total thickness δa in this estimate; this is because the layer δa does not become solid ice in time t_1, but only a mixture of ice crystals and water.) Using observed values for G for growth parallel to the basal plane of ice (see Section 16.4), and assuming that growth proceeds perpendicular to the water layer, Macklin and Payne determined that for $T = -20\,^\circ\text{C}$, $t_1 \approx 1 \times 10^{-6}$, 1×10^{-5}, 7.5×10^{-5}, 2×10^{-4} sec, for $\delta a = 0.2$, 2, 15, 40 μm, respectively.

To determine t_2 Macklin and Payne (1967, 1968) considered first the heat flow into the spherical ice substrate of radius a. Assuming that during t_1 the initial temperature T_d of the ice substrate is raised instantaneously to $T_0 = 273\,^\circ\text{K}$ (0 °C), and using expressions derived by Carslaw and Jaeger (1959), Macklin and Payne found the amount of heat conducted into the sphere $(T_d < T_0)$ as a function of time to be given by

$$q(t) \approx \frac{4\,\pi}{3} a^3 \rho_i c_i \left[\frac{6\,(T_0 - T_d)(\kappa_i t)^{1/2}}{a \pi^{1/2}} - \frac{3\,(T_0 - T_d)\kappa_i t}{a^2} \right], \tag{16-5}$$

where κ_i is the thermal diffusivity of ice, which was assumed constant in order to obtain (16-5). (For the purpose of the calculation this is a valid assumption, though it should be noted that $\kappa_i = k_i/\rho_i c_i$ actually depends on temperature through the temperature dependence of ρ_i and c_i (recall Chapter 4), and also of the thermal conductivity k_i, which for the range $-50 \leqslant T \leqslant 0\,^\circ\text{C}$ obeys the empirical relation

$$k_i = (537.4 - 1.48\,T + 0.028\,T^2) \times 10^{-5}, \tag{16-6}$$

with k_i in cal cm^{-1} sec^{-1} °C^{-1}. This expression for k_i fits the experimental data of Ratcliff (1962) to an accuracy of $\pm 0.5 \times 10^{-5}$ cal cm^{-1} sec^{-1} °C^{-1}.) The amount of heat $\hat{q}_{in}$ conducted per unit area into the sphere during the freezing time t_2 is then

$$\hat{q}_{in} = \rho_i c_i (T_0 - T_d) \left[2 \left(\frac{\kappa_i t_2}{\pi} \right)^{1/2} - \frac{\kappa_i t_2}{a} \right]. \tag{16-7}$$

Secondly, Macklin and Payne considered the amount of heat transferred by forced convection and evaporation to the environmental air. For a smooth spherical ice substrate of surface temperature T_0, the average amount of heat $\hat{q}_{out}$ transferred per unit area during time t_2 by these two mechanisms is (from

(13-64))

$$\hat{q}_{out} = t_2[k_a(T_0 - T_\infty)\bar{f}_h + L_e D_v(\rho_{v,a} - \rho_{v,\infty})\bar{f}_v]/a, \qquad (16\text{-}8)$$

where $\bar{f}_h$ and $\bar{f}_v$ are the mean ventilation coefficients for heat and vapor transport in air (see Section 13.2.3), D_v is the diffusivity of water vapor in air, k_a is the heat conductivity of air, T_∞, $\rho_{v,\infty}$ are the temperature and water vapor density in air far away from the ice sphere, and T_a $(=T_0)$, $\rho_{v,a}$ are the temperature and water vapor density of air at the surface of the water layer. Considering that the amount of heat to be removed per unit area is $\rho_w \delta a [L_m - c_w(T_0 - T_\infty)]$, where ρ_w is the density of water, t_2 is found as the solution of the equation

$$\hat{q}_{out} + \hat{q}_{in} = \rho_w \delta a [L_m - c_w(T_0 - T_\infty)], \qquad (16\text{-}9)$$

where $\hat{q}_{out}$ and $\hat{q}_{in}$ are given by (16-7) and (16-8). For small values of t_2, i.e., at relatively low deposit temperatures, $\hat{q}_{out} \ll \hat{q}_{in}$ and $(\kappa_i t_2/a) \ll (\kappa_i t_2/\pi)^{1/2}$, so that approximately

$$t_2 \approx \frac{\pi \rho_w^2(\delta a)^2[L_m - c_w(T_0 - T_\infty)]^2}{4\, k_i \rho_i c_i(T_0 - T_d)^2}. \qquad (16\text{-}10)$$

From (16-10) Macklin and Payne determined that for $T_\infty = -20\,°C$ and $T_d = -10\,°C$, $t_2 \approx 1 \times 10^{-5}$, 1×10^{-3}, 5×10^{-2} sec, for $\delta a = 0.2, 2, 15\,\mu m$, respectively. Similarly, for $T_\infty = -10\,°C$ and $\delta a = 2\,\mu m$, $t_2 \approx 1$ and 10^{-3} sec, for $T_d = 0\,°C$ and $-10\,°C$, respectively. These results demonstrate that t_2 is strongly dependent on the thickness δa of the water layer and on the deposit temperature T_d, and that it is considerably longer than t_1, except for very thin water layers.

Observations show that the model used above is highly idealized since drops colliding with the ice surface do not spread such as to form a liquid layer over the entire underlying ice particle, except perhaps at large accretion rates when the surface temperature of the ice surface is raised to near $0\,°C$. At other deposit temperatures the observations described in Section 16.1 show that drops freeze onto the ice deposit in a more or less 'chain-like' manner. In order to estimate the freezing time of stage two for more realistic conditions, Macklin and Payne assumed that the drop and substrate could be considered as a half hemisphere attached to the end of an ice rod. For this model the total amount of heat conducted from the ventilated hemisphere of area $2\,\pi a^2$ and temperature T_0 to air of T_∞ is then approximately

$$q_{out} = 2\,\pi a\, t_2[k_a(T_0 - T_\infty)\bar{f}_h + L_e D_v(\rho_{v,a} - \rho_{v,\infty})\bar{f}_v], \qquad (16\text{-}11)$$

and the total amount of heat conducted into the rod is (from Carslaw and Jaeger (1959))

$$q_{in} = (T_0 - T_d)k_i a^2 \left(\frac{\pi t_2}{\kappa_i}\right)^{1/2}. \qquad (16\text{-}12)$$

Considering that the heat to be removed is $(2\,\pi/3)a^3\rho_w[L_m - c_w(T_0 - T_\infty)]$, t_2 is found as the solution of the equation

$$q_{out} + q_{in} = \frac{2\,\pi}{3} a^3\rho_w[L_m - c_w(T_0 - T_\infty)], \qquad (16\text{-}13)$$

where q_{out} and q_{in} are given by (16-11) and (16-12). For $T_\infty = -20\,°C$ and $a = 10\,\mu m$, Macklin and Payne found $t_2 \approx 5 \times 10^{-3}\,\text{sec}$ and $5 \times 10^{-2}\,\text{sec}$, for $T_d = -10\,°C$ and $-2\,°C$, respectively. These results suggest again that the subsequent freezing time t_2 is considerably longer than the initial freezing time t_1. Macklin and Payne also showed that, for the same values of T_∞, a, and T_d, the ratio of t_2, computed by assuming no loss of heat to the ice surface, to t_2 computed by assuming no loss of heat to the air, is 8.2 and 1.4, respectively. This indicates that the rate of heat conduction into the ice substrate plays a much more important role in the drop freezing process than does heat flow into the air, substantiating the suggestions of Brownscombe and Hallett (1967), and of Macklin and Payne (1967).

It is instructive to compare the freezing times computed above for drops in contact with an ice surface with those for drops freely falling in air. Due to the relative ineffectiveness of air in removing the latent heat released during freezing, we expect that these latter freezing times will be considerably longer. This expectation has been verified by Dye and Hobbs (1968), Johnson and Hallett (1968), and Murray and List (1972) in independent treatments. The essentials of these are summarized below.

Consider a water drop supercooled to the temperature T_∞ of the environment. When the drop is nucleated at time $t = 0$, ice crystals grow rapidly through the drop, completing the initial growth in a relatively short time, to be neglected here, while heating the drop to a temperature close to $0\,°C$. At this stage the drop consists of a mixture of water and ice, with a fraction $(1 - c_w \Delta T / L_m)$ of the drop volume yet to freeze. Subsequent freezing occurs at a much slower rate through the transfer of heat by conduction and evaporation to the environment. Assuming that the heat transfer is spherically symmetric, a spherical shell of ice will form with internal radius r at time t. According to Hallett (1964), the ice shell thickens at the small rate of $\approx 10^{-2}\,\text{cm sec}^{-1}$, and the local supercooling at the ice-water interface is about $0.2\,°C$. For the present computations we shall neglect this supercooling and assume that the ice-water interface is at $T_0 = 273\,°K$ $(0\,°C)$. Then for a quasi steady state the rate of release of latent heat of freezing is equal to the rate of heat conduction through the ice shell, which in turn is equal to the rate of heat loss by evaporation and conduction to the environmental air. These balance conditions are expressed by the relations

$$4\,\pi\rho_w L_m r^2 \frac{dr}{dt}\left(1 - \frac{c_w \Delta T}{L_m}\right) = \frac{4\,\pi k_i a r (T_0 - T_a(r))}{a - r}, \qquad (16\text{-}14)$$

$$\frac{4\,\pi k_i a r (T_0 - T_a(r))}{a - r} = 4\,\pi a k_a (T_a(r) - T_\infty)\bar{f}_h + 4\,\pi a L_s D_v (\rho_{v,a} - \rho_{v,\infty})\bar{f}_v, \qquad (16\text{-}15)$$

where from Section 13.2.3, we have

$$\rho_{v,a} - \rho_{v,\infty} = \rho_{v,\text{sat}}(T_a) - \rho_{v,\infty} = (1 - \phi_v)\rho_{v,\text{sat}}(T_\infty) + (T_a(r) - T_\infty)\left(\overline{\frac{d\rho_v}{dT}}\right)_{\text{sat,i}}, \qquad (16\text{-}16)$$

and where $T_a(r)$ is the surface temperature of the freezing drop, L_s is the latent heat of sublimation, ϕ_v is the fractional relative humidity of the air, and $(d\rho_v/dT)_{sat,i}$ is the mean slope of the ice saturation vapor density curve over the interval from $T_a(r)$ to T_∞. From (16-14) we find the freezing time t_f may be expressed as

$$t_f = \frac{\rho_w L_n (1 - c_w \Delta T/L_m)}{a k_i} \int\limits_a^0 \frac{r(a-r)}{T_0 - T_a(r)}\, dr, \tag{16-17}$$

where $T_a(r)$ is given by (16-15).

We may obtain an approximate solution by assuming for simplicity that the air has a relative humidity of 100%, i.e., $\phi_v = 1$, and that $\bar{f}_h \approx \bar{f}_v = \bar{f}$ (see Section 12.7.2); then (16-15) reduces to

$$\frac{k_i r(T_0 - T_a(r))}{a - r} = \bar{f}(T_a(r) - T_\infty)\left[k_a + L_s D_v \left(\frac{\overline{d\rho_v}}{dT}\right)_{sat,i}\right]. \tag{16-18}$$

Eliminating $T_a(r)$ between (16-18) and (16-14), we find

$$3\, t_0 \frac{dy}{dt} = -\frac{1}{(1-m)y^2 + my}, \tag{16-19}$$

where $y = r/a$, and

$$t_0 = \frac{\rho_w L_m a^2 [1 - (T_0 - T_\infty)c_w/L_m]}{3\,\bar{f}(T_0 - T_\infty)\left[k_a + L_s D_v \left(\dfrac{\overline{d\rho_v}}{dT}\right)_{sat,i}\right]} \tag{16-20}$$

$$m = \frac{\bar{f}\left[k_a + L_s D_v \left(\dfrac{\overline{d\rho_v}}{dT}\right)_{sat,i}\right]}{k_i}. \tag{16-21}$$

Integrating (16-19) together with (16-20) and (16-21) from $t = 0$, $y = r/a = 1$ to $t = t_f$, $y = 0$ ($r = 0$), we find

$$t_f = t_0\left(1 + \frac{m}{2}\right). \tag{16-22}$$

Since for water drops in air $m/2 \ll 1$,

$$t_f \approx t_0, \tag{16-23}$$

with t_0 given by (16-20). For $(T_0 - T_\infty) = 10\,°C$, $L_m = 74.5\ cal\ g^{-1}$, $L_s = 677.5\ cal\ g^{-1}$, $k_a = 5.63 \times 10^{-5}\ cal\ cm^{-1}\ sec^{-1}\,°C^{-1}$, $D_v = 0.196\ cm^2\ sec^{-1}$, and $(d\rho_v/dT)_{sat,i} = 1.8 \times 10^{-7}\ g\ cm^{-3}$, we find for a drop of 500 μm radius ($\bar{f} = 5$) a freezing time of $t_f \approx 13$ sec. Similarly, for a drop of 0.2 mm equivalent radius ($\bar{f} = 14$), we obtain a freezing time of $t_f \approx 80$ sec. Both of these estimates are in good agreement with values determined by Murray and List (1972) from laboratory observations.

16.3 Structure and Growth Mode of Ice in Supercooled Water

The structure and growth mode of ice in supercooled water has been studied experimentally by Kumai and Itagaki (1953), Lindenmeyer (1959), Hallett (1960, 1964), Macklin and Ryan (1965, 1966), Knight (1966), and Pruppacher (1967a, b). Most experiments were carried out with relatively large supercooled water bodies nucleated by single ice crystals which had a temperature close to 0 °C. It was found that at supercoolings less than 0.9 °C ice crystals, nucleated with their c-axis normal to the surface of the water, develop as thin, almost circular disks. On the other hand, crystals with their c-axis parallel to the water surface grow as long surface needles. Each of these consists of a dendritic portion which is co-planar with the seed crystal and grows into the water, and of a rib-like portion which grows along the water surface boundary. At supercoolings between about 0.9 and 2.5 °C ice crystals grow as plane stellar dendrites or dendritic sheets co-planar with the seed crystal, i.e., parallel to the seed crystal's basal plane. At supercoolings larger than about 2.5 °C ice crystals no longer grow co-planar with the seed crystal, but rather split into two symmetrical, hollow, hexagonal, pyramidal segments joined together at their apices (Figure 16-2 and Plate 9). With increasing supercooling the angle between these segments increases. According to Macklin and Ryan (1966), the angle φ between a primary growth segment and the basal plane of ice (which is one half of the angle between the two primary growth segments) increases from $\varphi \approx 4°$ at $\Delta T = 2\,°C$ to $\varphi \approx 20°$ at $\Delta T = 7\,°C$. At supercoolings larger than about 5.5 °C secondary and higher order splitting takes place on the major growth planes, causing the formation of complex, three-dimensional structures favored by the presence of salts dissolved in water (Pruppacher, 1967b). Primary, secondary, and higher order splitting leads to non-rational growth, i.e., growth in directions which cannot be described by rational crystallographic indices (see Chapter 3). Lindenmeyer (1959), Hallett (1964), and Macklin and Ryan (1965, 1966) suggested that this non-rational growth of ice in supercooled water can be explained, as illustrated in Figure 16-3, on the basis of a step growth mechanism analogous to the explanation of the hopper structure of ice crystals grown from the vapor (see Section 5.7.3). Here two growth components, G_a and G_c, are involved. The component G_a is the growth velocity of ice parallel to the crystallographic a-axis of ice (i.e., the growth velocity of the crystallographic prism plane of ice), while G_c is the growth velocity of ice parallel to the crystallographic c-axis of ice (i.e., the growth velocity of the basal plane of ice). Observations discussed in Section 16.4 show that $G_a > G_c$ at all water supercoolings. Note from Figure 16-3 that the

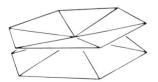

Fig. 16-2. Schematic representation of ice structure formed in water supercooled by about 5 °C. (From Macklin and Ryan, 1965; by courtesy of Amer. Meteor. Soc., and the authors.)

Plate 9. Ice structure formed in supercooled water at −5.2 °C. (From Macklin and Ryan, 1965; by courtesy of Amer. Meteor. Soc., and the authors.)

angle of splitting is determined by the ratio of the height of the growth step to the distance between the steps, i.e., by the ratio of the growth velocities parallel and perpendicular to the c-axis. Since observations show that G_c increases more rapidly with supercooling than G_a, the angle φ between a primary growth segment and the basal plane must increase with increasing supercooling – as has been observed. Another implication of such a mechanism is that one would expect the non-rational ice structures to be single crystals. This expectation was confirmed by Lindenmeyer (1959), who used X-rays to analyze the ice structures grown in water, and by Macklin and Ryan (1965, 1966) who noted that all the secondary growth features are aligned parallel to the primary growth features.

Additional support for this conclusion was given by Hallett (1963, 1964), who found that when millimeter size water drops were frozen by contact with an ice single crystal of a temperature near 0 °C, they developed into single crystals with

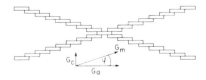

Fig. 16-3. Schematic representation of the stepped growth mechanism considered to be responsible for explaining the ice structures formed in supercooled water. (From Macklin and Ryan, 1965, with changes.)

the crystal orientation of the substrate, irrespective of the temperature of the drop. This occurred also when the temperatures of both the drop and ice surface were warmer than −5 °C. If the temperatures of the drop and the surface were below −5 °C the drop froze into polycrystalline ice, the polycrystallinity apparently being due to the existence of more than one point of nucleation.

The crystalline structure of water drops nucleated while freely suspended in the vertical air stream of a wind tunnel was studied by Pitter and Pruppacher (1973). They found that whether or not a drop became a polycrystalline ice particle depended critically on the size and supercooling of the drop before freezing. The critical size was found to be related to the drop's supercooling by

$$a_c = \left(\frac{\beta}{\Delta T}\right)^3, \quad \text{with } \beta = Ak^B, \tag{16-24}$$

where $A = 23$ and $B = -1/8$ are constants, k $(\text{cal cm}^{-1}\text{sec}^{-1}\text{°C}^{-1})$ is the heat conductivity of the medium in contact with the drop, and a_c is in μm. This relation together with observational evidence is plotted in Figure 16-4 for drops frozen while freely falling in air, and for drops frozen by contact with an ice surface. Note from this figure that at a drop supercooling of as large as 20 °C drops up to 80 μm radius become single crystals on freezing. If we now recall Figure 2-1, which suggests that most atmospheric clouds glaciate at temperatures warmer than −20 °C, and Figure 2-4b, which suggests that typically cloud drops have radii less than 80 μm, we find that under atmospheric conditions most cloud drops freeze as single crystals, each of which can grow into a hexagonal shaped ice crystal after sufficient time for growth by vapor diffusion. Hexagonal shaped ice crystals may therefore be the result of both vapor deposition on ice forming nuclei and of drop freezing, in good agreement with the observational evidence cited in Chapter 2.

Experiments show that the polycrystallinity of ice grown in supercooled water is a result of c-axis reorientation during the epitaxial overgrowth of ice on ice in an environment of supercooled water. According to Hallett (1964) and Rye and Macklin (1975) a most common reorientation is one in which the c-axis of the

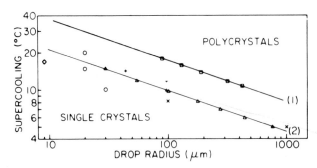

Fig. 16-4. Dependence of the polycrystallinity of drops frozen in air and on the surface of a single ice crystal on drop size and on supercooling; (1) Equation (16-24) for k = k$_{air}$, (2) Equation (16-24) for k = k$_{ice}$, × Hallett (1963b, 1964), ○ Magono and Aburukawa (1968), + Aufdermauer and Mayes (1965), △, □ Pitter and Pruppacher (1973), ⊥ Rye and Macklin (1975). (From Pitter and Pruppacher, 1973, with changes.)

nucleated ice makes an angle of 90° with the c-axis of the substrate ice. Under these conditions the crystallographic misfit δ (see Section 9.2.3.4) is -6.0%. Other common reorientations of small misfits to the basal plane of ice are: $\delta = -2.5\%$ for a $(20\bar{2}1)$ plane, causing an angle of 75.2° between the c-axes; $\delta = +6.7\%$ for a $(10\bar{1}1)$ plane, causing an angle of 62.1° between the c-axes; $\delta = -4.0\%$ for a $(12\bar{1}1)$ plane, causing an angle of 58.5° between the c-axes; $\delta = -2.0\%$ for a $(1\bar{2}10)$ plane, causing an angle of 90° between the c-axes; and $\delta = -2.3\%$ for a $(2\bar{4}21)$ plane, causing an angle of 72.8° between the c-axes (Higuchi and Yoshida, 1967). The basal matching between these selected crystallographic planes of ice is illustrated in Figure 16-5. The angles between the c-axes predicted from such matching techniques have been verified by the same authors, and by Magono and Suzuki (1967), Aburukawa and Magono (1972), and Lee (1972) through measurements on spatial crystals (see Section 2.2.1).

Following Levi (1970), Levi and Aufdermauer (1970), Rye and Macklin (1973), and Macklin and Rye (1974), we finally may suggest some reasons for the preferred orientation of ice crystallites in ice deposits grown by riming. We recall that the c-axis of these crystallites is preferentially oriented at an angle α of 0° and 90° to the radial growth direction of the riming ice deposit in the dry growth and wet growth regimes, respectively. This can be understood if we consider that in the dry growth regime in which the cloud drops collide with a relatively cold substrate the latent heat of freezing is preferentially lost to the substrate. Since heat is more efficiently conducted along the crystallographic c-axis of ice than along other directions (Hobbs, 1974a; the effect arises from the greater lineal density of molecules along the c-axis, making this the preferred direction for the propagation of thermal fluctuations; see Section 5.71 and Figure 3-3), growth of those ice dendrites which nucleate by chance with their a-axis parallel to the surface of the accreting ice deposit is facilitated as compared to dendrites which nucleate in other orientations. Thus in this growth regime ice crystallites assume preferentially an

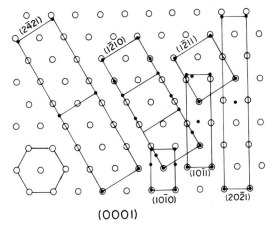

Fig. 16-5. Matching of various ice crystal planes on the basal plane of ice. (From Higuchi and Yoshida, 1967; by courtesy of Institute for Low Temperature Science, Hokkaido University.)

orientation in which the c-axis of the ice crystallites is oriented perpendicular to the accreting ice surface, i.e., parallel to the radial growth direction of the surface. On the other hand, in the wet growth regime the ice substrate is relatively warm, and latent heat of freezing is dissipated mainly by conduction and convection into the surrounding air. Consequently, the growth of those ice dendrites which nucleate by chance with their a-axes perpendicular to the accreting ice surface is facilitated as compared to growth of ice dendrites nucleated in other directions. Thus, in this growth regime the ice crystallites assume preferentially an orientation in which their c-axis is oriented parallel to the accreting ice surface, i.e., perpendicular to the radial growth direction of the surface.

The temperature dependence of the angular orientation of the ice crystallites is a result of the non-rational growth of ice dendrites in supercooled water, as illustrated in Figure 16-1. Thus, since the non-rational growth angle φ increases with increasing supercooling (see Section 16.1), while in the dry growth regime G_m tends to be oriented parallel to the growing ice surface, the orientation angle α also will increase with increasing supercooling. An analogous argument holds for the wet growth regime in which G_m tends to be oriented perpendicular to the growing ice surface.

16.4   Growth Rate of Ice in Supercooled Water

Most studies of the growth rate of ice crystals in supercooled water have dealt with ice growth out in narrow tubes and capillaries (for a review of some of the earlier work see Pruppacher, 1967c). It is obvious that such measurements have very limited application to the growth rate of ice in supercooled water drops, since these are not affected by the proximity of heat conducting walls. The so-called free growth rate G_m of ice crystals in supercooled bulk water and in supercooled drops has been studied by Lindenmeyer (1959), Hallett (1964), Pruppacher (1967d), Macklin and Ryan (1968), Ryan and Macklin (1968), Ryan (1969), and Gokhale and Lewinter (1971). Their results are summarized in Table 16-1 and Figure 16-6. Note from this figure and table that the growth rate G_m of ice in supercooled water varies nearly as the square of the bath supercooling $\Delta T_\infty = T_0 - T_\infty$ for $0 \leqslant \Delta T_\infty \leqslant 10\,°C$, where $T_0 = 273\,°K$ ($0\,°C$) and where T_∞ is the temperature far away from the growing ice crystal surface. Note also that the component growth rate of ice parallel to its crystallographic a-axis, $G_a = G_m \cos \varphi$ (see Figure 16-3), shows a similar dependence on supercooling because of the smallness of angle φ. On the other hand, the component growth rate of ice parallel to the crystallographic c-axis, $G_c = G_m \sin \varphi$, is considerably smaller than G_a and varies approximately as the third power of ΔT_∞. At $9 \leqslant \Delta T_\infty \leqslant 12\,°C$ a discontinuity in G is observed which is characterized by little change in growth rate with decrease in ΔT_∞. From Figure 16-6 this discontinuity is evidently a manifestation of a transition from a higher to a lower power growth law, as at $\Delta T_\infty > 12\,°C$ the growth rate appears to vary nearly linearly with ΔT_∞.

The effect of water soluble salts on the growth rate of ice has been studied by Pruppacher (1967d), Ryan and Macklin (1968), Macklin and Ryan (1968), and Ryan (1969). These studies show that the growth rate of ice in supercooled

TABLE 16-1

Experimentally derived laws for the variation of the growth rate G of ice in water supercooled by ΔT_∞ degrees. G_m is the non-rational dendrite growth velocity, G_a is the component growth velocity along the basal plane of ice, and G_c is the component growth velocity perpendicular to the basal plane of ice.

Lindenmeyer (1959)	$G_m = 0.023\,(\Delta T_\infty)^{2.39}$	$2 \leqslant \Delta T_\infty \leqslant 6.5\,°C$
Hallett (1964), max. values	$G_m = 0.08\,(\Delta T_\infty)^{1.9}$	$1 \leqslant \Delta T_\infty \leqslant 10\,°C$
Pruppacher (1967d)	$G_m = 0.035\,(\Delta T_\infty)^{2.22}$	$0.5 \leqslant \Delta T_\infty \leqslant 9\,°C$
Macklin and Ryan (1968)	$G_a = 0.015\,(\Delta T_\infty)^{2.49}$	$2 \leqslant \Delta T_\infty \leqslant 10\,°C$
	$G_c = 9 \times 10^{-4}\,(\Delta T_\infty)^{3.35}$	$2 \leqslant \Delta T_\infty \leqslant 10\,°C$
Gokhale and Lewinter (1971)	$G_m = 0.044\,(\Delta T_\infty)^{2.3}$	$5 \leqslant \Delta T_\infty \leqslant 10\,°C$

aqueous solutions remains unaffected until the salt concentration becomes greater than about $10^{-2}\,\ell^{-1}$. At larger concentrations the growth rate is progressively retarded by the salt. Fluorides, however, behave exceptionally in that they invariably enhance the growth rate at concentrations between 10^{-3} and $10^{-2}\,\text{mol}\,\ell^{-1}$. Such growth rate enhancement was also noted by Michaels *et al.* (1966) and Pruppacher (1967c). The effects of dissolved salts on the growth rate of ice were attributed by the above authors: (1) to a change of the thermal conductivity of water by the presence of salt ions, thus affecting the rate at which latent heat is dissipated by conduction from the ice water interface; (2) to a concentration buildup of salt rejected at the ice-water interface, thus lowering the local equilibrium freezing temperature; (3) to adsorption of salt ions at growth steps on the ice surface, thus inhibiting the incorporation of water molecules at these locations; (4) to a change of the mobility and therefore also the diffusivity of water molecules through water by the presence of salt ions, thus affecting the rate at which water molecules can reach the ice-water interface and

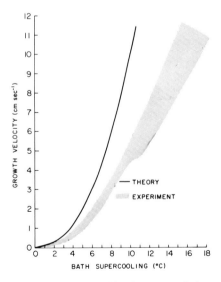

Fig. 16-6. Variation of the free growth velocity of ice in supercooled water. Theory is due to Bolling and Tiller (1961) for $\Delta T_S = 0$. Experimental results are due to Lindenmeyer (1959), Hallett (1964), Pruppacher (1967a, d), and Macklin and Ryan (1968). (From Pruppacher, 1967e, with changes.)

the rate at which the salt ions can diffuse away from the surface into the solution under the effect of the concentration gradient at the ice-water interface; (5) to a formation of additional dislocations in the growing ice crystal lattice by ions incorporated into the ice lattice, thus causing additional strain and stresses in the lattice; and (6) to the development of relatively large, local electric fields at the ice-water interface due to differential incorporation of ions into the ice crystal lattice (see Section 5.8), thus affecting the local structure of water.

From our discussions in Section 9.2.3.1 it is obvious that mechanisms (2) and (3) will impede the rate of ice crystal propagation. Also, mechanisms (1) and (4) will tend to impede ice crystal propagation, since the experiments of Eigen (1952) and Kauptinskii and Razanin (1955) show that most salts decrease the heat conductivity of water, and since the experiments of Wang and Miller (1952), Wang (1954), and McCall and Douglas (1965) show that most salts reduce the self diffusion coefficient of water, increasingly so with increasing salt concentration. Little is known about the effects of mechanisms (5) and (6). However, from our discussions of the effect of surface dislocations on ice nucleation (see Section 9.2.3.5), and on the growth rate of ice from the vapor (see Section 13.3), one would expect that an increase in the number density of dislocations and, consequently, an increase in the number density of surface steps on a growing ice crystal would facilitate the incorporation of water molecules into the ice crystal lattice. Also, Ryan (1969) has argued that the electric fields which tend to build up at the ice-aqueous solution interface would also facilitate the incorporation of water molecules into the ice crystal lattice. Thus, while mechanisms (1) to (4) appear to explain the generally observed growth rate retardation of ice crystals in salt solutions, mechanisms (5) and (6) may conceivably be responsible for the growth rate enhancement of ice crystals in the aqueous solutions of fluoride salts.

In order to interpret the experimentally found variation of the growth rate of ice in supercooled water as a function of supercooling, we shall consider first the simple case of a planar ice face propagating into supercooled water. We shall assume that the face contains no steps due to emerging dislocations, but nevertheless is molecularly rough. The latter assumption implies that any site at the surface is a potential site for attachment of water molecules. Such a case has been studied by Wilson (1900) and Frenkel (1932) through use of the simple theory of reaction rates. If μ_i is the chemical potential of ice and $\mu_{w,\ell}$ is that of water one mean molecular jump distance ℓ away from the ice-water interface, the thermodynamic driving force responsible for the advance of the crystal face is $(\mu_{w,\ell} - \mu_i)/\ell$. The rate of advance of the ice crystal face is then given by the product of this driving force and the mobility $D_w/\mathscr{R}T_S$ of the water molecules, where T_S is the temperature of the ice-water interface (Glasstone et al., 1941). Thus,

$$G = \frac{D_w}{\mathscr{R}T_S} \frac{(\mu_{w,\ell} - \mu_i)}{\ell}, \qquad (16\text{-}25)$$

where D_w is the diffusivity of water molecules in water, given from the experimental data of Mills (1971, 1973), Gillen et al. (1972), and Pruppacher

(1973) to $\pm 0.002 \times 10^{-5} \, cm^2 \, sec^{-1}$ by the expression

$$D_w = 1.076 \times 10^{-5} \exp (4.14 \times 10^{-2} \, T + 2.048 \times 10^{-4} \, T^2 + 2.713 \times 10^{-5} \, T^3),$$
(16-26a)

for temperatures below 0 °C, and by

$$D_w = (1.075 + 4.26 \times 10^{-2} \, T + 2.667 \times 10^{-4} \, T^2 - 2.667 \times 10^{-6} \, T^3) \times 10^{-5}$$
(16-26b)

for temperatures above 0 °C, with T (°C) and D_w in $cm^2 \, sec^{-1}$. Since $(\partial F/\partial T)_p = -S$, where S is the entropy and F the Helmholtz free energy of the system (Section 4.1), we may write for the volume free energy change of the system $\Delta F_{vol} = -\int_0^{\Delta T_S} S_{vol} \, d(\Delta T) \approx \overline{-\Delta S_{vol}} \Delta T_S$, where $\Delta T_S = T_0 - T_S$. From (7-17) we have, on the other hand, $\Delta F_{vol} = n_i(\mu_i - \mu_w)$. We therefore find that

$$\mu_{w,\ell} - \mu_i = \frac{\overline{\Delta S_{vol} \Delta T}}{n_i} = \overline{\Delta S_{vol}} \Delta T = \overline{\frac{\mathcal{L}_m}{T}} \Delta T_S,$$
(16-27)

where $\mathcal{L}_m$ is the molar latent heat of fusion taken at the mean temperature $\bar{T} \approx (T_S + T_0)/2$. On combining (16-25) with (16-27) we find that the rate of advance of the molecularly rough ice crystal face is given by

$$G = \frac{D_w \mathcal{L}_m}{\ell \mathcal{R} T_S \bar{T}} \Delta T_S \approx \frac{D_w \bar{\mathcal{L}}_m}{a_0 \mathcal{R} T_0^2} \Delta T_S,$$
(16-28)

where we have now assumed ℓ to be of the order of the molecular spacing a_0 in ice, and $T_S \bar{T} \approx T_0^2$. Note that the given treatment predicts a growth rate which is linear in ΔT_S. Crystals exhibiting such a growth law are said to grow in the 'continuous growth regime', as their faces advance continuously without involving a lateral spreading of surface steps.

Frank (1949) and Burton et al. (1951) pointed out that many crystal surfaces are not perfect on a microscopic scale (i.e., molecularly rough), but contain steps due to emerging lattice dislocations. By assuming an abrupt step which advances laterally by addition of single molecules from the liquid through a diffusion jump mechanism, Hillig and Turnbull (1956) derived an expression for the growth rate of such a crystal in its supercooled melt. The starting point for their development is a modified form of the simple Wilson-Frenkel growth law. Instead of assuming that every site on the ice surface is available for molecular attachment, they assumed that only sites on the growth spirals of emerging screw dislocations can be used for molecular attachment. An approximate equation for such a growth spiral, considered nearly Archimedian, was given by Burton et al. (1951). Applied to a growth spiral on an ice surface in supercooled water, the equation can be expressed as

$$r = 2 \, \theta a_g,$$
(16-29)

where r and θ are polar coordinates describing the spiral, and a_g is the radius of a two-dimensional ice germ on an ice surface in supercooled water at temperature T_s. To determine an expression for a_g we rewrite (9-10) for the energy of i-mer formation of a cylindrical ice embryo of molecular height a_0 on an ice substrate, including a contribution from line tension, interpreted here as the step energy per unit length, as

$$\Delta F_{i,S} = \pi a_i^2 a_0 \Delta f_{vol} + \sigma_{i,i} \Omega_{i,i} + 2\pi\lambda a_i. \tag{16-30}$$

From (7-17) and (7-18), $\Delta f_{vol} = -(\mu_w - \mu_i)/v_i$, where v_i is the mole volume of ice. Considering that the ice embryo forms on an ice substrate, the second term on the right side of (16-30) is zero. Instead of (9-15) we then obtain for a_g (by setting $\partial(\Delta F_{i,S})/\partial a_i = 0$):

$$a_g = -\frac{\lambda}{a_0 \Delta f_{vol}} = -\frac{\lambda v_i}{a_0(\mu_w - \mu_i)}, \tag{16-31}$$

where λ is the step energy per unit length. With (16-31) one finds from (16-29) and (16-27) that

$$r = 2\theta \frac{\lambda v_i \overline{T}}{a_0 \overline{\mathcal{L}}_m \Delta T_S} \approx 2\theta \frac{\lambda v_i T_0}{a_0 \overline{\mathcal{L}}_m \Delta T_S}. \tag{16-32}$$

The change in r as θ advances by one revolution, i.e., by 2π, is thus

$$\Delta r = \frac{4\pi\lambda v_i T_0}{a_0 \overline{\mathcal{L}}_m \Delta T_S}. \tag{16-33}$$

The fraction $\alpha = a_0/\Delta r$ of surface sites available for molecular attachment is therefore

$$\alpha = \frac{a_0^2 \overline{\mathcal{L}}_m \Delta T_S}{4\pi\lambda v_i T_0}, \tag{16-34}$$

if attachment of molecules to the step occurs only within a distance a_0 of the step. The growth rate of a surface advancing by lateral spreading of steps is therefore

$$G = \frac{3\alpha D_w \overline{\mathcal{L}}_m \Delta T_S}{a_0 \mathcal{R} T_0^2} = \frac{3 D_w \overline{\mathcal{L}}_m^2 a_0}{4\pi\lambda v_i \mathcal{R} T_0^3}(\Delta T_S)^2, \tag{16-35}$$

in place of (16-28), assuming that 3 molecules are available for attachment at a given step site. (Note that this number is somewhat smaller than the number of nearest neighbors (namely 4.4) to a given water molecule in bulk water – see Section 4.7.) Assuming further that λ may be estimated from $\lambda = a_0 \sigma_{i/w}$ (e.g., Cahn et al., 1964), Hillig and Turnbull (1956) found

$$G = \frac{3 D_w \mathcal{L}_m^2}{4\pi\sigma_{i/w} v_i \mathcal{R} T_0^3}(\Delta T_S)^2. \tag{16-36}$$

Note that (16-36) predicts a parabolic growth law ('classical growth regime').

In deriving (16-36) Hillig and Turnbull assumed that the ice-water interface is sharp, i.e., that ordering of the water molecules at the growth front occurs within a distance $\ell \approx a_0$ of a step by a monomolecular transport process characterized by a diffusion constant D_w, with 3 molecules available for attachment at a given step. Cahn (1960) and Cahn et al. (1964) criticized this assumption by pointing out that the interface is not sharp but diffuse, the transition from the liquid to the solid phase occurring over several molecular layers. To correct for this feature they introduced a parameter g which is a measure of the diffuseness of the interface, and which depends on the number n of molecular layers comprising the transition zone. From a theoretical study of surface energy they found $g \approx (\pi^4/8)n^3 \exp(-\pi^2 n/2)$. For a sharp interface $g \approx 1$; g decreases rapidly toward zero with increasing interface diffuseness ($g \approx 0.01$ for $n = 2$). Instead of the traditionally assumed estimate for the step energy per unit length, namely that $\lambda = a_0 \sigma_{i/w}$, Cahn et al. argued that $\lambda = a_0 \sigma_{i/w} g^{1/2}$. They also conjectured that the number of molecules in position to jump into a growth site is given by $2 + g^{-1/2}$, which reduces to 3 (the value assumed by Hillig and Turnbull), when g approaches 1. For $g \approx 0.01$ the number of molecules in position to jump into a growth site is ~ 10. In addition, Cahn et al. proposed that an accommodation coefficient β should be introduced as a measure for the difficulty molecules have in moving to a step and in assuming an orientation suitable for incorporation into the ice lattice. With these corrections introduced into (16-35), Cahn et al. estimated the rate at which a crystal grows by a screw dislocation mechanism to be

$$G = \frac{\beta(1 + 2 g^{1/2})}{4 \pi g} \frac{D_w \mathscr{L}_m^2}{\sigma_{i/w} v_i \mathscr{R} T_0^3} (\Delta T_S)^2,$$ (16-37)

and that the corresponding rate at which a crystal grows in the continuous growth regime is

$$G = \frac{\beta D_w \mathscr{L}_m}{a_0 \mathscr{R} T_0^2} \Delta T_S.$$ (16-38)

Obviously, the growth rate of ice in supercooled water is not exclusively controlled by kinetic processes at the ice-water interface, as was assumed above, but by heat transport processes as well. Two distinct points of view commonly exist regarding the relative effect of these two mechanisms which influence crystal growth. The first point of view assumes that the growth rate is an interface-controlled, material-transport process determined mainly by the molecular mobility at the interface and by the temperature, crystallographic perfection, and crystallographic orientation of the interface. The second point of view assumes that there is no effective transport barrier at the interface between the crystal and its melt, and that the growth rate of the crystal is therefore predominantly controlled by the rate at which latent heat of solidification is removed from the growing interface.

In actuality, of course, these two points of view are only limiting cases since the temperature T_S of the interface is not exactly the bath temperature T_∞, or the heat of solidification could obviously not be extracted; nor is it exactly the equilibrium temperature T_0 or T_e (the latter being the equilibrium temperature of a curved ice-water interface), or there would be no driving force for freezing. Rather, in steady state growth T_S adjusts itself such that the rate of liberation of heat corresponding to the rate of molecular incorporation just balances the rate of heat dissipation from the interface. Thus, the supercooling $T_e - T_S$ of the curved ice interface controls the rate of deposition of water molecules, while the supercooling $T_S - T_\infty$ controls the rate of heat dissipation from the ice-water interface (note that $T_\infty < T_S < T_e < T_0 = 273\ °K = 0\ °C$). Experiments show that $T_e - T_\infty$ for ice growing in supercooling water is only a small fraction of $T_S - T_\infty$. This implies that the growth rate of ice in supercooled water is heat dissipation limited (as is the growth rate of ice in supersaturated vapor (Section 13.3)). This result was confirmed by Lindenmeyer *et al.* (1957), who found that the larger the heat conductivity of the walls encasing a sample of supercooled water, the larger the growth rate of ice in the water sample.

The rate of heat dissipation from an ice dendrite tip growing in supercooled water has been studied theoretically by Bolling and Tiller (1961), Horvay and Cahn (1961), and Holzman (1970), all of whom idealized the growing dendrite tip by an isothermal prolate spheroid and assumed $\Delta T_S = 0$. Values for the growth rate determined by Bolling and Tiller are plotted in Figure 16-6. The plotted curve fits the equation $G = 0.049\ \Delta T_\infty^{2.3}$ (for $\Delta T_S = 0$). On the other hand, Horvay and Cahn found $G = (\Delta T_\infty)^{1.3}/r$, where r is the curvature of the dendrite tip. According to Fisher (Hillig, 1959) and Bolling and Tiller (1961), however, r is proportional to $(\Delta T_\infty)^{-1}$. Therefore the result of Horvay and Cahn implies $G \propto (\Delta T_\infty)^{2.3}$, in good agreement with the result derived by Bolling and Tiller.

On comparing the observed values of G in Figure 16-6 with those computed for the case that $\Delta T_S = 0$, we notice that G is not completely thermally determined; i.e., for each value of G some value ΔT_S must be added to bring the theoretical prediction into agreement with experiment. Such comparison yields $G = G(\Delta T_S)$, i.e., the growth rate of ice is a function of interface supercooling. From the maximum observed values for G one finds approximately for $1 \leqslant \Delta T_\infty \leqslant 10\ °C$ that $G \approx 1.0\ (\Delta T_S)^2$, while for $\Delta T_\infty > 10\ °C$, $G \approx 2.3\ \Delta T_S$. From a comparison of the experimentally observed growth laws with (16-37) and (16-38) one finds that with $\delta = 3 \times 10^{-8}\ cm$, we have $\beta \approx 1$ and $g \approx 4 \times 10^{-3}$, which implies a diffuseness of about 3 molecular layers (Cahn *et al.*, 1964). Comparison of the experimental growth laws with (16-37) and (16-38) suggests further that growth of ice in supercooled water proceeds via the lateral spreading of steps at $1 \leqslant \Delta T_\infty \leqslant 10\ °C$, and via the continuous growth mechanism at $\Delta T_\infty > 10\ °C$, i.e., when $G > 5\ cm\ sec^{-1}$ (Pruppacher, 1967e; Macklin and Ryan, 1969). There is no obvious reason for the observed large scatter of data for $\Delta T_\infty > 10\ °C$.

In addition to the classical and continuous growth regimes, Cahn (1960) and Cahn *et al.* (1964) considered the existence of a transitional growth regime in which growth continues to proceed by lateral spreading of steps but which also progressively deviates from the simple parabolic growth law which characterizes the classical regime. Through theoretical arguments involving the free energy per unit volume ΔF_{vol} which acts as the driving force for crystal growth in a melt, they found that for the transitional regime $\sigma g/a_0 \leqslant -\Delta F_{vol} \leqslant \pi \sigma g/a_0$, where σ is the interface energy between the crystal and its melt. Thus, $\Delta F_{vol} < \sigma g/a_0$ characterizes the classical regime, while $-\Delta F_{vol} > \pi \sigma g/a_0$ characterizes the continuous growth regime. If we recall from Section 7.1.2 that ΔF_{vol} is proportional to the supercooling of the melt, we see that $\pi \Delta T_S^*$ will mark the supercooling onset of the continuous growth regime if ΔT_S^* marks the supercooling limit of the classical regime. From their experimental growth studies on ice Pruppacher (1967e) and Macklin and Ryan (1969) suggested that $\pi \Delta T_\infty^* \approx 9$ to

10 °C, from which it follows from Bolling and Tiller's theory (Figure 16-6) that $\pi \Delta T_S^* \approx 2$ °C and therefore $\Delta T_S^* \approx 0.6$ °C, giving $\Delta T_\infty^* \approx 2.7$ °C. This latter super-cooling has some physical significance in that it closely agrees with the super-cooling required for the onset of non-rational growth of ice in supercooled water, discussed in Section 16.3.

16.5   Growth Rate of Graupel and Hailstones

Assuming that a graupel or hailstone grows by collision with supercooled water drops in the dry growth regime (dr) where all the water collected is frozen on the growing ice particle, its growth can be described adequately by the continuous growth equation discussed in Section 15.1. For this regime the growth rate of a spherical ice particle due to collection of water drops is given approximately by

$$\left(\frac{dm}{dt}\right)_{dr} = E_c \pi (a_H + a_d)^2 (V_{\infty,H} - V_{\infty,d}) w_L, \tag{16-39}$$

where a_H and $V_{\infty,H}$, and a_d and $V_{\infty,d}$ are the radius and terminal velocity of the graupel or hailstone and the water drops, respectively, E_c is the collection efficiency of the ice particle, w_L is the liquid water content of the cloud, and m is the mass of the ice particle. Since $a_d \ll a_H$ and $V_d \ll V_H$, we may further reduce (16-39), after dropping subscripts, to the simple form

$$\left(\frac{dm}{dt}\right)_{dr} = E_c \pi a^2 V_\infty w_L. \tag{16-40}$$

Of course, the same equation applies when the ice particle grows in the wet growth regime if one assumes that none of the collected water is shed but rather is incorporated into the spongy structure of the ice particle.

On the other hand, if we assume that an ice particle grows in the wet growth regime (wr) and sheds all excess liquid water, its growth rate is given by the rate at which the collected cloud water can be frozen, i.e., by the rate at which the latent heat released is dissipated. The heat to be dissipated is

$$\left(\frac{dq}{dt}\right)_1 = \left(\frac{dm}{dt}\right)_{wr} [L_m - c_w(T_0 - T_\infty)], \tag{16-41}$$

where L_m is the latent heat of melting, c_w is the specific heat of water, and $(dm/dt)_{wr}$ is the rate of mass increase of the riming ice particle in the wet growth regime. The second term on the right side of (16-41) is due to the heating of the collected cloud water from its original temperature (the drops are assumed to be originally at the temperature T_∞ of the environment) to the temperature T_0 of the ice particle surface, assumed to remain at 0 °C. On the other hand, the rate at which heat is transferred to the environmental air by forced convection and by evaporation is given by

$$\left(\frac{dq}{dt}\right)_2 = 4 \pi a [k_a (T_0 - T_\infty) \bar{f}_h + L_e D_v (\rho_{v,a} - \rho_{v,\infty}) \bar{f}_v], \tag{16-42}$$

where k_a is the heat conductivity of air, L_e the latent heat of evaporation, D_v the diffusivity of water vapor in air, ρ_v the density of water vapor, and $\bar{f}$ the ventilation coefficient. For a steady state $(dq/dt)_1 = (dq/dt)_2$, and so we find the growth rate of the ice particle in the wet growth regime:

$$\left(\frac{dm}{dt}\right)_{wr} = \frac{4\,\pi a\,[k_a(T_0 - T_\infty)\bar{f}_h + L_e D_v(\rho_{v,a} - \rho_{v,\infty})\bar{f}_v]}{L_m - c_w(T_0 - T_\infty)}. \tag{16-43}$$

If the riming ice particle grows as a result of the collection of both ice crystals and drops the term $(dm/dt)_i c_i(T_0 - T_\infty)$, expressing the rate of sensible heat transfer from the relatively warm riming ice particles to the relatively cold ice crystals, has to be added to the numerator on the right side of (16-43), where $(dm/dt)_i$ is the rate of mass increase of the riming ice particle due to collision with ice crystals, and c_i is the specific heat of ice.

As a corollary to (16-43), the critical effective liquid water content of a cloud may be found by inserting (16-40) into (16-43):

$$E_c w_{L,c} = \frac{L_e D_v[\rho_v(T_0) - \rho_v(T_\infty)]\bar{f}_v + k_a(T_0 - T_\infty)\bar{f}_h}{a\,V_\infty[L_m - c_w(T_0 - T_\infty)]/4}, \tag{16-44}$$

where $w_{L,c}$ is the critical liquid water content of the cloud required for all the collected water to freeze while the ice particle maintains a surface temperature of $T_0 = 273\,°K$. For $w_L < w_{L,c}$ the riming ice particle grows in the dry growth regime, while for $w_L > w_{L,c}$ it grows in the wet growth regime.

In order to evaluate (16-43) or (16-44) it is necessary to estimate the ventilation coefficients $\bar{f}_h$ and $\bar{f}_v$ for heat and mass transfer for rimed ice particles. Unfortunately, only values for $\bar{f}_h$, determined from heat transfer studies on hailstones and hailstone models, are available in the literature. In order to estimate $\bar{f}_v$ one generally assumes $\bar{f}_v \approx \bar{f}_h$; the justification for this assumption has been given in Section 12.7.2. Relevant heat transfer studies have been carried out by List (1960b), Macklin (1963, 1964a, b), Bailey and Macklin (1968b), Schüepp and List (1969), Joss and Aufdermauer (1970), and Schüepp (1971). On expressing $\bar{f}_h = BN_{Re}^{1/2}N_{Sc,v}^{1/3}$, Macklin (1963) and Bailey and Macklin (1968b) found from wind tunnel experiments with smooth ice spheres that B increased from 0.37 to 0.46 as the Reynolds number N_{Re} increased from 1×10^4 to 2×10^5. Experiments with spherical hailstones in the same Reynolds number interval gave values for B which increased from $B \approx 0.37$ at $N_{Re} \approx 10^4$ to $B \approx 1.25$ at $N_{Re} \approx 2 \times 10^5$. The reason for the increase of B with N_{Re} lies in the fact that two physically distinct processes determine the heat transfer from a spherical particle, namely, transfer through the laminar boundary layer on the upstream side of the sphere, and transfer through the turbulent wake on the downstream side; B increases because the latter becomes increasingly important as the Reynolds number increases. Wind tunnel studies of Macklin (1964b) showed that rotational and oscillational motions typically exhibited by rimed ice particles did not significantly affect $\bar{f}_h$.

The findings of Macklin (1963) and Bailey and Macklin (1968b) for the dependence of heat transfer on Reynolds number agree quite well with the earlier studies of List (1960b) and with the more recent studies of Aufdermauer and Joss (1967), Schüepp and List (1969), Joss and Aufdermauer (1970), and

Schüepp (1971), all of whom made wind tunnel studies of the heat transfer rate from spherical hailstones and spherical hailstone models of various surface roughness. Thus, Joss and Aufdermauer found that for smooth spheres in a relatively quiet airflow with a turbulence level of 0.4%, $B \approx 0.35$ at $N_{Re} = 4 \times 10^3$, increasing to $B \approx 0.46$ at $N_{Re} = 7 \times 10^4$. For a spherical body with a roughness (ratio of height to diameter of roughness elements) of 8%, Joss and Aufdermauer found $B \approx 0.44$ at $N_{Re} = 4 \times 10^3$ and $B \approx 0.95$ at $N_{Re} = 7 \times 10^4$. With increasing turbulence of the air stream the value of B increased, reaching at a turbulence level of 5% and at $N_{Re} = 7 \times 10^4$ a value of $B \approx 0.63$ and 1.01 for a smooth sphere and a sphere of 8% roughness, respectively.

From their values for the ventilation coefficient, Bailey and Macklin (1968b) evaluated $E_c w_{L,c}$ for smooth spheres and hailstones from (16-44). Their findings are plotted in Figure 16-7 for different temperatures. Note from this comparison that for hailstones larger than 4 cm in diameter the effective critical water contents are about 1.2 to 3 times as large as those for smooth spheres.

Braham (1968), Jiusto (1968, 1971), Hindman and Johnson (1972), and Ryan (1972) computed the growth rate of a spherical ice particle growing by vapor diffusion and by the collection of supercooled drops in the dry growth regime. Ryan (1972) used the simple continuous growth model of (16-40), and assumed the collision efficiencies of Mason (1957), Hocking (1959), and Shafrir and Neiburger (1963), the choice depending on the size range. The change in mass due to diffusional growth was computed from (13-82) for the case of water saturated air at $T = -10\,°C$ and $p = 750$ mb. Two types of clouds were considered, namely a maritime type cloud and a modified continental type cloud, with drop spectra as shown in Figure 16-8. Updrafts were assumed negligible. The results of Ryan's calculations are shown in Figure 16-9, where the mass of the ice particle is given as a function of growth time. Note that, as expected, growth is faster in the maritime cloud, due to its broader drop size distribution. Note also

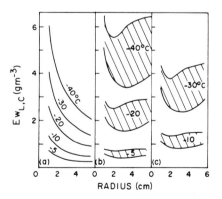

Fig. 16-7.   Effective critical liquid water content $E_c w_{L,c}$ as a function of radius in riming ice particle; (a) smooth ice sphere, (b) and (c) hailstones. Computed by Bailey and Macklin from an equation analogous to (16-44) using wind tunnel measurements for the ventilation coefficient of ice particles. (From Bailey and Macklin, 1968b; by courtesy of *Quart. J. Roy. Meteor. Soc.*)

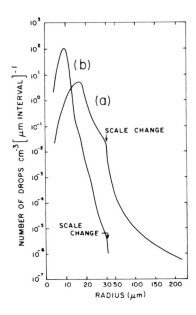

Fig. 16-8.   Model size distributions of water drops (a) in a maritime cumulus cloud ($w_L = 0.8\,\mathrm{g\,m^{-3}}$), and (b) in a modified continental cumulus cloud ($w_L = 1.5\,\mathrm{g\,m^{-3}}$), used in the computations displayed in Figure 16-9. (From Ryan, 1972; by courtesy of *J. de Rech. Atmos.*, and the author.)

that it is important to consider the whole drop size distribution rather than just the average drop size in a cloud.

The continuous growth equation for the case of an updraft W is (cf. (15-6)):

$$\frac{da_H}{dz} = \frac{V_\infty w_L E_c}{4\,\rho_H (W - V_\infty)}. \tag{16-45}$$

This equation was used by List *et al.* (1965, 1968), List and Dussault (1967), Musil (1970), Charlton and List (1972a, b), and Dennis and Musil (1973) in conjunction with an assumed updraft profile and a pseudo-adiabatic steady state parcel model (which provides values for w_L – see Chapter 11) to compute the growth rate of spherical hailstones. In the computations of Charlton and List (1972a, b), hail embryos of diameters d_0 and concentration N_0 ($\mathrm{m^{-3}}$) were assumed to be injected at the 0 °C level into the cloud updraft of strength W (m sec^{-1}), which obeyed the condition that $\rho_a W = $ constant. It was further assumed that the growing embryos always fell at their terminal velocity, had a collection efficiency of unity and the density of bulk ice, and that the accreted water remained on the growing ice particle. Various initial vertical velocities W_0 were considered. The temperature and humidity structure of the atmosphere in the environment of the rising air parcel was assumed to be that given by the soundings of Beckwith (1960). Thermal feedback through the release of latent heat was considered by means of (16-41) and (16-42). Charlton and List's computations are illustrated by Figure 16-10, which displays the growth rate of hailstones for the case that $W_0 = 20$ m sec^{-1} and $d_0 = 0.5$ cm. Note that for the given conditions spherical

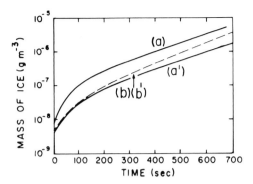

Fig. 16-9. Change with time of the mass of a spherical ice particle of initial radius 10 μm falling through a maritime cloud and a modified continental cloud. Curve (a) refers to the calculations using the drop size distribution (a) in Figure 16-8. Curve (a') refers to the calculation using only the average drop radius of the drop spectrum in that cloud. Curve (b) refers to the calculation using the drop size distribution (b) in Figure 16-8, and curve (b') refers to the calculation using only the average drop radius of the drop spectrum in that cloud. (From Ryan, 1972; by courtesy of *J. de Rech. Atmos.*, and the author.)

hailstones may grow to 3 cm in diameter within about 20 min. The computations also show, as expected, that relatively large hailstones are the result of relatively large embryo sizes and relatively low embryo concentrations. On the other hand, large updrafts were not found to lead necessarily to large hailstones, since under such conditions a stone may not spend a sufficiently long time in the cloud.

The above-mentioned studies demonstrate that hailstone growth depends strongly on cloud parameters such as: the number concentration of hailstone embryos and their size and level of formation, on the magnitude of the updraft velocity and its variation with height, on the collection and water-retention efficiency assumed for the growing hailstone, and on the drop sizes and liquid

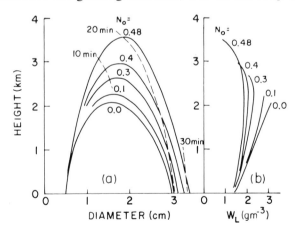

Fig. 16-10. Hailstone diameter (a), and cloud liquid water content (b), as a function of height during the ascent and descent of a hailstone, with different number concentrations N_0 (m^{-3}) of hailstone embryos of 0.5 cm diameter, injected at level z = 0, and for an updraft velocity of 20 m sec^{-1}. Dashed lines indicate hail growth times. (From Charlton and List, 1972; by courtesy of Amer. Meteor. Soc., and the authors.)

water content of the cloud. However, it has been observed, in addition, that the growth of a hailstone depends in a complicated manner also on cloud dynamic parameters, such as the tilt of the updraft system, the wind shear, and the existence of liquid water accumulation levels in the cloud. Unfortunately, little quantitative information is available on these effects (for a review, see Gokhale, 1975).

While it is justifiable to use the continuous growth equation to determine the growth of ice particles which are relatively large compared to the cloud drops, the growth problem must be treated as a stochastic process (see Chapter 15) if the ice particles and drops have similar sizes. The evolution of a size distribution of spherical ice particles growing by a stochastic riming process has been considered by Ryan (1973) and recently by Beheng (1976, 1978). Ryan used a simple stochastic model in which changes in the ice particle spectrum were studied without considering simultaneous changes in the drop size distribution, and without considering the feedback of these changes on ice particle riming. In addition, Ryan assumed relatively unrealistic collision efficiencies for the supercooled drops and ice crystals, and he further imposed the artificial condition that all the ice crystals appear in the cloud at one given time. In Beheng's more comprehensive study, the evolution of an ice particle size spectrum of originally plate-like ice crystals was determined for the case of growth by collision with supercooled water drops. The simultaneous evolution of the cloud drop spectrum resulting from the collision of drops with drops, and of drops with ice particles, was also followed. The initial cloud drop mass spectrum was that of Berry and Reinhardt (1974a) for a width parameter $p = 0$ (see (12-129)), an initial mean drop radius of $\bar{a}_d = 14 \, \mu$m, an initial drop concentration $N_d = 86 \, \text{cm}^{-3}$, and a liquid water content of $w_L = 1 \, \text{g m}^{-3}$. The drops were assumed to collide with efficiencies as given by Shafrir and Neiburger (1963), and by Davis and Sartor (1967), and to fall with terminal velocities as described by Berry and Pranger (1974). The initial ice crystal spectrum was assumed to be Gaussian. The total ice crystal concentration was assumed to be $N_i = 250 \, \text{m}^{-3}$ or $2500 \, \text{m}^{-3}$, and the mean radius of the plate-like crystals was considered to be 205, 272, 313, 361, or 478 μm. The plate-like ice crystals were assumed to follow the size relationship of Auer and Veal (1970), to fall with a terminal velocity as described by Heymsfield (1972), and to collide with supercooled cloud drops with an efficiency as computed by Pitter and Pruppacher (1974). The graupel particles, on the other hand, were assumed to have a mass-size relationship and terminal fall velocities as given by Locatelli and Hobbs (1974), and to have collision efficiencies like that of the drops. The ice crystals were introduced into the cloud gradually, with an injection time varying between 1 and 180 sec. The stochastic growth of the ice crystals and drops was modeled by means of a modified Berry-Reinhardt scheme (see Section 15.2.4.1).

Figure 16-11 gives some selected results obtained by Beheng. The figure illustrates the evolution of the assumed ice crystal spectrum (Figure 16-11a) and the simultaneous evolution of the assumed cloud drop spectrum (Figure 16-11b). We note that the drop spectrum rapidly broadens with time. However, this broadening is considerably less pronounced in clouds where ice particles are

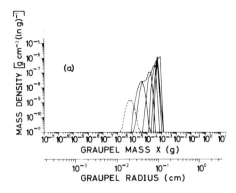

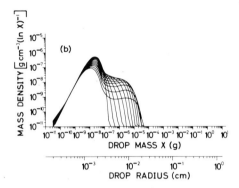

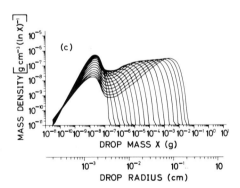

Fig. 16-11.   Evolution of a size spectrum of plate-like ice particles grown by riming, and simultaneous evolution of a size spectrum of cloud drops involved in the riming process; initial cloud drop mass distribution is that of Berry and Reinhardt (1974), with $(\bar{a}_d)_{initial} = 14\,\mu m$, $p = 0$, $(N_d)_{initial} = 86\,cm^{-3}$.

(a) Evolution of graupel spectrum for a Gaussian initial size distribution of plate-like ice crystals, with $\bar{a}_i = 272\,\mu m$, $N_i = 2500\,m^{-3}$, and an injection time of 180 sec; in 200 sec intervals up to 1300 sec; dashed line indicates initial ice crystal distribution.

(b) Simultaneous evolution of cloud drop spectrum; in 200 sec intervals up to 1800 sec.

(c) Evolution of cloud drop spectrum assuming no ice present in the cloud; in 100 sec intervals up to 1800 sec. (From Beheng, 1978; by courtesy of the Amer. Meteor. Soc., and the author.)

present which grow by riming than in warm clouds where ice particles are absent and where growth proceeds only by the collision and coalescence of drops (Figure 16-11c). This behavior is obviously a result of the depletion of drops by the riming process. Also note from Figure 16-11a that the drop depletion causes the graupel size distribution to narrow with time and to shift less and less on the mass axis.

16.6 Ice Particle Multiplication Processes

As might be expected from the immediately preceding discussion, presently available models for the cloud glaciation process are far from complete. For example, the model just discussed should be extended to include ice-ice as well as drop-drop interactions, and for these and the ice-drop interactions more realistic collision efficiencies should be used. In addition, some allowance should be made for various ice particle multiplication processes which are known to occur in clouds (recall Section 9.2.1). In the present section we shall briefly summarize our present knowledge of these processes.

The various known ice crystal multiplication mechanisms have been critically discussed by Mossop (1970, 1971) and Mossop *et al.* (1970, 1972). Of the enhancing mechanisms suggested, the three considered of greatest importance are: (1) mechanical fracture of fragile ice crystals such as dendrites, needles, sheaths, and capped columns caused by the collision of these crystals with graupel particles, other ice crystals, or large drops; (2) ice splinter formation during the riming of ice particles; and (3) fragmentation of relatively large individual cloud drops during freezing.

The efficiency of mechanical fracturing of delicate ice crystals was studied by Hobbs (1969, 1972, 1974b) and Hobbs and Farber (1972) in clouds over the Cascade Mts. These studies showed that the mechanical fracturing mechanism may, indeed, significantly enhance the ice particle concentration in these clouds. This conclusion was supported by the large number of ice crystal fragments collected in these clouds, and by the observation that over 50% of all stellar ice crystals collected at the ground had pieces (branches and portions of branches) missing. A similar observation was made on sector type crystals. Also, needles were often found fractured, especially at their tips. Ice crystals with spatial extensions and radiating assemblages of various crystal types often had portions missing. Similar observations were reported by Grant (1968) and by Vardiman and Grant (1972a, b) in clouds over the Colorado Rockies.

Several observations have implied that the presence of large drops leads to efficient ice multiplication. Thus, observations by Ono (1972) in Japan, Mossop (1970, 1972), Mossop *et al.* (1968, 1970, 1972) in Australia, and Koenig (1963) and Braham (1964) in Missouri are consistent with the interpretation that drops of diameter $\geq 250\,\mu m$ are required. These observations have been interpreted in terms of an ice multiplication mechanism involving the shattering or partial fragmentation of relatively large, freezing cloud drops.

Although earlier studies by Mason and Maybank (1960) suggested copious splinter formation during the freezing of supercooled drops, Dye and Hobbs

(1966, 1968) demonstrated that Mason and Maybank's results were biased by the presence of abnormally large concentrations of CO_2. They showed that drops freezing at rest in ordinary air and in thermal equilibrium with their surroundings did not shatter, although the formation of cracks, spikes, and proturberances was frequent. This result was essentially confirmed by Johnson and Hallett (1968) and Pena et al. (1969).

However, Johnson and Hallett found that whether or not a freezing drop shattered also depended crucially on whether or not the drop rotated during freezing, thus allowing a more or less radially uniform dissipation of the latent heat. This finding raised some questions as to the realism of drop splintering and shattering experiments which are carried out with drops that freeze while being suspended in a fixed orientation on fibers and other mounts. Under such conditions the lower or upstream side of a drop is colder than the upper or downstream side, since the former is much more effectively ventilated (see Section 13.2.3). Consequently, one would expect from the studies of Johnson and Hallett that the resulting asymmetric freezing would tend to inhibit shattering. Pitter and Pruppacher (1973), however, demonstrated by wind tunnel experiments that immediately after nucleation a freezing supercooled water drop which is freely falling at its terminal velocity in air begins to tumble and spin as it falls along a helical path, thus providing for a nearly radially symmetric heat loss from the drop to the environment during the initial stages of freezing. As expected from the findings of Johnson and Hallett, Hobbs and Alkezweeney (1968) found during experiments in which drops were allowed to fall freely in a long shaft that a small but measurable fraction of drops with diameters between 50 and 100 μm shattered at temperatures between -20 and $-32\,°C$; similar results were obtained at $-8\,°C$. However, none shattered if the drop diameter was less than 50 μm. Similarly, Kuhns (1968) found no shattering for 10 to 50 μm diameter drops of very pure water frozen between -36 and $-38\,°C$, and Brownscombe and Goldsmith (1972) found no shattering for drops of 20 to 50 μm diameter frozen at -10 to $-15\,°C$. On the other hand, Brownscombe and Thorndyke (1968) observed that a small fraction of 40 to 120 μm diameter drops, nucleated internally at temperatures between -5 and $-15\,°C$, shattered with 2 to 3 ice splinters being produced per drop. Similarly, a small fraction of 120 to 240 μm diameter drops shattered when nucleated at -5 to $-15\,°C$ by contact with ice crystals.

From their observations, Brownscombe and Thorndyke deduced that the ice particle enhancement factor (defined by Mossop et al. (1970) as the ratio of ice particles produced to drops frozen) was finite but rather small, ranging between values of 1.12 to 1.30. Only on one occasion did they find an enhancement factor of 2.45. Similarly, Bader et al. (1974) observed, for single free falling drops of 30 to 84 μm in diameter, an enhancement factor < 10. Takahashi and Yamashita (1969, 1970) deduced an enhancement factor of 1.74 near $-15\,°C$ for drops of 74 to 350 μm in diameter. Both at colder and warmer temperatures the enhancement factor was less. Pruppacher and Schlamp (1975) carried out wind tunnel studies with drops of 410 μm in diameter freely suspended in the air stream and nucleated by contact or freezing nuclei. They found an enhancement factor of

1.22 to 1.72 at temperatures between -7 and $-23\,°C$, the maximum enhancement occurring at temperatures between -11 and $-15\,°C$.

These observations suggest that shattering and splintering of freezing drops freely falling in the atmosphere results in an ice multiplication factor which at times surpasses a value of 2 but rarely, if ever, exceeds a value of 10. Further evidence that freezing drops at times do fragment in clouds has recently been provided by Knight and Knight (1974), who deduced from a photographic study of frozen drops preserved as hailstone embryos that drop break-up during freezing is a fairly common occurrence.

Until recently, considerable controversy has existed in the literature regarding the efficiency with which the ice crystal concentration in clouds is enhanced by the ice splinter formation which accompanies the riming of an ice particle. Experiments by Macklin (1960), and Latham and Mason (1961a) suggested that copious splintering accompanies the impaction of supercooled drops on an ice surface. However, Hobbs and Burrows (1966), Aufdermauer and Johnson (1972), and Brownscombe and Goldsmith (1972) failed to substantiate these findings. Recently, however, Hallett and Mossop (1974), Mossop and Hallett (1974), and Mossop (1976) appear to have resolved this quandary by demonstrating that the process depends rather sensitively on several factors, such as the drop size distribution, the liquid water content, the velocity of the drops impacting on a riming ice particle, the air temperature, and the surface temperature of the riming ice particle. They made observations on a cloud characterized by a liquid water content of approximately $1\,g\,m^{-3}$, with drop diameters from 5 to $45\,\mu m$ and total drop concentration of roughly $500\,cm^{-3}$, and found that ice splinter formation during riming was significant only for air temperatures between -3 and $-8\,°C$, drop diameters larger than $24\,\mu m$, and drop impact velocities between 1.4 and $3\,m\,sec^{-1}$. A pronounced maximum for splinter formation was found at a cloud air temperature of $-5\,°C$ and a drop impact velocity of $2.5\,m\,sec^{-1}$. At these optimal conditions, 1 secondary ice splinter was produced per 250 drops of diameter larger than $24\,\mu m$ impacting on the riming ice surface at $-5\,°C$, which was found to be equivalent to about 350 splinters produced per milligram of rime deposited (1 mg of rime is equivalent to a spherical graupel particle 2 mm in diameter and with a density of $0.24\,g\,cm^{-3}$).

Hallett and Mossop pointed out that, at temperatures at which riming is accompanied by copious splintering, ice crystals grow preferentially along the c-axis as needles or hollow columns. They therefore suggested that drop interactions with these extreme crystalline forms may be the cause of splinter formation. This explanation is supported by an earlier study of Macklin (1960), who investigated the structure of rime formed by drops of diameters up to $140\,\mu m$, impacting at $2.5\,m\,sec^{-1}$ on an ice surface at $-3\,°C$ in air at $-11\,°C$. Macklin found that at these warm temperatures and low impact velocities, the deposited rime was of very low density and consisted at the surface of very fragile columnar spicules which readily broke away on drop impact. Other possible splintering mechanisms during riming have been discussed by Mossop (1976). However, at the present time, none of these seems quantitatively satisfactory.

In support of these laboratory studies, the observations of Mossop (1970, 1972) and Mossop *et al.* (1968, 1970, 1972) in clouds over Australia, by Ono (1972) in Japan, and by Koenig (1963) and Braham (1964) in Missouri demonstrated that rimed ice crystals and graupel particles were invariably present before the ice particle concentration rose above that expected on the basis of the IN concentration which corresponded to the cloud top temperature. In this connection we note that Gagin's (1971) observation of negligible ice multiplication in winter cumuli over Israel appears consistent with the Hallett-Mossop temperature criterion. Such clouds have top temperatures considerably below $-10°C$. Thus it is probable that the initial riming growth of graupel particles in these clouds also occurs at temperatures colder than $-10°C$ at which, according to the Hallett-Mossop criterion, the formation of secondary ice particle is negligible.

As we have already indicated, theoretical models for assessing the rate of cloud glaciation must necessarily be rather complex, since a variety of cloud particle interactions must be included. Furthermore, the individual ice crystal production rates can generally be described only in terms of probability distributions, and the possibly significant coupling between the various growth processes is difficult to account for without making the overall formulation unwieldy. It is clear, for example, that a complete description of the glaciation of a cloud by a splintering avalanche process must include not only a description of the splintering mechanism itself, but also a simultaneous description of the growth of drops by collision and coalescence, of the growth of ice crystals by vapor diffusion and riming, and of the capture of ice splinters by supercooled drops. Unfortunately, neither the ice splinter production rates nor most of the relevant collection efficiencies are accurately known.

A few attempts have been made to develop models for glaciation by splintering (Koenig, 1966; Chisnell and Latham, 1974, 1976). Although these models attempt to take into account the probabilities of various cloud particle interactions, they nevertheless contain some serious deficiencies. For example, the numerical model of Koenig assumes that ice splinters are produced exclusively by the freezing of individual drops and that the number of splinters produced is that given by Mason and Maybank (1960) which is, due to the reasons mentioned earlier, much too large for atmospheric conditions. It is therefore not surprising that Koenig found from his calculations that the drop freezing and splintering mechanism converts cloud water rather rapidly to ice, even at temperatures as warm as $-10°C$.

The analytical model of Chisnell and Latham (1974) assumes, as does Koenig's, that ice splinters are produced only by the freezing of individual cloud drops which are assumed to eject a rather large number of splinters. In addition, the model neglects the evolution of the cloud drop size distribution during the glaciation process. In order to eliminate some of these deficiencies, Chisnell and Latham (1976) extended their analytical model to include ice-splintering by the Hallett-Mossop riming mechanism and to include the growth of ice crystals by diffusion and riming. However, the model is still far from complete. For example, the relevant growth processes are incorporated as though they proceeded independently. For this purpose the parameterized growth models of

Koenig (1972) are invoked. Also, collection efficiencies of unity are assumed for the interactions between drop and ice splinters, and between riming ice particles and supercooled drops. Also, all riming particles are assumed, independently of their size, to eject the same number of splinters per unit mass of accreted rime. Finally, the drop distribution is assumed stationary during the glaciation process. Since the collection efficiencies may be overestimated by a factor of as much as 10^2 (see Section 14.6), and no depletion in the concentration of supercooled drops is allowed, it is likely that splinter production rates are strongly over-estimated by this model.

All presently available theoretical descriptions of the glaciation mechanism in clouds with relatively warm temperatures assume that the initial number of ice crystals is that measured by contemporary IN counters. Thus, at the temperature at which the Hallett-Mossop splinter mechanism works most efficiently, one finds that the concentration of IN, i.e., ice crystals, is only about 1 to $10 \, cm^{-3}$. This represents a rather low initial ice crystal concentration, even for a powerful ice multiplication mechanism. In looking for conditions in the cloud which might provide a good running start for any ice multiplication mechanism, one could argue that the cloud drops at the cloud periphery are somewhat colder than the ambient air as a result of evaporative cooling. For example, from wet bulb temperatures derived from dew points observed on days when relatively warm cloud glaciation was observed in clouds over Missouri, Koenig (1965b) deter-mined that drops at the periphery of clouds could have cooled by as much as $6 \, °C$ below the ambient temperature. Similarly, Mossop et al. (1968) determined a maximum wet bulb depression of $4 \, °C$ for drops at the periphery of a cumulus cloud, located off the southern coast of Australia, which glaciated at $-4 \, °C$. However, these temperature depressions, considered together with the typical temperature variation of IN concentrations, suggest that evaporative cooling of drops at the edges of clouds can, at most, account for an increase of one order of magnitude in the ice crystal concentration, which is insufficient to explain the observed ice particle enhancement.

Recently, Young (1974) has suggested that the initiation of the ice phase in clouds which exhibit warm temperature glaciation may be the result of contact nucleation of supercooled cloud drops at the cloud edges. This suggestion seems to be supported by observations that the air around the top portion of the glaciating Australian and Missouri clouds was quite dry, and that dry aerosol particles serve as excellent ice-forming contact nuclei. Evidence for the effectiveness of ice formation by contact nucleation has been given by the wind tunnel studies of Gokhale and Spengler (1972) and Pitter and Pruppacher (1973), who showed that clay and soil particles of diameters between 0.1 and $30 \, \mu m$ nucleate supercooled water drops at temperatures as warm as -3 to $-4 \, °C$.

The number of ice crystals produced by contact nucleation may be estimated from the study of Young (1974), who determined the relative collection rate, i.e., the number of collection events per cubic centimeter per second, for drops capturing aerosol particles by Brownian diffusion, diffusiophoresis, and ther-mophoresis. On the basis of Young's computations, one finds that, in a cloud at $600 \, mb$ where the temperature is assumed to be $-5 \, °C$ and where the relative humidity is 98%, an ice crystal concentration of 30 to 3000 m^{-3} will be produced within 5 min if 10 μm radius drops are present in a concentration of 100 cm^{-3}, assuming that the drops are nucleated by contact nuclei with a radius of 0.03 μm and that the concentration of contact nuclei at −5 °C ranges between 1 and 10 ℓ^{-1}.

Assuming further that the primary ice particles thus formed will need about 12 to 20 min to grow by vapor diffusion and riming to graupel particles of 1 to 2 mm in diameter, and that during its growth each graupel produces about 350 secondary ice particles, the cloud may contain graupel particles in concentrations of 30 to 300 m^{-3}, and have a total ice particle concentration between 10 to 100 ℓ^{-1} within 20 min. Although this result is reasonable when compared with the observed concentrations, it must be remembered that the foregoing estimate is based on very approximate values for the concentration of contact nuclei in the atmosphere. Unfortunately, at present few quantitative studies on these concentrations are available, and among them there is little consensus. (For example, Blanchard (1957) arrived at the estimate of 100 ℓ^{-1} at −5 °C, while Vali (1974) estimated the concentration at 2 ℓ^{-1} at −16 °C.)

16.7 Melting of Ice Particles

When ice particles fall through the 0 °C level in the atmosphere they commence melting. Obviously, such melting is not instantaneous due to the finite rate at which heat can be supplied to provide for the necessary latent heat of melting. Due to their smaller mass-to-area ratio, smaller sized ice particles melt quicker than do larger ones. Because of this and the difference in fall speeds, generally only very large particles of the size of hailstones can survive the fall from a cloud base above 0 °C to ground. In this section we shall briefly discuss the quantitative details of such behavior, following the work of Mason (1956), Macklin (1963), Drake and Mason (1966), and Bailey and Macklin (1968b).

Let us consider a spherical, relatively large ice particle of radius a falling at terminal velocity in air of constant humidity and of constant temperature $T_\infty > T_0$, where $T_0 = 273$ °K. Since $T_\infty > T_0$, the ice particle melts and therefore consists at time t of an ice core of radius r surrounded by a layer of water of thickness $(a - r)$, assumed uniform and symmetrical. We shall also assume a = constant; strictly, this implies no water shedding or evaporation, but we shall nevertheless include the cooling effect of evaporation. Assuming a steady state, the rate of release of latent heat of melting must then be balanced by the rate at which heat is transferred through the water layer, so that

$$4 \pi \rho_i L_m r^2 \frac{dr}{dt} = \frac{4 \pi a r k_w (T_0 - T_a(r))}{a - r}, \tag{16-46}$$

where k_w is the thermal conductivity of water. (An empirical expression for k_w on the range $-20 \leqslant T °C \leqslant 35$, which agrees with the experimental data of Haywood (1966) to within $\pm 2 \times 10^{-6}$ cal cm^{-1} sec^{-1} °C^{-1} is given by

$$k_w = 135.8 \times 10^{-5} \exp(3.473 \times 10^{-3} T - 3.823 \times 10^{-5} T^2 + 1.087 \times 10^{-6} T^3), \tag{16-47}$$

with T in °C and k_w in cal cm^{-1} sec^{-1} °C^{-1}. In (16-46) T_a is the temperature at the surface of the liquid layer and T_0 is the temperature at the ice-water interface.

From (16-46) we thus find the time t_m for complete melting of the ice sphere to be given by

$$t_m = \frac{L_m \rho_i}{k_w a} \int_a^0 \frac{r(a-r)\,dr}{T_0 - T_a(r)}, \tag{16-48}$$

with T in °K. Since for a steady state the rate of heat transfer through the water layer is balanced by the rate of heat dissipation by forced convection and evaporation at the surface of the falling ice particle, $T_a(r)$ in (16-48) is found from

$$\frac{4\pi k_w a r(T_0 - T_a(r))}{a - r} = -4\pi a k_a(T_\infty - T_a(r))\bar{f}_h - 4\pi a L_e D_v(\rho_{v,\infty} - \rho_{v,a})\bar{f}_v, \tag{16-49}$$

where $\rho_{v,\infty} = \phi_v \rho_{v,sat}(T_\infty)$ and $\rho_{v,a} = \rho_{v,sat}(T_a)$, ϕ_v being the relative humidity of air.

Mason (1956) and Mason and Drake (1966) determined the melting time of ice particles from (16-48) and (16-49) for a variety of initial conditions. Mason's results are shown in Figure 16-12, and exhibit the qualitatively expected behavior that t_m should decrease, i.e., the fall distance for complete melting should decrease, with increasing humidity or with decreasing ice particle density. The dependence on density is such that a low density graupel particle survives about half the fall distance of a solid ice particle of the same initial radius, for comparable conditions. It is also interesting to note that the fall distance for complete melting varies nearly linearly with

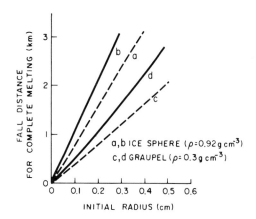

Fig. 16-12. Distance which a spherical ice particle has to fall from 700 mb level (0 °C) in the atmosphere in order to melt completely. Temperature lapse rate: 6.5 °C km^{-1}. —— no condensation or evaporation on ice particle surface; – – – ice particle falling in a water saturated atmosphere. (From Mason, 1956; by courtesy of *Quart. J. Roy. Meteor. Soc.*)

the initial particle radius. Wind tunnel studies by Drake and Mason (1966) have essentially confirmed the values of t_m predicted by (16-48) and (16-49) for $0 \leqslant T_\infty \leqslant 7\,°C$. However, at warmer temperatures they found that ice melts more slowly than is indicated by this simple theory. The discrepancies can probably be attributed to the neglect in the theory of the heat needed to raise the temperature of melt water above $0\,°C$.

CHAPTER 17

THE ELECTRICAL STATE OF THE ATMOSPHERE
AND ITS EFFECTS ON CLOUD MICROPHYSICS

The subject of cloud electricity is quite massive and complex in its own right, and has in addition many controversial aspects. Consequently, we shall make no attempt here to provide a comprehensive treatment of the entire subject; rather, we shall restrict ourselves primarily to summaries of some observed electrical properties of clouds and the particles contained within them, and to a consideration of the effects these electrical properties have on various microphysical processes. To a lesser extent we shall also consider concurrently the question of how the various types of clouds and cloud particles acquire their characteristic states of electrification. The topics of lightning, current budgets, and measurement techniques are omitted altogether.

Useful general references on atmospheric and cloud electricity include Israel (1970, 1973), Coroniti and Hughes (1969), and Chalmers (1967). Lightning is treated in detail by Uman (1969) and Golde (1977). Cloud charging mechanisms, and, to some extent, lighting phenomena are surveyed by Mason (1971, 1972).

17.1 Electrical State of the Cloudless Atmosphere

We shall first briefly summarize the main features of the 'fair weather' electrical state in order to establish the background conditions to which an isolated cloud is exposed, at least in its early stages of development. Under clear sky conditions, flat portions of the conducting Earth carry a negative surface charge density σ_0 which is approximately $\sigma_0 = -3.4 \times 10^{-4}$ e.s.u. cm^{-2}, or -1.1×10^{-9} G m^{-2} (1 e.s.u. of charge (1 statcoulomb) $\approx 1/3 \times 10^{-9}$ C(coulombs $\approx 2 \times 10^9$ e (elementary charges)). By Gauss's law there is a corresponding downward-directed surface electric field of magnitude $E_0 = 4 \pi\sigma_0$(e.s.u.) $= \sigma_0/\varepsilon_0$(m.k.s.) $\approx$ 130 V m^{-1}, where ε_0(m.k.s.) $= 8.85 \times 10^{-12}$ farad m^{-1} (1 e.s.u. of field strength (1 statvolt cm^{-1}) $\approx 3 \times 10^4$ V m^{-1}).

Given the existence of σ_0, one might expect the field strength in the atmosphere to vary with height in accordance with Coulomb's law (i.e., as d^{-2}, where d denotes distance from the center of the Earth). However, the rate of attenuation is much greater than this, owing to the existence of positive space charge which rapidly 'screens out' the surface field with increasing height. The main cause of this space charge is cosmic radiation, which provides an essentially continuous source of oppositely charged small ion pairs. Roughly speaking, the positive ions drift in the field toward the ground and encounter increasing resistance (decreasing atmospheric conductivity) as the air and aerosol density increase with decreasing height. This conductivity gradient leads to a 'traffic-jam' effect of space charge buildup with decreasing height.

An approximate empirical description of the variation of electric field strength with height is

$$E(V\ m^{-1}) = 81.8\ e^{-4.52z} + 38.6\ e^{-0.375z} + 10.27\ e^{-1.21z}, \tag{17-1}$$

where z is in km (Gish, 1944). From Gauss's law,

$$\nabla \cdot \vec{E} = \rho/\varepsilon_0, \tag{17-2}$$

the corresponding positive space charge density ρ (elementary charges cm^{-3}) is

$$\rho = 20.4\ e^{-4.52z} + 0.8\ e^{-0.375z} + 0.069\ e^{-0.121z}. \tag{17-3}$$

Thus at ground level the charge density is about $21\ e\ cm^{-3}$, while the average over the first kilometer of the atmosphere is about $5\ e\ cm^{-3}$.

The conduction current density or flux vector of charge due to ionic drift in the electric field is

$$\vec{j}_{q,cond} = \sum_i n_i q_i \vec{v}_i, \tag{17-4}$$

where n_i, q_i, and $\vec{v}_i$ respectively denote the number density, charge, and drift velocity of ion species i. The drift velocity due to the electric field is customarily expressed in the form

$$\vec{v}_i = \frac{q_i}{|q_i|}\ B_i \vec{E}, \tag{17-5}$$

where B_i is the *ionic mobility*. (Note this definition for mobility differs slightly from the one (B_i') introduced in (12-19): Since the force on the ion is $q_i\vec{E}$, (12-19) becomes $\vec{v}_i = q_i B_i'\vec{E}$, so that $B_i = |q_i|B_i'$.) From (12-21) the corresponding ionic diffusion coefficient is

$$D_i = \frac{B_i kT}{|q_i|}. \tag{17-6}$$

The small ions produced by cosmic rays and radioactive substances are generally singly charged molecules. They may become 'large' ions by attaching themselves to much larger aerosol particles. Since the mobilities of the large ions are usually 10^{-2} to 10^{-4} times less than those of the small ions, they contribute relatively little to the total ion current. Measurements show that the representative mobilities B_- of the small negative ions are greater than the corresponding mobilities B_+ of the positive ions; according to Bricard (1965), $B_+ = 1.4\ cm^2\ V^{-1}\ sec^{-1}$ and $B_- = 1.9\ cm^2\ V^{-1}\ sec^{-1}$ at STP. An extrapolation of these values to elevations $z \leqslant 10\ km$ in the Standard Atmosphere has been computed by Shreve (1970):

$$B_+ = 1.4\ e^{0.14z}, \quad B_- = 1.9\ e^{0.14z}, \tag{17-7}$$

where z is in km. Shreve also obtained expressions for the variation with altitude of the corresponding diffusivities (in $cm^2 sec^{-1}$), viz.,

$$D_+ = 3.6 \times 10^{-2} e^{0.092z}, \quad D_- = 4.8 \times 10^{-2} e^{0.092z}. \tag{17-8}$$

The values for $z = 0$ were obtained by substituting the corresponding mobilities and $T = 273\,°K$ into (17-6).

On substituting (17-5) into (17-4) for the case of small ions present in concentrations n_+ and n_-, we have

$$\vec{j}_{q,cond} = \Lambda \vec{E} = (\Lambda_+ + \Lambda_-)\vec{E}, \tag{17-9}$$

where Λ is the *total conductivity*, and

$$\Lambda_+ = en_+ B_+, \quad \Lambda_- = en_- B_- \tag{17-10}$$

are the *polar conductivities*. The conductivities $\Lambda_\pm \geq 0$ both contribute in the same way to Λ, since oppositely charged ions move in opposite directions in the electric field. Since the constant air-to-Earth conduction current density normally observed in fair weather is $j_{q,cond} = 2.7 \times 10^{-12}\,A\,m^{-2}$ (Gish, 1944), we find from (17-9) that the sea level value of conductivity is $\Lambda_0 \approx 2.7 \times 10^{-12}/130 \approx 2.1 \times 10^{-14}\,ohm^{-1}\,m^{-1} = 1.9 \times 10^{-4}\,e.s.u.$ (1 $ohm^{-1}\,m^{-1} = 9 \times 10^9\,e.s.u.$). The conductivity at other levels may be estimated from the relation $\Lambda = \Lambda_0 E/130$, where E is given by (17-1).

An interesting exercise is to calculate the time which would be required for $j_{q,cond}$ to reduce the fair weather surface charge density σ_0 to zero, if no charging mechanisms existed to maintain σ_0. The governing equation for this situation is

$$\frac{\partial \sigma_0}{\partial t} = -j_{q,cond} = -\Lambda_0 E_0 = -\frac{\Lambda_0}{\varepsilon_0}\sigma_0, \tag{17-11}$$

so that $\sigma_0(t) = \sigma_0(0) \exp(-t/\tau_c)$, where $\tau_c = \varepsilon_0/\Lambda_0 \approx 450\,sec \approx 7\frac{1}{2}\,min$. This short discharge time implies the existence of a very active Earth-charging mechanism. Observations indicate that thunderstorms world-wide collectively fulfill this role of serving as the 'battery' which maintains σ_0 near a constant value.

We may also estimate the fair weather small ion concentration from Λ. Since the ions are produced in pairs, it is reasonable to assume $n_- \approx n_+ = n \gg (n_+ - n_-)$ (this is justified *a posteriori*). Thus we may write $n \approx \Lambda/2\,eB_+$; at $z = 0$, 3, and 8 km this gives $n \approx 5 \times 10^2\,e$, $6 \times 10^2\,e$, and $8 \times 10^2\,e\,cm^{-3}$, respectively. Although in principle similar estimates may be made by solving directly for the steady state balance between ion production and loss rates (see also Section 17.3.3), some of the relevant quantities, and especially the ionization rate, have not been measured or calculated to a high degree of accuracy (see, for example, §43 of Israel (1970)).

17.2   Electrical State of the Atmospheric Aerosol

Aerosol particles acquire charge through Brownian deposition of ions. In turn, such charged particles experience Brownian coagulation which is enhanced or suppressed by electrostatic forces. We may formulate this problem using a slight generalization of the model of Section 12.5. Consider the relative motion of charged particles of volume v_2 toward a charged particle of volume v_1. We write the v_2-particle flux vector toward the v_1-particle as the sum of contributions due to diffusion and conduction, ignoring gas kinetic (finite mean free path) effects:

$$\vec{j}_q = -D_{12}\nabla n_2 + \frac{q_2}{|q_2|} B_{12}\vec{E}_{12}n_2, \tag{17-12}$$

where q_2 is the v_2-particle charge, and $\vec{E}_{12}$ is the electric field acting on a v_2-particle due to particle v_1. On using (17-6) and writing $q_2\vec{E}_{12} = -\nabla\Phi_{12}$, where Φ_{12} is the electrostatic interaction potential, (17-12) may also be expressed in the form

$$\vec{j}_q = -D_{12}e^{-\Phi_{12}/kT}\nabla(n_2e^{\Phi_{12}/kT}). \tag{17-13}$$

If we now make the reasonable assumptions (1) that Φ_{12} is a function only of the center-to-center distance $r \geq r_{12}$ and (2) that there is a steady state, then the constant v_2-particle flux J_q into any spherical surface $S = 4\pi r^2$ concentric with the v_1-particle is

$$J_q = -\int_S \vec{j}_q \cdot d\vec{S} = 4\pi D_{12}r^2e^{-\Phi_{12}/kT}\frac{d}{dr}(n_2e^{\Phi_{12}/kT}). \tag{17-14}$$

On integrating this equation and imposing the conditions $n_2(r_{12}) = 0$ and $\Phi_{12}(\infty) = 0$, we find

$$J_q = 4\pi D_{12}n_2(\infty)\left[\int_{r_{12}}^{\infty} \frac{e^{\Phi_{12}/kT}}{r^2}dr\right]^{-1}. \tag{17-15}$$

Therefore, the generalization of the Brownian collection kernel K_{ij} of (12-46) which includes the effects of particle charge is

$$K'_{ij} = K_{ij}Q_{ij}, \quad Q_{ij} = \left[r_{ij}\int_{r_{ij}}^{\infty} \frac{e^{\Phi_{ij}/kT}}{r^2}dr\right]^{-1} \tag{17-16}$$

(Fuchs, 1934).

Let us now use (17-16) to estimate the effect of charging on the coagulation rate. Under fair weather conditions aerosol charging is approximately symmetrical, since small ions are created in pairs and have roughly equal mobilities. Therefore, in equilibrium particles of any size have a charge distribution approximately symmetrical about zero charge, so that for every v_j-particle bearing charge $q_j > 0$, another has charge $-q_j$. For such particles coagulating with a v_i-particle of charge $q_i > 0$, the appropriate (Coulomb) interaction potential is

$$\Phi_{ij}(e.s.u.) = \pm\frac{q_iq_i}{r}. \tag{17-17}$$

This expression ignores induced (image) charges, which are insignificant in the present context (e.g., see Fuchs (1964), p. 307). On substituting (17-17) into (17-16) we find the effect of electrostatic repulsion is a decrease in the coagulation rate of v_i- and v_j-particles measured by the factor

$$Q_{ij}^+ = \frac{\tau}{e^\tau - 1}, \quad \tau = \frac{q_i q_j}{r_{ij} kT}. \tag{17-18a}$$

Similarly, electrostatic attraction enhances the coagulation rate by the factor

$$Q_{ij}^- = \frac{\tau}{1 - e^{-\tau}}. \tag{17-18b}$$

Because of the symmetry of the charging process, the overall effect may be estimated by the arithmetic mean of Q_{ij}^+ and Q_{ij}^-, viz.,

$$\bar{Q}_{ij} = \frac{\bar{\tau}(e^{\bar{\tau}} + 1)}{2(e^{\bar{\tau}} - 1)} \geq 1, \tag{17-19}$$

where $\bar{\tau}$ is the average value of τ. The coagulation rate of a symmetrically charged aerosol is thus higher than that of a neutral one. However, the effect turns out to be quite small: If we regard particle charge as providing an extra degree of freedom in the particle's energy budget, then by the equipartition theorem we have $\overline{q_i q_i}/r_{ij} = kT/2$ (e.g., Keefe $et\ al.$, 1959). Then $\bar{\tau} = 1/2$ and $\bar{Q}_{ij} \approx 1.02$, implying the charge carried by natural aerosols should have a negligible influence on their coagulation rate. This conclusion has been borne out by many experiments (e.g., see Devir, 1967).

Besides acquiring charge through ionic diffusion, aerosol particles polarize in the fair weather electric field, and thereby receive an additional ionic drift current. This problem is discussed in Section 17.3.1 for the case of charged drops. It can be dismissed in the present context in the following way: the energy acquired by an ion moving in the background field E over the characteristic length (the radius) r of an aerosol particle is small compared to the thermal energy kT. For example, for $r = 1\ \mu$m, $E = 1$ V cm^{-1}, and $T = 273$ °K, we have $eEr/kT \approx 4 \times 10^{-3}$. Consequently, Brownian diffusion of ions completely dominates under ordinary circumstances. Then the estimate $\bar{\tau} = 1/2$ remains good, which says that aerosol particles carry few, if any, charges on the average. Therefore, particle mobility in the field E remains low, so that the relative particle motion (and hence the coagulation rate) is changed only slightly by the presence of the field. For example, drift velocities of particles of 0.1 and 1 μm radius carrying one elementary charge in a field of 1 V cm^{-1} are roughly 10^{-4} and 10^{-5} cm sec^{-1}, respectively. Such small ordered velocities are insignificant compared to ambient air motions and sedimentation velocities.

The argument above which led to $\bar{\tau} = 1/2$ also implies a normal (Boltzmann) charge distribution centered about zero charge. Thus, if n(r, p) denotes the concentration of aerosol particles of radius r bearing p elementary charges, then $n(r, p) \sim n(r, 0) \exp(-pe^2/rkT)$, and the width (standard deviation) of the distribution is $\sigma = \sqrt{rkT}/e$. Also, the average number of charges, irrespective of sign, is given by

$$\overline{|p(r)|} = 2 \int_0^\infty pn(r, p)\, dp \Big/ \int_0^\infty n(r, p)\, dp = \frac{1}{e}\left(\frac{2rkT}{\pi}\right)^{1/2} \approx 300\, r^{1/2}, \tag{17-20}$$

for $0\,°C$ and r in cm (see also Section 17.3.2). These predictions are generally consistent with observations: For example, Whitby and Liu (1966) found that aerosol particles of $1\,\mu m$ diameter carry an average absolute charge of about $e/2$. Similarly, Israel (1973) quotes measurements which showed that $|p| \approx 7$ and 14 for particles of radius 5 and $20\,\mu m$, respectively; also, the distributions were found to broaden with increasing r, as expected.

17.3 Electrical Conductivity in Clouds

Cloud charging mechanisms are opposed by field-driven leakage currents. Hence the degree of electrification achieved will depend in part on the cloud conductivity, Λ_c. In this section we shall explore the dependence of Λ_c on field strength, ionization rates, and microphysical properties such as cloud particle concentrations, sizes, and charge states.

17.3.1 DIFFUSION AND CONDUCTION OF IONS TO CLOUD DROPS

The electrical conductivity in clouds is controlled by the local balance of ion sources and sinks. The dominant sinks are the cloud particles, which efficiently absorb ions through diffusion and conduction. The rate of ion attachment by diffusion alone has been treated in the previous section. Thus from (17-16)–(17-18) we see that a stationary drop of radius a and charge $pe > 0$, suspended in a region of negligible electric field and where the ambient concentrations of positive and negative ions are $n_+(\infty)$ and $n_-(\infty)$, absorbs positive ions at the rate

$$J_D^+(p) = \frac{4\,\pi a\,D_+ n_+(\infty)pc}{(e^{pc} - 1)},\tag{17-21a}$$

and negative ions at the rate

$$J_D^-(p) = \frac{4\,\pi a\,D_- n_-(\infty)pc}{(1 - e^{-pc})},\tag{17-21b}$$

where $c = e^2/akT \approx 6 \times 10^{-6}/a$ for $0\,°C$ and a in cm (Fuchs, 1934; Pluvinage, 1946; Bricard, 1949; Gunn, 1954).

If the magnitude of the background electric field is E, the relative strengths of the conduction and diffusion ion currents are measured by the ratio

$$\frac{|B_\pm \nabla n_\pm \cdot \vec{E}|}{|D_\pm \nabla^2 n_\pm|} \propto \frac{eEa}{kT} \equiv \gamma,\tag{17-22}$$

using (17-6) and applying the usual scaling arguments (cf. Section 12.7.1). From the analogy between (12-84) and (17-22), the parameter γ can be described as an 'electric' Peclet number. Conduction is clearly negligible when $\gamma \ll 1$, and it is for this case that (17-21) applies. Since $\gamma(\text{m.k.s.}) = 42.5\,Ea$ at $0\,°C$, we see that for a typical fair weather situation ($a = 10\,\mu m$, $E = 10^2\,V\,m^{-1}$, $\gamma \approx 0.04$), (17-21) provides a good estimate for the ion attachment rates.

Unfortunately, a general solution for simultaneous diffusion and conduction at higher field strengths is not available. However, a solution to first order in γ has been obtained by the method of matched asymptotic expansions (Klett, 1971b). (A description of this method is given in example 4 of Section 10.2.4.) The resulting modified ion attachment rates are given by

$$J^{\pm}_{D+C}(p) = \left(1 \pm \frac{\gamma pc}{2(1 - e^{\pm pc})}\right) J^{\pm}_D(p), \qquad (17\text{-}23)$$

where $J^{\pm}_D(p)$ are the rates expressed in (17-21).

For highly electrified clouds with $\gamma \gg 1$, the ion loss rate due to conduction overwhelms that from diffusion. This situation can be modeled by calculating the conduction currents to the polarized drop, assuming the ambient ion concentrations prevail right up to the drop surface. Let us proceed with this formulation by considering a drop of radius a and charge $Q > 0$ in a region where the background electric field is $\vec{E} = E\hat{e}_z$. Then the electric field in the vicinity of the drop may be expressed in spherical coordinates (and in e.s.u.) as

$$\vec{E} = \frac{Q}{r^2}\hat{e}_r + E\hat{e}_z + \frac{Ea^3}{r^3}(2 \cos \theta\hat{e}_r + \sin \theta\hat{e}_\theta), \qquad (17\text{-}24)$$

where θ is the polar angle measured from the direction of $\hat{e}_z$. The last term in (17-24) is the dipole field induced by the conducting drop. The radial field at the drop surface is therefore $(E_r)_{r=a} = 3 E \cos \theta + Q/a^2$, and can be seen to switch sign as θ passes through $\theta_0 = \cos^{-1}(-Q/3 Ea^2)$. Consequently, negative ions are conducted to the drop for $0 \leq \theta \leq \theta_0$, while positive ions flow to the surface where $\theta_0 < \theta \leq \pi$. Therefore, the positive and negative ion currents to the drop are

$$J^+_c = 2 \pi a^2 \Lambda_+ \int_{\theta_0}^{\pi} (E_r)_{r=a} \sin \theta \, d\theta = \frac{\pi \Lambda_+}{3 Ea^2}(Q - 3 Ea^2)^2, \qquad (17\text{-}25a)$$

and

$$J^-_c = 2 \pi a^2 \Lambda_- \int_0^{\theta_0} (E_r)_{r=a} \sin \theta \, d\theta = \frac{\pi \Lambda_-}{3 Ea^2}(Q_2 + 3 Ea^2). \qquad (17\text{-}25b)$$

In equilibrium $J^+_c = J^-_c$, which results in an expression for the drop charge as a function of radius, field strength, and polar conductivity ratio:

$$Q(\text{e.s.u.}) = 3 Ea^2 \left[\frac{(\Lambda_+/\Lambda_-)^{1/2} - 1}{(\Lambda_+/\Lambda_-)^{1/2} + 1}\right] \qquad (17\text{-}26)$$

(Pauthenier and Moreau-Hanot, 1931; Gunn, 1956). We thus find that the maximum charge attainable through conduction charging is $Q_{max} = 3 Ea^2$.

17.3.2 CONDUCTIVITY IN WEAKLY ELECTRIFIED CLOUDS

For the case $\gamma \ll 1$, we may assume the ion attachment rates per drop are described adequately by (17-21). However, to obtain the overall ion loss rate using this formulation we evidently need to know the distribution of drop charge which arises from diffusion charging. Fortunately, it turns out that only a small error is introduced if we assume that every drop carries the average charge of the distribution (assuming equal sized drops). To show this simplification is possible, we must first determine the stationary drop charge distribution function. Although, as we have indicated in Section 17.2, this is given to a good approximation by a Boltzmann distribution, it is of some interest to determine it more exactly. As emphasized by Fuchs (1964), the actual distribution differs in principle from a Boltzmann distribution since ions, once captured by a drop, are unable to leave it.

Let the fractional concentration of equal sized drops bearing charge pe be denoted by $n(p)$. We can determine $n(p)$ by the principle of detailed balancing, according to which the steady state rate at which drops of charge pe capture positive ions is equal to the rate at which drops of charge $(p + 1)e$ capture negative ions. Thus we have

$$n(p)J_D^+(p) = n(p + 1)J_D^-(p + 1), \qquad (17\text{-}27)$$

or

$$n(p) = \frac{(p - 1)(1 - e^{-pc})\Lambda_+}{p(e^{(p-1)c} - 1)\Lambda_-} n(p - 1), \qquad (17\text{-}28)$$

from (17-21). This recursion relation is easily solved by writing $n(p - 1)$ in terms of $n(p - 2)$, etc., and noting that $1 + 2 + \cdots + (p - 1) = p(p - 1)/2$; the result is

$$n(p) = n(0)\left(\frac{\Lambda_+}{\Lambda_-}\right)^p \frac{\sinh\left(\dfrac{pc}{2}\right)}{\dfrac{pc}{2}} e^{-p^2c/2} \qquad (17\text{-}29)$$

(Pluvinage, 1946; Sal'm, 1971).

In order to explore the implications of (17-29), let us first consider the special situation for which $\Lambda_+ = \Lambda_-$. Then if the total drop concentration is $N = \Sigma_p n(p)$, the average positive ion flux per drop, $\overline{J_D^+}$, may be expressed as

$$N\overline{J_D^+} = \sum_{p=-\infty}^{\infty} J_D^+(p)n(p) = 2 \sum_{p=0}^{\infty} J_D^+(p)n(p), \qquad (17\text{-}30)$$

using (17-27) and the symmetry conditions $n(-p) = n(p)$ and $J_D^+(-p) = J_D^-(p)$. On substituting (17-21a) and (17-29) with $\Lambda_+ = \Lambda_-$ into (17-30), we obtain

$$N\overline{J_D^+} = 2\,n(0)J_D^+(0) \sum_{p=0}^{\infty} e^{[p(p+1)c]/2} \approx \left(\frac{\pi}{2\,c}\right)^{1/2} J_D^+(0)n(0)e^{c/8}\left[1 - \mathrm{erf}\left(\frac{1}{2}\sqrt{\frac{c}{2}}\right)\right]. \qquad (17\text{-}31)$$

The last form on the right side has been obtained by approximating the sum over p by an integral. Since $c = e^2/akT < 10^{-2}$ for $a > 1\,\mu\mathrm{m}$ at $T = 273\,°\mathrm{K}$, (17-29) reduces approximately to $n(p) = n(0)\exp(-p^2c/2)$, so that $N \approx \sqrt{\pi/2}\,cn(0)$. Therefore, (17-31) becomes

$$\overline{J_D^+} \approx J_D^+(0)\left[1 - \sqrt{\frac{c}{2\,\pi}} + \frac{c}{8} + O(c^{3/2})\right], \qquad (17\text{-}32)$$

which is the desired result, namely that to within an error of a few percent we may calculate the total ion flux to the drops by assuming each carries the average (zero) charge. Furthermore, for this case the charge distribution is close to a Gaussian (Boltzmann) form, symmetric about zero charge.

Now let us suppose the average drop charge $\bar{p}e$ is not zero. From the nature of diffusion charging we might well expect the physics to be largely unchanged in this generalization, the only difference being a new reference level of average charge. Thus we would expect n(p) to be given approximately by a Gaussian distribution centered about $\bar{p}$, viz.,

$$n(p) \approx \frac{N}{(2\pi)^{1/2}\sigma} e^{-[(p-\bar{p})^2]/2\sigma^2}, \quad \sigma^2 = \frac{1}{c} = \frac{akT}{e}, \tag{17-33}$$

and that the total ion flux to the drops may be determined with sufficient accuracy by assuming each carries the average charge.

It is easy to show that these expectations are consistent with the more rigorous form of n(p) given by (17-29). Thus if we assume the average positive ion flux per drop is now $\overline{J_D^+} = J_D^+(\bar{p})$ (cf. (17-32)), and that similarly the average negative ion flux is $\overline{J_D^-} = J_D^-(\bar{p})$, then on setting these two rates equal for a steady state we find from (17-21) that

$$\bar{p} = \sigma^2 \ln \frac{\Lambda_+}{\Lambda_-} = \frac{akT}{e^2} \ln \frac{\Lambda_+}{\Lambda_-} \tag{17-34}$$

(Gunn, 1955). Now from (17-34) we find $\Lambda_+/\Lambda_- = \exp(\bar{p}c)$, which on substitution into (17-29) produces a charge distribution of the form

$$n(p) = n(0)e^{\bar{p}^2c/2} \frac{\sinh \dfrac{pc}{2}}{\dfrac{pc}{2}} e^{-[(p-\bar{p})^2c]/2} \approx n(0)e^{-[(p-\bar{p})^2c]/2}, \tag{17-35}$$

which is equivalent to (17-33).

With an electric field present the assumption of a Gaussian charge distribution is still reasonable if the energy acquired by an ion in moving a mean free path λ_a in the field direction is small compared with its thermal energy; i.e., if $eE\lambda_a \ll kT$ or $E \ll 2 \times 10^3$ V cm^{-1} for typical cloud conditions. The corresponding constraint on γ for 10 μm radius drops is $\gamma \ll 80$. Hence for $\gamma \lesssim 1$ the expression (17-23) may be used (with p replaced by $\bar{p}$) in conjunction with (17-33) to determine the first order effect of γ on $\bar{p}$:

$$\bar{p} = \left(1 + \frac{\gamma}{2}\right)\frac{akT}{e^2} \ln \frac{\Lambda_+}{\Lambda_-}. \tag{17-36}$$

Let us now use these results to explore the behavior of conductivity and the disposition of space charge in weakly electrified clouds. Within the cloud the attachment rates of ions to drops are much greater than the loss rates arising from ionic recombination. Therefore, if the local ionization rate is I (= number per unit volume of ions of either sign created per unit time), then for a steady state balance between the generation and loss rates we have to a good approximation from (17-21) and (17-23):

$$I = 4 \, \pi a \, ND_+ n_+ \left(1 - \frac{\bar{p}c}{2} + \frac{\gamma}{2}\right) = 4 \, \pi a \, ND_- n_- \left(1 + \frac{\bar{p}c}{2} + \frac{\gamma}{2}\right), \tag{17-37}$$

where we have made the assumption, justified above, that each of the N drops per unit volume of radius a carries the average charge $\bar{p}e$. To obtain (17-37) we have also retained just the first order term in $\bar{p}c$ from (17-21) and (17-23). From the preceding discussions we may expect this simplification to yield sufficient accuracy for $\bar{p} \lesssim 10^2$ under typical conditions (with $a \approx 10 \, \mu$m).

We see from (17-6), (17-10), and (17-37) that the total cloud conductivity Λ_c has, to $O(\bar{p}c)$ and $O(\gamma)$, the form

$$\Lambda_c \approx \Lambda_{co}\left(1 - \frac{\gamma}{2}\right), \quad \Lambda_{co}(\text{e.s.u.}) = \frac{Ic}{2 \, \pi N} = \frac{eIB}{2 \, \pi a ND}. \tag{17-38}$$

The quantity Λ_{co} is the estimate of total conductivity, due to Pluvinage (1946), for the case of uncharged drops (here B and D may be regarded as the average of the mobilities and diffusivities of the positive and negative ions). Equation (17-38) shows that the ambient field lowers the total conductivity, but that drop charge has no first order effect. Numerical evaluation with $I = 10 \, \text{cm}^{-3} \, \text{sec}^{-1}$ shows that Λ_c is typically 1/40 to 1/3 the clear air value at the same level, in good agreement with observations (e.g., Pluvinage (1946), Israel and Kasemir (1952), Allee and Phillips (1959)).

We may also use (17-37) to estimate the net ionic space charge in a region of charged drops. The ionic space charge density is $(n_+ - n_-)e$, while the drop charge density is $N\bar{p}e$; from (17-37), the ratio of these densities is

$$\frac{n_+ - n_-}{N\bar{p}} \approx \frac{Ic}{4 \, \pi a N^2 D}. \tag{17-39}$$

For typical conditions with $I = 10 \, \text{cm}^{-3} \, \text{sec}^{-1}$, $a = 10 \, \mu$m, and $N = 10^2 \, \text{cm}^{-3}$, we find the drop charge density is about 10^2 times greater than the ionic space charge density. Thus for practical purposes the charge in a region of cloud may be assumed to reside on the drops. Equation (17-39) also justifies a posteriori an assumption we have implicitly made in all of our models for calculating ion attachment rates to drops, namely that the ionic space charge is too dilute to have any noticeable screening effect on the Coulomb field of the charged drops. Finally, we see from (17-34) or (17-37) that the polar conductivity ratio is near unity ($\Lambda_+/\Lambda_- \approx 1 + \bar{p}c$) in weakly electrified clouds; hence the inequality $n_+ - n_- \ll n_+ \approx n_-$ ($\approx I/4 \, \pi a ND$) holds, just as for the case of a clear atmosphere.

17.3.3 CONDUCTIVITY IN STRONGLY ELECTRIFIED CLOUDS

The results of the previous section indicate that for a monodisperse cloud with average drop charge $\bar{Q}$, a good approximation to the equilibrium concentration of positive ions n_+ may be obtained by solving the following equation:

$$I = \alpha n_+ n_- + 4 \, \pi a ND_+ \, n_+ + \frac{\pi B_+ n_+ N}{3 \, Ea^2}(3 \, Ea^2 - \bar{Q})^2. \tag{17-40}$$

We have included in the first term on the right side an expression for the loss rate due to ionic recombination, the quantity $\alpha (\approx 1.6 \times 10^{-6} \, \text{cm}^3 \, \text{sec}^{-1})$ being the *recombination coefficient* for small ions. Outside the cloud we thus have $n_+ \approx n_- \approx \sqrt{I/\alpha}$. To first order in γ and $\bar{p}c$ ($\bar{Q} = \bar{p}e$) the diffusion and conduction terms in (17-40) reduce to $4 \, \pi a N D_+ n_+ (1 - \bar{p}c/2 + 3 \, \gamma/4)$, which is fairly close to (17-37); for $\gamma \gg 1$ the ion loss rate will be controlled by the last (conduction) term on the right side of (17-40). Hence (17-40) is reasonably accurate in both the low and high field limits.

Phillips (1967) used (17-40), along with its obvious counterpart for the negative ion concentration, to show that cloud conductivity is reduced sharply with increasing field strength and liquid water content, and more moderately with decreasing drop size for a given water content. The explanation for this last mentioned trend is simply that a given quantity of water presents an absorbing surface area which increases as it achieves a more finely divided state. (Incidentally, it is for this reason that we may safely ignore the effects of drop motion on ion capture rates. Ventilation effects are significant only for $a > 50 \, \mu\text{m}$ (see Section 13.2.3), and the drops in that fraction of the spectrum contribute negligibly to the total surface area of the cloud water, at least in the early stages of cloud development when electrification is weak. In the later stages, an increase in electrification will generally accompany any shift in the spectrum to larger drop sizes, so that ion drift velocities will grow faster than representative drop terminal velocities. Hence in-cloud drop motions probably never exert a significant influence on ion capture rates.) For example, for a cloud with $w_L = 1 \, \text{g} \, \text{m}^{-3}$ and $E = 0$, the cloud conductivity is less than the fair weather value at the same altitude by factors of about 40 and 10 for $a = 10$ and $20 \, \mu\text{m}$, respectively. If $E = 300 \, \text{V/cm}$ (1 e.s.u.) and the other condition remains unchanged, the reduction factors become 500 and 200. Changes in w_L alone produce proportionate changes in the reduction factor.

Since the mean spacing of drops is very large compared to their radii (see Section 14.1), it is reasonable to regard a cloud with a drop size distribution as a superposition of partial clouds, each of which absorbs ions in accordance with the diffusion and conduction terms of (17-40). This approach was adopted by Griffiths *et al.* (1974), who considered the extension of (17-40) to a cloud with a drop size distribution of the form of (2-1). They also considered the effects of ion emission (corona discharge) from hydrometeors for the case of very high electric fields. In addition, they investigated the dependence of cloud conductivity on the drop charge distribution by assuming several different forms for it and then calculating the response.

Some representative results for the case of no secondary ion emission by corona are shown in Table 17-1. The normalized conductivity values given are computed as the ratio of the cloud conductivity to the free air value at the same level, i.e., as $\hat{\Lambda}_c = \Lambda_c / 2 \, eB(I/\alpha)^{1/2}$. It is seen that $\hat{\Lambda}_c$ is sensitive only to changes in w_L and E, in agreement with the earlier results of Phillips. The effect of drop charge and the manner in which it is distributed over the drop spectrum are both quite small. The only significant effect of $\bar{Q}$ is to alter the relative ambient ion concentrations by about a factor of two for the case of drop charge levels near

TABLE 17-1

ormalized conductivity $\hat{\Lambda}_c$ and polar conductivity ratio Λ_+/Λ_- within a stratocumulus cloud as a function of eld strength E and the way in which a constant amount of positive charge is distributed over the droplet ▸ectrum; the three charge distributions are given by $Q = 2 \times 10^{-4} \, Ea$, $Q = 0.5 \, Ea^2$, $Q = 60 \, Ea^3$; $w_L = 0.5 \, g \, m^{-3}$, $z = 2 \, km$, $\Lambda_0 = 1.30 \times 10^{-3}$ e.s.u. (From Griffiths *et al.*, 1974; by courtesy of *Quart. J. Roy. Meteor. Soc.*)

E (kV m^{-1})	$\hat{\Lambda}_c (\times 10^4)$			Λ_+/Λ_-		
	$Q = 2 \times 10^{-4} \, Ea$	$Q = 0.5 \, Ea^2$	$Q = 60 \, Ea^3$	$Q = 2 \times 10^{-4} \, Ea$	$Q = 0.5 \, Ea^2$	$Q = 60 \, Ea^3$
0	84.5	84.5	84.5	0.98	0.98	0.98
3	60.4	60.4	60.4	1.2	1.2	1.2
9	38.9	39.1	39.1	1.4	1.4	1.4
30	17.7	17.8	17.7	1.7	1.7	1.7
300	2.22	2.24	2.23	1.9	1.9	1.9

the largest values observed in thunderstorms (cf. Table 17-1 and Figure 17-2). Thus we see that $(\Lambda_+/\Lambda_-)_{max} \approx 2$; for the case of pure conduction charging, we see from (17-26) that this corresponds to $\bar{Q}_{max} \approx 0.5 \, Ea^2$.

The experiments of Griffiths and Latham (1974) on the occurrence of corona from hydrometeors (Section 17.5.3) showed that typically corona onset requires $E_0 > 4 \, kV \, cm^{-1}$, and that for a reasonable concentration of corona sites in clouds the ionization rate due to cosmic rays must be negligible in comparison to that from corona discharge. Also, we see from Table 17-1 that for such large fields conduction completely dominates the other ion loss mechanisms, and further that drop charge has a negligible effect on the ion concentrations. Therefore, if we denote the corona ionization rate by I_c, we see that an adequate simplification of (17-40) for the purpose of estimating $n_+ \approx n_- = n_c$ when corona discharge occurs is just $I_c \approx 3 \, \pi B N E a^2 n_c$, or

$$n_c \approx \frac{I_c}{3 \, \pi B N E a^2}. \tag{17-41}$$

Griffiths *et al.* obtained the field-dependent values of I_c from the equation

$$I_c = 5 \times 10^5 [1 + 1.5 \times 10^{-4}(E - E_c)], \tag{17-42}$$

where I_c is measured in cm^{-3} sec^{-1} and $E > E_c$ is in V m^{-1}. The expression (17-42) is based on the experiments of Griffiths and Latham (1974), and the assumption that the number of active corona sites is 1 m^{-3} (this appears reasonable since the minimum size of an ice particle that can produce corona is about 2 to 3 mm diameter). The results of these computations are shown in Figure 17-1 for the case of a cloud of drops of radius 20 μm located at an altitude of 2.5 km (0 °C and 746 mb). The results for $E < E_c$ follow from (17-40) with $\bar{Q} = 0$. The figure shows that for 1 kV cm$^{-1} \lesssim E < E_c$ the ion concentration is typically about three orders of magnitude less than the clear air value at the same altitude, n_0 (≈ 1160 cm^{-3}). On the other hand, with the onset of corona ($E_c \approx 5.4 \, kV \, cm^{-1}$ for snowflakes and 7.2 kV cm^{-1} for hailstones), the value of n rises sharply to about an order of magnitude greater than the clear air value. Also, with corona the altitude z plays a more important role since E_c decreases with increasing z

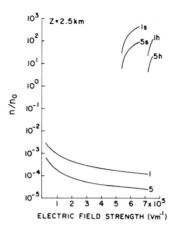

Fig. 17-1. Predicted variation of the normalized ionic concentration n/n_0 with electric field strength within a highly electrified cloud at an altitude $z = 2.5$ km; the symbols 1, 5, h, and s refer, respectively, to a liquid water content $1\,\mathrm{g\,m^{-3}}$, $5\,\mathrm{g\,m^{-3}}$, corona from hailstones, and corona from snowflakes; $n_0 = 1160\,\mathrm{cm^{-3}}$, number of active corona sites $= 1\,\mathrm{m^{-3}}$, air pressure $= 746$ mb, air temperature $= 0\,°\mathrm{C}$. (From Griffiths *et al.*, 1974; by courtesy of *Quart. J. Roy. Meteor. Soc.*)

(decreasing pressure); e.g., at $z = 5.5$ km, $E_c \approx 3.9\,\mathrm{kV\,cm^{-1}}$ for snowflakes and $5.4\,\mathrm{kV\,cm^{-1}}$ for hailstones.

Since E_c as measured by Griffiths and Latham is generally larger than observed values of electric fields in clouds (see Section 17.4.2.1), we may expect that cloud conductivities are generally very small, and that it is probably unnecessary to account for ionic leakage currents in calculations of electric field growth in clouds. Some workers have argued that effective cloud conductivities are often quite large, perhaps as much as 20 times the fair weather value, on the basis of the observed rapid recovery times of sudden electric field changes associated with thunderstorms (e.g., Freier, 1962; Colgate, 1969). However, a more convincing explanation for this behavior is that there is a rearrangement of charge over a large region of the atmosphere outside the cloud in response to the abrupt field change, so that recovery times (generally measured at the ground) are close to the short clear air relaxation times at higher elevations (Illingworth, 1971).

The difference in clear air and cloud conductivities causes a layer of space charge to form at cloud boundary regions, if a component of the electric field is normal to the boundary. For example, if we consider an idealized layer-type cloud with a boundary location given by $z = z_B$, then in the steady state the vertical conduction current must be continuous across the boundary, which leads to the relation $\Lambda_c E_c = \Lambda_0 E_0$ at $z = z_B$, the subscripts 0 and c referring to the values outside and within the cloud, respectively. Then from Gauss's law, (17-2), we find the boundary charge density is $\rho(\mathrm{m.k.s.}) \approx \varepsilon_0(E_c - E_0)/\Delta z = \varepsilon_0 E_c(1 - \Lambda_c/\Lambda_0)/\Delta z$, where Δz is the characteristic depth over which the charge is distributed. A crude estimate for Δz for the case of large fields ($\gamma \gg 1$) is that it is given by the conduction mean free path of ions, ℓ, in a cloud of uncharged drops. From (17-40) or (17-41) we see that the rate of depletion of ions by conductive attachment to uncharged drops is measured by $dn/dt = v_{\text{drift}} \, dn/dz = -3 \pi B N E_c a^2 n$; since $v_{\text{drift}} = B E_c$ we obtain $\Delta z \approx \ell = (3 \pi N a^2)^{-1}$ (typically, $5 \leqslant \ell \leqslant 50$ m), and so $\rho \approx \varepsilon_0 E_c / \ell$ with $\Lambda_c / \Lambda_0 \ll 1$. The corresponding charge per drop is

$$\bar{Q}(\text{boundary layer}) \approx 3 \pi \varepsilon_0 E_c a^2 (\text{m.k.s.}) = \frac{3 E_c a^2}{4} (\text{e.s.u.}). \qquad (17\text{-}43)$$

This generally large value of drop charge suggests our estimate for Δz is not very accurate. (Another source of error we have completely ignored is turbulent mixing at the cloud boundary, which can establish a characteristic cloud conductivity gradient, and hence a different characteristic value for Δz.) Nevertheless, our simple calculation suffices to illustrate how large drop charge densities can form from ion currents which enter clouds. (More detailed studies of the formation and structure of charge screening layers around clouds may be found in Brown et al. (1971), Hoppel and Phillips (1971, 1972), and Klett (1972).)

17.4   Cloud Electrification

17.4.1   WEAKLY ELECTRIFIED CLOUDS

Let us consider first the extent to which the preceding descriptions of charging by ion attachment can provide at least a qualitative basis for understanding the charge distributions observed in fair weather clouds. Since clouds are poor conductors, we would expect the fair weather electric field to deposit negative ions in the base of a newly formed cloud. We would further expect that this charge, essentially all of which will reside on the cloud particles, to be carried along in any updraft, so that the cloud should tend to develop a negatively charged core. On the other hand, the regions of the cloud near the upper surface should be positively charged from the positive ion current entering the cloud from above.

The above picture is in accord with most observations of fair weather clouds. For example, the field studies of Reiter (1964, 1969) showed that electrified stratocumulus, altocumulus, and convective cumulus clouds which contain neither ice nor precipitation-sized particles are electrically bi-polar with a pronounced negatively charged base and a positively charged upper portion. Also, Takahashi (1972) observed predominantly negatively charged drops in the bases of warm clouds in Hawaii. On the other hand, Twomey (1956) found a net positive charge in the bases of stratus and stratocumulus clouds. Krasnogorskaya (1969) observed predominantly negative charge in cumuli and stratified cumuli generally less than 1 km thick, but the location of the measurements relative to the cloud boundaries was not reported.

Although the observed overall pattern of charges in fair weather clouds is usually in agreement with what we would expect from the action of ion capture processes, the same cannot be said for the quantitative details of the distribution of charge among individual cloud particles. In particular, the observed particle charges are usually much larger than what would follow from diffusion and conduction charging alone. The best case of agreement with the theory of diffusion charging was provided by Phillips and Kinzer (1958), who measured the size and charge of drops in stratocumulus clouds with 'near normal' fair weather electric fields. They found the observed charge distributions were represented moderately well by (17-33) with $\bar{p} = 0$. For this case the mean absolute charge is given by (17-20); for example a $10 \, \mu$m radius drop carried about 10 elementary charges on the average, irrespective of sign. However, much larger drop charges were reported by Twomey (1956) under apparently similar conditions (unfortunately, the field strength was not stated). As noted above, Twomey found mainly positively charged drops; for these the charge versus size relation could be expressed as $\bar{Q}$(e.s.u.) $\approx 0.3 \, a^2$(cm). According to this result, for example, a $10 \, \mu$m radius drop usually carried about 6×10^2 positive elementary charges, in striking contrast to the observations of Kinzer and Phillips. (As we have seen, conduction charging produces a quadratic dependence of charge on radius also. However, under fair weather conditions we would generally expect the magnitude of charge to be somewhat less than that found by Twomey. If, for example, we were to suppose that he measured drop charges acquired through conduction charging in the boundary layer, then from (17-43) we see that the cloud electric field would have to have been about 10^2 V cm^{-1}, which implies $\Lambda_c/\Lambda_0 \approx 10^{-2}$. But for stratocumulus it is doubtful that the liquid water content would be large enough to achieve such a low cloud conductivity.) Krasnogorskaya (1969) reported two characteristic types of distributions in stratocumuli, a near Gaussian distribution centered about zero mean charge, and an asymmetrical distribution displaced toward negative charge values. In general the mean absolute charge could be expressed as $\overline{|Q|} = ca^2$, with $0.28 \leqslant c$(e.s.u. cm^{-2}) $\leqslant 1.6$, the exact value of c depending on the cloud shape and stage of development. Electric field values up to 30 V cm^{-1} were present. Colgate and Romero (1970) measured charges on small drops ($a \leqslant 12 \, \mu$m) "in the lower few hundred meters and at an early stage of a forming thunderstorm." They expressed their results in e.s.u. as $\overline{|Q|} = 1.72 \, a^2$, and reported in addition an average negative charge of about 4×10^2 elementary charges "at each of several drop sizes." The electric field was not measured.

It should be emphasized that although Twomey, Krasnogorskaya, and Colgate and Romero were all able to fit their data to a quadratic charge versus size relation, their respective measurements are really quite different, since Twomey found an average strong positive drop charge, Krasnogorskaya an average weak negative charge, and Colgate and Romero an average strong negative charge. In the literature such differences have often been minimized or overlooked altogether, with the result that it has often been considered reasonable, for purposes of modeling cloud electrification processes, to assume that all small cloud drops carry positive charge in proportion to their surface area (e.g., Colgate (1972), Pringle et al. (1973)). Further discussion of this point may be found in Illingworth (1973).

17.4.2   STRONGLY ELECTRIFIED CLOUDS

The sampling of experimental results given in the previous section illustrates the present unsettled state of the subject of cloud electrification, even for nearly fair weather conditions. It is obvious that further comprehensive measurements are needed, and that these should include more information on the prevailing cloud conditions. We can appreciate, for example, that particle charge measurements can be properly interpreted only if we also know the location of the measurements in the cloud, the type of cloud and its stage of development, and the ambient electric field and liquid water content (not to mention cloud turbulence levels and airflow patterns). Ideally, we should have available the full particle charge and size distributions, for both the liquid water and ice phases, in order to assess the role played by various particle-interaction charging mechanisms. Finally, this information should be obtained for several locations within the cloud simultaneously, since there is evidence that different cloud regions contribute in different characteristic ways to the overall growth of electrification. Of course, such a comprehensive observational program is not likely to be forthcoming in the near future, owing to massive logistical and instrumentation problems.

Unfortunately, the theoretical side of the subject is also in some disarray. An embarrassingly large number of possible charge separating mechanisms involving drops and/or ice particles have been conjectured and/or identified, and with few exceptions it is not very clear which combinations of them, if any, may be of significance in real clouds. A large part of the problem here stems from an insufficient knowledge of the quantitative details of the proposed charging mechanisms, especially in the context of a natural cloud environment. Also, of course, the often contradictory results of observations, illustrated above by the example of drop charge measurements, provides an inadequate foundation for the construction of theoretical models.

In view of these many difficulties, and also because it is too large a subject to be dealt with effectively here, we shall not attempt to provide a detailed treatment of the microphysics of cloud electrification. At this point we shall instead merely extend our previous description of charge and electric field observations to include the case of strongly electrified clouds, and then discuss very briefly as illustrative examples a few currently prominent theoretical models of electrification. More comprehensive surveys of possible cloud charging mechanisms may be found in Israel (1973), Mason (1971, 1972), Stow (1969), Chalmers (1967), and Reiter (1964).

17.4.2.1 *Observed Charges and Fields*

A summary of measured values of absolute charge on cloud drops up through precipitation size for situations of both strong and weak electrification has been compiled by Takahashi (1973), and is shown in Figure 17-2. Various empirical fits can be made to such data. A fair approximation for the larger values of charge (in e.s.u.) is given by curve 3 in the figure, which is a plot of the

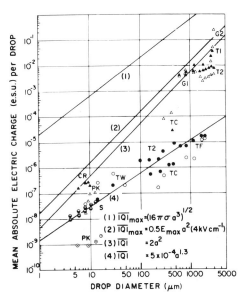

Fig. 17-2. Mean absolute electric charge on cloud and raindrops. Round symbols indicate warm cloud cases, triangular symbols indicate thunderstorm cases; solid symbols indicate negative charge, open symbols indicate positive charge. PK Phillips and Kinzer (1958), S Sergieva (1959), CR Colgate and Romero (1970), TW Twomey (1956), TC Takahashi and Craig (1973), T1 Takahashi (1965), T2 Takahashi (1972), TF Takahashi and Fullerton (1972), G1 Gunn (1949), G2 Gunn (1950). (Adapted with changes from Takahashi (1973).)

relationship

$$\overline{|Q|} = 2\,a_0^2, \tag{17-44}$$

where a_0(cm) is the equivalent drop radius. Similarly, an approximate fit of the data over the full drop spectrum for warm clouds (with no ice phase, and generally having relatively weak electrification) is

$$\overline{|Q|} = 5 \times 10^{-4}\,a_0^{1.3}, \tag{17-45}$$

which is plotted as curve 4 in Figure 17-2.

We have already mentioned the fact that for the case of small drops different workers have reported different prevailing charge signs. The same variability extends to larger drops as well. Thus, according to Takahashi (1972) and Takahashi and Craig (1973), drizzle drops falling from warm clouds appear to carry predominantly negative charge, while from thunderstorm clouds they are predominantly positively charged. Also, the charge sign for rain may depend on drop size. For example, Takahashi and Fullerton (1972) found that raindrops from warm clouds were predominantly negatively charged if $a_0 \lesssim 850\ \mu$m, and mainly positively charged for larger sizes. Just the opposite charge-sign versus size relationship has been observed for thunderstorms by Smith (1955) and Takahashi (1972). Their data appear consistent with a predominance of negatively charged drops for $a_0 \gtrsim 800\ \mu$m, and positively charged drops for smaller sizes. Other trends have been reported also (see Takahashi, 1973), so that overall there appears not to be a great deal of consistency in the measurements.

Numerous field observations of electric field changes associated with lightning discharges (see, for example, Chalmers (1967) and Mason (1971)) indicate that the charge distribution in a typical thunderstorm is roughly like that shown in Figure 17-3. This figure implies field strengths as large as $(4\pi\varepsilon_0)^{-1} \times 80 \times (3 \times 10^3)^{-2} \approx 10^5 \text{ V m}^{-1}$ may occur between the main positive and negative charge centers. Field strengths required to produce local dielectric breakdown of air, and hence lightning, may be perhaps one order of magnitude larger than this representative average maximum value. Thus, from aircraft Gunn (1948) measured mean maximum storm field intensities of $1.3 \times 10^5 \text{ V m}^{-1}$. On one occasion a field strength of $3.4 \times 10^5 \text{ V m}^{-1}$ was observed just before lightning struck the aircraft. Fitzgerald and Byers (1962) observed fields up to $2.3 \times 10^5 \text{ V m}^{-1}$, while Kasemir and Holitza (1972) reported fields up to $3 \times 10^5 \text{ V m}^{-1}$. Using instrumented rockets, Winn et al. (1974) observed a field of $4 \times 10^5 \text{ V m}^{-1}$ on one occasion, while peak values of 10^5 V m^{-1} were encountered 10% of the time.

It is interesting to note that none of these values is close to the so-called 'dielectric strength of air,' $E_s = 3 \times 10^7 \text{ V m}^{-1}$, which represents the approximate field strength required to initiate breakdown between plane parallel electrodes in dry air at STP. In clouds, the large surface curvature of some cloud particles can apparently cause sufficient local field enhancement and ion emission to initiate large scale breakdown when the ambient field is only about one percent of E_s (see also Section 17.5.3).

It is of some interest to use the observed maximum values of electric field to estimate the corresponding maximum drop charge which would occur by conduction charging. As we discussed in Section 17.3.3, this is given approximately by $0.5 E_{max}a^2$, assuming $(\Lambda_+/\Lambda_-)_{max} \approx 2$ in accordance with the

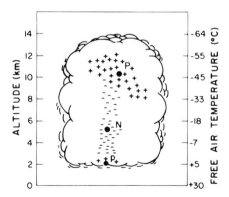

Fig. 17-3. Probable distributions of the thundercloud charges, P, N, and p for a South African thundercloud according to Malan (1952, 1963); solid black circles indicate locations of effective point charges, typically P = +40 C, N = −40 C, and p = +10 C, to give observed electric field intensity in the vicinity of the thundercloud. (From *Lightning* by M. A. Uman, copyrighted by McGraw-Hill Book Co., 1969.)

conductivity study of Griffiths *et al.* (1974). On substituting the value $E_{max} \approx$ 4×10^5 V m^{-1} (Winn *et al.*, 1974), we thus estimate $\overline{|Q|}_{max} \approx 7\,a^2$. This result, plotted as curve 2 in Figure 17-2, lies only slightly higher than many observed drop charge values under thunderstorm conditions. It therefore seems likely that the average magnitude of charge carried by particles of a given size in highly electrified clouds is fairly close to the equilibrium value arising from conduction charging in the ambient electric field.

The maximum charge that it is physically possible for a drop to carry is given by the *Rayleigh limit* Q_R for disruption (Rayleigh, 1882). This expresses the condition that mechanical instability occurs when the surface electrostatic stress equals the surface tension stress, i.e., when $E^2/8\,\pi = 2\,\sigma/a$; with $E = Q_R/a^2$, we thus have

$$Q_R\text{(e.s.u.)} = (16\,\pi\sigma a^3)^{1/2}. \tag{17-46}$$

This result is plotted as curve 1 in Figure 17-2, and indicates that drops in clouds are generally far from the Rayleigh limit. On the other hand, the Rayleigh limit has been reached in the laboratory by the controlled evaporation of charged drops on the size interval $30 \leqslant a \leqslant 170\ \mu$m (Doyle *et al.* (1964), Abbas and Latham (1967), Roulleau and Desbois (1972), and Dawson (1973)). (On reaching the Rayleigh limit, the drops were observed to undergo explosive mass disjection, the extent of which varied in the experiments from a mass loss of only a few percent (Roulleau and Desbois, Dawson) to up to 30% (Abbas and Latham, Doyle *et al.*)). It is therefore conceivable that the Rayleigh limit may be reached on occasion by evaporating drops at the edges of clouds, assuming that ionic leakage currents are relatively ineffective.

Figure 17-3 also implies the extensive presence of the ice phase under typical conditions of strong cloud electrification, and so it is of interest to study the charges on ice particles as well as water drops. Unfortunately, most early studies on the sign and magnitude of the charge carried by ice particles showed little consistency, due to experimental difficulties in measuring the charge and also because of the different conditions under which the observations were made (see Chalmers, 1967). However, the more recent studies of Isono *et al.* (1966), Magono and Orikasa (1966), Burrows and Hobbs (1970), and Kikuchi (1973, 1975) were carried out under better controlled conditions and with more sophisticated instruments, and consequently give a somewhat more unified picture. Burrows and Hobbs found that 99% of the ice particles studied carried a charge less than $Q_{99\%}$, where $10^{-3} \leqslant Q_{99\%} \leqslant 10^{-2}$ e.s.u., while 50% of the particles carried a charge less than $Q_{50\%}$, where $2 \times 10^{-4} \leqslant Q_{50\%} \leqslant 5 \times 10^{-3}$ e.s.u. These results are also consistent with the observations of Magono and Kikuchi (1961) and the other authors cited above.

Most of the field studies mentioned above were made at ground level in snowstorms which may or may not have contained convective cells of a structure similar to the one shown in Figure 17-3. Unfortunately, neither detailed electric field measurements nor detailed charge versus ice particle size measurements were made. However, Burrows and Hobbs observed that most particles which carried a relatively small magnitude of charge were negatively charged, while most particles which carried a relatively large magnitude of charge were positively charged. On comparing this outcome with the observations of Magono and Kikuchi, who found that the charge on ice crystals increased with increasing crystal mass, it was concluded that small ice crystals are more likely to carry negative charge, and large crystals positive charge.

A noticeable correlation between ice crystal shape and charge sign was found by Isono *et al.* (1966), Magono and Orikasa (1966), and Kikuchi (1973, 1975). Their observations showed that crystal plates, dendritic crystals, sector plates, and small snowflakes of a few dendrites predominantly carry negative charge, while bullets, columns, combined columns and side planes, and large snowflakes are mainly positively charged. Rimed ice crystals and soft hail or graupel particles were also observed to carry predominantly positive charge. To some extent these results appear consistent with the distribution of cloud-charge sign versus temperature shown in Figure 17-3, and the temperature dependence of ice crystal growth habits discussed in Section 2.2.1. However, the apparent correlation depends on the assumption that the observed ice particles actually fell from a cloud system with a charge versus temperature structure resembling that shown in Figure 17-3, and also on the assumption that little change in the charge state of the particles occurred as they fell to Earth.

17.4.2.2 *Models for Cloud Electrification*

The construction of theoretical models of cloud electrification is certainly one of the more interesting, although controversial, areas of cloud physics. Therefore, even though we must forego a detailed treatment for the reasons stated at the beginning of Section 14.4.2, we would like now to discuss briefly and in a largely qualitative manner a sampling of four prominent models. This will also provide an opportunity to identify particularly relevant microphysical processes which are in need of further study. General references for a more comprehensive approach to this topic have already been provided (just before Section 17.4.2.1); others of a more specific nature will be given below.

1. Convection Charging. Grenet (1974), and, independently, Vonnegut (1955) proposed that a convective cloud may operate as an electrostatic energy generator according to the following scenario: Initially, an updraft carries positive space charge from the lowest levels of the troposphere into the growing cloud; the electrified cloud soon acquires a negative charge screening layer at its edges due to cloud particle capture of ions drifting from clear air to cloud under the influence of the main positive charge; finally, downdrafts carry the negative charge closer to the ground, thereby increasing the (reversed) electric field at the Earth sufficiently to initiate positive point discharge which enhances the positive charge entering the cloud via the updraft. The continuance of this positive feedback cycle can thus provide for a strong buildup of electrostatic energy at the expense of the organized cloud convective motions.

We see that according to the above picture the cloud in its early stages is predicted to have a positively charged base and core, which is in conflict with the reasoning and the bulk of the evidence presented in Section 17.4.1. The cause of the disagreement is evident: In Section 17.4.1, we ignored the expected upward convection current of positive space charge; on the other hand, the convection model ignores the net upward conduction current of negative charge. However, a simple estimate shows that the latter should dominate the former. Thus, if we assume a fair weather electric field of $E \approx 10^2 \, \text{V m}^{-1}$, a negative ion concentration $n_- \approx 5 \times 10^2 \, \text{cm}^{-3}$ (see the paragraph following (17-11)), a negative ion mobility $B_- \approx 2 \, \text{cm}^2 \, \text{V}^{-1} \, \text{sec}^{-1}$ from (17-7), a reasonable updraft speed of $W \approx 1 \, \text{m sec}^{-1}$, and an average positive space charge density at cloud base level of $\rho_+ \approx 5 \, e \, \text{cm}^{-3}$ (recall from Section 17.1 that this is the average density over the first kilometer of the atmosphere), we find that the ratio of the convection and conduction currents is $\rho_+ W / e n_- B_- E \approx 5 \times 10^{-3}$. Hence it appears the neglect of the field-driven deposition of negative ions into the base of a young cloud constitutes a very serious flaw in the convection charging model.

This conclusion is supported by numerical simulations of the convective electrification process (Ruhnke (1970, 1972), Chiu and Klett (1976)). These studies show that for usual conditions of cloud formation a charge distribution in general qualitative agreement with that discussed in Section 17.4.1 is produced, namely a negatively charged cloud core capped by a relatively thin positively charged upper layer. However, Chiu and Klett also found that convective transport of positive charge may dominate the conductive transport of negative charge, and thus produce a cloud of polarity opposite to the usual case, if the cloud forms near ground level. This happens primarily because higher concentrations of positive space charge are then available to be carried aloft into the cloud. (Incidentally, this last result may conceivably have some bearing on the fact that drop charge measurements such as Twomey's (1956), which were taken on a mountain summit imbedded in the base of stratocumulus clouds, have often revealed a predominance of positive charge.)

For a lively debate on other aspects of the convective electrification model, and its merits relative to those of the various precipitation charging mechanisms, see Moore (1976a, b) and Mason (1976a, b), and the references given therein.

2. Particle Charging by Selective Ion Capture. Wilson (1929) described how an electrically polarized cloud particle may selectively capture ions of one sign as it falls. This happens because while the lower surface of the particle may easily attract and capture ions which carry a sign opposite to the local surface charge, the upper surface is not as effective in this respect, since ions attracted to it must first catch up with it in order to be captured. The net effect of this selective process is a large scale separation of charge, due to the sedimentation of the charged cloud particles. This reinforces the existing field, so that its occurrence in clouds would cause a field enhancement in qualitative agreement with what is expected for thunderstorms (recall Figure 17-3).

A mathematical model for the 'Wilson process,' based on spherical particles in Stokes flow (and thus of restricted validity), has been worked out by Whipple and Chalmers (1944). As might be expected, the equilibrium charge for this process is proportional to the ambient field strength and particle surface area. Also, the process becomes ineffective for large fields, since then the ionic drift velocities greatly exceed drop terminal velocities. A simple estimate by Mason (1971) suggests the mechanism may produce fields only as large as 10^2 V cm^{-1}, which is still three orders of magnitude below characteristic thunderstorm values.

3. Particle Charging by Thermoelectric Effects. Reynolds *et al.* (1957) obtained laboratory evidence that a hail pellet may become charged as a result of collisions with ice crystals having a temperature different from that of the pellet. The physical basis of the charge transfer was suggested as being due to the diffusion of hydrogen ions down the temperature gradient existing in the region of momentary contact (Brook, 1958). Thus, since H$^+$ ions have a greater mobility in the ice lattice than OH$^-$ ions, a temperature gradient maintained across a piece of ice will result in an excess of positive charge on the colder portion. Mason (in Latham and Mason, 1961b) formulated a one-dimensional model for this process on the basis of an ideal ice structure and found that the magnitude of equivalent surface charge density σ on the ends of an ice rod having a prescribed steady state temperature gradient dT/dx (°C cm^{-1}) is given approximately by $\sigma \approx 5 \times 10^{-5}$ dT/dx e.s.u. cm^{-2}. Similar results follow from a more complete theory of the process in real ice having orientational or Bjerrum defects (see Section 3.3), worked out by Jaccard (1963).

Since ice crystals and graupel or soft hail particles are often present under conditions of strong cloud electrification (e.g., Kuettner, 1950), and since the graupel particles are generally slightly warmer than the environment due to the latent heat released from accreted supercooled drops, it seems reasonable to expect that cloud conditions often favor the occurrence of thermoelectric charging. Although this mechanism would lead to a charge distribution qualitatively like that normally observed in thunderclouds, since by relative sedimentation under gravity the more massive and negatively charged graupel particles would predominate in the lower cloud levels (and vice versa for the lighter and positively charged ice crystals), the amount of charge which can be separated under natural conditions remains in question, since different investigators have obtained quite different results. Apparently the outcome depends sensitively on such factors as the time of contact (little charge transport can occur if this is too short, whereas the local temperature gradient is decreased by long contact times), the microtopography of the areas of contact, the presence of impurities in the ice, and the relative velocity of impact (which, for example, may control the amount of local frictional heating, and affect the contact time). All of these factors are difficult to control in an experimental situation, and hence their effects are not easily evaluated. There evidently remains a need for further studies of the microphysics of thermoelectric charging (see Stow (1969) and Mason (1971, 1972), and the references given therein for more information on this subject).

4. Particle Charging by Induction. From (17-24) we see that an uncharged drop polarized in the fair weather electric field carries a surface charge density σ e.s.u. $= (4 \pi)^{-1}(E_r)_{r=a} = 3 E \cos \theta/4 \pi$, θ being the polar angle measured from the lowest point on the drop; thus the lower hemisphere is positively charged, and the upper hemisphere negatively charged. Therefore, if such a drop were to experience on its lower hemisphere a momentary electrical contact with, and subsequent separation from, a similarly polarized smaller drop, there would result a net negative charge on the larger drop and a positive charge of equal magnitude on the smaller drop. Elster and Geitel (1913) were the first to point out that such a process of inductive charge transfer, occurring throughout a cloud and followed by the large scale separation of charge through relative sedimentation under gravity, would serve to increase the in-cloud electric field in the sense normally observed in thunderstorms.

Elster and Geitel also provided a simple estimate of the maximum charge transfer that could occur for the case of a sphere of radius a_2 which contacts the lowest point ($\theta = 0$) of a sphere of radius $a_1 \gg a_2$. In this case the smaller sphere will acquire a charge with average density equal to $\pi^2/6$ times the density of the charge on the larger sphere at the point of contact, since the curvature of the larger sphere can be ignored (the capacitance of the small sphere in contact with a conducting plane is $C_2 = a_2(\pi^2/6)$; see also Appendix A-17.5.5). Thus the maximum charge that can be acquired by the small sphere is approximately $(\pi^2/6)(4\pi a_2^2) \times (3E/4\pi) = \pi^2 E a_2^2/2 \approx 5 E a_2^2$. This is generally a very large charge magnitude (for example, it is ten times larger than the equilibrium charge expected from ion attachment by conduction under conditions of strong electrification – recall Section 17.3.3 and Figure 17-2), and suggests that induction charging may be powerful enough to produce strong cloud electrification.

However, any quantitative assessment of the efficacy of induction charging (also referred to as polarization or influence charging) must take into account several other factors as well. For example, let us first consider the more realistic situation where contact occurs for $\theta \neq 0$, and the drops initially carry charges Q_1 and Q_2. Then by a simple extension of the arguments given above the charges after contact and separation will be Q_1' and Q_2', where $Q_2' = (\pi^2/6)a_2^2(3E\cos\theta + Q_1'/a_1^2)$. On writing $Q_2' = Q_2 + \Delta Q$ and $Q_1' = Q_1 - \Delta Q$, we thus find that the charge transferred is given by

$$\Delta Q = \frac{\pi^2}{2} E a_2^2 \cos\theta + \frac{\dfrac{\pi^2}{6}p^2 Q_1}{\left(1 + \dfrac{\pi^2}{6}p^2\right)} - \frac{Q_2}{\left(+\dfrac{\pi^2}{6}p^2\right)}, \qquad (17\text{-}47)$$

where $p \equiv a_2/a_1 \ll 1$ (the problem of finding ΔQ for arbitrary p on the interval $0 \leqslant p \leqslant 1$ is discussed in Appendix A-17.5.5). This expression shows that in-duction *dis*charging may occur also; i.e., ΔQ may be negative for sufficiently large θ and $Q_2 > 0$. Since θ represents the polar angle between the point of contact and the electric field, and since the latter may have a large horizontal component in some cloud regions (recall Figure 17-3), we see that even collisions restricted to the lower hemisphere of the large drop may result in $\Delta Q < 0$.

On the other hand, if we use (17-47) to estimate the equilibrium charge $\langle Q_1 \rangle$ to be expected on the larger sphere after many induction charging events with smaller spheres of radius a_2, we find by setting $\Delta Q = 0$ for equilibrium and neglecting Q_2 relative to $\langle Q_1 \rangle$ that $\langle Q_1 \rangle \approx -3\, a_1^2 \langle E \cos \theta \rangle$ (here $\langle E \cos \theta \rangle$ denotes the expected value of $E \cos \theta$ under the assumed equilibrium conditions). This rough calculation implies the equilibrium charge on the large drop will have a large negative value, in spite of possibly frequent discharging events, if $\langle E \cos \theta \rangle$ is sufficiently large.

Consideration of the expected value of the electrical contact angle brings several additional problems into focus (assuming they have not been obvious from the outset!) Thus, to find $\langle E \cos \theta \rangle$ we must in effect solve the fairly difficult collision efficiency problem with the inclusion of electrostatic forces (see Section 17.5.5). Furthermore, we must determine the θ-dependence of the probability that separation actually occurs after contact is made. Since from our discussions in Section 14.8.1 we expect that coalescence generally follows a collision between drops under natural conditions unless they are larger than a few hundred microns in radius, we begin to suspect the effectiveness of inductive charge transfer must be quite small for water drops. Finally, since colliding drops are likely to coalesce, and the presence of charge only tends to increase the coalescence probability (see Section 17.5.5), we see that there appears to be a tendency for this charge separation mechanism to be 'shorted out' as large drops rapidly accrete smaller, oppositely charged droplets (e.g., Colgate (1972), Moore (1975)).

Some of these difficulties can be avoided by applying the inductive charging mechanism to ice-ice or ice-water particle interactions, since for these the separation probabilities are generally higher. On the other hand, for such interactions an additional physical barrier to charge separation may enter in, namely the relatively long relaxation time for charge conduction through ice ($\approx 10^{-2}$ sec at $-10\,°C$). In addition, the more complex geometry and dynamics for these cases greatly complicates the quantitative assessment of the relevant processes and parameters.

It has been known for some time that, with the aid of several favorable assumptions, the induction charging mechanism is capable of predicting a rapid growth rate of electric field commensurate with those deduced from observations of thunderstorms. Because of this apparent success, it is easily the most popular electrification mechanism, and has inspired a lengthy series of increasingly sophisticated modeling efforts. We shall now close this section with a brief summary of this evolutionary development, in order to provide some indication of the 'state of the art,' and to relate some additional problems which are common to all the precipitation charging mechanisms.

The first detailed estimate of the electric field growth rate to be expected from induction charging was made by Müller-Hillebrand (1954, 1955); similar calculations including leakage currents were also performed by Latham and Mason (1962). Müller-Hillebrand studied the charging and subsequent separation of soft hail- and ice crystals in a one-dimensional cloud. If we assume spherical particles as before, then from (17-47) we may write approximately, with $p \ll 1$ and $Q_2/Q_1 \ll 1$,

$$\Delta Q = \varepsilon(\theta) \left(\frac{\pi^2}{2} E a_2^2 \cos \theta + \frac{\pi^2}{6} p^2 Q_1 \right), \qquad (17\text{-}48)$$

where we have now included the separation probability for angle θ, $\varepsilon(\theta)$. Then if the hailstone falls with speed U relative to the ice crystals present in concentration n, its charging rate will be given approximately by

$$\frac{dQ_1}{dt} = -\pi a_1^2 E_c U n \left(\frac{\pi^2}{2} Ea_2^2 \overline{\varepsilon \cos\theta} + \frac{\pi^2}{6} p^2 Q_1 \overline{\varepsilon} \right), \qquad (17\text{-}49)$$

where E_c is the collision efficiency and $\overline{\varepsilon(\theta) \cos\theta}$ and $\overline{\varepsilon(\theta)}$ denote average values over the interval $0 \leqslant \theta \leqslant \theta_c$, θ_c being the electrical contact angle corresponding to the grazing collision which determines E_c. Müller-Hillebrand obtained the solution to (17-49) under the assumptions of constant E, ε, and $\cos\theta$, viz.,

$$Q_1 = -3\,Ea_1^2 \cos\theta (1 - e^{-t/\tau}), \quad \tau = (\pi^3 E_c \varepsilon U n a_2^2/6)^{-1}. \qquad (17\text{-}50)$$

Assuming $E_c = \varepsilon = 1$, $U = 800$ cm sec^{-1}, $a_2 = 50\,\mu$m, and $n = 10^5$ m^{-3} (all fairly realistic values except possibly for ε, which may be seriously overestimated), we obtain a hailstone charging time of $\tau \approx 10^2$ sec.

Assuming further that the ice crystals have negligible fall speeds, the magnitude of the local current density due to the charged cloud particles is just $j_Q = NUQ_1$, where N is the assumed concentration of hailstones of radius a_1 and charge Q_1. Therefore, for a one-dimensional 'parallel plate capacitor' cloud, the local change in electric field (in e.s.u.) due to j_Q will be $\partial E/\partial t = -4\,\pi j_Q$, or

$$\frac{\partial E}{\partial t} = 12\,\pi NU E_0 a_1^2 \cos\theta(1 - e^{-t/\tau}). \qquad (17\text{-}51)$$

The solution to this equation is

$$E(t) = E(0)\exp\left\{ \frac{9\,R'\cos\theta}{a_1\rho_1}[t - \tau(1 - e^{-t/\tau})] \right\}, \qquad (17\text{-}52)$$

where ρ_1 is the density of the hailstones and $R' = 4\,\pi NU a_1^3\rho_1/3$ is the precipitation rate expressed as a mass flux of hail. For $t > \tau$ we see from (17-52) that the e-folding time t_e for the field is approximately $t_e = a_1\rho_1/9\,R'\cos\theta$; assuming $\rho_1 = 0.5$ g cm^{-3}, $a_1 = 0.2$ cm, $\theta = 45°$ (this value, chosen by Müller-Hillebrand and many others, may be too small; i.e., it may overestimate the efficiency of charge transfer (e.g., see Moore, 1975)), and a precipitation rate of 10 mm hr^{-1} ($R'/\rho_1 = 2/(3.6 \times 10^3)$ cm sec^{-1}), we have $t_e \approx 50$ sec. Thus with these various assumptions we find that the field will increase by a factor of nearly 10^4 within a total time of about 8 min; presumably, for longer times the field growth would be interrupted by corona, lightning, or other discharge processes not included in the model.

These rather crude calculations suggest that induction charging is indeed a powerful electrification mechanism. As it turns out, many further refinements of the process have left that impression essentially intact, at least in the minds of most investigators. Among these refinements are the following noteworthy contributions: (1) the inclusion of ionic leakage currents and particle size distributions (Sartor, 1967) (however, charge transfer was determined as though the particles were uncharged); (2) the inclusion of electrical forces on the particle velocities (e.g., Kamra (1970), Mason (1972)) (electrostatic forces become significant only for fields within a few orders of magnitude of those which are apparently sufficient for lightning – see Section 17.5.4); (3) the inclusion of a more complete time history of the electric field (Paluch (1973), Paluch and Sartor (1973)) (note, for example, that E was assumed constant in the integration of (17-50), and that its time dependence was then considered separately and hence artificially via (17-51)); (4) the inclusion of effects arising from the stochastic evolution of mass and charge distributions over a full particle size spectrum (Ziv and Levin (1974), Scott and Levin (1975b), Levin (1976)).

All of the above studies assumed the one-dimensional or parallel plate capacitor cloud geometry. Illingworth and Latham (1975) pointed out that real clouds of finite extent could not be as easily electrified. Recently, Chiu and Orville (1976) have managed to incorporate the induction charging mechanism, along with the processes of ion attachment by diffusion and conduction (the latter including the bias arising from the 'Wilson process' for precipitation-sized particles – see item 2 above), into a two-dimensional, axially symmetric, time-dependent, numerical cloud model. This model is the most realistic overall to date, and so it is worth taking a closer look at some of the computational details. (Chiu, 1977, personal communication, has provided the information on which the following discussion is based.)

Chiu and Orville have studied the case of a warm cloud, with a partition of the water into characteristic populations of cloud droplets and precipitation drops: (1) The cloud droplets are assumed to be of 10 μm radius, and to be present in concentrations depending on the local cloud water content. (2) The raindrops, at a given location and time, all have a radius and concentration which are determined from the spectral characteristics of the Marshall-Palmer distribution, (2-12), and the local rainwater content (which varies due to accretion, autoconversion, and evaporation). For the drop-droplet interactions, the crucial parameters E_c, ε, and θ are assigned values, in the absence of reliable computed or measured values; a typical assignment is $E_c = 1$, $\varepsilon = 0.04$ and $\cos \theta = 2/3$ ($\theta \approx 48°$). From the charge which is separated in a given time step, the resulting electric field change is computed by solving Poisson's equation for the entire cloud. The influence of the electric field on drop velocities is included, as is the effect of its direction on the induction charge transfer process (recall the discussion after (17-47)). (The computations for a cloud with an ice phase should be similar in most important respects to that given above; in fact, the very important choice of separation probability, $\varepsilon = 0.04$, is likely to be more representative of ice-ice than water-water interactions.)

The results of the computations are encouragingly realistic. For example, a maximum vertical field of 7×10^5 V m^{-1} is achieved in about 28 min, which is well within the range of observed growth rates. Furthermore, the cloud charge distribution after 28 min is in general accord with the typical distribution shown in Figure 17-3. The main dipole structure is created by the induction charging mechanism (the Wilson process having little effect), while the smaller positive charge near cloud base is explained as a screening layer of positive charge due to the positive ion current into cloud base under the influence of the strong reversed field.

From this work it appears that at least a tentative understanding of some important aspects of strong cloud electrification may be at hand, in spite of the many experimental and theoretical problems we have discussed above.

17.5 Effect of Electric Fields and Charges on Microphysical Processes

We shall now describe some observed and/or predicted modifications in the behavior of isolated and interacting cloud particles in consequence of the presence of ions, particle charges, and ambient electric fields.

17.5.1 DROP AND ICE CRYSTAL NUCLEATION

Many expansion chamber experiments (for a summary see Mason (1971) and Rathje and Stranski (1955)) have shown that in the presence of singly charged, negative small ions the onset of an appreciable rate of drop formation ($J = 1$ in the notation of Section 7.2.2) in otherwise clean moist air requires critical saturation ratios $(S_{v,w})_c$ of 3.7 to 4.2, i.e., critical supersaturations $(S_{v,w})_c$ of 270 to 320%. For positive ions the corresponding critical saturation ratio is $(S_{v,w})_c \approx 6$. Both results apply to temperatures between -5 and $-8\,°C$. Note by comparison with Figure 7-4 that for the same temperature interval homogeneous nucleation at the rate $J = 1$ occurs for $4.6 \leqslant (S_{v,w})_c \leqslant 4.9$. Thus we see that the presence of negative ions decreases the supersaturation required for drop nucleation below the value required under homogeneous conditions, while positive ions raise the critical supersaturation. This electric sign effect was qualitatively explained by Loeb *et al.* (1938). They argued that embryonic water drops are in a pseudo-crystalline state in which the molecules assume a definite structural arrangement. In order to minimize the surface energy of such an arrangement, the water molecules at the surface tend to be oriented with their oxygen atoms outward, since the polarizability of an oxygen atom is considerably larger than that of a hydrogen atom. Thus, the capture of negatively charged ions will enhance the original tendency of the embryonic drops to orient approaching water molecules with their hydrogen atoms inward and thereby facilitate drop nucleation; by the same argument, positive ions will hinder nucleation. However, the small differences due to sign aside, the main point to be made is that since supersaturations in clouds rarely exceed a few percent (see Figure 2-2), small ions cannot affect the nucleation rate under natural conditions. Studies of the effect of net electric charges on aerosol particles and of external electric fields on heterogeneous drop nucleation are not available in the literature.

Work carried out prior to 1963 suggested qualitatively that electric fields and charges enhance ice nucleation (Pruppacher, 1963b). Subsequent more quantitative experimental studies confirmed these effects, but indicated that under atmospheric conditions only charges are likely to affect ice nucleation. Gabarashvili and Gliki (1967) and Gabarashvili and Kartsivadze (1968, 1969) found that supercooled drops containing particles of quartz or napthalene were nucleated to ice at significantly warmer temperatures when the particles carried a net negative charge than when they were neutral or carried a net positive charge. Abbas and Latham (1969a) and Morgan and Langer (1973) observed that charged nuclei produced during corona discharges or by sparks promoted ice nucleation of supercooled drops. The effect of charge sign was not reported. Pruppacher (1973b) found that the freezing temperature of supercooled water drops of 100 to 350 μm radius, freely suspended in the air stream of a wind tunnel, was considerably raised when contacted by predominantly negatively charged amorphous sulfur particles which, when uncharged are known to be poor ice forming nuclei.

The effect of external electric fields on ice nucleation has been studied under essentially two different experimental conditions. Pruppacher (1963c) observed in laboratory experiments that millimeter sized drops, forced in an external electric field to deform and thus to rapidly spread over a solid surface, froze at temperatures up to 10 °C warmer than when the drops were unaffected by the field. These results are consistent with the observations of Doolittle and Vali (1975), who found that an electric field had no effect on the freezing of supercooled drops if they remained motionless with respect to the supporting surface. Other studies have dealt with drops in free fall. For example, Dawson and Cardell (1973) observed millimeter sized drops suspended in the air stream of a wind tunnel at temperatures of -8 to -15 °C, and detected no electrofreezing effect for external fields up to 4 kV cm^{-1}. Coalescence between drops did not alter this outcome. In contrast to these studies, however, Abbas and Latham (1969a) and Smith *et al.* (1971) found that millimeter sized drops falling in air of -5 to -12 °C through intense fields froze if disruptions caused small filaments to be drawn out from or between drops.

17.5.2   DIFFUSIONAL GROWTH OF ICE CRYSTALS

Numerous laboratory studies suggest that both the growth mode and the growth rate of ice crystals are significantly affected by external electric fields. Studies prior to 1973 have critically been reviewed by Evans (1973) and by Crowther and Saunders (1973). From their summaries it appears that in intense electric fields ice crystals tend to assume the shape of a needle or spike, independently of temperature, which is oriented in the direction of the field lines. To understand this behavior we may consider a corner of an ice crystal pointing in the direction of the field lines: The local field enhancement caused by the large curvature of the corner causes the formation of ions in its neighborhood. Some of these ions are captured by the polar water molecules, which then move in the field to enhance the vapor flux to the crystals corner. As the corner grows into a needle or spike the process is self propagating as side branches which might compete for water vapor are suppressed. The growth rate of the crystal is further increased by the field enhancement of the migration distance of water molecules at the ice surface. Both effects have been observed to enhance the growth rate of an ice crystal by factors of 10 to 1000 over the rate in the absence of electric fields. For such significant effects to occur the field strength has to reach several kV cm^{-1}. No growth enhancement has been observed if $E \leqslant 500$ V cm^{-1}.

17.5.3 DROP DISRUPTION AND CORONA PRODUCTION

It is well known that a drop in an electric field will elongate along the field direction, due to the interaction of the field and the polarization charge induced on the drop. For situations in which gravitational and aerodynamic forces are negligible, the distorted shape is approximately prolate spheroidal, with an eccentricity which increases with increasing field strength until finally disruption occurs at a critical field value. Taylor (1964) modeled this situation theoretically for an uncharged drop by assuming a spheroidal shape, and then setting up equations of equilibrium between surface tension and electrical stresses at the poles and equator of the drop. As a result of this analysis he estimated the critical field for disruption (in e.s.u.) to be given by

$$E_{cr} = 1.625 \left(\frac{\sigma}{a_0}\right)^{1/2}, \tag{17-53}$$

where a_0 is the undeformed radius of the drop in cm and σ is its surface tension. This result is in excellent agreement with experiment (e.g., Nolan (1926), Macky (1931), Mathews (1967)).

The applicability of (17-53) to drops in clouds is of course very limited, since they will also be subjected to gravitational (i.e., hydrostatic) and aerodynamic forces. For example, in their experiments on the disintegration of drops falling in strong electric fields, Ausman and Brook (1967) observed discrepancies from (17-53) of as much as 15%. Furthermore, drops in strong fields will generally carry net charges, which can be expected to affect their stability as well. A combined theoretical and experimental study of the more general problem of charged drops falling through an electric field was first carried out by Abbas and Latham (1969b), who attempted to extend Taylor's analysis by including drop charge and some terms representing hydrostatic and aerodynamic effects. From their computations (also reported in Latham and Meyers (1970)), they obtained the following relationship between the critical field for disruption and the drop charge Q (in e.s.u.):

$$E_{cr}\left(\frac{a_0}{\sigma}\right)^{1/2} = 1.6\left(1 - \frac{1.5\,Q}{Q_R}\right), \tag{17-54}$$

where Q_R is the Rayleigh charge limit given by (17-46). They also found that (17-54), their experimental results, and the experiments of Ausman and Brook showed agreement to within 3% for $0.1 \le a_0 \le 0.17$ cm and $0 \le Q/Q_R \le 0.15$.

It is interesting to note that (17-54) does not yield the Rayleigh limit $Q = Q_R$ for $E_{cr} = 0$; however, (17-54) was not intended to be applicable for $Q/Q_R > 0.15$. On the other hand, (17-54) does closely approximate the Taylor limit, (17-53), for $Q = 0$. Levine (1971) obtained a result similar to (17-54), but with the coefficient 1.5 multiplying Q/Q_R replaced by 1.0, in order to achieve the Rayleigh limit for zero field.

Later experiments by Dawson and Richards (1970) and Richards and Dawson (1971), in which a vertical wind tunnel was used to ensure that the drops under study were falling at their terminal velocities, have raised serious doubts about the validity of (17-54) and Levine's similar results. They found that drops falling at terminal velocity become unstable at their upper surface, whereas (17-54) is based on the assumption that electrical instability ensues at the lower surface (the equilibrium conditions were set up for the equator and lower pole of the drop). The most likely explanation for the fact that (17-54) agrees well with the experiments of Abbas and Latham is that the drops in those experiments had not achieved terminal velocity (the drops studied by Ausman and Brook were nearly at terminal velocity, but they were subjected to horizontal rather than vertical electric fields). Also, it has been pointed out by Dawson and Richards (1970) and Foote and Brazier-Smith (1972) that the theoretical analysis of Abbas and Latham which led to (17-54) suffers from some additional errors, even granting the reasonableness of the spheroidal assumption.

The results of Richards and Dawson (1971) for the critical electric field for onset of instability in uncharged water drops falling in air at terminal velocity are shown in Figure 17-4. We see that for $a_0 \lesssim 1.5$ mm the wind tunnel results are in good agreement with the simple Taylor limit of (17-53), plotted as curve (1) in the figure. However, for larger sizes (17-53) (or (17-54) with $Q = 0$) progressively underestimates E_{cr}. The principal reason for this growing difference is that aerodynamic forces tend to flatten the drops into shapes resembling oblate spheroids, the more so as size increases (recall Section 10.3.2), thereby stabilizing the drops against the vertical field.

It is also interesting to note from Figure 17-4 that for drops of $a_0 \gtrsim 2.5$ mm the critical field for breakup remains nearly constant at about 9 kV cm^{-1}. This result is consistent with the observations of Griffiths and Latham (1972), who showed that at 1000 mb the onset of corona discharge (which may be expected to accompany instability) from water drops of $a_0 = 2.7$ mm in air at 1000 mb occurs in a vertical field of (9 ± 0.5) kV cm^{-1}, decreasing to about 5.5 kV cm^{-1} at 500 mb.

Unfortunately, no adequate wind tunnel observations are yet available for the corresponding case of charged drops falling in vertical fields. It is clear,

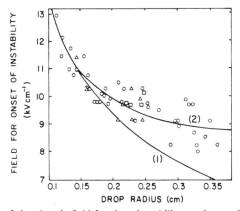

Fig. 17-4. Dependence of the electric field for drop instability on drop radius, for uncharged water drops in air; (1) Theory of Taylor (1964), (2) best fit to experiment, □ slow increase of field, pos.; ○ slow increase of field, neg.; △ impulse, pos. (From Richards and Dawson, 1971; by courtesy of the authors; copyrighted by American Geophys. Union.)

however, that drop charge lowers the critical field for instability. For example, from (17-54) we find that $E_{cr} \approx 6 \, kV \, cm^{-1}$ for $a_0 = 2 \, mm$, $Q = 2 \, e.s.u.$, and $\sigma = 76 \, erg \, cm^{-2}$. A more refined numerical study of Brazier-Smith (1972), in which the constraining assumption of a spheroidal shape is abandoned, yields $E_{cr} \approx 5.5 \, kV \, cm^{-1}$ for the same conditions. However, it must be pointed out that Brazier-Smith's analysis does not include the very important contribution of aerodynamic forces, and so must be regarded primarily as a refinement of the Taylor model. On the other hand, although no truly definitive experimental or theoretical values of $E_{cr} = E_{cr}(Q)$ are available, the above estimates are still considerably larger than the maximum fields typically found in thunderstorms, even though we have used unrealistically large values of Q in obtaining them. Consequently, it appears unlikely that typical isolated precipitation drops experience purely electrical disruption in clouds, though electrical forces may help trigger hydrodynamic instability for very large drops.

One would expect the behavior of pairs of drops in close proximity to be quite different. In this case a strong enhancement of the field between the drops is expected due to the mutual interaction of the polarization charges. This effect will be particularly strong if the field is nearly parallel to the line of centers of the drops. A theoretical and experimental study of this problem was performed by Latham and Roxburgh (1966). They employed the basic calculational procedure of Taylor (1964), described above, along with the theoretical values of Davis (1964a, b) for the local field enhancement in the gap between two spheres situated in an electric field. (Davis's analysis, confirmed experimentally by Latham et al. (1966), is discussed in Appendix A-17.5.5.) The computations of Latham and Roxburgh showed that the field required to initiate instability in one of a pair of closely separated drops may be several orders of magnitude less than that needed to disrupt either drop in isolation (see Table 17-2). The calculated

TABLE 17-2
Critical electric field E_{cr} required for disintegration of a water drop of undistorted radius $a_0 = 0.01 \, cm$ separated from an identical drop by a distance x_0; a/a_0 represents the elongation of the drop at the moment of disintegration. (From Latham and Roxburgh, 1966; by permission of the Royal Society, and the authors.)

x_0/a_0	$E_{cr}(a_0/\sigma)^{1/2}$	$E_{cr}(V \, cm^{-1})$	(a/a_0)
0.5	5.690×10^{-1}	18 340	1.098
0.3	3.287×10^{-1}	8 416	1.054
0.1	7.887×10^{-2}	2 019	1.019
0.08	5.911×10^{-2}	1 514	1.015
0.05	3.211×10^{-2}	1 035	1.009
0.03	1.647×10^{-2}	422	1.0053
0.01	3.910×10^{-3}	100	1.0019
0.005	1.574×10^{-3}	40	1.0009
0.002	4.716×10^{-4}	12	1.0004
0.001	1.898×10^{-4}	6	1.0002

critical fields for $a_0 \approx 1$ mm were also found to agree well with their experimental values.

On the other hand, Brazier-Smith (1971) applied the same numerical method used previously for single drops to this problem, and obtained critical field values consistently higher (by about a factor of 2 for close separations of equal drops, decreasing to a factor of about 1.1 for a separation of 2 a_0) than those found by Latham and Roxburgh. Brazier-Smith explained the agreement between the theoretical and experimental results of Latham and Roxburgh by noting that both actually dealt with the problem of supported drops, rather than of free drops as proposed and as studied by Brazier-Smith.

The relevance of these studies to pairs of drops falling under natural conditions may be somewhat limited, since hydrodynamic forces were ignored. By analogy with the previous case of studies on isolated drops, we might expect that hydrodynamic forces would tend to suppress the onset of electrical instability for interacting drop pairs. However, the effect is probably not as great in the present instance, since the deformations and instabilities occur in the vicinity of closely adjacent surfaces where air-flow effects may be relatively unimportant. In any case, in view of the very small values given in Table 17-2 for the critical field for close separations, it is reasonable to expect that local near-surface disruptions and charge exchange can occur between drops of perhaps $a_0 \gtrsim 100 \, \mu$m in momentary very close proximity. Photographic evidence of such behavior has in fact been obtained by Atkinson and Abbot (in Sartor, 1967).

Crabb and Latham (1974) have made measurements on the critical field required to produce corona from a pair of water drops, of radii 2.7 mm and 0.65 mm, colliding with a relative velocity of 5.8 m sec^{-1}, which is similar to their difference in terminal velocities. The critical field ranged from about 5 kV cm^{-1} for head on collisions to about 2.5 kV cm^{-1} for glancing collisions; these values are of course considerably less than those required to produce corona from single drops, and suggest that corona from colliding raindrops may be capable of triggering lightning. Similar results and conclusions follow from the experiments of Griffiths and Latham (1974) on corona production from ice crystals or hailstones. They found that the critical field for sustained positive or negative corona depends on the size, shape, and surface features of the particle, but may be as low as 4 kV cm^{-1} at pressures corresponding to those where lightning is initiated.

17.5.4 DROP TERMINAL VELOCITIES

A drop of mass m bearing charge of magnitude Q and falling in a vertical electric field E will experience a combined gravitational and electrical force of mg $\pm$ QE. If the drop is small enough to retain a spherical shape, the computational schemes of Section 10.3.5 for spherical drops may be used to determine its terminal velocity, the only modification required being merely the replacement of mg by mg $\pm$ QE. Table 17-3 gives some selected values for drop terminal velocities computed in this manner by Gay et al. (1974) for water drops in air. Note the pronounced increase in terminal velocity which charged drops

TABLE 17-3

Terminal velocities (cm sec^{-1}) of water drops of radius a and $Q = 2\,Ea^2$ (e.s.u.) falling in air in a vertical positive electric field of field strength E. Drops of radius less than 50 μm are assumed to carry positive charges while larger ones carry negative charges. The minus sign signifies upward motion of the drop. For 20 °C and 1013 mb. (Based on data given by Gay et al., 1974, in their Table 2.)

E (kV/cm)	$a(\mu$m)		$a(\mu$m)		
	10	20	50	100	300
0	1.2	4.6	25.6	71.0	245
0.5	2.8	7.7	19.3	63.4	237
1.0	7.4	16.8	−2.4	38.4	214
1.5	15.1	31.2	−35.1	−20.8	173
2.0	25.9	49.7	−70.8	−79.2	102
2.5	39.2	72.1	−110	−133	−57.8
3.0	55.1	97.8	−151	−189	−178
3.4	73.2	125	−193	−248	−278
4.0	92.8	155	−238	−309	−373

experience in an external electric field. Experiments carried out by Gay et al. for the same range of drop sizes, drop charges, and external electric fields yielded terminal velocities which were in good agreement with those theoretically predicted.

For deformed drops the computational schemes of Section 10.3.5 cannot be used unless the drop deformation is known as a function of drop size, charge, and field strength. Unfortunately, there are at present no satisfactory experimental or theoretical descriptions of this functional relationship. However, some information concerning the electric field dependence of terminal velocities of large uncharged drops falling in vertical fields has been obtained by Dawson and Warrender (1973). As we know from the previous section, a vertical field tends to counter the oblate spheroidal deformation of large falling drops, induced by hydrodynamic forces. Therefore, we would expect to see an increase in terminal velocity with increasing field strength, owing to the resulting decrease in the drop cross section presented to the flow. The extent of the increase found by Dawson and Warrender for millimeter sized drops was rather small: typically, $\delta V_\infty / \delta E = 0.1$ m sec^{-1} (kV cm^{-1})$^{-1}$; for example, for a drop of $a_0 = 2.1$ mm they found $\Delta V_\infty = 1.3$ m sec^{-1} if E was increased from 0 to 10 kV cm^{-1}.

17.5.5 COLLISIONAL GROWTH RATE OF CLOUD PARTICLES

We would also expect to see a significant change in the collision and coalescence (or sticking) efficiencies of cloud particles which are subjected to electrostatic forces of magnitude comparable to the acting hydrodynamic and gravitational forces. Several theoretical and experimental investigations of the possible electrostatic influences have been carried out. For example, in a field experiment

Latham (1969) found that the growth rate of 130 μm radius drops colliding with 15 μm diameter drops (all carrying negligible charge) was enhanced if the external electric field strength exceeded 150 V cm^{-1}. At field strengths of 500 V cm^{-1} the growth rate was about 20% higher than for the case of no field. No further enhancement was noted until the field reached 1200 V cm^{-1}; in larger fields the growth rate rapidly increased again, achieving for 1600 V cm^{-1} a 100% increase over the rate for zero field. Qualitatively similar results were obtained in laboratory investigations by Phan-Cong and Dinh-Van (1973). They showed that the number of collection events for a given time period in a cloud of uncharged drops in air was affected only by field strengths exceeding 400 V cm^{-1}. The collection rate increased linearly with further increasing field strength.

Some effects of drop charge on growth by collision and coalescence have been studied by Woods (1965). In the absence of an external electric field, oppositely charged drops of $a \leqslant 40$ μm required a charge $Q > 5 \times 10^{-5}$ e.s.u. for a discernible increase in growth. Above this threshold the number of collection events increased approximately linearly with increasing charge. Drops of $a > 40$ μm exhibited a similar trend, but no clear threshold for the onset of a charge effect could be detected. Charges of equal sign reduced the number of collection events below that occurring for uncharged drops if $a > 40$ μm, and inhibited collection completely if $a \leqslant 40$ μm.

Similar electrostatic effects on the growth rate of ice crystals colliding with ice crystals have been observed by Latham (1969), Latham and Saunders (1970), Crowther and Saunders (1973), and Saunders and Wahab (1975). In these experiments threshold fields of 100 to 600 V cm^{-1} were required for a discernible effect. For larger fields the growth rate increased rapidly, reaching values 80 to 100% larger than for the zero field case if $E \approx 1500$ V cm^{-1}. In the absence of a field only about 10% of the ice crystals were aggregated, the aggregates consisting of up to 6 component crystals. In fields of near 1500 V cm^{-1}, 100% of the ice crystals were aggregated, each consisting of up to 10 crystals. In addition, Saunders and Wahab (1975) found that aggregation in electric fields was most efficient at temperatures near -8 °C, i.e., in the temperature region where columnar crystals are the favored growth habit (see Section 2.2.1). At this temperature the crystal aggregates consisted of short columns joined at their basal or prism planes.

The effect of electric fields on the growth rate of ice crystals growing by collision with supercooled drops was studied by Latham (1969). He found a threshold field of $\sim$900 V cm^{-1} for a discernible effect. In the presence of larger fields the growth rate increased rapidly, reaching about 150% at 1500 V cm^{-1} and about 200% near 2000 V cm^{-1} of the growth rate in the absence of a field. No studies on the effect of charges on the growth rate of ice crystals are available in the literature.

Several theoretical studies of the effect of electrostatic forces on the collision efficiency of water drops in air have been carried out. For most cases of interest very complex hydrodynamic and electrostatic interactions are involved, and the solutions can be obtained only through numerical integration procedures. However, a few approximate analytical solutions are also available for special combinations of drop mass, charge, and electric fields. Two of these asymptotic estimates of collision efficiencies are especially simple and helpful in providing some additional physical insight into the more complex situations, and so we shall now turn to a discussion of them, following Atkinson and Paluch (1968) (similar results appear in Paluch (1970)).

First let us consider the case of large drops bearing large charges. If a pair of such drops has a sufficiently large initial relative velocity, the relative motion can be described adequately within the theoretical framework of the classical two-body problem, in which two particles approach each other in a frictionless medium and are subjected to an inverse square law mutual attraction. The solution of this problem is well known (e.g., see Goldstein (1965)), and may be expressed as $(y_p/r_a)^2 = 1 + (E_e/E_k)$, where y_p is the impact parameter of the encounter, r_a is the apsidal (orbital turning point) distance, E_k is the kinetic energy of the relative motion for large separations, and E_e is the decrease in electrical energy that occurs as the particles are brought from infinity to the minimal separation r_a at the turning point of the orbit. Considering the definition of the collision efficiency E (Section 14.4.2), we may view $(y_p/r_a)^2$ for $r_a = a_1 + a_2$ and $y_p = y_c$ as the collision efficiency for two point particles which approach to within the drop collision distance $a_1 + a_2$. In this manner we obtain the estimate

$$E \approx 1 + z, \quad z = \frac{-Q_1Q_2}{(a_1 + a_2)} \Big/ \frac{m_1m_2}{(m_1 + m_2)} \frac{\Delta V_\infty^2}{2}, \tag{17-55}$$

where ΔV_∞ is the initial relative velocity, Q_i is the drop charge in e.s.u., and $z \geq 0$. Numerical computations of the trajectories which evolve when the effects of polarization (image) charges and drag are included produce results which can be fitted within a few percent to the relationship

$$E = 1 + 1.4 z, \tag{17-56}$$

for the ranges $0 \leq z \leq 5$, $100 \leq a_1 \leq 1000 \ \mu m$, and $K \geq 4$, where $K = -Q_1Q_2/a_1^2a_2^2$. Thus we see that the simple two-body model and (17-55) describe the essential behavior for this class of interactions.

The other approximate analytical result applies to the case of highly (and oppositely) charged small drops in the Stokes regime ($a_i \leq 30 \ \mu m$). The model in this case involves the assumptions of negligible drop inertia and hydrodynamic interaction, so that the Stokes drag on each drop just balances the applied forces from gravity, the Coulomb interaction, and the external electric field $\vec{E}$; thus for drop 1 we write

$$6 \pi a_1 \eta \vec{v}_1 = m_1 \vec{g}^* + Q_1 \vec{E} + \frac{Q_1Q_2}{r^2} \hat{e}_r, \tag{17-57}$$

where r is the center to center distance between the drops and $\hat{e}_r$ is the unit vector along the line of centers from drop 1 to drop 2 ($Q_1Q_2 < 0$). On subtracting (17-57) from the similar equation for drop 2 and letting both gravity and the external field act in the positive x-direction, we find the velocity of drop 2 relative to drop 1 has the components

$$\frac{dx}{dt} = \Delta V_\infty + \frac{(B_1 + B_2)Q_1Q_2x}{r^3}, \quad \frac{dy}{dt} = \frac{(B_1 + B_2)Q_1Q_2y}{r^3}, \tag{17-58}$$

where $r = (x^2 + y^2)^{1/2}$ and $B_i = (6\pi a_1 \eta)^{-1}$ is the mobility of drop i (cf. (17-57) and (12-19)). On eliminating time from (17-58), the governing equation for the relative trajectories is obtained:

$$\frac{dx}{dy} = \frac{x}{y} + \frac{C(x^2 + y^2)^{3/2}}{y}, \quad C = -\frac{\Delta V_\infty}{Q_1 Q_2 (B_1 + B_2)}. \tag{17-59}$$

Integration of this equation yields the family of trajectories $x/r = Cy^2/2 + C_1$, where C_1 is the integration constant. We may determine C_1 in terms of the impact parameter y_p by noting that $y = y_p$ for $x \to \infty$. Furthermore, inspection of the resulting trajectories reveals that collisions ($y_p = y_c$) will occur only for those trajectories which approach the origin from the $-x$ direction; i.e., the criterion for $y_p = y_c$ is that $x/r \to -1$ as x and r approach zero. Therefore, we obtain $y_c^2 = 4/C$, or

$$E = \frac{4}{C(a_1 + a_2)^2} = -\frac{2 Q_1 Q_2}{3\pi\eta a_1 a_2 (a_1 + a_2)\Delta V_\infty}. \tag{17-60}$$

For highly charged drops (17-60) predicts collision efficiencies two to three orders of magnitude larger than the geometric value of unity. For example, (17-60) gives $E = 1.3 \times 10^3$ for $a_1 = 25\ \mu m$, $a_2 = 20\ \mu m$, $Q_1 = 3.4 \times 10^{-4}$ e.s.u., and $Q_2 = -1.9 \times 10^{-4}$ e.s.u.; and $E = 3.3 \times 10^2$ for $a_1 = 14\ \mu m$, $a_2 = 12\ \mu m$, $Q_1 = 2.6 \times 10^{-5}$ e.s.u., $Q_2 = -2.9 \times 10^{-5}$ e.s.u. Laboratory experiments by Krasnogorskay and Neizvestnyy (1973) and Abbott (1975) have shown (17-60) to be within the range of experimental scatter for $E > 2 \times 10^2$. However, for smaller E Abbott found that (17-60) progressively underestimates the experimental values. From the numerical examples given above we see that $E > 2 \times 10^2$ typically corresponds to drop charges much larger than those found in clouds (recall Figure 17-2). Therefore, for small drops and realistic charges it is necessary to employ a more complete physical-mathematical model and numerical methods of solution.

A complete description of the electrostatic forces acting on a pair of drops is available from the work of Davis (1964a, b), who solved the boundary value problem of two charged conducting spheres subjected to a background electric field. (A description of this solution is given in Appendix A-17.5.5.) For the hydrodynamic forces the various flow fields and flow-interaction descriptions discussed in Chapter 14 may be used. The two types of forces may then be superposed and the drop trajectories integrated numerically to determine the extent of electrostatic influences on the collision efficiency problem.

Such studies have been carried out by Sartor (1960), Davis (1965), and Krasnogorskaya (1965) using Stokes flow, and by Lindblad and Semonin (1963), Plumlee and Semonin (1965), and Semonin and Plumlee (1966) using Proudman and Pearson flow (see Section 10.2.4). In all studies except that of Krasnogorskaya the electric force expressions of Davis were used. The results of these studies, applicable to collector drops of low Reynolds number only, show that for a given charge or field the effect on the collision efficiency increases with decreasing size of the collector drop, and that for a given collector drop the effect increases with decreasing size of the collected drop. Furthermore, for a given drop pair a critical charge and/or field is generally found to be necessary for electrostatic effects to be noticeable. For example, Davis (1965) concluded that in the absence of electric fields and for $a_1 \geqslant 20\,\mu$m, the threshold charge is Q_c(e.s.u.)$\geqslant 0.1\,a^2$, with a in cm. Similarly, Krasnogorskaya found $Q_c \geqslant 10^{-7}$ e.s.u. for $a_1 \geqslant 10\,\mu$m. Semonin and Plumlee found $3 \times 10^{-1} \leqslant Q_c \leqslant 3 \times 10^{-6}$ e.s.u. for $30 \leqslant a_1 \leqslant 50\,\mu$m and $5 \leqslant a_2 \leqslant 10\,\mu$m. Lindblad and Semonin and Semonin and Plumlee found that the threshold vertical field for a noticeable enhancement of the collision efficiency for uncharged drops was $E_{cr} = 200$ V cm^{-1} for $a_1 = 30\,\mu$m and $a_2 = 5\,\mu$m, $E_{cr} \approx 500$ V cm^{-1} for $a_1 = 30\,\mu$m and $a_2 = 15\,\mu$m, $E_{cr} = 400$ V cm^{-1} for $a_1 = 40\,\mu$m and $a_2 = 5\,\mu$m, and $E_{cr} = 1$ kV cm^{-1} for $a_1 = 40\,\mu$m and $a_2 = 15\,\mu$m.

In order to determine the effect of electric fields and charges on the collision efficiency of larger drops, Schlamp et al. (1976) employed the superposition scheme (see Section 14.3) in conjunction with numerically determined flow fields around drops, and the Davis expressions for the electrostatic forces. The equation of motion for the a_1-drop was thus written as

$$m_1 \frac{d\vec{v}_1}{dt} = m_1 \vec{g}^* - \frac{\pi}{4} C_{D,1} N_{Re,1} a_1 \eta_a (\vec{v}_1 - \vec{u}_2) + \vec{F}_{e,1}, \qquad (17\text{-}61)$$

with an analogous equation for the a_2-drops. The expression for $\vec{F}_e$ is given by (A.17-3) and (A.17-4). The charge on the drops was assumed to be given by a relation of the form $Q = Aa^2$, with $0.05 \leqslant A \leqslant 2.0$ (see Equation (17-44)), in order to include most of the drop-charge range found in clouds (see Figure 17-2). In the computations the collector (a_1) drop was assumed positively charged, and the a_2-drop negatively charged. In all computations the electric field vector was assumed to be parallel to the gravity vector, pointing downward from a positive charge center in the upper portion of the cloud to a negative charge center in the lower portion of the cloud. The magnitude of the electric field was varied between 50 and 4000 V cm^{-1}. Selected results from the collision-efficiency study of Schlamp et al. (1976) for the case of a positively charged a_1-drop and a negatively charged a_2-drop are shown in Figures 17-5 to 17-7.

From these figures we note that the collision efficiencies of small highly charged drops in thunderstorm fields may be up to 10^2 times larger than the corresponding efficiencies for the same drop pairs in weakly electrified clouds. Therefore, studies of cloud electrification via precipitation charging mechanisms should obviously account for the electrostatic influence on collision efficiencies. On the other hand, the figures also suggest that the weak charges and fields present in young clouds will probably not significantly promote the early stages of drop spectral evolution by collision and coalescence. However, because of the complex nonlinear nature of the collection growth problem, the actual quantitative importance of any colloidal destabilization induced by electrostatic forces can only be determined by comprehensive simulations of the collection process which include the full coupling between changes in electrification and in the drop spectrum. To date no such studies are available. However, recent simulations of cloud electrification which take some aspects of the simultaneous

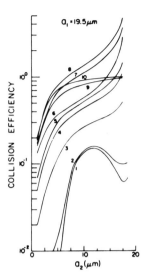

Fig. 17-5. Effect of electric fields and charges on the collision efficiency of a 19.5 μm water drop in air at 800 mb and 10 °C. Curves 1, 3, 4, 5, 7, 8: $\hat{Q}_1 = 0$, $\hat{Q}_2 = 0$, $E = 0$, 500, 1000, 1236, 2504, 3000 V cm^{-1}, respectively. Curves 2, 6: $E = 0$, $\hat{Q}_1 = +0.2, +2.0$ e.s.u. cm^{-2}, $\hat{Q}_2 = -0.2, -2.0$ e.s.u. cm^{-2}, respectively; Curves 9, 10: $\hat{Q}_1 = +2.0$ e.s.u. cm^{-2}, $\hat{Q}_2 = -2.0$ e.s.u. cm^{-2}, $E = 1236$, 2504 V cm^{-1}, respectively; where $\hat{Q} = Q/a^2$. (From Schlamp et al., 1976; by courtesy of Amer. Meteor. Soc., and the authors.)

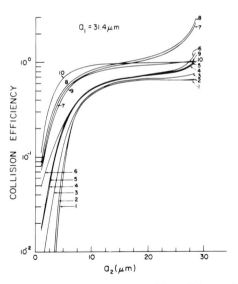

Fig. 17-6. Effect of electric fields and charges on the collision efficiency of a 31.4 μm water drop in air at 800 mb and 10 °C. Curves 1, 3, 4, 5, 7, 8: $\hat{Q}_1 = 0$, $\hat{Q}_2 = 0$, $E = 0$, 500, 1000, 1038, 3000, 3235 V cm^{-1}, respectively; Curves 2, 6: $E = 0$, $\hat{Q}_1 = +0.2, +2.0$ e.s.u. cm^{-2}, $\hat{Q}_2 = -0.2, -2.0$ e.s.u. cm^{-2}, respectively; Curves 9, 10: $\hat{Q}_1 = +2.0$ e.s.u. cm^{-2}, $\hat{Q}_2 = -2.0$ e.s.u. cm^{-1}, $E = 1038$, 3235 V cm^{-1}, respectively; where $\hat{Q} = Q/a^2$. (From Schlamp et al., 1976; by courtesy of Amer. Meteor. Soc., and the authors.)

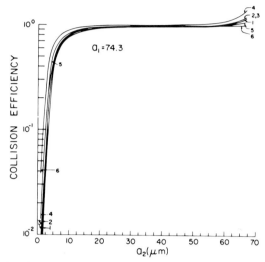

Fig. 17-7. Effect of electric fields and charges on the collision efficiency of a 74.3 μm water drop in
air at 800 mb and 10 °C. Curves 1, 2, 4: $\hat{Q}_1 = 0$, $\hat{Q}_2 = 0$, $E = 0$ and 500, 705 and 1000, 2960 and
3000 V cm^{-1}, respectively; Curve 3: $E = 0$, $\hat{Q}_1 = +2.0$ e.s.u. cm^{-2}, $\hat{Q}_2 = -2.0$ e.s.u. cm^{-2}; Curves 5, 6:
$\hat{Q}_1 = +2.0$ e.s.u. cm^{-2}, $\hat{Q}_2 = -2.0$ e.s.u. cm^{-2}, $E = 705$, 2960 V cm^{-1}, respectively; where $\hat{Q} = Q/a^2$.
(From Schlamp et al., 1976; by courtesy of Amer. Meteor. Soc., and the authors.)

growth of precipitation and electrification into account (but which do not include
the electrostatic enhancement of collision efficiencies) suggest that the coupling
between the two processes is very strong indeed (e.g., Scott and Levin, 1975b).

Figures 17-5 to 17-7 show further that the influence of electrostatic effects
decreases with increasing a_1, becoming negligible even in the presence of the
highest electric fields and charges observed in clouds if $a_1 \gtrsim 70 \mu$m. Also, as
expected, for a given drop pair the effect of electric fields and charges on the
collision efficiency generally increases with increasing drop charge in the absence of an external field, and with increasing field strength in the absence of
charges. For a given a_1 the effect on the collision efficiency of either electric
charges or an external electric field depends on the drop size ratio p = a_2/a_1, the
effect being largest for p ≪ 1, least for intermediate p, and increasing again for
p ≈ 1. We also note that for a given a_1 the collision efficiency generally increases
with increasing radius of the a_2-drop, i.e. with increasing p, and thus decreasing
relative velocity between the drops. The figures also show qualitative agreement
with the threshold values given above of charges and fields required for a
noticeable effect. Finally we should point out that a switch in choice of charge
for the drops may result in different collision efficiency values in some cases,
since the relative velocity difference of the drops and hence their interaction
time could be significantly altered by such a charge. Thus, it would be desirable
to extend the study of Schlamp et al. to the case of negatively charged drops
which collect smaller drops bearing positive charge.

In addition to increasing the collision efficiency, electric fields and charges
may raise the growth rate of cloud drops by increasing the fraction of 'colliding'

drops which coalesce. In Section 14.8.1 we pointed out that drops of $a > 100 \, \mu m$ are likely to rebound from each other rather than coalesce, due to the presence of an air film temporarily trapped between their deformed surfaces as they collide. We would expect that this barrier to coalescence would be weakened by the forces of electrostatic attraction, and also by the local surface deformations which may result (recall Section 17.5.3). These expectations have been verified by Lindblad (1964) and Semonin (1966), who found that for millimeter sized drops the apparent delay time between collision and coalesce was strongly reduced if the potential difference across the colliding drops was raised to several volts. Also, Goyer *et al.* (1960) observed that ~34% of 50 μm radius drops colliding with 300 to 400 μm radius drops coalesced if the electric field across them was raised to ~3 V cm^{-1}, while 95% of the colliding drops coalesced if the field was raised to ~40 V cm^{-1}. Jayaratne and Mason (1964) found that a critical electric field of ~100 V cm^{-1} was required to affect the coalescence of 150 μm radius drops impacting on a large essentially flat water target. Freier (1960) observed that the probability of coalescence between 2.5 mm drops began to be affected when the electric field exceeded ~300 V cm^{-1}. Note from these studies that the larger the colliding drops, the higher the field required to enhance the efficiency with which the drops coalesce. This is expected from our discussion in Section 14.8.1, where we pointed out that the larger the colliding drops, the more pronounced is the deformation which the drops experience on collision, and thus the more extensive the air film trapped between the drops.

The effect of charges of opposite sign on the coalescence of water drops in air has been studied by Jayaratne and Mason (1964), Park (1970), Whelpdale and List (1971), and by Brazier-Smith *et al.* (1972). From these studies it appears that for drops of $150 \lesssim a \lesssim 2000 \, \mu$m coalescence is affected only if the charge on the drops is larger than ~10^{-4} e.s.u. For charges of magnitude 10^{-2} to 10^{-1} e.s.u. the two colliding drops were found to always coalesce. A glance at Figure 17-2 shows us that the charge requirement for 100% coalescence is met only in highly electrified thunderclouds. In all other clouds the charge residing on drops of $a > 100 \, \mu$m is probably less than that required for no rebound.

Unfortunately, no theoretical studies of the electrostatic enhancement of ice particle collision efficiencies are available in the literature. However, some related studies have been carried out. For example, the force expressions of Davis (1964a, b) for the case of two electrically conducting spheres have been extended by Hall and Beard (1975) and by Grover (1976a) to include the case of electrical interaction between a conducting sphere and a dielectric sphere, and by Davis (1969) for the case of electrical interaction between a conducting plate and a dielectric sphere. The results of these studies show that the calculated force constants are not sensitive to the dielectric constant ε of the sphere if $\varepsilon > 80$, and agree essentially with the values found for $\varepsilon = 10^8$, taken to represent a conducting sphere. Only for sphere separations smaller than $0.01 \, a_1$ are there noticeable deviations from the conducting case. Since for water (ε_w) and ice (ε_i) we have $\varepsilon_w = 87.7$ (0 °C) (Malmberg and Mariott) and $\varepsilon_i = 105$ (0 °C, perpendicular to c-axis), $\varepsilon_i = 92$ (0 °C, parallel to c-axis) (Gränicher, 1963), the conducting sphere assumption and thus the solution of Davis (1964a, b) may be taken to represent reasonably well the forces of interaction between either charged water spheres or charged ice spheres in an external electric field.

The effect of ice crystal shape on the force of interaction has been studied experimentally by Latham et al. (1965, 1966), and Latham and Saunders (1970) for the case of two uncharged metallic ice crystal models of various shapes arranged in various configurations. The measured forces could be made to agree closely with the force expressions of Davis merely by introducing a simple shape factor γ, which was essentially independent of the direction and strength of the background field, and of the separation of the crystals. Depending on the combination of crystal shapes, γ was found to range between ~ 0.5 and ~ 6; thus in most cases the forces between ice crystals are greater than between drops of the same volume as the crystals and separated by the same distance. Shape factors for the electrical interaction between a sphere and an ice crystal-shaped body were not determined. If charge resided on the crystal model no corresponding simple modification of the Davis expressions was found to be adequate, since then the shape factor depended in an undetermined manner on the amount of charge present.

17.5.6 SCAVENGING OF AEROSOL PARTICLES

As would be expected from the discussions of the previous sections, the efficiency with which aerosol particles are scavenged by cloud and raindrops can be strongly enhanced by the presence of drop charge and external electric fields. Grover and Beard (1975) and Grover (1976b) demonstrated this behavior through theoretical studies of the effect of charges and fields on the inertial impaction of aerosols. More recently, Grover et al. (1977) have carried out similar computations including also the effect of phoretic forces. They studied the efficiency of scavenging of aerosol particles by water drops in air by means of the particle trajectory method discussed in Section 12.7.4, using an expanded version of (12-103), viz.,

$$m \frac{d\vec{v}}{dt} = m\vec{g} - \frac{6 \pi \eta_a r}{1 + \alpha N_{Kn}} (\vec{v} - \vec{u}) + \vec{F}_e + \vec{F}_{Th} + \vec{F}_{Df}, \qquad (17\text{-}62)$$

where the thermophoretic force $\vec{F}_{Th}$ and the diffusiophoretic force $\vec{F}_{Df}$ were computed from (12-93) and (12-99), respectively, using numerically determined vapor density and temperature fields around evaporating water drops in air, and where the electric force $\vec{F}_e$ was assumed to be given by (A.17-3) and (A.17-4). The drop charges were assumed to be $Q_a(e.s.u.) = 2 a^2$ or $0.2 a^2$, while similarly the particle charges were set equal to $Q_r = 2 r^2$ or $0.2 r^2$. The results for drop radii of 72 and 309 μm are shown in Figures 17-8 and 17-9, respectively. As would be expected, for fixed a, r, and relative humidity ϕ_v, the collision efficiency E increases with field strength or drop and particle charge (for drops and particles bearing charge of opposite sign). Note also that for given r and ϕ_v, the electrostatic influence on E increases with decreasing a, and that similarly for given a and ϕ_v it increases with decreasing r.

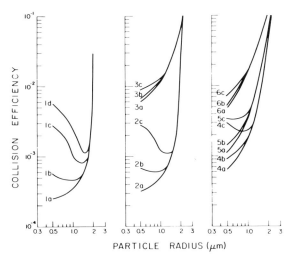

Fig. 17-8.   Effect of electric fields and charges on the efficiency with which a 72 μm radius water drop collides with aerosol particles of $\rho_P = 2 \, g \, cm^{-3}$ in air of various relative humidities ϕ_v, and at 10 °C and 900 mb. Curves 1a, b, c, d: $\hat{Q}_a = 0$, $\hat{Q}_r = 0$, $E = 0$, $\phi_v = 100\%$, 95%, 75%, 50%, respectively; Curves 2a, b, c: $\hat{Q}_a = \mp0.2 \, e.s.u. \, cm^{-2}$, $\hat{Q}_r = \pm0.2 \, e.s.u. \, cm^{-2}$, $E = 0$, $\phi_v = 100\%$, 95%, 75%, respectively; Curves 3a, b, c: $\hat{Q}_a = \mp2.0 \, e.s.u. \, cm^{-2}$, $\hat{Q}_r = \pm2.0 \, e.s.u. \, cm^{-2}$, $E = 0$, $\phi_v = 100\%$, 95%, 75%, respectively; Curves 4a, b, c: $\hat{Q}_a = 0$, $\hat{Q}_r = 0$, $E = 500 \, V \, cm^{-1}$, $\phi_v = 100\%$, 95%, 75%, respectively; Curves 5a, b, c: $\hat{Q}_a = 0$, $\hat{Q}_r = 0$, $E = 1000 \, V \, cm^{-1}$, $\phi_v = 100\%$, 95%, 75%, respectively; Curves 6a, b, c: $\hat{Q}_a = 0$, $\hat{Q}_r = 0$, $E = 2886 \, V \, cm^{-1}$, $\phi_v = 100\%$, 95%, 75%, respectively; where $\hat{Q}_a = Q_a/a^2$ and $\hat{Q}_r = Q_r/r^2$. (From Grover et al., 1977; by courtesy of Amer. Meteor. Soc., and the authors.)

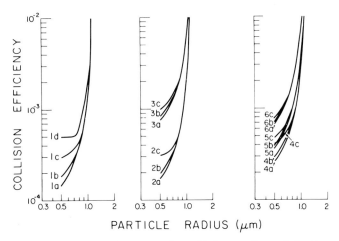

Fig. 17-9.   Same as Figure 17-8, except for a water drop of 309 μm radius, and that for case 6a, b, c the parameters are now: $\hat{Q}_a = 0$, $\hat{Q}_r = 0$, $E = 3000 \, V \, cm^{-1}$, and $\phi_v = 100\%$, 95%, 75%, respectively; where $\hat{Q}_a = Q_a/a^2$ and $\hat{Q}_r = Q_r/r^2$. (From Grover et al., 1977; by courtesy of Amer. Meteor. Soc., and the authors.)

Unfortunately, the trajectory method can only be used to determine the efficiency with which relatively large particles ($r \geq 0.5 \, \mu m$) are captured by cloud drops, since capture effects due to Brownian diffusion are not included in this method. A 'particle flux method' applying to smaller aerosol particles ($0.001 \leq r \leq 0.5 \, \mu m$) has recently been worked out by Wang (1978) and Wang *et al.* (1978). This method considers particle capture due to Brownian diffusion, and phoretic and electric forces, but neglects particle capture by inertial impaction. Although the flux method makes several restrictive assumptions, Wang *et al.* found good agreement between the particle capture efficiencies derived from this method and experimental values, and values derived from the more accurate trajectory method. This suggests that the flux method predicts reasonable collision efficiency values. The flux method and some results derived from it shall therefore be described briefly below.

Using (12-23) together with (12-19), the current density of aerosol particles moving toward a stationary water drop because of Brownian diffusion and the simultaneous influence of phoretic and electric forces can be written as

$$\vec{j}_P = n_P B_P (\vec{F}_e + \vec{F}_{Th} + \vec{F}_{Df}) - D_P \nabla n_P, \tag{17-63}$$

where n_P, D_P, and B_P are the number concentration, diffusivity and mobility of the aerosol particles, respectively, and where $\vec{F}_e$, $\vec{F}_{Th}$, and $\vec{F}_{Df}$ are given by (A.17-3) and (A.17-4), and by (12-93) and (12-99), respectively. The resulting flux of particles to the drop is easily determined for the case of a steady state and for simple inverse square law phoretic and electric forces. Under these simplifying conditions (which include the assumption that the background electric field has a negligible influence) the flux has the form determined earlier in Section 17.2, where we considered the problem of Brownian motion of charged aerosol particles. To adapt the results of that problem to our present situation we need only recognize that the effective interaction potential Φ between the drop and an aerosol particle must now include both phoretic and electric contributions, i.e., instead of (17-17) we now have $\Phi = C/r$, where

$$C = C_e + C_{Th} + C_{Df}, \tag{17-64a}$$

$$C_e = Q_r Q_a, \tag{17-64b}$$

$$C_{Th} = -\frac{12 \, \pi \eta_a r (k_a + c_t k_P N_{Kn}) k_a a (T_\infty - T_a)}{5 (1 + 3 \, c_m N_{Kn})(k_P + 2 \, k_a + 2 \, c_t k_P N_{Kn}) p}, \tag{17-64c}$$

from (12-93), and

$$C_{Df} = -\frac{6 \, \pi \eta_a r (1 + \sigma_{va} x_a) D_v M_a a (\rho_{v,\infty} - \rho_{v,a})}{(1 + \alpha N_{Kn}) M_w \rho_a}, \tag{17-64d}$$

from (12-99). Then by analogy with (17-16) and (17-18a), the collision kernel is

$$K_P = \frac{4 \, \pi a D_P \tau_c}{(e^{\tau_c} - 1)} \tag{17-65a}$$

with

$$\tau_c = \frac{C}{a k T}. \tag{17-65b}$$

An approximate extension of these results to the case of a falling drop was achieved by Wang *et al.* by including various ventilation factors. In particular, their final expression for the collision kernel is given by (17-65) with the replacements $D_P \rightarrow \bar{f}_P D_P$, $C_{Th} \rightarrow \bar{f}_h C_{Th}$, and $C_{Df} \rightarrow \bar{f}_v C_{Df}$. The factors $\bar{f}_P$, $\bar{f}_h$, and $\bar{f}_v$ were assumed to be approximately given by the results of Beard and Pruppacher (1971b) (see Equations (13-57) and (13-58)).

From (17-65) one may infer collision efficiencies using the relationship expressed by (12-77). This has been done for a wide range of conditions by Wang (1978) and Wang *et al.* (1978). The resulting efficiencies were also compared with those computed by the trajectory method for the same conditions. As an example of these computations, Figure 17-10 gives collision efficiency curves of E vs. r for a water drop of $310 \, \mu m$ collecting aerosol particles of $0.001 \leqslant r \leqslant 5 \, \mu m$ in air of $10 \,°C$ and 900 mb, and for various electric charges on the drop and aerosol particles. We note from Figure 17-10 that the flux and trajectory methods combined give a consistent description of the Greenfield gap and the sensitivity of the depth, width, and position of this gap to phoretic and electric forces. We note in particular that the presence of opposite electric charges on drop and aerosol particles causes a pronounced filling-in of the gap. This is a consequence of the negligible effects of electric charges on particle capture due to Brownian motion in the range $r < 0.01 \, \mu m$ and due to

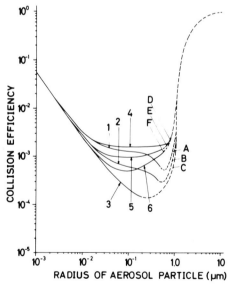

Fig. 17-10.   Effect of electric charges on the efficiency with which a $310 \, \mu m$ radius water drop collides with aerosol particles of $\rho_P = 2 \, g \, cm^{-3}$ in air of various relative humidities ϕ_v, and of $10 \,°C$ and 900 mb. Composite results combining the particle trajectory method of Grover *et al.* (1977) (dashed lines) with the particle flux model of Wang (1978), and Wang and Pruppacher (1978) (solid lines). Curves 1, 2, 3, i.e. A, B, C are for $\hat{Q}_a = 0$, $\hat{Q}_r = 0$ and $\phi_v = 50\%$, 75%, and 95%, respectively. Curves 4, 5, 6, i.e. D, E, F are for $\hat{Q}_a = \pm 2.0$, $\hat{Q}_r = \mp 2.0$ and $\phi_v = 50\%$, 75%, and 95%, respectively; where $\hat{Q}_a = Q_a/a^2$ and $\hat{Q}_r = Q_r/r^2$. (From Wang, 1978, with changes.)

inertial impaction in the range $r > 2\,\mu$m, contrasted by the rather strong enhancement of particle capture by electric charges in the range $0.01 \leqslant r \leqslant 2\,\mu$m. We also note from Figure 17-10 that a complete 'bridging' of the Greenfield gap is achieved neither through the effect of phoretic forces at typical atmospheric relative humidities, nor through the effect of electric forces due to electric charges as they are found under thunderstorm conditions.

For a survey of some other approaches to the problem of determining electrical influences on scavenging processes, see Rosenkilde (1976).

APPENDICES

APPENDIX TO CHAPTER 4

A-4.9 Convenient Formulations for Determining the Saturation Vapor Pressure Over Water and Ice (Lowe and Ficke, 1974)

$$e_{sat} = a_0 + T(a_1 + T(a_2 + T(a_3 + T(a_4 + T(a_5 + a_6 T))))),$$ (A.4-1)

with T (°C).

for water	for ice
$a_0 = 6.107799961$	$a_0 = 6.109177956$
$a_1 = 4.436518521 \times 10^{-1}$	$a_1 = 5.034698970 \times 10^{-1}$
$a_2 = 1.428945805 \times 10^{-2}$	$a_2 = 1.886013408 \times 10^{-2}$
$a_3 = 2.650648471 \times 10^{-4}$	$a_3 = 4.176223716 \times 10^{-4}$
$a_4 = 3.031240396 \times 10^{-6}$	$a_4 = 5.824720280 \times 10^{-6}$
$a_5 = 2.034080948 \times 10^{-8}$	$a_5 = 4.838803174 \times 10^{-8}$
$a_6 = 6.136820929 \times 10^{-11}$	$a_6 = 1.838826904 \times 10^{-10}$

Range of validity: $-50\,°C$ to $+50\,°C$ for water, $-50\,°C$ to $+0\,°C$ for ice.

$$\frac{de_{sat}}{dT} = b_0 + T(b_1 + T(b_2 + T(b_3 + T(b_4 + T(b_5 + b_6 T))))),$$ (A.4-2)

with T (°C).

for water	for ice
$b_0 = 4.438099984 \times 10^{-1}$	$b_0 = 5.030305237 \times 10^{-1}$
$b_1 = 2.857002636 \times 10^{-2}$	$b_1 = 3.773255020 \times 10^{-2}$
$b_2 = 7.938054040 \times 10^{-4}$	$b_2 = 1.267995369 \times 10^{-3}$
$b_3 = 1.215215065 \times 10^{-5}$	$b_3 = 2.477563108 \times 10^{-5}$
$b_4 = 1.036561403 \times 10^{-7}$	$b_4 = 3.005693132 \times 10^{-7}$
$b_5 = 3.532421810 \times 10^{-10}$	$b_5 = 2.158542548 \times 10^{-9}$
$b_6 = -7.090244804 \times 10^{-13}$	$b_6 = 7.131097725 \times 10^{-12}$

Range of validity: $-50\,°C$ to $+50\,°C$ for water, $-50\,°C$ to $0\,°C$ for ice.

APPENDIX TO CHAPTER 7

A-7.1 Relations From Statistical Mechanics

Our purpose here is primarily to summarize relations from statistical mechanics which are used in this chapter. Where possible in the limited space available, we have also sketched in relevant derivations. Of course, detailed treatments are available in any good reference on the subject (e.g., Landau and Lifshitz, 1958; Huang, 1963, Morse, 1969).

We need connecting links between thermodynamics and statistical mechanics. The most fundamental of these is the basic postulate, first stated by Boltzmann, which relates the entropy S for a thermodynamic macrostate of a system to the microstate distribution function f_ν:

$$S = -k \sum_\nu f_\nu \ln f_\nu. \tag{A.7-1}$$

The macrostate corresponds to an ensemble of systems with different microstates, and f_ν measures the probability that a system selected at random from the ensemble is in the microstate characterized by quantum numbers $(\nu_1, \nu_2, \ldots, \nu_\phi) \equiv \nu$, where ϕ is the number of degrees of freedom of the system. Of course, f_ν is subject to the normalization condition

$$\sum_\nu f_\nu = 1. \tag{A.7-2}$$

Canonical Ensemble

Consider a thermodynamic macrostate described by a system of N particles held at constant temperature T and volume V. The ensemble corresponding to this thermodynamic situation must be comprised of systems which also have constant T, V, and N. The energy E_ν of each system need not be constant, but the mean energy for the ensemble must satisfy the constraint

$$\sum_\nu f_\nu E_\nu = U = F + TS, \tag{A.7-3}$$

where U and F are respectively the internal energy and Helmholtz function for the thermodynamic system.

By the use of Lagrange multipliers one may readily solve for the f_ν which satisfies (A.7-1) and is subject to the constraints (A.7-2) and (A.7-3). On realizing that F is the thermodynamic function which is minimized for equilibrium at constant T and V (see (4-7)), and that S is maximized, the proper identification of the multipliers is straightforward and the following results are obtained:

$$f_\nu = Q^{-1} e^{-E_\nu/kT}, \tag{A.7-4}$$

where

$$Q = \sum_\nu e^{-E_\nu/kT} = e^{-F/kT}. \tag{A.7-5}$$

The ensemble corresponding to this distribution is called the *canonical ensemble*, and Q is called the *canonical partition function*. The result (A.7-5) provides another important link between thermodynamics and statistical mechanics.

If the interparticle forces in the system are negligible, the system energy will then be the sum of the separate energies of the individual particles, each depending only on its own set of quantum numbers. Then the sum in (A.7-5) will become a product of N factors, each of which will be a single-particle partition function. However, if the particles are indistinguishable, we must be careful to avoid counting the system microstates which are the same except for a reshuffling of the quantum numbers among the particles; rather, these must be regarded as the same microstate, which therefore is to be counted only once. For the cases of interest to us this bookkeeping problem can be resolved merely by dividing the product of N factors by N!. Thus, for example, the canonical partition function for a perfect gas of N molecules has the form

$$Q = q^N/N!, \tag{A.7-6}$$

where q is the partition function per molecule, i.e.,

$$q = \sum_j e^{-\varepsilon_j/kT} = e^{-\dot{F}/kT}, \tag{A.7-7}$$

where the sum runs over the allowed energies ε_j of the molecule, and $\dot{F}$ is the free energy per molecule.

The same process of factorization may be continued if the molecular energy can be written as the sum of separate contributions due to translation, rotation, etc.; then we can also write, in obvious notation,

$$q = q_T q_R \ldots . \tag{A.7-8}$$

Let us calculate the translational partition function q_T for a molecule in the volume V. For the classical context of interest to us, the energy levels are closely spaced and we may, for the sake of computational convenience, replace the indicated sum over energy levels by an integral over the Hamiltonian $H = p^2/2\,m$, where p is the momentum of the molecule, and m is its mass. For this purpose we must also take into account the fact that there are $h^{-\phi}(dV)_{ph}$ microstates corresponding to the phase space volume element $(dV)_{ph}$ where h is Planck's constant; this follows directly from the Heisenberg uncertainty principle. Thus we have, using Cartesian coordinates,

$$q_T = \frac{1}{h^3} \int_V dx\, dy\, dz \int_{-\infty}^{+\infty} dp_x\, dp_y\, dp_z e^{-p^2/2\,mkT} = V(2\,\pi mkT/h^2)^{3/2}. \tag{A.7-9}$$

In the same way one may obtain the rotational partition function for a molecule having the principal moments of inertia I_1, I_2, I_3:

$$q_R = \delta^{-1}(\pi I_1 I_2 I_3)^{1/2}(8\,\pi^2 kT/h)^{3/2}, \tag{A.7-10}$$

where δ is called the rotation symmetry number, and counts the number of physically indistinguishable orientations of the molecule.

Grand Canonical Ensemble

Let us now devise an ensemble (the *grand canonical ensemble*) in which we drop the requirement that the number of particles be fixed; instead, we require only that the average number of particles be specified as $\bar{N}$. In other respects the ensemble is to be chosen as before. A microstate for this ensemble is then specified by both the number of particles N the selected sample system has, and the quantum numbers $\nu_1, \nu_2, \ldots, \nu_{3N}$ which will determine its energy $E_{N,\nu}$. Thus for an equilibrium macrostate the distribution function $f_{N,\nu}$ must satisfy the following conditions:

$$S = -k \sum_{N,\nu} f_{N,\nu} \ln f_{N,\nu} \quad \text{is maximum}$$

$$\sum_{N,\nu} f_{N,\nu} = 1; \quad \sum_{N,\nu} N f_{N\nu} = \bar{N} \tag{A.7-11}$$

$$\sum_{N,\nu} E_{N,\nu} f_{N,\nu} = U = TS + \bar{N}\mu + \Omega,$$

where μ is the chemical potential per particle, and $\Omega \equiv F - \mu\bar{N}$ is the *grand potential*, whose differential is

$$d\Omega = -S\, dT - \bar{N}\, d\mu - p\, dV. \tag{A.7-12}$$

The solution to (A.7-11) may be obtained via Lagrange multipliers as in the previous case; the results are:

$$f_{N,\nu} = \mathcal{2}^{-1} \exp\left(\frac{\mu N - E_{N,\nu}}{kT}\right), \tag{A.7-13}$$

where the *grand canonical partition function* $\mathcal{2}$ is

$$\mathcal{2} = \sum_{N,\nu} \exp\left(\frac{\mu N - E_{N,\nu}}{kT}\right) = e^{-\Omega/kT}. \tag{A.7-14}$$

We see that $\mathcal{2}$ is the sum of the canonical partition functions Q(N) for ensembles with different values of N, each weighted by the factor $e^{\mu N/kT}$:

$$\mathcal{Q} = \sum_{N=0}^{\infty} e^{\mu N/kT} Q(N),$$

$$Q(N) = \sum_{\nu} e^{-E_{N,\nu}/kT}.$$

(A.7-15)

The relation (A.7-14) provides still another very useful connection between thermodynamics and statistical mechanics.

APPENDICES TO CHAPTER 10

A-10.1   Equations of Fluid Flow

The following is a somewhat descriptive development of the pertinent governing equations of fluid flow. More complete derivations may be found elsewhere (e.g., Landau and Lifshitz, 1959; Batchelor, 1967; White, 1974).

1. *Continuity Equation*

Consider a fluid of density ρ moving with velocity $\check{u}$, both functions of position $\check{r}$ and time t. If we follow a small fluid element of volume δV and mass $\delta m = \rho \delta V$ in its motion, the principle of conservation of mass tells us that

$$\frac{d\delta m}{dt} = \frac{d\rho}{dt}\delta V + \rho\frac{d\delta V}{dt} = 0,$$

where d/dt denotes differentiation following the moving element. The rate of change of δV is evidently given by the flux of $\check{u}$ out of the surface δS enclosing δV. Therefore, from the divergence theorem we have to first order in δV (or, more concisely, simply from the definition of the divergence),

$$\frac{d\delta V}{dt} = \int_{\delta S} \check{u}\cdot\hat{n}\,dS = \nabla\cdot\check{u}\delta V,$$

where $\hat{n}$ is the unit outward normal associated with the surface element dS. The principle of mass conservation therefore becomes

$$\frac{d\rho}{dt} + \rho\nabla\cdot\check{u} = 0,$$

(A.10-1)

which is known as the *continuity equation*. If $d\rho/dt = 0$, then $\nabla\cdot\check{u} = 0$ and the flow is termed *incompressible*.

The first order change in any function $f(\check{r}, t)$ due to changes in time dt and position $d\check{r} = \check{u}\,dt$ is

$$df = f(\check{r} + \check{u}\,dt, t + dt) - f(\check{r}, t) = \check{u}\cdot\nabla f\,dt + \frac{\partial f}{\partial t}dt,$$

which allows us to express d/dt as

$$\frac{d}{dt} + \frac{\partial}{\partial t} + \check{u}\cdot\nabla.$$

(A.10-2)

Using (A.10-2), we can obtain another form of the continuity equation, involving the local time derivative $\partial/\partial t$:

$$\frac{\partial\rho}{\partial t} + \nabla\cdot(\rho\check{u}) = 0.$$

(A.10-3)

2. Navier-Stokes Equation

An interior volume V of fluid with boundary surface S may be accelerated by external body forces, and by surface forces exerted on S by the fluid outside V. If we denote the external force per unit mass acting on volume element dV by $\vec{f}$, and the surface stress acting on surface element dS with outward normal $\hat{n}$ by $\vec{T}(\hat{n})$, the equation of motion of the fluid volume is

$$\frac{d}{dt} \int_V \rho \vec{u} \, dV = \int_V \rho \vec{f} \, dV + \int_S \vec{T}(\hat{n}) \, dS. \tag{A.10-4}$$

In terms of the usual suffix notation, where suffices i, j take the values 1, 2, 3 corresponding to the components of vectors and second rank tensors along the x, y, z axes, respectively, the equation of motion is

$$\frac{d}{dt} \int_V \rho u_i \, dV = \int_V \rho f_i \, dV + \int_S T(\hat{n})_i \, dS. \tag{A.10-5}$$

If the volume V has a characteristic dimension ℓ, then $V \sim \ell^3$ and $S \sim \ell^2$. Consequently, if we let $\ell \to 0$ while preserving the shape of the volume, (A.10-5) reduces to the condition of local equilibrium:

$$\lim_{\ell \to 0} \left[(1/\ell^2) \int_S T(\hat{n})_i \, dS \right] = 0. \tag{A.10-6}$$

If (A.10-6) is applied to a small tetrahedron whose slant face has unit normal n_i and whose other faces are parallel to the coordinate planes, we easily find

$$T(\hat{n})_i = T_{ij} n_j, \tag{A.10-7}$$

where T_{ij} is the stress in the x_i direction acting on a surface element whose normal is in the x_j direction, and we have used the usual convention of summation over a double index. Since $T(\hat{n})_i$ and n_i are vectors, the quantities T_{ij} are components of a second rank tensor $\bar{T}$, called the *stress tensor*, which describes completely the system of stresses in a fluid.

It is also easy to show that $\bar{T}$ is symmetric, i.e.,

$$T_{ji} = T_{ij}. \tag{A.10-8}$$

This is an immediate consequence of applying the condition that there must be zero net torque on a small cube of fluid whose faces are parallel to the coordinate planes.

The effect of pressure $p(\vec{r}, t)$ in the fluid is to contribute an inward (compressional) normal stress on any surface element. Thus we may write

$$T_{ij} = -p\delta_{ij} + T''_{ij}, \tag{A.10-9}$$

where δ_{ij} (Kronecker delta symbol) is unity for $i = j$ and zero otherwise, and where T''_{ij} includes the stress contribution due to the viscosity of the fluid. For ordinary (Newtonian) fluids, of which air and water are examples, the viscous stress is proportional to the amount of relative motion taking place near the point in question. Thus, at least for small velocity gradients, T''_{ij} must be a linear function of the derivatives $\partial v_i / \partial x_j$. By also imposing the conditions that T''_{ij} must be symmetric and that $T''_{ij} = 0$ for pure rotation, in which case no internal friction can arise in the fluid, one obtains

$$T''_{ij} = \eta \left(\frac{\partial u_i}{\partial x_j} + \frac{\partial u_j}{\partial x_i} \right), \tag{A.10-10}$$

for an incompressible fluid. The factor η is called the *dynamic viscosity*. For the problems of interest to us η may be assumed constant.

On substituting (A.10-7) into (A.10-5), and using the divergence theorem to transform the surface integral into a volume integral, we obtain

$$\frac{d}{dt} \int_V \rho u_i \, dV = \int_V \left(\rho f_i + \frac{\partial T_{ij}}{\partial x_j} \right) dV. \tag{A.10-11}$$

Furthermore, if we use the mass continuity principle in the form $d\delta m/dt = 0$, we can write

$$\frac{d}{dt}\int_V u_i(\rho\ dV) = \int_V \frac{du_i}{dt}(\rho\ dV).$$

If this last result is substituted into (A.10-11) we obtain, in view of the arbitrariness of the choice of volume V,

$$\rho\frac{du_i}{dt} = \rho f_i + \frac{\partial T_{ij}}{\partial x_j}.\qquad(A.10\text{-}12)$$

In vector notation, this (*Navier-Stokes*) equation of motion for the fluid is

$$\frac{\partial\vec{u}}{\partial t} + \vec{u}\cdot\nabla\vec{u} = -\frac{\nabla p}{\rho} + \vec{f} + \nu\nabla^2\vec{u},\qquad(A.10\text{-}13)$$

where $\nu = \mu/\rho$ is the *kinematic viscosity*, and we have assumed the flow is incompressible.

A-10.2.2 STREAM FUNCTION FORMULATION FOR AXISYMMETRIC, INCOMPRESSIBLE FLOW

Consider steady incompressible flow past a fixed axisymmetric body. Let us describe the flow in terms of a right-handed, orthogonal, curvilinear coordinate system with its origin at the center of the body, and with coordinates (q_1, q_2, ϕ), where q_1 and q_2 describe position in any meridian plane. We may express the velocity field in this system as $\vec{v} = (u_1, u_2, 0)$. The condition $\nabla\cdot\vec{u} = 0$ then becomes

$$\frac{\partial}{\partial q_1}(h_1 h_3 u_1) + \frac{\partial}{\partial q_2}(h_1 h_3 u_2) = 0,\qquad(A.10\text{-}14)$$

where the h's are the scale factors associated with the coordinates; i.e., $h_i\ dq_i$ is the element of arc length associated with dq_i. This equation is satisfied by choosing

$$h_2 h_3 u_1 = -\frac{\partial\psi}{\partial q_2},\quad h_1 h_2 u_2 = \frac{\partial\psi}{\partial q_1},\qquad(A.10\text{-}15)$$

where ψ is the *stream function*. For example, in cylindrical coordinates $(z, \bar{\omega}, \phi)$ we have $h_1 = 1$, $h_2 = 1$, $h_3 = \bar{\omega}$, so that

$$u_z = -\frac{1}{\bar{\omega}}\frac{\partial\psi}{\partial\bar{\omega}},\quad u_{\bar{\omega}} = \frac{1}{\omega}\frac{\partial\psi}{\partial z}.\qquad(A.10\text{-}16)$$

Similarly, in spherical coordinates (r, θ, ϕ) we have $h_1 = 1$, $h_2 = r$, $h_3 = r\sin\theta$, so that

$$u_r = -\frac{1}{r^2\sin\theta}\frac{\partial\psi}{\partial\theta},\quad u_\theta = \frac{1}{r\sin\theta}\frac{\partial\psi}{\partial r}.\qquad(A.10\text{-}17)$$

It is easy to see that $\psi = $ constant along a streamline by writing (A.10-15) in the form

$$\nabla\psi = -h_3\hat{e}_\phi\times\vec{u},\qquad(A.10\text{-}18)$$

where $\hat{e}_\phi$ is the unit vector in the ϕ direction. Therefore, we have $\nabla\psi\cdot\vec{u} = \vec{u}\cdot(\hat{e}_\phi\times\vec{u}) = \hat{e}_\phi\cdot(\vec{u}\times\vec{u}) = 0$. Since $\nabla\psi$ is normal to the line $\psi = $ constant in any meridian plane, this line must be parallel to u everywhere.

Using the vector identities $\nabla^2\vec{u} = \nabla(\nabla\cdot\vec{u}) - \nabla\times(\nabla\times\vec{u})$ and $\vec{u}\cdot\nabla\vec{u} = 1/2\ \nabla u^2 - \vec{u}\times(\nabla\times\vec{u})$, the steady state Navier-Stokes equation becomes

$$(\nabla\times\vec{u})\times\vec{u} = -\nabla\left(\frac{u^2}{2} + \frac{p}{\rho}\right) - \nu\nabla\times(\nabla\times\vec{u}).\qquad(A.10\text{-}19)$$

On taking the curl of this equation, we obtain

$$\nabla\times[(\nabla\times\vec{u})\times\vec{u}] = -\nu\nabla\times[\nabla\times(\nabla\times\vec{u})].\qquad(A.10\text{-}20)$$

In the system (q_1, q_2, ϕ) we have

$$\nabla \times \vec{u} = \frac{\hat{e}_\phi}{h_1 h_2} \left[\frac{\partial}{\partial q_1} (h_2 u_2) - \frac{\partial}{\partial q_2} (h_1 u_1) \right]. \tag{A.10-21}$$

Using (A.10-15) this becomes

$$\nabla \times \vec{u} = \frac{\hat{e}_\phi}{h_3} E^2 \psi, \tag{A.10-22}$$

where

$$E^2 \equiv \frac{h_3}{h_1 h_2} \left[\frac{\partial}{\partial q_1} \left(\frac{h_2}{h_1 h_3} \frac{\partial}{\partial q_1} \right) + \frac{\partial}{\partial q_2} \left(\frac{h_1}{h_2 h_3} \frac{\partial}{\partial q_2} \right) \right]. \tag{A.10-23}$$

After carrying out all the indicated differentiations in (A.10-20), using (A.10-18) and (A.10-21) to (A.10-23), we obtain the governing equation for the flow in terms of the stream function:

$$\left[\nabla \psi \times \nabla \left(\frac{E^2 \psi}{h_3^2} \right) \right] \cdot \hat{e}_\phi = \frac{\nu E^4}{h_3}, \tag{A.10-24}$$

where $E^4 \equiv E^2(E^2)$. In spherical coordinates, (A.10-24) becomes

$$\sin \theta \left[\frac{\partial \psi}{\partial r} \frac{\partial}{\partial \theta} - \frac{\partial \psi}{\partial \theta} \frac{\partial}{\partial r} \right] \left(\frac{E^2 \psi}{r^2 \sin^2 \theta} \right) = \nu E^4 \psi, \tag{A.10-25}$$

where

$$E^2 = \frac{\partial^2}{\partial r^2} + \frac{\sin \theta}{r^2} \frac{\partial}{\partial \theta} \left(\frac{1}{\sin \theta} \frac{\partial}{\partial \theta} \right). \tag{A.10-26}$$

For Stokes flow, the pressure may be easily recovered from the stream function by noting that

$$\nabla p = \eta \nabla^2 \vec{u} = -\nabla \times (\nabla \times \vec{u}) = \frac{\hat{e}_\phi \times \nabla(E^2 \psi)}{\bar{\omega}}, \tag{A.10-27}$$

where $\bar{\omega}$ denotes radial distance from the symmetry axis.

A-10.3.3 DROP OSCILLATIONS

We follow here the treatment given in §61 of Landau and Lifshitz (1959). Assuming potential flow within the drop, the velocity potential Φ satisfies $\nabla^2 \Phi = 0$. To solve this equation, the shape of the deformed drop must be specified. Since the shape is governed by (10-80), we must determine the sum $R_1^{-1} + R_2^{-1}$ for a slightly deformed sphere. To accomplish this, first note that the area of a surface $r = r(\theta, \phi)$ given in spherical coordinates (r, θ, ϕ) is

$$S = \int_0^{2\pi} \int_0^\pi \left[r^2 + \left(\frac{\partial r}{\partial \theta} \right)^2 + \frac{1}{\sin^2 \theta} \left(\frac{\partial r}{\partial \phi} \right)^2 \right]^{1/2} r \sin \theta \, d\theta \, d\phi. \tag{A.10-28}$$

Assuming $r = a_0 + \xi$ with $\xi \ll a_0$, (A.10-28) becomes

$$S = \int_0^{2\pi} \int_0^\pi \left\{ (a_0 + \xi)^2 + \frac{1}{2} \left[\left(\frac{\partial \xi}{\partial \theta} \right)^2 + \frac{1}{\sin^2 \theta} \left(\frac{\partial \xi}{\partial \phi} \right)^2 \right] \right\} \sin \theta \, d\theta \, d\phi. \tag{A.10-29}$$

Therefore, the variation in S due to a variation $\delta \xi$ is

$$\delta S = \int_0^{2\pi} \int_0^\pi \left\{ 2(a_0 + \xi) \delta \xi + \frac{\partial \xi}{\partial \theta} \frac{\partial (\delta \xi)}{\partial \theta} + \frac{1}{\sin^2 \theta} \frac{\partial \xi}{\partial \phi} \frac{\partial (\delta \xi)}{\partial \phi} \right\} \sin \theta \, d\theta \, d\phi \tag{A.10-30}$$

$$= \int_0^{2\pi} \int_0^\pi \left\{ 2(a_0 + \xi) - \frac{1}{\sin \theta} \frac{\partial}{\partial \theta} \left(\sin \theta \frac{\partial \xi}{\partial \theta} \right) - \frac{1}{\sin^2 \theta} \frac{\partial^2 \xi}{\partial \phi^2} \right\} \delta \xi \sin \theta \, d\theta \, d\phi, \tag{A.10-31}$$

This last result has been obtained by integrating the second and third terms in (A.10-30) by parts with respect to θ and ϕ, respectively.

Another expression for δS is available from (10-78) and (10-79), viz.,

$$\delta S = \int \delta \, dS = -\int \frac{(p_e - p_i)}{\sigma} \delta \xi \, dS = \int \left(\frac{1}{R_1} + \frac{1}{R_2}\right) \delta \xi \, dS, \tag{A.10-32}$$

where the integral is taken over the closed surface. In spherical coordinates, $dS = r^2 \sin \theta \, d\theta \, d\phi = a_0(a_0 + 2\xi) \sin \theta \, d\theta \, d\phi$ to first order in ξ. Therefore, on comparing (A.10-31) and (A.10-32) we obtain the result

$$\frac{1}{R_1} + \frac{1}{R_2} = \frac{2}{a_0} - \frac{2\xi}{a_0^2} - \frac{1}{a_0^2}\left[\frac{1}{\sin^2 \theta}\frac{\partial^2 \xi}{\partial \phi^2} + \frac{1}{\sin \theta}\frac{\partial}{\partial \theta}\left(\sin \theta \frac{\partial \xi}{\partial \theta}\right)\right]. \tag{A.10-33}$$

For small amplitude potential flow, the Navier-Stokes equation reduces to $\partial \tilde{u}/\partial t = -\nabla p/\rho_w$. Therefore, in terms of the velocity potential

$$\rho_w \frac{\partial \Phi}{\partial t} + p = p_e, \tag{A.10-34}$$

where p_e is the constant exterior pressure (gravity effects are ignored). Then from (10-80), (A.10-33), and (A.10-34), the boundary condition at $r = a_0$ for Φ is

$$\rho_w \frac{\partial \Phi}{\partial t} + \sigma \left\{\frac{2}{a_0} - \frac{2\xi}{a_0^2} - \frac{1}{a_0^2}\left[\frac{1}{\sin^2 \theta}\frac{\partial^2 \xi}{\partial \phi^2} + \frac{1}{\sin \theta}\frac{\partial}{\partial \theta}\left(\sin \theta \frac{\partial \xi}{\partial \theta}\right)\right]\right\} = 0. \tag{A.10-35}$$

To eliminate reference to ξ, we differentiate (A.10-35) with respect to time, and note that $\partial \xi/\partial t = u_r = \partial \Phi/\partial r$. The result at $r = a_0$ is

$$\rho_w \frac{\partial^2 \Phi}{\partial t^2} - \frac{\sigma}{a_0^2}\left\{2\frac{\partial \Phi}{\partial r} + \frac{\partial}{\partial r}\left[\frac{1}{\sin^2 \theta}\frac{\partial^2 \Phi}{\partial \phi^2} + \frac{1}{\sin \theta}\frac{\partial}{\partial \theta}\left(\sin \theta \frac{\partial \Phi}{\partial \theta}\right)\right]\right\}. \tag{A.10-36}$$

If we seek a solution in the form of a stationary wave, then $\Phi = e^{i\omega t}\psi$, where $\nabla^2 \psi = 0$. In spherical coordinates, therefore, the velocity potential is of the form

$$\Phi = A e^{i\omega t} r^n P_n^m(\cos \theta) e^{im\phi}, \tag{A.10-37}$$

where $P_n^m(\cos \theta)$ is the associated Legendre function. Finally, on substituting (A.10-37) into (A.10-36), and using the fact that $Y_{nm} \equiv P_n^m(\cos \theta)e^{im\phi}$ satisfies

$$\frac{1}{\sin \theta}\frac{\partial}{\partial \theta}\left(\sin \theta \frac{\partial Y_{nm}}{\partial \theta}\right) + \frac{1}{\sin^2 \theta}\frac{\partial^2 Y_{nm}}{\partial \phi^2} + n(n+1)Y_{nm} = 0,$$

we find a discrete set of allowed angular frequencies, namely

$$\omega_n^2 = \frac{n(n-1)(n+2)\sigma}{\rho_w a_0^3}. \tag{A.10-38}$$

Since ω_n is independent of m, there are $2n+1$ different oscillatory modes ($m = 0, \pm 1, \ldots, \pm n$) corresponding to each ω_n. Thus the asymmetric modes have the same frequencies as the axisymmetric modes of the same n.

A-10.3.4 RAYLEIGH-TAYLOR INSTABILITY OF TWO SUPERPOSED FLUIDS

We consider the stability of two superposed inviscid fluids in a vertical circular cylinder, following the treatment given in Chapter 4 of Yih (1965).

Denote the lower fluid density by ρ and the upper by ρ' ($\rho' > \rho$). Let the axis of the cylinder be along the vertical z-axis. Assume potential flow, and let the velocity potentials be Φ and Φ'.

The boundary conditions at the cylinder wall are that the radial velocities must vanish, so that in cylindrical coordinates (r, θ, z) we have

$$\frac{\partial \Phi}{\partial r} = \frac{\partial \Phi'}{\partial r} = 0 \quad \text{at} \quad r = a. \tag{A.10-39}$$

The kinematic condition at the fluid interface is that the vertical velocity is continuous, or

$$\frac{\partial \Phi}{\partial z} = \frac{\partial \Phi'}{\partial z} = \frac{\partial \xi}{\partial t} \quad \text{at} \quad z = \xi, \tag{A.10-40}$$

where ξ is the displacement of the interface. (When there is no motion, the interface is at $z = 0$.) If p and p' denote the perturbation pressures associated with the potential flow, then from arguments parallel to those used in deriving (A.10-33) we have the following dynamic boundary condition at the interface:

$$p' - p = \sigma \left(\frac{\partial^2 \xi}{\partial r^2} + \frac{1}{r} \frac{\partial \xi}{\partial r} + \frac{1}{r^2} \frac{\partial^2 \xi}{\partial \theta^2} \right). \tag{A.10-41}$$

Assuming the upper fluid extends to $z \to +\infty$ and the lower fluid to $z \to -\infty$, and recognizing that $\nabla^2 \Phi = \nabla^2 \Phi' = 0$, suitable particular solution forms for the velocity potentials are

$$\Phi = c \exp(\alpha t + k_m z) \cos n\theta J_n(k_m r),$$
$$\Phi' = c' \exp(\alpha t - k_m z) \cos n\theta \, J_n(k_m r), \tag{A.10-42}$$

where J_n is the Bessel function of the nth order. The eigenvalues k_m are roots of the equation

$$J_n'(kr) = dJ_n(kr)/dr = 0 \quad \text{at} \quad r = a.$$

These solutions thus satisfy the conditions (A.10-39).

To satisfy (A.10-40), we set

$$\xi = a e^{\alpha t} \cos n\theta \, J_n(kr), \tag{A.10-43}$$

and find

$$ck = -c'k = \alpha a, \tag{A.10-44}$$

by imposing (A.10-40) at $z = 0$ (ξ is regarded as an infinitesimal disturbance).

For small amplitude inviscid flow, the Navier-Stokes equation in the presence of gravity reduces to $\partial \tilde{u}/\partial t = -\rho^{-1} \nabla p - \nabla(gz)$, so that the pressures can be found from the (Bernoulli) equations

$$\frac{p}{\rho} = -\frac{\partial \Phi}{\partial t} - g\xi \quad \text{and} \quad \frac{p'}{\rho'} = -\frac{\partial \Phi}{\partial t} - g\xi. \tag{A.10-45}$$

On substituting (A.10-42), (A.10-43), and (A.10-45) into (A.10-41) for $z = 0$, we obtain

$$\rho(-\alpha c + ga) - \rho'(-\alpha c' + ga) = \sigma k^2 a. \tag{A.10-46}$$

Finally, by eliminating c, c', and a from (A.10-44) and (A.10-46), the value of α is determined in terms of the physical parameters ρ, ρ', and σ, and the eigenvalue k:

$$\alpha^2 = \frac{g(\rho' - \rho)k}{\rho + \rho'} - \frac{\sigma k^3}{\rho + \rho'}. \tag{A.10-47}$$

Therefore, if

$$k < k_c = \left[\frac{g(\rho' - \rho)}{\sigma} \right]^{1/2}, \tag{A.10-48}$$

the fluid is unstable, since then the amplitude of the disturbance will grow exponentially in time. It is noteworthy that this result is valid for any cross-sectional shape, which only affects the eigenvalues of k. In particular, (A.10-48) holds for the case $a \to \infty$, i.e., two semi-infinite fluids with an interface.

APPENDIX TO CHAPTER 12

A-12.4 Mutual Sedimentation and Diffusion of Aerosol Particles

The solution to (12-24) is facilitated by making the change of variable from n to n', where

$$n(z, t) = n'(z, t) \exp \left(-\frac{V_s z}{2D} - \frac{V_s^2 t}{4D} \right). \tag{A.12-1}$$

Substitution of (A.12-1) into (12-24) reduces the latter to the standard diffusion equation for n':

$$\frac{\partial n'}{\partial t} = D\nabla^2 n'. \tag{A.12-2}$$

From the initial and boundary conditions specified in Section 12.4, namely that $n(z, 0) = n_0 =$ constant and $n(0, t) = 0$, we find the corresponding conditions on n': $n'(z, 0) = n_0 \exp(V_s z/2 D)$ and $n'(0, t) = 0$. This problem may be reduced to one without boundary conditions by extending the solution to include the region $z < 0$, where we assume an initial distribution which simulates the actual boundary condition at $z = 0$ for all $t > 0$. This can be accomplished by choosing $n'(z, 0) = -n'(-z, 0)$, so that for the extended region the initial condition is

$$n'(z, 0) = \begin{cases} n_0 \exp(V_s z/2 D), & z > 0 \\ -n_0 \exp(-V_s z/2 D), & z < 0 \end{cases}. \tag{A.12-3}$$

Our problem is now one of solving the diffusion equation for a specified initial distribution in an unbounded medium. This can be accomplished by expanding the desired solution as a Fourier integral:

$$n'(z, t) = \int_{-\infty}^{+\infty} \Phi(k, t)e^{ikz} \, dk, \tag{A.12-4}$$

$$\Phi(k, t) = \frac{1}{2\pi} \int_{-\infty}^{+\infty} n'(z', t)e^{-ikz'} \, dz'. \tag{A.12-5}$$

On substituting (A.12-4) into (A.12-2) we find $\partial\Phi/\partial t + k^2 D\Phi = 0$; the solution is $\Phi(k, t) = \Phi(k, 0)e^{-k^2 Dt}$, where $\Phi(k, 0)$ is obtained by substituting the initial distribution for n' into (A.12-5). Therefore, (A.12-4) may be expressed as an integral over the specified initial distribution as follows:

$$n'(z, t) = \frac{1}{2\pi} \int_{-\infty}^{+\infty} \int_{-\infty}^{+\infty} n'(z', 0)e^{-k^2 Dt}e^{ik(z-z')} \, dz' \, dk. \tag{A.12-6}$$

The integration over k is easily carried out, whence the solution for n' becomes

$$n'(z, t) = \frac{1}{(4\pi Dt)^{1/2}} \int_{-\infty}^{+\infty} n'(z', 0) \exp[-(z - z')^2/4 Dt] \, dz'. \tag{A.12-7}$$

The solution may now be completed by substituting (A.12-3) into (A.12-7) and integrating. The result in terms of $n(z, t)$ is given by (12-25), where $\text{erf}(x) \equiv (2/\pi^{1/2}) \int_0^x e^{-x^2} \, dx$.

APPENDICES TO CHAPTER 14

A-14.1 Nearest Neighbor Distance Between Cloud Drops

Let $p(r) \, dr$ be the probability that the nearest neighbor to a given drop is located between a distance r and $r + dr$ from it. This probability is simply the product of the probability of no nearest neighbor between 0 and r, which is $1 - \int_0^r p(r) \, dr$, and the probability of a drop occurring between r and $r + dr$, which is $4\pi r^2 n \, dr$, where n is the average number concentration of drops. Therefore, we obtain the relation

$$p(r) = 4\pi r^2 n \left(1 - \int_0^r p(r) \, dr\right). \tag{A.14-1}$$

From this we find $dg/g = -4\pi r^2 n \, dr$, where $g = p/4\pi r^2 n$; hence $g = g_0 \exp(-4\pi r^3 n/3)$. From the condition $\int_0^\infty p(r)\, dr = 1$, we see that $g_0 = 1$, and so

$$p(r) = 4\,\pi r^2 n \exp(-4\,\pi r^3 n/3). \tag{A.14-2}$$

From (A.14-2) the nearest neighbor distance S is

$$S = \int_0^\infty rp(r)\, dr = \left(\frac{3}{4\,\pi n}\right)^{1/3} \int_0^\infty x^{1/3} e^{-x}\, dx, \tag{A.14-3}$$

in which we have set $x = 4\,\pi r^3 n/3$. The integral is $\Gamma(4/3)$, where Γ is the gamma function, so that finally we obtain

$$S = \frac{0.554}{n^{1/3}} \tag{A.14-4}$$

(Hertz, 1909). In terms of the liquid water content, $w_L = 4\,\pi a^3 n/3$, where a is the characteristic drop radius, the result is

$$\frac{S}{a} = \frac{\Gamma(4/3)}{w_L^{1/3}} = \frac{0.893}{w_L^{1/3}}. \tag{A.14-5}$$

The simple estimate of (14-1) is thus seen to be quite accurate.

A-14.3 Superposition Method for Stokes Flow

As an illustration of the use of the superposition method, we shall consider the special case of two spheres falling in the x-direction in Stokes flow. The situation is depicted in Figure A.14-1. For the velocity field $\tilde{u}_1$ generated by sphere-1 we write $\tilde{u}_1 = \tilde{u}_1(r_1, \theta_1)$, where θ is the polar angle between $\tilde{r}_1$ and the x_1-axis. Then according to (14-7) the hydrodynamic forces on the two spheres are

$$\tilde{F}_1 = -6\,\pi a_1 \eta[\tilde{v}_1 - \tilde{u}_2(d, \Theta + \pi)], \tag{A.14-6a}$$

$$\tilde{F}_2 = -6\,\pi a_2 \eta[\tilde{v}_2 - \tilde{u}_1(d, \Theta)], \tag{A.14-6b}$$

where $\tilde{v}_1 = v_1 \hat{e}_x$, $\tilde{v}_2 = v_2 \hat{e}_x$, d is the separation of the sphere centers, and Θ is the angle between the line of centers and the direction of fall. The unit vectors $\hat{e}_r$, $\hat{e}_\theta$, $\hat{e}_x$, and $\hat{e}_y$ satisfy the following relations: (a) along the line of centers, $\hat{e}_{r,1} = -\hat{e}_{r,2}$, $\hat{e}_{\theta,1} = -\hat{e}_{\theta,2}$; (b) $\hat{e}_{r,1} = \hat{e}_x \cos\Theta + \hat{e}_y \sin\Theta$, $\hat{e}_{\theta,1} = -\hat{e}_x \sin\Theta + \hat{e}_y \cos\Theta$. Therefore the forces may be expressed as

$$-\frac{\tilde{F}_1}{6\,\pi a_1 \eta v_1} = \Lambda_1 \hat{e}_x + \Gamma_1 \hat{e}_y, \tag{A.14-7a}$$

$$-\frac{\tilde{F}_2}{6\,\pi a_2 \eta v_2} = \Lambda_2 \hat{e}_x + \Gamma_2 \hat{e}_y, \tag{A.14-7b}$$

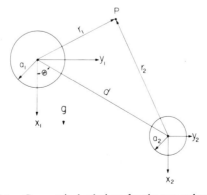

Fig. A.14-1. Geometrical relations for the two sphere problem.

where

$$\Lambda_1 = 1 + \frac{u_{r,2}(d, \Theta + \pi) \cos \Theta - u_{\theta,1}(d, \Theta + \pi) \sin \Theta}{v_1},$$ (A.14-8)

$$\Gamma_1 = \frac{u_{r,2}(d, \Theta + \pi) \sin \Theta + u_{\theta,2}(d, \Theta + \pi) \cos \Theta}{v_1},$$ (A.14-9)

$$\Lambda_2 = 1 + \frac{u_{\theta,1}(d, \Theta) \sin \Theta - u_{r,1}(d, \Theta) \cos \Theta}{v_2},$$ (A.14-10)

$$\Gamma_2 = \frac{-u_{r,1}(d, \Theta) \sin \Theta - u_{\theta,1}(d, \Theta) \cos \Theta}{v_2}.$$ (A.14-11)

Assuming Stokes flow, from (10-35) and (A.10-17) we have

$$\frac{\tilde{u}_1}{v_1} = \frac{\cos \theta_1}{2} \left(\frac{3 a_1}{d} - \frac{a_1^3}{d^3} \right) \hat{e}_{r,1} - \frac{\sin \theta}{4} \left(\frac{3 a_1}{d} + \frac{a_1^3}{d^3} \right) \hat{e}_{\theta,1},$$ (A.14-12)

and similarly for $\tilde{u}_2$. As a result the dimensionless forces are

$$\Lambda_1 = 1 - \frac{v_2}{v_1} \left[(3 \beta - \beta^3) \frac{\cos^2 \Theta}{2} + (3 \beta + \beta^3) \frac{\sin^2 \Theta}{4} \right],$$ (A.14-13)

$$\Gamma_1 = -\frac{3}{4} \frac{v_2}{v_1} \beta(1 + \beta^2) \sin \Theta \cos \Theta,$$ (A.14-14)

$$\Lambda_2 = 1 - \frac{v_1}{v_2} \left[(3 \alpha - \alpha^3) \frac{\cos^2 \Theta}{2} + (3 \alpha + \alpha^3) \frac{\sin^2 \Theta}{4} \right],$$ (A.14-15)

$$\Gamma_2 = -\frac{3}{4} \frac{v_1}{v_2} \alpha(1 + \alpha^2) \sin \Theta \cos \Theta,$$ (A.14-16)

where $\alpha = a_1/d$ and $\beta = a_2/d$.

According to (A.14-13) and (A.14-15), the dimensionless drag (x-component) force acting on either sphere of an equal pair falling side by side ($\Theta = \pi/2$) is

$$\Lambda_1 = \Lambda_2 = 1 - \frac{3}{4} \alpha - \frac{\alpha^3}{4}.$$ (A.14-17)

For motion along the line of centers ($\Theta = 0$), the result is

$$\Lambda_1 = \Lambda_2 = 1 - \frac{3}{2} \alpha + \frac{\alpha^3}{2}.$$ (A.14-18)

For touching spheres ($\alpha = 1/2$) we have $\Lambda = 0.313$ for $\Theta = 0$ and $\Lambda = 0.594$ for $\Theta = \pi/2$. These values compare rather poorly with the known rigorous results $\Lambda = 0.645$ for $\Theta = 0$ and $\Lambda = 0.725$ for $\Theta = \pi/2$ (see Appendix A-14.4.2), showing the superposition scheme to be indeed rather inaccurate for spheres of comparable size in close proximity. (If the spheres are free to rotate as they fall side by side, a slight reduction in drag results: $\Lambda = 0.715$.) Note spheres falling side by side experience a greater resistance than when falling along their line of centers, in agreement with intuition since the latter configuration is the more streamlined of the two.

We also see from (A.14-4) and (A.14-16) that there are no transverse (lift) forces acting on spheres falling side by side in Stokes flow: $\Gamma_1 = \Gamma_2 = 0$ for $\Theta = \pi/2$. This result happens to be true rigorously for Stokes flow.

A-14.4.2 SPECIAL PROBLEMS 1, 2, 3 FOR TWO SPHERES IN STEADY STATE STOKES FLOW

1. *Problem 1: Axisymmetric Motion*

For the axisymmetric problem we may use the stream function. The equation to be solved is

$$E^4 \psi = 0.$$ (A.14-19)

In cylindrical coordinates we have, from (A.10-23),

$$E^2 = \bar{\omega} \frac{\partial}{\partial \bar{\omega}} \left(\frac{1}{\bar{\omega}} \frac{\partial}{\partial \bar{\omega}} \right) + \frac{\partial^2}{\partial z^2}. \tag{A.14-20}$$

Since E^2 is linear in z, a solution to (A.14-19) is

$$\psi = \psi_1 + z\psi_2, \tag{A.14-21}$$

where $E^2\psi_{1,2} = 0$.

The principal difficulty in all of the 2-sphere flow problems is that of satisfying the no-slip boundary conditions on both spheres simultaneously. Best suited for this purpose are the bispherical coordinates, (ξ, η), defined by

$$\bar{\omega} = \frac{c \sin \eta}{\cosh \xi - \cos \eta}, \quad z = \frac{c \sinh \xi}{\cosh \xi - \cos \eta} \quad \begin{array}{l} 0 \le \eta \le \pi \\ c = \text{constant} > 0. \end{array} \tag{A.14-22}$$

The surface defined by $\xi = \xi_0$ is a sphere of radius $c|\text{cosech } \xi_0|$ with its center at $\bar{\omega} = 0$, $z = c \coth \xi_0$; if $\xi_0 > 0$ ($\xi_0 < 0$) the sphere lies completely in the half space $z > 0$ ($z < 0$). Thus if the two spheres are of radii a_1, a_2 and have their centers at distances d_1, d_2 from the origin, on opposite sides, then

$$\begin{aligned} a_1 &= c \text{ cosech } \alpha, \quad a_2 = -c \text{ cosech } \beta, \\ d_1 &= c \coth \alpha, \quad d_2 = -c \coth \beta, \end{aligned} \tag{A.14-23}$$

where $\alpha > 0$ and $\beta < 0$.

Writing $\cos \eta \equiv \tau$, we have in bispherical coordinates

$$E^2 = \frac{(\cosh \xi - \tau)}{c^2} \left\{ \frac{\partial}{\partial \xi} \left[(\cosh \xi - \tau) \frac{\partial}{\partial \xi} \right] + (1 - \tau^2) \frac{\partial}{\partial \tau} \left[(\cosh \xi - \tau) \frac{\partial}{\partial \tau} \right] \right\}. \tag{A.14-24}$$

A solution of $E^2\psi = 0$ has been given by Jeffery (1912) in the form

$$\psi = (\cosh \xi - \tau)^{-1/2} \sum_{n=0}^{\infty} [a_n \cosh (n + \tfrac{1}{2})\xi + b_n \sinh (n + \tfrac{1}{2})\xi]V_n(\tau), \tag{A.14-25}$$

where $V_n(\tau)$ is a simple combination of Legendre polynomials:

$$V_n(\tau) \equiv P_{n-1}(\tau) - P_{n+1}(\tau). \tag{A.14-26}$$

On combining two expressions of the form (A.14-25) with (A.14-21) the solution of (A.14-19) results:

$$\psi = (\cosh \xi - \tau)^{-3/2} \sum_{n=0}^{\infty} U_n(\xi)V_n(\tau), \tag{A.14-27}$$

where

$$U_n = A_n \cosh (n - \tfrac{1}{2})\xi + B_n \sinh (n - \tfrac{1}{2})\xi + C_n \cosh (n + \tfrac{3}{2})\xi + D_n \sinh (n + \tfrac{3}{2})\xi. \tag{A.14-28}$$

The constants $A_n, \ldots, D_n$ are to be determined from the no-slip boundary conditions, $\psi|_a = (\partial\psi/\partial n)|_a = 0$, where n measures distance along the outward normal $\hat{n}$ (cf. (10-15)). Then by the straightforward but tedious integration of the stress tensor over the sphere surfaces, the drag forces may be obtained. For details the reader may refer to the paper by Stimson and Jeffery (1926). For the special case of identical spheres moving with equal velocities, Stimson and Jeffery obtained the following expression for the drag D on either sphere:

$$\frac{D}{D_S} = \frac{4}{3} \sinh \alpha \sum_{n=1}^{\infty} \frac{n(n+1)}{(2n-1)(2n+3)} \left\{ 1 - \frac{4 \sinh^2 (n + \tfrac{1}{2})\alpha - (2n+1)^2 \sinh^2 \alpha}{2 \sinh (2n+1)\alpha + (2n+1) \sinh 2\alpha} \right\}, \tag{A.14-29}$$

where D_S is the corresponding single sphere Stokes drag given by (10-39). For touching spheres the value of this function is approximately 0.645.

2. Problems 2 and 3: Sphere Rotation, and Motion Perpendicular to the Line of Centers

We now proceed to outline the solutions to problems 2 and 3. These asymmetrical problems are much more difficult, and it was only relatively recently that an elegant method of solution was discovered by Dean and O'Neil (1963). Our presentation closely follows the application of their method by O'Neil and Majumdar (1970a, b). Another treatment may be found in Davis (1972).

As in problem 1 we choose a system of cylindrical coordinates $(z, \bar{\omega}, \phi)$ such that the line of centers of the spheres is along the z axis, with the centers of spheres 1 and 2 at $(d_1, 0, 0)$ and $(-d_2, 0, 0)$, respectively. For both problems the velocity is required to vanish at infinity and on the surface of sphere 2. In problem 2 sphere 1 rotates with angular velocity $\Omega \hat{e}_y$, so that the cylindrical components of the velocity field satisfy

$$u_z = -\Omega \bar{\omega} \cos \phi, \quad u_{\bar{\omega}} = \Omega(z - d_1) \cos \phi, \quad u_\phi = -\Omega(z - d_1) \sin \phi, \qquad (A.14\text{-}30)$$

on sphere 1. For problem 3 sphere 1 moves with velocity $U\hat{e}_x$, so that the cylindrical components of the velocity field in this case satisfy

$$u_z = 0, \quad u_{\bar{\omega}} = U \cos \phi, \quad u_\phi = U \sin \phi, \qquad (A.14\text{-}31)$$

on sphere 1.

Basic solutions of the governing equations (14-19) and (14-20) (with dynamic viscosity denoted by μ) having symmetry about the plane $y = 0$ yield the following expressions for the pressure and velocity fields:

$$p = \mu\, A P \cos \phi$$

$$u_{\bar{\omega}} = (A/2)[\bar{\omega} P + c(W + \Pi)] \cos \phi$$

$$u_\phi = (Ac/2)(W - \Pi) \sin \phi$$

$$u_z = (A/2)(zP + 2\, c\Phi) \cos \phi, \qquad (A.14\text{-}32)$$

where $A = \Omega$ for problem 2 and U/c for problem 3, c is the same positive constant defined by (A.14-22) and (A.14-23), and P, Φ, W, and Π are functions of $\bar{\omega}$ and z only satisfying

$$L_1^2 P = L_1^2 \Phi = L_2^2 W = L_0^2 \Pi = 0, \qquad (A.14\text{-}33)$$

the operators being defined by

$$L_m^2 \equiv \frac{\partial^2}{\partial \bar{\omega}^2} + \frac{1}{\bar{\omega}} \frac{\partial}{\partial \bar{\omega}} - \frac{m^2}{\bar{\omega}^2} + \frac{\partial^2}{\partial z^2}. \qquad (A.14\text{-}34)$$

The condition of incompressibility provides a constraint among the four functions:

$$\left[3 + \bar{\omega} \frac{\partial}{\partial \bar{\omega}} + z \frac{\partial}{\partial z} \right] P + c \left[\frac{\partial \Pi}{\partial \bar{\omega}} + \left(\frac{\partial}{\partial \bar{\omega}} + \frac{2}{\bar{\omega}} \right) W + \frac{2}{\partial z} \frac{\partial \Phi}{\partial z} \right] = 0. \qquad (A.14\text{-}35)$$

By combining (A.14-30) and (A.14-32) we find the problem 2 boundary conditions for sphere 1 are satisfied if

$$P = -\frac{2(\bar{\omega} + c\Phi)}{z}, \quad W = \frac{\bar{\omega}^2}{cz} + \frac{\bar{\omega}\Phi}{z}, \quad \Pi = W + \frac{2(z + d_1)}{c}, \qquad (A.14\text{-}36)$$

everywhere on sphere 1. Similarly, from (A.14-31) and (A.14-32) the problem 3 boundary conditions for sphere 1 are satisfied if

$$P = -\frac{2 c\Phi}{z}, \quad W = 1 - \frac{\bar{\omega} P}{2 c}, \quad \Pi = W - 1, \qquad (A.14\text{-}37)$$

everywhere on sphere 1. For both problems the flow vanishes on sphere 2, so that

$$P = -\frac{2 c\Phi}{z}, \quad W = \Pi = \frac{\bar{\omega}\Phi}{z}, \qquad (A.14\text{-}38)$$

everywhere on sphere 2.

The differential equations (A.14-33) must next be solved in terms of the bispherical coordinates introduced in (A.14-22). According to Jeffery (1912) and O'Neil (1964), suitable solutions are

TABLE A.14-1

Values of f_{ij} and g_{ij} for various values of $p = a_2/a_1$ and $s' = s/a$. (Based on data given by O'Neil and Majumdar, 1970a.)

s/a_2	$a_2/a_1 = 1.000$				$a_2/a_1 = 0.5$			
	$f_{ij}(A)$	$f_{ij}(B)$	$g_{ij}(A)$	$g_{ij}(B)$	$f_{ij}(A)$	$f_{ij}(B)$	$g_{ij}(A)$	$g_{ij}(B)$
10.0	-0.0004	0.0070	1.0000	-0.0003	-0.0002	0.0152	1.0000	-0.0003
	0.0070	-0.0004	-0.0003	1.0000	0.0009	-0.0001	-0.0000	1.0000
	1.0040	-0.0630	-0.0003	0.0052	1.0022	-0.0656	-0.0002	0.0028
	-0.0630	1.0040	0.0052	-0.0003	-0.0382	1.0022	0.0028	-0.0001
1.0	-0.0378	0.1236	1.0180	-0.0217	-0.0289	0.3437	1.0126	-0.0358
	0.1236	-0.0378	-0.0217	1.0180	0.0214	-0.0160	-0.0045	1.0058
	1.0921	-0.2972	-0.0284	0.0927	1.0577	-0.3429	-0.0216	0.0642
	-0.2972	1.0921	0.0927	-0.0284	-0.1714	1.0679	0.0644	-0.0120
0.1	-0.2630	0.4114	1.2330	-0.1017	-0.2163	1.2813	1.1625	-0.2122
	0.4114	-0.2630	-1.1017	1.2330	0.0814	-0.1754	-0.0265	1.1820
	1.3930	-0.6599	-0.1972	0.3086	1.2673	-0.8267	-0.1622	0.2442
	-0.6599	1.3930	0.3086	-0.1972	-0.4133	1.4140	0.2402	-0.1316

$$\Phi = (\cosh \xi - \tau)^{1/2} \sum_{n=1}^{\infty} [A_n \cosh (n + \tfrac{1}{2})\xi + B_n \sinh (n + \tfrac{1}{2})\xi] P_n^1(\tau),$$

$$P = (\cosh \xi - \tau)^{1/2} \sum_{n=1}^{\infty} [C_n \cosh (n + \tfrac{1}{2})\xi + D_n \sinh (n + \tfrac{1}{2})\xi] P_n^1(\tau),$$

$$W = (\cosh \xi - \tau)^{1/2} \sum_{n=1}^{\infty} [E_n \cosh (n + \tfrac{1}{2})\xi + F_n \sinh (n + \tfrac{1}{2})\xi] P_n(\tau),$$

$$\Pi = (\cosh \xi - \tau)^{1/2} \sum_{n=1}^{\infty} [G \cosh (n + \tfrac{1}{2})\xi + H_n \sinh (n + \tfrac{1}{2})\xi] P_n^2(\tau),$$

(A.14-39)

where the $P_n^m(\tau) \equiv (1 - \tau^2)^{m/2} \, d^m P_n(\tau)/d\tau^m$ are associated Legendre polynomials.

Substitution of these solutions into the boundary condition equations yields determining equations for the coefficients $C_n, D_n, \ldots, H_n$ in terms of A_n and B_n. The determination of A_n, B_n is then affected through the use of (A.14-35). The resulting lengthy expressions may be found in O'Neil and Majumdar (1970a), together with a discussion of the method of successive approximations used to obtain explicit results for the coefficients. The forces are found by integrating the surface stress and its moment over the sphere surfaces. The results may be superposed to describe the forces in more general circumstances. For example, suppose sphere 1 translates with velocity $U_1\hat{e}_x$ and rotates with angular velocity $\Omega_1\hat{e}_y$, while sphere 2 moves with $U_2\hat{e}_x$ and $\Omega_2\hat{e}_y$. Then the resultant forces $\vec{F}_1 = F_1\hat{e}_x$ and $\vec{F}_2 = F_2\hat{e}_x$ may be expressed as follows:

$$F_1 = -6 \pi a_1 \mu \{\Omega_1 a_1 f_{11}(p, s') - \Omega_2 a_1 f_{12}(p^{-1}, s'p^{-1}) + U_1 f_{21}(p, s') + U_2 f_{22}(p^{-1}, s'p^{-1})\},$$
$$F_2 = -6 \pi a_2 \mu \{\Omega_1 a_2 f_{12}(p, s') - \Omega_2 a_2 f_{11}(p^{-1}, s'p^{-1}) + U_1 f_{22}(p, s') + U_2 f_{21}(p^{-1}, s'p^{-1})\}.$$

(A.14-40)

Similarly, the torques on spheres 1 and 2 are $\vec{\Gamma}_1 = \Gamma_1\hat{e}_y$ and $\vec{\Gamma}_2 = \Gamma_2\hat{e}_y$, respectively, where

$$\Gamma_1 = -8 \pi a_1^2 \mu \{\Omega_1 a_1 g_{11}(p, s') - \Omega_2 a_1 g_{12}(p^{-1}, s'p^{-1}) + U_1 g_{21}(p, s') + U_2 g_{22}(p^{-1}, s'p^{-1})\},$$

(A.14-41)

$$\Gamma_2 = -8 \pi a_2^2 \mu \{-\Omega_1 a_2 g_{12}(p, s') + \Omega_2 a_2 g_{11}(p^{-1}, s'p^{-1}) - U_1 g_{22}(p, s) - U_2 g_{21}(p^{-1}, s'p^{-1})\}.$$

In these expressions the coefficients f_{ij} and g_{ij} are functions only of p and s'.

In Table A.14-1, taken from O'Neil and Majumdar (1970a), values are given for f_{ij} and g_{ij} for

$p = a_2/a_1 = 1, 0.5$, and for $s' = 10, 1, 0.1$. For each pair of p and s', the f_{ij}, g_{ij} are grouped as follows:

$$\begin{bmatrix} f_{11}(p, s') \\ f_{12}(p^{-1}, s'p^{-1}) \\ f_{21}(p, s') \\ f_{22}(p^{-1}, s'p^{-1}) \end{bmatrix} \begin{bmatrix} f_{12}(p, s') \\ f_{11}(p^{-1}, s'p^{-1}) \\ f_{22}(p, s') \\ f_{21}(p^{-1}, s'p^{-1}) \end{bmatrix} \begin{bmatrix} g_{11}(p, s') \\ g_{12}(p^{-1}, s'p^{-1}) \\ g_{21}(p, s') \\ g_{22}(p^{-1}, s'p^{-1}) \end{bmatrix} \begin{bmatrix} g_{12}(p, s') \\ g_{11}(p^{-1}, s'p^{-1}) \\ g_{22}(p, s') \\ g_{21}(p^{-1}, s'p^{-1}) \end{bmatrix}.$$

(A.14-42)

A-14.4.3 DETAILS OF THE SLIP-FLOW CORRECTION

As shown in the previous appendix, in bispherical coordinates (ξ, η) a sphere surface is defined by ξ = constant. Thus in modifying the formulation of two spheres in Stokes flow to take slip-flow into account, the boundary condition of no-slip is replaced by the following condition, expressed in bispherical coordinates:

$$\beta(u_\eta - u_\eta^0) = \pm T_{\xi\eta}, \quad \xi = \text{constant.}$$

(A.14-43)

In this expression $T_{\xi\eta}$ is the tangential stress component, β the coefficient of external friction, u_η^0 the η velocity component of the body surface, and u_η the fluid tangential velocity component at the surface. The upper sign is to be used for $\xi = \xi_1$ (sphere 1 of radius c cosech ξ_1 with center at $\bar{\omega} = 0$, $z = c \coth \xi_1$), and the lower sign for $\xi = -\xi_2$ (sphere 2 of radius c cosech ξ_2 with center at $\bar{\omega} = 0$, $z = -c \coth \xi_2$). The tangential stress component $T_{\xi\eta}$ has the form (μ is the dynamic viscosity):

$$T_{\xi\eta} = \frac{\mu}{c^3}(\cosh \xi - \tau)^2 \left\{ (\cosh \xi - \tau) \left[\sin \eta \frac{\partial^2 \psi}{\partial \tau^2} - \csc \eta \frac{\partial^2 \psi}{\partial \xi^2} \right] \right.$$
$$\left. - 3 \left[\sin \eta \frac{\partial \psi}{\partial \tau} + \sin \xi \csc \eta \frac{\partial \psi}{\partial \xi} \right] \right\}.$$

(A.14-44)

The boundary condition expressed by (A.14-43) was employed by Davis (1972) to obtain his slip-flow modification of the Stokes flow problem presented in Appendix A-14.4.2 (Problem 1). In brief, the procedure is as follows: The expression for u_η in terms of the stream function ψ is inserted into (A.14-43); by using the general form for ψ given in (A.14-27)–(A.14-28) and recursion relations for P_n, the boundary condition (A.14-43)–(A.14-44) results finally in a recursion relation connecting U_n', U_n'', U_{n+1}'', and U_{n-1}'', where the primes denote differentiation with respect to ξ (Equation 12.II of Davis); the other boundary condition specifying the equality of the normal velocity components of fluid and body is unchanged. The expansion coefficients $A_n, \ldots, D_n$ may be determined now as before, by invoking the orthogonality and completeness properties of the V_n.

A-14.4.4 FLOW FIELD AND FORCES FOR TWO SPHERES IN MODIFIED OSEEN FLOW

Consider the case of the spheres moving in the x direction (see Figure A.14-1). The total velocity and pressure fields in the fluid are written as

$$\tilde{u} = \tilde{u}_1(\tilde{r}_1) + \tilde{u}_2(\tilde{r}_2)$$
$$p = p_1(\tilde{r}_1) + p_2(\tilde{r}_2),$$

(A.14-45)

where

$$U_\ell \frac{\partial \tilde{u}_\ell}{\partial x_\ell} = -\frac{\nabla p_\ell}{\rho} + \nu \nabla^2 \tilde{u}_\ell$$
$$\nabla \cdot \tilde{u}_\ell = 0, \quad \ell = 1, 2.$$

(A.14-46)

$U_\ell \hat{e}_x$ is the velocity of sphere ℓ, and ρ is the density of the fluid (unit vectors in the x and y directions are denoted by $\hat{e}_x$ and $\hat{e}_y$, respectively). Solutions of (A.14-46) which satisfy the boundary condition of undisturbed fluid at infinity can be expressed in terms of potentials ϕ_ℓ, χ_ℓ, π_ℓ, ψ_ℓ following Lamb (1932) and Goldstein (1929, 1931):

$$\tilde{u}_\ell = \nabla \phi_\ell - \frac{1}{2 k_\ell} \nabla \chi_\ell - \chi_\ell \hat{e}_x + \nabla \pi_\ell - \frac{1}{2 k_\ell} \nabla \frac{\partial \psi_\ell}{\partial y_\ell} - \frac{\partial \psi}{\partial x_\ell} \hat{e}_y$$
$$p_\ell = \rho U_\ell \left(\frac{\partial \phi_\ell}{\partial \chi_\ell} + \frac{\partial \pi_\ell}{\partial \chi_\ell} \right),$$

(A.14-47)

provided that

$$\nabla^2 \left\{ \begin{matrix} \phi_\ell \\ \pi_\ell \end{matrix} \right\} = 0, \quad \left(\nabla^2 - 2 k_\ell \frac{\partial}{\partial x_\ell} \right) \left\{ \begin{matrix} \chi_\ell \\ \psi_\ell \end{matrix} \right\} = 0, \tag{A.14-48}$$

and

$$k_\ell = \frac{U_\ell}{2\nu}. \tag{A.14-49}$$

Letting $r_\ell = (x_\ell^2 + y_\ell^2 + z_\ell^2)^{1/2}$, we have

$$\phi_\ell = \sum_{n=0}^{\infty} A_{\ell n} \frac{\partial^n}{\partial x_\ell^n} \left(\frac{1}{r_\ell} \right)$$

$$\chi_\ell = \sum_{n=0}^{\infty} B_{\ell n} \frac{\partial^n}{\partial x_\ell^n} \left[\frac{\exp\left[-k_\ell(r_\ell + x_\ell) \right]}{r_\ell} \right]$$

$$\pi_\ell = \sum_{n=0}^{\infty} C_{\ell n} \frac{\partial^{n+1}}{\partial y_\ell^{n+1}} \left\{ \log \left[k_\ell(r_\ell + x_\ell) \right] \right\} \tag{A.14-50}$$

$$\psi_\ell = \sum_{n=0}^{\infty} D_{\ell n} \frac{\partial^n}{\partial y_\ell^n} \left[\int_{r_\ell + x_\ell}^{\infty} (e^{-k_\ell}/s) \, ds \right].$$

In order to avoid singularities along the negative x_ℓ-axis, we must impose a constraint between $C_{\ell n}$ and $D_{\ell n}$:

$$C_{\ell n} = -\frac{D_{\ell n}}{2 k_\ell}. \tag{A.14-51}$$

The boundary conditions which are needed along with (A.14-51) to determine the coefficients appearing in (14-50) are

$$\tilde{u}|_{r_\ell = a_\ell} = U_\ell \hat{e}_x. \tag{A.14-52}$$

Because the Oseen equations can provide only the correct first-order contribution of the fluid inertia, the boundary condition equations are set up in a form involving the expansion of a part of the flow field in the neighborhood of each sphere in a particular way. Near a given sphere, the portion of the total field generated by it is expanded to first order in the Reynolds number based on its radius and velocity. Thus, the resultant drag expressions are given essentially to first order in the Reynolds number, and reduce to the well known first-order Oseen formula (10-42) for the drag on a single sphere in the limit when either sphere becomes isolated.

The complexity of the problem necessitates working with a highly truncated version of (A.14-50); the retained terms include the four potentials corresponding to $n = 0$, the $n = 1$ term for ϕ_ℓ, and a term of the form $b_\ell y_\ell / r_\ell^3$, with b_ℓ a constant. The latter is thus a member of the set of potentials given either by $\partial \pi_\ell / \partial x_\ell$ or $\partial \phi_\ell / \partial y_\ell$. Its inclusion is dictated by the need to preserve the symmetry properties of Stokes flow in the limit of zero Reynolds number.

Completion of the analysis thus requires a determination of the coefficients $A_{\ell 0}$, $A_{\ell 1}$, $B_{\ell 0}$, $C_{\ell 0}$, $D_{\ell 0}$, and b_ℓ. Condition (A.14-51) connects $C_{\ell 0}$ and $D_{\ell 0}$, while the symmetry condition in Stokes flow referred to above leads to the constraints

$$A_{\ell 1} = -\frac{\alpha_\ell^2 B_{\ell 0}}{6}, \quad b_\ell = -\frac{\alpha_\ell^2 D_{\ell 0}}{6}, \quad \alpha_\ell \equiv \frac{a_\ell}{d}. \tag{A.14-53}$$

Also, satisfaction of the z-component of the boundary condition (A.14-52) leads to the additional constraint

$$A_{\ell 0} = \frac{B_{\ell 0}}{2 k_\ell}. \tag{A.14-54}$$

The remaining coefficients $B_{\ell 0}$ and $D_{\ell 0}$ are determined by requiring the x and y components of the boundary condition (A.14-52) to hold on the average over the sphere surfaces. When part of the flow field is expanded in the manner described just after (A.14-52), the boundary conditions produce the following set of linear equations for $B_{\ell 0}$ and $D_{\ell 0}$:

$$B_{10}(\tfrac{2}{3}-\tfrac{1}{4}R_1)+B_{20}\alpha_1G_1+D_{20}\alpha_1G_3=-a_1U_1$$

$$B_{10}\alpha_2G_2+B_{20}(\tfrac{2}{3}-\tfrac{1}{4}R_2)+D_{10}\alpha_2G_4=-a_2U_2$$

$$B_{20}\alpha_1G_3+D_{10}(\tfrac{2}{3}-\tfrac{3}{8}R_1)+D_{20}\alpha_1G_5=0 \tag{A.14-55}$$

$$B_{10}\alpha_2G_4+D_{10}\alpha_2G_6+D_{20}(\tfrac{2}{3}-\tfrac{3}{8}R_2)=0,$$

where the G's are functions of α_ℓ, R_ℓ', and Θ, the angle between the direction of fall and the line connecting the centers of the spheres (see Figure A.14-1). Finally, integration of the fluid stresses over the surface of sphere ℓ produces the following simple expressions for the force components in terms of the coefficients:

$$F_{ex}=-4\,\pi\eta\,B_{\ell 0}$$

$$F_{ey}=-4\,\pi\eta\,D_{\ell 0}. \tag{A.14-56}$$

This completes the solution in principle. For further details the reader is referred to Klett and Davis (1973).

In the collision problem the spheres generally have small components of motion along the y axis, at right angles to the direction of gravity. This component of motion is treated in the same manner as the principal x component; in fact, the force expressions already obtained are adapted for this situation with only minor adjustments being necessary: U_1 and U_2 are replaced by V_1 and V_2, the new velocities of the spheres along the y axis; the directional roles of the forces are interchanged; and Θ is replaced by $\pi/2-\Theta$. The total hydrodynamic force acting on a sphere moving with a general off-axis velocity is then given by the sum of the forces corresponding to motion along each of the axes. In this final representation for the force, one further adjustment is made: the Reynolds numbers for each sphere are now based on the total velocity; i.e., we now have $R_\ell=a_\ell(U_\ell^2+V_\ell^2)^{1/2}/\nu$ and similarly for R_ℓ'. If the new force components due to motion along the y axis are distinguished by primes from the original set for motion along the x axis, the total hydrodynamic force on sphere ℓ moving with velocity $u_\ell\hat{e}_x+v_\ell\hat{e}_y$ may be expressed as

$$F_{tot,\ell}=-6\,\pi a_\ell\eta\,[u_\ell\Lambda_\ell-v_\ell\Gamma_\ell')\hat{e}_x+(v_\ell\Lambda_\ell'-u_\ell\Gamma_\ell)\hat{e}_y], \tag{A.14-57}$$

where if we indicate the dependence of the dimensionless force components Λ_ℓ and Γ_ℓ by writing

$$\Lambda_\ell=\Lambda_\ell(u_1,u_2,R_\ell,R_\ell',\alpha_\ell,\theta),\quad \Gamma_\ell=\Gamma_\ell(u_1,u_2,R_\ell,R_\ell',\alpha_\ell,\theta), \tag{A.14-58}$$

then for Λ_ℓ' and Γ_ℓ' we have

$$\Lambda_\ell'=\Lambda_\ell\left(v_1,v_2,R_\ell,R_\ell',\alpha_\ell,\frac{\pi}{2}-\theta\right),\quad \Gamma_\ell'=\Gamma_\ell\left(v_1,v_2,R_\ell,R_\ell',\alpha_\ell,\frac{\pi}{2}-\theta\right). \tag{A.14-59}$$

It is now a simple matter to write down the equations of motion of two spheres of density ρ', falling under the influence of gravity and the hydrodynamic forces of (A.14-57). In nondimensional form the equations are

$$\frac{du_\ell}{dt}=\frac{p^4}{N_S^*}\left[1-\frac{(\Lambda_\ell u_\ell-\Gamma_\ell'v_\ell)}{p_\ell^2}\right],\quad \frac{dv_\ell}{dt}=\frac{p^4}{N_S^*}\left[\frac{\Gamma_\ell u_\ell-\Lambda_\ell'v_\ell}{p_\ell^2}\right]$$

$$\frac{dx_\ell}{dt}=u_\ell,\quad \frac{dy_\ell}{dt}=v_\ell. \tag{A.14-60}$$

In these equations sphere ℓ has velocity (u_ℓ, v_ℓ) and center coordinates (x_ℓ, y_ℓ), where now the unit of length is a_1 and the unit of time is $a_1/V_{s,1}$, $V_{s,1}$ being the Stokes terminal velocity of the large sphere in isolation. Also,

$$p=\frac{a_2}{a_1},\quad p_\ell=\begin{cases}1,\ell=1\\p,\ell=2\end{cases},\quad N_S^*=\frac{2\rho'p^4V_{s,1}a_1}{9\,\eta}. \tag{A.14-61}$$

The physical significance of the parameter N_S^* is discussed in Section 14.3.1.

A-14.5   Drop Interactions in Turbulent Air (de Almeida, 1975)

1. *Equations of Motion*

Since a particle in a turbulent environment is subject to random accelerations, its motion can be described with great accuracy only by abandoning the assumption of steady flow which has been used up to this point. The effects of flow unsteadiness may be assessed by inspection of the following solution for the force on a sphere moving in the Stokes regime with velocity $\vec{v}(t)$, arbitrary in time but constant in direction, through an otherwise undisturbed fluid (Basset, 1910):

$$\vec{F} = -6\,\pi\eta a\left[\vec{v} + \frac{a^2}{9\,\nu}\frac{d\vec{v}}{dt} + \frac{a}{(\pi\nu)^{1/2}}\int_{-\infty}^{t}\frac{d\vec{v}}{dt}\frac{d\tau}{(t-\tau)^{1/2}}\right]. \tag{A.14-62}$$

The first term in this equation is the familiar steady state Stokes drag. The other two terms are corrections due to the unsteady motion of the sphere. The first correction term,

$$-\frac{2}{3}\pi\rho a^3\frac{d\vec{v}}{dt} = -\frac{m_f}{2}\frac{d\vec{v}}{dt}, \tag{A.14-63}$$

is the pressure-gradient resistive force experienced by a sphere which accelerates in an ideal fluid (see, for example, §11 of Landau and Lifschitz (1959)). Its effect is seen to be equivalent to an increase in the mass of the sphere by half the mass m_f of displaced fluid. Thus, the coefficient $m_f/2$ is referred to as the 'induced mass.' The second correction term, the 'Basset term,' describes the manner in which the history of the sphere's motion is connected through the viscosity of the fluid with the frictional resistance experienced by the sphere at the present time t. In this sense viscous fluids are seen to have an effective memory.

A generalization of (A.14-62) to the case of a non-uniformly moving medium was achieved by Tchen (1947). If $\vec{u}$ denotes the velocity of the fluid in the neighborhood of the spherical particle, but far enough away to be unaffected by it, Tchen's equation for the motion of the particle may be expressed as follows:

$$m_p\frac{d\vec{v}}{dt} = -\frac{m_f}{2}\left(\frac{d\vec{v}}{dt} - \frac{d\vec{u}}{dt}\right) - 6\,\pi a\eta(\vec{v} - \vec{u}) + m_f\frac{d\vec{u}}{dt} - 6(\pi\nu)^{1/2}\rho a^2\int_{-\infty}^{t}\frac{\left(\dfrac{d\vec{v}}{dt'} - \dfrac{d\vec{u}}{dt'}\right)dt'}{(t-t')^{1/2}}. \tag{A.14-64}$$

This equation was obtained by application of (A.14-62) to a particle of mass m_p moving with velocity $(\vec{v} - \vec{u})$ in a fluid at rest, followed by imposition of the velocity $\vec{u}$ on the entire system (fluid plus particle). For this 'derivation' to be reasonable, it follows that during the particle's motion its neighborhood must be formed by the same fluid. As discussed at great length by Hinze (1959), this assumption requires not only that the particle density cannot be much larger than that of the turbulent medium, but also that the velocity of the particle relative to the medium must be small compared to the characteristic microscale velocity $(\varepsilon\nu)^{1/4}$. Unfortunately, these restrictions appear to make (A.14-64) of limited quantitative value as a basis for predicting the motion of drops in turbulent air.

It can be shown (e.g., Fuchs, 1964) that the effect of the Basset term in (A.14-64) is to cause a small reduction in the particle velocity, proportional to $\sqrt{\rho/\rho'}$, where ρ' is the particle density. This result, plus the fact that $|\vec{v} - \vec{u}|$ is usually not close to zero in the collision problem, means that the integral term in (A.14-64) is probably of negligible importance.

We are now in a position to write down de Almeida's equations for the motion of a pair of spherical droplets interacting hydrodynamically while falling under gravity in turbulent air. The equation set is obtained through the following modifications of (A.14-64): i) the Basset term is dropped, ii) a term for buoyancy-corrected gravity is included, iii) the Stokes resistance term $-6\,\pi a\eta(\vec{v} - \vec{u})$ for a single sphere is replaced by the corresponding Klett-Davis (1973) forces for relative motion $(\vec{v} - \vec{u})$. Thus, in terms of the normalization and notation employed in the previous Appendix A-14.4.4, the dimensionless equations of motion for sphere ℓ are (denoting the drop velocities by $\vec{v}_\ell = (v_{x,\ell}, v_{y,\ell})$ and the air velocity by $\vec{u}_\ell = (u_{x,\ell}, u_{y,\ell})$):

$$\frac{dv_{x\ell}}{dt} = \frac{1.5\,\sigma}{(1+0.5\,\sigma)}\frac{du_{x,\ell}}{dt} + \frac{p^4}{N'_S}\left\{1 + \frac{[\Gamma'_\ell(v_{x,\ell}-u_{x,\ell}) - \Lambda_\ell(v_{y,\ell}-u_{y,\ell})]}{p_\ell^2}\right\},$$

$$\frac{dv_{y,\ell}}{dt} = \frac{1.5\,\sigma}{1+0.5\,\sigma}\frac{du_{y,\ell}}{dt} + \frac{p^4}{N'_S}\left\{\frac{\Gamma_\ell(v_{x,\ell}-u_{x,\ell}) - \Lambda'_\ell(v_{y,\ell}-u_{y,\ell})}{p_\ell^2}\right\},$$

(A.14-65)

where $\sigma \equiv \rho/\rho'$ and $N'_S \equiv (1+0.5\,\sigma)N^*_S$ (cf. (A.14-60)). In view of the neglect of the Basset term, the distinction between 1 and $1+0.5\,\sigma$ in (A.14-65) appears superfluous.

2. Turbulent Velocity Field Simulation

A suitable simulation of $\vec{u}$ and $d\vec{u}/dt$ must be provided in order to achieve the numerical integration of (A.14-65). de Almeida's procedure is to use a Monte Carlo method, described briefly below, to simulate the random vector $\vec{u}$ in such a way as to satisfy known statistical constraints. Then a numerical differentiation operator is used to obtain $d\vec{u}/dt$.

2a. Velocity Correlation

From our discussion of turbulence in Section 12.6.2, we recall that for eddies of scale size λ in the inertial subrange ($\lambda_K \ll \lambda \ll \ell$), there is a simple correlation of velocities between points a distance λ apart: $u_\lambda \sim (\varepsilon\lambda)^{1/3}$, where u_λ is the magnitude of the velocity fluctuation over length λ. de Almeida employs similar correlations in his simulation of turbulence. We shall describe them briefly now, following the treatment of this subject by Landau and Lifschitz (1959).

Consider the two-point correlation tensor B_{ik}:

$$B_{ik} \equiv \overline{(u_{2i} - u_{1i})(u_{2k} - u_{1k})},$$

(A.14-66)

where in suffix notation u_{1i} and u_{2i} are the velocity vectors at points 1 and 2, and the bar denotes some appropriate time average. Denote the radius vector from point 1 to point 2 by $\vec{r}$ and assume $r \ll \ell$. For such r the turbulence is isotropic, so that B_{ik} cannot depend on any direction in space except $\hat{n}$, where $\vec{r} = \hat{n}r$. Thus, the form of B_{ik} must be

$$B_{ik} = A(r)\delta_{ik} + B(r)n_i n_k.$$

(A.14-67)

Now orient the coordinate axes so that one of them is in the direction of $\hat{n}$, and denote the velocity components along and perpendicular to this axis by u_r and u_n, respectively. The tensor components B_{rr} and B_{nn} are of particular interest. They are given by

$$B_{rr} = \overline{[u_r(\vec{x}_1 + \vec{r}) - u_r(\vec{x}_1)]^2} = A(r) + B(r),$$
$$B_{nn} = \overline{[u_n(\vec{x}_1 + \vec{r}) - u_n(\vec{x}_1)]^2} = A(r),$$

(A.14-68)

since $n_r = 1$, $n_n = 0$.

A simple relationship exists between B_{rr} and B_{nn}, due to the assumptions that the turbulence is isotropic, homogeneous, and incompressible. First of all, isotropy and homogeneity imply that $\overline{u_{1i}u_{1k}} = \overline{u_{2i}u_{2k}}$ and $\overline{u_{1i}u_{2i}} = \overline{u_{1k}u_{2i}}$. Then from (A.14-66) we have $B_{ik} = 2\overline{u_{1i}u_{1k}} - 2\overline{u_{1i}u_{2k}}$, and the derivative of this with respect to the coordinates of point 2 is $\partial B_{1k}/\partial x_{2k} = -2\,\overline{u_{1i}\,\partial u_{2k}/\partial x_{2k}}$. But this is identically zero because of the incompressible condition, $\partial u_{2k}/\partial x_{2k} = 0$ (we are of course using the convention of summation over a double index).

Since B_{ik} is a function only of the components $x_i = x_{2i} - x_{1i}$ of the vector $\vec{r}$, differentiation with respect to x_{2k} is equivalent to differentiation with respect to x_k. Then from (A.14-67) we have

$$\frac{\partial B_{ik}}{\partial x_{2k}} = \frac{\partial B_{ik}}{\partial x_k} = \frac{\partial}{\partial x_k}\left[A(r)\delta_{ik} + B(r)\frac{x_i x_k}{r^2}\right]$$

$$= \frac{A'(r)x_k \delta_{ik}}{r} + \frac{B'(r)x_i x_k x_k}{r^3} + B(r)\left[\frac{\delta_{ik}x_k}{r^2} + \frac{3\,x_i}{r^2} - \frac{2\,x_i x_k x_k}{r^4}\right]$$

$$= \left[A'(r) + B'(r) + \frac{2\,B(r)}{r}\right]n_i = 0,$$

where the prime denotes differentiation with respect to r. Then on substitution from (A.14-68) we obtain the relationship $B'_{rr} + 2(B_{rr} - B_{nn})/r = 0$, or

$$B_{nn}(\vec{r}) = \frac{1}{2r}\frac{d}{dr}[r^2 B_{rr}(\vec{r})].$$
(A.14-69)

For r on the inertial subrange ($\lambda_k \ll r \ll \ell$), we know that both B_{rr} and B_{nn} must be proportional to $(\varepsilon r)^{2/3}$ (see the remarks at the beginning of this section). Then from (A.14-69) we may write

$$B_{rr}(\vec{r}) = c\varepsilon^{2/3} r^{2/3},$$
$$B_{nn}(\vec{r}) = \frac{4}{3}c\varepsilon^{2/3} r^{2/3}, \quad \lambda_k \ll r \ll \ell,$$
(A.14-70)

where c is a positive number of order unity. These are the constraints imposed by de Almeida on the random vector components u_r and u_n.

Similarly, for $\lambda \ll \lambda_k$ we have the result $u_\lambda \sim (\varepsilon/\nu)^{1/2}\lambda$ (see Section 12.6.2) so that for $r \ll \lambda_k$ both B_{rr} and B_{nn} must be proportional to $(\varepsilon/\nu)r^2$. Then from (A.14-69) we find

$$B_{rr}(\vec{r}) = \frac{c_1 \varepsilon r^2}{\nu},$$
$$B_{nn}(\vec{r}) = \frac{2c_1 \varepsilon r^2}{\nu},$$
(A.14-71)

where c_1 is another positive number. Actually, for this range of r the numerical coefficient can be determined: $c_1 = 1/15$. This is possible because the mean energy dissipation rate is $\varepsilon = 0.5\nu \overline{(\partial u_i/\partial x_k + \partial u_k/\partial x_i)^2}$, and the various average square gradient terms can be related to terms appearing in B_{rr} (or B_{nn}) in the limit of $r \to 0$ (see, for example, §33 of Landau and Lifschitz, 1959).

It is interesting to note that since the scale size of cloud droplets is generally smaller by at least an order of magnitude than the Kolmogorov length scale λ_k of the small eddies (Section 12.6.2), the effect of turbulence on the interaction of neighboring droplets should be governed primarily by motion on a scale $r < \lambda_k$. Accordingly, it appears that de Almeida should have used the statistical constraints of (A.14-71) rather than those of (A.14-70). It is difficult to estimate the quantitative differences that would result from these two different constraints.

de Almeida attempts to justify the use of (A.14-70) in the following manner. He notes that the '2/3 law' for the correlation functions B_{rr} and B_{nn} corresponds to a '$-5/3$ law' for the spectral distribution of energy $E(k)$, where $E(k)\,dk$ is the energy/mass on the wave number interval k to $k + dk$. (This can be seen to be true by noting B_{rr} and $E(k)$ are essentially Fourier transforms of each other, so that $E(k) \sim k^{-5/3}$ is equivalent to $B_{rr} \sim r^{2/3}$, i.e., the exponents must sum to -1.) He then quotes experimental evidence for the $-5/3$ law (Ackerman (1967, 1968), for hurricanes; Voyt et al. (1971) for cumulus clouds). However, their measured values of ε (from 5 to 700 cm^2 sec^{-3}) were not so large as to indicate the 2/3 law should hold down to distances as small as those corresponding to strong interactions between droplets (say 10 radii or less). Thus the experimental evidence does not support the use of (A.14-70).

Another relatively minor consideration is the question of the applicability of results from the theory of turbulence in a homogeneous fluid to the inhomogeneous two phase flow of cloudy air. Fortunately, here there is no problem. It is easy to show (e.g., de Almeida, 1975) that: (i) the presence of cloud particles does not significantly alter either the viscosity of the air or the energy dissipation rate, (ii) thermal stratification (buoyancy force) has a negligible influence on the small length scales of interest here.

Monte Carlo Method

The Monte Carlo method, or method of statistical trials, can be defined as any procedure which involves the use of statistical sampling techniques to simulate the solution of a mathematical or physical problem. A general review of the method may be found in references such as Brown (1956) and Halton (1970). Here we merely summarize very briefly the Monte Carlo model constructed by de Almeida for the purpose of simulating the turbulent velocity field.

The goal is to simulate the velocity components $u_r(\vec{x} + \vec{r})$ and $u_n(\vec{x} + \vec{r})$. In order to simplify the notation, denote either u_r or u_n by u_s. Also, denote the vector location $\vec{x} + \vec{r}$ by $\vec{x} + i\hat{e}_r$, so that at the discrete set of points corresponding to $i = 1, 2, \ldots, N$, the turbulent velocity components may be designated as $u_{i,s}$. These random variables are related by de Almeida to another set $\eta_{i,s}$ as follows:

$$u_{i,s} = \eta_{i,s} + \beta_s. \tag{A.14-72}$$

The constants β_s are assumed independent of position. They are just the mean or expected values of $u_{i,s}(\equiv \langle u_{i,s} \rangle)$, as the $\eta_{i,s}$ will be seen shortly to have zero mean. The $\eta_{i,s}$ are given in turn as a linear combination of another set of random variables ζ_k as follows:

$$\eta_{i,s} = \sum_{k=1}^{N} a_{i,k}(s)\zeta_k; \quad i = 1, 2, \ldots, N, \tag{A.14-73}$$

where the $a_{i,k}$ are components of an $N \times N$ matrix to be specified.

The ζ_k are chosen such that $\langle \zeta_k \rangle = 0$ and $\langle \zeta_k^2 \rangle = 1$; i.e., they have zero mean and unit variance. They are also assumed independent, so that $\langle \zeta_k \zeta_{k+j} \rangle = \langle \zeta_k \rangle \langle \zeta_{k+j} \rangle = 0$. In consequence, we find from (A.14-72) and (A.14-73) that

$$\langle u_s(\tilde{x}) \rangle = \beta_s,$$

$$\langle [u_s(\tilde{x} + \tilde{r}) - u_s(\tilde{x})]^2 \rangle = \sum_{k=1}^{i} [a_{i+\ell,k}(s) - a_{i,k}(s)]^2 + \sum_{k=i+1}^{i+\ell} a_{i+\ell,k}^2(s), \tag{A.14-74}$$

$$\langle u_s^2(\tilde{x}) \rangle = \sum_{k=1}^{i} a_{i,k}^2(s) + \beta_s^2.$$

In arriving at these results the upper limit N in (A.14-73) has been set equal to i, and $\tilde{r} = \ell \hat{e}_r$.

It is now possible to solve for the matrix elements in terms of two specified statistical quantities. de Almeida takes one of these as

$$T_{\ell,s} = c_s \varepsilon^{2/3} (\ell r)^{2/3}, \tag{A.14-75}$$

which is seen to represent the velocity correlation between points separated by a distance ℓr, for ℓr on the inertial subrange. Here r represents the sampling distance between consecutive simulation points, and c_s is a positive number of order unity. The other quantity chosen by de Almeida is the variance, $S_{i,s}$, or second central moment:

$$S_{i,s} = \langle [u_s(\tilde{x} + \tilde{r}) - \langle u_s(\tilde{x} + \tilde{r}) \rangle]^2 \rangle = S_s. \tag{A.14-76}$$

The subscript i may be dropped because of the assumption of homogeneity. Letting $j = i + \ell$, the solution for the matrix elements is then expressed as

$$a_{j,i}(s) = \begin{cases} 0, & i > j \\ a_{i,j}^{-1}(s) \left[\dfrac{S_s - T_{\ell s}}{2} - \displaystyle\sum_{k=1}^{i-1} a_{j,k}(s) a_{i,k}(s) \right], & i \leq j \,. \end{cases} \tag{A.14-77}$$

The quantity S_s is specified by the further assumption that the components of the turbulent velocity field have a normal distribution at any point. Experiments show this to be an excellent approximation (Batchelor, 1953). Another consequence of this assumption is that the set of random variables ζ_k must have a normal distribution. This requirement is satisfied approximately in the computer simulation in the following manner: (i) a sequence of pseudorandom numbers uniformly distributed on the interval $[0, 1]$ is generated by means of a composite congruential-type algorithm (see Marsaglia et al., 1964), (ii) a sequence of pseudorandom numbers Z, normally distributed with zero mean and unit variance, is generated by the relation

$$Z = (-2 \log u_1)^{1/2} \sin (2 \pi u_2), \tag{A.14-78}$$

where u_1 and u_2 are independent pseudorandom numbers obtained from step (i). It can be shown (Box and Muller, 1958) that resulting distribution function $F(Z)$ is approximately normal as desired; i.e.,

$$F(Z) \approx \frac{1}{(2\pi)^{1/2}} \int_{-\infty}^{Z} \exp(-Z^2/2) \, dZ. \tag{A.14-79}$$

The Monte Carlo simulation of the turbulent velocity field can now be described according to the following steps: (i) the quantities (A.14-75) and (A.14-76) are specified, (ii) the matrix elements are determined by (A.14-77), (iii) the turbulent velocity field is simulated by using (A.14-72) and (A.14-73):

$$u_{i,s} = \sum_{k=1}^{i} a_{i,k}(s)\zeta_k. \tag{A.14-80}$$

In order to carry out step (iii), the random variables ζ_k are replaced by the numbers Z generated according to (A.14-78).

3. Numerical Differentiation Operator

A suitable representation for $d\bar{u}/dt$ has to be found in order to integrate the equations of motion (A.14-65). The numerical time differentiation scheme chosen by de Almeida on the basis of accuracy and speed is as follows:

$$\frac{df}{dt}\bigg|_{t=t_0+2\Delta t} = \Delta t^{-1}\{\tfrac{2}{3}[f(t_0+3\,\Delta t) - f(t_0+\Delta t)] + \tfrac{1}{12}[f(t_0) - f(t_0+4\,\Delta t)]\}. \tag{A.14-81}$$

APPENDICES TO CHAPTER 15

A-15.2.1.2 Correlations in a Stochastic Coalescence Process (Bayewitz et al., 1974)

1. Formulation of the Stochastic Model

Suppose that at time $t+dt$ the aerosol contains N_k droplets of size k, for $k = 1, 2, \ldots, L$. Suppose also that the aerosol is well-mixed, and that the probability per unit time of coalescences between any pair of droplets is a constant, $A_{ik} = A$. Let N denote the total number of droplets present: $N = \Sigma_k N_k$. For notational purposes it is convenient to describe the state as an ordered list, N, N_1, $N_2, \ldots, N_L$. This facilitates the description of neighboring states; e.g., N, N_1, $N_2 + 1, \ldots, N_L$ is meant to denote another state, identical to the first except for an additional double droplet.

Now we pose the following question: From what states could the distribution at $t+dt$ have evolved in time dt, assuming that at most one coalescence could occur in the entire system during the interval? The answer is that there are three possible routes: (1) a coalescence occurs between two droplets of the same size; (2) a coalescence occurs between two droplets of different size; (3) no coalescence occurs.

We now consider the first route. If the system is to reach the desired distribution at $t+dt$ through a coalescence of two droplets of size m, it must be, at time t, in the state $N + 1$, N_1, $N_2, \ldots, N_m + 2, \ldots, N_{2m} - 1, \ldots, N_L$. Let the probability of this state be denoted by $v(N + 1$, N_1, $N_2, \ldots, N_m + 2, \ldots, N_{2m} - 1, \ldots, N_L$; t). Since the probability that any two droplets will coalesce in dt is A dt, the probability that any two droplets of size m will coalesce in the interval $(t, t+dt)$ is just $\binom{N_m+2}{2}$A dt. Accordingly, the product of these probabilities, summed over m to account for all the possible mutually exclusive outcomes, gives the total transition probability in dt for route 1):

$$\sum_m v(N + 1, N_1, N_2, \ldots, N_m + 2, \ldots, N_{2m} - 1, \ldots, N_L; t)\binom{N_m + 2}{2}A\,dt. \tag{A.15-1}$$

Similarly, if the system is to reach the desired distribution through a coalescence of droplets of size m and n, it must be, at time t, in the state $N + 1$, N_1, $N_2, \ldots, N_m + 1, \ldots, N_n + 1, \ldots, N_{m+n} - 1, \ldots, N_L$. The probability of coalescence of two particles of sizes m and n in the interval $(t, t + dt)$ is $(N_m + 1)(N_n + 1)$A dt. Hence the desired transition probability for route 2) is

$$\sum_{m<n} v(N + 1, N_1, N_2, \ldots, N_m + 1, \ldots, N_n + 1, \ldots, N_{m+n} - 1, \ldots, N_L; t)$$
$$\times (N_m + 1)(N_n + 1)A\,dt. \tag{A.15-2}$$

The condition $m < n$ is included to avoid counting the same coalescence event twice.

Finally, the transition probability for having the desired distribution at time t and no coalescence in the interval $(t, t + dt)$ is

$$v(N, N, N_2, \ldots ; t) \left[1 - \binom{N}{2} A \, dt \right].$$ (A.15-3)

The probability $v(N, N_1, N_2, \ldots, t + dt)$ of having the desired distribution at time $t + dt$ is just the sum of the three mutually exclusive probabilities assembled above. From this mathematical statement one immediately obtains the fundamental equation for the stochastic description:

$$\frac{dv}{dt} (N, N_1, N_2, \ldots ; t) = A \sum_{m < n} \sum v(N + 1, N_1, N_2, \ldots, N_m + 1, \ldots, N_n + 1, \ldots,$$

$$N_{m+n} - 1, \ldots ; t)(N_m + 1)(N_n + 1) + A \sum_m v(N + 1, N_1, N_2, \ldots,$$

$$N_{m+2}, \ldots, N_{2m} - 1, \ldots, t) \binom{N_m + 2}{2} - Av(N, N_1, N_2, \ldots ; t) \binom{N}{2}.$$ (A.15-4)

2. Total Number of Particles

The probability $v_N(t)$ of having N droplets of arbitrary size at time t is evidently just

$$v_N(t) = \sum_{\substack{\text{over all } N_m}} \cdots \sum v(N, N_1, N_2, \ldots ; t).$$ (A.15-5)

Therefore, summing over (A.15-4) leads to the equation governing v_N:

$$\frac{dv_N}{dt} = A \binom{N + 1}{2} v_{N+1} - A \binom{N}{2} v_N.$$ (A.15-6)

[This equation may also be arrived at more directly. Thus, the probability of having $N + 1$ droplets of arbitrary size is v_{N+1}, and the number of ways of producing a set of N droplets of arbitrary size from the set of $N + 1$ arbitrary droplets via binary coalescences is $\binom{N+1}{2}$. Therefore, the 'rate of production' of the probability v_N associated with the set of N droplets of arbitrary size is just the first term on the right side of (A.15-6). Similar arguments account also for the second (loss rate) term on the right side.]

The coupled, linear equation set contained in (A.15-6) may be solved sequentially. For the initial conditions $v_N(0) = 0$ for $N \neq N_0$ and $v_N(0) = 1$ for $N = N_0$, the solution to (A.15-6) is

$$v_N(t) = \sum_{k=N}^{N_0} \frac{(-1)^{k+N}(2k - 1)(k + N + 2)!}{N!(N - 1)!(k - N)!} \prod_{\nu=0}^{k-1} \left(\frac{N_0 - \nu}{N_0 + \nu} \right) e^{-A\binom{k}{2}t}.$$ (A.15-7)

Therefore, the mean number of droplets at time t is

$$\langle N \rangle = \sum_N N v_N = \sum_{k=N}^{N_0} (2k - 1) \prod_{\nu=0}^{k-1} \left(\frac{N_0 - \nu}{N_0 + \nu} \right) e^{-A\binom{k}{2}t}.$$ (A.15-8)

It is of interest to compare (A.15-8) with the corresponding result from the SCE. As we have seen, the SCE arises from the quasi-stochastic model in which N is uniquely determined. For the present situation in which every pair of the N droplets has the same coalescence probability, the equation governing N is evidently given by $dN/dt = -A\binom{N}{2} = -AN(N - 1)/2$. The solution for $N(0) = N_0$ is

$$N(t) = \frac{N_0}{N_0 - (N_0 - 1)e^{-At/2}}.$$ (A.15-9)

[This is the result obtained by Bayewitz et al. (though in a slightly different manner), and they refer to it as following from the SCE. However, this is not quite so: on summing the discrete SCE over the index k, we get $dN/dt = -AN^2/2$; thus the solution for $N(0) = N_0$ is $N(t) = N_0(1 + N_0At/2)^{-1}$. However, the two expressions for N differ very little except in the limit of large t (i.e., $t \gg A^{-1}$); then (A.15-9) predicts $N \to 1$, while the SCE yields $N \to 0$. In general, the SCE approximates the coalescence rate of droplets of size k, $A_{kk}N_k(N_k - 1)/2$, by the expression $A_{kk}N_k^2/2$. Fortunately, the resulting error is unimportant.]

Evaluation of (A.15-8) and (A.15-9) shows that N(t) according to the SCE is an excellent approximation to the true mean $\langle N \rangle$, even for droplet counts as small as $N_0 = 10$. Therefore, it appears the SCE theory produces total particle counts in excellent agreement with the true stochastic averages, even for very small initial populations, at least for collection kernels which are not strongly size dependent.

3. Size Spectrum

For an initial population of N_0 droplets of unit size, there is a surprisingly simple solution for (A.15-4) in terms of $v_N(t)$:

$$v(N, N_1, N_2, \ldots ; t) = \phi(N_1, N_2, \ldots, N_{N_0}/N) v_N(t), \qquad (A.15\text{-}10)$$

where ϕ is the probability of having a specified distribution given that there are N droplets. The expression for ϕ is

$$\phi(N_1, N_2, \ldots, N_{N_0}/N) = \binom{N}{N_1, N_2, \ldots, N_0} \Big/ \binom{N_0 - 1}{N - 1}. \qquad (A.15\text{-}11)$$

The numerator is the multinomial coefficient which counts the number of ways a specified $(N_1, N_2,$ etc.) distribution of N_0 units of mass in N droplets can be obtained, and the denominator counts the total number of ways of distributing N_0 units of mass in N droplets. (The form of the denominator may be explained in the following manner: Imagine the N_0 units of mass in a row, so that there are $N_0 - 1$ spaces between them. A particular distribution of N droplets may be achieved by partitioning the row in $N - 1$ places. At least one mass unit is to go in each segment of the row $(N_0 \geq N)$. Therefore, the number of possible distributions is the number of ways of selecting any $N - 1$ of the $N_0 - 1$ spaces as partition locations.)

Bayewitz *et al.* demonstrate that (A.15-11) is the solution to (A.15-4) by direct substitution into the differential equation, followed by numerical evaluation in simple cases which verifies uniqueness of the solution.

It is now possible to determine the function P(n, m; t) defined by (15-16):

$$P(n, m, t) = \sum_N \sum_{N_1} \sum_{N_2} \cdots \sum v(N, N_1, N_2, \ldots, N_m = n, \ldots ; t). \qquad (A.15\text{-}12)$$
$$\text{except } N_m$$

Numerical evaluation in a few cases (m = 2, 5; $N_0 = 20$) indicates that as time increases P(n, m; t) approaches a Poisson distribution. As we discussed in Section 15.2.1.1, Gillespie (1972) showed this to be true also for a variable collection coefficient in the absence of correlations.

4. Correlations in Poorly-Mixed Systems

As we have noted, real clouds are not well-mixed, and this can be expected to give rise to correlation problems. Bayewitz *et al.* have considered the consequences of poor mixing by partitioning the cloud into small, isolated compartments of volume V_0. This approach has the merit of dictating only slight changes in the mathematics.

Let us now compare the descriptions of droplet growth in each compartment as provided by the SCE and the full stochastic equations. From (A.15-9) with A replaced by KV_0^{-1}, where K is the constant collection kernel, one may obtain the mean stochastic droplet concentration in any compartment, viz. $\langle f \rangle \equiv \langle N \rangle V_0^{-1}$. Similarly, from (A.15-9) we have the corresponding droplet concentration according to the SCE: $f \equiv N V_0^{-1}$. Computations show that for N_0 as small as 10, f exceeds $\langle f \rangle$ only slightly. Thus the results indicate the total droplet concentration is rather insensitive to poor mixing as simulated by the partitioning model.

Now let us consider the size spectrum in the partitioned system. Let $\langle \gamma_m(t) \rangle$ be the mean fraction of the total mass of any compartment consisting of particles of mass m. With N_0 being the total mass in each compartment (recall $N(0) = N_0$ droplets of unit mass), we have

$$\langle \gamma_m(t) \rangle = \frac{m \langle N_m(t) \rangle}{N_0}, \qquad (A.15\text{-}13)$$

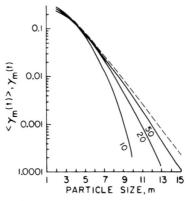

Fig. A.15-1.   Mass fraction distribution at $Kt = 0.002$. —— $\langle \gamma_m(t) \rangle$ for $N_0 = 10, 20, 50$; $-\cdot-\cdot-$ $\gamma_m(t)$. (From Bayewitz *et al.*, 1974; by courtesy of Amer. Meteor. Soc., and the authors.)

where

$$\langle N_m(t) \rangle = \sum_n nP(n, m; t). \tag{A.15-14}$$

The corresponding mass fraction from the SCE is $\gamma_m(t) = mN_m(t)/N_0$. In the present case of an initially homogeneous aerosol and a constant collision kernel, $N_m(t)$ is just the Smoluchowski solution (12-43); therefore,

$$\gamma_m(t) = m \left(\frac{f_0 Kt}{2} \right)^{m-1} \bigg/ \left(\frac{1 + f_0 Kt}{2} \right)^{m+1}. \tag{A.15-15}$$

In this expression f_0 is the concentration of droplets of unit mass at $t = 0$.

Figure A.15-1 compares $\langle \gamma_m(t) \rangle$ and $\gamma_m(t)$ at $Kt = 0.02$ and with $f_0 = 100$. For small N_0 (small V_0), there are seen to be substantial differences between the spectra predicted by the full stochastic model and the SCE, especially in the tails of the spectra. These differences decrease sharply with increasing V_0, and more moderately with time.

A-15.2.2 PARTICULAR SOLUTIONS TO THE SCE

Our purpose here is to evaluate the general solutions (15-40) and (15-47) for two choices of the initial spectrum. First, let us assume $g(x, 0) = \delta(x - 1)$. Then $G(s) = e^{-s}$, so that

$$L^{-1}[G^{k+1}(s)] = \frac{1}{2\pi i} \int_{\gamma - i\infty}^{\gamma + i\infty} e^{sx} e^{-s(k+1)} \, ds = \frac{1}{2\pi} \int_{-\infty}^{+\infty} e^{iy(x-k-1)} \, dy = \delta(x - k - 1). \tag{A.15-16}$$

Substitution of this result into (15-40) and (15-47) leads to the solutions (15-48) and (15-49), respectively.

For our second choice of the initial spectrum we choose the family of gamma distributions given by (12-129). The Laplace transform of the initial spectrum is therefore

$$G(s) = \frac{(p + 1)^{p+1}}{\Gamma(p + 1)} \int_0^\infty x^p e^{-(s+p+1)x} \, dx = \frac{(p + 1)^{p+1}}{(s + p + 1)^{p+1}}, \tag{A.15-17}$$

using the defining equation

$$\Gamma(p) \equiv \int_0^\infty e^{-x} x^{p-1} \, dx, \quad p > 0 \tag{A.15-18}$$

for $\Gamma(P)$. Therefore,

$$L^{-1}[G^{k+1}(s)] = \frac{(p+1)^{(p+1)(k+1)}}{2\pi i} \int_{\gamma-i\infty}^{\gamma+i\infty} \frac{e^{sx}\,ds}{(s+p+1)^{(p+1)(k+1)}}. \tag{A.15-19}$$

This integral may be evaluated by means of the well known residue theorem (e.g., Morse and Feshbach, vol. I, 1953), according to which the integral of a function of the complex variable z around a closed contour C is equal to $2\pi i$ times the sum of the residues of the function at its singular points within C. It is assumed that the function is analytic except at the singular points. The residue at an isolated singular point z_0 is just the coefficient of $(z - z_0)^{-1}$ in an expansion of the function in powers of $(z - z_0)$. From (A.15-19) we thus see that what is called for is an expansion of e^{sx} about $s_0 = -(p+1)$, so that the integrand may be expressed as

$$e^{-(p+1)x} \sum_{\ell=0}^{\infty} \frac{x^{\ell}}{\ell!}(s+p+1)^{\ell-(p+1)(k+1)}.$$

Assuming integral p (the result holds for all p > 0), the residue occurs for $\ell = (p+1)(k+1)-1$, and the contour may be closed in the left half of the s plane, for negative real s. Therefore, we obtain

$$L^{-1}[G^{k+1}(s)] = \frac{(p+1)^{(p+1)(k+1)}e^{-(p+1)x}x^{p+k(p+1)}}{\Gamma[(p+1)(k+1)]}. \tag{A.15-20}$$

Substitution of this result into (15-40) and (15-47) leads to the solutions (15-53) and (15-54), respectively.

A-15.2.4.3   *A Monte Carlo Algorithm for Stochastic Coalescence*

Our purpose here is to present Gillespie's (1975b) exact simulation algorithm. This requires focusing not on the average drop volume spectrum function which appears in the SCE, but rather on a quantity $P(\tau, i, j)$ which Gillespie calls the 'coalescence probability density function.' An explicit expression for $P(\tau, i, j)$ in terms of the basic probability of coalescence of a pair of drops with labels i and j in time dt can be derived, and the derivation is rigorous in that it is free of any no-correlation assumptions. A Monte Carlo simulation procedure is then based on $P(\tau, i, j)$ to calculate the stochastic evolution of a set of drops.

1.   *Coalescence Probability Density Function*

Consider a well-mixed cloud containing N drops at time t. Label these drops in any convenient way by the index i (i = 1, 2, ..., N), and let V_i denote the volume of drop i. We define the 'coalescence probability density function' $P(\tau, i, j)$ by the following statement:

$P(\tau, i, j)\,d\tau \equiv$ probability at time t that the next coalescence
  will occur in the time interval $(t + \tau, t + \tau + d\tau)$,
  and will be the coalescence of droplets i and j (i < j).   (A.15-21)

Now let us define also the set of numbers $A_{k\ell}$ as follows:

$A_{k\ell}\,d\tau \equiv$ probability that droplets k and ℓ will coalesce
  in the next infinitesimal time interval $d\tau$.   (A.15-22)

(As in Section 15.2.1, $A_{k\ell} = K_{k\ell}V^{-1}$, where $K_{k\ell}$ is the collection kernel and V is the volume of the well-mixed cloud.) An expression for (A.15-21) in terms of the quantities (A.15-22) can be derived as follows: The probability in (A.15-21) is the product of: (1) the probability that none of the droplets coalesce in the time interval $(t, t + \tau)$; times (2) the probability that droplets i and j coalesce in the next differential time interval $(t + \tau, t + \tau + d\tau)$; times (3) the probability that no other droplets coalesce in that same differential time interval.

To calculate the probability (1), imagine the time interval $(t, t + \tau)$ to be divided into m subintervals of equal length $\delta = \tau/m$. For small δ, the probability that drops k and ℓ (k < ℓ) will not coalesce in the first δ-interval is $(1 - A_{k\ell}\delta)$; hence the probability that all drop pairs will not coalesce in any of the m δ-intervals is

$$\left[\prod_{k=1}^{N-1}\prod_{\ell=k+1}^{N}(1-A_{k\ell}\delta)\right]^{m}.$$

We now obtain (1) by taking the limit of this expression as our subdivision of $(t, t + \tau)$ becomes infinitesimally fine:

$$
(1) \qquad = \lim_{m \to \infty} \left[\prod_{k=1}^{N-1} \prod_{\ell=k+1}^{N} (1 - A_{k\ell}\delta) \right]^m = \prod_{k=1}^{N-1} \prod_{\ell=k+1}^{N} \lim_{m \to \infty} \left[1 - \frac{A_{k\ell}\tau}{m} \right]^m
$$

$$
= \prod_{k=1}^{N-1} \prod_{\ell=k+1}^{N} \exp(-A_{k\ell}\tau) = \exp\left[-\sum_{k=1}^{N-1} \sum_{\ell=k+1}^{N} A_{k\ell}\tau \right]. \qquad (A.15\text{-}23)
$$

From (A.15-22) the probabilities (2) and (3) immediately follow:

$$
(2) \qquad = A_{ij}\, d\tau, \qquad\qquad\qquad\qquad\qquad\qquad\qquad\qquad\qquad (A.15\text{-}24)
$$

and

$$
(3) \qquad = \prod_{\substack{k=1 \\ k\ell \neq ij}}^{N-1} \prod_{\ell=k+1}^{N} (1 - A_{k\ell}\, d\tau). \qquad\qquad\qquad\qquad (A.15\text{-}25)
$$

Therefore, on setting (A.15-21) equal to the product of (A.15-23)–(A.15-25) and dividing through by $d\tau$, we obtain

$$
P(\tau, i, j) = A_{ij} \exp\left[-\sum_{k=1}^{N-1} \sum_{\ell=k+1}^{N} A_{k\ell}\tau \right]. \qquad (A.15\text{-}26)
$$

This result holds for $0 \leqslant \tau < \infty$ and $1 \leqslant i < j \leqslant N$; otherwise, $P(\tau, i, j) = 0$.

2. Basic Simulation Algorithm

The simulation algorithm based on $P(\tau, i, j)$ is as follows:

Step 0:   Set $t = 0$. Specify initial values for the N drop volumes $V_1, V_2, \ldots, V_N$, and calculate the corresponding $N(N-1)/2$ matrix elements $A_{k\ell}$. Specify a series of sampling times $t_1 < t_2 < \cdots$, and also a stopping time t_{stop}.

Step 1:   By employing suitable Monte Carlo techniques (see next section), generate a random triplet (τ, i, j) distributed according to the joint probability density function $P(\tau, i, j)$.

Step 2:   Using the values τ, i, and j obtained in Step 1, advance the time variable t by τ, remove drops i and j, and insert a new drop of volume $(V_i + V_j)$. Adjust the drop numbering scheme in any convenient way to reflect the fact that the cloud now contains one less drop than before, and make whatever rearrangements and recalculations of the matrix elements $A_{k\ell}$ are required.

Step 3:   If t has just been advanced through one of the sample times t_i, display the drop volume spectrum at time t_i as a frequency histogram of the current V_i values. If $t \geqslant t_{\text{stop}}$ (or if only one drop remains), terminate the calculation: otherwise, return to Step 1.

By carrying out this algorithm from time 0 to time t, one realization of the stochastic collection process is obtained. Several such realizations, starting from the same initial data, must be carried out to get a statistically complete picture. Let

$$
N_k(v, \delta v; t) = \text{number of drops found in run k at time t with}
$$
$$
\text{volumes between v and } v + \delta v. \qquad (A.15\text{-}27)
$$

Then for k runs the average number of drops at time t with volumes between v and $v + \delta v$ is

$$
\bar{N}(v, \delta v; t) \approx \frac{1}{K} \sum_{k=1}^{K} N_k. \qquad (A.15\text{-}28)
$$

(The expression becomes exact in the limit $k \to \infty$.) Similarly, the rms fluctuation about this average is

$$
\Delta(v, \delta v; t) \approx \left[\frac{1}{K} \sum_{k=1}^{K} N_k^2 - \left(\frac{1}{K} \sum_{k=1}^{N} N_k \right)^2 \right]^{1/2}. \qquad (A.15\text{-}29)
$$

If $\Delta/\bar{N} \ll 1$, then the results found in separate runs will be nearly identical. If $\Delta/\bar{N} \geqslant 1$, then an accurate estimate of $\bar{N}$ is not really necessary; of more interest is the scatter of N_k for several trials. Probably $k \leqslant 10$ should provide an adequate picture of the state of the spectrum in general. Finally, the connection between $\bar{N}(v, \delta v; t)$ and the usual average spectrum function $n(v, t)$ belonging to the

SCE is as follows: Let

$$N(v, t) \equiv Vn(v, t). \tag{A.15-30}$$

Then

$$N(v, t) = \lim_{\substack{\delta v \to 0 \\ k \to \infty}} \frac{\bar{N}(v, \delta v, t)}{\delta v}. \tag{A.15-31}$$

3. Implementing the Monte Carlo Step

Gillespie provides three Monte Carlo methods for implementing Step 1 of the simulation algorithm. Here we shall summarize only one of these, the 'first-coalescence' method, because of its intuitive appeal and relative simplicity. However, it is probably not as efficient as the other two methods.

Consider any two cloud drops k and ℓ (k < ℓ) at time t. From (A.15-22) it is easy to show that

$$\mathscr{P}_{k\ell}(\tau) \, d\tau = e^{-A_{k\ell}\tau} A_{k\ell} \, d\tau \tag{A.15-32}$$

would be the probability for k and ℓ to coalescence in the time interval $(t + \tau, t + \tau + d\tau)$, were it not for the fact that k or ℓ might coalesce with some other drop prior to time $t + \tau$. This being the case, Gillespie generates a *tentative* coalescence time $\tau_{k\ell}$ for the drop pair (k, ℓ) according to (A.15-32) (in a manner described below). This is done for all (k, ℓ) pairs, and of these $N(N - 1)/2$ tentative next coalescences, the one which occurs first is chosen as the actual next coalescence. Thus, we put

$$\tau = \text{smallest } \tau_{k\ell} \text{ for all } (k, \ell) \text{ pairs,}$$
$$(i, j) = (k, \ell) \text{ for which } \tau_{k\ell} \text{ is smallest.} \tag{A.15-33}$$

This procedure is a physically plausible way to pick values for τ, i, and j. Gillespie proves that it is, in fact, the correct way; i.e., if we let $\mathscr{P}'(\tau, i, j) \, d\tau$ be the probability that the procedure just described will result in the next coalescence being between drops i and j and occurring in the time interval $(t + \tau, t + \tau + d\tau)$, then it can be shown that $\mathscr{P}'(\tau, i, j) = \mathscr{P}(\tau, i, j)$.

We now consider the problem of generating the numbers $\tau_{k\ell}$. The goal is to generate a sequence of random numbers $\{\tau\}$ distributed according to some given probability density function $\mathscr{P}(\tau)$; in the present case $\mathscr{P}(\tau)$ has the exponential form (A.15-32). One way to accomplish this by means of the so-called 'method of inversion' (e.g., Gillespie, 1975c), which can be described in the following manner: For $\mathscr{P}(\tau)$ defined on the interval [a, b], consider the corresponding probability distribution function

$$F(\tau) \equiv \int_a^\tau \mathscr{P}(\tau') \, d\tau', \quad a \leq \tau \leq b. \tag{A.15-34}$$

Then to generate τ distributed according to $\mathscr{P}(\tau)$, we simply draw a random number r from a uniform random number generator (a source of pseudorandom numbers distributed uniformly on [0, 1]), and then choose for τ that value which satisfies $F(\tau) = r$; i.e., we take

$$\tau = F^{-1}(r). \tag{A.15-35}$$

As applied to (A.15-32), this procedure yields the sequence of numbers

$$\tau_{k\ell} = A_{k\ell}^{-1} \ln (1/r_{k\ell}), \tag{A.15-36}$$

where $r_{k\ell}$ is on the interval [0, 1] and k = 1, ..., N – 1; ℓ = k + 1, ..., N.

A-15.2.5 PARAMETERIZATION OF ACCRETION AND HYDROMETEOR SELF-COLLECTION

Here we follow Berry and Reinhardt (1974c). Let us consider accretion first. In this growth mode, large drops of volume v_2 continuously collect small droplets of volume $v_1 \ll v_2$. Accretion does not destroy the identity of the large drops, so that their total concentration N_2 remains constant; i.e.,

$$\left. \frac{dN_2}{dt} \right|_c = \int_{S2} \left. \frac{\partial n(v_2, t)}{\partial t} \right|_c dv_2 = 0, \tag{A.15-37}$$

where the subscript c denotes accretion. Now consider a plot of $n(v_2, t)$ versus v_2. The 'velocity' with which a v_2 drop moves along the volume axis is dv_2/dt. Therefore, the current density of such drops is $j_2 = n(v_2, t) \, dv_2/dt$. Since there are no sources or sinks in the accretion mode, the equation of continuity becomes $\partial n_2/\partial t = -\partial j_2/\partial v_2$; i.e.,

$$\left. \frac{\partial n(v_2, t)}{\partial t} \right|_c = -\frac{\partial}{\partial v_2}\left(n(v_2, t)\frac{dv_2}{dt} \right). \tag{A.15-38}$$

It is of interest to determine the change in shape of S2 due to accretion. One measure for shape is $\mathrm{var}\, v = v_g/v_f - 1$, and so Berry and Reinhardt proceed to calculate the rates of change of v_{f2} and v_{g2}. From (15-106) and (A.15-37) we see that

$$\frac{1}{v_{f2}}\frac{dv_{f2}}{dt} = \frac{1}{w_{L2}}\frac{dw_{L2}}{dt} \tag{A.15-39}$$

and

$$\frac{1}{v_{g2}}\frac{dv_{g2}}{dt} = \frac{1}{(\bar{M}_2)_2}\frac{d(\bar{M}_2)_2}{dt} - \frac{1}{w_{L2}}\frac{dw_{L2}}{dt}. \tag{A.15-40}$$

If we assume that $n(v_2, t)$ vanishes on the boundaries of S2, then from (A.15-38) we find

$$\frac{dw_{L2}}{dt} = \int_{S2} v_2 \frac{\partial n(v_2, t)}{\partial t}\, dv_2 = \int_{S2} n(v_2, t)\frac{dv_2}{dt}\, dv_2. \tag{A.15-41}$$

If we further assume S1 is sufficiently narrow, then

$$\frac{dv_2}{dt} = \int_{S1} K(v_2, v_1)v_1 n(v_1, t)\, dv_1 \approx K(v_2, v_1)w_{L1}. \tag{A.15-42}$$

Therefore, from (A.15-39), (A.15-41), and (A.15-42) it follows that

$$\left. \frac{1}{v_{f2}}\frac{dv_{f2}}{dt} \right|_c \approx \frac{w_{L1}}{w_{L2}} \int_{S2} K(v_2, v_1)n(v_2, t)\, dv_2. \tag{A.15-43}$$

Similarly, the fractional rate of change of v_{g2} is easily found to be given approximately by the expression

$$\left. \frac{1}{v_{g2}}\frac{dv_{g2}}{dt} \right|_c \approx \frac{w_{L1}}{w_{L2}} \int_{S2} K(v_2, v_1)\left(\frac{2v_2}{v_{g2}} - 1 \right) dv_2. \tag{A.15-44}$$

Now let us turn to hydrometeor self-collection. Directly from the moment equation (15-57) and from (15-106) we can write

$$\left. \frac{1}{v_{f2}}\frac{dv_{f2}}{dt} \right|_s = \frac{1}{2\,N_2} \int_{S2}\int_{S2} K(v, v')n(v, t)n(v', t)\, dv\, dv', \tag{A.15-45}$$

and

$$\left. \frac{1}{v_{g2}}\frac{dv_{g2}}{dt} \right|_s = \frac{1}{(\bar{M}_2)_2} \int_{S2}\int_{S2} vv' K(v, v')n(v, t)n(v', t)\, dv\, dv', \tag{A.15-46}$$

where the subscript S denotes the self-collection mode. Recalling again that the kernel $K(v, v')$ is given approximately by the Golovin sum-of-volumes form for $r \geq 50\ \mu\mathrm{m}$, we may obtain an estimate for the integral terms by setting $K = K_0(v + v')$ where K_0 is a constant. This yields

$$\left. \frac{1}{v_{f2}}\frac{dv_{f2}}{dt} \right|_s = \left. \frac{1}{2\,v_{g2}}\frac{dv_{g2}}{dt} \right|_s = K_0 w_{L2}, \quad K = K_0(v + v'). \tag{A.15-47}$$

These results lead Berry and Reinhardt to introduce a rate coefficient, b_c, defined as follows:

$$b_c(v_2, v_1) \equiv K(v_2, v_1)/(v_2 + v_1).$$ (A.15-48)

They further write the self-collection growth rates in the form

$$\frac{1}{v_{f2}}\frac{dv_{f2}}{dt}\bigg|_S \equiv b_{sf}w_{L2}, \quad \frac{1}{v_{g2}}\frac{dv_{g2}}{dt}\bigg|_S \equiv b_{sg}w_{L2},$$ (A.15-49)

which defines the coefficients b_{sf} and b_{sg} in terms of the integrals in (A.15-45) and (A.15-46). From (A.15-47) and (A.15-48), we might expect that b_{sg} is approximately twice the value of b_c evaluated for drop volumes comparable to v_{g2}. The expression found empirically through study of the various numerical solutions is as follows:

$$b_{sg} \approx 2\, b_c(3.4\, v_{g2}, v_{g2}).$$ (A.15-50)

This means that drop self-collection is reasonably well represented by the capture of the 'predominant' sized drop by one of 1.5 times its diameter. The similar approximation found for b_{sf} is

$$b_{sf} \approx b_c(v_{g2}, v_{f2}).$$ (A.15-51)

We now proceed to describe the accretion mode in analogous fashion; i.e., we write

$$\frac{1}{v_{f2}}\frac{dv_{f2}}{dt}\bigg|_c \equiv b_{cf}w_{L1}, \quad \frac{1}{v_{g2}}\frac{dv_{g2}}{dt}\bigg|_c \equiv b_{cg}w_{L1}.$$ (A.15-52)

Then from (A.15-43), (A.15-48), and (A.15-52) we obtain

$$b_{cf} \approx \frac{1}{w_{L2}} \int_{S2} v_2 b_c(v_2, v_{f1})\, dv_2 \approx b_c(v_{g2}, v_{f1}).$$ (A.15-53)

Similarly, from (A.15-44), (A.15-48), and (A.15-52) we obtain the same approximate result for b_{cg}:

$$b_{cg} \approx b_c(v_{g2}, v_{f1}).$$ (A.15-54)

APPENDIX TO CHAPTER 17

A-17.5.5 TWO CHARGED CONDUCTING SPHERES IN A BACKGROUND ELECTRIC FIELD

Here we summarize the work of Davis (1964a, b), who solved the boundary value problem of two electrically conducting spheres of given radii and charges, separated by a given distance and situated in an external electric field of given uniform strength far away from the spheres. Figure A.14-1 depicts the geometry of this problem, if we now assume that Θ describes the angle between the background field $\vec{E}_0$ and the line of centers of the spheres (i.e., if we replace $\vec{g}$ by $\vec{E}_0$ in the figure). To solve this problem the electrostatic potential function Φ must be determined which satisfies the boundary conditions (1) that at large distances from the spheres the potential corresponds to that of the uniform field, and (2) that the two spheres are equipotential surfaces carrying charges Q_1 and Q_2, respectively. The governing equation for Φ is Laplace's equation, $\nabla^2\Phi = 0$, which Davis solved in bispherical coordinates (a description of this system is given in Appendix A-14.4.2).

Given the solution for Φ, the force $\vec{F}_e$ (in e.s.u.) acting on either sphere may then be obtained by integrating the electrical stress over its surface:

$$\vec{F}_e \cdot \hat{\ell} = \frac{\varepsilon}{8\pi} \int_S \left(\frac{\partial\Phi}{\partial n}\right)^2 \hat{\ell} \cdot \hat{n}\, dS,$$ (A.17-1)

where $\hat{\ell}$ is an arbitrary unit vector, $\hat{n}$ the unit normal to the surface element dS, and ε is the dielectric constant of the medium in which the conducting spheres are imbedded. The forces on the two spheres must of course satisfy the relation

$$\vec{F}_e(a_1) + \vec{F}_e(a_2) = (Q_1 + Q_2)\vec{E}_0,$$ (A.17-2)

so that the integration in (A.17-1) need be carried out for only one of the spheres. Referring to Figure A.14-1, the components of the electrical force on sphere 2 along and at right angles to the line of centers were found by Davis to be given respectively by

$$F_{e,r}(a_2) = \varepsilon a_2^2 E_0^2 (F_1 \cos^2 \Theta + F_2 \sin^2 \Theta) + E_0 \cos \Theta (F_3 Q_1 + F_4 Q_2)$$

$$+ \frac{1}{\varepsilon a_2^2} (F_5 Q_1^2 + F_6 Q_1 Q_2 + F_7 Q_2^2) + E_0 Q_2 \cos \Theta, \qquad (A.17-3)$$

and

$$F_{e,\theta}(a_2) = \varepsilon a_2^2 E_0^2 F_8 \sin 2\Theta + E_0 \sin \Theta (F_9 Q_1 + F_{10} Q_2) + E_0 Q_2 \sin \Theta, \qquad (A.17-4)$$

where the coefficients $F_1, F_2, \ldots, F_{10}$ are complicated functions, tabulated in Davis (1964a, b), of the distance between the spheres and of $p = a_2/a_1$. The corresponding force components in the x and y directions of Figure A.14-1 are then given by

$$F_{e,x}(a_2) = F_{e,r}(a_2) \cos \Theta - F_{e,\theta}(a_2) \sin \Theta,$$
$$F_{e,y}(a_2) = F_{e,r}(a_2) \sin \Theta + F_{e,\theta}(a_2) \cos \Theta. \qquad (A.17-5)$$

Davis's solution may also be used to calculate the charge transferred between the two spheres if they make electrical contact. Most convenient for this purpose are the following expressions from Davis (his Eq. (24) from 1964b):

$$V_1 = P_{11}(Q_1 - Q_1^*) + P_{12}(Q_2 - Q_2^*),$$
$$V_2 = P_{12}(Q_1 - Q_1^*) + P_{22}(Q_2 - Q_2^*), \qquad (A.17-6)$$

where $V_1 = \Phi(a_1)$ and $V_2 = \Phi(a_2)$ are the potentials of the two spheres, the P_{ij} are the calculated coefficients of induction ($P_{12} = P_{21}$ by the Reciprocation Theorem; e.g., see §2.12 of Smythe (1950)), and Q_1^*, Q_2^* are the calculated charges which the spheres would have if they were at zero potential (i.e., grounded). Note that the only place $\vec{E}_0$ enters into (A.17-6) is in the functional dependence of the effective charges Q_1^* and Q_2^*.

Now suppose the spheres touch, so that $V_1 = V_2 = V$, and the original charges Q_1, Q_2 become Q_1', Q_2'. Then from (A.17-6) we have

$$Q_1' = \frac{-(P_{11} - P_{12})Q_1^* + (P_{12} - P_{22})(Q_T - Q_2^*)}{2 P_{12} - P_{11} - P_{22}}, \qquad (A.17-7)$$

where $Q_T = Q_1 + Q_2 = Q_1' + Q_2'$. A similar expression may be written down for Q_2'. The charge transferred is then given by $\Delta Q = Q_1' - Q_1 = Q_2 - Q_2'$.

Of course, another way to evaluate ΔQ is by setting the electric field equal to zero at the point of contact. From Eq. (31) of Davis, the field strength at the near surface point of sphere 2 is

$$E_{2,r} = \frac{(E_1 Q_1 + E_2 Q_2)}{\varepsilon a_2^2} + E_3 E_0 \cos \Theta, \qquad (A.17-8)$$

where E_1, E_2, and E_3 are functions of separation distance and a_2/a_1, tabulated by Davis. From (A.17-8) with $E_{2,r} = 0$ we therefore immediately find also that

$$\Delta Q = Q_1' - Q_1 = \frac{E_1 Q_1 + E_2 Q_2 + \varepsilon a_2^2 E_3 E_0 \cos \Theta}{(E_2 - E_1)}. \qquad (A.17-9)$$

REFERENCES

Aarons, A. B., and Keith, C. H., 1954: *J. Meteor.* **11**, 173.

Abbas, M. A., and Latham, J., 1967: *J. Fluid. Mech.* **30**, 663.

Abbas, M. A., and Latham, J., 1969a: *J. Meteor. Soc., Japan* **47**, 65.

Abbas, M. A., and Latham, J., 1969b: *Quart. J. Roy. Meteor. Soc.* **95**, 63.

Abbott, C. E., 1974: *J. Geophys. Res.* **79**, 3098.

Abbott, C. E., 1975: *J. Appl. Meteor.* **14**, 87.

Abel, N., Jaenicke, R., Junge, C., Kanter, H., Rodriguez, P., and Seiler, W., 1969: *Meteor. Rundschau* **22**, 158.

Abraham, F. F., 1968: *J. Atmos. Sci.* **25**, 47.

Abraham, F. F., 1974a: *J. Chem. Phys.* **61**, 1221.

Abraham, F. F., 1974b: *Homogeneous Nucleation Theory*, Academic Press, New York.

Aburukawa, H., and Magono, C., 1972: *J. Meteor. Soc., Japan* **50**, 166.

Achenbach, E., 1974: *J. Fluid Mech.* **62**, 209.

Ackerman, B., 1959: *J. Meteor.* **16**, 191.

Ackerman, B., 1963: *J. Atmos. Sci.* **20**, 288.

Ackerman, B., 1967: *J. Appl. Meteor.* **6**, 61.

Ackerman, B., 1968: *Proc. Cloud Phys. Conf.*, Toronto, Canada, Aug. 1968, p. 564, Univ. of Toronto Press, Toronto, Canada.

Acrivos, A., and Taylor, T. D., 1962: *Phys. Fluids* **5**, 387.

Adam, J. R., Lindblad, N. R., and Hendricks, C. D., 1968: *J. Appl. Phys.* **39**, 5173.

Adam, J. R., and Semonin, R. G., 1970: *Precipitation Scavenging*, 1970, AEC Symposium, Richland, Wash., June 1970, p. 151, U.S. Dept. of Commerce, Springfield, Va.

Adamson, A. W., 1960: *Physical Chemistry of Surface*, Interscience Publ., Inc., New York.

Adamson, A. W., Dormant, L. M., and Orem, M. W., 1967: *J. Coll. Interface Sci.* **25**, 206.

Admirat, P., 1962: *Bull. Obs. Puy de Dome*, No. 2, 87.

Admirat, P., and Grenier, J. C., 1975: *J. de Rech. Atmos.* **9**, 97.

Alkezweeney, A. J., and Lockhard, T. J., 1972: *Atmos. Environ.* **6**, 481.

Allard, E. F., and Kassner, J. L., 1965: *J. Chem. Phys.* **42**, 1401.

Allee, P. A., 1970: *Proc. Conf. on Weather Mod.*, Santa Barbara, Calif., April 1970, p. 244, Amer. Meteor. Soc., Boston, Mass.

Allee, P. A., and Phillips, B. B., 1959: *J. Meteor.* **16**, 405.

Allen, L. B., and Kassner, J. L., 1969: *J. Coll. and Interface Sci.* **30**, 81.

Alty, T., 1931: *Proc. Roy. Soc.* **A131**, 554.

Alty, T., 1933: *Phil. Mag.* **15**, 82.

Alty, T., and Mackay, C. A., 1935: *Proc. Roy. Soc., London* **A149**, 104.

Alty, T., and Nicole, F. H., 1931: *Can. J. Res.* **4**, 547.

Ames, W. F., 1965: *Nonlinear Partial Differential Equations in Engineering*, Academic Press, New York.

Anderson, J., and Hallett, J., 1976: *J. Atmos. Sci.* **33**, 822.

Annis, B. K., and Mason, E. A., 1975: *Aerosol Science* **6**, 105.

Antonoff, G., 1907: *J. Chem. Phys.* **5**, 372.

Aoi, T., 1955: *J. Phys. Soc., Japan* **10**, 119.

Arenberg, D. L., 1941: *Bull. Am. Meteor. Soc.* **22**, 113.

Aris, R., 1962: *Vectors, Tensors, and the Basic Equations of Fluid Mechanics*, Prentice-Hall, Englewood Cliffs, N.J.

Artemov, I. S., 1946: *J. Phys. Chem., U.S.S.R.* **20**, 553.

Atkinson, W. R., and Paluch, I., 1968 *J. Geophys. Res.* **73**, 2035.

Atlas, D., and Donaldson, R. J., 1955: *Air Force Survey in Geophys.* No. 65, p. 91, U.S. Air Force, Cambridge Res. Center, Cambridge, Mass.

Atlas, D., and Ludlam, F. H., 1961: *Quart. J. Roy. Meteor. Soc.* **87**, 523.

Auer, A. H., 1966: *J. de Rech. Atmos.* **3**, 289.

Auer, A. H., 1970: *J. de Rech. Atmos.* **4**, 145.

Auer, A. H., 1971: *J. Atmos. Sci.* **28**, 285.

Auer, A. H., 1972a: *J. Atmos. Sci.* **29**, 311.

Auer, A. H., 1972b: *Monthly Weath. Rev.* **100**, 325.

Auer, A. H., and Veal, D. L., 1970: *J. Atmos. Sci.* **27**, 919.

Auer, A. H., Veal, D. L., and Marwitz, J. D., 1969: *J. Atmos. Sci.* **26**, 1342.

Aufdermauer, A., 1963: *Schweiz. Z. f. Obst und Weinbau* **72**, 434.

Aufdermauer, and Johnson, D. A., 1972: *Quart. J. Roy. Meteor. Soc.* **98**, 369.

Aufdermauer, and Joss, J., 1967: *Z. Angew. Math. und Phys.* **18**, 852.

Aufdermauer, List, R., and Mayes, W. C., 1963: *Z. Angew. Math. und. Phys.* **14**, 574.

Aufdermauer, and Mayes, W. C., 1965: *Proc. Cloud Phys. Conf.*, Tokyo, Japan, May 1965, p. 281, Meteor. Soc., Japan, Tokyo.

Aufm Kampe, B. J., Weickmann, H. K., and Kelley, J. J., 1951: *J. Meteor.* **8**, 168.

Ausman, E. L., and Brook, M., 1967: *J. Geophys. Res.* **72**, 6131.

Austin, J. M., and Fleisher, A., 1948: *J. Meteor.* **5**, 240.

Bader, M., Goster, J., Brownscombe, J. L., and Goldsmith, P., 1974: *Quart. J. Roy. Meteor. Soc.* **100**, 420.

Bader, R. F., and Jones, G. A., 1963: *Canad. J. Chem.* **41**, 586.

Bagnold, R. A., 1965: *The Physics of Blown Sand Desert Dunes*, Methuen, London.

Bailey, I. H., and Macklin, W. C., 1968a: *Quart. J. Roy. Meteor. Soc.* **94**, 1.

Bailey, I. H., and Macklin, W. C., 1968b: *Quart. J. Roy. Meteor. Soc.*, **94**, 93.

Baird, M. H., Hamielec, A. E., 1962: *Canad. J. Chem. Eng.* **40**, 119.

Baramaev, M. K., 1939: *Zhurnal Fiz. Khim.* **13**, 1635.

Barchet, W. R., 1971: *Ph.D. Thesis*, Colorado State University, Fort Collins, Colo. (also: Atmos. Sci. Paper No. 168, Dept. of Atmos. Sci., Colorado State University).

Barchet, W. R., and Corrin, M. L., 1972: *J. Phys. Chem.* **76**, 2280.

Barker, J. A., and Watts, R. O., 1969: *Chem. Phys. Lett.* **3**, 144.

Barklie, R. H. D., and Gokhale, N. R., 1959: *Scientific Report*, MW-30, Stormy Weather Group, McGill University, Montreal.

Barnard, A. J., 1953: *Proc. Roy. Soc., London* **A220**, 132.

Barnard, A. J., 1954: *Ph.D. Thesis*, Glasgow University, Glasgow, England.

Barrie, L. A., 1975: *Berichte des Instit. f. Meteor. und Geophys.*, University of Frankfurt, Frankfurt.

Barrie, L. A., and Georgii, H. W., 1976: *Atmospheric Environment* **10**, 743.

Barthakur, N., and Maybank, J., 1963: *Nature* **200**, 866.

Barthakur, N., and Maybank, J., 1966: *J. de Rech. Atmos.* **2**, 475.

Bartlett, J. T., 1966: *Quart. J. Roy. Meteor. Soc.* **92**, 93.

Bartlett, J. T., 1968: *Proc. Cloud Phys. Conf.*, Toronto, Aug. 1968, p. 515, Amer. Meteor. Soc., Boston, Mass.

Bartlett, J. T., 1970: *Quart. J. Roy. Meteor. Soc.* **96**, 730.

Bartlett, J. T., and Jonas, P. R., 1972: *Quart. J. Roy. Meteor. Soc.* **98**, 150.

Bashkirova, G. M., and Pershina, T. A., 1956: *Trudy Glav. Geofiz. Observ.* No. 57, 19.

Bashkirova, G. M., and Pershina, T. A., 1964a: *Trudy Glav. Geofiz. Observ.* No. 165, 83.

Bashkirova, G. M., and Pershina, T. A., 1964b: *Trudy Glav. Geofiz. Observ.* No. 165, 90.

Basset, A. B., 1910: *Quart. J. Pure Appl. Math.* **41**, 369.

Batchelor, G. K., 1953: *The Theory of Homogeneous Turbulence*, Cambridge Univ. Press, London.

Batchelor, G. K., 1967: *Fluid Dynamics*, Cambridge University Press, London.

Battan, L. J., 1963: *J. Meteor.* **7**, 333.

Battan, L. J., 1973: *Radar Observations of Atmospheric Clouds*, University of Chicago Press, Chicago, Ill.

Battan, L. J., and Braham, R. R., 1956: *J. Meteor.* **13**, 587.

Battan, L. J., and Reitan, C. H., 1957: *Artificial Stimulation of Rain*, p. 184, Pergamon Press, New York.

Bayardelle, M., 1955: *C. R. Acad. Sci.*, **240**, 2553.

Bayewitz, M. H., Yerushalmi, J., Katz, S., and Shinnar, R., 1974: *J. Atmos. Sci.* **31**, 1604.

Beard, K. V., 1974a: *Precipitation Scavenging*, 1974, AEC Symposium, October 1974, Champagne, Ill., U.S. Dept. of Commerce, Springfield, Va.

Beard, K. V., 1974b: *J. Atmos. Sci.* **31**, 1595.

Beard, K. V., 1976: *J. Atmos. Sci.* **33**, 851.

Beard, K. V., and Grover, S. N., 1974: *J. Atmos. Sci.* **31**, 543.

Beard, K. V., and Pruppacher, H. R., 1968: *J. Geophysics* **73**, 6407.

Beard, K. V., and Pruppacher, H. R., 1969: *J. Atmos. Sci.* **26**, 1060.

Beard, K. V., and Pruppacher, H. R., 1971a: *J. Atmos. Sci.* **28**, 1455.

Beard, K. V., and Pruppacher, H. R., 1971b: *Quart. J. Roy. Meteor. Soc.* **97**, 242.

Becker, R., and Döring, W., 1935: *Ann. d. Phys.* **24**, 719.

Beckwith, W. B., 1960: *Geophys. Monogr.* No. 5, 348, Amer. Geophys. Union, Washington, D.C.

Beheng, K. D., 1976: *Ph.D. Thesis*, Dept. Meteor. and Geophys., University of Köln, Köln.

Beheng, K. D., 1978: *J. Atmos. Sci.*, April issue.

Belvayev, V. I., 1961: *Izv. Akad. Nauk.*, U.S.S.R., Geofis. Ser. **8**, 1209.

Ben-Naim, A., 1972: *Water*, F. Franks, Ed., p. 413, Plenum Press, New York.

Ben-Naim, A., 1974: *Water and Aqueous Solutions*, Plenum Press, New York.

Ben-Naim, A., and Stillinger, F. H., 1972: *Water and Aqueous Solutions*, R. A. Horne, Ed., p. 295, Wiley-Interscience, New York.

Benson, G. C., and Shuttleworth, R., 1951: *J. Chem. Phys.* **19**, 130.

Bentley, W. A., and Humphreys, W. J., 1931: *Snow Crystals*, McGraw-Hill, New York (Dover Publ. Inc., New York, 1962).

Berman, L. D., 1961: *Khim. Mashinost.* **36**, 66.

Bernal, J. D., and Fowler, R. H., 1933: *J. Chem. Phys.* **1**, 515.

Berry, E. X., 1965: *Ph.D. Thesis*, University of Nevada, Reno, Nev.

Berry, E. X., 1967: *J. Atmos. Sci.* **24**, 688.

Berry, E. X., 1968: *J. Atmos. Sci.* **25**, 151.

Berry, E. X., and Pranger, M. R., 1974: *J. Appl. Meteor.* **13**, 108.

Berry, E. X., and Reinhardt, R. L., 1973: *Techn. Rept.*, Ser. P, No. 16, Desert Research Institute, University of Nevada, Reno, Nev.

Berry, E. X., and Reinhardt, R. L., 1974a: *J. Atmos. Sci.* **31**, 1814.

Berry, E. X., and Reinhardt, R. L., 1974b: *J. Atmos. Sci.* **31**, 2118.

Berry, E. X., and Reinhardt, R. L., 1974c: *J. Atmos. Sci.* **31**, 1825.

Berry, E. X., and Reinhardt, R. L., 1974d: *J. Atmos. Sci.* **31**, 2127.

Best, A. C., 1947: *Meteor. Res. Comm.*, Air Ministry, Great Britain, M.R.P. 330.

Best, A. C., 1950a: *Quart. J. Roy. Meteor. Soc.* **76**, 16.

Best, A. C., 1950b: *Quart. J. Roy. Meteor. Soc.* **76**, 302.

Best, A. C., 1951a: *Quart. J. Roy. Meteor. Soc.* **77**, 418.

Best, A. C., 1951b: *Quart. J. Roy. Meteor. Soc.* **77**, 241.

Best, A. C., 1952: *Quart. J. Roy. Meteor. Soc.* **78**, 28.

Bigg, E. K., 1953a: *Quart. J. Roy. Meteor. Soc.* **79**, 510.

Bigg, E. K., 1953b: *Proc. Phys. Soc.*, B. **66**, 688.

Bigg, E. K., 1955: *Quart. J. Roy. Meteor. Soc.* **81**, 478.

Bigg, E. K., 1963: *Nature* **197**, 172.

Bigg, E. K., 1965: *Proc. Cloud Phys. Conf.*, Tokyo, May 1965, p. 137, Meteor. Soc. of Japan, Tokyo.

Bigg, E. K., 1967: *J. Atmos. Sci.* **24**, 226.

Bigg, E. K., 1975: *J. Atmos. Sci.* **32**, 910.

Bigg, E. K., and Giutronich, J., 1967: *J. Atmos. Sci.* **24**, 46.

Bigg, E. K., and Hopwood, S. C., 1963: *J. Atmos. Sci.* **20**, 185.

Bigg, E. K., Kriz, Z., and Thompson, W. J., 1971: *Tellus* **23**, 247.

Bigg, E. K., Kriz, Z., and Thompson, W. J., 1972: *J. Geophys. Res.* **77**, 3916.

Bigg, E. K., and Miles, G. T., 1963: *Tellus* **15**, 162.

Bigg, E. K., and Miles, G. T., 1964: *J. Atmos. Sci.* **21**, 396.

Bigg, E. K., Miles, G. T., and Hefferman, K. J., 1961: *J. Meteor.* **18**, 804.
Bigg, E. K., Ono, A., and Thompson, W. J., 1970: *Tellus* **22**, 550.
Bigg, E. K., and Stevenson, C., 1970: *J. de Rech. Atmos.* **4**, 41.
Bikerman, J. J., 1970: *Physical Surfaces*, Academic Press, New York.
Bilham, E. G., and Relf, E. F., 1937: *Quart. J. Roy. Meteor. Soc.* **63**, 149.
Bird, R. B., Stewart, W. E., and Lightfoot, E. L., 1960: *Transport Phenomena*, J. Wiley, New York.
Birstein, S. J., 1954: *Geophys. Res. Papers*, No. 32, Dec. 1954, Atmos. Phys. Lab., Geophys. Res. Dir., Air Force Cambridge Res. Center, Bedford, Mass.
Birstein, S. J., 1955: *J. Meteor.* **12**, 324.
Birstein, S. J., 1956: *J. Meteor.* **13**, 395.
Birstein, S. J., and Anderson, C. E., 1955: *J. Meteor.* **12**, 68.
Bjerrum, N., 1951: *K. Danske Vidensk., Selsk., Skr.* **27**, 1.
Blackman, M., and Lisgarten, N. D., 1957: *Proc. Roy. Soc., London* **A239**, 93.
Blair, T. A., 1928: *Monthly Weather Rev.* **56**, 313.
Blanchard, D. C., 1948: *Occas. Rept.*, No. 7, Project Cirrus, General Electric Res. Labs. Schenectady, N.Y.
Blanchard, D. C., 1949: *Occas. Rept.*, No. 17, Project Cirrus, General Electric Res. Labs. Schenectady, N.Y.
Blanchard, D. C., 1950: *Trans. Amer. Geophys. Union* **31**, 836.
Blanchard, D. C., 1953: *J. Meteor.* **10**, 457.
Blanchard, D. C., 1954: *Nature* **174**, 470.
Blanchard, D. C., 1955: *J. Meteor.* **12**, 91.
Blanchard, D. C., 1957: *Artificial Stimulation of Rain*, p. 233, Pergamon Press, New York.
Blanchard, D. C., 1962: *J. Atmos. Sci.* **19**, 119.
Blanchard, D. C., 1963: *Progr. in Oceanogr.* **1**, 71.
Blanchard, D. C., 1964: *Science* **146**, 396.
Blanchard, D. C., 1968: *Proc. Cloud Phys. Conf.*, Toronto, Aug. 1968, p. 25. University of Toronto Press, Toronto.
Blanchard, D. C., 1969: *J. de Rech. Atmos.* **4**, 1.
Blanchard, D. C., 1971: *J. Atmos. Sci.* **28**, 645.
Blanchard, D. C., and Spencer, A. T., 1970: *J. Atmos. Sci.* **27**, 101.
Blanchard, D. C., and Woodcock, A. H., 1957: *Tellus* **9**, 145.
Blottner, F. G., and Ellis, M. A., 1973: *Computers and Fluids* **1**, 133.
Bolling, G. F., and Tiller, W. A., 1961: *J. Appl. Phys.* **32**, 2587.
Borovikov, A. M., Gaivoronskii, I. I., Zak, E. G., Kostarev, V. V., Mazin, I. P., Minervin, V. E., Khragian, A. Kh, and Simeter, S. M., 1963: *Cloud Physics*, p. 65. Transl. by Israel Program f. Scientific Translation, U.S. Dept. Commerce, Washington, D.C.
Borovikov, A. M., and Nevzorov, A. N., 1971: *Izvest. Acad. Sci.*, U.S.S.R. Atm. and Ocean Phys. **7**, 235.
Bourot, J. M., 1969: *C. R. Acad. Sci. de Paris* **A269**, 1017.
Bowen, E. G., 1950: *Austral. J. Sci. Res.* **A3**, 193.
Bowen, E. G., 1953: *Austral. J. Phys.* **6**, 490.
Bowen, E. G., 1956a: *J. Meteor.* **13**, 142.
Bowen, E. G., 1956b: *Tellus* **8**, 394.
Bowers, R., 1953: *Phil Mag.* **44**, 467.
Box, G. E. P., and Muller, M. E., 1958: *Ann. Math. Stat.* **29**, 610.
Braham, R. R., 1963: *J. Appl. Meteor.* **2**, 498.
Braham, R. R., 1964: *J. Atmos. Sci.* **21**, 640.
Braham, R. R., 1967: *J. Atmos. Sci.* **24**, 311.
Braham, R. R., 1968: *Bull. Am. Meteor. Soc.* **49**, 343.
Braham, R. R., 1974: *Bull. Am. Meteor. Soc.* **55**, 100.
Braham, R. R., and Spyers-Duran, P., 1967: *J. Appl. Meteor.* **5**, 1058.
Braham, R. R., and Spyers-Duran, P., 1975: *J. Appl. Meteor.* **13**, 940.
Braham, R. R., and Squires, P., 1974: *Bull. Am. Meteor. Soc.* **55**, 543.
Brazier-Smith, P. R., 1971: *Phys. of Fluids* **14**, 1.

Brazier-Smith, P. R., 1972: *Quart. J. Roy. Meteor. Soc.* **98**, 434.

Brazier-Smith, P. R., Jennings, S. G., and Latham, J., 1972: *Proc. Roy. Soc., London* **A326**, 393.

Brazier-Smith, Jennings, S. G., and Latham, J., 1973: *Quart. J. Roy. Meteor. Soc.* **99**, 260.

Breach, D. R., 1961: *J. Fluid Mech.* **10**, 306.

Brenner, H., 1961: *Chem. Eng. Sci.* **16**, 242.

Brenner, H., 1963: *Chem. Eng. Sci.* **18**, 109.

Brenner, H., and Cox, R. G., 1962: *J. Fluid Mech.* **13**, 561.

Bricard, J., 1949: *J. Geophys. Res.* **54**, 39.

Bricard, J., 1965: *Problems of Atmospheric and Space Electricity*, S. C. Coronity, Ed., p. 82, Elsevier, Amsterdam.

Bricard, J., Billard, F., and Madelaine, G., 1968: *J. Geophys. Res.* **73**, 4487.

Bricard, J., Cabane, M., and Madelaine, G., 1971: *J. Aerosol. Sci.* **2**, 275.

Bridgeman, P. W., 1929: *Int. Crit. Tables* **4**, 11.

Bridgeman, P. W., 1931: *Dimensional Analysis*, Yale Univ. Press, New Haven, Conn.

Brill, R., and Tippe, A., 1967: *Acta Cryst.* **23**, 343.

Brock, J. R., 1962: *J. Coll. Sci.* **17**, 768.

Brock, J. R., 1971: *Atmos. Environment* **5**, 833.

Brook, M., 1958: *Recent Advances in Thunderstorm Electricity*, L. G. Smith, Ed., p. 383, Pergamon Press, New York.

Brook, M., and Latham, J. D., 1968: *J. Geophys. Res.* **73**, 7137.

Brown, G. W., 1956: *Modern Mathematics for the Engineer*, E. F. Beckenbach, Ed., p. 279, McGraw-Hill, New York.

Brown, S., 1970: *Atmospheric Sciences Paper*, No. 170, Dec. 1970, Dept. Atmos. Sci., Colorado State University, Fort Collins, Colo.

Brown, E. N., and Braham, R. R., 1959: *J. Meteor.* **16**, 609.

Brown, K. A., Krehbiel, P. R., Moore, C. B., and Sargent, G. N., 1971: *J. Geophys. Res.* **76**, 2825.

Browning, K. A., 1966: *Quart. J. Roy. Meteor. Soc.* **92**, 1.

Browning, K. A., and Beimers, J. G., 1967: *J. Appl. Meteor.* **6**, 1075.

Browning, K. A., Hallett, J., Harrold, T. W., and Johnson, D., 1968: *J. Appl. Meteor.* **7**, 603.

Brownscombe, J. L., and Goldsmith, P., 1972: *Proc. Cloud Physics Conf.*, London, Aug. 1972, p. 27, Roy, Meteor. Soc., London.

Brownscombe, J. L., and Hallett, J., 1967: *Quart. J. Roy. Meteor. Soc.* **93**, 455.

Brownscombe, J. L., and Thorndyke, N. S. C., 1968: *Nature* **220**, 687.

Brunauer, S., Copeland, L. E., and Cantro, D. L., 1967: *The Solid-Gas Interface*, E. A. Flood, Ed., p. 71, Marcel Dekker, New York.

Bryant, G. W., Hallett, J., and Mason, B. J., 1959: *J. Phys. Chem. Solids* **12**, 189.

Bryant, G. W., and Mason, B. J., 1960: *Phil. Mag.* **5**, 1221.

Bucaro, J. A., and Litovitz, T. A., 1971: *J. Chem. Phys.* **54**, 3846.

Buff, F. P., 1951: *J. Chem. Phys.* **19**, 1591.

Buff, F. P., 1955: *J. Chem. Phys.* **23**, 419.

Bullemer, B., and Riehl, N., 1966: *Solid State Communic.* **4**, 447.

Bullrich, K., Eiden, R., Jaenicke, R., and Nowak, W., 1966: *Final Techn. Rept.*, Contract No. DA-91-591-EUC-3458, Joh. Gutenberg University, Mainz.

Burley, G., 1963: *J. Chem. Phys.* **38**, 2807.

Burley, G., 1964: *J. Phys. Chem. Solids* **25**, 629.

Burrows, D. A., and Hobbs, P. V., 1970: *J. Geophys. Res.* **75**, 4499.

Burton, W. K., Cabrera, N., and Frank, F. C., 1951: *Phil. Trans. Roy. Soc., London* **243A**, 299.

Butcher, S. S., and Charlson, R. J., 1972: *An Introduction to Air Chemistry*, Academic Press, New York.

Byers, H. R., 1965: *Elements of Cloud Physics*, University of Chicago Press, Chicago, Ill.

Byers, H. R., Sievers, J. R., and Tufts, B., 1957: *Artificial Stimulation of Rain*, p. 47, Pergamon Press, New York.

Cadle, R. D., Ed., 1961: *Proc. Int. Symposium on Chemical Reactions in the Lower and Upper Atmosphere*, San Francisco, April 1961, Interscience Publ., New York.

Cadle, R. D., 1965: *Tellus* **18**, 176.

Cadle, R. D., 1966: *Particles in the Atmosphere and Space*, Reinhold Publ. Co., New York.

Cadle, R. D., 1972: *Trans. Amer. Geophys. Union* **53**, 812.

Cadle, R. D., 1973: *Chemistry of the Lower Atmosphere*, S. I. Rasool, Ed., p. 69, Plenum Press, New York.

Cadle, R. D., and Allen, E. R., 1970: *Science* **167**, 243.

Cadle, R. D., and Langer, G., 1975: *Geophys. Res. Letters* **2**, 329.

Cadle, R. D., and Powers, J. W., 1966: *Tellus* **18**, 176.

Cahn, J. W., 1960: *Acta Metallurgica* **8**, 554.

Cahn, J. W., Hillig, W. B., and Sears, G. W., 1964: *Acta Metallurgica* **12**, 1421.

Campbell, E. S., 1952: *J. Chem. Phys.* **20**, 1411.

Carras, J. N., and Macklin, W. C., 1973: *Quart. J. Roy. Meteor. Soc.* **99**, 639.

Carras, J. N., and Macklin, W. C., 1975: *Quart. J. Roy. Meteor. Soc.* **101**, 127.

Carrier, G. F., 1953: *Final Rept.*, Office of Naval Res., Contract Nonr-653-00/1.

Carslaw, H. S., and Jaeger, J. C., 1959: *Heat and Mass Transfer*, McGraw-Hill, New York.

Carstens, J. C., 1972: *J. Atmos. Sci.* **29**, 588.

Carstens, J. C., Grayson, M. A., Kassner, J. L., Rivers, J. L., and Lund, L. H., 1966: *J. de Rech. Atmos.* **2**, 95.

Carstens, J. C., and Kassner, J. L., 1968: *J. de Rech. Atmos.* **3**, 33.

Carstens, J. C., Podzimek, J., and Saad, A., 1974: *J. Atmos. Sci.* **31**, 592.

Carte, A. E., 1956: *Proc. Phys. Soc., London* **69**, 1028.

Carte, A. E., 1959: *Proc. Phys. Soc., London* **73**, 324.

Carte, A. E., and Kidder, R. E., 1966: *Quart. J. Roy. Meteor. Soc.* **92**, 382.

Carte, A. E., and Mossop, S. C., 1960: *Bull. Bos. Puy de Dome*, No. 4, 137.

Cataneo, R., Adam, J. R., and Semonin, R. G., 1971: *J. Atmos. Sci.* **28**, 416.

Caton, P. G., 1966: *Quart. J. Roy. Meteor. Soc.* **92**, 15.

Chagnon, C. W., and Junge, C. E., 1961: *J. Meteor.* **18**, 746.

Chappell, C. F., Magzainer, E. L., and Fritsch, J. M., 1974: *J. Appl. Meteor.* **13**, 726.

Chalmers, B., 1964: *Principles of Solidification*, p. 62, J. Wiley, New York.

Chalmers, J. A., 1967: *Atmospheric Electricity*, 2nd ed., Pergamon Press, New York.

Chamberlain, A. C., 1953: *British Rept. Atomic Energy Res. Establ.*, AERE-HP/12-1261.

Chandrasekhar, S., 1943: *Rev. Mod. Phys.* **15**, 1.

Chandrasekhar, S., 1961: *Hydrodynamic and Hydromagnetic Stability*, Oxford University Press, Oxford.

Charlton, R. B., and List, R., 1972a: *J. de Rech. Atmos.* **6**, Nos. 1-3, 55.

Charlton, R. B., and List, R., 1972b: *J. Atmos. Sci.* **29**, 1182.

Chen, C. S., 1971: *Ph.D. Thesis*, Dept. Meteor., Univ. of Calif., Los Angeles, Calif.

Chepil, W. S., 1951: *Soil Sci.* **72**, 387.

Chepil, W. S., 1957: *Am. J. Sci.* **225**, 12, 206.

Chepil, W. S., 1965: *Meteor. Monographs* **6**, 123.

Chepil, W. S., and Woodruff, N. P., 1957: *Am. J. Sci.* **225**, 104.

Chepil, W. S., and Woodruff, N. P., 1963: *Adv. Agronomy* **15**, 211.

Chernov, A. A., and Melnikova, A. M., 1971: *Soviet Physics (Crystallography)* **16**, 404.

Chester, R., and Johnson, L. R., 1971: *Nature* **229**, 105.

Chester, W., and Breach, D. R., 1969: *J. Fluid Mech.* **37**, 751.

Chidambaram, R., 1961: *Acta Cryst.* **14**, 467.

Chin, E. H. C., 1970: *Ph.D. Thesis*, Dept. Meteor., Univ. of California, Los Angeles, Calif.

Chin, E. H. C., and Neiburger, M., 1972: *J. Atmos. Sci.* **29**, 718.

Chisnell, R. F., and Latham, J., 1974: *Quart. J. Roy. Meteor. Soc.* **100**, 296.

Chisnell, R. F., and Latham, J., 1976: *Quart. J. Roy. Meteor. Soc.* **102**, 133.

Chiu, C. S., and Klett, J. D., 1976: *J. Geophys. Res.* **81**, 1111.

Chiu, C. S., and Orville, H. D., 1975: *EOS Abstract* **56**, 992.

Chodes, N., Warner, J., and Gagin, A., 1974: *J. Atmos. Sci.* **31**, 1351.

Clark, A., 1970: *The Theory of Adsorption and Catalysis*, Academic Press, New York.

Clementi, E., Kistenmacher, H., and Popkie, H., 1973: *J. Chem. Phys.* **58**, 2460.

Cobb, A. W., and Gross, G. W., 1969: *J. Electrochem. Soc.* **116**, 796.

Cohen, N. V., Cotti, M., Iribarne, J. V., and Weissmann, M., 1962: *Trans. Farad. Soc.* **58**, 490.

Cohen, E. R., and Vaughn, E. U., 1971: *J. Coll. Interface Sci.* **35**, 612.

Cohen, R. E., 1970: *J. Stat. Phys.* **2**, 147.

Colgate, S. A., 1969: *Planetary Electrodynamics*, Vol. 2, S. Coroniti and J. Hughes, Eds., p. 143.

Colgate, S. A., 1972: *J. Geophys. Res.* **72**, 479.

Colgate, S. A., and Romero, J. M., 1970: *J. Geophys. Res.* **75**, 5873.

Collatz, L., 1960: *The Numerical Treatment of Differential Equations*, 3rd ed., Springer-Verlag, Berlin.

Cooper, W. A., 1974: *J. Atmos. Sci.* **31**, 1832.

Coriel, S. R., Hardy, S. C., and Sekerka, R. F., 1971: *J. Crystal Growth* **11**, 53.

Corn, M., 1966: *Aerosol Science*, C. N. Davies, Ed., p. 359, Academic Press, New York.

Coroniti, S. C., and Hughes, J., Eds., 1969: *Planetary Electrodynamics*, Gordon and Breach, New York.

Corrin, M. L., 1975: *Proc. Symposium on Hail*, Section IV, Estes Park, Col., Sept. 1975, NCAR, Boulder, Colo.

Corrin, M. L., Edwards, H. W., and Nelson, J. A., 1964: *J. Atmos. Sci.* **21**, 565.

Corrin, M. L., Moulik, S. P., and Cooley, B., 1967: *J. Atmos. Sci.* **24**, 530.

Corrin, M. L., and Nelson, J. A., 1968: *J. Phys. Chem.* **72**, 643.

Cotton, W. R., 1971: *J. Atmos. Sci.* **28**, 647.

Cotton, W. R., 1972: *Monthly Weath. Rev.* **100**, 764.

Cotton, W. R., 1975: *Rev. Geophys. and Space Phys.* **13**, 419.

Cotton, W. R., and Gokhale, N. R., 1967: *J. Geophys. Res.* **72**, 4041.

Coulter, L. V., and Candela, G. A., 1952: *Z. Elektrochem.* **56**, 449.

Cox, R. A., and Penkett, S. A., 1970: *Atmos. Environment* **4**, 425.

Crabb, J. A., and Latham, J., 1974: *Quart. J. Roy. Meteor. Soc.* **100**, 191.

Craig, R. A., 1965: *The Upper Atmosphere*, Academic Press, New York.

Cram, D. J., and Hammond, G. S., 1955: *Organic Chemistry*, McGraw-Hill Publ. Co., New York.

Crandall, W. K., Molenkamp, C. R., Williams, A. L., Fulk, M. M., Lange, R., and Knox, J. B., 1973: *Res. Rept.*, University of California, UCRL-75896, Lawrence Livermore Lab., Livermore, Calif.

Crowther, A. G., and Saunders, C. P. R., 1973: *J. Meteor. Soc., Japan* **51**, 490.

Cunningham, E., 1910: *Proc. Roy. Soc., London* **A83**, 357.

Cwilong, B. M., 1947: *Proc. Roy. Soc., London* **A190**, 137.

Czerwinski, N., and Pfisterer, W., 1972: *J. de Rech. Atm.* **6**, 89.

Dalal, N., 1947: *Ph.D. Thesis*, University of Heidelberg, Heidelberg.

Dana, M. T., 1970: *Precipitation Scavenging 1970*, AEC Symposium, Dec. 1970, p. 137, U.S. Dept. of Commerce, Springfield, Va.

Danford, M. D., and Levy, 1962: *J. Am. Chem. Soc.* **84**, 3965.

Davies, C. N., 1945: *Proc. Phys. Soc., London* **57**, 259.

Davies, C. N., 1966: *Aerosol Science*, Academic Press, New York.

Davies, C. N., and Peetz, C. V., 1956: *Proc. Roy. Soc., London* **A234**, 269.

Davies, J. T., and Rideal, E. K., 1961: *Interfacial Phenomena*, Academic Press, New York.

Davis, B. L., Johnson, L. R., and Moeng, F. J., 1975: *J. Appl. Meteor.* **14**, 891.

Davis, C. I., 1974: *Ph.D. Thesis*, Dept. Environ. Sci., University of Wyoming, Laramie, Wyoming.

Davis, C. I., and Auer, A. H., 1974: *Proc. Cloud Phys. Conf.*, Aug. 1974, Tucson, Ariz., p. 141, Amer. Meteor. Soc., Boston, Mass.

Davis, C., and Litovitz, T., 1965: *J. Chem. Phys.* **42**, 2563.

Davis, D., 1973: *Proceedings Conf. on Environmental Impact of Aerospace Operations*, June 1973, Amer. Instit. of Aeron. and Astron., Denver, Col.

Davis, M. H., 1964a: *Res. Rept.*, RM-3860-PR, Jan. 1964, The RAND Corp., Santa Monica, Calif.

Davis, M. H., 1964b: *Quart. J. Mech. and Appl. Math.* **17**, 499.

Davis, M. H., 1965: *Proc. Cloud Phys. Conf.*, Tokyo and Sapporo, May 1965, p. 118, Meteor. Soc. Japan, Tokyo.

Davis, M. H., 1966: *J. Geophys. Res.* **71**, 3101.

Davis, M. H., 1969: *Amer. J. Phys.* **37**, 26.

Davis, M. H., 1972: *J. Atmos. Sci.* **29**, 911.

Davis, and Sartor, J. D., 1967: *Nature* **215**, 1371.

Davy, J. G., and Somorjai, G. A., 1971: *J. Chem. Phys.* **55**, 3624.

Dawson, G. A., 1973: *J. Geophys. Res.* **78**, 6364.

Dawson, G. A., and Cardell, S. R., 1973: *J. Geophys. Res.* **78**, 8864.

Dawson, G. A., and Richards, C. N., 1970: *J. Geophys. Res.* **75**, 4589.

Dawson, G. A., and Warrender, R. A., 1973: *J. Geophys. Res.* **78**, 3619.

Day, G. J., 1955: *Geofys. Pura e Appl.* **31**, 169.

Day, J. A., 1958: *J. Meteor.* **15**, 226.

Day, J. A., 1964: *Quart. J. Roy. Meteor. Soc.* **90**, 72.

Day, J. A., 1965: *Natural History*, Jan. issue.

de Almeida, F. C., 1975: *Research Rept.* 75-2, Dept. Meteor., University of Wisconsin, Madison, Wisc.

de Almeida, F. C., 1976: *J. Atmos. Sci.* **33**, 571.

de Almeida, F. C., 1977: *J. Atmos. Sci.* **34**, 1286.

Dean, W. R., and O'Neill, M. E., 1963: *Mathematika* **10**, 13.

Defay, R., Prigogine, I., Bellermans, A., and Everett, D. H., 1966: *Surface Tension and Adsorption*, Wiley Inc., New York.

Deirmendjian, D., 1969: *Electromagnetic Scattering on Spherical Polydispersions*, American Elsevier Publ. Co., New York.

Delaney, L. J., Houston, R. W., and Eagleton, L. C., 1964: *Chem. Engr. Sci.* **19**, 105.

Delany, A. C., Parkin, D. W., Griffin, J. J., Goldberg, E. D., and Reimann, B. E., 1967: *Geochim, Cosmo. Chim. Acta* **31**, 885.

de Micheli, S. M., and Iribarne, J. V., 1963: *J. de Chim. Phys.*, No. 48, 767.

Dennis, A. S., and Musil, D. J., 1973: *J. Atmos. Sci.* **30**, 278.

Dennis, S. C. R., and Shimshoni, M., 1964: *Aero. Res. Counc.*, Current Paper No. 797.

Derjaguin, B. V., and Yamalov, Yu. I., 1972: *Reviews in Aerosol Phys. and Chem.* **3**, G. M. Hidy and J. R. Brock, Eds., p. 3, Pergamon Press, New York.

Devir, S. E., 1963: *J. Colloid. Sci.* **18**, 744.

Devir, S. E., 1966: *J. Coll. Interface Sci.* **21**, 9.

Diem, M., 1968: *Arch. Meteor. Geophys. Biokl.* **B16**, 347.

Diem, M., and Strantz, R., 1971: *Meteor. Rundschau* **24**, 23.

Dinger, J., Howell, H. B., and Wojciechowski, T. H., 1970: *J. Atmos. Sci.* **27**, 1791.

Dingle, A. N., 1975: *Progress Res. Rept.*, No. 11, C00-1407-58, Univ. of Michigan, Ann Arbor, Mich.

Dingle, A. N., and Lee, Y., 1972: *J. Appl. Meteor.* **11**, 877.

Dingle, A. N., and Lee, Y., 1973: *J. Appl. Meteor* **12**, 1295.

Dobrowolski, A. B., 1903: *Rapports Scientifiques* (Resultats du Voyage de S. Y. Belgica, 1897–1899), 3/4 (Meteorology), pt. 3: La neige et le givre, J. E. Buschman, Anvers (Belgium).

Dodd, K. N., 1960: *J. Fluid Mech.* **9**, 175.

Donaldson, R. J., Chmela, A. C., and Shackford, C. R., 1960: *Geophys. Monographs*, No. 5, 354, Amer. Geophys. Union, Washington, D.C.

Doolittle, J. B., and Vali, G., 1975: *J. Atmos. Sci.* **32**, 375.

Dorsch, R. G., and Boyd, B., 1951: *NACA, Techn. Note*, 2532.

Dorsch, R. G., and Hacker, P., 1950: *NACA, Techn. Note*, 2142.

Dorsch, R. G., and Hacker, P., 1951: *NACA, Techn. Note*, 2510.

Dorsey, N. E., 1940: *Properties of Ordinary Water Substance*, Reinhold Publ. Co., New York.

Douglas, R. H., 1960: *Nubila* **3**, No. 1, 5.

Doyle, A., Moffet, D. R., and Vonnegut, B., 1964: *J. Coll. Sci.* **19**, 136.

Draginis, M., 1958: *J. Meteor.* **15**, 481.

Drake, J. C., and Mason, B. J., 1966: *Quart. J. Roy. Meteor. Soc.* **92**, 500.

Drake, R. L., 1972a: *J. Atmos. Sci.* **29**, 537.

Drake, R. L., 1972b: *Topics in Current Aerosol Research*, G. M. Hidy and G. R. Brock, Eds., p. 201, Pergamon Press, New York.

Drake, R. L., and Wright, T. J., 1972: *J. Atmos. Sci.* **29**, 548.

Droessler, E. G., 1964: *J. Atmos. Sci.* **21**, 201.

Droessler, E. G., and Heffernan, K. J., 1965: *J. Appl. Meteor.* **4**, 442.

Drost-Hansen, W., 1967: *J. Coll. Interface Sci.* **25**, 131.

Dufour, L., 1965a: *Report*, Publ. Ser. B, No. 46, Inst. Roy. Meteor. de Belgique, Brussels.

Dufour, L., 1965b: *Report*, Bull. de l'Acad. Royale de Belgique (Classes des Sciences) Ser. 5, **60**, 298.

Dufour, L., and Defay, R., 1963: *Thermodynamics of Clouds*, Academic Press, New York.

Duncan, A. B., and Pople, J. A., 1953: *Trans. Farad. Soc.* **49**, 217.

Dundon, M. L., and Mack, E., 1923: *J. Am. Chem. Soc.* **45**, 2479, 2650.

Dunham, S. B., 1966: *J. de Rech. Atmos.* **2**, 231.

Dunning, W. J., 1955: *Chemistry of the Solid State*, W. E. Garner, Ed., p. 159, Butterworths, London.

Dunning, W. J., 1967: *The solid-gas interface*, E. A. Flood, Ed., Vol. 1, p. 271, Marcel Dekker, New York.

Dunning, W. J., 1969: *Nucleation*, A. C. Zettlemoyer, Ed., Marcel Dekker, Inc., New York.

Durbin, N. G., 1959: *Tellus* **11**, 202.

Dye, J. E., and Hobbs, P. V., 1966: *Nature* **209**, 464.

Dye, J. E., and Hobbs, P. V., 1968: *J. Atmos. Sci.* **25**, 82.

Eadie, W. J., 1971: *Ph.D. Thesis*, Dept. of Geophys. Sciences, University of Chicago, June 1971. (Also Techn. Note No. 40, Cloud Phys. Lab., Univ. of Chicago.)

East, T. W., 1957: *Quart. J. Roy. Meteor. Soc.* **83**, 61.

Easter, R. C., and Hobbs, P. V., 1974: *J. Atmos. Sci.* **31**, 1586.

Eaton, L., 1970: *Proc. Cloud Phys. Conf.*, Fort Collins, Col., Aug. 1970, p. 137, Amer. Meteor. Soc., Boston, Mass.

Edwards, G. R., and Evans, L. F., 1968: *J. Atmos. Sci.* **25**, 249.

Edwards, G. R., and Evans, L. F., 1971: *J. Atmos. Sci.* **28**, 1443.

Edwards, G. R., and Evans, L. F., and La Mer, V. K., 1962: *J. Coll. Sci.* **17**, 749.

Edwards, G. R., and Evans, L. F., and Zipper, A. F., 1970: *Trans. Farad. Soc.* **66**, 220.

Ehhalt, D. H., 1973: *J. Geophys. Res.* **78**, 7076.

Eigen, M., 1952: *Z. f. Elektrochemie* **56**, 176.

Einstein, A., 1905: *Ann. d. Physik* **17**, 549.

Eisenberg, D., Kauzmann, W., 1969: *The Structure and Properties of Water*, Oxford Press, Oxford.

Eldrige, R. G., 1957: *J. Meteor.* **14**, 55.

Elster, J., and Geitel, H., 1913: *Phys. Z.* **14**, 1287.

Enukashvili, I. M., 1964a: *Bull. Acad. Sci., USSR*, Geophys. Ser., No. 10, 944.

Enukashvili, I. M., 1964b: *Bull. Acad. Sci., USSR*, Geophys. Ser., No. 11, 1043.

Erikson, E., 1959: *Tellus* **11**, 375.

Evans, L. F., 1965: *Nature* **206**, 822.

Evans, L. F., 1966: *Nature* **211**, 281.

Evans, L. F., 1967a: *Nature* **213**, 384.

Evans, L. F., 1967b: *Trans. Farad. Soc.* **63**, 1.

Evans, L. F., 1970: *Proc. Conference Cloud Phys.*, Fort Collins, Col., Aug. 1970, p. 14, Amer. Meteor. Soc., Boston, Mass.

Evans, L. F., 1973: *J. Atmos. Sci.* **30**, 1657.

Evans, L. F., and Lane, J. E., 1973: *J. Atmos. Sci.* **30**, 326.

Eyring, H., and Jhon, M. S., 1969: *Significant Liquid Structures*, J. Wiley, New York.

Fabian, P., and Junge, C. E., 1970: *Arch. Meteor. Geophys. and Biokl.* **A19**, 161.

Facy, L., 1955: *Arch. Meteor. Geophys. and Biokl.* **A8**, 229.

Facy, L., 1958: *C. R. Acad. Sci.*, Paris, **246**, 3161.

Facy, L., 1960: *Geofys. Pura e Appl.* **46**, 201.

Fage, A., 1934: *Proc. Roy. Soc.*, London **A144**, 381

Fage, A., 1937a: *Aero Res. Comm., Repts. and Mem.*, No. 1765.

Fage, A., 1937b: *Aero Res. Comm., Repts. and Mem.*, No. 1766.

Faraday, M., 1860: *Proc. Roy. Soc.*, London **10**, 440.

Farkas, L., 1927: *Z. Phys. Chem.* **A125**, 236.

Farley, F. J., 1952: *Proc. Roy. Soc.*, London **A212**, 530.

Federer, B., 1968: *Z. Angew. Math. Phys.* **19**, 637.

Federer, B., and Waldvogel, A., 1975: *J. Appl. Meteor.* **14**, 91.

Federer, B., and Waldvogel, A., 1978: *Quart. J. Roy. Meteor. Soc.* **104**, 69.

Feller, W., 1967: *An Introduction to Probability Theory and Its Application*, Vol. 1, J. Wiley, New York.

Ferguson, W. S., Griffin, J. J., and Goldberg, E. D., 1970: *J. Geophys. Res.* **75**, 1137.

Finn, R. K., 1953: *J. Appl. Phys.* **24**, 771.

Fischer, K., and Hänel, G., 1972: '*Meteor.*', *Forschg. Ergebn.*, Ser. B., No. 8, p. 59.

Fitzgerald, D., and Byers, H. R., 1962: *Res. Rept.*, AF 19(604)2189, Dept. Meteorology, University of Chicago, Chicago, Ill.

Fitzgerald, J. W., 1970: *J. Atmos. Sci.* **27**, 70.

Fitzgerald, J. W., 1972: *Ph.D. Thesis*, Dept. of Geophys. Sci., Univ. of Chicago. (Also Techn. Note, No. 44, Cloud Phys. Lab., Univ. of Chicago.)

Fitzgerald, J. W., 1973: *J. Atmos. Sci.* **30**, 628.

Fitzgerald, J. W., 1974: *J. Atmos. Sci.* **31**, 1358.

Fitzgerald, J. W., and Spyers-Duran, P. A., 1973: *J. Appl. Meteor.* **12**, 511.

Flachsbart, O., 1927: *Phys. Z.* **28**, 461.

Fletcher, A. N., 1972: *J. Appl. Meteor.* **11**, 988.

Fletcher, N. H., 1958: *J. Chem. Phys.* **29**, 572.

Fletcher, N. H., 1959a: *J. Meteor.* **16**, 173.

Fletcher, N. H., 1959b: *J. Chem. Phys.* **31**, 1136.

Fletcher, N. H., 1959c: *J. Chem. Phys.* **30**, 1476.

Fletcher, N. H., 1960: *Austral. J. Phys.* **13**, 408.

Fletcher, N. H., 1962a: *Physics of Rain Clouds*, Cambridge University Press, London.

Fletcher, N. H., 1962b: *Phil. Mag.* **7**, 255.

Fletcher, N. H., 1963: *Phil. Mag.* **8**, 1425.

Fletcher, N. H., 1968: *Phil. Mag.* **18**, 1287.

Fletcher, N. H., 1969: *J. Atmos. Sci.* **26**, 1266.

Fletcher, N. H., 1970a: *The Chemical Physics of Ice*, Cambridge University Press, London.

Fletcher, N. H., 1970b: *J. Atmos. Sci.* **27**, 1098.

Fletcher, N. H., 1973: *Proc. Conf. Phys. and Chem. of Ice*, Ottawa, Aug. 1972, p. 132, Roy. Soc, of Canada, Ottawa.

Flood, E. A., Ed., 1967: *The Solid-Gas Interface*, p. 11, Marcel Dekker, New York.

Flower, W. I., 1928: *Proc. Phys. Soc.* **A40**, 167.

Flyger, H., 1973: *J. Appl. Meteor.* **12**, 161.

Fonselius, S., 1954: *Tellus* **6**, 90.

Foote, G. B., 1971: *Ph.D. Thesis*, Dept. Atmos. Sci., University of Arizona, Tucson, Ariz.

Foote, G. B., and Brazier-Smith, P. R., 1972: *J. Geophys. Res.* **77**, 1695.

Foote, G. B., and du Toit, P. S., 1969: *J. Appl. Meteor.* **8**, 249.

Foote, G. B., and Fankhauser, J. C., 1973: *J. Appl. Meteor.* **12**, 1330.

Forslind, E., 1952: *Acta Polytechn. Scand.* **3**, No. 5.

Fournier d'Albe, E. M., 1949: *Quart. J. Roy. Meteor. Soc.* **75**, 1.

Fournier d'Albe, E. M., and Hidayetulla, M. S., 1955: *Quart. J. Roy. Meteor. Soc.* **81**, 610.

Frank, F. C., 1949: *Disc. Farad. Soc.* **5**, 45.

Frank, F. C., 1958: *Growth and Perfection of Crystals*, R. H. Doremus *et al.*, Eds., p. 411, J. Wiley, New York.

Frank, H. S., and Wen, W. Y., 1957: *Disc. Farad. Soc.* **24**, 133.

Frank, H. S., 1958: *Proc. Roy. Soc.* **A247**, 481.

Frank, H. S., and Quist, A. S., 1961: *J. Chem. Phys.* **34**, 604.

Franks, F., Ed., 1972: *Water*, Vol. 1, Plenum Press, New York.

Franks, F., Ed., 1973: *Water*, Vol. 3, Plenum Press, New York.

Freier, G., 1960: *J. Geophys. Res.* **65**, 3979.

Freier, G., 1962: *J. Geophys. Res.* **67**, 4683.

Frenkel, J., 1932: *Phys. Z., Swietunion* **1**, 498.

Frenkel, J., 1946: *Kinetic Theory of Liquids*, Oxford University Press, Oxford.

Fresch, R. W., 1973: *Res. Rept.*, AR-106, Dept. of Atmos. Resources, University of Wyoming, Laramie, Wyoming.

Freundlich, H., 1926: *Colloid and Capillary Chemistry*, Methuen and Co., Ltd., London.

Frey, F., 1941: *Z. f. Phys. Chemie* **B49**, 83.

Friedlander, S. K., 1957: *A.I.Ch.E. Journal* **3**, 43.

Friedlander, S. K., 1960a: *J. Meteor.* **17**, 373.

Friedlander, S. K., 1960b: *J. Meteor.* **17**, 479.

Friedlander, S. K., 1961: *J. Meteor.* **18**, 753.

Friedlander, S. K., 1965: *Proc. Conf. on Aerosols*, Liblice, Oct. 1962, p. 115, K. Spurny, Ed., Czechoslovak Acad. Sci., Prague.

Friedlander, S. K., and Hidy, G. M., 1969: *Proc. Int. Conf. on Condensation and Ice Nuclei*, Sept. 1969, Prague and Vienna, p. 21, Czechoslovak Acad. of Sci., Prague.

Friedlander, S. K., and Wang, C. S., 1966: *J. Colloid. Interface Sci.* **22**, 16.

Friend, J. P., 1966: *Tellus* **18**, 465.

Friend, J. P., Leifer, R., and Trichon, M., 1973: *J. Atmos. Sci.* **30**, 465.

Fromm, J., Clementi, E., and Watts, R. O., 1975: *J. Chem. Phys.* **62**, 1388.

Frössling, N., 1938: *Gerlands Beitr. z. Geophysik* **52**, 170.

Fuchs, N., 1934: *Z. f. Physik* **89**, 736.

Fuchs, N, 1951: *Dokl. Akad. Nauk, SSSR* **81**, 1043.

Fuchs, N., 1959: *Evaporation and Droplet Growth in Gaseous Media*, Pergamon Press, New York.

Fuchs, N., 1964: *The Mechanics of Aerosols*, Pergamon Press, New York.

Fukuta, N., 1958: *J. Meteor.* **15**, 17.

Fukuta, N., 1963: *Nature* **199**, 475.

Fukuta, N., 1966: *J. Atmos. Sci.* **23**, 191.

Fukuta, N., 1969: *J. Atmos. Sci.* **26**, 522.

Fukta, N., 1975a: *J. Atmos. Sci.* **32**, 1597.

Fukuta, N., 1975b: *J. Atmos. Sci.* **32**, 2371.

Fukuta, N., and Armstrong, P. A., 1974: *Proc. Cloud Phys. Conf.*, Tucson, Arizona, Oct. 1974, p. 46, Amer. Meteor. Soc., Boston, Mass.

Fukuta, N., and Mason, B. J., 1963: *J. Phys. Chem. Solids* **24**, 715.

Fukuta, N., Sano, I., and Asaoka, M., 1959: *J. Meteor. Soc., Japan* **37**, 274.

Fukuta, N., and Walter, L. A., 1970: *J. Atmos. Sci.* **27**, 1160.

Gabarashvili, T. G., and Gliki, N. V., 1967: *Izv. Acad. Sci., U.S.S.R.*, Atmos. and Ocean. Phys. **3**, 324.

Gabarashvili, T. G., and Kartsivadze, A. I., 1968: *Proc. Conf. Cloud Phys.*, Toronto, Aug. 1968, p. 188, Univ. of Toronto Press, Toronto.

Gabarashvili, T. G., and Kartsivadze, A. I., 1969: *Proc. Conf. on Conden. and Ice Nucl.*, Prague-Vienna, Sept. 1969, p. 220, Czechoslovak. Acad. Sci., Prague.

Gagin, A., 1965: *Proc. Cloud Phys. Conf.*, Tokyo, May 1965, p. 155, Meteor. Soc. of Japan, Tokyo.

Gagin, A., 1971: *Proc. Weather Mod. Conf.*, Canberra, Sept. 1971, p. 5, Amer. Meteor. Soc., Boston, Mass.

Gagin, A., 1972: *J. de Rech. Atmos.* **6**, 175.

Gallily, I., and Michaeli, G., 1976: *Nature* **259**, 110.

Garland, J. A., 1971: *Quart. J. Roy. Meteor. Soc.* **97**, 483.

Garner, F. H., and Lane, J. J., 1959: *Trans. Inst. Chem. Eng.* **37**, 167.

Garner, F. H., and Lihou, D. A., 1965: *DECHEMA Monographien* **55**, 155.

Garner, F. H., Skelland, A. H., and Haycook, P. J., 1954: *Nature* **173**, 1239.

Garret, W. D., 1965: *Linnology and Oceanography* **10**, 602.

Garret, W. D., 1967: *Deep-Sea Res.* **14**, 221.

Garret, W. D., 1969: *Ann. d. Meteor. N.F.*, **4**, 25.

Garret, W. D., 1970: *Proc. Cloud Phys. Conf.*, Fort Collins, Aug. 1970, Col., p. 131, Amer. Meteor. Soc., Boston, Mass.

Garten, V. H., and Head, R. B., 1964: *Nature* **201**, 573.

Gay, M. J., Griffiths, R. F., Latham, J., and Saunders, C. P. R., 1974: *Quart. J. Roy. Meteor. Soc.* **100**, 682.

Gentile, A. L., and Drost-Hansen, W., 1956: *Naturwissenschaften* **43**, 274.

Georgii, H. W., 1959a: *Ber. Deutsch. Wetter Dienst, U.S. Zone*, 8, No. 58.

Georgii, H. W., 1959b: *Geophys. Pura e Appl.* **44**, 249.

Georgii, H. W., 1963: *Z. Angew. Math. Phys.* **14**, 503.

Georgii, H. W., 1965: *Ber. Deutsch Wetter Dienst, U.S. Zone*, No. 100.

Georgii, H. W., 1975: *Promet* **5**, 21.

Georgii, H. W., and Gravenhorst, G., 1972: *Meteor. Rundschau* **25**, 180.

Georgii, H. W., Jost, D., and Vitze, W., 1971: *Ber. Inst. Meteor. u. Geophys.*, No. 23, University of Frankfurt, Frankfurt.

Georgii, H. W., Jost, D., and Meszaros, E., 1974: *Acta Geologica Acad. Scient. Hungaricae* **18**, 79.

Georgii, H. W., and Kaller, 1970: *Ber. Inst. Meteor. u. Geophys.*, No. 21, University of Frankfurt, Frankfurt.

Georgii, H. W., and Kleinjung, E., 1967: *J. de Rech. Atmos.* **3**, 145.

Georgii, H. W., and Kleinjung, E., 1968: *Pure and Appl. Geophys.* **71**, 181.

Georgii, H. W., and Metnieks, A. L., 1958: *Geophys. Pura e Appl.* **41**, 159.

Georgii, H. W., and Müller, W. J., 1974: *Tellus* **26**, 180.

Georgii, H. W., and Wötzel, D., 1970: *J. Geophys. Res.* **75**, 1727.

Gerber, H., 1972: *J. Atmos. Sci.* **29**, 391.

Gerhard, E. R., and Johnstone, H. R., 1955: *Ind. Eng. Chem.* **47**, 972.

Gilbert, H., Couderc, J. P., and Angelina, H., 1972: *Chem. Eng. Sci.* **27**, 45.

Gillen, K. T., Douglass, D. C., and Hoch, M. J. R., 1972: *J. Chem. Phys.* **57**, 5117.

Gillespie, D. T., 1972: *J. Atmos. Sci.* **29**, 1496.

Gillespie, D. T., 1975a: *J. Atmos. Sci.* **32**, 600.

Gillespie, D. T., 1975b: *J. Atmos. Sci.* **32**, 1977.

Gillette, D. A., Blifford, I. H., and Fenster, C. R., 1972: *J. Appl. Meteor.* **11**, 977.

Gillette, D. A., Blifford, I. H., and Fryrear, D. W., 1974: *J. Geophys. Res.* **79**, 4068.

Ginnings, D. C., and Corrucini, R. J., 1947: *J. Res. Nat. Bur. Stand.* **38**, 583.

Gish, O. H., 1944: *Terr. Magn. Atmos. Electr.* **49**, 159.

Gitlin, S. N., Goyer, G. G., and Henderson, T. J., 1968: *J. Atmos. Sci.* **25**, 97.

Gittens, G. J., 1969: *J. Coll. Interface Sci.* **30**, 406.

Glasstone, S., 1959: *Physical Chemistry*, Van Nostrand, Co., Inc., New York.

Glasstone, S., Laidler, K. J., and Eyring, H., 1941: *The Theory of Rate Processes*, McGraw-Hill, New York.

Goetz, A., 1965: *Proc. Cloud Phys. Conf.*, Tokyo, May 1965, p. 42, Meteor. Soc. of Japan, Tokyo.

Goff, J. A., 1942: *Trans. Am. Soc. Heating and Vent. Engr.* **48**, 299.

Goff, J. A., 1949: *Trans. Am. Soc. Mech. Engs.* **71**, 903.

Goff, J. A., 1957: *Trans. Am. Soc. Heat, Air Cond. Eng.* **63**, 347.

Goff, J. A., 1965: *Humidity and Moisture*, Vol. 3, A. Wexler, Ed., p. 289, Reinhold Publ. Co., New York.

Goff, J. A., and Gratch, S., 1945: *Trans. Am. Soc. Heating and Vent. Engr.* **51**, 125.

Goff, J. A., and Gratch, S., 1946: *Trans. Am. Soc. Heating and Vent. Engr.* **52**, 95.

Gokhale, N., 1975: *Hailstorms and Hailstone Growth*, State University of New York Press, Albany, N.Y.

Gokhale, N., and Goold, J., 1969: *J. Geophys. Res.* **74**, 5374.

Gokhale, N., and Lewinter, O., 1971: *J. Appl. Meteor.* **10**, 469.

Gokhale, N., and Spengler, J. I., 1972: *J. Appl. Meteor.* **11**, 157.

Gold, L. W., and Power, B. A., 1952: *J. Meteor.* **9**, 447.

Gold, L. W., and Power, B. A., 1954: *J. Meteor.* **11**, 35.

Goldburg, A., and Florsheim, B. H., 1966: *Phys. Fluids* **9**, 45.

Golde, R. H., Ed., 1977: *Lightning*, Academic Press, New York.

Goldsmith, P., Delafield, H. J., and Cox, L. C., 1963: *Quart. J. Roy. Meteor. Soc.* **89**, 43.

Goldsmith, P., and May, F. G., 1966: *Aerosol Science*, C. N. Davies, Ed., p. 163, Academic Press, New York.

Goldstein, H., 1965: *Classical Mechanics*, Addison Wesley, New York.

Goldstein, S., 1929: *Proc. Roy. Soc., London* **A123**, 216.

Goldstein, S., 1º31: *Proc. Roy. Soc., London* **A125**, 198.

Golovin, A. M., 1963a: *Izv. Geophys. Ser.*, No. 5, 482.

Golovin, A. M., 1963b: *Izv. Geophys. Ser.*, No. 9, 880.

Golovin, A. M., 1965: *Izv. Atmos. Oceanic Phys.* **1**, 422.

Goyer, G. G., McDonald, J. E., Baer, F., and Braham, R. R., 1960: *J. Meteor.* **17**, 442.

Graedel, T. E., and Franey, J. P., 1975: *Geophys. Res. Letters* **2**, 325.

Granat, L., 1972: *Tellus* **24**, 550.

Gränicher, H., 1958: *Z. f. Kristallogr.* **110**, 432.

Gränicher, H., 1963: *Phys. d. kondens, Materie* **1**, 1.

Gränicher, H., Jaccard, C., Scherrer, P., and Steinemann, A., 1957: *Disc. Farad. Soc.*, No. 23, 50.

Grant, L. O., 1968: *Proc. Cloud Phys. Conf.* Toronto, Aug. 1968, p. 305, University of Toronto Press, Toronto.

Gravenhorst, G., and Corrin, M. L., 1972: *J. de Rech. Atmos.* **6**, 205.

Green, A. W., 1975: *J. Appl. Meteor.* **14**, 1578.

Green, H. L. and Lane, R. L., 1964: *Particulate Clouds*, E.a.F.N. Spon Ltd., London.

Greenfield, S. M., 1957: *J. Meteor.* **14**, 115.

Gregory, P. H., 1961: *The Microbiology of the Atmosphere*, Leonard Hill Publ. Co., London.

Gregory, P. H., 1967: *Science Progr.* **55**, 613.

Grenet, G., 1947: *Extrait des Ann. Geophys.* **3**, 306.

Gretz, R. D., 1966a: *Surface Science* **5**, 255.

Gretz, R. D., 1966b: *Surface Science* **5**, 239.

Gretz, R. D., 1966c: *J. Chem. Phys.* **45**, 3160.

Griffin, F. O., 1972: *M.S. Thesis*, Dept. Chem. Eng., University of British Columbia, Vancouver.

Griffiths, R. F., and Latham, J., 1972: *J. Meteor. Soc., Japan* **50**, 416.

Griffiths, R. F., and Latham, J., 1974: *Quart. J. Roy. Meteor. Soc.* **100**, 163.

Griffiths, R. F., Latham, J., and Myers, V., 1974: *Quart. J. Roy. Meteor. Soc.* **100**, 181.

Grim, R. E., 1953: *Clay Mineralogy*, McGraw-Hill Co., New York.

Gross, B., 1954: *Phys. Rev.* **94**, 1545.

Gross, G. W., 1968: *Adv. in Chem. Series* **73**, 27.

Gross, G. W., 1971: *J. Atmos. Sci.* **28**, 1005.

Grosse, G., and Stix, E., 1968: *Ber. Deutsch. Bot. Gesellsch.* **81**, 528.

Grover, S. N., 1976a: *Pure and Appl. Geophys.* **114**, 521.

Grover, S. N., 1976b: *Pure and Applied Geophys.* **114**, 509.

Grover, S. N., 1978: *Ph.D. Thesis*, Dept. Atmos. Sci., University of California, Los Angeles, Calif.

Grover, S. N., and Beard, K. V., 1975: *J. Atmos. Sci.* **32**, 2156.

Grover, S. N., Pruppacher, H. R., and Hamielec, A. E., 1977: *J. Atmos. Sci.* **34**, 1655.

Grunow, J., 1960: *Geophys. Monograph* No. 5, p. 130, Amer. Geophys. Union, Washington, D.C.

Gudris, N., and L. Kulikova, L., 1924: *Z. Phys.* **24**, 121.

Guenadiev, N., 1970: *J. de Rech. Atmos.* **4**, 81.

Guenadiev, N., 1972: *Pure and Appl. Geophys.* **99**, 251.

Guilder, L., Johnson, D. P., and Jones, F. E., 1975: *Science* **191**, 1261.

Gunn R., 1948: *Phys. Rev.* **71**, 181.

Gunn, R., 1949: *J. Geophys. Res.* **54**, 57.

Gunn, R., 1950: *J. Geophys. Res.* **55**, 171.

Gunn, R., 1954: *J. Meteor.* **11**, 339.

Gunn, R., 1955: *J. Coll. Sci.* **10**, 107.

Gunn, R., 1956: *J. Meteor.* **13**, 283.

Gunn, R., 1965: *Science* **150**, 695.

Gunn, R., and Kinzer, G. D., 1949: *J. Meteor.* **6**, 243.

Gunn, R., and Marshall, J. S., 1958: *J. Meteor.* **15**, 452.

Gupalo, Y. P., and Ryazantsev, Y. S., 1972: *Chem. Eng. Sci.* **27**, 61.

Gurikov, Yu., V., 1960: *Zh. Strukt. Khim.* **1**, 286.

Gurikov, Yu., V., 1965: *Zh. Strukt. Khim.* **6**, 817.

Haagen-Smit, A. J., and Wayne, L. G., 1968: *Air Pollution*, 2nd ed., Vol. 1, A. C. Stern, Ed., p. 149, Academic Press, New York.

Hadamard, J., 1911: *C. R. Acad. Sci.*, Paris **152**, 1735.

Hagen, D. E., 1973: *Bull. Amer. Phys. Soc.* **18**, 609.

Haggis, G. H., Hasted, J. B., and Buchanan, T. J., 1952: *J. Chem. Phys.* **20**, 1452.

Hale, B. N., and Kiefer, J., 1975: *J. Statist. Phys.* **12**, 437.

Hale, B. N., and Plummer, P. L., 1974a: *J. Chem. Phys.* **61**, 4012.

Hale, B. N., and Plummer, P. L., 1974b: *J. Atmos. Sci.* **31**, 1015.

Hall, W. D., and Beard, K. V., 1975: *Pure and Appl. Geophys.* **113**, 515.

Hall, W. D., and Pruppacher, H. R., 1977: *J. Atmos. Sci.* **33**, 1995.

Haller, W., and Duecker, H. C., 1960: *Nat. Bur. Stand., J. Res.* **A64**, 527.

Hallett, J., 1960: *J. Glaciol.* **3**, 698.

Hallett, J., 1961: *Phil. Mag.* **6**, 1073.

Hallett, J., 1963a: *Proc. Phys. Soc., London* **82**, 1046.

Hallett, J., 1963b: *Meteor. Monogr.* **5**, No. 27, p. 168, Amer. Meteor. Soc., Boston, Mass.

Hallett, J., 1964: *J. Atmos. Sci.* **21**, 671.

Hallett, J., 1965: *J. de Rech. Atmos.* **2**, 81.

Hallett, J., and Mason, B. J., 1958: *Proc. Roy. Soc., London* **A247**, 440.

Hallett, J., and Mossop, S. C., 1974: *Nature* **249**, 26.

Hallett, J., and Shrivastava, S. K., 1972: *J. de Rech. Atmos.* **6**, 223.

Hallgren, R. E., and Hosler, C. L., 1960: *Geophys. Monogr.*, No. 5, p. 257, Amer. Geophys. Union, Washington, D.C.

Halsey, G. D., 1948: *J. Chem. Phys.* **16**, 931.

Halton, J. H., 1970: *Comp. Sci. Tech. Rept.*, No. 13, University of Wisconsin, Madison, Wisc.

Hamielec, A. E., Hoffmann, J. W., and Ross, L., 1967: *A.I.Ch.E. Journal* **13**, 212.

Hamielec, A. E., and Raal, J. D., 1969: *Phys. Fluids* **12**, 11.

Hammeke, K., and Kappler, E., 1953: *Z. Geophys.* **19**, 181.

Hampl, B., Kerker, M., Cooke, D. D., and Matijevic, E., 1971: *J. Atmos. Sci.*, **28**, 1211.

Handa, B. K., 1969: *Tellus* **21**, 95.

Hänel, G., 1966: *Diplom. Thesis*, Meteor. Inst., Univ. of Mainz, Mainz.

Hänel, G., 1968: *Tellus* **20**, 371.

Hänel, G., 1969: *Ann. Meteor. N.F.* **4**, 138.

Hänel, G., 1970: *Beitr. z. Phys. d. Atmos.* **43**, 119.

Hänel, G., 1972a: *Ph.D. Thesis*, Meteor. Inst., Univ. of Mainz, Mainz.

Hänel, G., 1972b: *Aerosol Science* **3**, 455.

Hänel, G., 1976: *Advances in Geophys.* **19**, 73.

Hänel, G., and Thudium, J., 1977: *Pure and Appl. Geophys.*, **115**, 799.

Happel, J., and Brenner, H., 1965: *Low Reynolds Number Hydrodynamics*, Prentice-Hall, Englewood Cliffs, N.J.

Harimaya, T., 1975: *J. Meteor. Soc., Japan* **52**, 384.

Harimaya, T., 1976: *J. Meteor. Soc., Japan* **54**, 42.

Harker, A. B., 1975: *J. Geophys. Res.* **80**, 3399.

Harries, J. E., 1976: *Rev. Geophys. and Space Phys.* **14**, 565.

Harrison, H., and Larson, T., 1974: *J. Geophys. Res.* **79**, 3095.

Harrison, L. P., 1965a: *Humidity and Moisture*, Vol. 3, A. Wexler, Ed., p. 105, Reinhold Publ. Co., New York.

Harrison, L. P., 1965b: *Humidity and Moisture*, Vol. 3, A. Wexler, Ed., p. 3, Reinhold Publ. Co., New York.

Hartley, E. M., and Matteson, P. J., 1975: *Ind. Eng. Chem. Fund.* **14**, 67.

Head, R. B., 1961a: *Bull. Obs. de Puy de Dôme*, No. 1, 47.

Head, R. B., 1961b: *Nature*, **191**, 1058.

Head, R. B., 1962a: *J. Phys. Chem. Solids* **23**, 1371.

Head, R. B., 1962b: *Nature* **196**, 736.

Heath, D. F., and Linnett, J., 1948: *Trans. Farad. Soc.* **44**, 556.

Hedlin, C. P., and Trofimenkoff, F. N., 1965: *Humidity and Moisture*, Vol. 3, A. Wexler, Ed., p. 519, Reinhold Publ. Co., New York.

Heffernan, K. J., and Bracewell, R. N., 1959: *J. Meteor.* **16**, 337.

Heist, R. H., and Reiss, H., 1973: *J. Chem. Phys.* **59**, 665.

Herne, H., 1960: *Int. J. Air Poll.* **3**, 26.

Hertz, P., 1909: *Math. Ann.* **67**, 387.

Hess, S. L., 1959: *Introduction to Theoretical Meteorology*, H. Holt and Co., New York.

Heverly, J. R., 1949: *Trans. Am. Geophys. Union* **30**, 205.

Heymsfield, A., 1972: *J. Atmos. Sci.* **29**, 1348.

Heymsfield, A., 1973: *Ph.D. Thesis*, Dept. Geophys. Sci., University of Chicago, Chicago, Ill.

Heymsfield, A., 1975a, b, c: *J. Atmos. Sci.* **32**, 799, 809, 820.

Heymsfield, A., and Knollenberg, R. E., 1972: *J. Atmos. Sci.* **21**, 1358.

Hickman, K. C., 1954: *Ind. Engr. Chem.* **46**, 1442.

Hidy, G. M., 1965: *J. Colloid. Sci.* **20**, 123.

Hidy, G. M., Ed., 1972: *Aerosols and Atmospheric Chemistry*, Academic Press, New York.

Hidy, G. M., 1973: *Chemistry of the Lower Atmosphere*, S. I. Rasool, Ed., Plenum Press, New York.
Hidy, G. M., and Brock, J. R., 1970: *The Dynamics of Aerocolloidal Systems*, Internat. Reviews in Aerosol Phys. and Chem., Vol. 1, Pergamon Press, New York.
Hidy, G. M., and Brock, J. R., 1971: *Topics in Current Aerosol Research*, Internat. Rev. in Aerosol Phys. and Chem., Vol. 2, Pergamon Press, New York.
Hidy, G. M., and Brock, J. R., 1972: *Topics in Current Aerosol Research*, Internat. Rev. in Aerosol Phys. and Chem., Vol. 3, Pergamon Press, New York.
Higuchi, K., 1958: *Acta Metallurgica* **6**, 363.
Higuchi, K., 1961: *J. Meteor. Soc., Japan* **39**, 237.
Higuchi, K., 1962a, b, c: *J. Meteor. Soc., Japan*, **40**, 65, 73, 266.
Higuchi, K., and Wushiki, H., 1970: *J. Meteor. Soc., Japan* **48**, 248.
Higuchi, K., and Yoshida, T., 1967: *Proc. Conf. on Low Temp. Sci.*, Aug. 1966, Vol. 1, p. 79, Instit. of Low Temp. Sci., Hokkaido University, Sapporo, Japan.
Hill, M. A., 1894: *Phil. Trans. Roy. Soc., London* **A185**, 213.
Hill, T. L., 1946: *J. Chem. Phys.* **14**, 263.
Hill, T. L., 1947: *J. Chem. Phys.* **15**, 767.
Hill, T. L., 1949: *J. Chem. Phys.* **17**, 580, 668.
Hill, T. L., 1952: *Adv. in Catalysis*, **4**, Academic Press, New York.
Hillig, W. B., 1959: *Kinetics of High Temperature Processes*, W. D. Kingery, Ed., p. 131, J. Wiley, New York.
Hillig, W. B., and Turnbull, D., 1956: *J. Chem. Phys.* **24**, 1956.
Hindman, E. E., and Johnson, J. B., 1972: *J. Atmos. Sci.* **29**, 1313.
Hinze, J. O., 1959: *Turbulence*, McGraw-Hill Co., New York.
Hirschfelder, J. O., Curtiss, C. F., and Bird, R. B., 1954: *Molecular Theory of Gases and Liquids*, J. Wiley, New York.
Hirth, J. P., and Pound, G. M., 1963: *Condensation and Evaporation*, Pergamon Press, London.
Hobbs, P. V., 1965: *J. Atmos. Sci.* **22**, 296.
Hobbs, P. V., 1969: *J. Atmos. Sci.* **26**, 315.
Hobbs, P. V., 1971: *Quart. J. Roy. Meteor. Soc.* **97**, 263.
Hobbs, P. V., 1974: *Ice Physics*, Oxford University Press, Oxford.
Hobbs, P. V., and Alkezweeny, A. J., 1968: *J. Atmos. Sci.* **25**, 881.
Hobbs, P. V., Bluhm, G. C., and Ohtake, T., 1971b: *Tellus* **23**, 28.
Hobbs, P. V., and Burrows, D. A., 1966: *J. Atmos. Sci.* **23**, 757.
Hobbs, P. V., Chang, S., and Locatelli, J. D., 1974a: *J. Geophys. Res.* **79**, 2199.
Hobbs, P. V., and Farber, J., 1972: *J. de Rech. Atmos.* **6**, 245.
Hobbs, P. V., Fullerton, C. M., and Bluhm, G. C., 1971c: *Nature (Phys. Sci)* **230**, 90.
Hobbs, P. V., and Ketcham, W. M., 1969: *Proc. Conf. on Phys. of Ice*, Munich, Sept. 1968, p. 95, Plenum Press, New York.
Hobbs, P. V., and Locatelli, J. D., 1969: *J. Appl. Meteor.* **8**, 833.
Hobbs, P. V., and Locatelli, J. D., 1970: *J. Atmos. Sci.* **27**, 90.
Hobbs, P. V., and Mason, B. J., 1964: *Phil. Mag.* **9**, 181.
Hobbs, P. V., and Radke, L. F., 1967: *J. Glaciol.* **6**, 879.
Hobbs, P. V., and Radke, L. F., 1969: *Science* **1163**, 279.
Hobbs, P. V., and Radke, L. F., 1970: *J. Atmos. Sci.* **27**, 81.
Hobbs, P. V., Radke, L. F., Fraser, A. B., Locatelli, J. D., Robertson, C. E., Atkinson, D. G., Farber, R. J., Weiss, R. R., and Easter, R. C., 1971a: *Res. Rept. No. 6*, Dept. of Atmospheric Sciences, University of Washington, Seattle, Washington.
Hobbs, P. V., Radke, L. F., Locatelli, J. D., Atkinson, D. G., Robertson, C. E., Weiss, R. R., Turner, F. M., and Brown, R. R., 1972: *Res. Rept. VII*, Dec. 1972, Dept. Atmos. Sci., University of Washington, Seattle, Washington.
Hobbs, P. V., Radke, L. F., Weiss, R. R., Atkinson, D. G., Locatelli, J. D., Biswas, K. R., Turner, F. M., and Robertson, C. E., 1974b: *Res. Rept. No. VIII*, Dec. 1974, Cloud Physics Group, Dept. Atmospheric Sci., University of Washington, Seattle, Washington.
Hobbs, P. V., and Scott, W. D., 1965: *J. Geophys. Res.* **70**, 5025.
Hocking, L. M., 1959: *Quart. J. Roy. Meteor. Soc.* **85**, 44.

Hocking, L. M., 1973: *J. Engr. Math.* **7**, 207.

Hocking, L. M., and Jonas, P. R., 1970: *Quart. J. Roy. Meteor. Soc.* **96**, 722.

Hoffer, T., 1960: *Techn. Note*, No. 22, Cloud Phys. Lab., Dept. of Meteorology, University of Chicago, Chicago, Ill.

Hoffer, T., 1961: *J. Meteor.* **18**, 766.

Hoffer, T., and Warburton, J. H., 1970: *J. Atmos. Sci.* **27**, 1068.

Hofmann, D. J., Rosen, J. M., Pepin, T. J., and Pinnick, R. G., 1975: *J. Atmos. Sci.* **32**, 1446.

Hogan, A. W., 1968: *J. de Rech. Atmos.* **3**, 53.

Hogan, A. W., 1975: *J. Appl. Meteor.* **14**, 550.

Hollomon, J. H., and Turnbull, D., 1953: *Progr. Metal Phys.* **4**, B. Chalmers, Ed., p. 333, Interscience Publ., New York.

Holroyd, E. W., 1964: *J. Appl. Meteor.* **3**, 633.

Holzman, E. G., 1970: *J. Appl. Phys.* **41**, 1460, 4769.

Homan, F., 1936: *Forschg. auf d. Gebiete d. Ing. Wesens* **7**, 1.

Hoppel, W. A., Dinger, J. E., and Ruskin, R. E., 1973: *J. Atmos. Sci.* **30**, 1410.

Hoppel, W. A., and Phillips, B. B., 1971: *J. Atmos. Sci.* **28**, 1258.

Hoppel, W. A., and Phillips, B. B., 1972: *J. Atmos. Sci.* **29**, 1218.

Horguani, V. G., 1965: *Izv. Acad. Sci., SSSR*, Atmos. Ocean, Phys., Ser., **1**. No. 2, 208.

Horne, R. A., Ed., 1972: *Water and Aqueous Solutions*, Wiley-Interscience, New York.

Horvay, G., and Cahn, J. W., 1961: *Acta Metallurgica* **9**, 695.

Hosler, C. L., 1951: *J. Meteor.* **8**, 326.

Hosler, C. L., and Hallgren, R. E., 1960: *Trans. Farad. Soc.*, No. 30, 200.

Hosler, C. L., Jensen, D. C., and Goldshlak, 1957: *J. Meteor.* **14**, 415.

Hosler, C. L., and Spalding, G. R., 1957: *Artificial Stimulation of Rain*, p. 369, Pergamon Press, New York.

Houghton, H. G., 1933: *Physics* **4**, 419.

Houghton, H. G., 1950: *J. Meteor.* **7**, 363.

Houghton, H. G., 1953: *J. Meteor.* **12**, 355.

Houghton, H. G., 1972: *J. de Rech. Atmos.* **6**, 657.

Houghton, H. G., and Radford, W. H., 1938: *MIT Papers in Phys. Oceanogr. and Meteor.* **6**, No. 4.

Howell, W. E., 1949: *J. Meteor.* **6**, 134.

Huan, Mei-Yuan, 1963: *Izv. Akad. Nauk., SSSR*, Ser. Geofiz, No. 2, 362.

Huang, K., 1963: *Statistical Mechanics*, J. Wiley, New York.

Huffman, P. J., 1973a: *J. Appl. Meteor.* **12**, 1080.

Huffman, P. J., 1973b: *Res. Rept. No. AR 108*, Sept. 1973, Dept. of Atmospheric Resources, University of Wyoming, Laramie, Wyoming.

Humphreys, W. J., 1937: *Weather Rambles*, Williams and Wilkins, Baltimore, Md.

Humphreys, W. J., 1942: *Ways of the Weather*, A Cultural Survey of Meteorology, Lancaster, Pa.

Illingworth, A., 1971: *Quart. J. Roy. Meteor. Soc.* **97**, 440.

Illingworth, A., 1973: *J. Geophys. Res.* **78**, 3628.

Illingworth, A., and Latham, J., 1975: *J. Atmos. Sci.* **32**, 2206.

Imai, I., 1950: *Geophys. Mag.* (Tokyo), **21**, 244.

Inmann, R. L., 1969: *J. Appl. Meteor.* **8**, 155.

Iribarne, J. V., 1972: *J. de Rech. Atmos.* **6**, 265.

Iribarne, J. V., and Godson, W. L., 1973: *Atmospheric Thermodynamics*, D. Reidel Publ. Co., Dordrecht, Holland.

Isaka, C., 1972: *Ph.D. Thesis*, Université de Clermont-Ferrand, Dept. Nat. Sciences, Clermont Ferrand, France.

Ishizaka, Y., 1972: *J. Meteor. Soc., Japan* **50**, 362.

Ishizaka, Y., 1973: *J. Meteor. Soc., Japan* **51**, 325.

Isono, K., 1955: *J. Meteor.* **12**, 456.

Isono, K., 1957: *Geofys. Pura e Appl.* **36**, 156.

Isono, K., 1959a: *Japan J. Geophys.* **2**, No. 2.

Isono, K., 1959b: *Nature* **183**, 317.

Isono, K., 1965: *Proc. Conf. Cloud Phys.*, Tokyo, May 1965, p. 150, Meteor. Soc. of Japan, Tokyo.

Isono, K., and Ikebe, Y., 1960: *J. Meteor. Soc., Japan* **38**, 213.

Isono, K., and Ishizaka, Y., 1972: *J. de Rech. Atmos.* **6**, 283.

Isono, K., and Komabayasi, M., 1954: *J. Meteor. Soc., Japan* **32**, 29.

Isono, K., Komabayasi, M., and Ono, A., 1959: *J. Meteor. Soc., Japan* **37**, 211.

Isono, K., Komabayasi, M., and Takahashi, T., 1966: *J. Meteor. Soc., Japan* **44**, 227.

Isono, K., Komabayasi, M., Takeda, T., Tanaka, T., and Iwai, K., 1970: *NWAP Rept.*, 70-R Water Resources Lab., Nagoya University, Nagoya, Japan.

Isono, K, and Tanaka, T., 1966: *J. Meteor. Soc., Japan* **44**, 255.

Isono, K., Tanaka, T., and Iwai, K., 1966: *J. de Rech. Atm.* **3**, 341.

Israel, H., 1970: *Atmospheric Electricity*, Vol. 1, Akad. Verlags-Gesellsch., Leipzig.

Israel, H., 1973: *Atmospheric Electricity*, Vol. 2, Akad. Verlags-Gesellsch., Leipzig.

Israel, H., and Kasemir, H. W., 1952: *Arch. Meteor. Geophys. Biokl.* **A5**, 71.

Itagaki, K., 1967: *J. Coll. Interface Sci.* **25**, 218.

Itoo, K., 1957: *Tokyo Papers in Meteor. and Geophys.* **8**, 220.

Iwai, K., 1973: *J. Meteor. Soc., Japan* **51**, 458.

Jaccard, C., 1963: *Phys. d. Kondens. Materie* **2**, 143.

Jaccard, C., 1967: *Proc. Conf. on Phys. of Snow and Ice*, Aug. 1966, Sapporo, Vol. 1, p. 173, University of Hokkaido, Sapporo, Japan.

Jaccard, C., 1971: *Water and Aqueous Solutions*, R. A. Horne, Ed., p. 25, Wiley-Interscience, New York.

Jaccard, C., and Levi, L., 1961: *Z. Angew. Math. und Phys.* **12**, 70.

Jacobi, W., 1955a: *Z. f. Naturforschung* **10a**, 322.

Jacobi, W., 1955b: *J. Meteor.* **12**, 408.

Jacobson, B., 1953: *Nature* **172**, 666.

Jaenicke, R., and Junge C. E., 1967: *Beitr. z. Phys. d. Atmos.* **40**, 129.

Jaenicke, R., Junge C. E., and Kanter, H. J., 1971: *'Meteor.' Forschg. Ergebn.*, Ser. B, No. 7, p. 1.

Jamieson, D. T., 1965: *3rd Symposium on Thermophysical Properties*, Purdue Univ., p. 230, Amer. Soc. Mech. Engrs.

Jaw Jeou-Jang, 1966: *Tellus* **18**, 722.

Jayaratne, O. W., and Mason, B. J., 1964: *Proc. Roy. Soc.* **A280**, 545.

Jayaweera, K. O. L. F., 1971: *J. Atmos. Sci.* **28**, 728.

Jayaweera, K. O. L. F., 1972a: *Proc. Cloud Phys. Conf.*, Aug. 1972, p. 48, Roy. Meteor. Soc., London.

Jayaweera, K. O. L. F., 1972b: *J. Atmos. Sci.* **29**, 596.

Jayaweera, K. O. L. F., and Cottis, R. E., 1969: *Quart. J. Roy. Meteor. Soc.* **95**, 703.

Jayaweera, K. O. L. F., and Mason, B. J., 1965: *J. Fluid Mech.* **22**, 709.

Jayaweera, K. O. L. F., and Mason, B. J., 1966: *Quart. J. Roy. Meteor. Soc.* **92**, 151.

Jayaweera, K. O. L. F., and Ohtake, T., 1974: *J. Atmos. Sci.* **31**, 280.

Jayaweera, K. O. L. F., and Ryan, B. F., 1972: *Quart. J. Roy. Meteor. Soc.* **98**, 193.

Jeffrey, G. B., 1912: *Proc. Roy. Soc.* **A87**, 109.

Jeffreys, H., 1918: *Phil. Mag.* **35**, 270.

Jenson, V. G., 1959: *Proc. Roy. Soc., London* **A249**, 346.

Jhon, M. S., Grosh, J., Ree, T., and Eyring, H., 1966: *J. Chem. Phys.* **44**, 1465.

Jindal, B. K., and Tiller, W. A., 1972: *J. Coll. Interface Sci.* **39**, 339.

Jiusto, J. E., 1966: *J. de Rech. Atmos.* **2**, 245.

Jiusto, J. E., 1967: *Tellus* **19**, 359.

Jiusto, J. E., 1968: *Proc. Conference on Weather Modification*, Albany, N.Y., April 1968, p. 287, Amer. Meteor. Soc., Boston, Mass.

Jiusto, J. E., 1971: *J. de Rech. Atmos.* **5**, 69.

Jiusto, J. E., and Bosworth, G. E., 1971: *J. Appl. Meteor.* **10**, 1352.

Jiusto, J. E., and Kocmond, W. C., 1968: *J. de Rech. Atmos.* **3**, 101.

Jiusto, J. E., and Weickmann, H. K., 1973: *Bull. Am. Meteor. Soc.* **54**, 1148.

Joe, P. I., 1975: *M.S. Thesis*, Dept. Physics, University of Toronto, Toronto.

Johnson, D. A., and Hallett, J., 1968: *Quart. J. Roy. Meteor. Soc.* **94**, 468.

Jonas, P. R., 1972: *Quart. J. Roy. Meteor. Soc.* **98**, 681.

Jonas, P. R., and Barlett, J. T., 1972: *J. Comput. Phys.* **9**, 290.

Jonas, P. R., and Goldsmith, P., 1972: *J. Fluid Mech.* **52**, 593.

Jonas, P. R., and Mason, B. J., 1974: *Quart. J. Roy. Meteor. Soc.* **100**, 286.

Jones, D. M., 1959: *J. Meteor.* **16**, 504.

Jones, G., and Ray, W. A., 1937: *J. Am. Chem. Soc.* **59**, 187.

Joss, J., and Aufdermauer, A. N., 1970: *Internat. J. Heat and Mass. Transfer* **13**, 213.

Joss, J., Thams, J. C., and Waldvogel, A., 1968: *Proc. Cloud Physics Conference*, Toronto, Aug. 1968, p. 369, University of Toronto Press, Toronto.

Joss, J., and Waldvogel, A., 1969: *J. Atmos. Sci.* **26**, 566.

Junge, C. E., 1936: *Gerlands Beitr. z. Geophys.* **46**, 108.

Junge, C. E., 1950: *Ann. Meteor.* **3**, 128.

Junge, C. E., 1952a: *Ann. Meteor.* **5**, Beiheft.

Junge, C. E., 1952b: *Ber. Deutsch. Wett. D. U.S. Zone*, Nr. 35, 261.

Junge, C. E., 1952c: *Arch. Meteor. Geophys. Biokl.* **5**, 44.

Junge, C. E., 1953: *Tellus* **5**, 1.

Junge, C. E., 1954: *J. Meteor.* **11**, 323.

Junge, C. E., 1955: *J. Meteor.* **12**, 13.

Junge, C. E., 1956: *Tellus* **8**, 127.

Junge, C. E., 1957a: *Artificial Stimulation of Rain*, p. 24, Pergamon Press, New York.

Junge, C. E., 1957b: *Artificial Stimulation of Rain*, p. 3, Pergamon Press, New York.

Junge, C. E., 1961: *J. Meteor.* **18**, 501.

Junge, C. E., 1962: *Tellus* **14**, 364.

Junge, C. E., 1963a: *Air Chemistry and Radioactivity*, Academic Press, New York.

Junge, C. E., 1963b: *J. de Rech. Atmos.* **1**, 185.

Junge, C. E., 1969a: *Proc. Conference on Condensation and Ice Nuclei*, Prague and Vienna, Sept. 1969, p. 31, Czechoslovak Akad. of Sci., Prague.

Junge, C. E., 1969b: *J. Atmos. Sci.* **26**, 603.

Junge, C. E., 1971: *Yearbook of the Max Planck Society*, 150.

Junge, C. E., 1972a: *Quart. J. Roy. Meteor. Soc.* **98**, 711.

Junge, C. E., 1972b: *J. Geophys. Res.* **77**, 5183.

Junge, C. E., 1972c: *Koll. Z. and Z. Polymere* **250**, 628.

Junge, C. E., 1974: *Tellus* **26**, 477.

Junge, C. E., and Abel, N., 1965: *Final Tech. Rept.*, Univ. Mainz, No. DA91-591 EVC-3484.

Junge, C. E., Chagnon, C. W., and Manson, J. E., 1961a: *J. Meteor.* **18**, 81.

Junge, C. E., Chagnon, C. W., and Manson, J. E., 1961b: *Science* **133**, 1478.

Junge, C. E., and Jaenicke, R., 1971: *J. Aerosol Sci.* **2**, 305.

Junge, C. E., and Manson, J. E., 1961: *J. Geophys. Res.* **66**, 2163.

Junge, C. E., and McLaren, E., 1971: *J. Atmos. Sci.* **28**, 382.

Junge, C. E., Robinson, E., and Ludwig, F. L., 1969: *J. Appl. Meteor.* **8**, 340.

Junge, C. E., and Ryan, T. G., 1958: *Quart. J. Roy. Meteor. Soc.* **84**, 46.

Junge, C. E., and Werbey, R. T., 1958: *J. Meteor.* **15**, 417.

Kachurin, L. G., Zaiceva, N. A., and Lomanooa, S. I., 1956: *Izv. Akad. Nauk., SSSR*, Ser. Geog. **7**, 857.

Kajikawa, M., 1971: *J. Meteor. Soc., Japan* **49**, 367.

Kajikawa, M., 1972: *J. Meteor. Soc. Japan* **50**, 577.

Kajikawa, M., 1973: *J. Meteor. Soc., Japan* **51**, 263.

Kajikawa, M., 1974: *J. Meteor. Soc., Japan* **52**, 328.

Kajikawa, M., 1975a: *J. Meteor. Soc., Japan* **53**, 267.

Kajikawa, M., 1975b: *J. Meteor. Soc., Japan* **53**, 476.

Kamra, A. K., 1970: *J. Atmos. Sci.* **27**, 1182.

Käselau, K. H., 1975: *Staub-Reinhalt Luft* **35**, 101.

Kasemir, H. W., and Holitza, F. D., 1972: *EOS Trans.*, AGU **53**, 1001.

Kassner, J. L., Carstens, J. C., and Allen, L. B., 1968a: *J. Atmos. Sci.* **25**, 919.

Kassner, J. L., Carstens, J. C., and Vietti, M. A., 1968b: *J. de Rech. Atmos.* **3**, 45.

Kassner, J. L., Plummer, P. L., Hale, B. N., and Biermann, A. H., 1971: *Proc. Weather Mod. Conf.*, Canberra, Australia, Sept. 1971, p. 51, Amer. Meteor. Soc., Boston, Mass.

Katz, J. L., and Blander, M., 1972: *J. Stat. Phys.* **4**, 55.

Katz, J. L., and Ostermier, B. J., 1967: *J. Chem. Phys.* **47**, 478.

Katz, U., 1960: *Z. Angew. Math. Phys.* **11**, 237.

Katz, U., 1961: *Z. Angew. Math. Phys.* **12**, 76.

Katz, U., 1962: *Z. Angew. Math. Phys.* **13**, 333.

Katzoff, S., 1934: *J. Chem. Phys.* **2**, 1841.

Kauptinskii, A. F., and Razanin, I. I., 1955: *Zhur. Fiz. Khim.* **29**, 222.

Kavanau, J. L., 1964: *Water and Solute-Water Interactions*, Holden-Day, Inc., San Francisco, Calif.

Kawaguti, M., 1953: *J. Phys. Soc., Japan* **8**, 747.

Kawaguti, M., and Jain, P., 1966: *J. Phys. Soc., Japan* **21**, 2055.

Keefe, D., Noland, P. J., and Rich, T. A., 1959: *Proc. Roy. Irish Acad.* **A60**, 27.

Keith, C. H., and Aarons, A. B., 1954: *J. Meteor.* **11**, 173.

Kell, G. S., 1972a: *Water*, Vol. I, F. Franks, Ed., p. 363, Plenum Press, New York.

Kell, G. S., 1972b: *Water and Aqueous Solutions*, R. A. Horne, Ed., p. 331, Wiley-Interscience, New York.

Kell, G. S., McLauring, G. E., and Whalley, E., 1968, *J. Chem. Phys.* **48**, 3805.

Kennard, E. H., 1938: *Kinetic Theory of Gases*, McGraw-Hill, New York.

Kerker, M., and Hampl. B., 1974: *J. Atmos. Sci.* **31**, 1368.

Kern, C. W., and Karplus, M., 1972: *Water*, F. Franks, Ed., p. 21, Plenum Press, New York.

Kessler, E., 1969: *Meteor. Monograph* **10**, No. 32, Am. Meteor. Soc., Boston, Mass.

Kestin, J., and Whitelaw, J. H., 1965: *Humidity and Moisture*, A. Wexler, Ed., p. 301, Reinhold Pub. Co., New York.

Ketcham, W. M., and Hobbs, P. V., 1968: *Phil. Mag.* **18**, 659.

Ketcham, W. M., and Hobbs, P. V., 1969: *Phil. Mag.* **19**, 1161.

Ketseridis, G., Hahn, J., Jaenicke, R., and Junge, C. E., 1976: *Atmos. Environment* **10**, 603.

Kientzler, C. F., Aarons, A. B., Blanchard, D. C., and Woodcock, A. H., 1954: *Tellus* **6**, 1.

Kikuchi, K., 1968: *J. Meteor. Soc., Japan* **46**, 128.

Kikuchi, K., 1970: *J. Meteor. Soc., Japan* **48**, 243.

Kikuchi, K., 1971: *J. Meteor. Soc., Japan* **49**, 20.

Kikuchi, K., 1972a: *J. Meteor. Soc., Japan* **50**, 142.

Kikuchi, K., 1972b: *J. Meteor. Soc., Japan* **50**, 131.

Kikuchi, K., 1973: *J. Meteor. Soc., Japan* **51**, 337.

Kikuchi, K., 1975: *J. Meteor. Soc., Japan* **53**, 322.

Kikuchi, K., and Ishimoto, K., 1974: *J. Fac. Sci.*, Hokkaido University, Ser. 7, **4**, 69.

Kikuchi, K., and Yanai, K. 1971: *Antarctic Record*, No. 41, 34, Polar Res. Center, Tokyo.

Kikuchi, R., 1969: *J. Stat. Phys.* **2**, 351.

Kikuchi, R., 1971: *J. Stat. Phys.* **3**, 331.

Kingery, W. D., 1960a: *J. Glaciol.* **3**, 577.

Kingery, W. D., 1960b: *J. Appl. Phys.* **31**, 833.

Kinzer, G. D., and Cobb, W. E., 1958: *J. Meteor.* **15**, 138.

Kinzer, G. D., and Gunn, R., 1951: *J. Meteor.* **8**, 71.

Kirkwood, J. G., and Buff, F. P., 1949: *J. Chem. Phys.* **17**, 338.

Kirkwood, J. G., and Oppenheim, I., 1961: *Chemical Thermodynamics*, McGraw-Hill, New York.

Kiryukhin, B. V., and Pevzner, S. I., 1956: *Trudy Glavnoi Geofiz. Obs.* **57**, 101.

Kiryukhin, B. V., and Plaude, N. O., 1965: *Studies of Clouds, Precip. and Thunderstorm Electricity*, N. I. Vulfson and L. M. Levin, Eds., p. 29, Amer. Meteor. Soc., Boston, Mass.

Kistenmacher, H., Lie, G. C., Popkie, H., and Clementi, E., 1974a: *IBM Research Report*, RJ 1334, Jan. 1974. IBM Res. Lab., San Jose, Calif.

Kistenmacher, H., Popkie, H., Clementi, E., and Watts, R. O., 1974b: *IBM Research Report*, RJ 1335, Jan. 1974. IBM Res. Lab., San Jose, Calif.

Klazura, G. E., 1971: *J. Appl. Meteor.* **10**, 739.

Klett, J. D., 1971a: *J. Atmos. Sci.* **28**, 646.

Klett, J. D., 1971b: *J. Atmos. Sci.* **28**, 78.

Klett, J. D., 1972: *J. Geophys. Res.* **77**, 3187.

Klett, J. D., 1975: *J. Atmos. Sci.* **32**, 380.

Klett, J. D., 1976: *J. Atmos. Sci.* **33**, 870.

Klett, J. D., 1977: *J. Atmos. Sci.* **34**, 551.

Klett, J. D., and Davis, M. H., 1973: *J. Atmos. Sci.* **30**, 107.
Kline, D. W., 1963: *Monthly Weather Review* **91**, 681.
Kline, S. J., 1965: *Similitude and Approximation Theory*, McGraw-Hill, New York.
Knelman, F., Dombrowski, N., and Newitt, D. M., 1954: *Nature* **173**, 261.
Knight, C. A., and Knight, N. C., 1968a: *J. Atmos. Sci.* **25**, 445.
Knight, C. A., and Knight, N. C., 1968b: *J. Atmos. Sci.* **25**, 458.
Knight, C. A., and Knight, N. C., 1970a: *J. Atmos. Sci.* **27**, 667.
Knight, C. A., and Knight, N. C., 1970b: *J. Atmos. Sci.* **27**, 659.
Knight, C. A., and Knight, N. C., 1970c: *J. Atmos. Sci.* **27**, 672.
Knight, C. A., and Knight, N. C., 1973a: *J. Atmos. Sci.* **30**, 118.
Knight, C. A., and Knight, N. C., 1973b: *J. Atmos. Sci.* **30**, 1665.
Knight, C. A., and Knight, N. C., 1974: *J. Atmos. Sci.* **31**, 1174.
Knight, C. A., and Knight, N. C., 1976: *Proc. Cloud Phys. Conf.*, Boulder, Colo., Aug. 1976, p. 222, Amer. Meteor. Soc., Boston, Mass.
Knutson, E. O., Sood, S. K., and Stockham, J. D., 1976: *Atmospheric Environment* **10**, 395.
Kobayashi, T., 1957: *J. Meteor. Soc. Japan*, 75th Anniversary Volume, p. 38.
Kobayashi, T., 1958: *J. Meteor. Soc. Japan* **36**, 193.
Kobayashi, T., 1961: *Phil. Mag.* **6**, 1363.
Kobayashi, T., 1965a: *J. Meteor. Soc., Japan* **43**, 359.
Kobayashi, T., 1965b: *Contrib. of Inst. Low Temp. Sci.*, Ser. A, No. 20, Hokkaido University, Sapporo, Japan.
Kobayashi, T., 1967: *Proc. Conf. on Low Temp. Science*, Sapporo, Aug. 1966, Vol. 1, p. 95, Institute of Low Temp. Sci., Hokkaido University, Sapporo, Japan.
Kocmond, W. C., 1965: *Res. Dept. RM-1788-p9*, p. 36, Cornell Aeronaut. Lab., Buffalo, N.Y.
Kocmond, W. C., and Mack, E. J., 1972: *J. Appl. Meteor.* **11**, 141.
Koenig, F. O., 1950: *J. Chem. Phys.* **18**, 449.
Koenig, L. R., 1963: *J. Atmos. Sci.* **20**, 29.
Koenig, L. R., 1965a: *J. Atmos. Sci.* **22**, 448.
Koenig, L. R., 1965b: *Proc. Cloud Phys. Conf.*, Tokyo, Japan, May 1965, p. 242, Meteor. Soc. of Japan, Tokyo.
Koenig, L. R., 1966: *J. Atmos. Sci.* **23**, 726.
Koenig, L. R., 1971: *J. Atmos. Sci.* **28**, 226.
Koenig, L. R., 1972: *Monthly Weather Review* **100**, 417.
Köhler, H., 1921a: *Meteor. Z.* **38**, 168.
Köhler, H., 1921b: *Geofysiske Publ.* **2**, No. 1.
Köhler, H., 1922: *Geofysiske Publ.* **2**, No. 6.
Köhler, H., 1927: *Geofysiske Publ.* **5**, No. 1.
Köhler, H., 1936: *Trans. Farad. Soc.* **32**, 1152.
Kojima, K., Ono, T., and Yamaii, K., 1952: *Teion-Kagaku*, Hokkaido University **4**, p. 223.
Kolmogorov, A. N., 1941: *Doklady Acad. Sci., SSSR* **32**, 16.
Kolmogorov, A. N., 1949: *Doklady Akad. Nauk. SSSR* **66**, 825.
Komabayasi, M., Gonda, T., and Isono, K., 1964: *J. Meteor. Soc., Japan* **32**, 330.
Komabayasi, M., and Ikebe, Y., 1961: *J. Meteor. Soc., Japan*, Ser. 11, **39**, 82.
Kornfeld, P. Shafrir, V., and Davis, M. H., 1968: *Proc. Cloud Phys. Conf.*, Toronto, Aug. 1968, p. 107, University of Toronto Press, Toronto.
Koros, R. B., Dechers, J. M., Andes, R. P., and Boudert, M., 1966: *Chem. Engr. Sci.* **21**, 941.
Kortüm, G., 1972: *Einführung in die Chemische Thermodynamik*, 6th ed., Vandenhoeck and Rupprecht, Göttingen.
Kotler, G. R., and Tarshis, L. A., 1968: *J. Crystal Growth* **4**, 603.
Koutsky, J. A., Walton, A. G., and Baer, 1965: *Surface Science* **3**, 165.
Kovetz, A., and Olund, B., 1969: *J. Atmos. Sci.* **26**, 1060.
Kramers, H. A., and Kistenmacher, J., 1943: *Physica* **10**, 699.
Kramers, H. A., and Stemerding, S., 1951: *Appl. Sci. Res.* **A3**, 73.
Krasnogorskaya, N. V., 1965: *Izv. Acad. Sci., U.S.S.R.*, Atmos. and Oceanic Phys. **1**, 200.
Krasnogorskaya, N. V., 1969: *Planetary Electrodynamics*, Vol. 1, S. C. Coroniti and J. Hughes, Eds., p. 427, Gordon and Breach, New York.

Krasnogorskaya, N. V., and Neizvestnyy, A. I., 1973: *Izv. Atmos. Ocean Phys.* **9**, 220.

Krastanow, L., 1940: *Meteor. Z.* **58**, 37.

Krastanow, L., 1943: *Meteor. Z.* **60**, 15.

Krestov, G. A., 1964: *Zh. Struct. Khim.* **5**, 909.

Kubelka, P., 1932: *Z. Elektrochemie* **38**, 611.

Kubelka, P., and Prokscha, R., 1944: *Koll. Z.* **109**, 79.

Kuczynski, 1949: *J. Meteor.* **1**, 169.

Kuettner, J., 1950: *J. Meteor.* **7**, 322.

Kühme, H. W., 1968: *Ber. Inst. Meteor. u. Geophys.*, Nr. 15, Oct. 1968, University of Frankfurt, Frankfurt.

Kuhns, I., 1968: *J. Atmos. Sci.* **25**, 878.

Kuhns, I. E., and Mason, B. J., 1968: *Proc. Roy. Soc., London* **A302**, 437.

Kumai, M., 1951: *J. Meteor.* **8**, 151.

Kumai, M., 1957: *Geofys. Pura e Appl.* **36**, 169.

Kumai, M., 1961: *J. Meteor.* **18**, 139.

Kumai, M., 1965: *CRREL Rept.*, No. 150.

Kumai, M., 1966a: *J. Meteor. Soc., Japan* **44**, 185.

Kumai, M., 1966b: *J. Geophys. Res.* **71**, 3397.

Kumai, M., 1968: *J. Glaciol.* **7**, 95.

Kumai, M., 1969a: *CRREL Rept.*, No. 235.

Kumai, M., 1969b: *CRREL Rept.*, No. 245.

Kumai, M., 1976: *J. Atmos. Sci.* **33**, 833.

Kumai, M., and Francis, K. E., 1962a: *CRREL Rept.*, No. 100.

Kumai, M., and Francis, M., 1962b: *J. Atmos. Sci.* **19**, 434.

Kumai, M., and Itagaki, K., 1953: *J. Fac. Sci.*, Ser. 2, **4**, **235**, Hokkaido University.

Kumai, M., and Itagaki, K., 1954: *J. Meteor. Soc., Japan* **32**, 1.

Kuroiwa, D., 1951: *J. Meteor.* **8**, 157.

Kuroiwa, D., 1953: *Hokkaido University Rept.*, 349.

Kuroiwa, D., 1961: *Tellus* **13**, 252.

Kuroiwa, D., 1962: *CRREL Rept.*, No. 86.

Kvajuc, G., and Brajovic, V., 1971: *J. Crystal Growth* **11**, 73.

Kvlividze, V. I., Kiselev, V. F., and Ushakova, L. A., 1970: *Dok. Phys. Chem. U.S.S.R.* **191**, 307.

Lagford, J., 1961: *Bull. Obs. du Puy de Dome*, No. 2, 87.

Laktinov, A. G., 1972: *Izv. Atmos. Ocean Phys.* **8**, 443.

Lamb, H., 1881: *Proc. Math. Soc., London* **13**, 51.

Lamb, H., 1911: *Phil. Mag.* **21**, 112.

Lamb, H., 1932: *Hydrodynamics*, Dover Publications, New York.

Lamb, D., and Hobbs, P. V., 1911: *J. Atmos. Sci.* **28**, 1506.

Lamb, D., and Scott, W. D., 1974: *J. Atmos. Sci.* **31**, 570.

La Mer, V. K., and Gruen, R., 1952: *Trans. Farad. Soc.* **48**, 410.

Landau, L., and Lifshitz, E. M., 1958: *Statistical Physics*, Pergamon Press, New York.

Landau, L., and Lifshitz, E. M., 1959: *Fluid Mechanics*, Pergamon Press, New York.

Landry, C. R., and Hardy, K. R., 1970: *Proc. 14th Conference on Radar Meterol.* Tucson, Amer. Meteor. Soc., Boston, Mass.

Landsberg, H., 1938: *Ergebnisse d. Kosm. Physik* **3**, 155.

Lane, W. R., 1951: *Ind. Eng. Chem.* **43**, 1312.

Langham, E. J., and Mason, B. J., 1958: *Proc. Roy. Soc.* **A247**, 493.

Langer, G., 1968: *Proc. Conf. on Weather Mod.*, Albany, p. 220, Amer. Meteor. Soc., Boston, Mass.

Langer, G., and Rodgers, J., 1975: *J. Appl. Meteor.* **14**, 560.

Langer, G., and Rosinski, J., 1967: *J. Appl. Meteor.* **6**, 114.

Langer, G., Rosinki, J., and Bernsen, S., 1963: *J. Atmos. Sci.* **20**, 557.

Langham, E. J., and Mason, B. J., 1958: *Proc. Roy. Soc.* **A247**, 493.

Langleben, M. P., 1954: *Quart. J. Roy. Meteor. Soc.* **80**, 174.

Langmuir, I., 1918: *J. Amer. Chem. Soc.* **40**, 1361.

Langmuir, I., 1944: *Research Rept.*, RL-223, General Electric Labs., Schenectady, N.Y.

Langmuir, I., 1948: *J. Meteor.* **5**, 175.

Langmuir, I., and Blodgett, K. B., 1946: *Techn. Rept.*, No. 5418.40, U.S. Army and Air Force Command, Wright Field, Dayton, Ohio.

Lapidus, A., and Shafrir, U., 1972: *J. Atmos. Sci.* **29**, 1308.

La Placa, S., and Post, B., 1960: *Acta Cryst.* **13**, 503.

Laplace P., 1806: *Traité de Mécanique Céleste* **4**, Suppl. to book 10. Courcier, Paris.

Latham, J., 1969: *Quart. J. Roy. Meteor. Soc.* **95**, 349.

Latham, J., Freisen, R. B., and Mystrom, R. E., 1966: *Quart. J. Roy. Meteor. Soc.* **92**, 407.

Latham, J., and Mason, B. J., 1961a: *Proc. Roy. Soc., London* **A260**, 523.

Latham, J., and Mason, B. J., 1961b: *Proc. Roy. Soc., London* **A260**, 537.

Latham, J., and Mason, B. J., 1962: *Proc. Roy. Soc., London* **A266**, 387.

Latham, J., and Myers, V., 1970: *J. Geophys. Res.* **75**, 515.

Latham, J., Mystrom, R. E., and Sartor, J. D., 1965: *Nature* **206**, 1344.

Latham, J., and Roxburgh, I. W., 1966: *Proc. Roy. Soc., London* **A295**, 84.

Latham, J., and Saunders, C. P. R., 1970: *Quart. J. Roy. Meteor. Soc.* **96**, 257.

Laws, J. O., 1941: *Trans. Amer. Geophys. Union* **22**, 709.

Laws, J. O., and Parsons, D. A., 1943: *Trans. Am. Geophys. Union* **24**, 452.

Lazarev, P. P., 1912: *Annal. d. Physik* **37**, 233.

Lazrus, A. L., and Gandrud, B. W., 1974: *J. Geophys. Res.* **79**, 3424.

Lazrus, A. L., Gandrud, B. W., and Cadle, R. D., 1971: *J. Geophys. Res.* **75**, 8083.

Le Clair, B. P., 1970: *Ph.D. Thesis*, Dept. Chem. Eng., McMaster University, Hamilton, Canada.

Le Clair, B. P., Hamielec, A. E., and Pruppacher, H. R., 1970: *J. Atmos. Sci.* **27**, 308.

Le Clair, B. P., Hamielec, A. E., Pruppacher, H. R., and Hall, W. D., 1972: *J. Atmos. Sci.* **29**, 728.

Lee, C. W., 1972: *J. Meteor. Soc., Japan* **50**, 171.

Lee, C. W., and Magono, C., 1967: *J. Meteor. Soc., Japan* **45**, 343.

Lee, I. Y., and Pruppacher, H. R., 1977: *Pure and Appl. Geophys.*, **115**, 523.

Lee, J. K., Abraham, F. F., and Pound, G. M., 1973: *Surface Sci.* **34**, 745.

LeFebre, V., 1967: *J. Coll. Interface Sci.* **25**, 263.

Leighton, P. A., 1961: *Photochemistry of Air Pollution*, Academic Press, New York.

Leighton, H. G., and Rogers, R. R., 1974: *J. Atmos. Sci.* **31**, 271.

Lenard, P., 1904: *Meteor. Z.* **21**, 249.

Letey, J., Osborn, J., and Pelishek, R. E., 1962: *Soil Science* **93**, 149.

Levi, L., 1970: *J. Glaciol.* **9**, 109.

Levi, L., Achaval, E. M., and Lubart, L., 1970: *J. Cryst. Growth* **22**, 303.

Levi, L., Achaval, E. M., and Aufdermauer, A. N., 1970: *J. Atmos. Sci.* **27**, 512.

Levi, L., and Aufdermauer, A. N., 1970: *J. Atmos. Sci.* **27**, 443.

Levich, V., 1954a: *Dokl. Akad. Nauk, SSSR* **99**, 809.

Levich, V., 1954b: *Dokl. Akad. Nauk, SSSR* **99**, 1041.

Levich, V., 1962: *Physico-Chemical Hydrodynamics*, Prentice-Hall Inc., Englewood Cliffs, N.J.

Levin, L. M., 1954: *Dokl. Akad. Nauk, SSSR* **94**, 1045.

Levin, L. M., and Sedunov, Y. S., 1966: *J. de Rech. Atmos.* **2**, 425.

Levin, Z., 1976: *J. Atmos. Sci.* **33**, 1756.

Levin, Z., and Machnes, B., 1976: *Proc. Cloud Phys. Conf.*, Boulder, July 1976, p. 196, Amer. Meteor. Soc., Boston, Mass.

Levin, Z., Neiburger, M., and Rodriquez, L., 1973: *J. Atmos. Sci.* **30**, 944.

Levine, J., 1950: *NACA*, Techn. Note, 2234.

Levine, N. E., 1971: *J. Geophys. Res.* **76**, 5097.

Levkov, L., 1971: *J. de Rech. Atm.* **5**, 133.

Lewis, G. N., and Randall, M., 1961: *Thermodynamics*, revised by K. S. Spitzer and L. Brewer, 2nd ed., McGraw-Hill, New York.

Lin, C. L., and Lee, S. C., 1975: *J. Atmos. Sci.* **32**, 1412.

Lin, Sin-Shong, 1973: *Rev. Sci. Instruments* **44**, 516.

Lindblad, N. R., 1964: *J. Coll. Sci.* **19**, 729.

Lindblad, N. R., and Semonin, R. G., 1963: *J. Geophys. Res.* **68**, 1051.

Lindenmeyer, C. S., 1959: *Ph.D. Thesis*, Harvard University, Cambridge, Mass.

Lindenmeyer, C. S., Orrok, G. T., and Jackson, K. A., 1957: *J. Chem. Phys.* **27**, 822.

List, R., 1958a: *Z. Angew. Math. Phys.* **9a**, 180.

List, R., 1958b: *Z. Angew. Math. Phys.* **9a**, 217.

List, R., 1959: *Z. Angew. Math. Phys.* **10**, 143.

List, R., 1960a: *Geophys. Monograph*, No. 5, p. 317, Amer. Geophys. Union, Washington, D.C.

List, R., 1960b; *Z. Angew. Math. Phys.* **11**, 273.

List, R., 1961: *Bull. Am. Meteor. Soc.* **42**, 452.

List, R., 1965: *Proc. Cloud Phys. Conf.*, Tokyo, May 1965, p. 481, Meteor. Soc. of Japan, Tokyo.

List, R., and Agnew, T. A., 1973: *J. Atmos. Sci.* **30**, 1158.

List, R., Cantin, J., and Farland, M. G., 1970: *J. Atmos. Sci.* **27**, 1080.

List, R., Charlton, R. B., and Butterlus, P. I., 1968: *J. Atmos. Sci.* **25**, 1061.

List, R., and Dussault, J. G., 1967: *J. Atmos. Sci.* **24**, 522.

List, R., and Gillespie, J. R., 1976: *J. Atmos. Sci.* **33**, 2007.

List, R., and Hand, M. J., 1971: *Physics of Fluids* **14**, 1648.

List, R., Kry, P. R., deQuervein, M. R., Liu, P. Y., Stagg, P. W., McTaggart-Cowan, J. D., Lozowski, E. P., v. Niederhauser, J. Stewart, R. E., Freire, E., and Lesins, G., 1976: *Proc. Cloud Phys. Conf.*, Boulder, Aug. 1976, p. 264, Amer. Meteor. Soc., Boston, Mass.

List, R., MacNeil, C. F., and McTaggart-Cowan, J. D., 1970: *J. Geophys. Res.* **75**, 7573.

List, R., Murray, W. A., and Dyck, C., 1972: *J. Atmos. Sci.* **29**, 916.

List, R., Rentsch, U. W., Byram, A. C., and Lozowski, E. P., 1973: *J. Atmos. Sci.* **30**, 653.

List, R., Rentsch, U. W., and Schüepp, P. H., and McBuoney, M. W., 1969: *Proc. Int. Conference on Severe and Local Storms*, Chicago, April 1969, p. 267, Amer. Meteor. Soc., Boston, Mass.

List, R., and Schemenauer, R. S., 1971: *J. Atmos. Sci.* **28**, 110.

List, R., Schüepp, P. H., and Method, R. G., 1965: *J. Atmos. Sci.* **22**, 710.

List, R., and Whelpdale, D. M., 1969: *J. Atmos. Sci.* **26**, 305.

Litvinov, I. V., 1956: *Iz. Akad. Nauk., SSSR*, Ser. Geofiz., No. 7, 853.

Liu, B. Y., and Whitby, K., 1968: *J. Coll. Interface Sci.* **26**, 161.

Locatelli, J. D., and Hobbs, P. V., 1974: *J. Geophys. Res.* **79**, 2185.

Lodge, J. P., 1955: *J. Meteor.* **12**, 493.

Lodge, J. P., and Pate, J. B., 1966: *Science* **153**, 408.

Loeb, L. B., Kip, A. F., and Einarsson, A. W., 1938: *J. Chem. Phys.* **6**, 264.

Long, A. B., 1971: *J. Atmos. Sci.* **28**, 210.

Long, A. B., 1974: *J. Atmos. Sci.* **31**, 1040.

Long, A. B., and Manton, M. J., 1974: *J. Atmos. Sci.* **31**, 1053.

Lonsdale, K., 1958: *Proc. Roy. Soc., London* **A247**, 424.

Lothe, J., and Pound, G. M., 1962: *J. Chem. Phys.* **36**, 2080.

Lothe, J., and Pound, G. M., 1966: *J. Chem. Phys.* **45**, 630.

Lothe, J., and Pound G. M., 1968: *J. Chem. Phys.* **48**, 1849.

Lothe, J., and Pound, G. M., 1969: *Nucleation*, A. C. Zettlemoyer, Ed., p. 109. Marcel Dekker, New York.

Low, R. D., 1969a: *J. Atmos. Sci.* **26**, 608.

Low, R. D., 1969b: *J. de Rech. Atmos.* **4**, 65.

Low, R. D., 1969c: *J. Atmos. Sci.* **26**, 1345.

Low, R. D., 1969d: *Research Rept.*, AD 691700, ECOM-5249, May 1969, Atmos. Sci. Lab., White Sands Missile Range, N.M.

Low, R. D., 1971: *Research and Develop. Techn. Rept.*, ECOM-5358, Jan 1971, U.S. Army Electronics Command, Fort Monmouth, N.J.

Lowe, P. R., and Ficke, J. M., 1974: *Techn. Paper*, No. 4-74, Environmental Prediction Res. Facility, Naval Post Grad. School, Monterey, Calif.

Ludlam, F. H., 1950: *Quart. J. Roy. Meteor. Soc.* **76**, 52.

Ludlam, F. H., 1951: *Quart. J. Roy. Meteor. Soc.* **77**, 402.

Ludlam, F. H., 1958: *Nubila* **1**, 12.

Ludlam, F. H., and Macklin, W. C., 1959: *Nubila* **2**, No. 1, 38.

Ludwig, J. H., Morgan, G. B., and McMullen, T. B., 1971: *Man's Impact on Climate*, Matthews *et al.*, Eds., p. 321, MIT Press, Cambridge, Mass.

Macklin, W. C., 1960: *Nubila* **3**, 30.

Macklin, W. C., 1961: *Quart. J. Roy. Meteorol. Soc.* **87**, 413.

Macklin, W. C., 1962: *Quart. J. Roy. Meteor. Soc.* **88**, 30.

Macklin, W. C., 1963: *Quart. J. Roy. Meteor. Soc.* **89**, 360.

Macklin, W. C., 1964a: *J. Atmos. Sci.* **21**, 227.

Macklin, W. C., 1964b: *Quart. J. Roy. Meteor. Soc.* **90**, 84.

Macklin, W. C., and Bailey, I. H., 1966: *Quart. J. Roy. Meteor. Soc.* **92**, 297.

Macklin, W. C., and Bailey, I. H., 1968: *Quart. J. Roy. Meteor. Soc.* **94**, 393.

Macklin, W. C., Carras, J. N., and Rye, P. J., 1976: *Quart. J. Roy. Meteor. Soc.* **102**, 25.

Macklin, W. C., and Ludlam, F. H., 1961: *Quart. J. Roy. Meteor. Soc.* **87**, 72.

Macklin, W. C., Merlivat, L., and Stevenson, C. M., 1970: *Quart. J. Roy. Meteor. Soc.* **96**, 472.

Macklin, W. C., and Payne, G. S., 1967: *Quart. J. Roy. Meteor. Soc.* **93**, 195.

Macklin, W. C., and Payne, G. S., 1968: *Quart. J. Roy. Meteor. Soc.* **94**, 167.

Macklin, W. C., and Payne, G. S., 1969: *Quart. J. Roy. Meteor. Soc.* **95**, 724.

Macklin, W. C., and Ryan, B. F., 1965: *J. Atmos. Sci.* **22**, 452.

Macklin, W. C., and Ryan, B. F., 1966: *Phil. Mag.* **14**, 847.

Macklin, W. C., and Ryan, B. F., 1968: *Phil. Mag.* **17**, 83.

Macklin, W. C., and Ryan, B. F., 1969: *J. Chem. Phys.* **50**, 551.

Macklin, W. C., and Rye, P. J., 1974: *J. Atmos. Sci.* **31**, 849.

Macklin, W. C., Strauch, E., and Ludlam, F. H., 1960: *Nubila* **3**, 12.

Macky, W. A., 1931: *Proc. Roy. Soc., London* **A133**, 565.

Madonna, L. A., Sciulli, C. M., Canjar, L. N., and Pound, G. M., 1961: *Proc. Phys. Soc., London* **A78**, 1218.

Maeno, N., 1973: *Proc. Conf. Phys. Chem. of Ice*, Ottawa, Aug. 1972, p. 140, Roy. Soc., Canada, Ottawa.

Magarvey, R. H., and Taylor, B. H., 1956: *J. Appl. Phys.* **27**, 1129.

Magono, C., 1951: *J. Meteor.* **8**, 199.

Magono, C., 1953: *Science Rept.*, Yokohama Natl. University, Sect. I, No. 2, 18, Yokohama, Japan.

Magono, C., 1954a: *J. Meteor.* **11**, 77.

Magono, C., 1954b: *Science Rept.*, Yokohama Natl. University, Sect. I, No. 3, 33, Yokohama, Japan.

Magono, C., 1960: *Geophys. Monograph*, No. 5, p. 142, Am. Geophys. Union, Washington, D.C.

Magono, C., and Aburukawa, H., 1968: *J. Fac. Sci.*, Ser. 7, Geophys. **3**, 85, Hokkaido University, Sapporo, Japan.

Magono, C., Endoh, T., Harimaya, T., and Kubota, S., 1974: *J. Meteor. Soc., Japan* **52**, 407.

Magono, C., and Kikuchi, K., 1961: *J. Meteor. Soc., Japan* **39**, 258.

Magono, C., Kikuchi, K., and Kimura, T., 1959: *J. Fac. Sci.*, Hokkaido University, Ser. 7 (Geophys.) **1**, 195.

Magono, C., Kikuchi, K., and Kimura, T., 1960: *J. Fac. Sci.*, Hokkaido University, Ser. 7 (Geophys.) **1**, 267.

Magono, C., Kikuchi, K., and Kimura, T., 1962: *J. Fac. Sci.*, Hokkaido University, Ser. 7 (Geophys.) **1**, 373.

Magono, C., Kikuchi, K., and Kimura, T., 1963: *J. Fac. Sci.*, Hokkaido University, Ser. 7 (Geophys.) **2**, 49.

Magono, C., Kikuchi, K., and Kimura, T., 1965: *J. Fac. Sci.*, Hokkaido University, Ser. 7 (Geophys.) **2**, 123.

Magono, C., Kikuchi, K., and Kimura, T., 1966: *J. Fac. Sci.*, Hokkaido University, Ser. 7 (Geophys.) **2**, 287.

Magono, C., Kikuchi, K., and Yamami, N., 1971: *J. Meteor. Soc., Japan* **49**, 179.

Magono, C., and Lee, C. W., 1966: *J. Fac. Sci.*, Hokkaido University, Ser. 7, **2**, No. 4.

Magono, C., and Lee, C. W., 1973: *J. Meteor. Soc. Japan* **51**, 176.

Magono, C., and Nakamura, T., 1959: *J. Meteor. Soc., Japan* **37**, 124.

Magono, C., and Nakamura, T., 1965: *J. Meteor. Soc., Japan* **43**, 139.

Magono, C., and Orikasa, K., 1966: *J. Meteor. Soc., Japan* **44**, 260, 280.

Magono, C., and Suzuki, S., 1967: *J. Fac. Sci.*, Hokkaido University, Ser. 7, **3**, 27.

Mahata, P. C., and Alofs, D. J., 1975: *J. Atmos. Sci.* **32**, 116.

Mahrous, M. A., 1954: *Quart. J. Roy. Meteor. Soc.* **80**, 99.

Maidique, M. A., Hippel, A. V., and Westphal, W. B., 1971: *J. Chem. Phys.* **54**, 150.

Malan, D. J., 1952: *Ann. Geophys.* **8**, 385.

Malan, D. J., 1963: *Physics of Lightning*, The English University Press, Ltd., London.

Mallow, J. V., 1975: *J. Atmos. Sci.* **32**, 440.

Malmberg, C. G., and Mariott, A. A., 1956: *J. Res. Nat. Bur. Stand.* **56**, 1.

Manley, R., and Mason, S., 1952: *J. Colloid Sci.* **7**, 354.

Manton, M. J., 1974: *Geophys. Fluid Dyn.* **6**, 83.

Marreno, T. R., and Mason, E. A., 1972: *J. Phys. Chem., Ref. Data* **1**, 1.

Marsaglia, G., MacLaren, M. D., and Bray, T. A., 1964: *Comm. ACM* **7**, 4.

Marshall, J. S., and Palmer, W. M., 1948: *J. Meteor.* **5**, 165.

Martell, E. A., 1970: *Advances in Chem. Series*, No. 93, 138.

Martell, E. A., 1971: *Man's Impact on the Climate*, Mathews *et al.*, Eds., p. 421, MIT Press, Cambridge, Mass.

Martell, E. A., and Moore, H. E., 1974: *J. de Rech. Atmos.* **8**, 903.

Martin, R. T., 1959: *Clays and Clay Minerals*, Monograph No. 2, Earth Sci. Series, p. 259, Pergamon Press, New York.

Martynov, G. A., and Bakanov, S. P., 1961: *Surface Forces*, B. Derjaguin, Ed., p. 182.

Maruyama, M., and Kitagawa, T., 1967: *J. Meteor. Soc., Japan* **65**, 126.

Marwitz, J. D., 1972: *Proc. Conf. Weather Modific*, Rapid City, June 1972, p. 245, Amer. Meteor. Soc., Boston, Mass.

Marwitz, J. D., 1974: *J. Appl. Meteor.* **13**, 450.

Masliyah, J. H., 1972: *Phys. Fluids* **15**, 1144.

Masliyah, J. H., and Epstein, N., 1970: *J. Fluid. Mech.* **44**, 493.

Masliyah, J. H., and Epstein, N., 1971: *Proc. Internat. Symp. on Two Phase Systems*, Haifa, Israel.

Mason, B. J., 1952a: *Quart. J. Roy. Meteor. Soc.* **22**, 78.

Mason, B. J., 1952b: *Quart. J. Roy. Meteor. Soc.* **78**, 377.

Mason, B. J., 1953: *Quart. J. Roy. Meteor. Soc.* **79**, 104.

Mason, B. J., 1954a: *J. Meteor.* **11**, 514.

Mason, B. J., 1954b: *Nature* **174**, 470.

Mason, B. J., 1956: *Quart. J. Roy. Meteor. Soc.* **82**, 209.

Mason, B. J., 1957a: *Physics of Clouds*, 1st ed., Oxford University Press, London.

Mason, B. J., 1957b: *Geofysica Pura e Appl.* **36**, 148.

Mason, B. J., 1959: *Tellus* **2**, 216.

Mason, B. J., 1960a: *Quart. J. Roy. Meteor. Soc.* **86**, 552.

Mason, B. J., 1960b: *J. Meteor.* **17**, 459.

Mason, B. J., 1971: *The Physics of Clouds*, 2nd ed., Oxford University Press, London.

Mason, B. J., 1972: *Proc. Roy. Soc., London* **A327**, 433.

Mason, B. J., 1975: *Quart. J. Roy. Meteor. Soc.* **101**, 178.

Mason, B. J., 1976a: *Quart. J. Roy. Meteor. Soc.* **102**, 219.

Mason, B. J., 1976b: *Quart. J. Roy. Meteor. Soc.* **102**, 933.

Mason, B. J., and Andrews, J. B., 1960: *Quart. J. Roy. Meteor. Soc.* **86**, 346.

Mason, B. J., Bryant, G. W., and van den Heuvel, H. P., 1963: *Phil. Mag.* **8**, 505.

Mason, B. J., and Chien, C. W., 1962: *Quart. J. Roy. Meteor. Soc.* **88**, 136.

Mason, B. J., and Jonas, P. R., 1974: *J. Meteor. Soc.* **100**, 23.

Mason, B. J., and Maybank, J., 1958: *Quart. J. Roy. Meteor. Soc.* **84**, 235.

Mason, B. J., and Maybank, J., 1960: *Quart. J. Roy. Meteor. Soc.* **86**, 176.

Mason, B. J., and Shaw, D., 1955: *J. Meteor.* **12**, 93.

Mason, B. J., and van den Heuvel, H. P., 1959: *Proc. Phys. Soc., London* **A74**, 744.

Matsubara, K., 1973: *J. Meteor. Soc., Japan* **51**, 54.

Mathews, J. B., 1967: *J. Geophys. Res.* **72**, 3007.

Mathews, J. B., and Mason, B. J., 1964: *Quart. J. Roy. Meteor. Soc.* **90**, 275.

Matthews, W. H., Kellogg, W. W., and Robinson, G. D., Eds., 1971: *Man's Impact on the Climate*, MIT Press, Cambridge, Mass.

Maude, A. D., 1961: *Brit. J. Appl. Phys.* **12**, 293.

Maxwell, J. C., 1890: *The Scientific Papers of James Clerk Maxwell*, Vol. 2, p. 636, Dover Publ., New York.

Maxworthy, T., 1965: *J. Fluid Mech.* **23**, 369.

May, D. R., and Kolthoff, I. M., 1948: *J. Phys. Chem.* **52**, 836.

Maybank, J., and Mason, B. J., 1959: *Proc. Phys. Soc., London* **74**, 11.

Mazin, I. P., 1965: *Moscow Cent. Aerosl. Obs. Trudy, vyp.* **64**, 57.

Mazzega, E., Pennino, U., Loria, A., and Mantovani, S., 1976: *J. Chem. Phys.* **64**, 1028.

McCall, D. W., and Douglass, D. C., 1965: *J. Phys. Chem.* **69**, 2001.

McCarthy, J., 1975: *J. Atmos. Sci.* **32**, 997.

McCormac, B. M., Ed., 1971: *Introduction to the Scientific Study of Atmospheric Pollution*, D. Reidel Publ. Co., Dordrecht, Holland.

McDonald, J. E., 1953a: *J. Meteor.* **10**, 68.

McDonald, J. E., 1953b: *J. Meteor.* **10**, 416.

McDonald, J. E., 1954: *J. Meteor.* **11**, 478.

McDonald, J. E., 1958: *Advances in Geophys.* **5**, 223.

McDonald, J. E., 1962: *Am. J. Phys.* **30**, 870.

McDonald, J. E., 1963a: *Am. J. Phys.* **31**, 31.

McDonald, J. E., 1963b: *J. Geophys. Res.* **68**, 4993.

McDonald, J. E., 1963c: *Z. Angew. Math. Phys.* **14**, 610.

McDonald, J. E., 1964: *J. Atmos. Sci.* **21**, 109.

McKay, H. A. C., 1971: *Atmos. Environ.* **5**, 7.

McKinnon, D., Morewood, H. W., 1970: *J. Atmos. Sci.* **37**, 483.

McLeod, J. B., 1962: *Quart. J. Math.*, Oxford, **13**, 119.

McLeod, J. B., 1964: *Proc. London Math. Soc.* **14**, 445.

McTaggart-Cowan, J. D., and List, R., 1975: *J. Atmos. Sci.* **32**, 1401.

Mecke, K., 1933: *Z. Phys.* **81**, 313.

Meissner, F., 1920: *Z. Anorg. Allg. Chemie* **110**, 169.

Melzak, Z. A., 1953: *Quart. Appl. Math.* **11**, 231.

Meszaros, A., 1965: *J. de Rech. Atmos.* **2**, 53.

Meszaros, A., and Vissy, K., 1974: *J. Aerosol Sci.* **5**, 101.

Meszaros, E., 1968: *Tellus* **20**, 443.

Meszaros, E., 1969: *Ann. d. Meteor.*, Nr. 4, 132.

Meyer, J., and Pfaff, W., 1935: *Z. Anorg. Allg. Chem.* **224**, 305.

Meyer, L., 1958: *Z. Phys. Chem.* **16**, 333.

Michaeli, G., 1971: *Nature, Phys. Sci.* **230**, 117.

Michaels, A. S., Brian, P. L., and Sperry, P. R., 1966: *J. Appl. Phys.* **37**, 4649.

Middleton, J. R., 1971: *Research Report*, No. AR 101, Dec. 1971, Dept. of Atmos. Resources, University of Wyoming, Laramie, Wyoming.

Middleton, W. E., 1965: *A History of the Theories of Rain and Other Forms of Precipitation*, Oldbourne, London.

Millero, F. J., 1972: *Water and Aqueous Solutions*, R. A. Horne, Ed., p. 519, Wiley-Interscience, New York.

Mills, A. F., and Seban, R. A., 1967: *Int. J. of Heat, Mass Transfer* **10**, 1815.

Mills, R., 1971: *Ber. Bunsengesellschaft* **75**, 195.

Mills, R., 1973: *J. Phys. Chem.* **77**, 685.

Miyake, Y., 1948: *Geophys. Mag. (Japan)* **16**, 64.

Miyamoto, S., and Letey, J., 1971: *Proc. Soil Science Soc. of America* **35**, 856.

Mockros, L. F., Quon, J. E., and Hjelmfelt, A. T., 1967: *J. Coll. Sci.* **23**, 90.

Mohnen, V. A., 1970: *J. Geophys. Res.* **75**, 7117.

Mohnen, V. A., 1971: *Pure and Appl. Geophys.* **84**, 141.

Mohnen, V. A., and Lodge, J. P., 1969: *Proc. Conf. on Cloud Nuclei*, Prague and Vienna, Sept. 1969, p. 69, Czechoslovak Akad. Sci., Prague.

Möller, W., 1938: *Phys. Z.* **39**, 57.

Monahan, E. C., 1968: *J. Geophys. Res.* **73**, 1127.

Monahan, E. C., 1969: *J. Atmos. Sci.* **26**, 1026.

Monahan, E. C., 1971: *J. Phys. Oceanogr.* **1**, 139.

Monin, A. S., and Yaglom, A. M., 1971: *Statistical Fluid Mechanics*, MIT Press, Cambridge, Mass.

Montgomery, D. N., 1971: *J. Atmos. Sci.* **28**, 291.

Moore, C. B., 1975: *J. Atmos. Sci.* **32**, 608.

Moore, C. B., 1976a: *Quart. J. Roy. Meteor. Soc.* **102**, 225.

Moore, C. B., 1976b: *Quart. J. Roy. Meteor. Soc.* **102**, 935.

Moore, D. J., 1952: *Quart. J. Roy. Meteor. Soc.* **78**, 596.

Moore, D. J., and Mason, B. J., 1954: *Quart. J. Roy. Meteor. Soc.* **80**, 583.

Moore, H. E., Poet, S. E., and Martell, E. A., 1973: *J. Geophys. Res.* **78**, 7065.

Morariu, V. V., and Mills, R., 1972: *Z. Phys. Chem.* **79**, 1.

Mordy, W. A., 1959: *Tellus* **11**, 16.

Morgan, G. M., and Langer, G., 1973: *Quart. J. Roy. Meteor. Soc.* **99**, 387.

Morgan, J., and Warren, B. E., 1938: *J. Chem. Phys.* **6**, 666.

Morris, T. R., 1957: *J. Meteor.* **14**, 281.

Morris, T. R., and Braham, 1968: *Proc. Weather Modific. Conference*, Albany, New York, April 1968, p. 306, Am. Meteor. Soc., Boston, Mass.

Morse, P., 1969: *Thermal Physics*, W. A. Benjamin, Inc., New York.

Morse, P., and Feshbach, H., 1953: *Methods of Theoretical Physics*, Vol. 1, McGraw-Hill, New York.

Mossop, S. C., 1955: *Proc. Phys. Soc.* **B68**, 193.

Mossop, S. C., 1956: *Proc. Phys. Soc.* **B69**, 165.

Mossop, S. C., 1963a: *Nature* **199**, 325.

Mossop, S. C., 1963b: *Z. Angew. Math. Phys.* **14**, 456.

Mossop, S. C., 1965: *Geochim. et Cosmochim. Acta* **29**, 201.

Mossop, S. C., 1970: *Bull. Am. Meteor. Soc.* **51**, 474.

Mossop, S. C., 1971a: *Weather* **26**, 222.

Mossop, S. C., 1971b: *Proc. Conf. Weather Mod.*, Canberra, Sept. 1971, p. 1, Amer. Meteor. Soc., Boston, Mass.

Mossop, S. C., 1972: *J. de Rech. Atmos.* **6**, 377.

Mossop, S. C., 1976: *Quart. J. Roy. Meteor. Soc.* **102**, 45.

Mossop, S. C., Cottis, R. E., and Bartelett, B. M., 1972: *Quart. J. Roy. Meteor. Soc.* **98**, 105.

Mossop, S. C., and Hallett, J., 1974: *Science* **186**, 632.

Mossop, S. C., and Jayaweera, K. O. L. F., 1969: *J. Appl. Meteor.* **8**, 241.

Mossop, S. C., and Kidder, R. E., 1964: *Nubila* **6**, 74.

Mossop, S. C., and Ono, A., 1969: *J. Atmos. Sci.* **26**, 130.

Mossop, S. C., Ono, A., and Heffernan, K. J., 1967: *J. de Rech. Atmos.* **3**, 45.

Mossop, S. C., Ono, A., and Wishart, E. R., 1970: *Quart. J. Roy. Meteor. Soc.* **96**, 487.

Mossop, S. C., Ruskin, R. E., and Heffernan, K. J., 1968: *J. Atmos. Sci.* **25**, 889.

Mrose, H., 1966: *Tellus* **18**, 266.

Müller-Hildebrand, D., 1954: *Tellus* **6**, 367.

Müller-Hildebrand, D., 1955: *Ark. Geofis.* **2**, 395.

Munzah, T., 1960: *Tellus* **12**, 282.

Murray, F. W., 1967: *J. Appl. Meteor.* **6**, 203.

Murray, F. W., 1970: *Research Report*, D-19910-ARPA, Feb. 23, 1970, The RAND Corp. Santa Monica, California.

Murray, J. D., 1967: *J. Math. and Phys.* **46**, 1.

Murray, W. A., and List, R., 1972: *J. Glaciol.* **11**, 415.

Murty, A. S. R., and Murty, B. V. R., 1972: *Tellus* **24**, 581.

Murty, Bh. V., 1969: *J. Meteor. Soc., Japan* **47**, 219.

Musil, D. J., 1970: *J. Atmos. Sci.* **27**, 474.

Nabavian, K., and Bromley, L. A., 1963: *Chem. Engr. Sci.* **18**, 651.

Nagle, J. F., 1966: *J. Math. Phys.* **7**, 1484.

Nakamura, T., 1964: *J. Meteor. Soc., Japan* **42**, 65.

Nakaya, U., 1954: *Snow Crystals*, Harvard University Press, Harvard, Mass.

Nakaya, U., and Higuchi, K., 1960: *Geophys. Monograph* No. 5, p. 118, Amer. Geophys. Union, Washington, D.C.

Nakaya, U., and Matsumoto, A., 1954: *J. Coll. Sci.* **9**, 41.

Nakaya, U., and Tereda, T., 1934: *J. Fac. Sci.*, Hokkaido Univ., Ser. II, **1**, 191.

Nakaya, U., and Tereda, T., 1935: *J. Fac. Sci.*, Hokkaido Univ., Ser. II, **1**, 191.

Namiot, A. Yu., 1961: *Zh. Struct. Khim.* **2**, 476.

Narten, A. H., Danford, M. D., and Levy, H. A., 1967: *Disc. Farad. Soc.* **43**, 97.

Narten, A. H., and Levy, H. A., 1969: *Science* **165**, 447.

Narten, A. H., and Levy, H. A., 1970: *Science* **167**, 1521.

Narten, A. H., and Levy, H. A., 1971: *J. Chem. Phys.* **55**, 2263.

Narten, A. H., and Levy, H. A., 1972: *Water*, F. Franks, Ed., p. 311, Plenum Press, New York.

Narusawa, M., and Springer, G. S., 1975: *J. Coll. Interface Sci.* **50**, 392.

Naruse, H., and Maryama, H., 1971: *Papers in Meteor. and Geophys.* **22**, 1.

Neiburger, M., 1949: *J. Meteor.* **6**, 98.

Neiburger, M., 1967: *Monthly Weather Rev.* **95**, 917.

Neiburger, M., and Chien, C. W., 1960: *Meteor. Monographs*, No. 5, p. 191. Amer. Geophys. Union, Washington, D.C.

Neiburger, M., Levin, Z., and Rodriguez, L., 1972: *J. de Rech. Atmos.* **6**, 391.

Neiburger, M., and Wurtele, M. G., 1949: *Chem. Rev.* **44**, 21.

Nelson, A. R., and Gokhale, N. R., 1972: *J. Geophys. Res.* **77**, 2724.

Nelson, A. R., and Gokhale, N. R., 1973: *J. Geophys. Res.* **78**, 1472.

Nemethy, G. and Scheraga, H. A., 1962a: *J. Chem. Phys.* **36**, 3382.

Nemethy, G., and Scheraga, H. A., 1962b: *J. Phys. Chem.* **66**, 1773.

Nemethy, G., and Scheraga, H. A., 1964: *J. Chem. Phys.* **41**, 680.

Newmann, G. H., Fonselius, S., and Wahlman, 1959: *Intl. J. Air Poll.* **2**, 132.

Nishioka, M., and Sato, H., 1974: *J. Fluid Mech.* **65**, 97.

Nolan, J. J., 1926: *Proc. Roy. Irish Acad.* **37**, 28.

Noll, K. E., and Pilat, M. J., 1971: *Atmos. Environment* **5**, 527.

Oakes, B., 1960: *Intl. J. Air Poll.* **3**, 179.

Oberbeck, A., 1876: *Crelle's J. Math.* **81**, 62.

O'Brien, F. E., 1948: *J. Sci. Instruments* **25**, 73.

Ockendon, J. R., and Evans, G. A., 1972: *J. Aerosol Sci.* **3**, 237.

O'Connell, J. P., Gillespie, M. D., Kostek, W. D., and Prausnitz, J. M., 1969: *J. Phys. Chem.* **73**, 2000.

Ogiwara, S., and Okita, T., 1952: *Tellus* **4**, 233.

Ohta, S., 1951: *Bull. Am. Meteor. Soc.* **82**, 30.

Ohtake, T., 1967: *Proc. Conf. Low Temp. Sci.*, Sapporo, Aug. 1966, Vol. 1, p. 105, Institute of Low Temp. Sci., Hokkaido University, Sapporo, Japan.

Ohtake, T., 1968: *Proc. Cloud Phys. Conf.*, Toronto, Aug. 1968, p. 285, University of Toronto Press, Toronto.

Ohtake, T., 1969: *J. Atmos. Sci.* **26**, 545.

Ohtake, T., 1970a: *J. Atmos. Sci.* **27**, 509.

Ohtake, T., 1970b: *Final Rept.*, Ap-100449, June 1970, Geophys. Institute, University of Alaska, Fairbanks, Alaska.

Ohtake, T., 1970c: *J. Atmos. Sci.* **27**, 804.

Okita, T., 1955: *J. Meteor. Soc. Japan, Ser. II*, **33**, 220.

Okita, T., 1958: *Sci. Rept.*, Tohoku University, 5th Ser., Geofys., **10**, No. 1.

Okita, T., 1962a: *J. Meteor. Soc., Japan* **40**, 39.

Okita, T., 1962b: *J. Meteor. Soc., Japan* **40**, 163.

Okuyama, M., and Zung, J. T., 1967: *J. Chem. Phys.* **46**, 1580.

O'Neil, M. E., 1964: *Ph.D. Thesis*, University of London, London.

O'Neil, M. E., and Majundar, S. R., 1970a: *Z. Angew. Math. und Phys.* **21**, 164.

O'Neil, M. E., and Majundar, S. R., 1970b: *Z. Angew. Math. und Phys.* **21**, 180.

Ono, A., 1969: *J. Atmos. Sci.* **26**, 138.

Ono, A., 1970: *J. Atmos. Sci.* **27**, 649.

Ono, A., 1972: *J. de Rech. Atmos.* **6**, 399.

Ono, S., and Kondo, S., 1960: *Handbuch der Physik*, S. Flügge, Ed., Vol. 10, p. 134, Springer-Verlag, Berlin.

Orem, M. W., and Adamson, A. W., 1969: *J. Coll. Interface Sci.* **31**, 278.

Orr, C., Hurd, F. K., and Hendrix, W. P., 1958: *J. Meteor.* **15**, 240.

Oseen, C. W., 1910: *Ark. Math. Astron. Fys.* **6**, No. 29.

Oseen, C. W., 1915: *Archiv., Math. Phys., Ser. 3* **24**, 108.

Oseen, C. W., 1927: *Hydrodynamik*, Akad. Verlagsgellschaft, Leipzig.

Ostwald, W., 1900: *Z. Phys. Chem.* **34**, 495.

Ostwald, W., 1902: *Lehrbuch der Allgem. Chemie*, Verlag W. Engelmann, Leipzig.

Osipow, L. I., 1962: *Surface Chemistry*, Reinhold Publ. Co., New York.

Owen, P. R., 1964: *J. Fluid Mech.* **20**, 225.

Pai, S-I., 1956: *Viscous Flow Theory, Vol. I: Laminar Flow*, Van Nostrand, Publ. Co., New York.

Palmer, L. S., 1952: *Proc. Phys. Soc., London* **65**, 674.

Paluch, I., 1970: *J. Geophys. Res.* **75**, 1633.

Paluch, I., 1973: *Proc. Roy. Soc., London* **A328**, 395.

Paluch, I., and Sartor, J. D., 1973: *J. Atmos. Sci.* **30**, 1166.

Park, R. W., 1970: *Ph.D. Thesis*, Dept. Chem. Engr., University of Wisconsin, Madison, Wisconsin.

Parker, B. C., 1968: *Nature* **219**, 617.

Parkin, D. W., Phillips, D. R., Sullivan, R. A., and Johnson, L. R., 1970: *J. Geophys. Res.* **73**, 1782.

Parkin, D. W., Phillips, D. R., Sullivan, R. A., 1972: *Quart. J. Roy. Meteor. Soc.* **98**, 798.

Parkinson, W. C., 1952: *J. Geophys. Res.* **57**, 314.

Parungo, F., and Lodge, J. P., 1965: *J. Atmos. Sci.* **22**, 309.

Parungo, F., and Lodge, J. P., 1967a: *J. Atmos. Sci.* **24**, 274.

Parungo, F., and Lodge, J. P., 1967b: *J. Atmos. Sci.* **24**, 439.

Parungo, F., and Weickmann, H. K., 1973: *Beitr. z. Phys. der Atmosphäre* **46**, 289.

Parungo, F., and Wood, J., 1968: *J. Atmos. Sci.* **25**, 154.

Pasternak, I. S., and Gauvin, W. H., 1960: *Canad. J. Chem. Eng.* **38**, 35.

Patterson, M. P. and Spillane, K. T., 1967: *J. Atmos. Sci.* **24**, 50.

Patterson, M. P., and Spillane, K. T., 1969: *Quart. J. Roy. Meteor. Soc.* **95**, 526.

Patterson, H. S., and Cawood, W., 1932: *Proc. Roy. Soc., London* **A136**, 538.

Patton, A. J., and Springer, G. S., 1969: *Rarefied Gas Dynamics*, L. Trilling and H. Y. Wachman, Eds., Vol. 2, p. 1947, Academic Press, New York.

Pauling, L., 1935: *J. Am. Chem. Soc.* **57**, 2680.

Pauling, L., 1959: *Hydrogen Bonding*, D. Hadzi, H. W. Thompson, Eds., Pergamon Press, New York.

Pauling, L., 1960: *The Nature of the Chemical Bond*, 3rd ed., Cornell University Press, Ithaca, New York.

Pauling, L., 1962: *Nature of the Chemical Bond*, Cornell University Press, Ithaca, New York.

Pauthenier, M., and Moreau-Hanot, M., 1931: *C. R. Acad. Sci., Paris* **13**, 1068.

Pawlow, P., 1910: *Z. Phys. Chem.* **74**, 562.

Pearcey, T., and Hill, G. W., 1956: *Quart. J. Roy. Meteor. Soc.* **83**, 77.

Pearcey, T., and McHugh, B., 1955: *Phil. Mag.* **46**, Ser. 7, 783.

Pena, J. A., and Pena, R. G., 1970: *J. Geophys. Res.* **75**, 2831.

Pena, J. A., Pena, R. G., and Hosler, C. L., 1969: *J. Atmos. Sci.* **26**, 309.

Penndorf, R., 1954: *Geophys. Res. Papers*, **25**, 1, Air Force Cambridge Res. Center, Cambridge, Mass.

Peppler, W., 1940: *Forschg. u. Erfahrung. Reichsamt f. Wetterdienst.*, **B.**, No. 1.

Perry, J., 1950: *Chem. Eng. Handbook*, 3rd ed., p. 1017, McGraw-Hill, New York.

Peterson, J. T., and Junge, C. E., 1971: *Man's Impact on the Climate*, Matthews *et al.*, Eds., p. 310, MIT Press, Cambridge, Mass.

Peterson, S. W., and Levy, H. A., 1957: *Acta Cryst.* **10**, 70.

Petrenchuk, O. P., and Drozdova, V. M., 1966: *Tellus* **18**, 280.

Petrenchuk, O. P., and Selezneva, E. S., 1970: *J. Geophys. Res.* **75**, 3629.

Pflaum, J., Martin, J., and Pruppacher, H. R., 1977: *Quart. J. Roy. Meteor. Soc.* **104**, 179.

Pflaum, J., and Pruppacher, H. R., 1976: *Proceedings, Cloud Phys. Conf.*, Boulder, July 1976, p. 113, Amer. Meteor. Soc., Boston, Mass.

Phan-Cong, L., and Dinh-Van, P., 1973: *Tellus* **25**, 63.

Phillips, B. B., 1967: *Monthly Weather Review* **95**, 854.

Phillips, B. B., and Kinzer, G. D., 1958: *J. Meteor.* **15**, 369.

Pich, J., Friedlander, S. K., and Lai, F. S., 1970: *Aerosol Sci.* **1**, 115.

Pick, W. H., 1929: *Quart. J. Roy. Meteor. Soc.* **55**, 305.

Pick, W. H., 1931: *Quart. J. Roy. Meteor. Soc.* **57**, 288.

Picknett, R. G., 1960: *Int. J. Air Pollution* **3**, 160.

Pierce, C., 1960: *J. Phys. Chem.* **64**, 1184.

Pitter, R. L., 1977: *J. Atmos. Sci.* **34**, 684.

Pitter, R. L., and Pruppacher, H. R., 1973: *Quart. J. Roy. Meteor. Soc.* **99**, 540.

Pitter, R. L., and Pruppacher, H. R., 1974: *J. Atmos. Sci.* **31**, 551.

Pitter, R. L., Pruppacher, H. R., and Hamielec, A. E., 1973: *J. Atmos. Sci.* **30**, 125.

Pitter, R. L., Pruppacher, H. R., and Hamielec, A. E., 1974: *J. Atmos. Sci.* **31**, 1058.

Pitzer, K. S., and Polissar, J., 1956: *J. Phys. Chem.* **60**, 1140.

Plumlee, H. R., and Semonin, R. G., 1965: *Tellus* **17**, 356.

Plummer, P. L., and Hale, B. N., 1972: *J. Chem. Phys.* **56**, 4329.

Pluvinage, P., 1946: *Ann. d. Geophys.* **2**, 31.

Pocza, J. F., Barna, A., and Barna, P. B., 1969: *J. Vac. Sci. Techn.* **6**, 242.

Podzimek, J., 1965: *Proc. Cloud Phys. Conference*, Tokyo, May 1965, p. 224, Meteor. Soc. of Japan, Tokyo.

Podzimek, J., 1966: *Studia geophys. et geodet.* **10**, 235.

Podzimek, J., 1968: *Proc. Cloud Phys. Conf.*, Toronto, Aug. 1968, p. 295, University of Toronto Press, Toronto.

Podzimek, J., 1969: *Docent Thesis*, Charles University, Prague.

Podzimek, J., and Cernoch, I., 1961: *Geofysica Pura e Appl.* **50**, 96.

Podzimek, J., Haberl, J. B., and Sedlacek, W. A., 1975: *Proc., 4th Conference on Climatic Impact Assessment Program*, Feb. 1975, U.S. Dept. of Transportation, Cambridge, Mass.

Podzimek, J., Sedlacek, W. A., and Haberl, J. B., 1974: *Rept. of SANDS to U.S. Dept. of Transportation.*

Poet, S. E., Moure, H. E., and Martell, E. A., 1972: *J. Geophys. Res.* **77**, 6515.

Popkie, H., and Kistenmacher, H., and Clementi, E., 1973: *J. Chem. Phys.* **59**, 1325.

Pople, J. A., 1951: *Proc. Roy. Soc., London* **A205**, 163.

Pound, G. M., Madonna, L. A., and Peake, L., 1953: *J. Coll. Sci.* **8**, 187.

Pound, G. M., Madonna, L. A., and Sciulli, C. M., 1955: *Proc. Conf. on Interfacial Phenomena and Nucleation*, H. Reiss, Ed., Aug. 1951, Boston, p. 85, U.S. Dept. of Commerce, Washington, D.C.

Pound, G. M., Simnad, M. T., and Yang, L., 1954: *J. Chem. Phys.* **22**, 1215.

Prandtl, L., 1904: *Verh. 3d Int. Math. Kongr.* Heidelberg, p. 484 (transc. as NACA Tech. Mem. No. 452).

Price, S., and Pales, J., 1963: *Arch. Meteor. Geophys. Biokl., A.* **13**, 398.

Prigogine, I., and Defay, R., 1967: *Chemical Thermodynamics*, 4th Ed., Longmans and Green Co., London.

Pringle, J. E., Orville, H. D., and Stechmann, T. D., 1973: *J. Geophys. Res.* **78**, 4508.

Prodi, F., 1970: *J. Appl. Meteor.* **9**, 903.

Prokhorov, P. S., 1951: *Doklady Akad. Nauk SSSR* **81**, 637.

Prokhorov, P. S., 1954: *Disc. Farad. Soc.* No. 18, 41.

Prospero, J. M., 1968: *Bull. Am. Meteor. Soc.* **49**, 649.

Prospero, J. M., and Bonatti E., 1969: *J. Geophys. Res.* **74**, 3362.

Prospero, J. M., Bonatti, E., Schubert, C., and Carlson, T. N., 1970: *Earth and Planetary Sci. Lett.* **9**, 287.

Prospero, J. M., and Carlson, T. N., 1972: *J. Geophys. Res.* **77**, 5225.

Prosperetti, A., 1976: *Proc. Conf. on Drops and Bubbles*, California Institute of Technology, and J. P. L. Pasadena, Ca., Aug. 1974, Vol. 2, p. 357.

Proudman, I., 1969: *J. Fluid Mech.* **37**, 759.

Proudman, I., and Pearson, J. R. A., 1957: *J. Fluid Mech.* **2**, 237.

Pruger, W., 1940: *Z. Phys.* **115**, 202.

Pruppacher, H. R., 1962: *Ph.D. Thesis*, Dept. Meteor., University of California, Los Angeles, Calif.

Pruppacher, H. R., 1963a, *J. Chem. Phys.* **39**, 1586.

Pruppacher, H. R., 1963b: *Z. Angew Math. und Phys.* **14**, 590.

Pruppacher, H. R., 1963c: *J. Geophys. Res.* **68**, 4463.

Pruppacher, H. R., 1967a: *Pure and Appl. Geophys.* **68**, 186.

Pruppacher, H. R., 1967b: *J. Glaciol.* **6**, 651.

Pruppacher, H. R., 1967c: *Z. f. Naturforschg.* **22a**, 895.

Pruppacher, H. R., 1967d: *J. Coll. Interface Sci.* **25**, 285.

Pruppacher, H. R., 1967e: *J. Chem. Phys.* **47**, 1807.

Pruppacher, H. R., 1972: *J. Chem. Phys.* **56**, 101.

Pruppacher, H. R., 1973a: *Chemistry of the Lower Atmosphere*, S. I. Rasool, Ed., p. 1, Plenum Press, New York.

Pruppacher, H. R., 1973b: *Pure and Appl. Geophys.* **104**, 623.

Pruppacher, H. R., and Beard, K. V., 1970: *Quart. J. Roy. Meteor. Soc.* **96**, 247.

Pruppacher, H. R., Le Clair, B. P., and Hamielec, A. E., 1970: *J. Fluid Mech.* **44**, 781.

Pruppacher, H. R., and Neiburger, M., 1963: *J. Atmos. Sci.* **20**, 376.

Pruppacher, H. R., and Pflaum, J., 1975: *J. Coll. Interface Sci.* **52**, 543.

Pruppacher, H. R., and Pitter, R. L., 1971: *J. Atmos. Sci.* **28**, 86.

Pruppacher, H. R., and Sänger, R., 1955: *Z. Angew Math. Phys.* **6**, 407.

Pruppacher, H. R., and Schlamp, R. J., 1975: *J. Geophys. Res.* **80**, 380.

Pruppacher, H. R., and Steinberger, E. H., 1968: *J. Appl. Phys.* **39**, 4129.

Pruppacher, H. R., Steinberger, E. H., and Wang, T. L., 1968: *J. Geophys. Res.* **73**, 571.

Pshenay-Severin, S. V., 1957: *Izvestyia Geophys. Ser.* No. 8, 1054.

Pshenay-Severin, S. V., 1958: *Izvestyia Geophys. Ser.* No. 10, 1254.

Pueschel, R. F., and Langer, G., 1973: *J. Appl. Meteor.* **12**, 549.

Quon, J. E., and Mockros, L. F., 1965: *Internat. J. Air and Waste Pol.* **9**, 279.

Qureshi, M. M., and Maybank, J., 1966: *Nature* **211**, 508.

Radke, L. F., 1970: *Proc. Cloud Phys. Conf.*, Fort Collins, Col., Aug. 1970, p. 7, Amer. Meteor. Soc., Boston, Mass.

Radke, L. F., and Hobbs, P. V., 1969: *J. Atmos. Sci.* **26**, 281.

Rahman, A., and Stillinger, F. H., 1971: *J. Chem. Phys.* **55**, 3336.

Ranz, W. E., and Marshall, W. R., 1952: *Chem. Eng. Progr.* **48**, 141, 173.

Ranz, W. E., and Wong, J. B., 1952: *Ind. Eng. Chem.* **44**, 1371.

Rao, C. N. R., 1972: '*Water*,' F. Franks, Ed., p. 93, Plenum Press, New York.

Rasmussen, R. A., and Went, F. W., 1965: *Proc. Nat. Acad. Sci.* **53**, 215.

Rasool, S. I., Ed., 1973: *Chemistry of the Lower Atmosphere*, Plenum Press, New York.

Ratcliff, E. H., 1962: *Phil. Mag.* **7**, 1197.

Ratcliffe, J. A., Ed., 1960: *Physics of the Upper Atmosphere*, Academic Press, New York.

Rathje, W., and Stranski, I. N., 1955: *Proc. Conf. Internat. Phenomena and Nucleation*, Vol. 1, H. Reiss, Ed., p. 1, Air Force Cambridge Res. Center, Cambridge, Mass.

Rau, W., 1950: *Z. f. Naturforschung* **5a**, 667.

Rayleight, Lord, 1882: *Phil. Mag.* **14**, 184.

Reid, R. C., and Sherwood, T. K., 1966: *The Properties of Gases and Liquids*, Vol. 2, McGraw-Hill, New York.

Reinhardt, R. L., 1972: *Ph.D. Thesis*, University of Nevada, Reno, Nev.

Reinking, R. F., 1976: *Proc. Cloud Phys. Conf.* July 1976, Boulder, Col., p. 207, Amer. Meteor. Soc., Boston, Mass.

Reinking, R. F., and Lovill, J. E., 1971: *J. Atmos. Sci.* **28**, 812.

Reiquam, H., and Diamond, M., 1959: *CRREL Rept.* No. 52.

Reischel, M. T., 1972: *Research Rept.*, No. AR105, Dept. of Atmos. Resources, University of Wyoming, Laramie, Wyo.

Reischel, M. T., and Vali, G., 1975: *Tellus* **27**, 414.

Reiss, H., 1965: *Methods of Thermodynamics*, Blaisdell Publ. Co., Waltham, Mass.

Reiss, H., 1970: *J. Stat. Phys.* **1**, 83.

Reiss, H., and Katz, J. L., 1967: *J. Chem. Phys.* **46**, 2496.

Reiss, H., Katz, J. L., and Cohen, E. R., 1968: *J. Chem. Phys.* **48**, 5553.

Reiter, R., 1964: *Felder, Ströme und Aerosole*, Dietrich Steinkopff Verlag, Darmstadt.

Reiter, R., 1969: *Planetary Electrodynamics*, S. C. Coroniti and J. Hughes, Eds., Vol. 1, p. 59, Gordon and Breach, New York.

Relf, E. F., and Simmons, L. F., 1924: *Great Britain Aeronaut. Res. Council*, Repts. and Memoranda, No. 917.

Reuck, A. V. S., 1957: *Nature* **179**, 1119.

Rex, R. W., and Goldberg, E. D., 1958: *Tellus* **10**, 153.

Reynolds, O., 1876: *Nature* **15**, 163.

Reynolds, S. E., 1952: *J. Meteor.* **9**, 36.

Reynolds, S. E., Brook, M., and Gourley, M. F., 1957: *J. Meteor. Soc.* **14**, 426.

Richards, C. N., and Dawson, G. A., 1971: *J. Geophys. Res.* **76**, 3445.

Richards, T. W., and Carver, E. K., 1921: *J. Am. Chem. Soc.* **43**, 827.

Riehl, N., Bullemer, B., and Engelhardt, H., Eds., 1969: *Physics of Ice*, Plenum Press, New York.

Rimmer, P. L., 1969: *J. Fluid Mech.* **32**, 1.

Rimon, Y., and Cheng, S. I., 1969: *Phys. Fluids* **12**, 949.

Rimon, Y., and Lugt, H. J., 1969: *Phys. Fluids* **12**, 2465.

Robbins, R. C., Cadle, R. D., and Eckhard, D. L., 1959: *J. Meteor.* **16**, 53.

Roberts, P., and Hallett, J., 1968: *Quart. J. Roy. Meteor. Soc.* **94**, 25.

Robertson, D., 1974: *J. Atmos. Sci.* **31**, 1344.

Robinson, E., and Robbins, R. C., 1968: *Final Rept.*, SRI Project PR-6755, Feb. 1968, Stanford Res. Inst., Menlo Park, Calif.

Robinson, E., and Robbins, R. C., 1969: *Final Rept. Supplement*, SRI Project PR-6755, June 1969, Stanford Res. Inst., Menlo Park, Calif.

Robinson, E., and Robbins, R. C., 1971: *Final Rept.*, SRI Project SCC 8507, Stanford Research Institute, Menlo Park, Calif.

Robinson, R. A., and Stokes, R. H., 1970: *Electrolyte Solutions*, 2nd ed. Butterworths, London.

Rodgers, D. C., 1974a: *Proc. Cloud Phys. Conf.*, Tuscon, Ariz., Oct. 1974, p. 108, Amer. Meteor. Soc., Boston, Mass.

Rodgers, D. C., 1974b: Res. Rept. June 1974, No. AR110, Dept. of Atmos. Resources, University of Wyoming, Laramie, Wyo.

Rogers, C. P., and Squires, P., 1974: *Proc. Cloud Phys. Conf.*, Tucson, Ariz., Oct. 1974, p. 17, Am. Meteor. Soc., Boston, Mass.

Rogers, L. N., 1971: *Bull. Am. Meteor. Soc.* **52**, 994.

Roos, D. S., 1972: *J. Appl. Meteor.* **11**, 1008.

Rosen, J. M., 1974: *Final Rept.*, Contract N0001-70-A-0266-0011, University of Wyoming, Laramie, Wyo.

Rosen, J. M., Hofmann, G. J., and Laby, J., 1975: *J. Atmos. Sci.* **32**, 1457.

Rosenhead, L., Ed., 1963: *Laminar Boundary Layers*, Oxford Press, London.

Rosenkilde, C. E., 1976: *Res. Rept.*, Univ. of Calif., UCRL-52027, Lawrence Livermore Lab, Livermore, Calif.

Roshko, A., 1954: *NACA*, Rept. 1191.

Rosinski, J., 1966: *J. Appl. Meteor.* **5**, 481.

Rosinski, J., 1967a: *J. Appl. Meteor.* **6**, 1066.

Rosinski, J., 1967b: *J. Atmos. Terrestr. Phys.* **29**, 1201.

Rosinski, J., and Kerrington, T. C., 1969: *J. Atmos. Sci.* **26**, 695.

Rosinski, J., Nagamoto, C. T., Langer, G., and Parungo, F. P., 1970: *J. Geophys. Res.* **75**, 2961.

Rosinski, J., and Snow, R. H., 1961: *J. Meteor.* **18**, 736.

Rossknecht, G. F., Elliott, W. P., and Ramsey, F. L., 1973: *J. Appl. Meteor.* **12**, 825.

Rossmann, F., 1950: *Ber. Deutsch Wett. D.*, U.S. Zone, **15**, 1.

Rottner, D., and Vali, G., 1974: *J. Atmos. Sci.* **31**, 560.

Roulleau, M., and Desbois, M., 1972: *J. Atmos. Sci.* **29**, 565.

Roussel, J. C., 1968: *J. de Rech. Atmos.* **3**, 253.

Rowe, P. N., Claxton, K. T., and Lewis, J. B., 1965: *Trans. Inst. Chem. Engrs.* **43**, T 14.

Rowlinson, J. S., 1949: *Trans. Farad. Soc.* **45**, 974.

Rowlinson, J. S., 1951: *Trans. Farad. Soc.* **47**, 120.

Ruckenstein, E., 1964: *Chem. Eng. Sci.* **19**, 131.

Rucklidge, J., 1965: *J. Atmos. Sci.* **22**, 301.

Ruepp, R., and Käss, M., 1969: *Proc. Conf. of Phys. on Ice*, Munich, Sept. 1968, p. 555, Plenum Press, New York.

Ruhnke, L. H., 1970: *J. Appl. Meteor.* **9**, 947.

Ruhnke, L. H., 1972: *Meteor. Res.* **25**, 38.

Ryan, B. F., 1969: *J. Crystal Growth* **5**, 284.

Ryan, B. F., 1972: *J. de Rech. Atmos.* **6**, 673.

Ryan, B. F., 1973: *J. Atmos. Sci.* **30**, 824.

Ryan, B. F., 1974: *J. Atmos. Sci.* **31**, 1942.

Ryan, B. F., and Macklin, W. C., 1968: *J. Crystal Growth* **2**, 337.

Ryan, B. F., and Macklin, W. C., 1969: *J. Coll. Interface.* **31**, 566.

Ryan, B. F., and Scott, W. D., 1969: *J. Atmos. Sci.* **36**, 611.

Ryan, B. F., Wishart, E. R., and Holroyd, E. W., 1974: *J. Atmos. Sci.* **31**, 2136.

Ryan, B. F., Wishart, E. R., and Shaw, D. E., 1976: *J. Atmos. Sci.* **33**, 842.

Ryan, R. T., Blau, H. H., Thuna, P. C., and Cohen, M. L., 1972: *J. Appl. Meteor.* **11**, 149.

Rybczinski, W., 1911: *Bull. Acad. Cracovie, A,* **40**.

Rye, P. J., and Macklin, W. C., 1973: *J. Atmos. Sci.* **30**, 1421.

Rye, P. J., and Macklin, W. C., 1975: *Quart. J. Roy. Meteor. Soc.* **101**, 207.

Saffman, P. G., 1968: *Topics in Nonlinear Physics,* N. J. Zabusky, Ed., Springer-Verlag, New York.

Saffman, P. G., and Turner, J. S., 1956: *J. Fluid Mech.* **1**, 16.

Sal'm, Ya., I., 1971: *Bull. (Izv.) Acad. Sci., SSSR,* Atmos. Oceanic Phys. **7**, 306.

Sambles, J. R., 1971: *Proc. Roy. Soc., London* **A423**, 339.

Sambles, J. R., Skinner, L. M., and Lisgarten, N. D., 1970: *Proc. Roy. Soc., London* **A318**, 507.

Samoilov, O. Ya., 1946: *Zh. Fiz. Khim* **20**, 1411.

Samoilov, O. Ya., 1957: *Zh. Fiz. Khim.* **31**, 537.

Samorjai, G. A., 1972: *Principles of Surface Chemistry,* Prentice-Hall, Inc., Englewood Cliffs, New Jersey.

Sander, A., and Damköhler, G., 1943: *Naturwissenschaften* **3**, 460.

Sano, I., Fujitani, Y., and Maena, Y., 1960: *Memoirs of the Kobe Marine Obs.* **14**, 1.

Sano, T., 1972: *J. Eng. Math.* **6**, 217.

Sartor, J. D., 1960: *J. Geophys. Res.* **65**, 1953.

Sartor, J. D., 1967: *J. Atmos. Sci.* **24**, 601.

Sartor, J. D., and Abbott, C. E., 1975: *J. Appl. Meteor.* **14**, 232.

Sasyo, Y., 1971: *Papers in Meteor. and Geophys.* **22**, 69.

Sasyo, Y., and Tokune, H., 1973: *Papers in Meteor. and Geophys.* **24**, 1.

Saunders, C. P. R., and Wahab, N. M. A., 1975: *J. Meteor. Soc., Japan* **53**, 121.

Savic, P., 1953: *Res. Rept.* No. NRC-MT-22, Nat. Res. Council of Canada, Div. of Mech. Eng., Ottawa.

Sax, R. I., and Goldsmith, P., 1972: *Quart. J. Roy. Meteor. Soc.* **98**, 60.

Saxena, V. K., Burford, J. N., and Kassner, J. L., 1970: *J. Atmos. Sci.* **27**, 73.

SCEP, 1970: *Man's Impact on the Global Environment,* Rept. on the Study of Critical Environmental Problems, MIT Press, Cambridge, Mass.

Schaefer, V. J., 1947: *Trans. Amer. Geophys. Union* **28**, 587.

Schaefer, V. J., 1950: *Occasional Rept.,* No. 20, Project Cirrus, General Electric Res. Lab., Schenectady, N.Y.

Schäfer, K., 1932: *Z. f. Physik* **77**, 198.

Scharrer, L., 1939: *Ann. d. Physik* **35**, 619.

Schedlovsky, J. P., and Paisley, S., 1966: *Tellus* **18**, 499.

Schiff, L., 1968: *Quantum Mechanics,* 3rd ed., McGraw-Hill, New York.

Schlamp, R. J., Grover, S. N., Pruppacher, H. R., and Hamielec, A. E., 1976: *J. Atmos. Sci.* **33**, 1747.

Schlamp, R. J., and Pruppacher, H. R., 1977: *Pure and Appl. Geophys.* **115**, 805.

Schlamp, R. J., Pruppacher, H. R., and Hamielec, A. E., 1975: *J. Atmos. Sci.* **32**, 2330.

Schlichting, H., 1968: *Boundary Layer Theory,* McGraw-Hill, New York.

Schmiedel, J., 1928: *Phys. Z.* **29**, 593.

Schmidt, M., 1972: *Pure and Appl. Geophys.* **97**, 219.

Schmidtkunz, H., 1963: *Ph.D. Thesis,* Institute of Meteor. and Geophys., University of Frankfurt, Frankfurt.

Schmitt, K. H., 1961: *Z. f. Naturforschung* **16a**, 144.

Schmitt, K. H., and Waldmann, L., 1960: *Z. f. Naturforschung* **15a**, 843.

Schneider-Carius, K., 1955: *Wetterkunde und Wetterforschung*, Karl Alber Verlag, Munich.
Schneider-Carius, K., 1962: *Linkes Meteorologishes Taschenbuch*, Vol. 1, 2nd ed., F. Bauer, Ed., Akad. Verlagsgesellschaft, Leipzig.
Schnell, R. C., 1972: *Res. Rept.* AR-102, Dept. of Atmos. Resources, University of Wyoming, Laramie, Wyoming.
Schnell, R. C., 1974: *Res. Rept.*, AR-111, Dept. of Atmos. Resources, University of Wyoming, Laramie, Wyoming.
Schnell, R. C., and Vali, G., 1976: *J. Atmos. Sci.* **33**, 1554.
Schotland, R. M., 1957: *J. Meteor.* **14**, 381.
Schotland, R. M., 1960: *Discuss. Farad.*, *Soc.* No. 30, 72.
Schrage, R. W., 1953: *Interphase Mass Transfer*, Columbia University Press, New York.
Schreiber, B., 1977: *Diploma Thesis*, Institute of Meteor., University of Mainz, Mainz.
Schuman, T. E., 1938: *Quart. J. Roy. Meteor. Soc.* **64**, 3.
Schuman, T. E., 1940: *Quart. J. Roy. Meteor. Soc.* **66**, 195.
Schuepp, P. H., 1971: *J. Appl. Meteor.* **10**, 1018.
Schuepp, P. H., and List, R., 1969: *J. Appl. Meteor.* **8**, 254.
Schütz, L., and Jaenicke, R., 1974: *J. Appl. Meteor.* **13**, 863.
Schütz, K., Junge, C., Beck, R., and Albrecht, B., 1970: *J. Geophys. Res.* **75**, 2230.
Scorer, R., 1972: *Clouds of the World*, Lothian Publ. Co., Melbourne.
Scott, W. D., and Hobbs, P. V., 1967: *J. Atmos. Sci.* **24**, 54.
Scott, W. D., Lamb, D., and Duffy, D., 1969: *J. Atmos. Sci.* **26**, 727.
Scott, W. D., and Levin, Z., 1975a: *J. Atmos. Sci.* **32**, 843.
Scott, W. D., and Levin, Z., 1975b: *J. Atmos. Sci.* **32**, 1814.
Scott, W. T., 1965: *Techn. Rept.*, No. 9, Desert Res. Instit., University of Nevada, Reno, Nev.
Scott, W. T., 1967: *J. Atmos. Sci.* **24**, 221.
Scott, W. T., 1968: *J. Atmos. Sci.* **25**, 54.
Searcy, J. Q., and Fenn, J. B., 1974: *J. Chem. Phys.* **61**, 5282.
Sedunov, Yu. S., 1960: *Izv. Geophys. Ser.*, No. 5, 792.
Sedunov, Yu. S., 1963: *Izv. Geophys. Ser.*, No. 11, 1747.
Sedunov, Yu. S., 1964: *Izv. Geophys. Ser.*, No. 1, 150.
Sedunov, Yu. S., 1965: *Izv. Akad. Nauk.*, Atm. Ocean. Phys. **1**, 722.
Seiler, W., 1975: *Promet.* **5**, 14.
Seiler, W., and Schmidt, U., 1973: *The Sea*, Vol. 5, E. Goldberg, Ed., J. Wiley, New York.
Sekhorn, R. S., and R. C. Srivastava, 1971: *J. Atmos. Sci.* **28**, 983.
Selezneva, E. S., 1966: *Tellus* **18**, 525.
Semonin, R. G., 1966: *Proc. 47th Annual AAAS and AMS meeting* (Pacific Division), June 1966, Seattle, Washington.
Semonin, R. G., and Plumlee, H. R., 1966: *J. Geophys. Res.* **71**, 427.
Sergieva, A. P., 1959: *Izv. Akad. Nauk. SSSR* **7**, 721.
Serpolay, R., 1958: *Bull. de l'Obs. du Puy de Dome*, No. 3, 81.
Serpolay, R., 1959: *Bull. de l'Obs. du Puy de Dome*, No. 3, 81.
Shafrir, U., and Gal-Chen, T., 1971: *J. Atmos. Sci.* **28**, 741.
Shafrir, U., and Neiburger, M., 1963: *J. Geophys. Res.* **68**, 4141.
Shanks, D., 1955: *J. Math. Phys.* **34**, 1.
Shewchuk, S. R., and Iribarne, J. V., 1971: *Quart. J. Roy. Meteor. Soc.* **97**, 272.
Shiratori, K., 1934: *Mem. Fac. Sci. Agric.*, Taihoku Imperial Univ., 10, 175.
Shreve, E. L., 1970: *J. Atmos. Sci.* **27**, 1186.
Shutt, F. T., 1907: *Trans. Roy. Soc.*, *Canada*, Ser. 3 **1**, 35.
Simpson, J., 1971: *J. Atmos. Sci.* **28**, 449.
Simpson, J., 1972: *J. Atmos. Sci.* **29**, 220.
Simpson, J., 1976: *Advances in Geophys.* **19**, 1.
Simpson, J., Simpson, R. H., Andrews, D. A., and Eaton, M. A., 1965: *Rev. Geophys.* **3**, 387.
Simpson, J., and Wiggert, V. W., 1969: *Monthly Weather Review* **97**, 471.
Singer, S. F., Ed., 1970: *Global Effects of Environmental Pollution*, D. Reidel Publ. Co., Dordrecht, Holland.

Sinnarwalla, A. M., Alofs, D. J., and Carstens, D. J., 1975: *J. Atmos. Sci.* **32**, 592.

Skapski, A., 1959: *J. Chem. Phys.* **31**, 533.

Skapski, A., Billups, R., and Rooney, A., 1957: *J. Chem. Phys.* **26**, 1350.

Skatskii, V. I., 1965: *Izv. Atmos. Oceanic Phys.* **1**, 479.

Skelland, A. H., and Cornish, A. R., 1963: *A.I.Ch.E. Journal* **9**, 73.

Skinner, L. M., and Sambles, J. R., 1972: *Aerosol Science* **3**, 199.

Slinn, W. G., and Gibbs, A. G., 1971: *J. Atmos. Sci.* **28**, 973.

Slinn, W. G., and Hales, J. M., 1971: *J. Atmos. Sci.* **28**, 1465.

Slinn, W. G., and Shen, S. F., 1970: *J. Geophys. Res.* **75**, 2267.

SMIC, 1971: *Inadvertant Climate Modification*, Rept. on the Study of Man's Impact on Climates, MIT Press, Cambridge, Mass.

Smith, A. M. O., and Clutter, D. W., 1963: *Am. Inst. Aeron. and Astron.* **1**, 2062.

Smith, L. G., 1955: *Quart. J. Roy. Meteor. Soc.* **81**, 23.

Smith, M. H., Griffiths, R. R., and Latham, J., 1971: *Quart. J. Roy. Meteor. Soc.* **97**, 495.

Smith, P. L., Musil, D. J., Weber, S. F., Spahn, J. F., Johnson, G. N., and Sand, W. R., 1976: *Proc. Cloud Phys. Conf.*, Boulder, Colo., Aug. 1976, p. 252, Amer. Meteor. Soc., Boston, Mass.

Smoluchowski, M., 1916: *Physik Z.* **17**, 557.

Smoluchowski, M., 1917: *Z. Phys. Chem.* **92**, 129.

Smythe, W. R., 1950: *Static and Dynamic Electricity*, McGraw-Hill, New York.

Son, J. S., and Hanratty, T. J., 1969: *J. Fluid. Mech.* **35**, 369.

Sood, S. K., and Jackson, M. R., 1969: *Proc. Conf. on Condensation and Ice Nuclei*, Prague and Vienna, Sept. 1969, p. 299, Czechoslovak Acad. of Sci., Prague.

Sood, S. K., and Jackson, M. R., 1970: *Precipitation Scavenging*, 1970, AEC Symposium, Richland, June 1970, p. 121, U.S. Dept. of Commerce, Springfield, Va.

Soulage, G., 1955a: *Arch. Meteor. Geophys., Biokl.* **8**, 211.

Soulage, G., 1957: *Ann. Geophys.* **13**, 103.

Soulage, G., 1958: *Bull. Obs. Puy de Dome*, No. 4, 121.

Soulage, G., 1961: *Nubila* **4**, 43.

Soulage, G., 1964: *Nubila* **6**, 43.

Soulage, G., 1966: *J. de Rech. Atmos.* **2**, 219.

Spells, K. E., 1952: *Proc. Phys. Soc.* **B65**, 541.

Spengler, J. D., and Gokhale, N. R., 1972: *J. Appl. Meteor.* **11**, 1101.

Spengler, J. D., and Gokhale, N. R., 1973a: *J. Geophys. Res.* **78**, 497.

Spengler, J. D., and Gokhale, N. R., 1973b: *J. Atmos. Sci.* **12**, 316.

Spilhaus, A. F., 1948: *J. Meteor.* **5**, 108.

Squires, P., 1952: *Australian J. Sci. Res.* **5**, 59, 473.

Squires, P., 1958a: *Tellus* **10**, 256.

Squires, P., 1958b: *Tellus* **10**, 372.

Squires, P., and Turner, J. S., 1962: *Tellus* **14**, 422.

Squires, P., and Twomey, S., 1966: *J. Atmos. Sci.* **23**, 401.

Srivastava, R. C., 1971: *J. Atmos. Sci.* **28**, 410.

Starr, J. R., 1967: *Ann. Occup. Hyg.* **10**, 349.

Starr, J. R., and Mason, B. J., 1966: *Quart. J. Roy. Meteor. Soc.* **92**, 490.

Stansbury, E. J., and Vali, G., 1965: *Scientific Report*, MW-46, Stormy Weather Group, McGill University, Montreal.

Starr, J. R., 1967: *Ann. Occup. Hyg.* **10**, 349.

Starr, J. R., and Mason, B. J., 1966: *Quart. J. Roy. Meteor. Soc.* **92**, 490.

Stefan, J., 1873: *Wien Akad. Sci. Sitzungs Ber., Math. Naturw. Kl.* **68**, 2 Abt., 943.

Steinberger, E. H., Pruppacher, H. R., and Neiburger, M., 1968: *J. Fluid Mech.* **34**, 809.

Steinberger, R. L., and Treybal, R. E., 1960: *A.I.Ch.E. Journal* **6**, 227.

Stepanov, A. S., 1975: *Izv. Akad. Nauk., SSSR, FAO*, 11, 267.

Stepanov, A. S., 1976: *Proc. Cloud Phys. Conf.*, Boulder, Colo., July 1976, p. 27, Amer. Meteor. Soc., Boston, Mass.

Stern, A. C., Ed., 1968: *Air Pollution*, Vols. 1, 2, 3, Academic Press, New York.

Stevenson, C., 1968: *Quart. J. Roy. Meteor. Soc.* **94**, 35.

Stevenson, R. E., and Collier, A., 1962: *Lloydia* **25**, 89.

Stille, U., 1961: *Messen und Rechnen in der Physik*, Vieweg Verlag, Braunschweig.

Stillinger, F. H., 1970: *J. Phys. Chem.* **74**, 3677.

Stillinger, F. H., and Rahman, A., 1972: *J. Chem. Phys.* **57**, 1281.

Stimson, M., and Jeffrey, G. B., 1926: *Proc. Roy. Soc.* **A111**, 110.

Stix, E., 1969: *Umschau in Wiss. Techn.*, Nr. 19/69, 620.

Stix, E., and Grosse, G., 1970: *Flora* **159**, 1.

Stockmayer, W. H., 1941: *J. Chem. Phys.* **9**, 398.

Stokes, G. G., 1851: *Trans. Cambridge Phil. Soc.* **9**, pt. 2, 8.

Stokes, R. H., and Miles, R., 1965: *Viscosity of Electrolytes and Related Properties*, p. 74, Pergamon Press, New York.

Storebo, P. B., 1972: *Geofys. Publ.* **28**, 1.

Stow, C. D., 1969: *Rept. Progr. Phys.* **32**, 1.

Strantz, R., 1971: *Meteor. Rundschau* **24**, 19.

Stringham, G. E., Simons, D. B., and Guy, H. P., 1969: *Geological Survey Paper*, 562-C, Government Printing Office, Washington, D.C.

Summers, P. W., 1968: *Proc. Cloud Phys. Conf.*, Toronto, Aug. 1968, p. 455, University of Toronto Press, Toronto.

Sutton, O. G., 1932: *Proc. Roy. Soc., London* **A135**, 143.

Sutton, O. G., 1942: *Meteor. Res. Paper*, No. 40, Chem. Defense Exp. Station, Porton, England.

Sutugin, A. G., and Fuchs, N. A., 1968: *J. Coll. Interface Sci.* **27**, 216.

Sutugin, A. G., and Fuchs, N. A., 1970: *J. Aerosol Sci.* **1**, 287.

Sutugin, A. G., Fuchs, N. A., and Kotsev, E. I., 1971: *J. Aerosol Sci.* **2**, 301.

Swift, D. L., and Friedlander, S. K., 1964: *J. Colloid. Sci.* **19**, 621.

Takahashi, C., and Yamashita, A., 1969: *J. Meteor. Soc., Japan* **47**, 431.

Takahashi, C., and Yamashita, A., 1970: *J. Meteor. Soc., Japan* **48**, 373.

Takahashi, K., and Kasahara, M., 1968: *Atmos. Environment* **2**, 441.

Takahashi, T., 1963: *J. Meteor. Soc., Japan* **41**, 327.

Takahashi, T., 1965: *J. Meteor. Soc., Japan* **43**, 206.

Takahashi, T., 1972: *J. Geophys. Res.* **77**, 3869.

Takahashi, T., 1973: *Rev. of Geophys. and Space Phys.* **11**, 903.

Takahashi, T., and Craig, T., 1973: *J. Meteor. Soc., Japan* **51**, 191.

Takahashi, T., and Fullerton, C. M., 1972: *J. Geophys. Res.* **77**, 1630.

Takami, H., and Keller, H. B., 1969: *Phys. of Fluids*, Suppl. II, 12, 11.

Takeda, T., 1968: *J. Meteor. Soc., Japan* **46**, 255.

Tamir, A., and Hasson, D., 1971: *Chem. Engr. Journal* **2**, 200.

Tammann, G., 1910: *Z. Phys. Chem.* **72**, 609.

Tammann, G., 1920: *Z. Anorg. Allg. Chemie* **110**, 166.

Tanaka, M., 1966: *Special Contrib.*, Geophys. Inst., No. 6, 45, Kyoto University, Kyoto, Japan.

Taneda, S., 1956a: *Rept. Res. Inst. Appl. Mech.* **6**, 16, 99, Kyushu University, Hakozaki, Fukuoka, Japan.

Taneda, S., 1956b: *J. Phys. Soc., Japan* **11**, 302.

Taylor, G. I., 1964: *Proc. Roy. Soc., London* **A280**, 383.

Taylor, T. D., and Acrivos, A., 1964: *J. Fluid Mech.* **18**, 466.

Tazawa, S., and Magono, C., 1973: *J. Meteor. Soc., Japan* **51**, 168.

Tchen, C. M., 1947: *Ph.D. Thesis*, Den Haag, Holland.

Tcheurekdjian, N., Zettlemoyer, A. C., and Chessick, J.J., 1964: *J. Phys. Chem.* **68**, 773.

Telford, J. W., 1955: *J. Meteor.* **12**, 436.

Telford, J. W., 1960: *J. Meteor.* **17**, 86.

Telford, J. W., and Cottis, R. E., 1964: *J. Atmos. Sci.* **21**, 549.

Telford, J. W., and Thorndike, N. S., 1961: *J. Meteor.* **18**, 32.

Telford, J. W., Thorndike, N. S., and Bowen, E. G., 1955: *Quart. J. Roy. Meteor. Soc.* **81**, 241.

Tennekes, H., and Lumley, J. L., 1972: *A First Course in Turbulence*, MIT Press, Cambridge, Mass.

Tennekes, H., and Woods, J. D., 1973: *Quart. J. Roy. Meteor. Soc.* **99**, 758.

Terliuc, B., and Gagin, A., 1971: *Collected Papers*, Rept. No. 4, Dept. Meteor., The Hebrew University, Jerusalem.

Thom, A., 1933: *Proc. Roy. Soc.* **A141**, 651.

Thomson, J. J., 1888: *Application of Dynamics to Physics and Chemistry*, 1st ed., p. 165, Cambridge University Press, London.

Thomson, W. (Lord Kelvin), 1870: *Proc. Roy. Soc., Edinburgh* **7**, 63.

Thorpe, A. D., and Mason, B. J., 1966: *Brit. J. Appl. Phys.* **17**, 54.

Thudium, J., 1976: *J. Aerosol Sci.* **7**, 167.

Thudium, J., 1978: *Pure and Appl. Geophys.* **116**, 130.

Thuman, W. C., and Robinson, E., 1954: *J. Meteor.* **11**, 151.

Toba, Y., 1965a: *Tellus* **17**, 131.

Toba, Y., 1965b: *Tullus* **17**, 365.

Toba, Y., and Tanaka, M., 1968: *J. de Rech. Atmos.* **3**, 17.

Tolman, R., 1949a: *J. Chem. Phys.* **17**, 118.

Tolman, R., 1949b: *J. Chem. Phys.* **17**, 333.

Tomotika, S., 1935: Great Britain, *Aero Res. Comm. Repts. and Memoranda*, No. 1678, 86.

Toulcova, J., and Podzimek, J., 1968: *J. de Rech. Atmos.* **3**, 89.

Townsend, A. A., 1956: *The Structure of Turbulent Shear Flow*, Cambridge University Press.

Tritton, D. J., 1959: *J. Fluid Mech.* **37**, 95.

Truby, F. K., 1955a: *Science* **121**, 404.

Truby, F. K., 1955b: *J. Appl. Phys.* **26**, 1416.

Tschudin, K., 1945: *Helvetica Phys. Acta* **19**, 91.

Tsuji, M., 1950: *Geophys. Mag. (Tokyo)* **22**, 15.

Tsunogai, S., and Fukuda, K., 1974: *Geochem. J., Japan* **8**, 141.

Tunitskii, N., 1946: *Zh. fiz. Khim.* **20**, 1136.

Turnbull, D., 1950: *J. Chem. Phys.* **18**, 769.

Turnbull, D., 1956: *Solid State Physics* **3**, 225.

Turnbull, D., 1965: *Liquids: Structure, Properties, Solid Interactions*, Th. Hugel, Ed., p. 15, Elsevier Publ. Co., New York.

Turnbull, D., and Fischer, J. C., 1949: *J. Chem. Phys.* **17**, 71.

Turnbull, D., and Vonnegut, B., 1952: *Industr. Eng. Chem.* **44**, 1292.

Turner, J. S., 1955: *Quart. J. Roy. Meteor. Soc.* **81**, 418.

Turner, J. S., 1962: *J. Fluid Mech.* **13**, 356.

Turner, J. S., 1963: *J. Fluid Mech.* **16**, 1.

Turner, J. S., 1969: *Ann. Rev. of Fluid Mech.* **1**, 29.

Turner, J. S., 1973: *Buoyancy Effects in Fluids*, Cambridge University Press, London.

Twomey, S., 1953: *J. Appl. Phys.* **24**, 1099.

Twomey, S., 1954: *J. Meteor.* **11**, 334.

Twomey, S., 1955: *J. Meteor.* **12**, 81.

Twomey, S., 1956: *Tellus* **7**, 445.

Twomey, S., 1959a: *Geofys. Pura e Appl.* **43**, 227.

Twomey, S., 1959b: *J. Chem. Phys.* **30**, 941.

Twomey, S., 1959c: *Geofys. Pura e Appl.* **43**, 243.

Twomey, S., 1960: *Bull. Obs. de Puy de Dome*, No. 1, 1.

Twomey, S., 1963: *J. de Rech. Atmos.* **1**, 101.

Twomey, S., 1964: *J. Atmos. Sci.* **21**, 553.

Twomey, S., 1965: *J. de Rech. Atmos.* **2**, 113.

Twomey, S., 1966: *J. Atmos. Sci.* **23**, 405.

Twomey, S., 1968: *J. de Rech. Atmos.* **3**, 281.

Twomey, S., 1969: *J. de Rech. Atmos.* **4**, 179.

Twomey, S., 1971: *J. Atmos. Sci.* **28**, 377.

Twomey, S., 1972: *J. Atmos. Sci.* **29**, 318.

Twomey, S., 1976: *J. Atmos. Sci.* **33**, 720.

Twomey, S., and Davidson, K. A., 1970: *J. Atmos. Sci.* **27**, 1056.

Twomey, S., and Davidson, K. A., 1971: *J. Atmos. Sci.* **28**, 1295.

Twomey, S., and Severynse, G. T., 1964: *J. de Rech. Atmos.* **1**, 81.

Twomey, S., and Warner, J., 1967: *J. Atmos. Sci.* **24**, 702.

Twomey, S., and Wojciechowski, T. A., 1969: *J. Atmos. Sci.* **26**, 684.

Uman, M. A., 1969: *Lightning*, McGraw-Hill, New York.

Underwood, R. L., 1969: *J. Fluid Mech.* **37**, 95.

Valencia, M. J., 1967: *Amer. J. Sci.* **265**, 843.

Vali, G., 1968a: *Proc. Cloud Phys. Conf.*, Toronto, Aug. 1968, p. 232, University of Toronto Press, Toronto.

Vali, G., 1968b: *Sci. Rept.*, MW-58, Stormy Weather Group, McGill University, Montreal.

Vali, G., 1971: *J. Atmos. Sci.* **28**, 402.

Vali, G., 1974: *Proc. Cloud Phys. Conf.*, Tucson, Arizona, Oct. 1974, p. 34, Amer. Meteor. Soc., Boston, Mass.

Vali, G., Christensen, M., Fresh, R. W., Galyen, E. L., Malu, L. R., and Schnell, R. C., 1976: *J. Atmos. Sci.* **33**, 1565.

Vali, G., and Schnell, R. C., 1973: *Nature* **246**, 212.

Vali, G., and Stansbury, E. J., 1965: *Scientific Report*, MW-41, Stormy Weather Group, McGill University, Montreal.

Vali, G., and Stansbury, E. J., 1966: *Canad. J. Phys.* **44**, 477.

Vance, J. L., and Peters, L. K., 1976: *J. Atmos. Sci.* **33**, 1824.

Van den Heuvel, A. P., and Mason, B. J., 1963: *Quart. J. Roy. Meteor. Soc.* **89**, 271.

Van der Hage, J., 1972: *J. de Rech. Atmos.* **6**, 595.

Van Dyke, M., 1964: *Perturbation Methods in Fluid Mechanics*, Academic Press, New York.

Van Eck, C. L. P., Mendel, H., and Fahrenfort, 1958: *Proc. Roy. Soc.*, London **A247**, 472.

Van Mieghem, J., and Dufour, L., 1948: *Thermodynamique de l'atmosphère*, Memoires, Vol. 30, Instit. Roy. Meteor. de Belgique, Bouxelles.

Vardiman, L., and Grant, L. D., 1972a: *Proc. Cloud Phys. Conf.*, London, England, Aug. 1972, p. 22, Roy. Meteor. Soc., London.

Vardiman, L., and Grant, L. D., 1972b: *Proc. Weather Mod. Conf.*, Rapid City, June 1972, p. 113, Amer. Meteor. Soc., Boston, Mass.

Verwey, E. J., 1941: *Rec. Trav. Chim. Pays Bas, Belg.* **60**, 887.

Vittori, A., and Prodi, V., 1967: *J. Atmos. Sci.* **24**, 533.

Vohra, K. G., and Nair, P. V., 1970: *J. Aerosol Sci.* **1**, 127.

Vohra, K. G., Ramu, S., and Vasudevan, K. N., 1969: *Atmos. Environ.* **3**, 99.

Vohra, K. G., Vasudevan, K. N., and Nair, P. V., 1970: *J. Geophys. Res.* **75**, 2951.

Volmer, M., 1939: *Kinetik der Phasenbildung*, Verlag Th. Steinkopff, Dresden.

Volmer, M., and Flood, H., 1934: *Z. Phys. Chem.* **A170**, 273.

Volmer, M., and Weber, A., 1926: *Z. Phys. Chem.* **A119**, 277.

Vonnegut, B., 1955: *Geophys. Res. Paper*, No. 42, Air Force Cambridge Res. Center, Cambridge, Mass.

Voyt, F. Ya., Korniyenko, Ye., and Khusid, S. B., 1971: *Izv. Atmos. Oceanic Phys.* **7**, 1206.

Vulfson, N. I., Laktinov, A. G., and Skatskii, V. I., 1973: *J. Appl. Meteor.* **12**, 664.

Vulfson, N. I., and Levin, L. M., 1965: *Studies of Clouds, Precipitation, and Thunderstorm Electricity*, N. I. Vufson and L. M. Levin, Eds., p. 33, Amer. Meteor. Soc., Boston, Mass.

Waldmann, L., and Schmitt, K. H., 1966: *Aerosol Science*, C. N. Davies, Ed., p. 137, Academic Press, New York.

Waldvogel, A., 1974: *J. Atmos. Sci.* **31**, 1067.

Walrafen, G. E., 1966: *J. Chem. Phys.* **44**, 1546.

Walrafen, G. E., 1967: *J. Chem. Phys.* **47**, 114.

Walrafen, G. E., 1968a: *J. Chem. Phys.* **48**, 244.

Walrafen, G. E., 1968b: *Hydrogen Bonded Solvent Systems*, A. K. Covington and P. Jones, Eds., p. 9, Taylor and Francis, London.

Walrafen, G. E., 1972: *Water*, F. Franks, Ed., p. 151, Plenum Press, New York.

Walter, H., 1973: *J. Aerosol Sci.* **4**, 1.

Wang, C. S., 1966: *Ph.D. Dissertation*, California Inst. of Technology, Pasadena, Calif.

Wang, C. S., and Friedlander, S. K., 1967: *J. Colloid. Interface Sci.* **24**, 170.

Wang, J. H., 1954: *J. Phys. Chem.* **58**, 686.

Wang, J. H., and Miller, S., 1952: *J. Am. Chem. Soc.* **74**, 1611.

Wang, J. Y., 1974: *J. Atmos. Sci.* **31**, 513.

Wang, K. C., 1970: *J. Fluid Mech.* **43**, 187.

Wang, P. K., and Pruppacher, H. R., 1977a: *J. Appl. Meteor.* **16**, 275.

Wang, P. K., and Pruppacher, H. R., 1977b: *J. Atmos. Sci.* **34**, 1664.

Wang, P. K., Grover, S. N., and Pruppacher, H. R., 1978: *J. Atmos. Sci.* **35**, Sept. issue.

Wanlass, F. M., and Eyring, H., 1961: *Adv. in Chem.* **33**, 140.

Warner, J., 1955: *Tellus* **7**, 449.

Warner, J., 1968a: *J. de Rech. Atmos.* **3**, 233.

Warner, J., 1968b: *J. Appl. Meteor.* **7**, 247.

Warner, J., 1969a: *J. Atmos. Sci.* **26**, 1049.

Warner, J., 1969b: *J. Atmos. Sci.* **26**, 1272.

Warner, J., 1970: *J. Atmos. Sci.* **27**, 1035.

Warner, J., 1972: *J. Atmos. Sci.* **29**, 218.

Warner, J., 1973: *J. Atmos. Sci.* **30**, 256.

Warner, J., 1975a: *J. Atmos. Sci.* **32**, 995.

Warner, J., 1975b: *Quart. J. Roy. Meteor. Soc.* **101**, 176.

Warner, J., and Twomey, S., 1967: *J. Atmos. Sci.* **24**, 704.

Warshaw, M., 1967: *J. Atmos. Sci.* **24**, 278.

Warshaw, M., 1968: *J. Atmos. Sci.* **25**, 874.

Watson, C., and Allenson, R. E., 1977: *Res. Rept.*, Los Alamos Sci. Lab., Los Alamos, New Mexico.

Watson, J. D., and Crick, F. H., 1953: *Nature* **171**, 737.

Watts, R. G., and Farhi, I., 1975: *J. Atmos. Sci.* **32**, 1864.

Webb, W. L., 1966: *Structure of the Stratosphere and Mesosphere*, Academic Press, New York.

Weber, E., 1968: *Staub* **28**, 462.

Weber, E., 1969: *Staub* **29**, 272.

Weickman, H. K., 1945: *Beitr. Z. Phys. d. freien Atmosphäre* **28**, 12.

Weickman, H. K., 1949: *Ber. Deutsch. Wett. Dienst. U.S. Zone*, Nr. 6.

Weickman, H. K., 1953: *Thunderstorm Electricity*, H. R. Byers, Ed., p. 66, University of Chicago Press, Chicago, Ill.

Weickman, H. K., 1957a: *Artificial Stimulation of Rain*, p. 315, Pergamon Press, New York.

Weickman, H. K., 1957b: *Artificial Stimulation of Rain*, p. 81, Pergamon Press, New York.

Weickman, H. K., 1964: *Nubila* **6**, 7.

Weickman, H. K., 1972: *J. de Rech. Atmos.* **6**, 603.

Weickman, H. K., and Aufm Kampe, H. J., 1953: *J. Meteor.* **10**, 204.

Weickman, H. K., and Pueschel, R. F., 1973: *Beitr. z. Phys. d. Atmos.* **48**, 112.

Weinstein, A. I., 1971: *J. Atmos. Sci.* **28**, 648.

Weissweiler, W., 1969: *Z.f. Meteor.* **21**, 108.

Welander, P., 1954: *Arkiv f. Fysik* **7**, 507.

Welch, J. E., Harlow, F. H., Shannon, J. P., and Daly, B. J., 1966: *Rept. LA-3425*, Los Alamos Scientific Laboratory, Los Alamos, N.M.

Wen Ching-Sung, 1966: *Sci. Sinica* **15**, 870.

Went, F. W., Slemons, D. B., and Mozingo, H. N., 1967: *Proc. Nat. Acad. Sci.* **58**, 69.

Weyl, W. A., 1951: *J. Coll. Sci.* **6**, 389.

Whalley, E., Jones, S. J., and Gold, L. W., Eds., 1973: *Physics and Chemistry of Ice*, Roy. Soc. of Canada, Ottawa.

Whelpdale, D. M., 1970: *Ph.D. Thesis*, Dept. of Physics, University of Toronto, Toronto.

Whelpdale, D. M., and List, R., 1971: *J. Geophys. Res.* **76**, 2836.

Whipple, F. J. W., and Chalmers, J. A., 1944: *Quart. J. Roy. Meteor. Soc.* **70**, 103.

Whitby, K. T., and Liu, B. T., 1966: *Aerosol Science*, C. N. Davies, Ed., p. 59, Academic Press, New York.

White, F. M., 1974: *Viscous Flow*, McGraw-Hill, New York.

Whitehead, A. N., 1889: *Quart. J. Math.* **23**, 143.

Whitehead, H. C., and Feth, J. H., 1964: *J. Geophys. Res.* **69**, 3319.

Whitten, R. C., and Poppoff, I. G., 1971: *Fundamentals of Aeronomie*, John Wiley Inc., New York.

Whytlaw-Gray, R., 1935: *J. Chem. Soc., London*, p. 268.

Whytlaw-Gray, R., and Patterson, H. S., 1932: *Smoke*, Edward Arnold, London.

Wieland, W., 1956: *Z. Angew. Math. Phys.* **7**, 428.

Wieselberger, C., 1922: *Phys. Z.* **23**, 219.

Wilkins, R. D., and Auer, A. H., 1970: *Proc. Cloud Phys. Conf.*, Fort Collins, Colo., Aug. 1970, p. 81, Amer. Meteor. Soc., Boston, Mass.

Wilkniss, P. E., and Bressan, D. J., 1972: *J. Geophys. Res.* **77**, 5307.

Williams, A. L., 1974: *Res. Rept.*, Univ. of California, UCRL-75897, Lawrence Livermore Lab., Livermore, Calif.

Williams, G. L., 1970: *J. Atmos. Sci.* **27**, 1220.

Williamson, R. E., and McCready, P. B., 1968: *Final Report*, Contract No. 5-2-67, Meteor. Res. Inc., Altadena, Calif.

Willmarth, W. W., Hawk, N. E., and Harvey, R. L., 1964: *Phys. Fluids* **7**, 197.

Wilson, A. T., 1959: *Nature* **184**, 99.

Wilson, C. T. R., 1899: *Phil. Trans. Roy. Soc.* **A193**, 289.

Wilson, C. T. R., 1929: *J. Franklin Inst.* **208**, 1.

Wilson, H. A., 1900: *Phil. Mag.* **50**, Ser. 5, 238.

Winkler, P., 1967: *Diplom. Thesis*, Meteor. Instit., University of Mainz, Mainz.

Winkler, P., 1968: *Ann. Meteor., N.S.* **4**, 134.

Winkler, P., 1970: *Ph.D. Thesis*, Meteor. Inst., University of Mainz, Mainz.

Winkler, P., 1973: *Aerosol Sci.* **4**, 373.

Winkler, P., and Junge, C. E., 1971: *J. Appl. Meteor.* **10**, 159.

Winkler, P., and Junge, C. E., 1972: *J. de Rech. Atmos.* **6**, 617.

Winn, W. P., Schwede, G. W., and Moore, C. B., 1974: *J. Geophys. Res.* **79**, 176.

Wirth, E., 1966: *J. de Rech. Atmos.* **2**, 1.

Wobus, H. B., Murray, F. W., and Koenig, L. R., 1971: *J. Appl. Meteor.* **10**, 751.

Wolff, G. A., 1957: *Artificial Stimulation of Rain*, p. 332, Pergamon Press, New York.

Wollan, E. O., Davidson, W. L., and Shull, C. G., 1949: *Phys. Rev.* **75**, 1348.

Woo, S., 1971: *Ph.D. Thesis*, Dept. Chem., Eng., McMaster University, Hamilton, Canada.

Woo, S., and Hamielec, A. E., 1971: *J. Atmos. Sci.* **28**, 1448.

Wood, G. R., and Walton, H. G., 1970: *J. Appl. Phys.* **41**, 3027.

Woodcock, A. H., 1953: *J. Meteor.* **10**, 362.

Woodcock, A. H., 1957: *Tellus* **9**, 521.

Woodcock, A. H., 1972: *J. Geophys. Res.* **77**, 5316.

Woodcock, A. H., and Duce, A., 1972: *J. de Rech. Atmos.* **6**, 639.

Woodcock, A. H., and Jones, R. H., 1970: *J. Appl. Meteor.* **9**, 690.

Woodcock, A. H., Kientzler, C. F., Aarons, A. B., and Blanchard, D. C., 1953: *Nature* **173**, 1144.

Woods, J. D., 1965: *Quart. J. Roy. Meteor. Soc.* **91**, 353.

Woods, J. D., Drake, J. C., Goldsmith, P., 1972: *Quart. J. Roy. Meteor. Soc.* **98**, 135.

Woods, J. D., and Mason, B. J., 1964: *Quart. J. Roy. Meteor. Soc.* **90**, 373.

Woods, J. D., and Mason, B. J., 1965: *Quart. J. Roy. Meteor. Soc.* **91**, 35.

Woods, J. D., and Mason, B. J., 1966: *Quart. J. Roy. Meteor. Soc.* **92**, 171.

Worley, J. D., and Klotz, I. M., 1966: *J. Chem. Phys.* **45**, 2868.

Wronski, C. R., 1967: *Brit. J. Appl. Phys.* **18**, 1731.

Wu, T. H., 1964: *J. Geophys. Res.* **69**, 1083.

Wulff, G., 1901: *Z. Kristallogr.* **34**, 449.

Wyckoff, R. W. G., 1963: *Crystal Structures*, 2nd ed., Vol. 1, Interscience Publ., New York.

Wylie, R. G., 1953: *Proc. Roy. Soc., London* **B66**, 241.

Yamamoto, G., and Miura, A., 1949: *J. Meteor. Soc., Japan* **27**, 257.

Yamamoto, G., and Ohtake, T., 1953: *Sci. Rept.*, Tohoku University, Ser. 5, (Geophys.) **4**, 141.

Yamamoto, G., and Ohtake, T., 1955: *Sci. Rept.*, Tohoku University, Ser. 5, (Geophys.) **7**, 10.

Yamashita, A., 1969: *J. Meteor. Soc., Japan* **47**, 57.

Yamashita, A., 1971: *J. Meteor. Soc., Japan* **49**, 215.

Yamashita, A., 1973: *J. Meteor. Soc., Japan* **51**, 307.

Yih, C-S., 1965: *Dynamics of Nonhomogeneous Fluids*, Macmillan, Publ. Co., New York.

Young, K. C., 1974: *J. Atmos. Sci.* **31**, 768.

Young, K. C., 1975: *J. Atmos. Sci.* **32**, 965.

Young, R. G., and Browining, K. A., 1967: *J. Atmos. Sci.* **24**, 58.

Zaitsev, V. A., 1948: *Trudy Glavnoi Geofiz., Observ.* **13**, 75.

Zaitsev, V. A., 1950: *Trudy Glavnoi Geofiz. Observ.* **19**, 122.

Zebel, G., 1956: *Z. f. Aerosol Forschg. und Therapie* **5**, 263.

Zeldovich, J., 1942: *Zh. Eksp. Theor. Fiz.* **12**, 525.

Zettlemoyer, A. C., 1968: *J. Coll. Interface Sci.* **28**, 343.

Zettlemoyer, A. C., Ed., 1969: *Nucleation*, Marcel Dekker, Inc., New York.

Zettlemoyer, A. C., Tcheurekdjian, N., and Chessick, J. J., 1961: *Nature* **192**, 653.

Zettlemoyer, A. C., Tcheurekdjian, and Hosler, C. L., 1963: *Z. Angew. Math. Phys.* **14**, 496.

Zikmunda, J., 1970: *J. de Rech. Atmos.* **4**, 7.

Zikmunda, J., 1972: *J. Atmos. Sci.* **29**, 1511.

Zikmunda, and Vali, G., 1972: *J. Atmos. Sci.* **29**, 1334.

Zimin, A. G., 1964: *Problems in Nuclear Meteorology*, I. L. Karol and S. G. Malahov, Eds., p. 139, USAEC Report AEC-tr-6128, State Publ. House for Literature.

Ziv, A., and Levin, Z., 1974: *J. Atmos. Sci.* **31**, 1652.

Zoebell, C. E., and Matthews, H., 1936: *Proc. Nat. Acad. Sci.* **27**, 567.

LIST OF PRINCIPAL SYMBOLS

a	radius of spherical drop, radius of rigid sphere, radius of circle circumscribed to basal plane of ice crystal, semi-major axis of oblate spheroid, radius of curvature of interface, radius of circular cylinder
$\bar{a}$	mean drop radius
a_0	equivalent radius of deformed drop, crystallographic lattice parameter in a-axis direction
a_i, a_p	radius of spherical ice crystal, potential drop radius
a_i, a_g	radius of spherical embryo consisting of i molecules, of spherical germ consisting of g molecules
a_k	activity of component k in mixture
a_w, a_s	activity of water in aqueous solution, of solute in aqueous solution
A_c	cross sectional area of body oriented perpendicular to viscous flow
A_B	area of basal face of planar ice crystal
A	virial coefficient
b	semi-minor axis of oblate spheroid
B	mobility
$\dot{c}$	concentration of molecules
c	number of components
c_w, c_i	specific heat of water, of ice
c_{pv}, c_{pa}	specific heat of water vapor at constant pressure, of air at constant pressure
c_0	crystal lattice parameter in c-axis direction
c_i	equilibrium concentration of i-mers
C	electrostatic capacitance
C_D	hydrodynamic drag force coefficient
C_a, C_v, C_m	compressibility factor for dry air, water vapor, moist air
C_w, C_i, C_{pv}, C_{pa}	molar heat capacity of water, of ice, of water vapor at constant pressure, of air at constant pressure
c_w	$= N_w/V$ (Ch. 7)
$c_{sat,w}, c_{sat,i}$	$= N_{sat,w}/V, N_{sat,i}/V$
d	diameter of spherical drop, of rigid sphere, of circle circumscribed to basal plane of ice crystal, of graupel, of hailstone
d	separation between the centers of two spheres
d_S	mean migration distance of water molecules by surface diffusion on ice
D	hydrodynamic drag on body in viscous medium (Ch. 10), diffusivity of aerosol particles in air (Ch. 12)
D_v	diffusivity of water vapor in air
D_+, D_-	diffusivity of positive ions in air, of negative ions in air

D_0	equivalent diameter of raindrop
D_S	Stokes drag
D_w	self diffusivity of water molecules in water
e	partial pressure of water vapor in moist air, vapor pressure of pure water vapor
e_∞, e_a	water vapor pressure in environment, at drop surface
$e_{a,w}, e_{a,i}$	water vapor pressure over spherically curved water surface, over spherically curved ice surface
$e_{sat,w}, e_{sat,i}, e_{sat,s}$	saturation vapor pressure over plane water surface, over plane ice surface, over plane aqueous solution surface
e	elementary electric charge
$\hat{e}$	unit vector
$\vec{E}$	electric field vector
E	magnitude of electric field (electric field intensity)
E_ν	energy of system of quantum number ν
E	collision efficiency (14.2)
E_c	collection efficiency
E_{coa}	coalescence efficiency
E_m	energy per molecule for cleaving a crystal
E_H	hydrogen bond energy
E_L	lattice energy of ice
f	vibration frequency of drops, shedding frequency of vortices from rear of a rigid sphere
f_w, f_s	rational activity coefficient of water, of salt in aqueous solution
f_i', f_i	unsteady, steady concentration of i-mers
$\bar{f}_v, \bar{f}_h$	ventilation coefficient for vapor diffusion, for heat diffusion
Δf_{vol}	bulk Helmholtz free energy change per unit volume
F	Helmholtz free energy of system
$\vec{F}, \vec{F}_1, \vec{F}_2$	force on body, hydrodynamic force on sphere -1, on sphere -2
$\vec{F}_e$	electric force on body
$\vec{F}_{Th}, \vec{F}_{Df}$	thermophoretic, diffusiophoretic force on aerosol particle
ΔF^+	energy of activation for the diffusion of a water molecule across the ice-water interface
ΔF_{act}	activation energy for the displacement of water molecules in bulk water
$\Delta F_i, \Delta F_g$	energy of i-mer formation, of germ formation
$\Delta F_{ad}, \Delta F_{des}$	energy of adsorption, of desorption, per molecule
ΔF_{sd}	activation energy for surface diffusion per molecule
$\vec{g}, g$	acceleration of gravity, magnitude of acceleration of gravity
g_k	partial molar Gibbs free energy of component k in mixture
G	Gibbs free energy of system (Ch. 4), linear growth rate of ice crystal face (Chs. 13, 16)
h	thickness of planar ice crystal
h	Planck constant (Ch. 7), enthalpy per unit mass
h	molar enthalpy
h_k	partial molar enthalpy of component k in mixture
h_w, h_v	partial molar enthalpy of water in aqueous solution, of water vapor in air
h_i	molar enthalpy of ice
h_m	molar enthalpy of mixing of water in aqueous solution
H	enthalpy of system

i	Van't Hoff coefficient for non-ideal aqueous salt solution
i	number of water molecules per embryo
I	moment of inertia
j_{Th}, j_{Df}, j_{CD}	current density of particles due to thermophoresis, due to diffusiophoresis, due to convective diffusion
j_v, j_h, j_q	mass flux of water vapor, flux of heat, flux of electrical charge
$J, J_{Th}, J_{Df}, J_{CD}$	nucleation rate (Ch. 7), total particle flux due to thermophoresis, due to diffusiophoresis, due to convective diffusion (Ch. 12)
k	Boltzmann constant
k_v, k_a, k_w, k_i	heat conductivity of water vapor, of air, of water, of ice
$\bar{k}_v$	mean mass transfer coefficient for water vapor in air
K	collection kernel
$K_B, K_{i,j}$	collection kernel for Brownian coagulation
K_{Th}, K_{Df}	collection kernel for thermophoresis, for diffusiophoresis
K_G	collection kernel for gravitational collection
ℓ	length, liter (Chs. 4, 5, 6, 8)
$L, \mathscr{L}$	latent heat of phase change per unit mass, per mole
$L_{e,0}, L_{m,0}, L_s$	latent heat of evaporation of pure water, of melting of ice, of sublimation of ice, per unit mass
$\mathscr{L}_{e,0}, \mathscr{L}_{m,0}, \mathscr{L}_s$	latent heat of evaporation of pure water, of melting of ice, of sublimation of ice, per mole
$\mathscr{L}_e, \mathscr{L}_m$	molar latent heat of evaporation of water from an aqueous solution, of melting of ice in an aqueous solution
L	length of columnar ice crystal, angular momentum of coalesced drop pair (Ch. 14)
L*	$\equiv \dfrac{\Omega}{P}$ (13-76)
$\dot{m}, \dot{m}_w$	mass of molecule, mass of water molecule
m	mass of aerosol particle (Ch. 12), of water drop (Ch. 13, Ch. 15)
m_N	mass of nucleus
m_0	mass of aerosol deposit
m_s	mass of salt particle, of salt deposit
$m_{i/v}, m_{i/w}$	compatibility parameter for ice on a solid substrate (5-24a, 5-24b)
$m_{w/v} \equiv \cos \theta$	compatibility parameter for water on a solid substrate (5-23)
$m_{s,t}$	mass of water-soluble portion of aerosol deposit
m_u	mass of insoluble portion of aerosol particle, of aerosol deposit
m_c	mass of ice crystal
$m_{v,ad}$	mass of water vapor adsorbed on solid substrate
m_v, m_w, m_a, m_i	mass of water vapor, of water, of air, of ice
M	molecular weight
M_a, M_w, M_s, M_m	molecular weight of air, of water, of salt, of moist air
$\mathfrak{M}$	molality of aqueous solution
$\hat{M}$	$= M_w/1000$ (4-64)
n_k	number of moles of component k in mixture
$n(a), n_d(a)$	number of cloud drops of radius a per unit volume and per unit size interval
$n(D_0)$	number of raindrops of diameter D_0 per unit volume and per unit size interval
$n(r)$	number of aerosol particles of radius r per unit volume and per unit size interval
n_m	number of adsorption sites available on solid substrate

n_{ad}	number of vapor molecules adsorbed on solid substrate
n_v, n_w, n_s, n_a	number of moles of water vapor, of water, of salt, of air
$\hat{n}$	unit outward normal vector from surface element
n_+, n_-	number of positive ions, of negative ions, per unit volume of air
N	total number of particles in system
N_A	Avogadro number
N_{Re}	Reynolds number (10-9)
N_{Pe}	Peclet number (12-84)
N_{Sc}	Schmidt number (12-85b)
N_{Sh}	Sherwood number (13-53)
N_{Pr}	Prandtl number (Section 12.7.2)
N_{Nu}	Nusselt number (Section 13.2.3)
N_{Fr}	Froude number (Section 14.3)
N_{St}	Strouhal number (Section 10.2)
N_{We}	Weber number (10-96)
N_{Bo}	Bond number (Section 10.3.5)
N_S	Stokes number (14-11)
N_{CCN}, N_{IN}	number of cloud condensation nuclei, of ice forming nuclei, per unit volume of air
N_i	number of i-mers in volume V
N_1	total number of single water molecules in volume V
$N_{sat,w}, N_{sat,i}$	number of water molecules in gas phase of volume V, at water saturation, at ice saturation
N_c	number of water molecules in water contacting unit area of ice germ
N_w	number of water molecules in water of volume V
N_u	number of unfrozen drops of drop population
N_f	number of frozen drops of drop population
p	pressure, number of elementary charges (Ch. 17)
p_k	partial pressure of component k in mixture
p_a	partial pressure of dry air in gas mixture, pressure of dry air
p_w	pressure of water inside a motionless water drop
p_∞	free stream pressure
p_s	static pressure
p_i, p_e	pressure inside, outside of a water drop
p_m	melting pressure of ice
p_0	frontal stagnation pressure on a sphere
P	perimeter of a body
p	momentum of molecule
p-ratio	$= a_2/a_1$ (Section 14.3)
q	heat
q_v	specific humidity of moist air
q_i	electric charge of ion species i
q_i	partition function of i-mer
$Q(N_i)$	canonical partition function of gas of i-mers
$\mathcal{Q}_i$	grand canonical partition function of gas of i-mers
Q, Q_a, Q_r	electric charge, electric charge on drop of radius a, electric charge on aerosol particle of radius r
dQ	incremental heat change (an imperfect differential)
$\vec{r}, r$	position vector, radial distance
r	radius of aerosol particle
r_N	radius of dry nucleus, radius of curvature of solid substrate
r_u	radius of water-insoluble portion of aerosol particle

R	precipitation rate (Ch. 2), Reynolds number (Ch. 10, (10-8))
R_M	maximum ice enhancement ratio (Section 9.2.1)
R_v, R_a, R_m	specific gas constant for water vapor, for dry air, for moist air
$\mathscr{R}$	universal gas constant
R_1, R_2	principal radii of curvature
s	radius of aerosol particle (Ch. 12), separation between the surfaces of two spheres (Ch. 14)
s	molar entropy
$s_{v,w}, s_{v,i}$	supersaturation of moist air with respect to a plane water surface, with respect to a plane ice surface
S	surface area, entropy (Chs. 4, 5, 11, 16), nearest neighbor distance between drops (Ch. 14)
S_0	zero-point entropy
$S_{v,w}, S_{v,i}$	saturation ratio of moist air with respect to a plane water surface, with respect to a plane ice surface
t	time
T	absolute temperature
T_a	temperature at drop surface
T_∞	temperature of environment
T_e	equilibrium freezing temperature
T_v	virtual temperature (4-31)
T_m	median freezing temperature of population of drops
T_s	$= T_0 - T$
T_{cr}	critical temperature
T_0	melting temperature of ice
T_S	surface temperature of ice crystal
$\vec{u}$	flow velocity
$\vec{u}_v, \vec{u}_a$	velocity of water vapor, of dry air
u	magnitude of flow velocity
$u_{i,\theta}$	magnitude of internal velocity in falling drop
u	volume of aerosol particle (Ch. 12), volume of drop (Ch. 15)
$\hat{u}$	internal energy per unit mass
U	internal energy of system, interaction potential between molecules
U_∞	free stream velocity of viscous flow
U_S	Stokes velocity
$\vec{v}$	particle velocity
v	mole volume (Chs. 4–7), magnitude of particle velocity (Chs. 10–17)
v_k	mole volume of component k in mixture
$v_{w,0}, v_{s,0}$	mole volume of pure water, of pure salt
v_i	mole volume of ice
$v_{v,0}, v_{a,0}$	mole volume of pure water vapor, of dry air
v_m	mole volume of moist air
v_w, v_s	partial molar volume of water, of salt, in aqueous solution
v	volume of aerosol particle (Ch. 12), volume of drop (Ch. 15)
$\dot{v}$	volume of molecule
v_0	propagation speed of steps on crystal surface
$\bar{v}$	mean speed of gas molecules
$\hat{v}$	specific volume
V	volume of system
V_∞, U_∞	terminal fall velocity of particle
V_S	Stokes fall velocity of particle

V_{uc}	volume of crystallographic unit cell
V_s	volume of solution
$V_d^{(\ell)}$	liquid portion of drop condensed on mixed aerosol particle
V_c	volume of ice crystal
V_i	volume of i-mer
V_d	drop volume

w	variance of system
w_L	liquid water content of cloud
w_s	cloud ice content
w_v	mixing ratio of unsaturated moist air
$w_{v,sat}$	mixing ratio of moist air saturated with water vapor
$w^{\downarrow}, w^{\uparrow}$	flux of molecules to and from a surface
W_c.	work for cleaving a crystal
W	vertical velocity of air, number of distinguishable microstates in ice (Ch. 3)
dW	incremental work change (an imperfect differential)

x	length coordinate
x_k	mole fraction of component k in mixture
x_v, x_a	mole fraction of water vapor, of dry air, in moist air
x_w, x_s	mole fraction of water, of salt, in aqueous solution

| y | length coordinate |
| y_c | linear collision efficiency (14-3) |

| z | (vertical) length coordinate |
| Z | Zeldovich factor |

α	angle, phase (Ch. 4, Ch. 5)
α_c	condensation coefficient (5-55)
α_d	deposition coefficient (5-56)
α_T	thermal accommodation coefficient (5-58)

| $\beta_0, \beta_s, \beta_N$ | mass increase coefficients (6-34), (6-60), (6-57) |

γ_c	cooling rate
$\gamma_{\pm}$	mean activity coefficient of salt ions in aqueous solution
$\Gamma_k^{(\sigma)}, \Gamma_s^{(s/v)}, \Gamma_w^{(s/v)}, \Gamma_w^{(w/v)}$	Gibbs adsorption of component k at interface (σ), of salt solution/vapor interface, of water at solution/vapor interface, of water at water/vapor interface, (5-9)

δ	crystallographic misfit (9-32)
δ_u	momentum boundary layer thickness
δ_D	diffusion boundary layer thickness
$\bar{\delta}$	average distance of molecular diffusion step or jump

ε	$= M_w/M_a$ (Chs. 4, 6), average elastic strain produced inside ice germ (Ch. 9), turbulent energy dissipation rate (Chs. 12, 14, 17)
$\varepsilon_v, \varepsilon_m$	volume-, mass fraction, of water-soluble substance in mixed aerosol particle
$\varepsilon_w, \varepsilon_i$	dielectric constant of water, of ice

| $\vec{\zeta}, \zeta$ | vorticity vector, magnitude of vorticity |

η, η_a, η_w	dynamic viscosity of viscous medium, of air, of water
ϑ	angle
θ	contact angle for water on solid substrate (Chs. 5, 9), angular coordinate
Θ	angle between the line of centers of interacting spheres and direction of fall
κ	compressibility (Ch. 6); $\kappa = R/c_p$ (Ch. 11); thermal diffusivity (Chs. 12, 13, 16)
λ	wave length of waves on surface of water (Ch. 10), step energy per unit length (Ch. 16), line tension (Ch. 9)
λ_a	mean free path of air molecules
Λ	electric conductivity
μ	chemical potential
μ_k	chemical potential of component k in mixture
$\mu_{v,0}, \mu_{w,0}, \mu_i$	chemical potential of pure water vapor, of pure water, of ice
μ_v, μ_w, μ_s	chemical potential of water vapor in air, of water in aqueous solution, of salt in aqueous solution
ν	number of ions into which a salt molecule dissociates in water (Ch. 4), kinematic viscosity of viscous medium (Chs. 10, 14)
ν_a, ν_w	kinematic viscosity of air, of water
ν_s	frequency of vibration of water molecule adsorbed on solid substrate
Π	spreading pressure (5-29)
ρ	density of viscous medium, of moist air
$\rho_w, \rho_i, \rho_a, \rho_m, \rho_v$	density of water, of ice, of dry air, of moist air, of water vapor
ρ_c	bulk density of ice crystal
ρ_s''	density of aqueous salt solution
ρ_N, ρ_P	bulk density of nucleus, of aerosol particle
ρ_u, ρ_s	bulk density of water-insoluble, of water-soluble, portion of aerosol particle
σ	surface tension, surface tension of water against air (Chs. 10, 14, 17)
$\sigma_{w/v}, \sigma_{i/v}, \sigma_{i/w}$	surface tension of water against vapor, of ice against vapor, of ice against water
$\sigma_{w/a}, \sigma_{i/a}$	surface tension of water against air, of ice against air
$\sigma_{s/a}, \sigma_{i/s}$	surface tension of aqueous solution against air, of ice against aqueous solution
$\sigma_{N/v}, \sigma_{N/w}, \sigma_{N/i}$	surface tension of nucleus against water vapor, against water, against ice
τ_{AP}	residence time of aerosol particle
τ_s	mean residence time of water molecule adsorbed on water surface
τ_V	vibration period of water molecule
τ_R	rotation period of water molecule
τ_D	translational displacement period of water molecule
ϕ	angular coordinate
ϕ_v	relative humidity

φ	number of bulk phases (Chs. 4, 5), angle (Ch. 16)
Φ_s	osmotic coefficient for aqueous solution (Chs. 4, 6, (4-68))
Φ	potential function, velocity potential function (Ch. 10), electrostatic potential function (Chs. 13, 17)
χ	number of surface phases
ψ	stream function
$\dot{\omega}$	angular fluid velocity
$\bar{\omega}$	radial distance from symmetry axis
Ω	surface area

TABLE OF PHYSICAL CONSTANTS

Absolute temperature of ice point $T_0 = 273.15\ °K$

Gas constant for 1 mole of ideal gas $\mathscr{R} = 8.3144 \times 10^7\ \text{erg mol}^{-1}\ (°K)^{-1}$
$= 1.858\ \text{cal}_{IT}\ \text{mol}^{-1}\ (°K)^{-1}$

Gas constant for 1 g of dry air $R_a = 2.8704 \times 10^6\ \text{erg g}^{-1}\ (°K)^{-1}$
$= 6.8557 \times 10^{-2}\ \text{cal}_{IT}\ \text{g}^{-1}\ (°K)^{-1}$

Gas constant for 1 g of water vapor $R_v = 4.6150 \times 10^6\ \text{erg g}^{-1}\ (°K)^{-1}$
$= 1.1023 \times 10^{-1}\ \text{cal}_{IT}\ \text{g}^{-1}\ (°K)^{-1}$

Boltzmann's constant $k = 1.3804 \times 10^{-16}\ \text{erg}\ (°K)^{-1}$

Planck's constant $h = 6.6252 \times 10^{-27}\ \text{erg sec}$

Avogadro's number $N_A = 6.02257 \times 10^{23}\ \text{molecule mol}^{-1}$

Molecular weight of dry air $M_a = 28.9644$

Molecular weight of water $M_w = 18.0160$

$\varepsilon = (R_a/R_v) = (M_w/M_a)$ $\varepsilon = 0.6220$

NACA Standard Atmosphere $p_{st} = 1013.250\ \text{mb}$
(sea level values) $\rho_{st} = 1.2250 \times 10^{-3}\ \text{g cm}^{-3}$
$T_{st} = 288.15\ °K = 15\ °C$
$g_{st} = 980.665\ \text{cm sec}^{-2}$
$n_{st} = 2.5471 \times 10^{19}\ \text{air molecule cm}^{-3}$

electronic charge $e = 4.80298 \times 10^{-10}\ \text{e.s.u.} = 1.60210 \times 10^{-19}\ \text{C}$

INDEX OF SUBJECTS